P9-CRQ-569

PHYSICS

PHYSICS A WINDO

THIRD EDITION

Jay Bolemo

The Universit
of Central Florid

Prentice Hall, Englewood Cliffs, NJ 076

Library of Congress Cataloging-in-Publication Data

Bolemon, Jay
Physics: a window on your world / Jay Bolemon—3rd ed.
p. cm.
Includes index.
ISBN 0-13-014309-X
1. Physics. I. Title
QC21.2.B65 1995
530—dc20 94-37112
CIP

For the curious and enthusiastic students who inspired this work

Acquisition Editor: Ray Henderson
Editorial Director: Tim Bozik
Director of Production and Manufacturing: David W. Riccardi
Development Editor: Irene Nunes
Production Editor: Susan Fisher
Marketing Manager: Gary June
Cover and Interior Design: Jeannette Jacobs
Cover Photo: Dale Nichols
Manufacturing Buyer: Trudy Pisciotti
Editorial Assistant: Pam Holland-Moritz
Design Director: Paula Maylahn
Page Make-up: Meryl Poweski
Photo Editor: Lorinda Morris-Nantz
Photo Researcher: Rhoda Sidney
Assistant Editor: Wendy Rivers

A Simon & Schuster Company, Englewood Cliffs, New Jersey 07632

Printed in the United States of America

10 9 8 7 6 5 4 3 2 1

ISBN: 0-13-014309-X

Prentice-Hall International (UK) Limited, *London*
Prentice-Hall of Australia Pty. Limited, *Sydney*
Prentice-Hall Canada Inc., *Toronto*
Prentice-Hall Hispanoamericana, S.A., *Mexico*
Prentice-Hall of India Private Limited, *New Delhi*
Prentice-Hall of Japan, Inc., *Tokyo*
Simon & Schuster Asia Pte. Ltd., *Singapore*
Editora Prentice-Hall do Brasil, Ltda., *Rio de Janeiro*

BRIEF CONTENTS

CONTENTS

CHAPTER THREE: LAWS OF MOTION 34

CHAPTER FOUR: USING THE LAWS OF MOTION 49

CHAPTER FIVE: MOMENTUM 65

CHAPTER SIX: ENERGY 77

CHAPTER SEVEN: GRAVITY 94

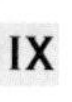

CHAPTER EIGHT: ATOMS AND MOLECULES 109

CHAPTER NINE: SOLIDS 125

CHAPTER TEN: LIQUIDS 140

CHAPTER ELEVEN: GASES 156

CHAPTER TWELVE: TEMPERATURE AND MATTER 170

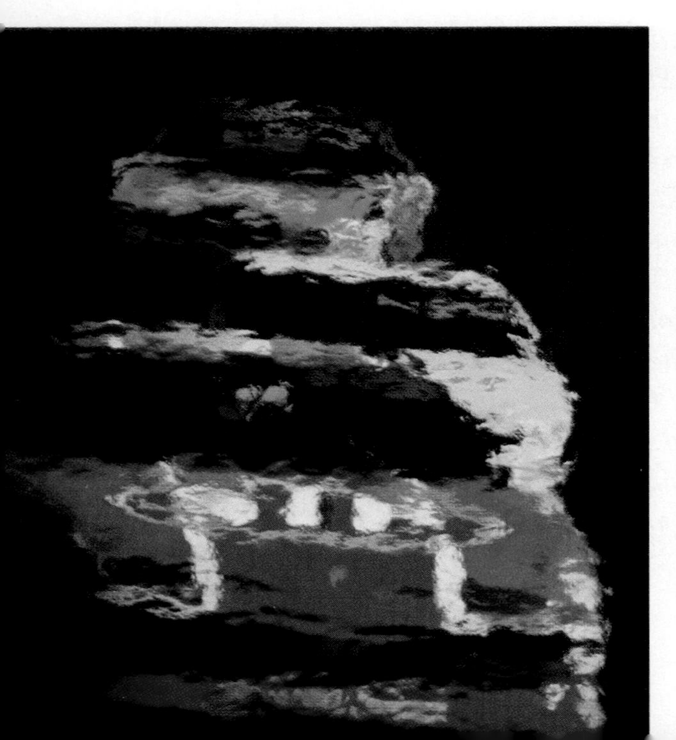

CHAPTER THIRTEEN: HEAT: ENERGY IN MOTION 185

CHAPTER FOURTEEN: HEAT AND THE EARTH 200

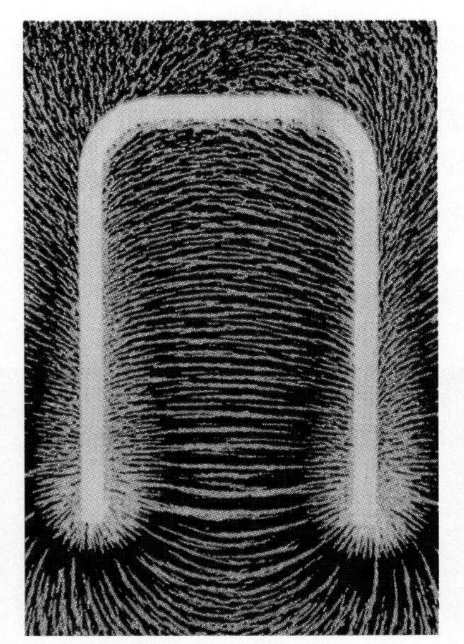

CHAPTER FIFTEEN: WAVES IN MATTER 217

CHAPTER SIXTEEN: SOUND 233

CHAPTER SEVENTEEN: ELECTRIC CHARGE 247

CHAPTER EIGHTEEN: MOVING CHARGES 264

CHAPTER NINETEEN: MAGNETISM 279

CHAPTER TWENTY: ELECTRICITY AND MAGNETISM 290

CHAPTER TWENTY-ONE: LIGHT: ELECTROMAGNETIC RADIATION 303

CHAPTER TWENTY-TWO: PROPERTIES OF LIGHT 318

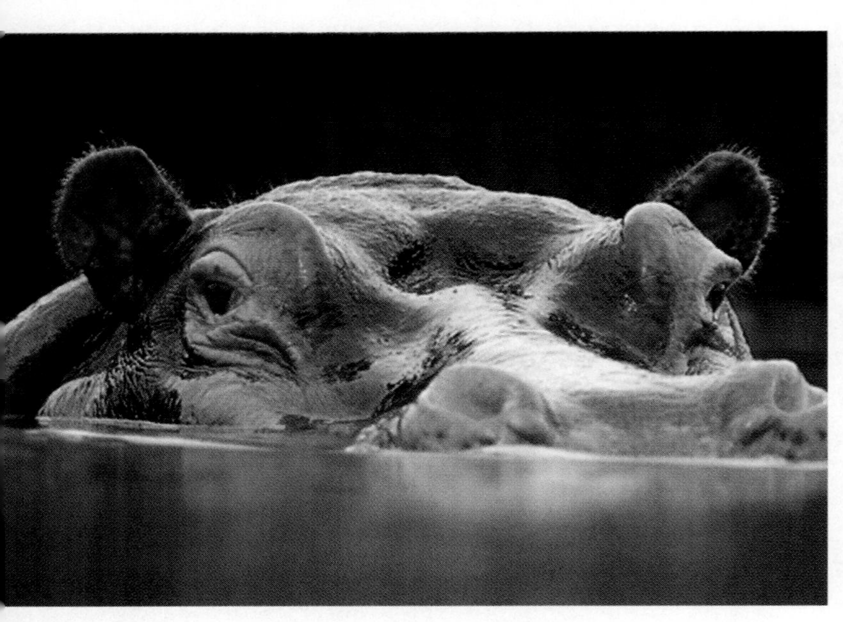

CHAPTER TWENTY-THREE: COLORS AND VISION 339

CHAPTER TWENTY-FOUR: LIGHT AND ATOMS 356

CHAPTER TWENTY-FIVE: THE WAVE NATURE OF PARTICLES 375

CHAPTER TWENTY-SIX: SPECIAL RELATIVITY 390

CHAPTER TWENTY-SEVEN: THE NUCLEUS 406

CHAPTER TWENTY-EIGHT: NUCLEAR ENERGY AND OUR ENERGY FUTURE 421

FOR THE STUDENT

Since you are almost certainly not a science major, you may be worried that physics will be really hard to understand. After all, topics which are candy for students of one major can be very sticky hurdles for others. But this book was written for you, with input from many students like yourself, and its principal goal is to help you understand the concepts and ideas of physics. It is about descriptions, how physics describes the things you see around you. (While physicists use equations to describe nature, the calculations here serve more to shows how physics works, how it makes predictions. Don't worry, this text won't try to turn you into a physics major!) Everyday experience makes you familiar with some of the ideas of physics, and we use your experiences as starting points whenever we can.

This book, then, is about appreciating the things you see around you in terms of what physicists have discovered over the years. Along the way we will discuss many topics with environmental applications, flagged by an earth icon. Physics gives you a unique perspective of your environment—a window on your world, as the title suggests—that is important to everyone today.

How to Learn from This Book

Here are a few tips on how to study. You'll want to read the chapters several times to be sure you get the vocabulary and the concepts down. Read the chapter quickly the first time, don't spend time with any formula, and stop after reading the summary. Then put the book down for an hour or even a day. The words, definitions, and examples will be more familiar the next time you read through. When you reread the chapter, go more slowly and read everything. After you feel comfortable with the chapter material, you are ready for the Concept Checks. The true-false answers for these are at the back of the book. If you can't answer them correctly, go back to the chapter for help. You'll retain more of the ideas each time you read it. Next, try some of the exercises under Applying the Concepts. Exercises with numbers in orange have answers at the back of the book. Sometimes these are related to things not in your background, and you'll learn from reading the answers. If you are asked to do calculation exercises, save for last those marked with a calculator icon. If the calculation uses one of the equations in the chapter, the icon color sends you to the equation with the icon of matching color. The problems in this text use only addition, subtraction, multiplication, and division, the math you use to balance a checkbook or figure the miles per gallon a car gets during a trip. But before you do these problems, *study Appendix I of this text,* which has examples with powers of ten, significant figures, and changing units of measurement.

The Learning Guide for this text, which comes with computer simulation software, is also valuable for reviewing and preparing for exams. It contains a quick review of each section of the text, with extra conceptual examples and extra calculation examples as well. In addition, it has a practice exam of conceptual questions and another with calculation problems, with detailed answers provided for both. In this guide you'll also find narrative answers for the Concepts Checks.

A last tip: The sooner you start, the better, because last minute studying doesn't work well in physics courses. Begin sooner and you'll learn more. Happy studying!

About the 3D Drawings and Photos

You'll find approximately forty pairs of stereo images throughout the book. Sometimes these will be photographic, and others will be illustrations. The stereo views offer an extra way to visualize the most difficult spatial topics in the book, and we hope you think they're fun to look at! Every copy of the book has a Stereopticon viewer attached to the inside back cover. Remove the viewer from the envelope, and follow the directions printed on it to assemble it (this is really a snap). By placing the viewer between the stereo pairs with the solid side of the viewer nearest you, the viewer will create a 3D image. If you lose your stereo viewer, you can also try a different technique. Hold the book close to you face with your nose between the image pair (close enough so the images are blurry). Then, relax your eyes by pretending to look across the room while you slowly move the book away from your face. This is tricky—but with practice you should be able to see the image in 3D about arms length away from your face. If you enjoy looking at these, your professor may have other stereo pairs that you can view.

TO THE INSTRUCTOR

The new name of this text, *Physics: A Window on Our World,* reflects the major changes in its third edition. As before, its goal is to teach students about physics principally through explanations of topics they can relate to from their own experiences. Its shortened and refined coverage makes it easier for students to study and understand, and its new, colorful format makes it easier to stay with during study sessions. The revision began more than four years ago with comments solicited from hundreds of students who used the second edition. Becky Jakubcin and Shannon Pack, study group leaders in one of the author's classes who became familiar first-hand with where students had difficulty understanding topics, critiqued the second edition for level of presentation and interest. Tino Wright acted as the ultimate "student advocate" who not only criticized the second edition but two versions of manuscript produced in the first two years of the revision. Because of her valued advice and editorial skill, the level of presentation is more even and understandable than before. During the second two years of the revision my editor, Ray Henderson, gathered the comments of many instructors on the manuscript, improving the accuracy of the physics descriptions on many subtle and some not-so-subtle points. Ray also sculpted the new format, fine-tuning its appearance and delivery even during production for interest and ease of understanding. Irene Nunes was the developmental editor during the last year, and she strove for clarity in every sentence. Mike Steffancin helped with the research for this edition, reviewed all the galleys line-by-line, and wrote the excellent Learning Guide for the text. Thomas Kimble drafted the 3D line drawings, often improving on the author's ideas. Dale Nichols furnished many beautiful photographs and suggested others, and Rhoda Sidney tracked down the author's photo requests, not an easy job. Andy Graham of Appalachian State University helped with some of the technical photos. Further assistance was given by Paula Adams, Tim Bandy, Mildred Bolemon, Faylla Chapman, Frank Diefenderfer, Paul Doherty, Freddie Flake, Ralph Llewellyn, Allen Mathews, Bob Miller, Gary Patterson, Wendy Rivers, Mary Louise Schmid, Barry Schroder, and Cynthia Turbeville. Susan Fisher expertly took the text through an extremely complicated production, maintaining a sense of humor under difficult conditions—none of which were the author's fault, of course! The labors of these people have improved the value of the text for students enormously. In the last stages a sabbatical granted by the University of Central Florida saw the book through to completion. Inspirational sustenance for the author was provided by the Louisiana Purchase in Banner Elk, North Carolina and by Pebbles and Cordon Bleu, both of Winter Park, Florida. Finally the author is immensely grateful to Sally, partner *extraordinaire* and patron of science education. Her support, encouragement, and help in improving the manuscript were essential.

Reviewers

Larry Bigelow, *Seminole Community College*
Edward Borchardt, *Mankato State University*
Michael J. Bozack, *Auburn University*
Dr. Dieter Brill, *University of Maryland*
David Buckley, *East Strondsburg University*
Tai L. Chow, *California State University, Stanislaus*
Robert Beck Clark, *Texas A&M University*
Robert Cole, *University of Southern California*
Roger D. Crelf, *Lamar University*
R. Coker, *University of Texas at Austin*
Roger C. Crawford, *Pierce College, CA.*
James R. Crawford, *Southwest Texas State University*
Doug Davis, *Eastern Illinois University*
Lawrence Day, *Utica College of Syracuse University*
Paul D'Alessandris, *Monroe Community College*
Alex Dickison, *Seminole Community College*
John R. Dunning Jr., *Sonoma State University*
Michael Hones, *Villanova*
Sanford Kern, *Colorado State University*
Robert L. Kernell, *Old Dominion University*
N. Franklin Kolp, *Trenton State College*
David Kruegar, *Colorado State*
Forrest Newman, *Sacramento City College*
Robert Packard, *Baylor University*
Don Rathjen, *Pleasant High School*
Ray Renken, *University of Central Florida*
Charles R. Rhyner, *University of Wisconsin—Green Bay*
Ronald Royer, *Minot State University*
Ajit Rupaal, *Western Washington University*
Joseph Scanio, *University of Cincinnati*
Terry Schalk, *University of California—Santa Cruz*
Larry Shepley, *University of Texas at Austin*
William S. Smith, *Boise State University*
Stanley Sobolewski, *Indiana University of PA.*
Mike Steffancin, *Valencia Community College*
Frank D. Stekel, *University of Wisconcin—Whitewater*
Gerald Taylor, *James Madison University*
Paul Varlashkin, *East Carolina University*
James Watson, *Ball State University*
Arthur Wiggins, *Oakland Community College*
Dong Ho Wu, *University of Maryland*

My thanks to you all!
Jay Bolemon
Syzygy, North Carolina

ONE

Science and Physics

A multiple exposure of (left to right) Jupiter, Venus and the sun rising behind Manhattan taken at about ten-minute intervals. The motions of the sun, the moon and the planets played a large role in the beginnings of science.

STUDY TIP: BEFORE YOU START THIS CHAPTER, YOU MIGHT WANT TO READ THE *PREFACE, FOR THE STUDENT.* IT HAS TIPS ON STUDYING THAT COME LARGELY FROM STUDENTS WHO USED THIS TEXTBOOK IN EARLIER EDITIONS.

Welcome to your study of physics! Physics, as you will see, is the study of how nature behaves. From your own experiences, many topics we take up will be familiar to you—rainbows, for instance, or the path a football or baseball takes through the air. You will also read about things that you cannot see because physics explores nature down to its smallest details. For example, physics can explain a spark that jumps between your finger and a doorknob in terms of the tiny atoms and subatomic particles that make up all matter, from the tips of your fingers to the most distant stars.

This book is about our present picture of the world as described by physics. For an introduction, this chapter takes a look at the ways science came into being and at what science is today. It discusses physics, how it relates to other sciences, and how physicists go about doing physics.

THE PREHISTORIC ORIGINS OF SCIENCE

Science, our body of knowledge of natural phenomena, springs entirely from human curiosity and imagination. It began when our early ancestors realized that some natural phenomena are repetitive and therefore *predictable.* A simple example of such a phenomenon is the cycle of night and day. Every day, the sun rises in the east, moves across the sky, and sets in the west (Figure 1-1). We all have seen enough days and nights to believe that the rising and setting of the sun are cyclical, predictable actions.

40,000–35,000 B.C.

For another example, also from the sky, we can turn to the first calendars. During the last Ice Age, 35,000 years ago, early people carved "moon calendars" in mammoth ivory and bones. A series of 29 notches and then another series of 30 notches were followed by a symbol that probably meant "repeat."

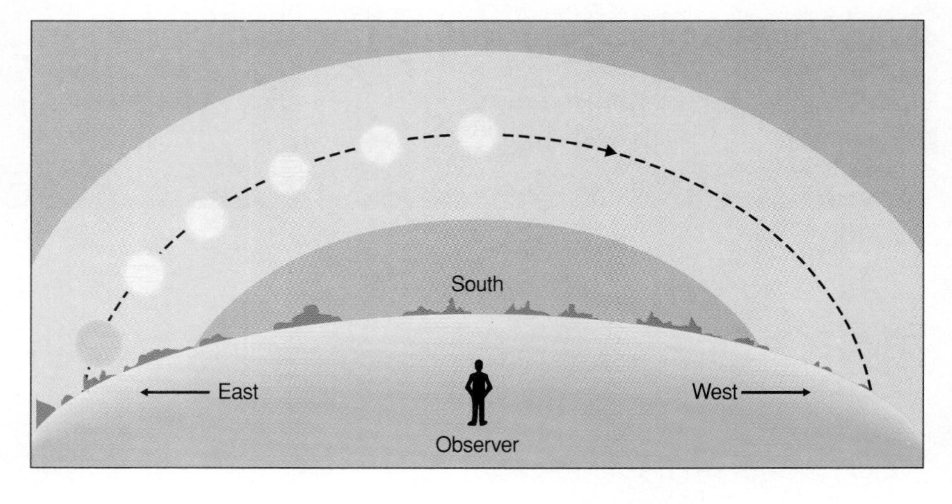

FIGURE 1-1 THE PATH OF THE SUN ACROSS THE SKY DURING ONE DAY IN THE U.S. THE SUN RISES IN THE EAST, ARCS ACROSS THE SKY, AND SETS IN THE WEST. AT NIGHT IN THIS SAME PORTION OF THE SKY, AN OBSERVER WILL SEE STARS MOVING FROM EAST TO WEST LIKE THE SUN DOES DURING THE DAY. IN SUMMER THE SUN'S DAILY PATH IS HIGHER IN THE SKY, AND IN WINTER IT IS LOWER, AS INDICATED BY THE HIGHLIGHTS IN THIS DRAWING.

FIGURE 1-2 THE VARYING APPEARANCE OF THE MOON IN THE SKY THROUGH ONE LUNAR CYCLE OF 29.5 DAYS.

These portable calendars marked the moon's changing appearance in the sky, which repeats itself every 29.5 days (Figure 1-2). (Today we refer to these changes in appearance as the "phases" of the moon.) Early hunters probably kept such moon calendars because bright moonlight, during and near full moons, could help them hunt at night, and keeping track of months could help them anticipate seasonal movements of their game.

After the last Ice Age, but still long before written languages were invented, people painted moon calendars and other objects on cave walls in Europe. Those early observers, by counting and keeping records, pursued what could be called pre-science. Careful observing of natural phenomenon is the first step in the process of science.

THIS ARTWORK FROM A CAVE IN FRANCE WAS DONE ABOUT 15,000 B.C., THOUSANDS OF YEARS BEFORE THE FIRST WRITTEN LANGUAGES WERE DEVELOPED BUT THOUSANDS OF YEARS AFTER THE FIRST KNOWN LUNAR CALENDARS.

THE BABYLONIANS

6000–5000 B.C.

Written languages appeared about 6000 B.C., and before 5000 B.C. the ancient Babylonians (who lived in what we today call southern Iraq and Iran) had developed agriculture to an extent never realized before. In this "cradle of civilization," communities flourished, commerce was vigorous, and scholars began keeping written records of motions and events in the skies. They estimated that the sun passed across the sky 360 times before the seasons, along with the patterns of the stars in the nighttime sky, repeated themselves. They used that cycle when they chose to divide the circle into 360 equal angular parts called *degrees,* a measure we still use today. Continued observations led them to add a few days to their calendar from time to time to adjust for the fact that there are 365.25 days in a year (Figure 1-3). Today we take care of the extra fraction of a day with a "leap" year every fourth year, when the short month of February gets an extra day.

600–500 B.C.

With their records from generations of observers, the Babylonians eventually discovered, in the sixth century B.C., a 56-year pattern of repetition of *lunar eclipses*—the name of the event when the moon passes through the

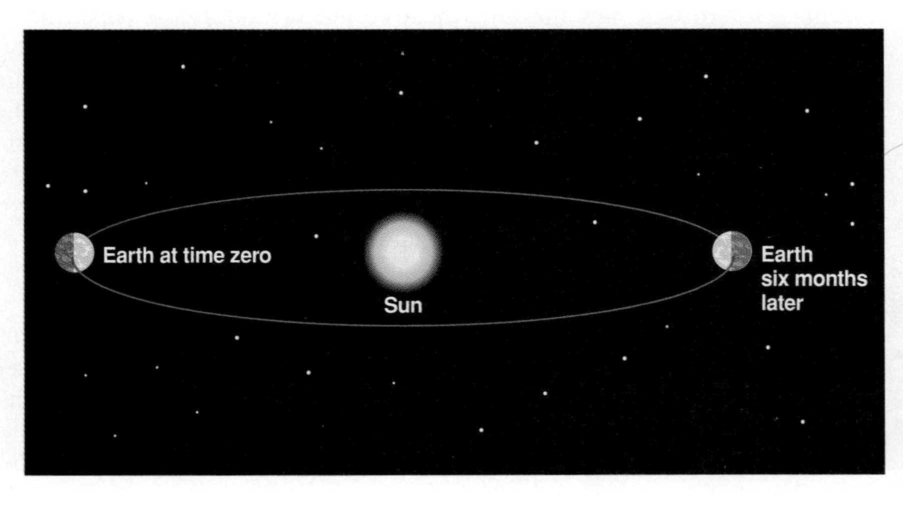

FIGURE 1-3 AS EARTH SWINGS AROUND THE SUN, WE SEE DIFFERENT STARS IN THE NIGHTTIME SKY. THE MIDNIGHT SKY AT THE POSITION ON THE LEFT IS TOTALLY DIFFERENT FROM THE MIDNIGHT SKY SIX MONTHS LATER AT THE POSITION ON THE RIGHT. BUT IN 6 MORE MONTHS, WHEN EARTH IS BACK AT THE POSITION ON THE LEFT, THE STARS IN THE NIGHTTIME SKY WILL LOOK EXACTLY THE SAME AS THEY DID 12 MONTHS EARLIER. DURING THIS YEAR-LONG JOURNEY, THE EARTH ALSO PERFORMS 365.25 COMPLETE ROTATIONS ON ITS AXIS, TURNING LIKE A ROTISSERIE IN THE SUNLIGHT TO GIVE US "DAYTIME" IN THE SUNLIGHT AND "NIGHTTIME" IN EARTH'S SHADOW.

FIGURE 1-4 THE MOON IN THE EARTH'S SHADOW DURING A TOTAL LUNAR ECLIPSE. THE MOON IS RED BECAUSE RED LIGHT REACHES IT FROM EARTH. IF AN ASTRONAUT WERE ON THE EARTH SIDE OF THE MOON DURING A LUNAR ECLIPSE, SHE OR HE WOULD SEE A RED RING OF SUNSET LIGHT AROUND THE EARTH!

GEO = EARTH
HELIO = SUN
CENTRIC = CENTER

AROUND 2000 B.C. THE INHABITANTS OF ANCIENT BRITAIN BUILT STONEHENGE, AN OBSERVATORY WHERE THE EXTREME POSITIONS OF THE SUN AND MOON ON THE FAR HORIZON WERE MARKED BY HUGE COLUMN-STONES. CIRCLING THE EDIFICE WERE 56 CHALK-FILLED HOLES, PROBABLY INDICATING THAT THE BUILDERS WERE AWARE OF THE 56-YEAR CYCLE OF LUNAR ECLIPSES 14 CENTURIES BEFORE THE BABYLONIANS DISCOVERED IT. NO WRITTEN RECORDS SURVIVE FROM THE PEOPLE OF STONEHENGE, HOWEVER, AND THEIR KNOWLEDGE WAS LOST LONG BEFORE THE ROMANS CAME TO THE ISLAND.

shadow of the earth (Figure 1-4). This was an amazing feat—suddenly humans could predict all the *future* eclipses of the moon! Even more important, this discovery showed the value of accumulating and passing on knowledge.

THE ANCIENT GREEKS

The rise of ancient Greek civilization began about 600 B.C., and a long line of Greek philosopher-scholars brought curiosity, logic, and mathematics together to explain their observations of nature. (The word *philosopher* stems from the Greek words *philos,* loving, and *sophos,* wise, so a philosopher is a *lover of wisdom.*) These scholars of natural phenomena came to be called **natural philosophers.**

About 400 B.C. Democritus and Leucippus speculated about the invisible nature of matter. They knew that a drop of water evaporates invisibly into the air, just as smoke vanishes into the air over a fire and tiny ashes from that fire disappear if rubbed between two fingers. From everyday observations like these, they concluded that matter was made of basic units too small to be seen. These tiny units, they decided, could not be broken or divided, explaining how a large collection of them could have the permanence seen in objects such as rocks. The Greeks called these invisible units *atoms,* from the Greek word *atomos,* which means "indivisible" or "uncuttable." But more than twenty centuries would pass before science could prove the existence of atoms.

Aristotle (384–322 B.C.), a teacher of Alexander the Great, wrote about logic, physics, astronomy, politics, zoology, and poetry. He was convinced that the earth was round in an age when most educated people believed it to be flat, and he developed a model to explain the phases of the moon by assuming the moon shines by reflected sunlight (Figure 1-5). He came to believe that nature could be understood *just by reasoning alone,* and his successes led many to accept his ideas. As late as 1600 A.D., educators in England could be fined if they questioned the correctness of Aristotle's writings, almost 20 centuries after he lived.

To explain the 24-hour cycle of day and night and the yearly cycle of the nighttime sky, the Greeks debated two ideas. These centered on two possible arrangements for the *solar system,* which is what we call our sun and the planets that revolve around it. Was the sun revolving around a stationary earth as proposed in the **geocentric model** or was the earth revolving around the sun as proposed in the **heliocentric model,** rotating on an axis each day as it did? Aristotle's followers argued that the earth must be stationary, or, for example, a bird that flew upward would be left behind as the earth moved onward. (For an even more compelling argument by Aristotle, see Demonstration 1.) The geocentric model, with earth at the center of the circling sun, moon, and planets, eventually won acceptance in Greek philosophy.

Later Greek philosophers developed geometry, and they wrote of the principles of levers, the properties of curved mirrors, and the buoyancy of objects immersed in water. They managed to estimate the diameter of the earth, the moon, and the sun, and they even calculated the distance to the moon to within a few percent of its actual value. Ptolemy (85–165 A.D.) lived and worked in Alexandria, Egypt, where the center of Greek science had moved around 250 B.C. to escape the invading Romans. There he brought the Greek ideas of the geocentric solar system to their highest point with a model for the sun, moon, and the five planets that are visible to the naked eye. Unlike the stars, which appear as if they are "fixed," or painted, on some distant sphere, these five planets move

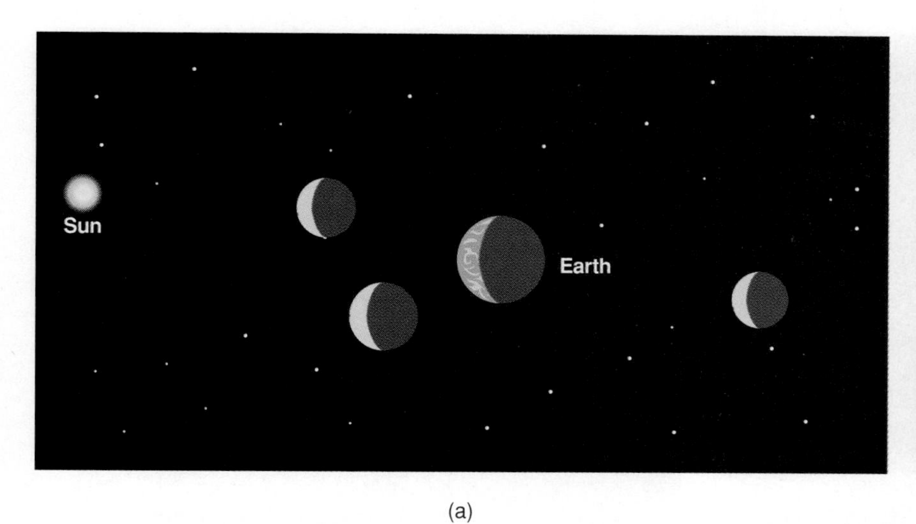

(a)

(b)

FIGURE 1-5 (A) ARISTOTLE REALIZED THAT JUST AS FOR EARTH, ONLY HALF OF THE MOON IS SUNLIT AT ONE TIME AND THAT THE DIFFERENT PHASES OF THE MOON APPEAR BECAUSE WE SEE DIFFERENT PORTIONS OF ITS SUN-LIT PART AS IT ORBITS EARTH. AT THE MOON'S LOWEST POSITION IN THIS DRAWING, FROM EARTH WE WOULD SEE THE FIRST QUARTER MOON (SECOND FROM LEFT IN FIGURE 1-2). WHEN THE MOON IS AT THE FAR RIGHT POSITION SHOWN HERE, FROM EARTH WE SEE THE FULL SUNLIT FACE, OR THE FULL MOON. AT ITS HIGHEST POSITION IN THIS DRAWING WE SEE A CRESCENT MOON. (B) A PHOTOGRAPH FROM A SPACECRAFT (GALILEO) OUTSIDE THE MOON'S ORBIT THAT SHOWS THE EARTH AND THE MOON TOGETHER IN ONE PICTURE.

among the stars (Figure 1-6). Over time they trace out peculiar paths, as shown in Figure 1-7. They move principally toward the east but sometimes back up toward the west for a while. This reverse motion is called *retrograde* motion. With his model (Figure 1-8) Ptolemy was able to predict planetary paths with fair accuracy months in advance. And so another repetitive action in nature was modeled successfully, and it would not be seriously challenged for some seventeen hundred years.

PHYSICS IN 3D/PHASES OF THE MOON

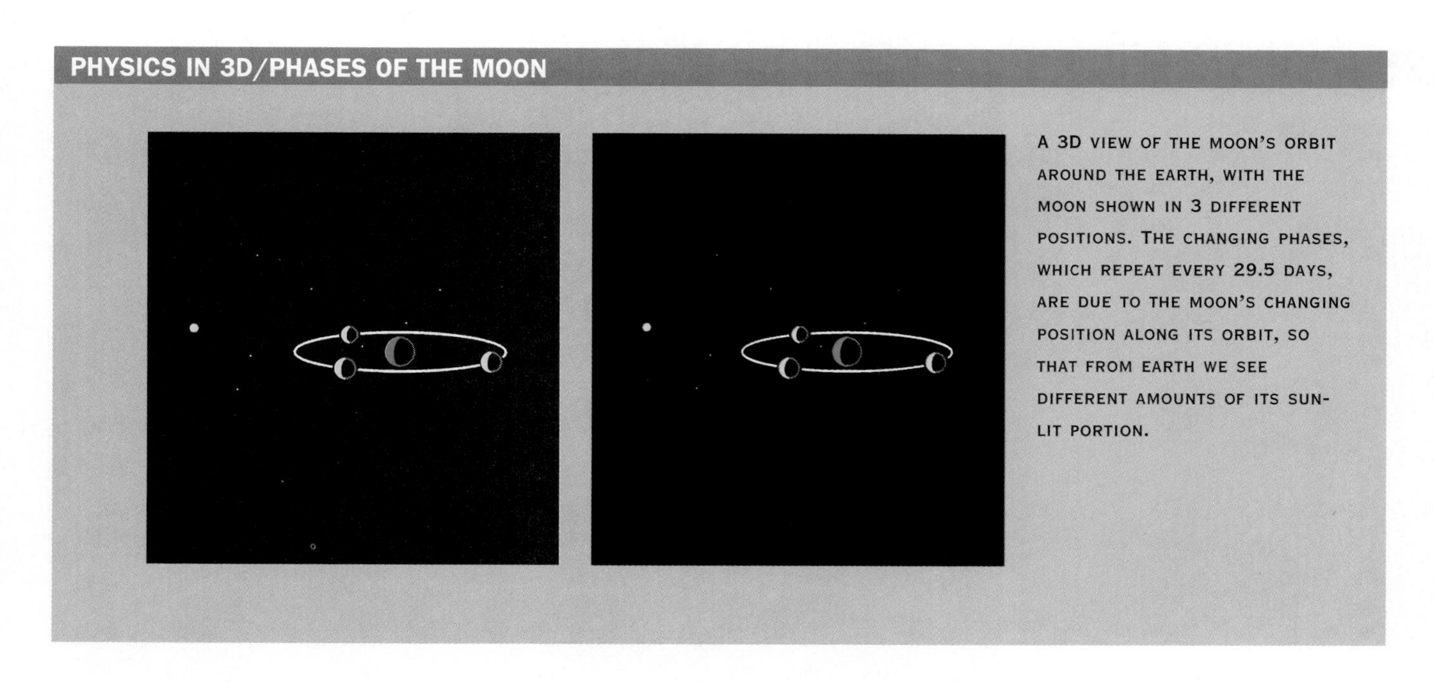

A 3D VIEW OF THE MOON'S ORBIT AROUND THE EARTH, WITH THE MOON SHOWN IN 3 DIFFERENT POSITIONS. THE CHANGING PHASES, WHICH REPEAT EVERY 29.5 DAYS, ARE DUE TO THE MOON'S CHANGING POSITION ALONG ITS ORBIT, SO THAT FROM EARTH WE SEE DIFFERENT AMOUNTS OF ITS SUN-LIT PORTION.

FIGURE 1-6 PHOTOGRAPHS TAKEN FROM THE SAME SPOT 24 HOURS APART SHOW THE RELATIVE MOTION OF VENUS AND JUPITER IN THE SKY. EARLY ASTRONOMERS CHARTED THESE MOTIONS WITH GREAT CARE AGAINST THE "FIXED" BACKGROUND OF THE STARS, WHICH DO NOT SEEM TO MOVE WITH RESPECT TO EACH OTHER OVER A HUMAN LIFETIME. (VENUS IS THE BRIGHTER OF THE TWO SEEN HERE.)

THE ROMANS

The Romans brought centuries of stability to Western Europe, North Africa, and the Middle East. They excelled as fighters and administrators, and as builders of roads, cities, and ports (Figure 1-9). In contrast to the Greeks before them, however, they showed little interest in the concepts of natural philosophy, which seemed to them to have no practical applications. In the fourth century A.D., Roman leaders zealously set out to destroy everything they considered pagan in the light of their religious beliefs. Feeling threatened by Greek culture, in 390 A.D. they burned the great library in Alexandria. Even

FIGURE 1-7 A TIME EXPOSURE OF PLANETARY MOTIONS AS SEEN IN A PLANETARIUM. WHILE THE PLANETS MOVE PRINCIPALLY TOWARD THE EAST (PROGRADE MOTION), THEY ALL BACK UP OCCASIONALLY TOWARD THE WEST (RETROGRADE MOTION). EAST IS TO THE LEFT HERE.

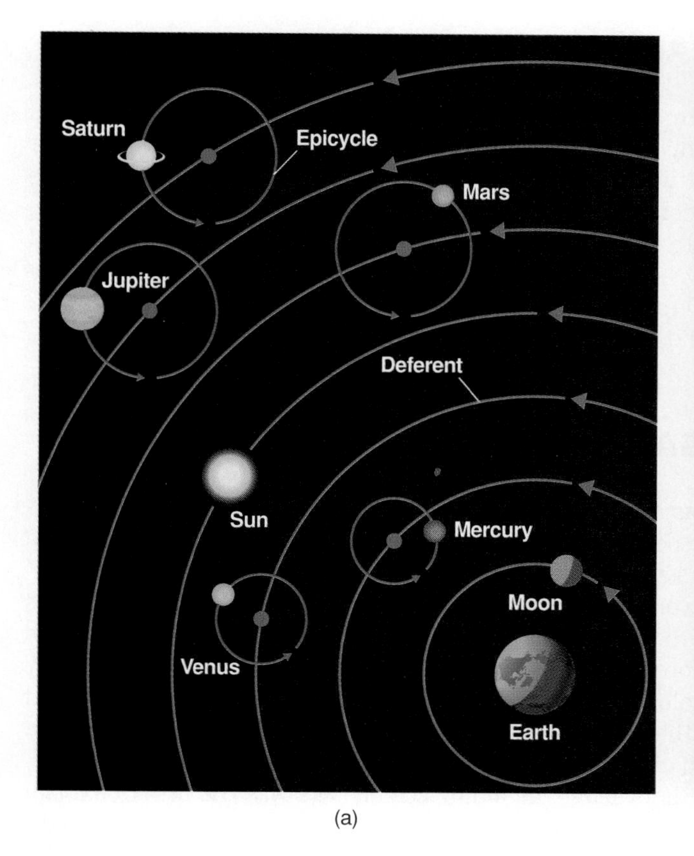

(a)

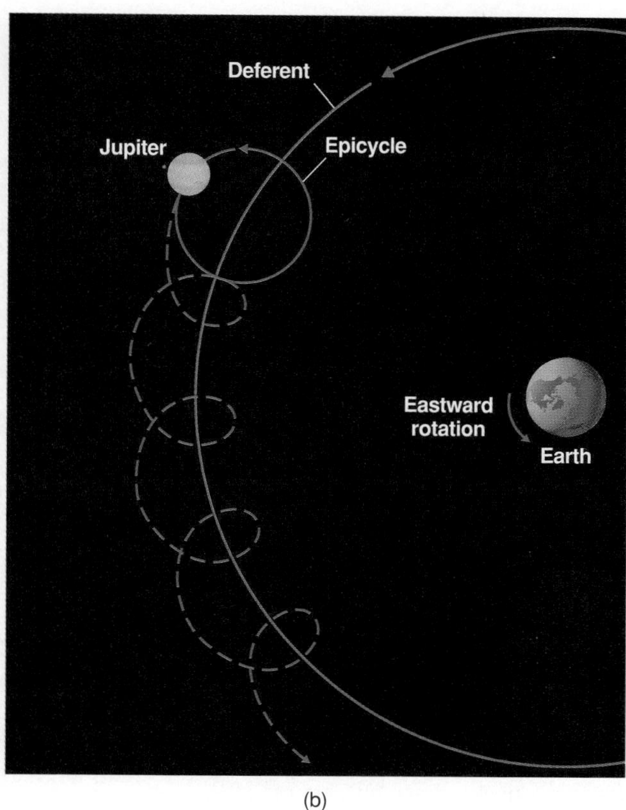

(b)

FIGURE 1-8 (A) PTOLEMY'S GEOCENTRIC MODEL OF THE SOLAR SYSTEM. EACH PRINCIPAL ORBIT AROUND THE EARTH WAS CALLED A DEFERENT. WHILE A PLANET CIRCLED EARTH, IT ALSO MADE A SMALLER ORBIT AROUND A POINT ON THE DEFERENT. THIS SMALLER ORBIT WAS CALLED AN EPICYCLE. THE EPICYCLE CENTER MOVED ALONG THE DEFERENT AS THE PLANET MOVED ON THE EPICYCLE. BECAUSE A PLANET'S MOTION ON ITS EPICYCLE WAS FASTER THAN THE EPICYCLE'S MOTION ALONG THE DEFERENT, THE PLANET APPEARED TO MOVE BACKWARDS (WESTWARD) WHEN IT WAS CLOSEST TO EARTH. (B) DETAILS OF JUPITER'S SUPPOSED MOTION AROUND EARTH, SHOWING THE LOOPS WHERE IT WOULD BE IN RETROGRADE MOTION.

FIGURE 1-9 HERE IS AN AQUEDUCT, STILL STANDING AT PONT DU GARD, FRANCE, BUILT BY THE ROMANS IN THE 1ST CENTURY A.D. MASTER BUILDERS, ADMINISTRATORS, AND WARRIORS, THE ROMANS PAID LITTLE ATTENTION TO SCIENCE.

the scholars there became targets; in 415 A.D. they murdered the first well-known woman scholar of natural philosophy, Hypatia. Greek and Egyptian philosophers fled, carrying hand-copied manuscripts that would find their way to non-Roman Arabian regions. It would be many centuries before this knowledge would reappear in Europe.

THE MIDDLE AGES

400–1400 A.D.

Rome was overrun by the Visigoths (410 A.D.) and the Vandals (455 A.D.), and soon the Huns, led by Attila, followed. In the 700s the Arabs controlled about half of the western world, and in 846 A.D. they too conquered Rome. The pursuit of science continued in the realm of Islam and in India and China, but Europe sank into feudalism as waves of invaders kept life uncertain and civilization at a low ebb. The 1000-year period between the decline of the Romans (about 400 A.D.) and the blooming of the Renaissance (about 1400 A.D.) is called the Middle Ages.

FIGURE 1-10 **DURING THE CRESCENT PHASES OF THE MOON, WE SEE THE REST OF THE MOON'S FACE DIMLY BECAUSE SUNLIGHT REFLECTS FROM EARTH TO THE MOON AND THEN IS REFLECTED FROM THE MOON BACK TO EARTH. (THE PLANET VENUS ALSO APPEARS HERE.)**

THE RENAISSANCE

The Renaissance was a period of rebirth in the arts and scholarship in Europe as the feudalism of the Middle Ages waned. Europe emerged from the Middle Ages gradually over several centuries. In the eleventh century, European scholars translated the works of Euclid, Aristotle, and other natural philosophers from 1000 years earlier from Arabic into Latin. Universities appeared in the 1200s at Paris, Oxford, Naples, and Cambridge, to name a few. At first they were schools of theology, law, and medicine, but in later centuries they would become centers of the sciences as well.

Leonardo da Vinci (1452–1519) was a Renaissance artist who was interested in all fields of knowledge, including science. He was the first to explain that we see the "dark" side of the moon in its crescent phases because of light reflected to the moon from earth (Figure 1-10). His notebooks are also the first written records that contain the idea of forces (which we take up in Chapter 3).

The printing press and movable type were perfected in Europe during the Renaissance, and in 1456 the Bible was set into print. Soon after, the early Greek manuscripts that had been translated into Latin in the eleventh century were also printed. Books by Ptolemy and Aristotle appeared in 1475 and 1476. Euclid's geometry text was printed in 1482 and has gone through perhaps a thousand editions since that time. The knowledge of the Greeks, inaccessible to Europeans during the Middle Ages, finally became available. By the time Columbus sailed to America in 1492, over 20 million books had been printed. Later Magellan's expedition sailed around the world, proving the earlier Greek idea that earth is spherical.

AN EARLY GROUND RULE FOR MAKING MODELS OF NATURAL PHENOMENA, OR THEORIES, WAS THAT THEY SHOULD BE AS SIMPLE AS POSSIBLE. CALLED OCKHAM'S RAZOR (AFTER WILLIAM OF OCKHAM, 1300–1349?), A MODERN TRANSLATION OF THIS RULE GOES LIKE THIS: "IF THERE ARE TWO OR MORE POSSIBLE MODELS THAT EXPLAIN THE SAME FACTS, THE ONE MOST LIKELY TO BE TRUE IS THE SIMPLEST ONE." THIS PRINCIPLE IS VALUABLE BECAUSE THE MORE ASSUMPTIONS MADE, THE GREATER THE POSSIBILITY OF ERROR. SINCE ITS EXPLANATION OF RETROGRADE MOTION WAS SIMPLER, COPERNICUS'S MODEL WAS PREFERABLE TO PTOLEMY'S.

THE COPERNICAN REVOLUTION

Copernicus (1473–1543) was born in Poland about 100 years after the Renaissance began in Italy. He challenged the geocentric model of the solar system perfected by Ptolemy, pointing out a simpler way to explain planetary motion. Copernicus's model had the earth and the other planets going in circles around the sun, with the planets nearer to the sun traveling *faster* than earth and those farther from the sun moving *more slowly* than earth. With this model, the retrograde motion of the planets occurs without any epicycles (Figure 1-11). Copernicus did use small circles on his circular orbits, just as Ptolemy did, in order to match Ptolemy's accuracy. Nevertheless, this heliocentric model explained retrograde motion more simply than Ptolemy's model and for that reason it soon drew attention from other perceptive scholars.

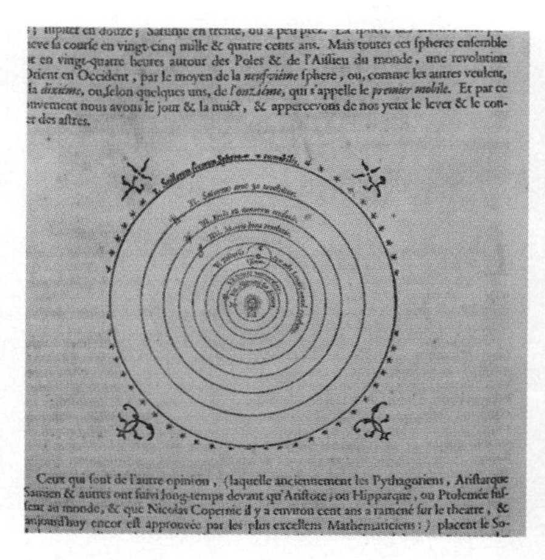

eve sa course en vingt-cinq mille & quatre cents ans. Mais toutes ces spheres ensemble
t en vingt-quatre heures autour des Poles & de l'Aissieu du monde, une revolution
Drient en Occident, par le moyen de la *neufviéme* sphere, ou, comme les autres veulent,
la *dixiéme*, ou, selon quelques uns, de *l'onziéme*, qui s'appelle le *premier mobile*. Et par ce
uvement nous avons le jour & la nuict, & appercevons de nos yeux le lever & le con-
r des astres.

Ceux qui sont de l'autre opinion, (laquelle anciennement les Pythagoriens, Aristarque
Samien & autres ont suivi long-temps devant qu'Aristote, ou Hipparque, ou Ptolemée fus-
sent au monde, & que Nicolas Copernic il y a environ cent ans a ramené sur le theatre, &
aujourd'huy encor est approuvée par les plus excellens Mathematiciens :) placent le So-

(a)

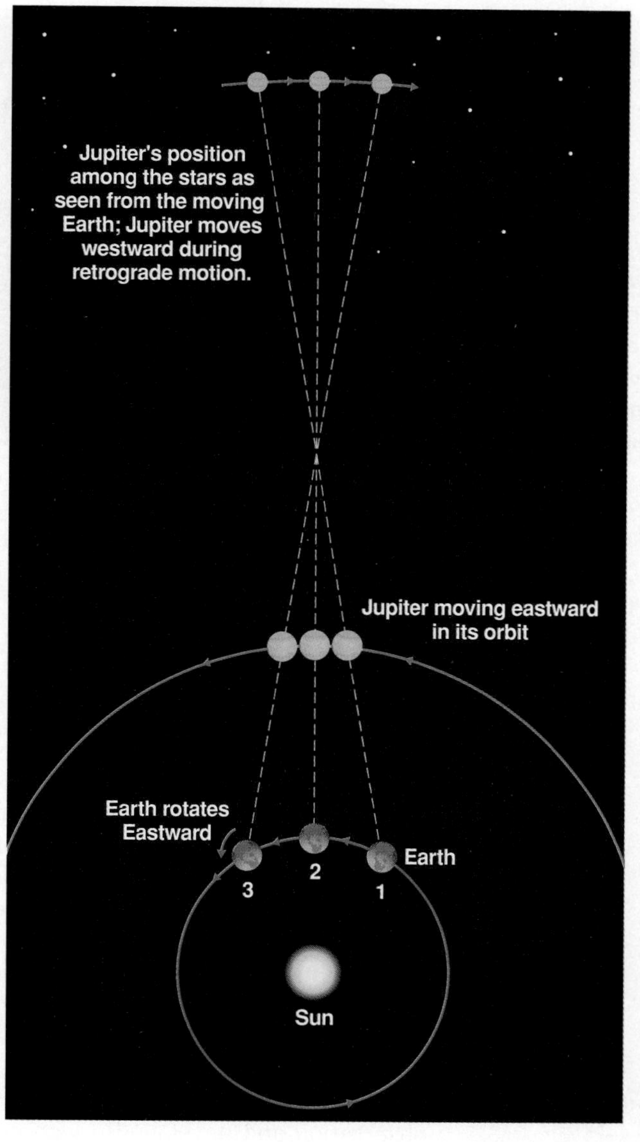

(b)

FIGURE 1-11 (A) A DRAWING CIRCA 1667 OF COPERNICUS'S SUN-CENTERED MODEL OF THE SOLAR SYSTEM. (B) COPERNICUS EXPLAINED THE RETROGRADE MOTION OF THE PLANETS MORE SIMPLY THAN PTOLEMY. EARTH GOES FASTER IN ITS ORBIT AROUND THE SUN THAN JUPITER DOES IN *ITS* ORBIT. AS EARTH PASSES BETWEEN JUPITER AND THE SUN, JUPITER *APPEARS TO* BACK UP (WESTWARD) IN THE SKY RELATIVE TO THE STARS BUT IN ACTUALITY IS STILL MOVING EASTWARD, AS IS THE EARTH.

Because Copernicus's views contradicted the authoritative religious views of that time, he delayed publishing his ideas until just before he died in 1543. Later, Johannes Kepler (1571–1630) found that if he assumed the planets move around the sun in *elliptical* paths rather than circular paths, with speeds that *changed* as they moved along their orbits, he could match their observed positions with impressive accuracy (Figure 1-12). Kepler's findings gave extremely accurate predictions of the planets' locations years in advance. Besides being more accurate, this theory was much simpler; the need for epicycles was over. However, this revelation was not instantly accepted either. In fact, Galileo, one of the first supporters of the findings of Copernicus and Kepler, soon angered the religious authorities.

1564–1727 A.D.
GALILEO AND NEWTON

Galileo was born in 1564, the year of the death of Michelangelo and the birth of William Shakespeare. In his *Dialogue on the Great World Systems*, Galileo argued persuasively that earth orbits the sun. For this he was tried by the Inquisition, forced to recant his beliefs, and died under house arrest in

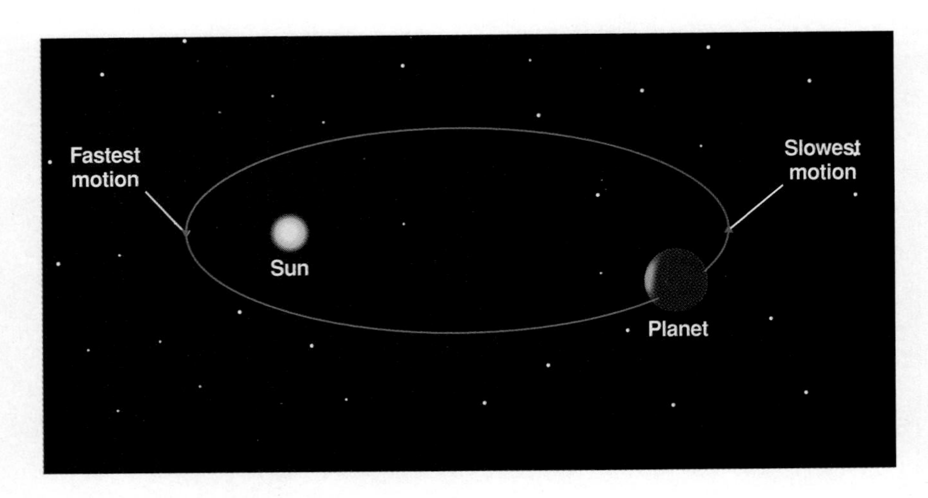

FIGURE 1-12 In Kepler's model, each planet orbits the Sun in an elliptical path, moving faster when it is closer to the Sun and slower when it is farther from the Sun. Kepler predicted the future positions of the planets as accurately as could be observed with the naked eye. These orbits are discussed in detail in Chapter 7.

Galileo and others introduced telescopes and microscopes during the Renaissance, opening the world beyond our eyesight to the scrutiny of scholars. This is one of Galileo's telescopes.

1642, the year Isaac Newton was born. It was Galileo and Newton who put science, and especially physics, on its modern course, as we shall see in later chapters.

SO, WHAT EXACTLY IS SCIENCE?

To begin with, *measurements* are the basis for the sciences. The sciences deal only with measurable quantities, which are also called **physical quantities.** Science deals with nature in **quantitative** terms (which means to *quantify,* or to express in terms of numbers or a formula), whereas experiences such as love, happiness, and philosophical convictions are described in **qualitative** terms (which means to *qualify,* or express attributes with words). Science deals only with the things we can observe and quantify.

Science is our accumulated understanding of the natural world. Though its roots go back to prehistory, its traditions were established only over the last several centuries, beginning with the technique of experimentation introduced by Galileo in the late 1500s. Starting with the assumption that the universe is orderly enough to be understood, scholars established methods of observation and developed rules for reasoning and making predictions. This framework is called the **scientific process** or the **scientific method.**

Briefly, the scientific process most often begins with a careful observation of some phenomenon followed by the formation of a **hypothesis** (which is an educated guess, or an untested theory) to explain what is going on. The hypothesis is then tested by **experiments** that show to some degree whether the hypothesis is supported. If the hypothesis correctly predicts the results of the experiments, *and if the hypothesis is general enough to make new predictions that also prove to be correct,* it may be accepted by scientists as a **theory.** More rarely, an insightful scientist might make a creative leap and formulate a correct theory without much or any experimental evidence as a guide. (We'll see in Chapter 26 that Albert Einstein did just that.) Those theories, too, are accepted only if later experiments show the predictions they make are accurate.

Theories are models of how things behave; they are a way of looking at nature so that we can understand it. A theory of the most general kind, with an extremely wide range of proven validity and application, may be called a **physical law.** These laws are the basic findings of science; they give science its predictive powers. However, even the laws of science are not expected to be the last word on the behavior of nature. *Laws and theories are only our* *best descriptions to date of the things we have observed.* Most scientists presume there will always be a next step to be taken once nature has been understood at a certain level. Writing in the first century A.D., Seneca put it well—"Our universe is a sorry little affair unless it has in it something for every age to investigate . . . Nature does not reveal her mysteries once and for all."

As you can see from all this, there are two essential sides to the pursuit of science: experiments and theories. A scientist (or often a team of scientists) does experiments to carefully check some aspect of how nature behaves, and other scientists accept the results of these experiments only if other independent workers set up experiments and confirm the findings. Modern science accepts no "authorities"; that is, no scientist's work is believed without independent confirmation, as Aristotle's was for so long. (Aristotle, remember, believed that *thinking* about what he observed revealed the truth about nature; he never tested his ideas with experiments.)

IN THE GRAND CANYON EROSION HAS EXPOSED LAYERS OF ROCKS WHICH FORM A GEOLOGICAL RECORD OF THIS AREA. TODAY GEOPHYSICISTS CAN ESTIMATE THE TIMES SUCH LAYERS WERE DEPOSITED WITH RADIOACTIVITY AND OTHER MEANS.

Today's sciences have a highly developed view of the universe. They are **quantitative** (expressed in terms of numbers), **empirical** (their assertions can be checked numerically with observations or experiments), and **theoretical** (models are built to explain what is observed and allow the prediction of other events not yet observed).

SCIENTIFIC DEBATES

Debates are a part of the scientific process, since it often takes time to understand results and come to valid conclusions. Opposing views or even great controversy in the scientific community usually means not enough evidence is available, and more needs to be understood before the answers are certain. Scientific debates, however, even when they appear in newspapers, are not like debates in the national political arena, where opposing viewpoints often come from differences in philosophies, nor like international affairs, where controversy often has to do with cultural or religious differences. Scientific debate is aimed at uncovering truths of nature that can be verified with experiments, and scientists learn from discussing and debating new ideas or experimental results with each other.

A MODEL OF DNA. CHEMISTS UNDERSTAND THE PROPERTIES AND STRUCTURES OF MOLECULES USING THEORIES AND INSTRUMENTS DEVELOPED BY PHYSICISTS.

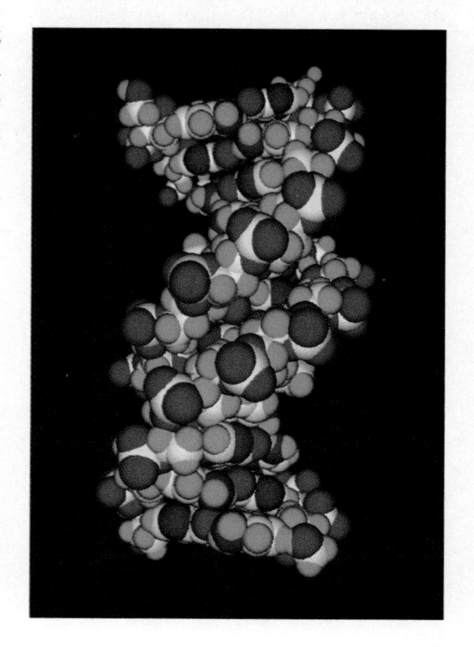

LIFE SCIENCES AND PHYSICAL SCIENCES

Science today consists of two broad areas, the life sciences and the physical sciences, although there is a growing overlap between the two. The **life sciences** include biology, medical science, agricultural science, and ecology, among others. The **physical sciences** include physics, chemistry, geology, astronomy, oceanography, the atmospheric sciences, and more. Both the physical sciences and the life sciences are grouped together under the name *natural sciences*. (The social sciences, economic sciences, and political science use

Using the laws of physics, present-day astronomers and astrophysicists make models that predict such things as the explosions of certain stars. The Crab Nebula is the remains of a star that exploded in 1054 A.D. and was observed in China and by native American Indians, but not, apparently, in Europe.

some of the techniques commonly used in the natural sciences, such as statistical analysis, but they deal with human nature and human activities rather than the nature of the physical universe.) Science is sometimes confused with technology. Science is our understanding of nature, whereas technology is concerned with applying the discoveries of science for practical purposes. Engineers are the *applied scientists* of technology and industry, and they develop both the techniques and the tools for the practical applications of science. Similarly, medical doctors are the applied scientists of medical science.

PHYSICS, THE BASIC SCIENCE

In simple terms, **physics** is *the study of matter and energy and space and time.* Physics is the most basic of the physical sciences because its principles form the foundation of all the others. Consider the science of chemistry, which deals with the interactions of the atoms and molecules that are the building blocks of all matter. Physicists developed the modern theories describing the structure of atoms and how they behave, and these theories provide the framework to calculate what happens in chemical reactions. Geologists use the laws of physics to explain the processes taking place on earth's surface and beneath it. Present-day astronomy depends entirely on physics. Astronomers cannot experiment directly with the stars; instead, they use the principles of physics discovered here on earth to build models of stars, and computers to calculate the mathematical details. When the models make predictions that match the observations, they can be assumed to represent a portion of reality. Besides being the foundation for the other physical sciences, physics is increas-

Atmospheric scientists use physics to model the dynamics of the atmosphere. Together with chemists, some are studying the ozone hole that appears over Antarctica each spring, as seen here in a satellite image.

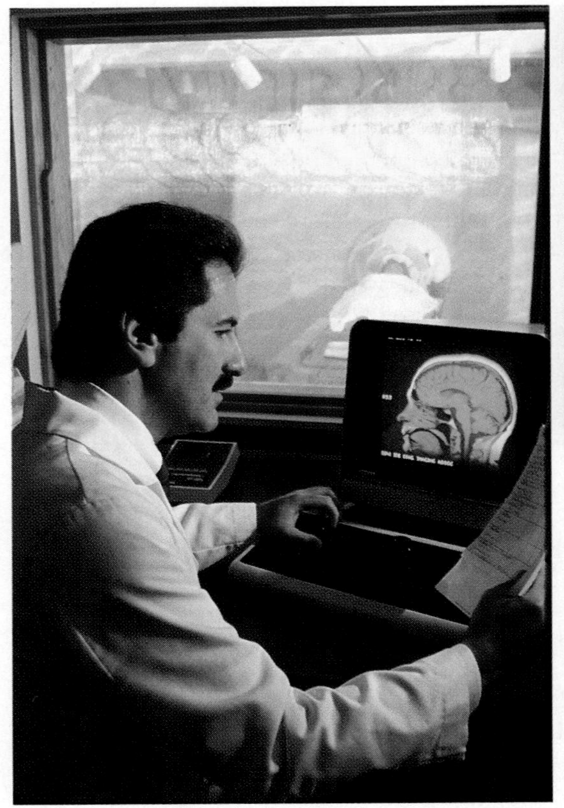

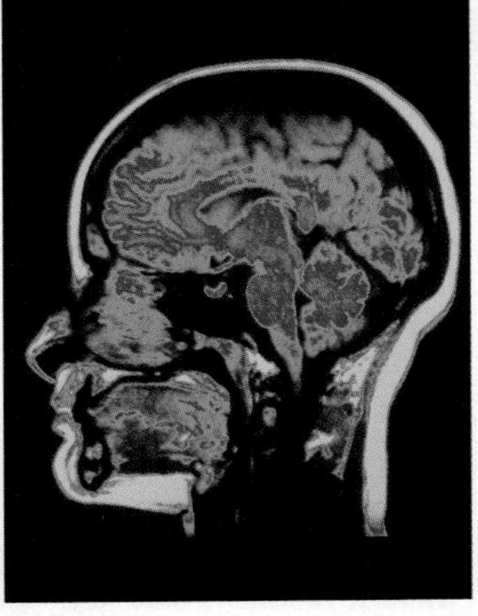

(A) A magnetic resonance imaging machine uses a strong magnetic field to help sense the differences in bones and tissues within the body. (B) A magnetic resonance image of a patient's head. The pituitary gland is shown in pink.

ingly important to the life sciences. Many of the modern instruments and techniques used in medicine and biology are designed by physicists who work in the fields of medical physics and biophysics.

The physics discovered in this century has both revolutionized our understanding of the world we live in and enlarged our view of the entire universe. Today physicists can speculate knowledgeably about how and when the physical universe came into being, as well as its ultimate fate. The science called physics, which is a human creation, encompasses all this!

ABOUT THE PHYSICS IN THIS BOOK

As a quantitative science, physics makes predictions with mathematics, but physics calculations are not the principal goal of this text. Just as music can be enjoyed by people who don't play musical instruments or even read music, physics can be appreciated at many levels. It is, after all, just a human invention, and you can grasp the explanations of physics without "playing the instruments" yourself. In this book the *ideas* and *concepts* are the most important things to learn, rather than the details of the predictions.

Often we will discuss points of physics that relate to the environment or the uses we make of our resources. Sensitivity to environmental issues is directly related to the quality of life we experience; these topics, passages, and exercises are indicated throughout the book by an earth icon as shown in the margin here.

Summary and Key Terms

THE KEY TERMS, SHOWN HERE IN COLOR, ARE DEFINED AGAIN IN THE GLOSSARY.

Science, our *accumulated understanding of the natural world*, began in prehistory when early people realized that some natural phenomena are *predictable*. The ancient Greeks made great strides in ***natural philosophy***, explaining many observations using *reasoning* alone. Ptolemy developed a ***geocentric*** (earth-centered) model for the solar system that accounted for the motions of the sun, moon, and the five planets visible with the naked eye.

In the 1500s, Copernicus developed a ***heliocentric*** (sun-centered) model that explained the retrograde motion of the planets more simply than the Greek model. A ground rule for making models of natural phenomena is ***Ockham's razor***, which states that *if there are two or more possible models that explain the same facts, the one most likely to be true is the simplest one.* Since Copernicus's model was simpler, it was preferable to the Greek model. In the 1600s Kepler's model for the planets was simpler still, and, even more important, much more accurate than the earlier models.

Modern science deals only with ***physical quantities***, which we can express in ***quantitative*** rather than ***qualitative*** terms. The ***scientific method*** begins with careful *observations* of some phenomenon, followed by the formation of a ***hypothesis***, an educated guess that tries to explain the observations. Scientists then test the hypothesis ***empirically*** with ***experiments***. If the hypothesis correctly predicts the results of the experiments and it is general enough to correctly predict behavior not previously observed, it may be accepted as a ***theory***. A theory with an extremely wide range of proven validity and application is called a ***physical law***.

science–2
natural philosophy–4
geocentric–4
heliocentric–4
Ockham's razor–8
physical quantities–10
quantitative–10
qualitative–10
scientific method–10
hypothesis–10
empirically–11
experiments–10
theory–10
physical law–10

life sciences–11
physical sciences–11
physics–12

The *natural sciences* today consists of two broad areas, the ***life sciences*** and the ***physical sciences***, although there is a growing overlap between them. ***Physics***, *the study of matter and energy and space and time*, is the most basic of the physical sciences since its principles are at the foundation of the others.

EXERCISES

Exercises with orange numerals have answers at the back of the text. In every set of exercises there are some where you may want to use a calculator to find the answer. These exercises have calculator icons in front of them for easy identification. If you are assigned these exercises, you will want to study the worked-out examples in Appendix 1 first.

Concept Checks: Answer true or false

1. A theory can be defined as "a possible explanation for observed phenomenon."

2. A theory is valid only so long as it fits all the observations.

3. Scientific knowledge is changeable, subject to revision.

4. A scientific law is developed on the basis of many observations of nature. If further observations reveal facts that do not follow that law, the law must be changed.

5. Scientists expect some theories to change as experiments probe deeper into the nature of things, just as they expect new theories to predict new things to look for.

6. (a) When a scientist publishes the results of a new study, other scientists assume that the results are correct. (b) When you read the results of a new scientific study in a newspaper, you should assume the results are correct.

Applying the Concepts

7. In the sixth century B.C., the famous fabulist Aesop said (in a poetic translation):

"What is discovered only serves to show
That nothing's known to what is yet to know."

Discuss what Aesop may have meant in a philosophical sense.

8. Discuss what the Greek mathematician Pythagoras must have meant when he said, "Everything, to be understood, must be a number."

9. In the fourth century B.C., the Greek historian Xenophon said, "What are called the mechanical arts [present-day engineering] carry a social stigma and are rightly dishonored in our cities." What attitudes held by the ancient Greek philosophers might have caused them to feel that way?

10. Long before the cause of tides was discovered, people learned to predict the tides in a port by recording the times of high and low tides and looking for patterns. Was this what we now call science? What steps did these people take to predict future tides that we now associate with science?

11. Suppose you come home one evening, turn your stereo on, and nothing happens. What would you do? You might think it wasn't getting electricity, and place the plug of its electrical cord into another outlet, or even connect a lamp to its present outlet to see if that's the case. Which of those steps is associated with *observation, hypothesis,* and *experiment?*

12. You turn the key to start your car. It doesn't start and not a sound comes from under the hood. That's your observation. What would be your first hypothesis? How can you experiment to check it out?

13. After a long hard weekend you go to class on Monday and no one is there. Think of three or four hypotheses you might quickly make. How would you try to gather support for each hypothesis?

14. We all function in the framework of previous experience. If you walk outside and find a leaf suspended in mid-air at eye level under a tree, it may go against your sense of what it should do because of your previous experience with things *falling*. What would be your first hypothesis on seeing such a suspended leaf? Would this hypothesis come from your previous experiences?

15. Which of these properly qualifies as a scientific hypothesis? (a) The closest parking lot to your home today has only four-wheeled vehicles in it. (b) Your room is filled with a substance that is undetectable. (c) The temperature at your school next Tuesday will at some time of the day pass through 73°F. (d) The matter around us is made from invisible atoms that are indestructible.

16. In the U.S. House and Senate in 1991, only eight of 535 members listed their professions as engineers, and none were scientists. (Only one of the engineers had a Ph.D, and he was defeated in 1992.) How do you think having scientists in Congress might affect national policy on scientific and technical matters?

17. Imhotep, the Egyptian who apparently designed the first of the Great Pyramids, was subsequently deified and worshipped as a God, in great contrast to the reaction noted in Exercise 9. Would you say that society today holds extreme views like these toward people who work in science and technology? Contrast your own feelings about scientists and engineers with those you hold for professional athletes, politicians, lawyers, and others.

18. How is the study of science different from, say, the study of history? How are they the same?

19. If humans did not exist, would there still be science?

20. Many statements made in astronomy textbooks in the 1960s and 1970s are contradicted by statements made in textbooks today. (The reason? The 80s and early 90s were times of great advances in the science of astronomy, due to space probes and improvements in telescopes and related instruments.) What would your reaction be if you had learned from the earlier texts and picked up a newer one today? Should you have expected that to happen? Should this affect how much you believe of what you read about science today?

21. Science assumes that the universe has a basic orderliness that lets us describe it. Does science also assume that all people perceive things basically in the same way? For example, suppose you met some people who thought that a red shirt was blue. Could you prove that it was the color that should be called red? Should they change their minds if you could produce a hundred other people to look at the shirt and tell them it is red?

22. Give examples to support this statement: How scientific knowledge is put to use through technology depends a great deal on the laws of the society using this knowledge.

23. Albert Einstein, perhaps in response to someone saying that one of his theories seemed too complicated to be true, once said, "Everything should be as simple as possible, but no simpler." Discuss this statement that seems to paraphrase Ockham's razor.

24. In March 1989, two U.S. scientists thought an experiment they did gave proof of "cold" fusion, a process that could have great promise as a new source of energy. However, no other scientists were able to confirm their results with independent experiments. What must other scientists assume about the original claims?

25. After an initial much-publicized report (in 1984) that oat bran reduces blood cholesterol levels, another study (1990) indicated that *any* kind of grain fiber did the same thing, and a third round of studies (1992) concluded that oat bran does have extra benefits. Could you explain to a friend that all three studies are part of the scientific process? (After the first study, Quaker Oats reported a six-fold increase in sales for their oatmeal, once the stuff of childhood tantrums!)

26. Aristotle advanced the idea that sunlight and shadows cause the phases of the moon, as seen in Figure 1-5. Would you say that this statement, which explained the observed phases very well, was a hypothesis or a theory? Why?

27. Early Chinese mathematicians divided a circle into $365\frac{1}{4}$ degrees. Discuss the significance of this fact.

28. From 1924 to 1988, Carlsbad Caverns in New Mexico had a sign above the entrance saying the caverns were 260 million years old. In 1988 the sign was changed to read 7–10 million years old, then in 1990 it was changed again to read 2 million years old. In 1992 the sign was gone entirely! Could you defend the science of geology from these estimates?

29. An Indian tribe of California explained the phases of the moon as being caused by an eagle in the sky spreading its wings, blocking parts of the moon's light. Was this a statement of scientific value? Would it qualify as a hypothesis?

30. How are legal laws or the rules of a basketball game different from laws of science? How are they the same?

31. Today we all still say that the sun "rises" in the east, moves across the sky and "sets" in the west, which is what the ancient Greeks said with their geocentric view of the solar system. Now we know the sun really doesn't go around the earth, that earth rotates and brings the sun up, across, and down in our local skies. Yet it isn't easy to think of a more scientifically correct way to describe "sunrise." Give it a try!

32. In the next chapter we shall discuss rates, which are *ratios* (i.e., $\frac{a}{b}$) that tell how fast something happens. The denominator of such a rate (b above) has units of time. This exercise will acquaint you with rates. (Keep only two significant figures.) The world record for peeling hard boiled eggs by a team of six people is 10,000 in eight hours. Find the average rate at which each person must peel, in eggs per minute. The record time for eggs peeled in one minute is held by a person in that group, 48 eggs/minute. If all six people could have sustained that rate for 8 hours, how many eggs would they have peeled?

33. A student types 66 words per minute on an exam. If the average word is 5.5 letters long, what is the student's average rate of letters typed per second?

34. Each year in the United States, about 2 million high school seniors go on to universities, colleges, and junior colleges, where about 80 percent take a course in the natural sciences. In 1992 there were an estimated 395,000 natural scientists in the United States. Roughly half of these scientists were employed in higher education, and roughly one in five of these taught a freshman course. Calculate what the average class size should be. How does that compare with your present class?

35. The U.S. Post Office projected that in the year 1995 about 116 million household, business, and post office boxes will receive about 185 billion pieces of mail. About how many pieces of mail per day would an average address receive? Would this number be close to your own experience?

36. In 1991 there were about 160 million registered drivers in the United States and about 110 billion gallons of gasoline sold at gas pumps. How many gallons of gasoline did an average driver use? If cars averaged 20 mpg in 1991, how many miles did an average driver drive per day? Per year?

37. U.S. highway statistics show 2.5 deaths in automobile collisions per 100 million vehicle miles traveled. Using numbers from the previous exercise, estimate how many people died in this country in collisions in 1991.

38. Twenty years ago, about 450 million cups of coffee were consumed per day in the United States, by 135 million coffee drinkers. What was the average drinker's daily consumption? In 1992 about 160 million coffee drinkers consumed 280 million cups of coffee per day. Compare this recent cups per day average with the number from 20 years ago. (Such consumption of coffee is not just a modern phenomenon; in 1715, before Isaac Newton died, there were already 2000 coffee houses in London.)

39. In the 1800s, the English magazine, "Punch," ran a cartoon of the reknowned English scientist Michael Faraday inspecting the polluted Thames river. (He was holding his nose!) Discuss why the cartoonist might have put a scientist in this position.

40. For many centuries, a span of about 20 years represented a human generation. That is, a baby, its parents, its grandparents, its great-grandparents, and so forth were all born about 20 years apart. About how many such generations separate you from Galileo (d. 1642) and Newton (b. 1642)? Today, 25 years is a more representative span for a generation. How many generations separate you from Galileo and Newton using this number?

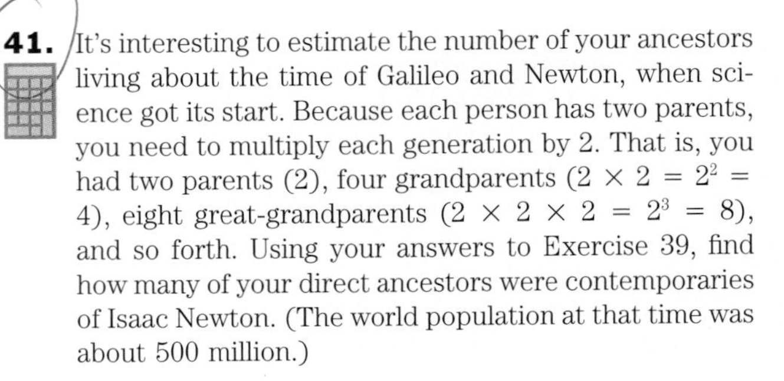

41. It's interesting to estimate the number of your ancestors living about the time of Galileo and Newton, when science got its start. Because each person has two parents, you need to multiply each generation by 2. That is, you had two parents (2), four grandparents ($2 \times 2 = 2^2 = 4$), eight great-grandparents ($2 \times 2 \times 2 = 2^3 = 8$), and so forth. Using your answers to Exercise 39, find how many of your direct ancestors were contemporaries of Isaac Newton. (The world population at that time was about 500 million.)

Demonstration

1. If you sit still and close one eye, what you see looks something like what a camera might record, a flat picture with no depth. With both eyes open, however, you sense depth. Because your eyes are some $2\frac{1}{2}$ inches apart, they see nearby points from slightly different angles. If you blink one eye and then the other, you will see the nearest objects shifting positions against the background (Figure 1-13a). That shift is called *parallax*. Our brains have learned to see things in three dimensions because of the slightly different view each eye sees.

Aristotle pointed out that if the earth circled the sun the nearby stars should gradually move as seen against stars farther away during the course of the year (Figure 1-13b). He argued that the earth could not be in motion around the sun because no one could see parallax with the stars. Today we know that this shifting does occur, although the stars are so distant it cannot be seen with the naked eye. In the 1800s photographs through telescopes revealed parallax for the closest stars, thus proving Aristotle's idea correct some 21 centuries later! Parallax is how you sense depth in the three-dimensional figures in this text.

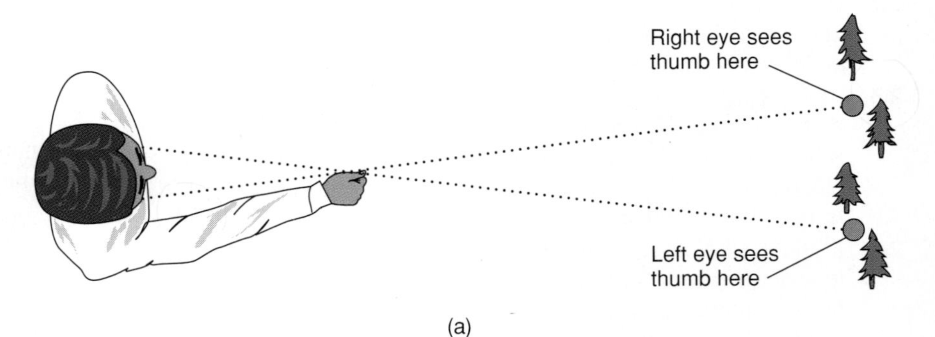

FIGURE 1-13 (A) OBSERVING PARALLAX BY USING THE THUMB AND SHUTTING ONE EYE AT A TIME. WHEN THE PERSON PULLS THE THUMB CLOSER, PARALLAX INCREASES; WHEN THE THUMB IS MOVED FARTHER AWAY, PARALLAX DECREASES. THIS SHIFTING, VIEWED WITH BOTH EYES OPEN, LETS THE PERSON SENSE THAT THE THUMB IS CLOSER THAN THE TREES. (B) PARALLAX OF A CLOSER STAR AS SEEN AGAINST FARTHER STARS FROM TWO POINTS SIX MONTHS APART IN EARTH'S ORBIT.

TWO

Motion

The motion of these flying geese is caught by a camera, which can "freeze" their motion better than our eyes can. Why are their wings usually blurred more than their bodies in this photograph?

The *description* and *prediction* of motion are fundamental to our understanding of nature, from the motions of the planets to our own motions to the motions of atoms and molecules. Aristotle wrote about motion, but he did not accurately describe it or discover how to predict it. So today the ancient Greek philosophers are remembered more for their studies of planetary motion than for their ideas about motions on earth. During the Renaissance Leonardo da Vinci would say, "To understand motion is to understand nature." You take part in motion every time you take a step, bat an eye, or take a breath, so you can relate to motion through your own experiences.

SPEED

We might describe a motion as "fast" or "slow," but to describe it more accurately, we use rates. A **rate** tells how fast something happens, or how much something changes in a certain amount of time. Speed, which is a distance divided by a time, is a rate. Suppose some students decide to jog for 2 miles. They might start out fast but slow down along the way or even stop if one needs to tighten a shoelace, which means they won't travel at the same rate all the time. If they finish in 20 minutes, however, their **average speed** is

$$\textbf{average speed} = \frac{\text{distance covered}}{\text{elapsed time}} =$$

2 miles/20 minutes = **0.1 mile/minute** (read as: point 1 mile per minute)

An average speed, however, tells us nothing about what happened at any point during the run. If we want to know how fast someone is moving at a specific instant or at a specific point in space, we want to know the person's **instantaneous speed.** For example, if you say, "At twelve noon my car was moving at 35 mph," or, "When the bicycle hit the bump, its speed was 12 mph," then you've specified an instantaneous speed.

WE USE *V* FOR SPEED SINCE SPEED IS A PART OF AN OBJECT'S VELOCITY, WHICH IS ITS SPEED AND DIRECTION TAKEN TOGETHER. VELOCITY IS DISCUSSED LATER IN THIS CHAPTER.

To make the definition of speed easy to use, we use d for the distance, t for the elapsed time, and v for the average speed. Now the definition for average speed becomes

PUTTING IT IN NUMBERS—AVERAGE SPEED

$$v = \frac{d}{t}$$

Whenever something moves, this formula gives the relationship between its average speed and the distance it moves in an amount of time. To use this formula to find the distance you'll travel in a certain time if you know your average speed, rearrange it to read $d = vt$. Or, to find the time you'll need to cover a certain distance at a certain average speed, use $t = d/v$.

Example: Suppose you know that you'll average 60 mph on a trip along an interstate highway and that your destination is 150 miles away. Calculate the time you'll need to drive that distance.

$$t = \frac{150 \text{ miles}}{60 \frac{\text{miles}}{\text{hour}}} = \frac{150}{60 \frac{1}{\text{hour}}} \cdot \frac{\text{hour}}{\text{hour}} = \textbf{2.5 hours}$$

UNITS OF LENGTH AND TIME

We measure length and time by comparing them to standard units that were defined with convenience in mind. Two systems of measurement, the British and the International System, are used in the United States today. The British system has a quaint history. An inch was originally the distance between the knuckles of the thumb of King Edgar (A.D. 959–975), and the yard was the distance between the outstretched fingers and nose of Henry I (A.D. 1100–1135). A system called the metric system was set up in France in the 1790s. After scholars carefully estimated the distance between the north pole and the equator, they divided that distance by 10 again and again until they came to a convenient length for everyday use. After seven divisions by 10, they reached a length equal to 39.37 inches, only a little over a yard in the British system, and that became the meter. (The European spelling is metre.) Modifications of the metric system in 1960 led to the International System, abbreviated SI from the French *Systeme International.* This system is used today in all countries for scientific purposes. SI units are related to each other by powers of 10, which is very convenient in many calculations. The meter is equal to 100 of the smaller units called centimeters, and for larger distances we use 1000 meters, the kilometer (pronounced "KILL ometer").

Both the British and the International systems use the second as the unit of time. Originally the second was defined in terms of an average day (24 hours, with 60 minutes in an hour and 60 seconds in a minute). Today the motions of electrons in atoms define the standard time unit for precise measurements. The modern second flits by in just 9,192,631,770 oscillations of a certain electron in a cesium atom, and atomic clocks tick off each second accurately to all those digits. Likewise, the meter is defined in terms of the wavelengths of orange light given off by an electron in a krypton atom. For everyday purposes, however, ordinary clocks, tape measures, and meter sticks do just fine. In this text, we use SI units much of the time, but we will often give quantities in the familiar British units as well.

A short list of the abbreviations of units of measurement:

second, s	meter, m	foot, ft
hour, h	kilometer (1000 m), km	inch (1/12 foot), in
year, yr	centimeter (1/100 m), cm	yard (3 feet), yd
	millimeter (1/10 cm), mm	mile (5280 feet), mi

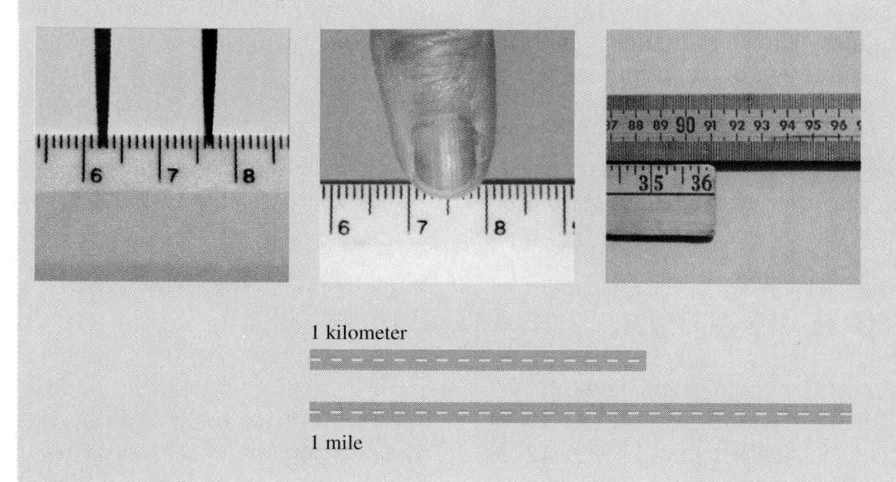

THE END OF A TOOTHPICK IS ABOUT A MILLIMETER ACROSS, WHILE ITS CENTER IS BETWEEN TWO AND THREE MILLIMETERS ACROSS.

A CENTIMETER IS ABOUT THE DISTANCE ACROSS THE FINGERNAIL OF YOUR SMALLEST FINGER.

THE YARDSTICK AND METER STICK ARE ALIGNED AT THEIR OTHER ENDS. YOU CAN SEE FROM THESE ENDS THAT A METER IS ABOUT 9 PERCENT LONGER THAN A YARD.

ONE MILE IS 1.609 KILOMETERS, AND A KILOMETER IS 0.6215 MILE. (THE RELATIVE LENGTHS ARE TO SCALE HERE, BUT NOT THE WIDTHS OF THE ROADS.)

ABOUT MEASUREMENTS AND ERRORS

Speed is a rate that tells how fast something moves, but to find a speed requires *measurements* of distance and time (Figure 2-1). Any measurement of a length or a time always has an *uncertainty* in it. Human judgment is sometimes a factor. Give a tape measure to a few of your friends and ask each of them to measure the length of a table very carefully. You'll get a variety of answers because each person must *estimate* where the table's edges fall between the finest marks on the tape. If you were timing how fast a swimmer covered a pool's length, you could use a stopwatch, but your reaction time would introduce an error. Your eyes detect the swimmer's finish, your brain signals your thumb, and your thumb contracts to stop the watch. Typically this takes at least 1/7

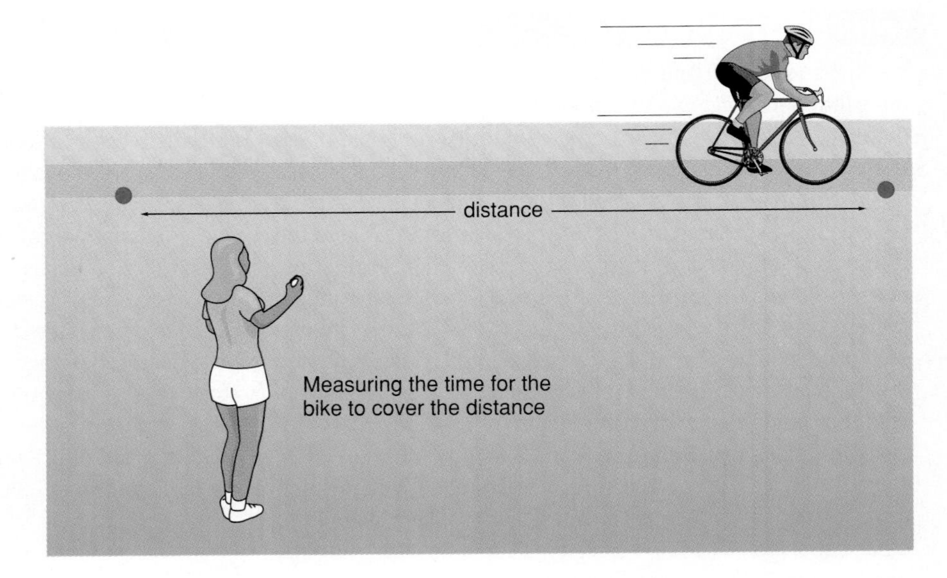

FIGURE 2-1 TO MEASURE SOMETHING'S AVERAGE SPEED, YOU MUST MEASURE BOTH THE DISTANCE IT COVERS AND THE TIME IT TAKES TO COVER THAT DISTANCE.

second, but the error can be less than that if you anticipate when the swimmer will touch the end of the pool. Musicians use a similar anticipation to perform—by listening to the beat of the music, they anticipate when to play each note.

Don't confuse "measuring" with "counting." You can count the pennies in a jar and have an exact answer because there is a definite number of pennies. In contrast, distance and time are quantities that (as far as we know) can be divided indefinitely, which means they don't come in definite amounts like pennies do. We call such quantities **continuous.** Errors are a part of any measurement of a continuous quantity, but the word "error" doesn't mean the experimenter has made a mistake or a blunder. Even when electronic equipment makes measurements of distances and times, the devices still must compare one continuous quantity with another, and no such comparison can be perfectly accurate. Though the errors may be very small, they cannot be zero.

MORE ABOUT INSTANTANEOUS SPEED

FIGURE 2-2 STARTING FROM REST, SPRINTERS ACCELERATE, GAINING SPEED.

A moving bike has a speed at every instant during its motion, but an *instant* is merely a point in time, such as 12 noon. It is no time interval at all, not even the smallest fraction of a second. At some instant, then, *there is no time for something to move,* to change its location. You cannot measure anything's speed at a single instant; there must be a time interval and a distance covered.

To *estimate* a bike's instantaneous speed at some point, you could draw two lines very close together around that point and find the bike's average speed between those marks. The bike's speed cannot change much over that small distance, so that its average speed is a good *approximation* of its instantaneous speed at that mark. Not even police radar can measure instantaneous speed exactly, as we'll see in a later chapter.

ACCELERATION

If a car you are riding in moves at a steady speed and the road is straight and smooth, your ride is very comfortable. As a passenger, you could pour a cup of coffee and drink it as easily as if you were sitting still beside the road; a baby could sleep in its carseat unaware of any motion. Even on a smooth, straight road, however, traffic and traffic signals cause drivers to change speeds. Any change from a constant speed affects you—you shift backward or forward in your seat, and if there is a cup of coffee aboard it sloshes about. If the speed changes slowly, you hardly notice it, but a quick change in speed is obvious. *How fast* speed changes is another rate—the rate of change of speed. We call this rate **acceleration.** The average acceleration along a straight line is the change in speed divided by the time required for that change, or

JUST AS FOR SPEED, THE INSTANTANEOUS ACCELERATION TELLS HOW FAST THE SPEED IS CHANGING AT AN INSTANT IN TIME OR A

PUTTING IT IN NUMBERS—AVERAGE ACCELERATION

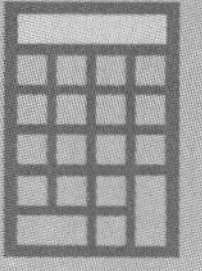

$$\textbf{average acceleration} = \frac{\text{change in speed}}{\text{time required}} = \frac{\text{final speed} - \text{initial speed}}{\text{time required}}$$

$$a = \frac{v_f - v_i}{t}$$

Example: A woman on a mountain bike steps hard on the pedals and goes from 2 meters per second (about 6 ft/s) to 4 meters per second (about 13 ft/s) in 2 seconds. The average acceleration is a = (4 m/s − 2 m/s)/2 s = (2 m/s)/2 s = (1 m/s)/s, which is abbreviated as **1 m/s²**.

Acceleration is a change in speed over time, so if something slows down it is also accelerating. Suppose a windsurfer slows from 10 m/s to 4 m/s in 4 seconds. The windsurfer's change in speed is the final speed (4 m/s) minus the initial speed (10 m/s), or −6 m/s. The average acceleration is then a = (−6 m/s)/4s = −1.5 m/s². A *negative* value of the acceleration means the speed is *decreasing*. Sometimes a negative acceleration is called a deceleration.

You know about acceleration from experiences you've had since childhood. A new baby picked up or put down too quickly stiffens and throws out its arms and legs, frightened by the sudden unfamiliar acceleration. If it is asleep in a carseat, a sudden change in the car's speed will wake it. As the baby grows, it learns to take such accelerations in stride. If you watch a movie filmed from a roller coaster or a race car, you will see how well you've been trained. At all the appropriate places, your mind *expects* accelerations, and you feel them in the pit of your stomach even though your stomach isn't actually accelerating.

THE ACCELERATION OF GRAVITY

As children we all learn that most things accelerate toward the ground if we let them go in midair, and we use the word "gravity" to describe this behavior. Galileo was the first to understand how gravity affects things near the surface of our planet. As mentioned in Chapter 1, in Galileo's time most scholars still accepted Aristotle's assumptions about motion, and Aristotle thought that a heavier object would fall toward the ground faster than a lighter object. Since heavier bodies make more noise and larger dents when they strike the ground, Aristotle's idea was easy to believe—and besides, we all know that some objects *do* behave that way. Feathers and small bits of paper accelerate more slowly toward the ground than books or baseballs.

Aristotle's ideas went untested until Galileo introduced experimental pro-

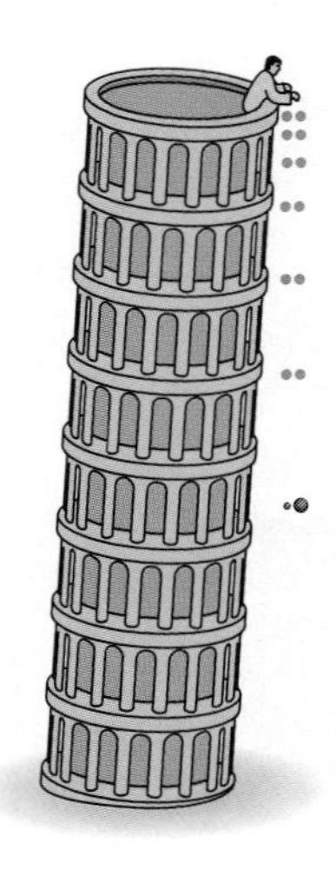

FIGURE 2-3 GALILEO WROTE ABOUT DROPPING TWO SIZES OF HEAVY METAL BALLS FROM THE LEANING TOWER OF PISA. INFLUENCED LITTLE BY AIR RESISTANCE, THEY WOULD REACH THE GROUND AT ALMOST THE SAME TIME, DESPITE THEIR DIFFERENT WEIGHTS.

cedures and actually measured accelerations. From his experiments (Figure 2-3), Galileo argued that if objects having different weights fell "totally devoid of resistance," that is, without air (or anything else) to interfere with their downward motion, they would fall with the *same* acceleration. A rock and a leaf would reach the same speed if they fell the same amount of time. Although Galileo didn't have the means to eliminate air resistance and prove his deduction, his conclusion was correct. This acceleration toward earth is given the symbol ***g*** ("gee"), and its value is about 9.8 meters/second/second, or about 32 feet/second/second. In other words, the speed of a freely falling object increases by 9.8 meters/second (or by 32 feet/second) for each second that it falls.

Galileo's experiments proved a formula he had derived from the definitions of average speed and acceleration (see Appendix 2). That is,

PUTTING IT IN NUMBERS—DISTANCE MOVED IN FREE FALL FROM REST

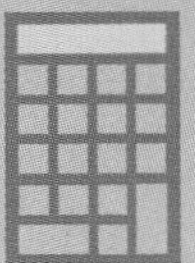

the distance *d* an object moves from rest in a time *t* with a constant acceleration *a* is equal to $\frac{1}{2}at^2$.
If the acceleration is due to gravity, then $a = g$, and

$$d = \frac{1}{2}gt^2$$

Example: An apple falls from its tree to the level ground in 0.7 s. How high from the ground was it growing? $d = 1/2\ (9.8\ \text{m/s}^2)(0.7\ \text{s})^2 = (4.9\ \text{m/s}^2)(0.7\ \text{s})(0.7\ \text{s}) = (4.9\ \text{m/s}^2)\ (0.49\ \text{s}^2) =$ **2.4 m,** or 7.9 ft.

Not long after Galileo's death, Isaac Newton used the newly invented vacuum pump to give the first visual proof of Galileo's deduction. After taking most of the air out of a tall glass tube, he released a gold coin and a feather at the top of the tube. They landed together. Figure 2-4 shows a modern version of Newton's experiment. Apollo astronauts demonstrated the same thing with a hammer and a feather on the surface of the moon. There, with no atmosphere to affect them, the hammer and feather accelerated side by side, only at a much lower rate than on earth because the gravitational acceleration at the moon's surface is only 1/6 of the value *g* at the earth's surface.

FIGURE 2-4 A FEATHER AND AN APPLE ACCELERATING SIDE-BY-SIDE IN A PARTIALLY EVACUATED GLASS TUBE. WITHOUT AIR RESISTANCE, ALL OBJECTS HAVE THE SAME ACCELERATION NEAR EARTH'S SURFACE.

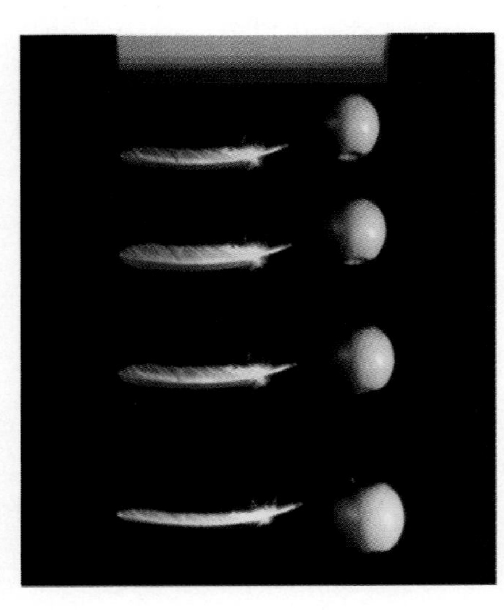

FREEZING MOTION

If you pass your hand quickly in front of your eyes, it sweeps past much too fast for your eyes and brain to "stop" the motion. However, cameras stop such motion by exposing film in a very short interval. An exposure time of 1/1000 of a second would catch a sharp image of your moving hand, and the shutters of many cameras blink open for such short exposures. To "freeze" faster motions requires even shorter exposure times, however. Special electronic flashes make very brief exposure times possible. The camera's shutter is left open, but the diaphragm, which lets the light in, is almost closed so that under normal lighting not enough light enters to expose the film. An electronic flash then floods the scene with a brief blaze of light. A typical camera flash lasts about 1/50,000 of a second, but custom electronic flashes deliver bursts of light in a millionth of a second. During those brief periods of intense illumination, the film is exposed, freezing almost any motion.

An electronic flash that triggers at regular time intervals is called a *stroboscope,* or *strobe*. When a strobe illuminates someone moving in a darkened room, a camera with its shutter open records the motion (Figure 2-5). Someone walking by at a constant speed appears as a row of images on the film, all the same distance apart. Something falling accelerates downward, and as it gains speed the successive images are farther and farther apart. On such photos you could measure the distances between the images with a ruler and calculate speeds and accelerations.

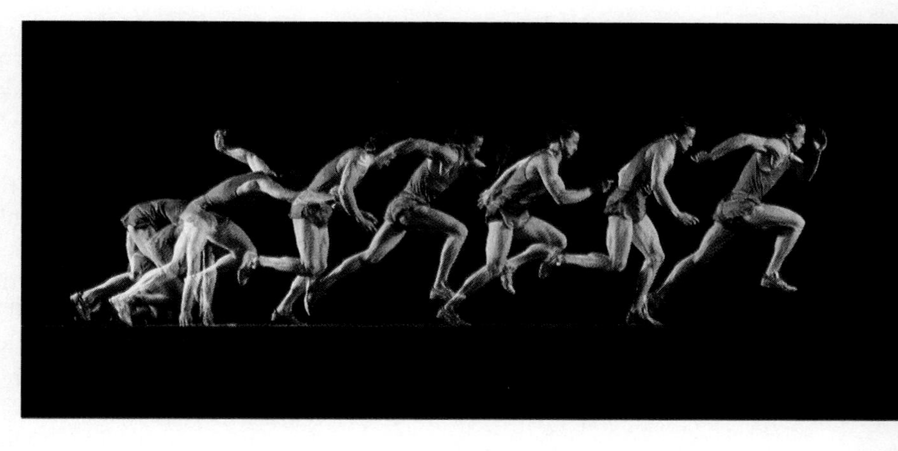

FIGURE 2-5 A STROBE PHOTO CATCHES A RUNNER ACCELERATING QUICKLY FROM THE STARTING BLOCKS.

A television is a strobe-like device. Thirty different "still" pictures ("frames") flash on the screen each second. Your eyes and brain cannot "stop" an individual frame; instead they interpret the quick sequence of pictures—which are not very different from one to the next—as a scene with real-time motion, a live performance. Movie theaters project only 24 frames per second, but each frame flashes up twice, so that the "motion" you see seems very smooth because 48 pictures per second pass before your eyes. Early movies showed from 16 to 12 or even fewer frames per second, and people saw a flickering motion on the screen—which is why they were called the "flicks."

AN EXAMPLE OF ACCELERATED MOTION DUE TO GRAVITY

To see how gravity affects motion, just take your keys and toss them straight up, as in Figure 2-6. You give them a certain initial speed and they move upward. As they do, gravity slows them down; it decelerates them. (At such slow speeds we can neglect the very small affect the air has on the keys' motion; their downward acceleration is g.) When the keys finally reach a speed of zero, they have come to the highest point in their path. How long is the speed zero? No time at all. Because the downward acceleration g is constant (earth's gravity never turns off), their speed *changes constantly*. Even at the top of their path, the keys accelerate downward; their speed is zero only instantaneously. They may *seem* to pause because they are moving so slowly there, and our eyes and our brains form a clearer image of them. Then the keys fall, gaining speed as they accelerate all the way down at the rate of g.

FIGURE 2-6 THE PATH OF KEYS IN FLIGHT. THE POINTS SHOW WHERE THE KEYS' MOTION WOULD BE FROZEN BY CONSECUTIVE FLASHES OF A STROBE. THE POINTS THAT ARE FARTHER APART SHOW THAT THE KEYS ARE MOVING FASTER AND COVERING MORE DISTANCE OVER THE SAME TIME INTERVAL THAN WHEN THE POINTS ARE CLOSE TOGETHER.

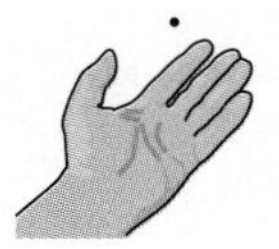

RELATIVE SPEED

Biking with a friend, you might ride side by side so you could talk. You would both move over the ground, but the distance between your bikes would change very little as you rode along. Since your friend and you weren't moving much toward or away from each other, you could say there is little *relative* motion between you. The **relative speed** between two objects measures *how fast the distance between those objects is changing*. Your friend's speed relative to a signpost beside the road might be 15 mph, but relative to you, your friend's speed is almost zero. Every motion of an object—a bike, a river, a planet—is relative to some other object.

The relative speed between two objects moving in the same direction is just the *difference* in their speeds over the ground. Suppose you are driving a car at 50 mph relative to the ground while a slower car ahead of you moves in the same direction at only 30 mph relative to the ground. You approach that car with a relative speed of 20 mph. Objects along the road zip by your car at 50 mph, but the motion between yourself and the slower car in front is as if it were standing still and you were approaching it at 20 mph.

The relative speed between cars traveling in *opposite* directions is the *sum* of their speeds over the ground. If your car is moving at 50 mph and an approaching car travels at 40 mph, the two approach each other with a relative speed of 90 mph. This is why head-on collisions between moving automobiles are so devastating, even when neither is exceeding the speed limit.

INDEPENDENCE OF MOTION IN PERPENDICULAR DIRECTIONS

If you toss your keys straight up, they rise and fall along a straight line, and you need only one number (or coordinate), the keys' elevation, to locate their position along that line. You need two numbers to locate a rowboat on a lake's surface, however. Notice that the rowboat in Figure 2-7 could move north or south on the lake without moving east or west at the same time, or it might go due east or west without changing its north–south coordinate. There are two *independent* directions of motion on a lake's surface, and motion in one direction does not mean any motion has to take place in the direction perpendicular to that one. A fish below that rowboat on the lake might rise or sink as well as travel north–south or east–west, so it has three independent directions of motion.

To see the independence of motion in perpendicular directions, look at

PHYSICS IN 3D/LONGITUDE, LATITUDE, AND ELEVATION

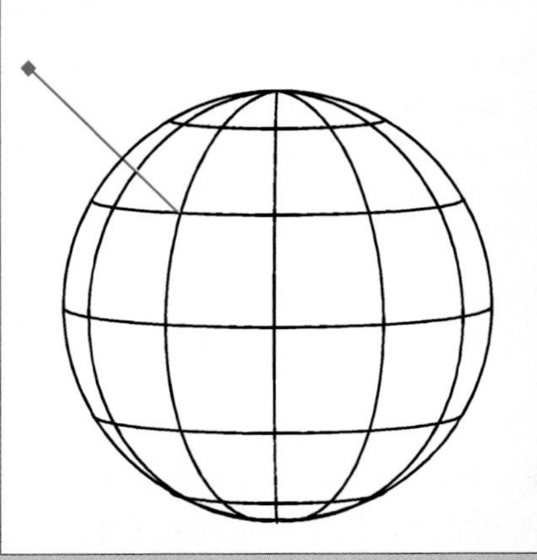

AN EARTH SATELLITE MUST BE LOCATED IN SPACE WITH 3 COORDINATES, SUCH AS ITS LONGITUDE AND LATITUDE AND ITS ELEVATION ABOVE EARTH'S SURFACE, AS SEEN HERE.

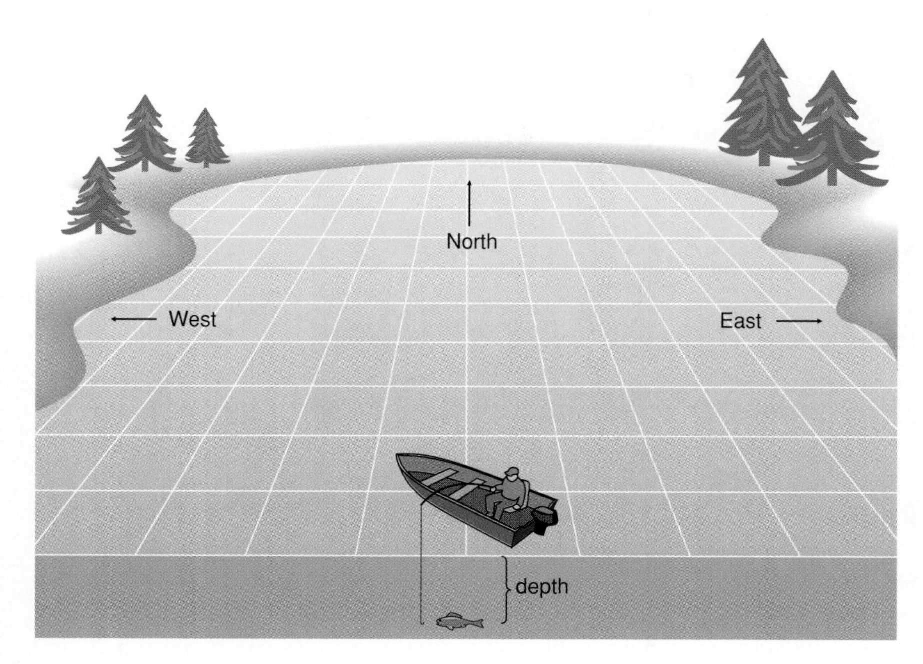

FIGURE 2-7 A GRID OF NORTH–SOUTH AND EAST–WEST LINES FURNISHES A WAY TO PINPOINT A BOAT'S LOCATION ON A LAKE. A THIRD COORDINATE, DEPTH, ESTABLISHES THE LOCATION OF A FISH.

the strobe photograph in Figure 2-8. The two balls start at precisely the same time from the same elevation. One simply falls from rest and moves straight down. The other has some initial speed in the horizontal direction so it moves to the side as it falls. Notice that, flash after flash, the balls have the same coordinate in the vertical direction. Why? Because gravity accelerates both balls downward at exactly the same rate. Notice how the ball with the horizontal motion moves to the side. Between flashes its horizontal movements are equal in length, meaning its speed in the horizontal direction is constant. Gravity accelerates both balls downward exactly the same, but it does *not* affect the speed this ball has in the horizontal direction. Experiments like this show us that *motions in perpendicular directions are independent of one another.*

Since there are three independent directions of motion in space, we say that space is three-dimensional, while a plane or other surface such as earth's surface is two-dimensional. (Any point on earth's surface can be located by two coordinates, its latitude and longitude.) Motion along a straight line takes place in only one dimension, however.

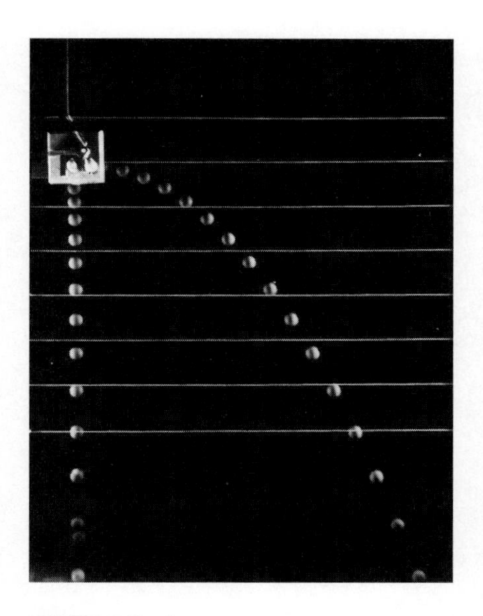

FIGURE 2-8 A STROBE PHOTOGRAPH OF TWO BALLS THAT BEGIN TO FALL AT THE SAME INSTANT. THE BALL ON THE RIGHT HAS AN INITIAL SPEED IN THE HORIZONTAL DIRECTION, WHICH REMAINS CONSTANT.

VELOCITY: SPEED AND DIRECTION

If you are jogging along a winding trail, you change directions as you move, taking the path's direction with each step. At any point you could draw an arrow at your position to show the direction of your motion, and the length of the arrow could represent your speed (Figure 2-9). Such an arrow is your **velocity,** and velocity is an example of a *vector quantity.* A **vector** is an arrow drawn to represent a quantity that has a direction. Vectors provide a useful way of looking at motion in more than one dimension.

The symbol for velocity is $\vec{v}$**,** where the arrow tells us velocity is a vector quantity. The velocity vector of a moving object points in its direction of travel at every instant, while the length of the arrow represents the speed v of the object. If you move along a curved path, your velocity vector points along that path, or is *tangent* to that path, at any position, as Figure 2-9 shows.

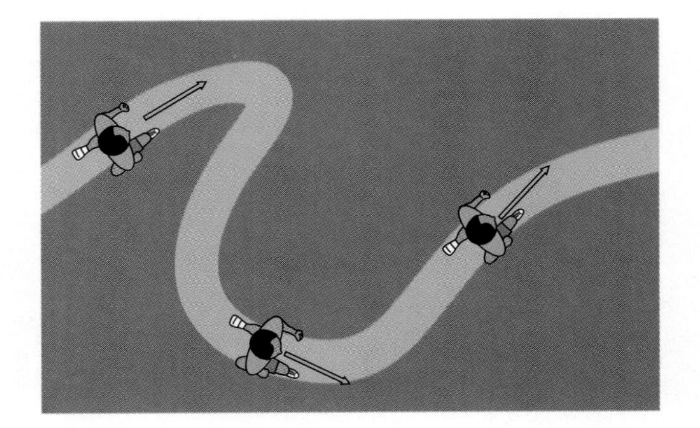

FIGURE 2-9 AS A RUNNER FOLLOWS THE WINDING TRAIL, HIS VELOCITY VECTOR POINTS ALONG (OR TANGENT TO) THE DIRECTION OF THE TRIAL WHEREVER HE IS. THE LENGTH OF THE VELOCITY VECTOR REPRESENTS HIS SPEED AT ANY INSTANT.

ADDING VELOCITY VECTORS: COMPONENTS

Suppose a woman paddles a canoe in a river, her steady strokes giving it a speed of 3 mph through the water directly toward the far bank, as in Figure 2-10a. The water she paddles through is also moving, however, perpendicular to the direction in which she is paddling. It carries her downstream at 1 mph, so that she has *two independent velocities.* Figure 2-10b shows what happens. Her actual, or *net,* velocity as seen from the bank can be found by sliding these two vectors together tail-to-tip without changing their directions or lengths. Her **net velocity** is a vector drawn from the tail of the first velocity vector in the sequence to the tip of the second velocity vector. We say the two independent velocities are *added* to find her net velocity. The length of this **net vector** represents her speed, and its direction shows her actual direction of motion. So she travels across diagonally with a speed that carries her both downstream at 1 mph and crossriver at 3 mph, and those two independent velocities are called **components** of her net velocity. A careful measurement on Figure 2-11 shows her speed is about 3.2 mph.

Suppose you saw a woman canoeing just like this in a river. How could you determine her velocity components? If you walked along the bank, as in

FIGURE 2-10 (A) THE CANOEIST GIVES HERSELF A VELOCITY OVER THE SURFACE TOWARD THE OPPOSITE BANK, BUT THE STREAM MOVES PARALLEL TO THE BANK. SHE MOVES ALONG THE DIRECTION OF THE NET VELOCITY, WITH A SPEED REPRESENTED BY THE LENGTH OF THE NET VELOCITY VECTOR. (B) ADDING THE TWO VELOCITIES TO FIND HER NET VELOCITY. THE VECTORS ARE PLACED TAIL TO TIP *IN ANY ORDER*, AND THE NET VECTOR GOES *FROM* THE TAIL OF THE FIRST VECTOR IN THE SEQUENCE *TO* THE TIP OF THE LAST VECTOR.

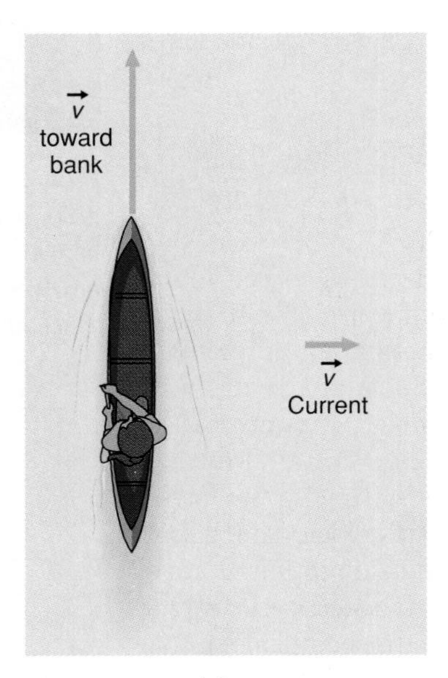

(a)

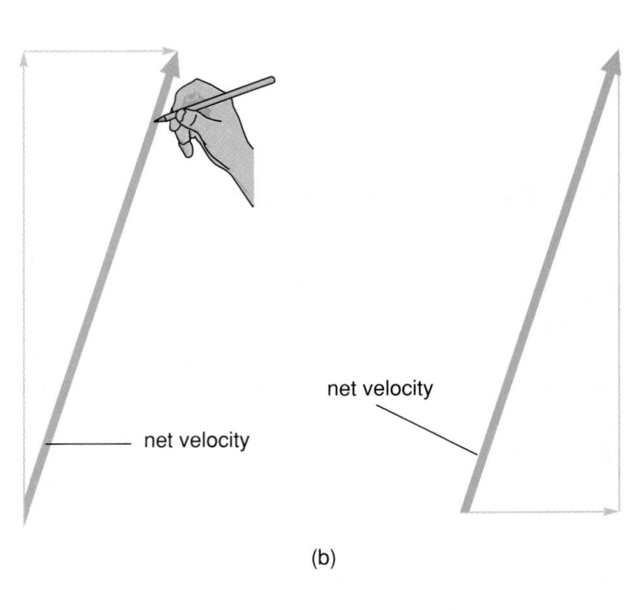

(b)

FIGURE 2-11 GRAPHICALLY FINDING THE NET VELOCITY. IF YOU DON'T HAVE A RULER, CUT AN EDGE FROM A SHEET OF GRAPH PAPER AND USE IT TO MEASURE ANY DIAGONAL VECTOR.

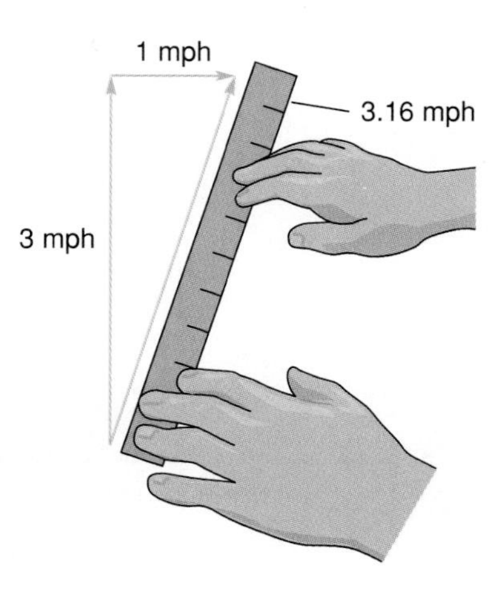

FIGURE 2-12 TWO PEOPLE FINDING THE PERPENDICULAR COMPONENTS OF A CANOEIST'S NET VELOCITY.

Figure 2-12, keeping pace with her as she moved downstream, your speed would match the downstream component of her velocity. A friend on a footbridge that goes straight across the river could keep pace with her along that perpendicular (and independent) direction to find her cross-current speed.

ACCELERATION IN TWO OR THREE DIMENSIONS

An **acceleration** can be either a change in speed, a change in direction, or a change in both speed and direction at the same time. That is, *a vector acceleration* is the *rate of change of the velocity vector.*

You can change a car's velocity in either of two ways: by changing its speed (which means you change the length of the velocity vector) or by changing its direction (which means you change the direction of the velocity vector). To gain speed, you press the accelerator pedal; to lose speed, you use the brakes. To change the velocity vector's direction, you turn the car to the right or the left. Because you can change its direction without changing its speed, these are *independent* ways to change its velocity.

Because acceleration is a vector, it can be separated into components as we did for the velocity vector. The component of acceleration that changes only the speed points in the car's direction of travel (to increase the speed) or opposite its direction of travel (to decrease the speed). The component of acceleration that turns the car left (or right) points to the left (or right) of the car's direction of travel, perpendicular to the velocity vector. Airplanes have *three* independent components of acceleration, since a pilot can also turn a plane's velocity vector upward or downward (Figure 2-13).

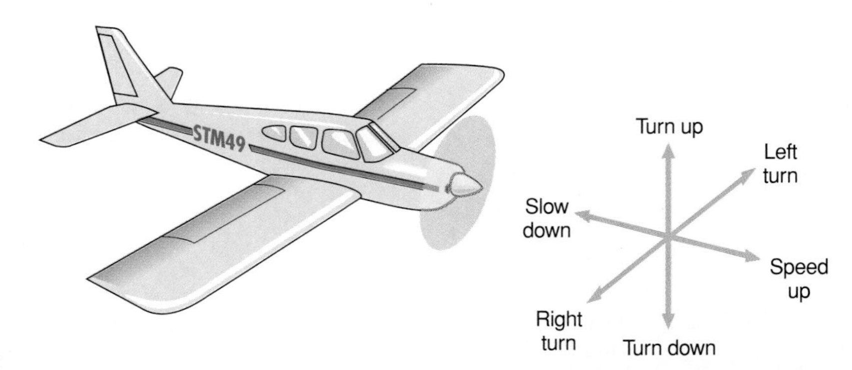

FIGURE 2-13 AN AIRPLANE CAN ACCELERATE IN THREE INDEPENDENT DIRECTIONS.

Summary and Key Terms

rates–18
average speed–18
instantaneous speed–18
continuous quantity–20

To describe motion accurately we use ***rates,*** which tell how fast something changes in a certain amount of time. Speed is a rate. ***Average speed*** is the total distance covered/time spent, or $v = d/t$. ***Instantaneous speed*** is the speed something has at a specific *instant* in time or point in space. Finding a speed requires measurements of distance and time, which are ***continuous*** (divisible without limit) ***quantities***. A measurement of continuous quantities always involves some inaccuracy.

relative speed–23

Every motion of an object is measured relative to some other object. The ***relative speed*** between two objects measures *how fast the distance between those objects changes.* If the objects move in the same direction, the relative speed is the difference in their speeds; if they move in opposite directions, the relative speed is the sum of their speeds.

acceleration–21
average acceleration–21
g–22

Acceleration along a single direction is the *rate at which speed changes.* The ***average acceleration*** along a straight line is the change in speed divided by the time required for that change, or $a = (v_f - v_i)/t$. Galileo argued, and Newton later proved, that, without air resistance, all objects fall toward the earth with the same constant acceleration, g, which has a value of about 9.8 m/s^2 or 32 ft/s^2. The distance an object moves if its acceleration is constant and it starts at rest is given by $\frac{1}{2}at^2$; if the acceleration is due to gravity, then $d = \frac{1}{2}gt^2$.

vectors–25
velocity–25

The motions of an object in perpendicular directions are independent of one another. Space is three-dimensional, a surface is two-dimensional, and a straight line is one-dimensional. We describe motion in more than one direction with ***vectors,*** arrows to represent quantities with a direction as well as a magnitude. ***Velocity,*** which is speed and direction together, is a vector quantity. Acceleration, a, is also a vector quantity, representing the rate of change of the velocity vector.

component–26
net velocity–26
net vector–26

If an object's motion has two or more ***component*** velocity vectors, we can find the ***net velocity*** by *adding* the vectors. To add two or more vectors to form a ***net vector,*** place the individual vectors tail to tip without changing their directions or length. The net vector is the one that connects the tail of the first vector in this sequence of joined vectors to the tip of the last vector in the sequence. The length of a net velocity represents the object's actual speed, and its direction shows the object's actual direction of motion.

EXERCISES

Concept Checks: Answer true or false

1. A rate shows how fast something happens; it is a quantity of something (such as a distance) divided by a time interval.

2. For ten minutes on an interstate highway a car's speed is absolutely constant. For that ten minutes, the car's *average* speed was equal to its *instantaneous* speed.

3. If you measure a continuous quantity, you compare it to a standard quantity.

4. Measurements of continuous quantities always have errors.

5. An instant in time can be said to be no time at all.

6. Since speed is a rate, an acceleration is the rate of change of a rate.

7. In a vacuum on earth, the speed of any freely falling object increases by 9.8 meters per second or by 32 feet per second for each second that it falls. The same is true for the moon.

8. On earth, air resistance accounts for the various accelerations we see as different things (such as a book and a sheet of paper) fall. Although Galileo understood this, Newton was the first to demonstrate it by dropping objects in a partial vacuum.

9. If you throw your keys straight up, their instantaneous speed at the very top of their path is zero and so is their acceleration.

10. The relative speed between two objects moving in the same direction is the sum of their speeds relative to the ground.

11. A vector may be built up from several vectors or separated into several component vectors.

12. To add two vectors, you must first, without changing their directions or length, place them tail to tip. The net vector then points from the tail of the first vector in the line of vectors to the tip of the last vector.

13. An acceleration of an object in one direction always influences the speed the object might have in a direction perpendicular to that acceleration.

Applying the Concepts

14. Label the following quantities as exact or continuous: (a) the number of paragraphs in this book, (b) the number of people you will meet in your lifetime, (c) the distance across a page of this book, (d) the number of square feet in a house, (e) the time of one orbit of the earth around the sun, (f) the number of leaves on a tree, (g) your height.

15. Two friends enter a 5-kilometer (3.1-mile) race. One is always a fast starter and the other starts slow but finishes fast. If they cross the finish line at the same time, you know that they both had the same (a) speed (b) instantaneous speed (c) average speed.

16. Late in the last century, George Littlewood of England ran 623.75 miles in six days at Madison Square Garden, a record that stood for almost 100 years. Calculate his average speed in mph.

17. In July 1991, 4 people survived when their small plane crashed in the ocean near the Bahamas. For 42 hours they drifted northward in a life raft, and then they were found about 100 miles from where they came down. What was their approximate average speed in the Gulf Stream in mph?

18. On a toll parkway, a driver takes a ticket at one toll booth and travels to another some miles away. The ticket is stamped by a time clock as the driver takes it at the entrance point. Can the toll collector at the exit point tell if the driver exceeded the speed limit for the parkway? Discuss.

19. A grizzly bear can run 500 yards in 30 seconds. What is the bear's average speed in yards/second? Compare that speed with the speed of a competitive sprinter who covers the 100-m dash in 10 seconds. (*Hint:* To make this easier, remember that a meter is about 9% longer than a yard.)

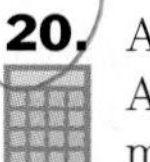

20. A football field in the United States is 100 yards long. About how long is that in meters? How long is a 100-meter track in yards?

21. When you walk, your average speed during one step is the distance of your step divided by the time between your footfalls. From this definition, think of two independent ways in which you could increase your speed.

22. If you throw your keys straight up, do they ever stop moving before you catch them as they come back down? Do they ever stop accelerating before you catch them? Discuss.

23. On the average, human hair grows about 1 centimeter in 3 weeks. If an actress bleaches her 1-foot-long hair for a play, she'll have to let her hair grow another foot before it can be back to its normal color and length. How long will that take? If you wanted to grow hair down to your waist, how long would it take?

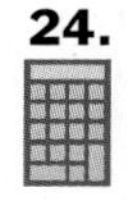

24. Measurements indicate that the continents of Europe and North America are moving away from each other at the rate of about 2 centimeters per year. If Columbus could have repeated his famous 1492 voyage in 1992, how many more meters would he have had to go?

25. Figure 2-14 shows the speed of a radio-controlled toy car. (a) In which time interval(s) was the car's speed constant? (b) In which time interval(s) was the car gaining speed? (c) In which time interval(s) was it losing speed? (d) In which interval does the car accelerate fastest?

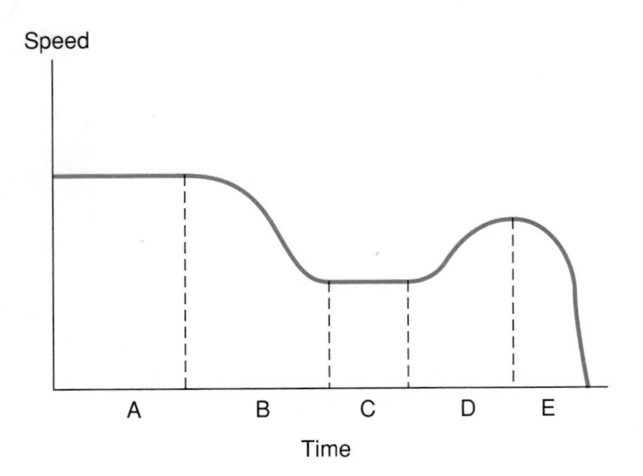

FIGURE 2-14

26. To accelerate a bicycle at a constant rate you would have to take exactly the same amount of time to go from 5 mph to 10 mph as you took to go from zero to 5 mph, and so on. True or false?

27. A schoolboy goes from rest on his bike to a speed of 10 mph in the same amount of time that his father takes to go from 50 mph to 60 mph on a motorcycle. Which has the greater (a) change in speed, (b) average speed, (c) average acceleration?

28. If you took 100 separate photos of honeybees in flight, you'd see their wings in many different positions. In which positions do you think their wings would appear most distinct (unblurred)?

29. In the strobelike illustration of a moving bug in Figure 2-15, assume the strobe flashed 3 times each second. (a) In which interval(s) did the bug have a constant speed? (b) In which interval(s) was the bug accelerating? (c) In which interval did it stay the longest time?

30. In 1992 three astronauts left the space shuttle to capture a broken satellite. All were moving at about 18,000 mph in orbits around the earth. Did their high speeds make this maneuver dangerous? Discuss.

31. If a car's speedometer registers 45 mph, what is that speed relative to?

32. As you drive along at 55 mph, another car overtakes and passes you at 70 mph. What is your speed relative to the passing car?

33. A fastball approaches the plate at 90 mph while a hitter swings his bat at 65 mph. How fast do the ball and the bat approach each other?

FIGURE 2-15

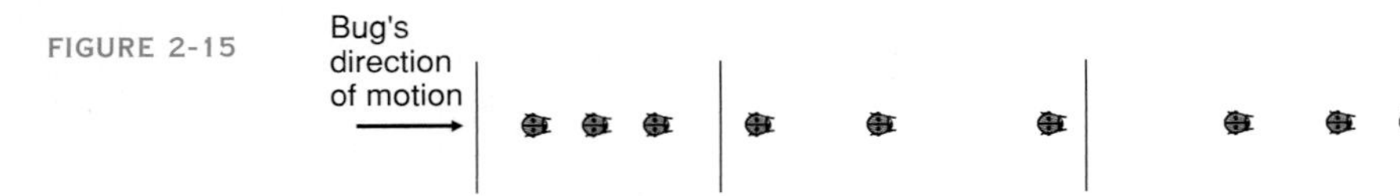

34. Figure 2-16 is a strobelike picture of a moving object. Identify (a) the regions of constant speed, (b) the regions of acceleration, (c) the regions of constant velocity.

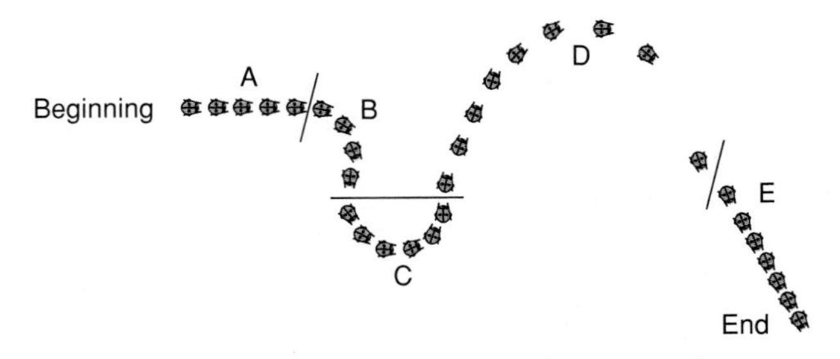

FIGURE 2-16

35. A boxer sees the glove of an opponent coming toward his own nose. To reduce the relative speed of those two objects, should he (a) stand as still as possible, (b) lean forward quickly, (c) lean backward quickly?

36. A car traveling at 80 km/hr collides with the rear of a large truck, but there is very little damage. What can you conclude about their relative speed?

37. Everything on this page can be pinpointed with ____ coordinates.

38. A pronghorn antelope can reach 60 mph (only a cheetah is faster) but it can average 40 mph for more than half an hour. At that lower speed, how long would it take to run a marathon?

39. Which quantities are not vectors: (a) acceleration, (b) temperature, (c) velocity, (d) speed?

40. If two snowmobiles have the same speed, does this mean they have the same (a) acceleration (b) position (c) velocity (d) direction of motion?

41. Draw the net vector of the two vectors shown in Figure 2-17.

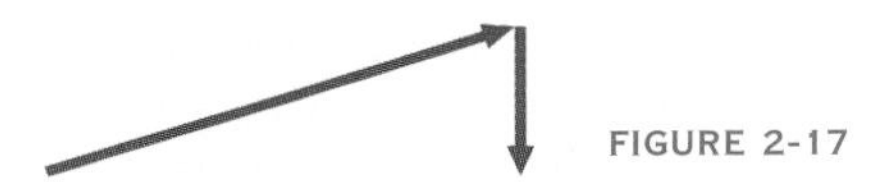

FIGURE 2-17

42. On a clear and windless day a plane flies overhead and the villains inside push James Bond out the door without a parachute. Would you say he moves in (a) 1, (b) 2 or, (c) 3 dimensions as he falls?

43. When a car moves along the road, does every point on the car's metal body have the same velocity vector? When you walk across the room, does every point on your body have the same velocity vector?

44. The ocean's level is now rising between 1.5 and 2.0 mm/yr. At this rate, how many years will it be before sea level is 10 meters (about a 3-story building) higher?

45. Three astronauts stand on a level plain on the moon. Acting together, one drops a feather, one throws a baseball horizontally, and one shoots an arrow horizontally from a bow, each from the same height. Which of these statements is correct? (a) All three objects hit the surface at the same time. (b) The baseball hits first, followed by the arrow and then the feather. (c) The arrow hits first, followed by the baseball and then the feather.

46. If you walk along a trail all day to a campsite and then walk back during the same hours the next day, will it happen at least once that you will be at some point on the trail at the same time on both days? Using the graph in Figure 2-18, draw several paths and find out.

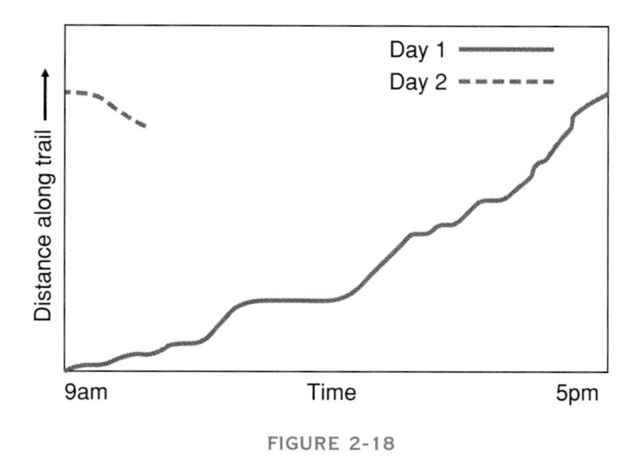

FIGURE 2-18

47. A sports car accelerates from rest to 60 mph in 6.8 seconds. What is its acceleration in mph/s? In ft/s^2? (Use 60 mph = 88 ft/s for an easy conversion.) Compare this second answer to g, the acceleration due to gravity.

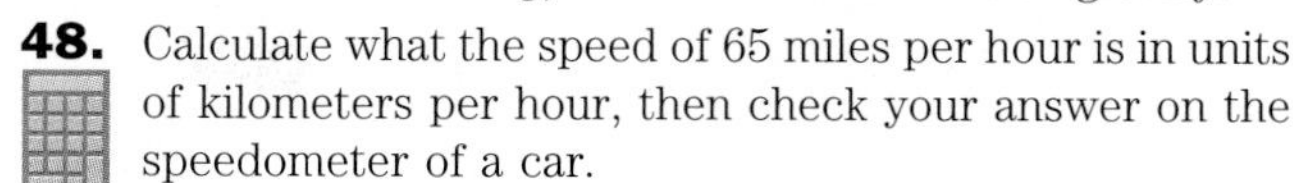

48. Calculate what the speed of 65 miles per hour is in units of kilometers per hour, then check your answer on the speedometer of a car.

49. A student drops her physics book from the edge of a canyon. In two seconds, how far will it fall (in meters).

50. A left-hand turn in a car accelerates the car (a) toward the left (b) toward the right (c) in neither direction, right or left.

51. A certain canoeist can paddle at 4 mph over still water. Would you expect her to be successful at paddling upstream in a deep creek that flows at a uniform 5 mph? Discuss.

53. In a physics lab, two students must find the speed of a cart rolling along the floor. Using a digital wristwatch, they time its passage between two lines. One says to the other, "It took exactly 1.73 seconds." It is very likely that the last digit (the 3) is significant in this number? Why or why not?

Demonstrations

1. Your sense of rhythm is an excellent internal timer! Place yourself in front of a clock that has a second hand. Pronounce a four-syllable word, such as Mississippi or locomotive, or four one-syllable words, such as "I can do it," and name the second immediately afterward. Say it with a cadence.

I can do it ONE
I can do it TWO
I can do it THREE

Because 5 syllables are spoken during each second, you are splitting each second into 5 parts, each syllable marking the passage of about 0.2 second. With a friend watching the clock, look away and count to 10 seconds this way. You'll probably determine 10-second intervals with an error of only one- or two-fifths of a second. When you watch a baseball game, time the pitches. Most take about 0.6 seconds to get from the mound to home plate, so you'll be able to say only "I can do . . ." before the ball hits the catcher's mitt or the bat.

2. Find your eye-to-hand reaction time. Have someone hold a ruler and center the lower end between your thumb and forefinger. When your partner drops the ruler without any warning, catch it on its way down and see how many inches slipped through your fingers. Table 2-1 gives your reaction time. After a few tries, you'll see that 4 or more inches invariably slip past—no matter who does the catching—because our eye-to-hand reaction times are at least 1/7 second and usually more. That fact is useful in a great party trick. Instead of a ruler, use a dollar bill and have someone center their thumb and forefinger over George Washington's nose. Then say, "If you catch it, you keep it!" (Explain that moving the hand downward with the bill is illegal.) The secret? It's only 7.6 cm to 8.3 cm from the edge of the bill to George's nose, and 1/7 s = 0.143 s.

Table 2·1 Distances and times from $d = \frac{1}{2}gt^2$ to 2 significant figures

Distance	Time
8.3 cm (3.2 in.)	0.13 second
9.6 cm (3.8 in.)	0.14 second
12.5 cm (4.9 in.)	0.16 second
14.2 cm (5.5 in.)	0.17 second
16 cm (6.2 in.)	0.18 second
18 cm (6.9 in.)	0.19 second
20 cm (7.7 in.)	0.20 second

3. Using a cardboard tube from a roll of paper towels, tape a pen inside as in Figure 2-19a. Then place the tube on a flat surface with the pen's point resting against an upright pad of paper as in Figure 2-19b. Roll the tube along and the path the pen traces shows the curve that a point on the edge of a moving wheel follows. It moves over and down, makes contact with the ground, then abruptly rises up and curves over again. Just when the point touches the ground, it changes directions. At this turning point, as when a set of keys is tossed straight up and turns around, its speed is instantaneously zero.

4. Some dark night, find an isolated city street light. Although the lamp seems to send out a steady stream of light, it does not. The brightness of its light varies rapidly, hitting a maximum value 120 times each second as the alternating electric current from the power company peaks. In effect, then, a street light is a strobe. If there is no other lighting nearby, you can see the flashes of light when something beneath the lamp moves fast enough. Take a stick that is several feet long and swish it quickly back and forth at arm's length. *You'll see the stick appear and disappear as it moves a noticeable distance between flashes.* Or just spread your fingers wide apart and sling you hand rapidly back and forth. *You'll see many different images of your fingers all at once.* The lamp's light seems steady when you look at slower objects, however.

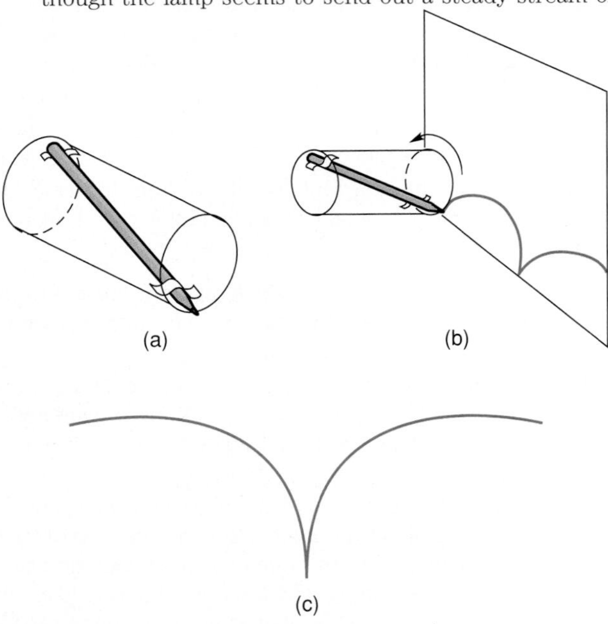

FIGURE 2-19

5. Estimate the thickness of one page of this book. It's easy! Measure a centimeter across the edge of the pages and then see which pages fall within that distance. Subtract the page number of the first page in that centimeter from the page number of the last page, divide by two (since pages are numbered on both sides), and divide 1 centimeter by that number to find the thickness of one page. Where will errors enter this measurement?

BEGINNINGS

Aristotle knew that if he pushed a plate across a table and then took his hand away, the motion of the plate would stop. To describe this, he wrote: "All that is moved is moved by something else." He reasoned that when the push from the "something else" stopped, so did the motion, and he decided that *rest* must be the natural state of any matter.

This description didn't explain how a spear continues in flight once it leaves the hand, however, or why an arrow keeps going once it leaves the bow. Aristotle decided that an arrow moving through air must compress the air with its front surface and leave the air in the space behind it rarefied, or thin. He argued that the air from the front must rush to the rear to fill the partial vacuum and that, as the air filled this space, it pushed the arrow along. To explain why an arrow in flight eventually slows, he said the transfer of air was never complete. This reasoning led to another wrong deduction, that motion must be impossible in the absence of air. "Nature," he said, "abhors a vacuum."

Aristotle's unproved ideas were still taught when Galileo Galilei (1654–1642) lived and worked. Galileo introduced experimental procedures, and he deduced that *force is not necessary to keep things moving,* that instead *forces of friction bring moving things to a halt.* Galileo knew that he had only begun to understand motion, however. He wrote that he had "opened up to this vast and most excellent science, of which my work is merely the beginning, ways and means by which other minds more acute than mine will explore its remote corners."

In addition to his work on motion, Galileo made many important astronomical discoveries. Using his newly designed telescope, he discovered four large moons orbiting Jupiter, and he saw that Venus was illuminated by the sun, having "phases" just as earth's moon does. Galileo's astronomical discoveries were there for anyone to see through a telescope, and his experiments on motion could be checked anywhere. Progressive scholars interested in experimental science soon formed groups such as the Royal Society of London for Improving Natural Knowledge. Now called the Royal Society, its present members are among the most renowned living scientists.

Isaac Newton made the next steps in the study of motion, and his contributions to physics are so immense that they may be unmatched in the whole history of science. Newton was born on Christmas Day 1642 in Lincolnshire, England. He was a premature baby, so tiny that his mother said she could have put him in a beer mug. As a schoolboy, he was very creative in making things, such as waterclocks, sundials, and even a wheeled chair. He carved his name in his desk at school, and one of his notebooks, still preserved, has an article he copied that told how to get birds drunk! He startled his neighbors one night by flying a kite that carried a paper lantern—they likely thought of witches rather than UFOs when they saw that dim light floating in the sky. Despite his creativity in building things and designing experiments, Newton did not excel at school. Being poor, he had to work at part-time jobs and graduated from Trinity College in Cambridge without any special distinction in 1665.

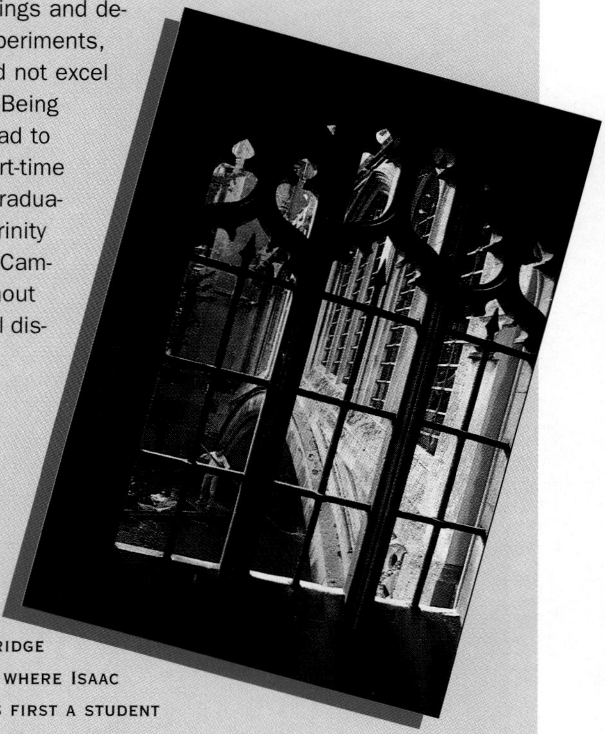

A SCENE FROM CAMBRIDGE UNIVERSITY, WHERE ISAAC NEWTON WAS FIRST A STUDENT AND LATER A PROFESSOR.

That summer the college closed, for the plague was raging in nearby London, killing over 10 percent of the city's people within 3 months. Newton returned to his home and in the peace and quiet of the countryside devoted his time to mathematics and natural philosophy. During 18 months of intense, uninterrupted study, he accomplished wonders. He discovered how to *predict* motion, extending and completing the work begun by Galileo. He began his investigations of gravity and the colors of light, and he invented the methods of calculus. In Chapters 3 and 4 we shall see how his laws of motion let us understand the motion around us and let us *predict* what will happen in situations where the forces are known.

THREE

Laws of Motion

This snow boarder and his dog have launched into the air, where gravity is the only substantial force acting on them. The forces around us are the subject of this chapter.

When you want to drink from a glass of water, you first *pull* on the glass with your hand to make it accelerate toward you, and then you *push* to stop it at your lips. Whenever anything accelerates, *a push or a pull* is causing the acceleration. Another word for push or pull is **force.** You can't see a force, but you can see the change it makes in something's motion. You use pushes and pulls whenever you move around. You can also *feel* pushes and pulls with your body, so you can relate to how these forces act on other objects. If you hold the string as a kite sails in a stiff breeze, you can feel how a soaring bird is pushed upward in a current of air. Likewise, if a friend's bicycle runs over your foot, you know what it means to say that a tire pushes against the road. You can also see that every push or pull has a direction; just like velocity and acceleration, force is a vector quantity.

ADDING FORCES: THE NET FORCE

Suppose you are standing at rest when two friends take your arms and pull on them equally, but in opposite directions. You won't go anywhere, and we say the two forces are *balanced.* If one friend pulls you northward as the other pulls you eastward, however, the forces won't balance. You will move off, and, as in Figure 3-1, your direction of motion will be at some angle between the directions of the two pulls. Equal pulls from your two friends will send you in a northeasterly direction; but if the person to the north tugs harder, you'll move more north than east.

You can represent these forces on a graph; draw vectors to show their directions, with the lengths of the vectors representing their strengths (Figure 3-2). Then place the two force vectors tail to tip and draw the *net* force. The **net force** describes the effect of the two forces acting together. Your body responds to both pulls at once, and it is as if *only the net force is present.*

FIGURE 3-1 WHEN MORE THAN ONE FORCE ACTS ON YOU, YOUR BODY RESPONDS TO THE VECTOR SUM OF THE FORCES, THE NET FORCE.

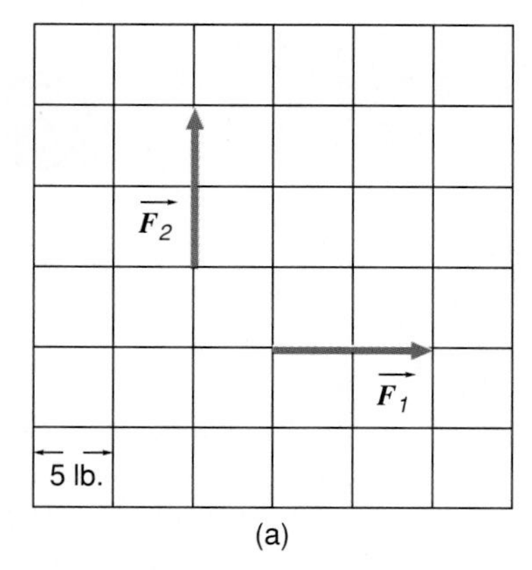

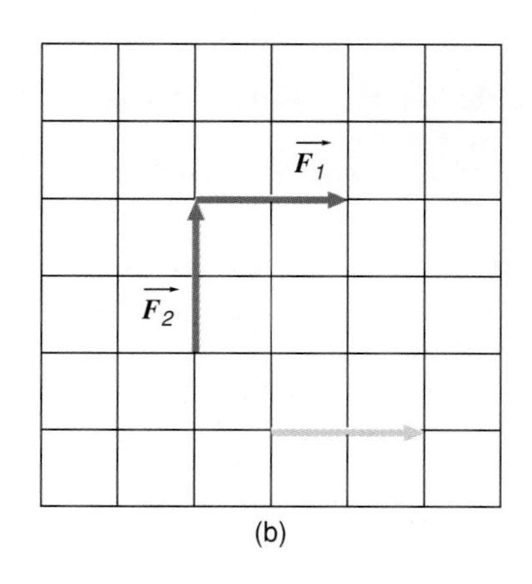

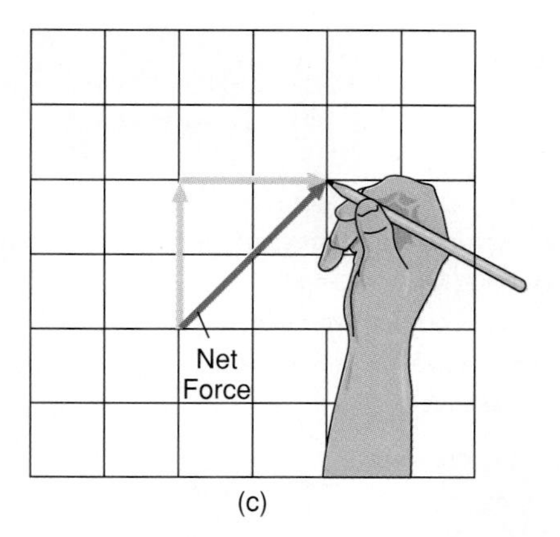

FIGURE 3-2 (A) REPRESENT EACH FORCE WITH AN ARROW POINTING IN THE DIRECTION OF THE FORCE, WITH A LENGTH THAT CORRESPONDS TO ITS STRENGTH, OR MAGNITUDE. (B) THEN SLIDE THE ARROWS TOGETHER TAIL TO TIP IN ANY ORDER. THE NET FORCE IS A VECTOR DRAWN FROM THE TAIL OF THE FIRST VECTOR IN THE SEQUENCE TO THE TIP OF THE LAST VECTOR. ITS LENGTH GIVES ITS SIZE, ABOUT 14 POUNDS HERE AS A MEASUREMENT WILL SHOW.

PHYSICS IN 3D/ADDING FORCES IN 3 DIMENSIONS

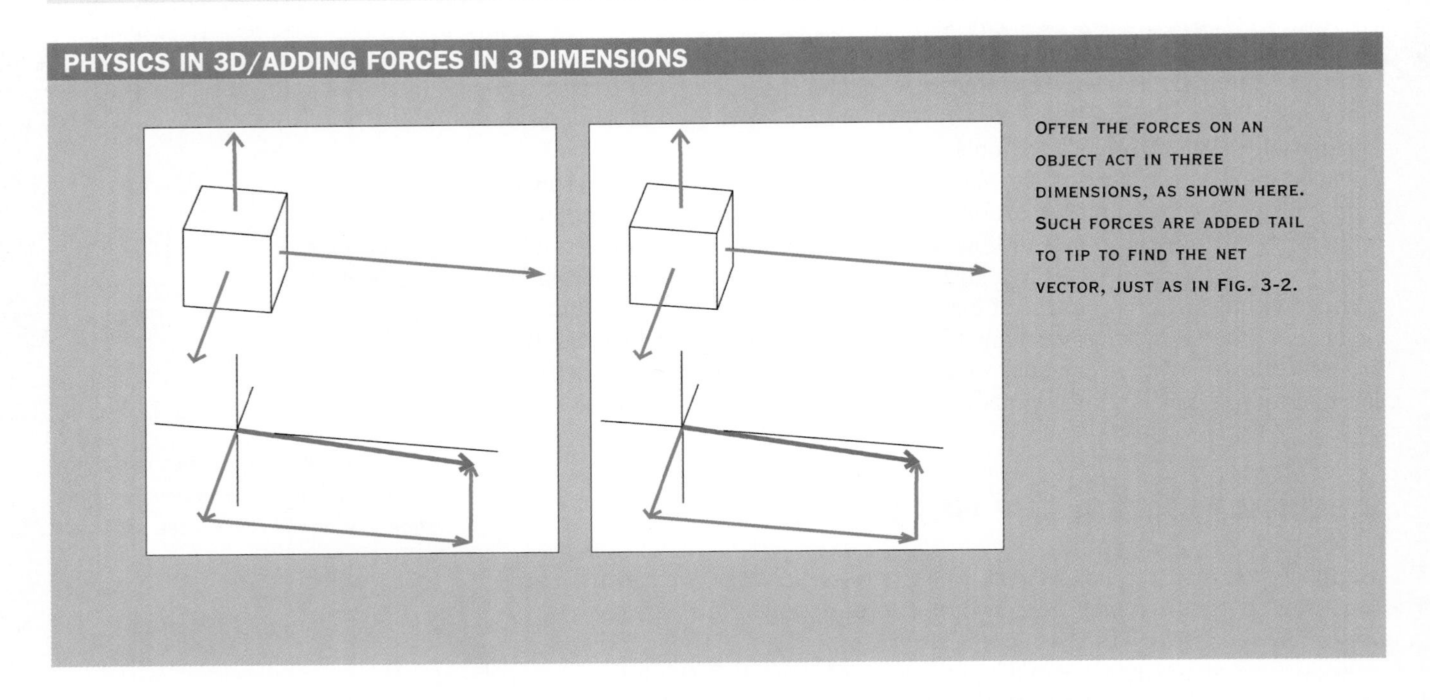

OFTEN THE FORCES ON AN OBJECT ACT IN THREE DIMENSIONS, AS SHOWN HERE. SUCH FORCES ARE ADDED TAIL TO TIP TO FIND THE NET VECTOR, JUST AS IN FIG. 3-2.

NEWTON'S FIRST LAW OF MOTION

To get a book to slide across the floor, you have to push it. When you stop pushing, the book stops moving. Aristotle described this action by saying that rest is the natural state of matter and, if no force acts, there is no motion, but he missed an important point. Forces can *stop* motion as well as start it, and in this case *a frictional force from the floor stops the book.*

Galileo realized that when there is little frictional resistance a rolling ball, a sliding book, or any other moving body travels much farther before stopping. He came to the conclusion that if there was no friction at all, an object in motion should continue to travel without any change in speed or direction, but he failed to state his idea in a convincing way. Isaac Newton, who was born the year Galileo died, made clear what Galileo had discovered in what is now called **Newton's first law of motion:**

Any object remains at rest or in motion along a straight line with constant speed unless acted upon by a net force.

Only a net push or pull can change something's speed or direction. This property of matter, that it will not accelerate unless a net force acts on it, is what we call **inertia.**

Because forces abound on the earth's surface, there are no examples of objects that move in perfectly straight lines with constant speeds. Astronauts go into an environment where objects seem to remain at rest or in motion with no net force (Figure 3-3). A flashlight thrown by an astronaut in the space shuttle will move along a straight path until it hits something in its way. Pencils and food containers left in midair will stay at rest there until small air currents in the spacecraft cause them to drift off. Even in the orbiting space shuttle, however, a force is present, as we'll see later in this chapter.

With a pair of in-line skates or a skateboard, you can push to get up some speed and then coast a long way. Your inertia opposes any change in your speed or direction. A fully-loaded supertanker (Figure 3-4) in motion acts the

FIGURE 3-3 ASTRONAUTS "JUGGLING" IN THE SPACE SHUTTLE; THE APPLES AND ORANGES STAY IN PLACE RATHER THAN FALL, MAKING IT MUCH EASIER TO KEEP THEM IN THE AIR SIMULTANEOUSLY.

same way, except it is much harder to change its speed or direction. (To slow it, its engines are cut off when it is still miles out of the harbor.) And that brings us to the subject of the second law of motion, one that takes into account the *amount* of inertia of an object.

FIGURE 3-4 SUPERTANKERS ILLUSTRATE THE PRINCIPAL OF INERTIA: THE ENGINES MAY BE CUT OFF A MILE OR MORE BEFORE THE TANKER ENTERS PORT, AND IT WILL COAST STRAIGHT ON IN.

NEWTON'S SECOND LAW OF MOTION

As children we all learned about inertia by pulling and pushing things around. Starting an empty wagon moving is easy, but not so when it is filled with sand. Because of inertia, it takes practice before our everyday motions become smooth. We handle forks and spoons with ease only because we know what sort of inertia they have before we pick them up and use them.

The point is this: From experience we all know there is a connection between the quantity of matter in a wagon or a spoon and how it will accelerate when we push it or pull it. Experiments can demonstrate this connection more plainly. For instance, suppose you pull an empty wagon with a steady force, say, of 10 pounds, as in Figure 3-5. With a strobe photograph of its motion, you could see how the wagon's speed changed—you could measure its acceleration. If you double the force to 20 pounds, you can see that the wagon's acceleration becomes twice as large as before. Or, suppose you load an identical wagon on top of the first one and pull with a 10-pound force again. This time, with twice as much matter to be moved, that force causes only half as much acceleration as before. Results much like these led Newton to his second law of motion, which predicts how motions change when forces act. Newton used the phrase "quantity of matter" to mean "amount of inertia," but today we use the term *inertial mass,* or more briefly, just **mass.**

PUTTING IT IN NUMBERS—NEWTON'S SECOND LAW

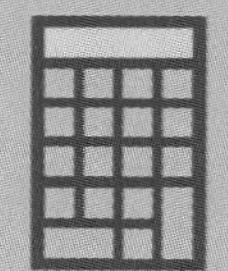

Newton's second law: the product of the mass (m) of any object times its acceleration ($\vec{a}$) is equal to the net force ($\vec{F}_{net}$) acting on the object

$$\vec{F}_{net} = m\vec{a}$$

The units of mass and force are discussed in the box on this page. Note that the acceleration is in the same direction as the net force.

Example: A friend wearing in-line skates has a mass of 60 kilograms and is standing at rest on level ground. You push him with a force of 45 newtons (about 10 pounds). What is his acceleration? The force you apply is the net force, since when he was at rest there were no unbalanced forces acting on him (Newton's first law) and the skates contribute almost no friction. Newton's second law gives $a = F_{net}/m = 45\ \text{N}/60\ \text{kg} = \mathbf{0.75\ m/s^2}$, and the acceleration is in the direction of your push.

UNITS OF MASS AND FORCE

In the SI system the basic unit of force is the newton (N) and the basic unit of mass is the kilogram (kg). One gram is the mass of 1 cm^3 of water, so 1 liter (1000 cm^3) of water has a mass of 1000 grams, or 1 kilogram. The mass of a quart of milk is also about one kilogram. The pound (lb) is the basic unit of force of the British system, where 1 pound = 4.45 newtons. If you hold a one-pound jar of peanut butter, it pushes down on your hand with the force of a pound. And a McDonald's Quarter-Pounder pushes down on your hand with about the force of one newton (next time ask for a "Newtoner!"). The British unit of mass is called a slug, since a larger mass responds more sluggishly to a force than a small mass does. This unit is seldom used, and we won't use it in this book.

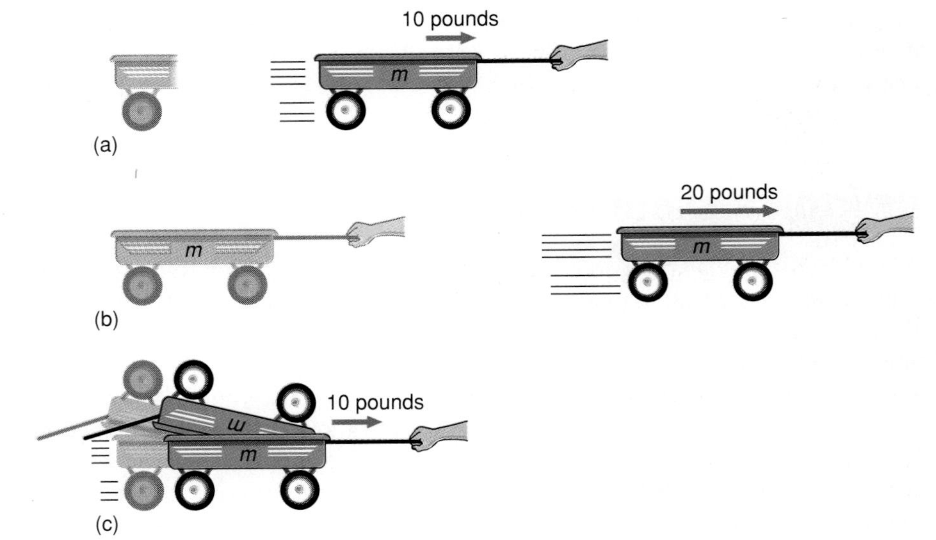

FIGURE 3-5 A STROBE PHOTO CAN REVEAL A WAGON'S ACCELERATION UNDER A STEADY FORCE. DOUBLING THE FORCE ON THE WAGON DOUBLES ITS ACCELERATION, WHILE DOUBLING THE MASS OF THE WAGON HALVES THE ACCELERATION IF THE FORCE REMAINS THE SAME.

Newton's second law works for baseballs and baby carriages, moons and planets, simple motions and complex motions. As long as you know all the forces acting on any object, you can add them tail to tip to find the net force. The direction of the net force gives you the direction of the object's acceleration, and the strength of the force F_{net} divided by the object's mass m tells you the acceleration a. In short, this law lets us *predict* changes in an object's motion.

NEWTON'S THIRD LAW OF MOTION

Any time something gets a push, something else gives that push. If you toss a book across the room to a friend, your hand exerts a force on the book. You can *feel* the book resist the change in its motion as you push it; that is, the book exerts a force back against your hand! Newton realized that forces always occur in pairs. When two children collide on a playground, *both* get a push, not just one. When a cue ball hits an eight ball on a billiard table, the motions of *both* change. Newton studied the actions of colliding objects and came to the conclusion now known as **Newton's third law of motion:**

For every force, or action, there is an equal but opposite force, or reaction.

The important thing to realize about this law is that the "action" force is on one object and the "reaction" force is on the other. These two forces always *act on different objects,* so they can never add and cancel. Only when equal and opposite forces act on the *same* object do they balance one another; in a playground collision, the force on one child can't cancel the force on the other.

If you lean over and push against a wall with your hands, the third law says the wall pushes back on you. That may sound strange, but imagine what would happen if the wall suddenly vanished while you were leaning on it. You would take a spill. The wall really does push back on you to hold you up, and you can feel that force on your hands. Why don't you accelerate in response to the wall's push? The friction between your feet and the floor balances the force from the wall. If you want to test this, try standing on a skateboard while you push against a wall (Figure 3-6).

Suppose you push a bowling ball. As the ball accelerates, you feel a resistance from it, and this resistance is its reaction force on you. As you push the ball, it presses against your hand with the same amount of force. The bowling ball's reaction force would cause you to accelerate backward except for the friction between the ground and your feet. If you stand on a frozen pond on ice skates and push that ball, you *will* accelerate backward (Figure 3-7). You won't accelerate as fast as the ball does because $a = F/m$, and if two objects get the same amount of force, the one with the greater mass will get the smaller acceleration.

Have you ever stepped ashore from a small boat and felt the boat slip backward beneath you? That's action and reaction. A motorboat pushes through the water by pushing water in the opposite direction with its propeller, just as an aircraft pushes forward through the air by forcing large amounts of air backward. Every time a force acts, there is a reaction force.

FIGURE 3-6 THE SKATEBOARDER PUSHES AGAINST THE WALL AND THE WALL PUSHES BACK ON THE SKATEBOARDER, WHO THEN ACCELERATES AWAY FROM THE WALL.

FIGURE 3-7 THE BOWLER AND THE BALL RECEIVE EQUAL BUT OPPOSITE FORCES. THE BALL, HAVING A SMALLER MASS, GETS A GREATER ACCELERATION FROM THAT FORCE THAN THE BOWLER DOES.

WEIGHT

While there are many kinds of forces all around us, perhaps nothing is so ingrained in our senses as the earth's pull on all of our surroundings. This pull is always there, never changing. Earth's gravity is built into our descriptions of our world with words like *up, down,* and *weight.*

Exactly what do we mean by weight? **Weight** is a force, it is the *pull of earth's gravity on an object.* To measure your body's weight, you can step onto a bathroom scale. Inside the scale is a spring that stretches when a force acts on it. If the spring is pulled, it exerts a reaction force in the other direction and this force is indicated by the pointer of the scale. When the scale supports your weight, the spring's force balances gravity's pull on your mass, and the force registered on the scale is equal to your weight. If you step down, drink two cups of coffee, and then step back up on the scale, you'll weigh about 1 pound more.

TWO CUPS OF COFFEE OR WATER WEIGH ONE POUND, WHICH GAVE RISE TO THE OLD SAYING, "A PINT'S A POUND THE WORLD AROUND." AND A MEDIUM-SIZED APPLE OR A STICK OF MARGARINE WEIGHS ABOUT ONE NEWTON.

The weight of a car is the pull of earth's gravity on the car. The greater the mass, the larger the attraction. For example, two identical pickup trucks weigh exactly twice as much as one. And, of course, there's twice as much matter, or mass, in two trucks as in one. Yet mass and weight are two very different things. Mass measures inertia, while the weight of an object is a force.

WEIGHT AT EARTH'S SURFACE: *mg*

In a vacuum near earth's surface, everything falls with acceleration g, as we saw in Chapter 2. We can use this fact and Newton's second law to find the weight of an object. If an object is falling in a vacuum, its weight is the only force acting, so its weight *is* the net force. The value of the acceleration a is just g, and substituting in the formula $\vec{F}_{\text{net}} = m\vec{a}$ we find

PUTTING IT IN NUMBERS—EQUATION FOR WEIGHT

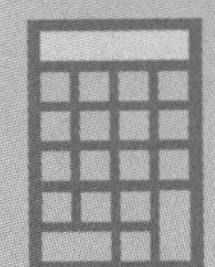

weight $= mg$

Example: If you weigh 600 newtons (about 135 pounds), what is your mass in kilograms?
$mg = 600$ N, so $m = 600 \text{ N}/g = 600 \text{ N}/(9.8 \text{ m/s}^2) =$ **61 kilograms.**

FIGURE 3-8 TESTING INERTIAS ON A FRICTIONLESS TABLE. THE BOOK WITH THE MOST INERTIA (MASS) REQUIRES THE GREATER FORCE.

The equation for weight points out the difference between mass and weight; mass measures an object's inertia while its weight is a force, the pull of gravity.

Suppose someone hands you two books and asks which is the more massive. Almost certainly you will "weigh" one in each hand and say the heavier book is the more massive one. That's correct, the heavier book does have more mass. If the two books were on a smooth, level table, however, as in Figure 3-8, you could just push them back and forth, one with each hand, to estimate which has the larger mass. What you are judging this time is not weight but *inertia*. The book with the greater mass requires the greater force when you accelerate the two books side by side.

An astronaut could pick up a large rock on the moon using much less force than required on earth, since its weight is so much less there ($g_{\text{moon}} = \frac{1}{6}\, g_{\text{earth}}$). If the astronaut shoved the rock in a horizontal direction, however, it would take just as much of a push to accelerate it at, say, 1 m/s^2, as it would take to accelerate it at 1 m/s^2 on earth. An object's weight changes if the pull of gravity changes, but an object's mass is the same everywhere.

FREE FALL

FIGURE 3-9 DURING A JUMP, THE JUMPERS EXPERIENCE "WEIGHTLESSNESS." EUNICE WINKLER IS NOT SUPPORTED BY HER HORSE AS THEY PLUNGE TOWARD THE (UNSEEN) WATER BARREL.

At the turn of the century, Eunice Winkler entertained crowds by jumping with her horse from a high tower into a large water barrel (Figure 3-9). As they fell, gravity pulled on them all the way down, so they both still had their weights, mg. While in the air, there was no force to support them, to oppose the gravitational pull. (Air resistance was negligible at their low speeds.) Because gravity was the only force acting, we say that they were in **free fall.**

If you strap on a backpack full of books and hop down from a chair, you and the pack are in free fall for the short time before you touch the floor. The pack's weight doesn't disappear—it causes the pack to accelerate (along with

THIS SURFER IS DOING AN "AERIAL," WHERE A WAVE CATAPULTS HIM FROM HIS SURFBOARD AND HE IS (ALMOST) IN FREE FALL. WHAT FORCE BESIDES GRAVITY IS ACTING ON HIM?

you) at the rate of g—but the pack's weight vanishes from your shoulder straps because you don't *support* this weight as you did before you jumped. Reporters sometimes say orbiting astronauts are "weightless," but gravity still pulls on them, so they do have weight. (Otherwise they would not constantly "fall" along a path that takes them around earth but rather would go off in a straight line courtesy of Newton's first law! We will look at this in more detail in Chapter 7). They are in free fall, and everything in the space shuttle falls *together* as it orbits the earth; nothing supports (or balances) anything else's weight. Everything in the orbiting shuttle seems to be floating, just like that pack on your back did if you jumped.

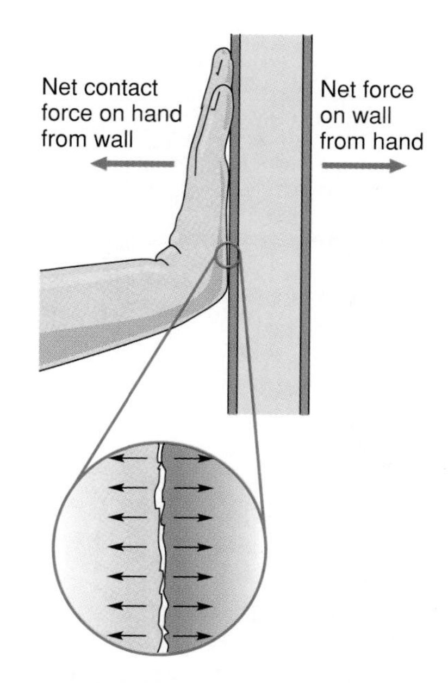

FIGURE 3-10 MANY SMALL CONTACT FORCES AT THE ATOMIC/MOLECULAR LEVEL ADD TO GIVE THE NET CONTACT FORCES BETWEEN THE HAND AND THE WALL. THE PUSH ON THE HAND BY THE WALL IS THE REACTION FORCE FROM THE HAND'S PUSH ON THE WALL.

CONTACT FORCES

When you push your hand against a door to open it, atoms and molecules collide and repel each other, keeping your hand from going through the door. As the ancient Greeks guessed, atoms and many of their combinations are almost indestructible, pushing off mightily when they are squeezed together. This repulsion is called the **contact force.** The contact force pushes atoms straight apart. (This force is often called the *normal* force, since it points normal, or perpendicular, to the surfaces that touch.) The sum of many such forces at the atomic/molecular level gives the contact force of the door on your hand, which is equal to the force your hand exerts on the door but pointed in the opposite direction (Figure 3-10).

The contact force is a reaction force, and it appears wherever things touch. If we place a lamp on a table, as in Figure 3-11, gravity pulls downward on the lamp and presses it against the tabletop. Because of this contact, the table reacts with an upward push—a contact force that supports the lamp. In fact, you can see that this contact force on the lamp exactly cancels the pull of gravity on it. How? The lamp doesn't accelerate down into the table, and the table doesn't throw the lamp upward; *the lamp doesn't accelerate, so the net force on it must be zero.* The upward contact force from the table wouldn't be there if the lamp were not pressing down on the tabletop. The contact force, a reaction force, is always equal in strength but opposite in direction

THIS CHICK GETS A CONTACT FORCE FROM THE GROUND THAT BALANCES ITS WEIGHT. SHOULD THE ELEPHANT'S FOOT DESCEND, IT WOULD ADD A MUCH LARGER CONTACT FORCE FROM ABOVE, WHICH IN TURN WOULD INCREASE THE CHICK'S CONTACT FORCE WITH THE GROUND.

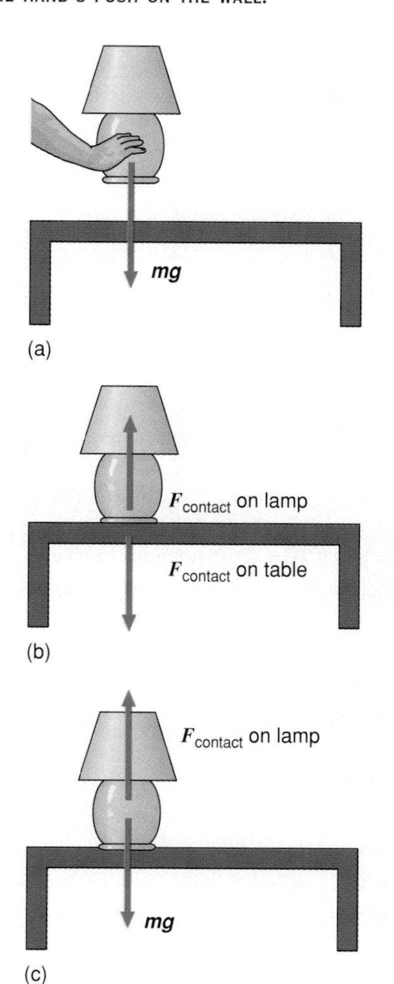

FIGURE 3-11 (A) A LAMP BEING PLACED ON A TABLE. (B) THE CONTACT FORCES APPEAR WHEN THE LAMP RESTS ON THE TABLE. THEY ARE EQUAL BUT OPPOSITE IN DIRECTION. ONE ACTS ON THE TABLE, AND THE OTHER ON THE LAMP. (C) THE FORCES ON THE LAMP ALONE ARE *MG* AND THE UPWARD CONTACT FORCE FROM THE TABLE. THESE TWO FORCES BALANCE, SO THE NET FORCE ON THE LAMP IS ZERO.

to the action. The contact force is always perpendicular to the surfaces in contact, as Figure 3-12 shows. When the table is tilted, the direction of the contact force changes.

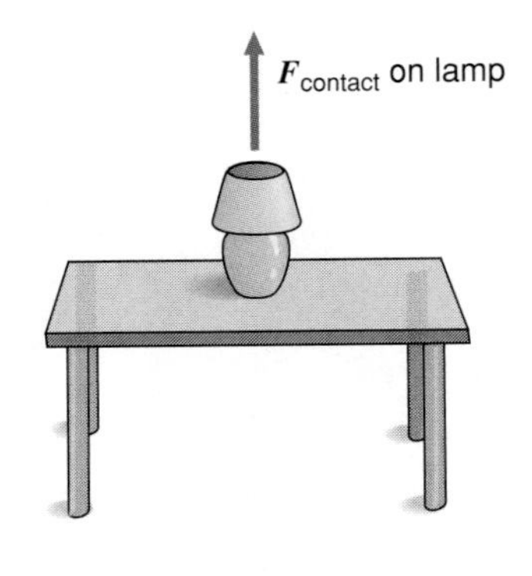

FIGURE 3-12 THE CONTACT FORCES ALWAYS PUSH THE OBJECTS IN CONTACT STRAIGHT APART, PERPENDICULAR TO THE TWO SURFACES THAT MEET. WHEN THE TABLE IS TILTED, SO IS THE CONTACT FORCE.

ELASTICITY

Besides the many forces around us, there are also forces that act within matter. When you squeeze a solid, its atoms press together and the contact force between them pushes back. When you stretch the solid, the **bonding forces** (**bonds** for short) connecting the atoms act to pull them back toward each other. These two actions, the contact forces and the bonding forces, cause the atoms in a solid to behave as if they were connected by tiny springs (Figure 3-13). We say a solid is **elastic** because it can normally recover its original size and shape after an external force is removed. Liquids, too, have bonding forces that keep their molecules close together, but these are weaker than a solid's bonds, so liquid molecules can easily move past each other, or flow.

No solid is perfectly rigid; all show some degree of springiness, or elasticity. Push in the side of an aluminum soft drink can slightly and it pops out when you let go. (If you squeeze too hard, however, you'll break some bonds, making a permanent dent. Elasticity has its limits.) Even your fingertips have elasticity. Press a pencil point against one. When you take away the pencil, your fingertip returns to its normal shape.

Place your hand on a countertop and press down. Even though you can't see it happen, the atoms of the counter's surface squeeze a little closer together as they push back on your hand. Below them, countless others share the force from your hand, shifting a tiny bit and passing the load along all the way to the floor below. When you lift your hand, all the molecules return to their former positions. However slight their shifting, that's elasticity.

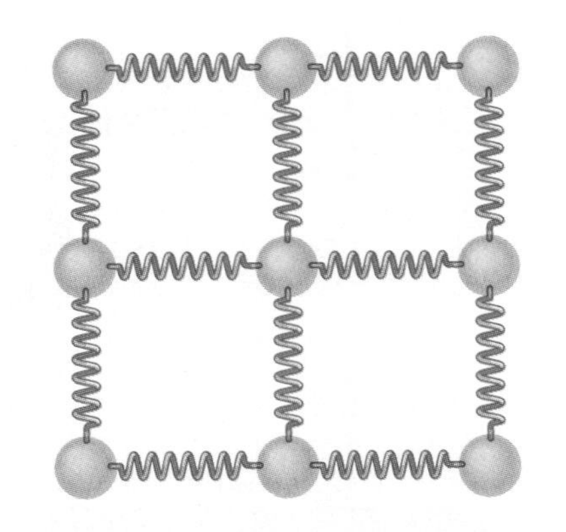

FIGURE 3-13 THE BONDING FORCES IN SOLIDS AND THE CONTACT FORCES COMBINE TO GIVE SOLIDS ELASTICITY. THESE FORCES CAUSE THE ATOMS TO ACT AS IF THEY WERE CONNECTED BY SPRINGS.

FRICTION

If you try to slide a book across a desktop by pushing gently, as in Figure 3-14a, the book won't budge. All solid surfaces, even the smoothest, have microscopic bumps and dips. Press two surfaces together and those irregularities catch on each other and resist any motion. If you push harder, the growing force from your hand will finally break the holds the irregularities have on each other and the book will begin to move (Figure 3-14c). Then there is less resistance—in other words, less *friction*—because the surfaces are skimming over each other and fewer of the bumps and dips catch and grab. Even if the surfaces are smooth at the atomic level, the individual atoms attract and cause friction. If two extremely flat surfaces of the same metal are pressed together, they "cold weld" and the two pieces of metal become one.

The force of friction between the desk and book before the book moves is called **static friction.** As you push harder, the static friction force grows, matching your push as you press harder—up to some maximum force, at which point the surfaces slip and begin to move past each other. The smaller frictional force that remains between them once motion begins is called **kinetic friction.**

STATIC FRICTION TAKES ITS NAME FROM THE GREEK WORD *STATIKOS*, "CAUSING TO STAND," WHILE KINETIC FRICTION COMES FROM THE GREEK WORD *KINETIKOS*, "OF MOTION."

Frictional forces, like all other forces, occur in pairs. When the desk exerts a frictional force on the book, resisting the book's motion, the book exerts an equal but opposite force on the desk. The direction of a frictional force is easy

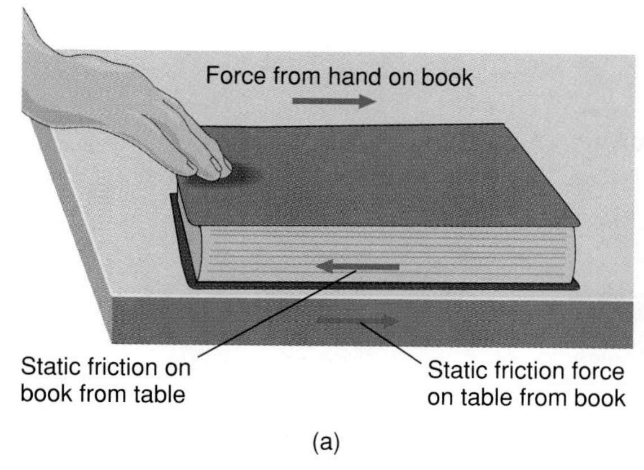

(a)

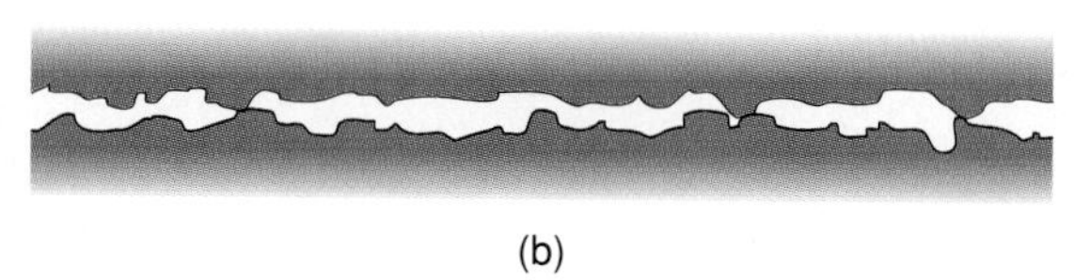

(b)

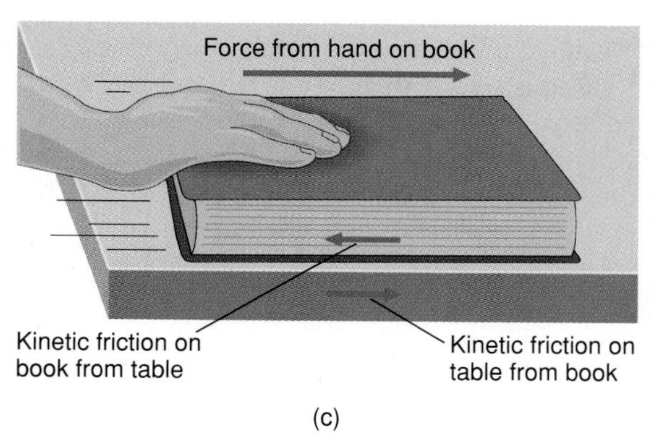

(c)

FIGURE 3-14 (A) IF THE FORCE FROM THE HAND IS TOO SMALL TO OVERCOME THE STATIC FRICTION FROM THE TABLE, THE STATIC FRICTION EXACTLY BALANCES THE FORCE FROM THE HAND. (B) AN ENLARGED DRAWING OF THE TWO ROUGH SURFACES IN CONTACT. MICROSCOPICALLY, ONLY A FEW POINTS ARE IN CONTACT. (C) IF THE FORCE FROM THE HAND OVERCOMES THE STATIC FRICTION FROM THE TABLE, THE BOOK ACCELERATES. IF THE FORCE APPLIED TO THE BOOK THEN EXCEEDS THE KINETIC FRICTIONAL FORCE FROM THE TABLE, THERE IS A NET FORCE ON THE BOOK AND IT WILL CONTINUE TO ACCELERATE.

to find: Kinetic frictional forces on objects always oppose the relative motion, and static frictional forces oppose the "pending" relative motion.

The strength of the frictional force between two surfaces also depends on how hard the surfaces press together. That is, *the frictional force is greater when the contact force is greater.* Put another book on top of the first one and push on that bottom book again (Figure 3-15). You'll find it is now much harder to break the hold of static friction between it and the desk; the maximum value of the static friction is greater than before because of the extra contact force due to the weight of that second book.

Our bodies are adapted to using friction in everyday activities. The fingerprint ridges on the elastic skin of our fingers and palms increase friction when we grasp things. The soles of our feet have similar patterns that aid in gripping. You use friction when you turn a doorknob, tie your shoelace, or scratch an itch.

A PAGE FROM LEONARDO DA VINCI'S NOTEBOOKS, SHOWING THE FORCES INVOLVED IN THE CONSTRUCTION OF AN ARCH.

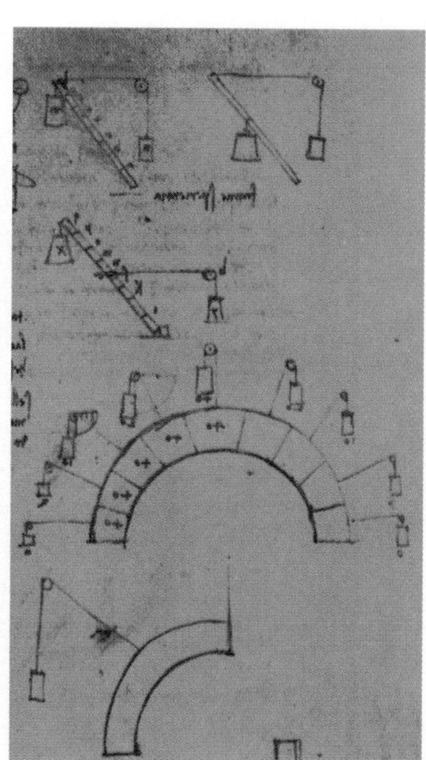

DRAG

Liquids and gases also resist the passage of a solid object, and this is friction too. Air exerts a force against any solid that moves through it; as air molecules collide with the oncoming surface, their impacts retard the object's motion.

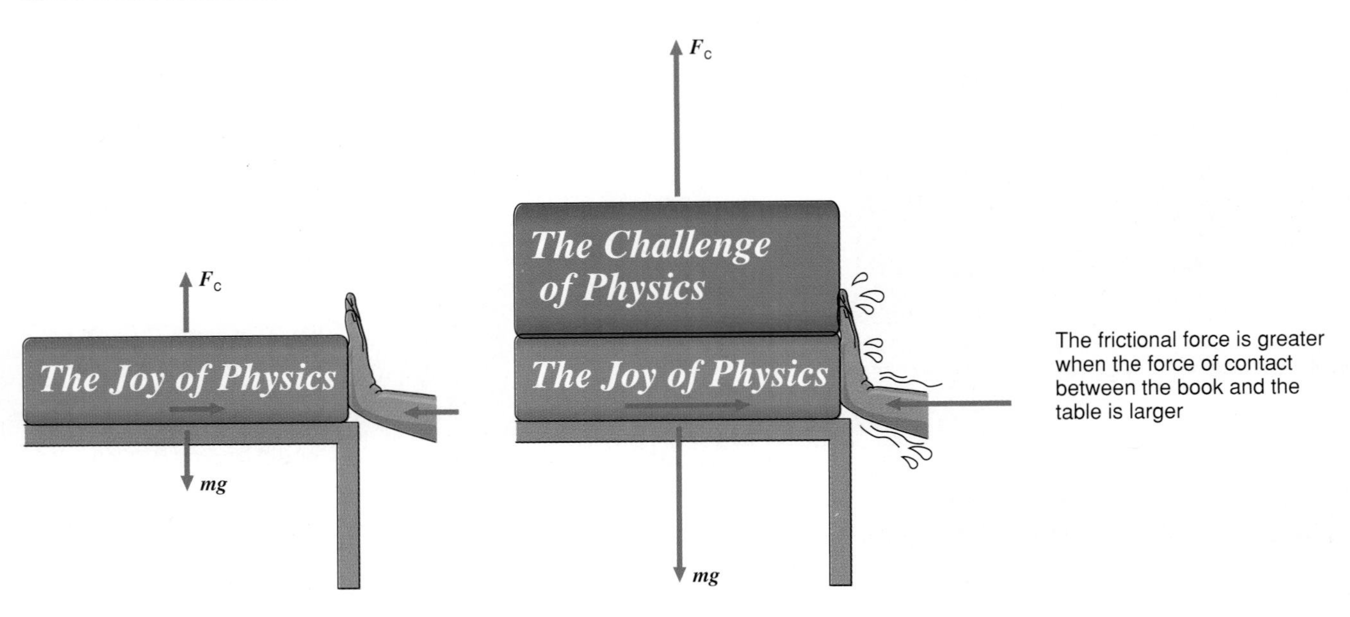

FIGURE 3-15 FRICTIONAL FORCE

Just wave your hand back and forth rapidly and you can feel the air with your hand. If you push your hand through water, there's much more resistance. Such friction from liquids and gases is called **drag.** Hold a hand out of the window of an accelerating car and you can feel the air drag increase rapidly with speed. Two reasons for this increase are easy to understand. The faster the object moves, the more molecules it will collide with per second, and the harder the collisions will be because of the increased relative speed.

DRAG, FRICTION, AND EROSION

Water moving along in streams or rivers exerts drag forces on any particles it encounters. Because drag depends on speed, the size and mass of the largest particles a stream can move is determined by the water's speed. Water-carried

PHYSICS IN 3D/EROSION CAUSED BY FRICTION

LONG-TERM EROSION BY WATER-CARRIED PARTICLES MADE THESE STRIKING FORMATIONS IN SOUTHWEST UTAH.

sand and gravel-size particles abrade any other solid they strike just as sandpaper would. This action is the principal way erosion cuts deep gorges through solid rock. Pebbles and larger rocks get trapped in depressions in the rock where they swirl in the eddies of moving water and enlarge the hole. In the process the pebbles and rocks become rounded themselves. When forests are clear-cut, rainwater moves, or erodes, the loose topsoil easily. Then, when rivers are dammed, the stilled waters behind the dam drop their solid particles, eventually filling the volume behind the dam with sediment.

PHYSICS OF EROSION

HOW ACCURATE ARE NEWTON'S LAWS?

Isaac Newton's discoveries in physics explained and predicted for the first time some of the grandest actions in nature. With his laws of motion and his law of gravity (Chapter 7), he explained the elliptical paths of the planets around the sun. He also explained the ocean tides caused by the moon and the sun, and he showed other unsuspected effects that were subsequently measured, such as the fact that the earth bulges somewhat at its equator because of its rotation. Newton discovered powerful, predictive laws, and the new science of physics quickly took an important, permanent place in human affairs. Newton's discoveries stood unaltered for more than 200 years, but he knew that his conclusions could not be the final word on how things behave. When he published these laws, he also wrote, ". . . I hope the principles here laid down will afford some light either to this or some truer method of philosophy."

In the early 1900s revolutionary discoveries found limits to the accuracy of Newton's laws of motion. Albert Einstein found that time and distance are not the same for moving objects as for those same objects at rest (Chapter 26). Since these quantities appear in the second law, $\vec{F} = m\vec{a}$, it was shown to be in error. However, differences between the predictions with $\vec{F} = m\vec{a}$ and Einstein's laws aren't normally measurable unless the speed of an object is a large fraction of the speed of light (about 186,000 miles per second). For the ordinary events that most of us deal with, speeds are never so great. In fact, when NASA scientists calculated the paths that took the astronauts to the moon, they used Newton's laws, not Einstein's.

Other discoveries in the early 1900s concerned the subatomic particles called electrons and protons, and even individual whole atoms. Even though these tiny particles make up the matter around us, they do not behave like ordinary objects behave. Physicists discovered that it is impossible to predict the motions of these particles precisely. A new theory quite unlike anything that Newton or Einstein discovered was developed to account for their behavior (Chapter 25). Yet even that theory predicted that when large numbers of atoms bond together to make everyday objects, Newton's laws do predict their actions with great accuracy. Physics evolves through discovery today as it has since Newton's time, and progress always presents new questions, ones that will need "some truer method of philosophy" for their answers.

Summary and Key Terms

Forces are the pushes and pulls that cause all accelerations. Force is a vector quantity. If more than one force acts on an object, you can draw vectors to show the direction and magnitude of each force. Then you can add them to find the ***net force;*** the object will accelerate as if the net force is the only force acting.

force–35

net force–35

Newton's laws explain how forces affect an object's motion, making it possible to understand and *predict* the motions in our everyday experience. ***Newton's first law*** says that when there is no net force, an object at rest remains at rest and an object in motion continues moving with a constant speed and direction. ***Newton's second law*** says that the net force on an object is equal to its mass times its acceleration, $\vec{F}_{net} = m\vec{a}$. An object's ***mass*** is its "quantity of matter," which measures its ***inertia***. ***Newton's third law*** tells us that for every force, or action, there is an equal and opposite force, or reaction.

Newton's first law–36

Newton's second law–37

mass–37

inertia–36

Newton's third law–38

There are many kinds of forces. An object's ***weight*** is the force of gravity's

weight–39

free fall–40

contact force–41
bonding force–42

elasticity–42

static friction–42

kinetic friction–42

drag–44

pull on that object. Ignoring air resistance, everything near earth's surface falls with acceleration g, so $weight = mg$. ***Free fall,*** often called "weightlessness," occurs when gravity is the only force acting. Objects in free fall near earth have weight, but since no object supports any other, they appear weightless.

There are also forces that act within matter. The ***contact force*** pushes atoms apart if they are squeezed together. ***Bonding forces*** keep a solid's atoms in place, so they don't move with respect to each other. The bonds are weaker in liquids, and liquids flow. The combined action of bonds and contact forces give solids the property of ***elasticity***.

Friction is a force that resists motion. When you try to begin sliding a heavy book across your desk, ***static friction*** between the two surfaces holds the book in place, matching your force up to some maximum point at which the surfaces slip. The frictional force that remains between the surfaces as you slide the book along is ***kinetic friction***. A stack of two books would be more difficult to set sliding than one because *the frictional force between solids is greater when the contact force between the surfaces is greater.* The friction that occurs when a solid moves through a liquid or gas is called ***drag***. The direction of a frictional force always opposes the relative (or pending) motion.

EXERCISES

Concept Checks: Answer true or false

1. All forces are pushes or pulls, but not all forces have directions.

2. A net force describes the effect of two or more forces acting on an object at the same time.

3. Newton's first law deals with objects that have no net force acting on them.

4. Mass can be described as the measure of something's inertia.

5. Newton's second law can predict an object's acceleration if the forces acting on the object are known.

6. The acceleration of an object is always in the direction of the net force on it.

7. The third law of motion tells us that forces always act in pairs.

8. The pull of earth's gravity on your body is what you call your weight. It is a force.

9. Mass and weight are measures of two different things, and an object's weight depends upon its mass. Rocks on the moon weigh six times more when they are brought to earth.

10. The contact force always pushes things straight apart when they are pressed together.

11. When you are in free fall, as when you jump into a swimming pool, you are without weight, since nothing supports your weight before you hit the water.

12. Bonding forces lock atoms or molecules together in solids, and combined with contact forces they are responsible for elasticity.

Applying the Concepts

13. We use many words for force in addition to *push* and *pull*. Some of these are jostle, nudge, butt, prod, punch, and jab. Can you think of others?

14. Humor aside, discuss how you can feel this book's inertia.

15. Is it more accurate to say someone diets to lose mass rather than weight?

16. Is inertia more closely related to weight or mass? Explain.

17. After stirring a stew with a spoon, you might tap the spoon against the side of the pot so that the excess liquid on the spoon falls back into the stew. What principle are you applying? After using a beach towel for relaxing on a sandy beach, you might hold it downwind and shake it sharply to get most of the sand off. What principle is being applied?

18. When you carry a large book in one hand, your hand is more likely to get bruised should it bump into something. What property of matter does this illustrate best?

19. When you hold your toothbrush in your hand, what is the net force on it? If you let it go, what is the net force on it just after you've dropped it? If it is at rest on the floor afterward, what is the net force on it?

20. Two youngsters have a shoving match on the playground, and one is finally pushed over. The victor shouts, "I pushed harder!" Is he right?

21. If you've ever had to push a stalled car, you know it is much easier to keep the car rolling than it is to accelerate it from rest. Why is that true?

22. Why is it easier to pull a paper towel from a roll if you jerk the towel quickly? Why is it sometimes easier to tear a towel from a full roll than a roll that is almost empty?

23. Cheetahs are bigger and faster than gazelles, but more often than not gazelles escape a pursuing cheetah by zigzagging. Why does this maneuver put the cheetah at a disadvantage?

24. When you watch something move, can you see its acceleration? Its velocity? Its inertia? The force that acts on it?

25. A vase being pushed across a table exerts a frictional force of 10 newtons on the table. What is the frictional force exerted on the vase by the table?

26. The head of someone who is pushed from behind seems to snap backwards. How does the first law of motion explain this whiplash motion? What force do headrests in cars apply to help guard against whiplash?

27. When you are standing up and the contact force from the floor is supporting your weight, the contact force is equal to mg, where m is your mass. True or false?

28. An astronaut in orbit shoves a flashlight away with a 1-newton push. If the flashlight accelerates at 5 m/s^2 while being pushed, what is its mass?

29. A new college grad in a Jaguar convertible accelerates along a level street at 3.4 m/s^2. If its mass is 1950 kg, what is the net force on the car?

30. Two wagons, one empty and one fully loaded, are parked side by side. If you grab the handles and begin to run, pulling the wagons behind you and keeping them perfectly even with each other, do both have the same (a) inertia? (b) force on them? (c) acceleration? (d) velocity? (e) speed? (f) mass?

31. An average-sized apple weighs about 1 newton. What is this weight in ounces? (There are 16 ounces in a pound.)

32. A car is towing another car of equal mass. It starts from rest, and as it accelerates the tow chain breaks. What happens to its acceleration?

33. When a large car and a small car collide, the forces exerted on them are equal but opposite, but what about their accelerations and the accelerations of the passengers?

34. If a rock is dropped on a future astronaut's toe while she is standing on the moon, should it hurt more or less or the same as it would on earth? If she then kicks the rock along the surface, should it hurt more or less or the same as it would on earth?

35. Bathroom scales often show kilograms as well as pounds. The kilogram reading indicates your mass, not your weight. Suppose the astronauts had taken such a scale to the moon to weigh themselves. Would the pound reading be correct? Would the kilogram reading be correct?

36. How much horizontal force does a car seat exert on a 55-kg person to accelerate that person at 2 m/s^2? Do you think you would feel such a force on your back? (*Hint:* Compare that force to mg.)

37. A 60-kg cowboy sits on a 500-kg bull at a rodeo. What is the contact force between the cowboy and the bull? Will it remain the same if the bull begins to buck? What is the total contact force between the bull's feet and the ground?

38. Can a 1-kg mass ever weigh more than a 2-kg mass? Explain. Can a 1-lb object ever have more mass than a 2-lb object? Explain.

39. Find your weight (on earth) in newtons. Find your mass (anywhere) in kilograms.

40. A 1-pound box of crackers generally has "454 grams" on its label. The pound is its weight, while the 454 grams is its mass. Use $F_{weight} = mg$ to prove that a mass of 0.454 kg weighs 1 pound on earth.

41. If something is traveling at a constant velocity, does this mean there are no forces acting on it? If not, then what does it mean?

42. If two things accelerate at the same rate, must they have the same mass? Must they have the same force acting on them?

43. Do you think perfectly frictionless shoelaces would stay tied? Explain.

44. At a roller rink a boy and girl push straight apart. The girl's mass is 40 kg, and during the push she accelerates at 1 m/s^2. If the boy's mass is 65 kg, what is his acceleration while they push off?

45. A certain rocket's engine gives a constant thrust (which is just a push) to the rocket during flight, and so in the equation $F = ma$, the force F is constant. As the rocket travels, however, it uses the fuel that initially accounts for most of its mass. What does the second law predict will happen during the flight?

46. When you clap your hands, what force stops them when they collide? When you chomp through a piece of celery, what stops your top teeth when they collide with your bottom teeth?

47. When a book lies at rest on a table, which of Newton's laws of motion tells us that the net force on the book is zero?

48. A rock placed on a scale registers 100 pounds. A girl tries to lift the rock unsuccessfully, but while she is exerting a steady pull, the scale registers only 60 pounds. How much force is she applying? If she stands on another scale while applying that steady force, what will that scale register? If both of these scales are on a third scale, what will it read?

49. When a book lies at rest on a level table, does the contact force exerted by the table arise only because the book presses against the table? Is the contact force equal to the force of gravity on the book? Does the contact force on the *book* point in the same direction as mg, the book's weight? Does the contact force on the *table* point in the same direction as mg, the book's weight?

50. If the tires of your car slip, skid, or spin over the road, you won't be able to accelerate or decelerate as quickly as you could otherwise. Why?

51. A ball rolls along a horizontal table and rolls off the edge. After losing contact with the edge of the table, only the force of gravity acts on it. Does it then fall straight down? Why or why not?

52. The fastest serve by a man at the 1992 U.S. Open was clocked by radar gun at 129 mph. Would the speed of that serve depend on where the radar gun clocked it? Discuss.

53. Suppose you wish to push a stack of books across a table and accelerate them at 1 m/s^2. If their mass is 12 kg and the frictional force between the books and the table is 49 newtons, what force would you have to apply?

54. You probably know to watch out for shaggy dogs coming out of a lake or just fresh from a bath. What principle does the dog use to rid its fur of excess water by shaking?

Demonstrations

1. Rest a paper cup in your hand while you fill it with water. It gets heavier in your hand because the weight of the water presses against the cup's bottom. Punch a small hole near the bottom and a stream of water will squirt straight out. Holding the cup by the rim, release it. Immediately the stream vanishes as cup and water fall freely toward the ground. Once the cup and water are in free fall, the cup no longer supports the water so the water no longer presses against the bottom or the sides of the cup—hence the water doesn't squirt out as it is falling.

2. A well-known magician's trick is to pull a tablecloth from underneath dishes and glasses without disturbing them. If the cloth is jerked away rapidly, the frictional force between it and the objects acts for such a short time that there can be little actual change in speed for the tableware. Use a sheet of paper and a dry glass to do this trick for yourself. (If the paper extends a few centimeters over the edge of the table, it is easier to grasp.) Of course, there is always some friction. In which direction does the glass move? How does its motion depend on where it sits on the paper?

3. Rub talcum powder on your fingertips. Then touch a mirror. Fingerprints appear where the talc-coated ridges of the skin pressed against the mirror. Natural oils in your skin do the same thing, which is how people leave almost invisible fingerprints on objects they've touched. A dusting with powder makes the "prints" visible as the powder sticks only to the oily marks. When a surface applies a force perpendicular to the ridges, they help to "grab" the surface, because each elastic ridge can stretch (much like a spring) into the groove next to it, and the more it stretches, the more it resists.

FOUR

Using the Laws of Motion

Whether in sports or everyday activities, we use forces to accelerate and decelerate and to keep our stability as we move around. How forces are applied is the subject of this chapter.

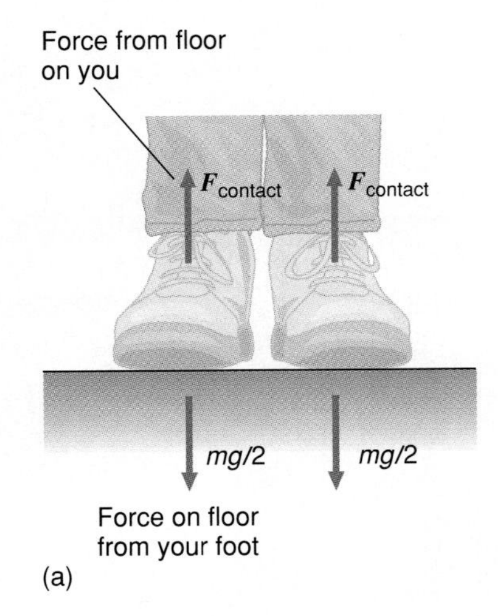

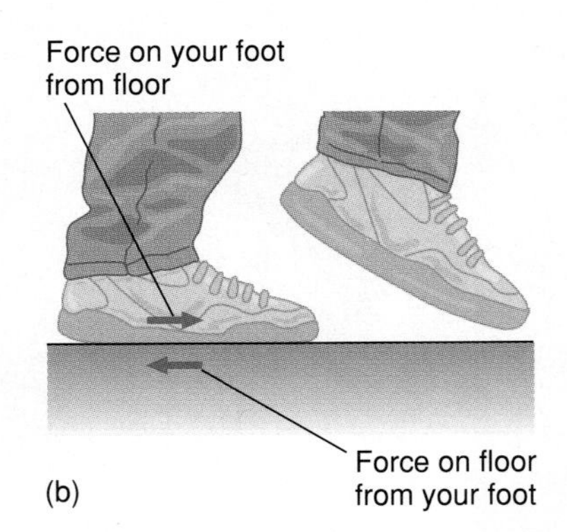

FIGURE 4-1 (A) STANDING AT REST, THE PULL OF GRAVITY CAUSES YOU TO PRESS AGAINST THE FLOOR WITH A FORCE EQUAL TO YOUR WEIGHT. EACH SQUARE CENTIMETER OF YOUR FEET THAT TOUCHES THE GROUND SUPPORTS SOME OF YOUR WEIGHT. THE FLOOR PRESSES BACK ON YOUR FEET WITH AN EQUAL FORCE, BALANCING THE FORCE OF GRAVITY. THE NET FORCE ON YOU IS ZERO, SO YOU DON'T ACCELERATE. (B) AS YOU TAKE A STEP, YOU PUSH OFF, MEANING YOUR PUSH AGAINST THE FLOOR HAS A HORIZONTAL COMPONENT. THE FRICTIONAL FORCE FROM THE FLOOR, WHICH PUSHES IN THE OPPOSITE DIRECTION FROM YOUR PUSH, ACCELERATES YOU IN THE HORIZONTAL DIRECTION. (YOUR WEIGHT AND THE CONTACT FORCE ARE NOT SHOWN HERE.)

This chapter shows how we can use Newton's laws of motion to understand everyday motions. Applying Newton's laws can make clear why things happen the way they do. To be sure of understanding the examples in this chapter, you should always *draw your own sketch* of the situation. Newton's laws explain the motion *only* if you get the forces and their directions right, and a simple (but correct!) drawing helps every time.

STANDING, WALKING, AND SLIPPING

When you stand on a level floor, the contact force balances your weight with each foot supporting about half (Figure 4-1a). You don't accelerate because the net force on you is zero. Then, to take a step, you must accelerate in the horizontal direction. That requires a horizontal force. If a foot pushes off against the floor, the floor pushes back on that foot (Figure 4-1b). That frictional force, the floor's reaction to your action, acts on your mass, and you accelerate horizontally. Now imagine trying to walk in a pool of grease on a smooth marble floor: No amount of body motion would propel you horizontally, because there is no frictional force. For you to accelerate, a force from outside your body must act on you. Such an outside force is called an *external* force.

When you stand on a slope, your weight still pulls straight down on you, but the contact force pushes straight out from (perpendicular to) the slope (Figure 4-2). The contact force, remember, is the surface's reaction to a force pressing against it. In this case the "action" force is not the full weight of your body but only that component of your weight that presses directly into (or perpendicular to) the surface. The other component of your weight is parallel to the slope, and it acts to pull your body down the slope. You can stand still on a slope only if the static friction between your feet and the surface balances the component of your weight that is parallel to the slope.

Skiers, of course, *want* to slip on an incline, so they eliminate as much friction as they can by waxing their skis. If they could remove all the friction between skis and snow, the net force down the slope would be the component of the skier's weight parallel to the slope minus air drag.

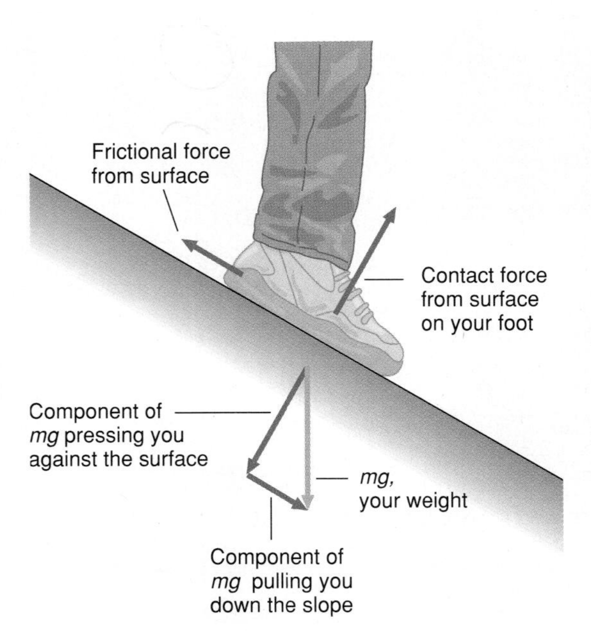

FIGURE 4-2 ON A SLOPE, NOT ALL OF YOUR WEIGHT PRESSES AGAINST THE SURFACE. A COMPONENT ACTS TO PULL YOU DOWN THE SLOPE. THE CONTACT FORCE, WHICH IS THE REACTION FORCE TO THE COMPONENT OF YOUR WEIGHT PRESSING AGAINST THE SURFACE, GIVES RISE TO A FRICTIONAL FORCE THAT OPPOSES THE COMPONENT OF YOUR WEIGHT POINTING DOWN THE SLOPE. (THE FORCE YOUR SHOES EXERT ON THE SLOPE IS NOT SHOWN HERE.)

RIDING UP AND DOWN WITH NEWTON'S SECOND LAW

Let's imagine that you take a bathroom scale into an elevator and stand on it. It will indicate your weight. If the elevator starts upward, the scale shows a reading higher than before, more than just your weight. How could you explain this? First, *identify the forces acting on you.* Your weight, mg, pulls downward on you. Then there's the contact force from the scale on the elevator's floor that pushes upward on you. Those are the only forces, and Newton's second law tells us the rest.

You accelerate upward only if there is an upward-pointing net force acting on you. The scale on the floor of the upwardly accelerating elevator pushes you upward with a contact force that is *greater than your weight,* as indicated by the higher scale reading. As you can see from the example in Figure 4-3a, the net force on you is upward as the elevator accelerates, and it is equal to the contact force from the scale minus your weight, mg.

An elevator's speed usually levels off after about a second of acceleration. Then you and the elevator travel up at a constant speed, as in Figure 4-3b. Newton's first law tells us that whenever anything moves with a constant speed and direction, the net force on it is zero. Now the contact force from the scale exactly balances your weight, and you no longer feel heavier than usual.

If the elevator stops and you press a button for a lower floor, the elevator

FIGURE 4-3 THREE CASES OF MOTION IN AN ELEVATOR. (A) THE CONTACT FORCE FROM THE FLOOR, WHICH IS THE SAME AS THE READING ON THE SCALE, IS GREATER THAN HER WEIGHT. THERE IS A NET FORCE UPWARD (IN THIS CASE 140 LB − 120 LB = 20 LB) AND SHE ACCELERATES UPWARD. (B) WHEN THE CONTACT FORCE BALANCES THE WEIGHT, THE ACCELERATION IS ZERO AND SHE FEELS NORMAL EVEN THOUGH SHE IS MOVING UP. (C) IF HER WEIGHT IS GREATER THAN THE CONTACT FORCE, THERE IS A NET FORCE DOWNWARD ON HER (20 POUNDS HERE) AND SHE ACCELERATES DOWNWARD.

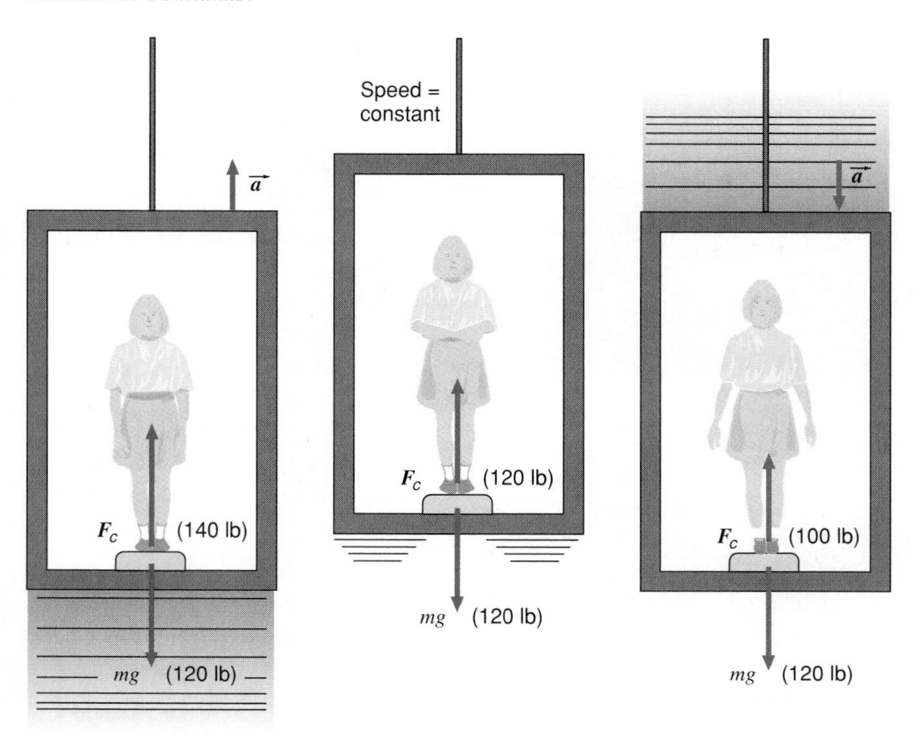

accelerates downward for a second or so, and as it does you'll feel lighter. For that short time the contact force, again shown by the reading on the scale, is less than your weight, meaning the net force on you is in the downward direction (Figure 4-3c).

USING COMPONENTS OF FORCE

Suppose you need to pull a heavy trunk across a floor. You can better understand what happens if you split your pulling force into horizontal and vertical components as in Figure 4-4. The upward component of your force opposes the weight of the trunk and therefore lessens the contact force the trunk exerts against the floor. But unless this vertical part of your pull exceeds the trunk's weight, the trunk remains on the floor. Even so, because the contact force between the trunk and the floor is now less, the friction from the floor is less and the trunk is easier to accelerate in the horizontal direction. If it is to accelerate, there must be a net force on the trunk in the horizontal direction. The trunk won't move at all unless the horizontal component of your pull is greater than the static frictional force from the floor. If you break the static friction's hold, the sliding (kinetic) friction will be less than the static friction was, and the trunk will accelerate—unless you decrease your pull to just *balance* the sliding frictional force. Then the trunk would move at a constant speed.

Instead of pulling, you could push this trunk as in Figure 4-5. That push has a downward component in the same direction as the trunk's weight. This extra downward force increases the contact force with the floor, which increases the frictional force with the floor. So the trunk is easier to pull along than to push along, as you might have guessed. From this we can see why it's easier to pull a lawnmower through high grass than it is to push it and why you would pull a wagon through soft sand rather than push it.

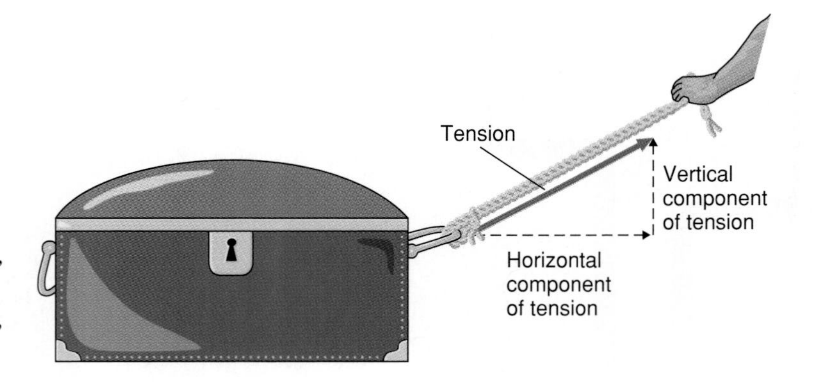

FIGURE 4-4 ONE COMPONENT OF THE APPLIED FORCE SUBTRACTS FROM THE WEIGHT, AND THE OTHER ACTS HORIZONTALLY. (NOT SHOWN ARE THE CONTACT FORCE, THE WEIGHT, AND THE FRICTIONAL FORCE.)

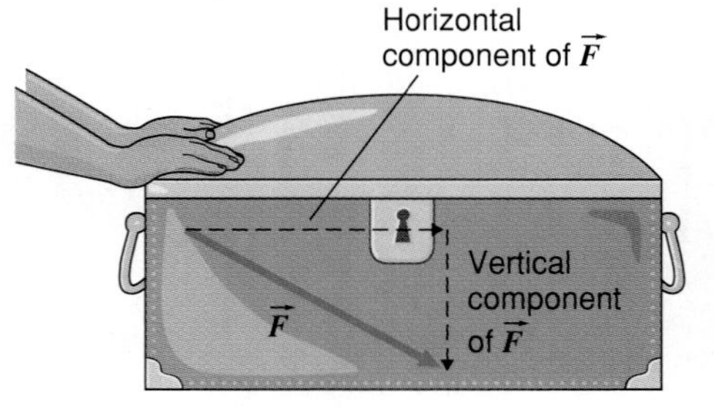

FIGURE 4-5 HERE THE DOWNWARD COMPONENT OF THE APPLIED FORCE ADDS TO THE WEIGHT OF THE TRUNK. THIS INCREASES THE CONTACT FORCE BETWEEN THE TRUNK AND THE FLOOR AND HENCE THE FRICTIONAL FORCE BETWEEN THEM.

AIR RESISTANCE

Air resistance comes from collisions with the molecules of the air, as we saw in Chapter 3, and the more surface area the air hits, the greater the drag will be. A small car, for example, will usually have less air drag than a larger car traveling at the same speed. An object's shape also plays a large role in the amount of air drag it receives. If air strikes a surface at a glancing angle, it exerts a much smaller force than if it strikes the surface straight on. That's one reason why "streamlining" is so important for cars, airplanes, and birds (Figure 4-6).

FIGURE 4-6 STREAMS OF SMOKE IN A WIND TUNNEL SHOW HOW SMOOTHLY THE AIR IS PARTED BY A MODERN CAR.

Autumn leaves float down lazily when they fall, but without air resistance, they would drop as if made of lead. Since they have little mass, the drag on their large surface areas really slows them down. A bed sheet billows and gently sinks when you make up your bed. Without air to resist its motion, it, too, would descend like a rock. Films of astronauts using the Lunar Rover show moon dust kicked up by the wheels. On that airless surface, those fine particles were sprayed outward for great distances. In the air of the earth, such tiny particles would slow to a stop in only a centimeter or two.

From the instant a skydiver jumps from an airplane, air drag affects the skydiver's motion. As the skydiver gains speed in the downward direction, the upward force from the air grows, opposing the downward pull of gravity and decreasing the net force on the skydiver, which decreases the skydiver's acceleration. When the skydiver falls fast enough, the air drag becomes as large as the pull of gravity, and the net force on the skydiver is *zero:* The skydiver gains no more speed. The skydiver's speed of fall when this happens is called the **terminal speed.** Skydivers can vary their terminal speeds. A skydiver who falls with arms and legs spread wide will catch a lot of air. The typical terminal speed for someone in this position is about 120 mph. By diving nose first, a more streamlined position, the same skydiver can plunge downward at speeds approaching 200 mph. Then, before landing, the parachute is opened and its extra drag lowers the skydiver's terminal speed, usually to 10 to 12 mph, a safe speed for landing.

If you put your hand out of the window of a moving car, you feel air resistance. Flatten your hand and hold it palm down. Then tilt the forward edge upward. Your hand gets pushed upward and backward by the air. The component of force opposing the hand's motion through the air is the drag, while the upward component of force on your hand is called **lift.** For an even more dramatic effect, hold a book out of the window and tilt it.

SKYDIVERS EXPERIENCE AIR DRAG AS WELL AS GRAVITY.

FORCES AND AIRPLANES

A moving airplane's wings get *lift* and *drag* from the air, the same as your upward-tilted hand when stuck outside a car's window. The body of the plane also gets drag, and the net drag force on the plane points in the direction opposite the plane's velocity. The plane's engine(s) exert a forward-pointing force called *thrust.* When the plane flies horizontally with a constant speed, there is no net force in the

AIRPLANES EXPERIENCE THRUST AND LIFT AS WELL AS AIR DRAG AND GRAVITY.

horizontal direction, so the drag balances the thrust. That plane has no vertical acceleration, so the lift on the wings balances the airplane's weight.

Passengers inside a commercial airplane are shielded from the outside rush of air. Only the pull of earth's gravity, their weight, has to be supported. The chairs and floor of the level-flying aircraft do that just as they would on earth. If the plane flies through an updraft or a downdraft, however, those aboard feel momentarily heavier or lighter as the plane accelerates up or down. In a severe downdraft, a plane can be accelerated downward at a rate *greater* than g. In the rare instances when this has happened, passengers and flight attendants who were not buckled into their seats have hit the ceiling as the plane fell faster than they were falling. Passengers buckled in at the time had the memorable experience of watching their drinks and ice cubes, which accelerated downward at only g, fly out of the cups and up to the ceiling.

An airplane that flies upward and then noses sharply downward will lose the upward lift on the wings. Then, for a short while, the airplane will fall toward earth with an acceleration of g and the passengers will be in free fall (Figure 4-7). This is the way NASA gives its new astronauts their first few moments of free-fall experience.

APPLYING FORCE: PRESSURE

You are probably sitting as you read this page. The contact force that supports your weight acts on the areas of your body that are in contact with the chair and the floor. Each square centimeter that touches the chair or floor receives a certain amount of upward force, and the sum of those forces balances your weight. When you stand, the smaller areas of the bottoms of your feet get all that force, so each square centimeter there supports a bigger share of your weight. And if you stand on your toes for a moment, you'll really feel the difference. The *force per unit of area* is called **pressure.** More precisely,

FIGURE 4-7 (A) WHEN THE PLANE IS IN FREE FALL, IT FOLLOWS A PATH THAT WOULD BE TAKEN BY A HEAVY STONE THROWN UPWARD FROM THE GROUND. WHILE ON THAT PATH, EVERYTHING INSIDE THE PLANE EXPERIENCES "WEIGHTLESSNESS." (B) ASTRONAUTS IN TRAINING, DURING A ZERO-G SESSION IN A NASA AIRCRAFT.

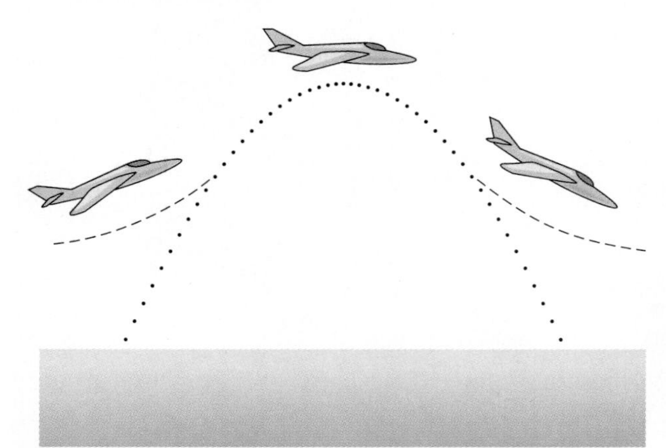

(a)

(b)

PUTTING IT IN NUMBERS—DEFINING PRESSURE

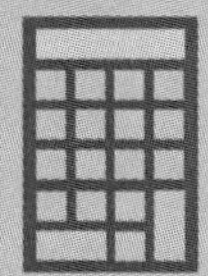

pressure = force on the area of application/area of application

$$P = \frac{F}{A}$$

where F is perpendicular to the surface area A. The metric unit of pressure is the pascal (Pa), which is 1 N/m^2, and the British unit is lb/in^2 (psi). Also, one atmosphere (atm) is 14.7 lb/in^2, the accepted standard atmospheric pressure at sea level.

Example: Sherlock Homes in on the footprints his wet feet leave on the bathroom tiles (1 inch on each side) after his shower. Counting the tiles his feet have wet, he sees that he supports his 180 pounds with an area of about 30 square inches. What is the average pressure on his feet in psi? $P = 180 \text{ lb}/30 \text{ in}^2 = \mathbf{6.0 \text{ lb/in}^2}$, or about 0.4 atm.

When you stand barefoot on a hard floor, the floor does not conform to your foot shape, so small areas on the balls and heels of your feet support your weight. Soft shoes and sand and plush carpets feel good to walk on because they conform to your feet, distributing your weight over more area. Likewise, a soft, cushiony chair conforms more to your body than a hard, straight chair, and the different pressures are easy to feel.

The definition of pressure shows us how sharp knives and razors cut so well. The slight force you apply with the tiny cutting area of the thin blade edge produces a large pressure. Naturally, the sharper the blade, the smaller the area (A) of contact. That is why it is easy to nick your finger on the edge of a piece of paper: If the area is very small, the applied force on it doesn't have to be large to apply enough pressure to break the skin.

TAKING A FALL IN THE JUDO CHAMPIONSHIP OF FRANCE.

During a fall it is important to maximize the force-bearing area in order to lessen the pressure on any one area. It looks dangerous when professional wrestlers are slammed to the mat, but they take care to land flat, on large area of their bodies. That minimizes the pressure, and they usually walk away unharmed.

PARTICLES IN THE AIR

High pressures are used in many industrial processes today. For one example, stone such as granite is crushed to make small gravel that is used as a component of construction concrete as well as road surfaces. When high pressure is applied to the stone, it cracks into all sizes of particles, and makes large amounts of rock dust. Carried by the wind away from the crushing sites, this is believed to be the second largest industrial pollutant in the form of suspended particulate matter, following the combustion products of fossil fuels.

THE FORCE THAT TURNS THINGS

Whirl a small object attached to a string in a circle on a tabletop. The object turns constantly because a force from the string, the *tension* in the string, is perpendicular to its direction of travel at any instant. This turning force is called a **centripetal force.** You feel this force with your hand. Because this force is perpendicular to the object's velocity at any point, it does not increase or decrease the object's speed. If you release the string, the turning force vanishes and the object continues moving *in a straight line along the direction it was moving at the instant the tension vanished.*

When an object follows a circular path, the turning force always points toward the center of the circle (Figure 4-8). What if the curve isn't a perfect

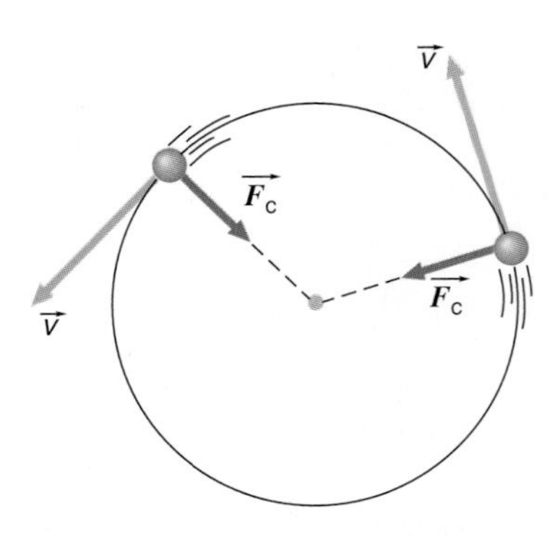

FIGURE 4-8 THE CENTRIPETAL FORCE, WHICH STAYS PERPENDICULAR TO THE OBJECT'S VELOCITY VECTOR, CONTINUOUSLY TURNS IT. SINCE THE CENTRIPETAL FORCE HAS NO COMPONENT IN THE OBJECT'S DIRECTION OF MOTION (SHOWN BY ITS VELOCITY), THE CENTRIPETAL FORCE DOES NOT CHANGE THE OBJECT'S SPEED.

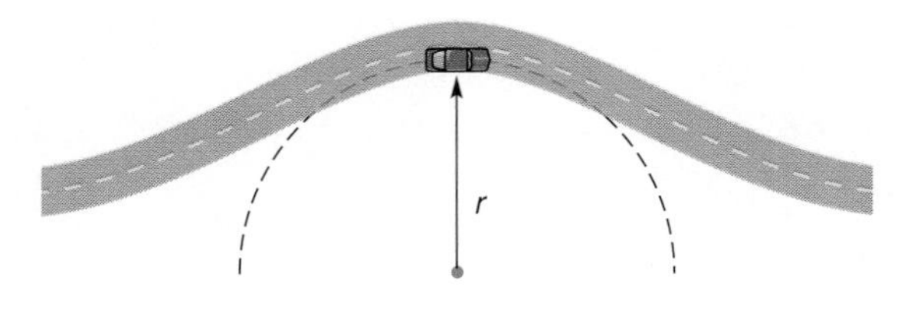

FIGURE 4-9 ON ANY CURVE, THERE IS A CIRCLE THAT MATCHES THE CURVE AT ANY POINT. THAT CIRCLE'S RADIUS, USED IN $\frac{MV^2}{R}$, GIVES THE TURNING FORCE ON THE CAR AT THAT POINT ON THE CURVE.

circle? The centripetal force (which means *center-seeking* force) then points toward the center of a circle that matches the curve perfectly wherever the object is on the curve (Figure 4-9). All kinds of forces can act as centripetal forces. For instance, the moon travels around the earth because the pull of the earth's gravity supplies a centripetal force.

When your car moves along a curve, so do you, and a centripetal force turns you. It comes from friction with the seat, or from the seatbelt, or even from contact with a door. At the same speed, a sharp turn (a small circle) causes a bigger push on you than a gentle turn (a large circle), so a smaller radius means a larger centripetal force. Likewise, on any given curve a low speed might be comfortable, but a high speed might require a stomach-wrenching force to turn you as your direction changes more quickly, meaning the centripetal force is larger. The relationship between the speed on the circle, the radius of the circle, and the centripetal force is

PUTTING IT IN NUMBERS—CENTRIPETAL FORCE

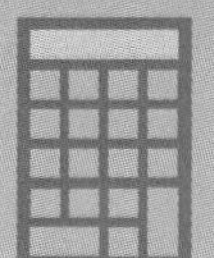

centripetal force = mass times speed times speed divided by the radius

$$F_c = \frac{mv^2}{r}$$

Example: A physics student and his bicycle have a combined mass of 60 kg. He tries to turn along a path of radius 4 meters when his speed is 7 m/s. What sideways force must act on the tires of his bike to turn them? $F_c = \frac{(60 \text{ kg}) \times (7 \text{ m/s})^2}{4 \text{ m}} =$ **735 N** (165 lb). That's a lot of force, so his tires may slip!

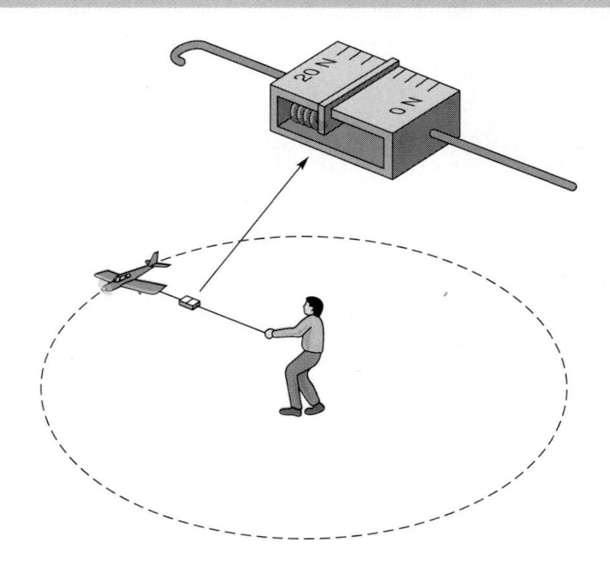

THE TENSION IN THE SPRING, MEASURED BY THE SCALE, IS $\frac{MV^2}{R}$ FOR THE MODEL AIRPLANE.

Notice the factor of v times v in the centripetal force. That means the centripetal force increases *very* rapidly if the speed increases. Perhaps as a child you played on a playground merry-go-round. When it turned about as fast as a small child could push it, hanging onto the edge was easy. The ride got much more exciting if an adult came over to spin the platform faster. Only a little extra speed made a great deal of difference in how firmly you needed to hang on. That was the effect of v^2 in the formula for the centripetal force.

AT THE RACE CALLED ROAD ATLANTA, THE TIRE TRACKS SHOW HOW THE DRIVERS TAKE EACH CURVE WITH THE LARGEST POSSIBLE RADIUS TO KEEP THE NEEDED CENTRIPETAL FORCE TO A MINIMUM.

ROTATIONS AND REVOLUTIONS: THE "FIRST LAW" OF ROTATIONAL MOTION

To this point, we've studied the motion of objects that move through a distance in space, a type of motion called a *displacement* or a *translation*. Now we will focus on *rotations,* such as when a wheel spins on its axle. When an object **rotates,** it spins about an **axis** (a straight line about which it rotates) within or on its body. If it turns about an axis outside its body, it is said to **revolve.** The earth spins around its polar axis about once per day, its rate of rotation. It revolves about the sun once per year, its rate of revolution. Sometimes such rates are measured in terms of angular speed. For example, the sun moves across earth's sky at a rate of about 15°/hr since the earth turns about 360° in 24 hours.

For translational motion, Newton's first law tells us an object's speed (or direction) is unchanged unless a net force acts to change it. A similar law applies to rotations. Toss a hammer into the air with a twist to set it spinning. While it is in the air, its spin rate stays the same until you catch it. *Once rotating, any rigid object continues to turn at the same rate if it is left alone*. This is the "first law" of rotational motion: Only an external force, applied in a definite manner as we'll see below, can change a rigid object's rate of rotation.

TORQUE: THE "FORCE" OF ROTATIONAL MOTION

A book at rest on a table can show you a lot about rotational motion. Press your forefinger on the book's center to hold it in place and spin it with the other hand, so that your forefinger becomes the axis of rotation. How do you start the book spinning? Intuitively you push at its corner, a point that's as far as possible from the axis of rotation. It's easier to spin or twist something if the force is applied *as far from the axis of rotation as possible*. Try other points on the book and compare. Notice that the *direction* of your push is just as important as the place it is applied. If you push inward toward the book's center, the book won't turn. Very likely you directed your push perpendicular to a line between where you push on the book and the axis of rotation (Figure 4-10). Experience tells you that a push in that direction, applied as far as possible from the axis of rotation, spins the book fastest.

FIGURE 4-10 IF YOU SPIN A BOOK WITH YOUR FINGERS LIKE THIS, YOU'LL SEE THAT YOU HAVE A FEELING FOR WHERE THE FORCE SHOULD BE APPLIED TO GIVE THE GREATEST INCREASE IN SPIN RATE.

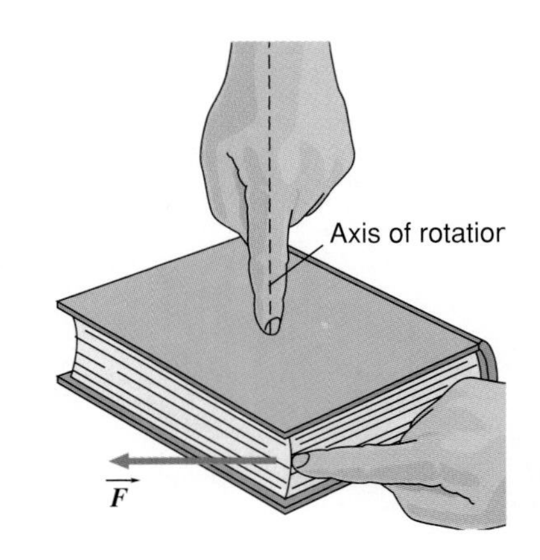

Likewise, when you push a door open, you push so that the force you exert makes a right angle with a line from where you are pushing to the hinges, the door's axis of rotation. And where should you push? Perhaps you've leaned

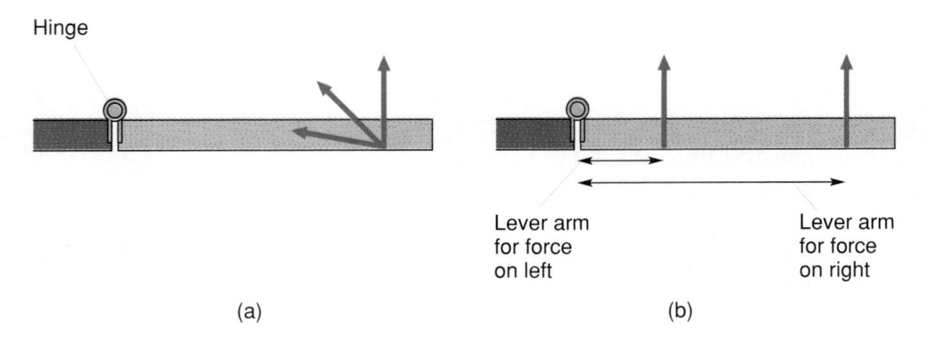

FIGURE 4-11 (A) THE BEST DIRECTION FOR YOUR PUSH IS PERPENDICULAR TO THE LINE FROM THE POINT WHERE YOU APPLY THE PUSH TO THE AXIS OF ROTATION. (B) THE LEVER ARM IS THE PERPENDICULAR DISTANCE FROM THE AXIS OF ROTATION TO THE LINE OF THE FORCE. THE LARGER THE LEVER ARM, THE GREATER THE TORQUE, AND THE LARGER THE FORCE, THE GREATER THE TORQUE.

against one edge of a heavy door, and when nothing happened you moved over to the other edge—farther from the axis of rotation, the hinges—to push the door open. We say you needed more *leverage* to get the door open. The **lever arm** of a force is the shortest distance from the axis of rotation to the line of the force, as defined in Figure 4-11. The lever arm is always perpendicular to the line of the force. Simple experiments show that an increase in the lever arm has the same effect on the rotation rate as an increase in the force. These two quantities multiplied together are what we call **torque**. Just as with forces, when more than one torque acts they must be added or subtracted to find the *net* torque.

torque = force × lever arm

Now we can restate the "first law" of rotational motion: *only a net external torque can change an object's rate of rotation.* You can think of a torque as a "rotational force." A torque can set something spinning, as when you push along the top edge of a tire to roll it on the ground. It can also stop that tire's rotation if the push from your hand is in the other direction. Two such opposing torques can balance each other if they have the same numerical value.

THE "SECOND LAW" OF ROTATIONAL MOTION

Every rigid object that rotates—book, wheel, or planet—requires a net torque to change its rate of rotation. How fast the rotational speed changes when a torque acts depends on the object's *rotational inertia,* which is *the measure of how difficult it is to change a body's rate of rotation*. The amount of an object's rotational inertia is called its **moment of inertia,** and the net torque equals its value times the rate of change of its rotation rate, or spin rate. So the "second law" of rotational motion is

net torque = moment of inertia × rate of change of spin

This law helps us explore the properties of the moment of inertia.

A yardstick (or meter stick), two cans of soup, and some tape let you experiment with rotational inertia. Tape the cans close to each side of the center point. Then, holding the stick at its center, make it rotate by twisting it back and forth. Note how easy it is to increase or decrease the rotation rate.

(a)

(b)

(A) CYNDEE TRIES TO ROTATE A STUBBORN LUG NUT ON A CAR'S WHEEL. (B) USING EXTRA LEVERAGE, WITH A PIPE, SHE INCREASES THE TORQUE TO BREAK THE LUG NUT LOOSE.

Next, tape those cans to each end of the yardstick and twirl it back and forth as before. Most of the mass is now much farther from the axis of rotation, and the difficulty of changing the system's rotation rate is impressive. This tells us the moment of inertia of an object depends strongly on how far the object's mass is from the axis of rotation. Measurements in such experiments reveal the formula for the moment of inertia I of a small object with mass m that is a distance d away from its axis of rotation: $\boldsymbol{I = md^2}$. To find the rotational inertia I for a solid object like a planet or a book, you'd need to calculate the sum of the values of md^2 for each particle in it. A slowly spinning planet would be very difficult to stop because it has a very large moment of inertia.

A POINT OF BALANCE: THE CENTER OF MASS

You can easily balance a book horizontally on your finger by finding a point near its middle where it won't rotate around your finger and fall off. This balance point lies under a point in the book called the **center of mass.** For symmetrical shapes, the center of mass is at the geometrical center. Gravity pulls on every particle in the book. On one side of the finger, torques try to rotate the book clockwise, while torques on the other side try to rotate it counterclockwise. The result is no rotation; it is as if the net weight of the book acts at the center of mass, right over your finger (Figure 4-12). (Actually, gravity acts as if its force is applied at a point called the *center of gravity,* but for our purposes the center of mass and the center of gravity can be taken to be exactly the same.) So the weight of the book presses straight down on your finger—*the weight has no lever arm and therefore cannot rotate the book*. Move your finger even 1 centimeter to the side, however, and the book's weight, acting at its center of mass, has a lever arm about your finger and exerts a torque. Your finger becomes an axis of rotation until the book falls off. Use a piece of tape to hold that same book closed and toss it in the air, giving it a twirl as you release it. Watch carefully and you'll see that the book spins naturally about this balance point. *Any freely rotating object always spins about its center of mass.*

FIGURE 4-12 THE NET EFFECT OF GRAVITY ON THE BOOK IS AS IF THE NET FORCE ACTED DIRECTLY ON THE CENTER OF MASS. IF THE PERSON MOVES HER FINGER OFF CENTER, THE BOOK WILL TWIST.

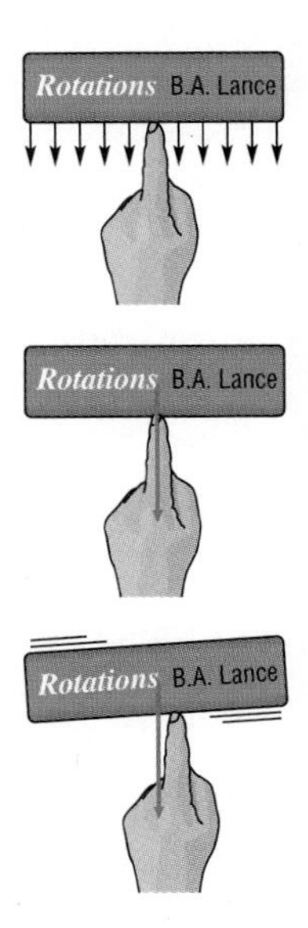

AN EVERYDAY EXAMPLE: STOPPING CARS

When a car stops quickly, it takes a nose dive, its front end dipping toward the road while the rear rises. That is, the car begins a rear-to-front rotation, and torques show why this happens. When a car stops, the friction forces on the tires point backward along the ground. These forces have lever arms with respect to the car's center of mass, and their torques begin to rotate the car. The direction of rotation is the same as you would get by lifting the rear bumper. This lightens the load on the rear tires and increases the weight on the front tires. That very action soon stops the car's rotation. When the front end dips, the car's front tires push harder against the ground, and the ground pushes back on the tires. The extra push up on the front wheels gives the car a torque that counters its rotation from the friction's torque, and the car stops rotating.

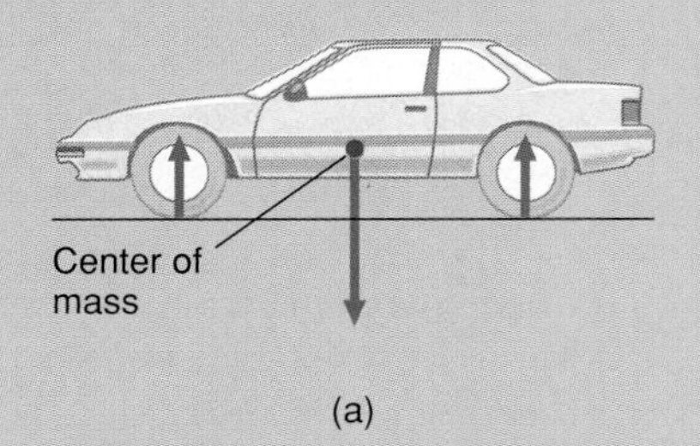

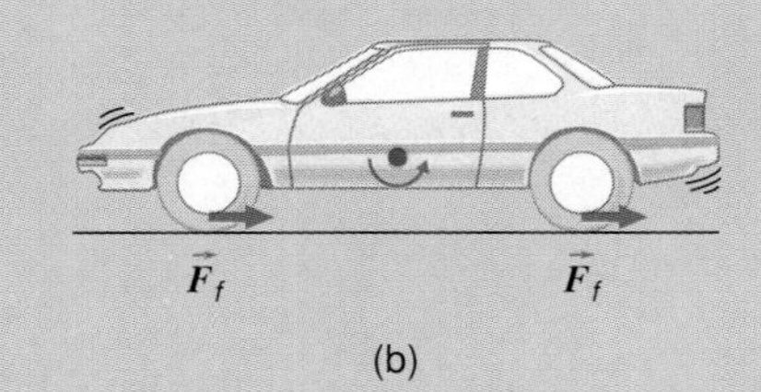

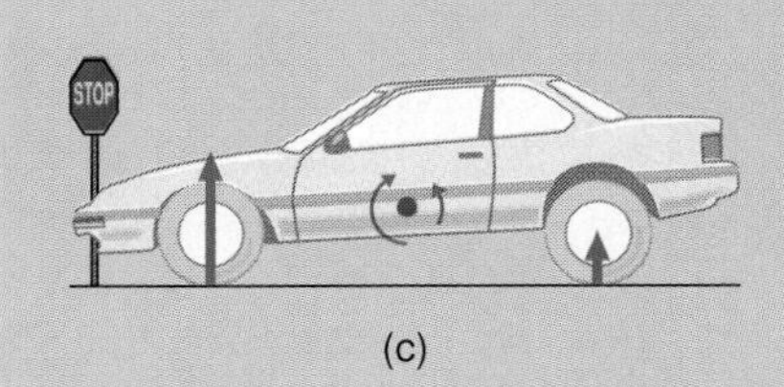

(A) APPROXIMATE VERTICAL FORCES ON A CAR DURING MOTION AT A CONSTANT VELOCITY. (B) THE FRICTIONAL FORCES FROM THE ROAD THAT STOP THE CAR WHEN THE BRAKES ARE APPLIED TEND TO ROTATE THE CAR COUNTER-CLOCKWISE ABOUT ITS CENTER OF MASS. (C) THE COUNTER-CLOCKWISE ROTATION PRESSES THE FRONT TIRES HARDER TO THE GROUND WHILE LESSENING THE PRESSURE OF THE REAR TIRES ON THE GROUND. THE GROUND PUSHES BACK MORE ON THE FRONT TIRES AND LESS ON THE REAR TIRES, AND THE NET TORQUE FROM THOSE FORCES SLOWS THE COUNTER-CLOCKWISE ROTATION AND STOPS IT.

An object's moment of inertia depends upon where its axis of rotation is. Find the balance point of a baseball bat, which is just under its center of mass. Hold the bat there with one hand and twist it back and forth. Then slide your hand in either direction and twist again. The bat rotates more easily about its center of mass than anywhere else because its smallest moment of inertia is about that axis.

STABILITY

Think of tipping over two bricks, one lying on its side and the other resting on one end, as in Figure 4-13. Grab the brick that's on its side at the bottom edge closest to you, lift it, and the brick pivots about its far edge where it meets the ground. That far edge is its axis of rotation, and you're exerting a torque. But your torque is opposed by another torque—the torque that the brick's weight exerts at the center of mass. *Unless your torque is greater than gravity's torque, the brick won't rotate.* As Figure 4-13(b) shows, if the brick is on its end, gravity's lever arm about the point of rotation is much shorter, making it easier for you to tip that brick over. Because of its relatively high center of mass, that brick also has a shorter angle to rotate through before its center of mass is beyond its base of support—in other words, it will topple over more easily than the brick that was resting on its side. For the same reasons, automobiles with wider wheelbases and lower centers of mass are more stable against turning over.

If you stand with your feet together, your base of support is small, and a slight push can move your center of gravity to one side of your feet. Then if

(a)

(b)

FIGURE 4-13 (A) WHICH IS HARDER TO ROTATE CLOCKWISE? (B) IT IS EASIER TO ROTATE THE BLOCK THAT IS STANDING, BECAUSE THE FORCE OF GRAVITY, WHICH ACTS AT THE CENTER OF MASS, HAS A SMALLER LEVER ARM AND SO EXERTS A SMALLER TORQUE TO OPPOSE THE ROTATION.

you don't shuffle your feet or grab something to steady yourself to oppose gravity's torque, you'll fall. Subway riders routinely stand with feet planted far apart for stability as the cars sway. Toddlers trying out inexperienced legs wobble as they learn to summon the small torques from back and leg muscles, and feet and toes, that keep them upright. A person who's had too much alcohol has slower-than-normal reaction times, and the delayed responses to correct his or her balance lead to "tipsiness."

Summary and Key Terms

Newton's laws tell us that a net force causes an acceleration, and if there is no acceleration, then there is no net force. For example, when you take a step, the backward push of your foot on the floor gives rise to the forward push of friction on you, giving a net force that accelerates your body forward.

Air drag is the frictional force air exerts on objects moving through it. The faster an object moves, the greater the drag becomes. If an object is falling through air, the drag grows until the object reaches its ***terminal speed,*** when the forces of air drag and gravity balance and the net force is zero. Air exerts an upward force on a moving airplane wing ***lift.***

terminal speed–53

lift–53

Pressure is the force applied to a unit of area: $P = F/A$. If the area of application is very small, such as the edge of a razor, the applied force on that area doesn't have to be large to create a large pressure. ***Centripetal force,*** the force that turns objects, points toward the center of the circle that matches the curve the object travels on. It changes only the object's direction, not its speed, because it points perpendicular to the object's direction of travel. Centripetal force grows very rapidly with increasing speed according to the equation $F_c = mv^2/r$.

pressure–54

centripetal force–55

Rotational motion occurs when an object ***rotates,*** turning about an ***axis*** within or on its body, or ***revolves,*** turning about an axis outside its body. To change an object's spin, you must apply a force with a ***lever arm,*** which is the perpendicular distance from the axis of rotation to the line of the force. The product of the lever arm and the force is called ***torque,*** and *only an external torque can change an object's spin.* An object's ***moment of inertia,*** its amount of rotational inertia, is a measure of its resistance to changing its spin rate. ***Net torque*** = moment of inertia × rate of change of spin. An object's moment of inertia can take on different values depending on the location of the object's axis of rotation. Any freely rotating body rotates around its ***center of mass,*** where its moment of inertia is the least.

rotate–57

axis–57

revolves– 57

lever arm–58

torque–58

moment of inertia–58

net torque–58

center of mass–59

EXERCISES

Concept Checks: Answer true or false

1. Only an external force can cause your body to accelerate; no amount of motion with your arms and legs alone will cause you to accelerate.
2. If you stand on an incline, the full force of your weight presses directly into the incline.
3. If you pull at an upward angle on a trunk, its contact force with the floor remains the same.
4. A falling object can have only one terminal speed.
5. When a solid object moves through air, its speed affects the amount of drag on it, as does its shape and its size.
6. When a plane travels horizontally with a constant speed, its drag balances its thrust.
7. Pressure is defined as force times area.
8. The centripetal force is always perpendicular to a turning object's direction at any instant.
9. The centripetal force on an object depends on its speed and on the radius of the circular path it moves on, and that's all.
10. Either friction, gravity, or a contact force can supply the centripetal force on an object if the circumstances are right.
11. The longest possible lever arm and the largest possible force produce the fastest change in rate of rotation.
12. Without a net torque, a rotating body's rate of rotation cannot change.
13. The measure of rotational inertia is called the moment of inertia.
14. The moment of inertia of a rotating object depends on how its mass is distributed about the axis of rotation.

Applying the Concepts

15. If a wagon is pulled at a constant velocity, the net force on it is zero. (a) true (b) false (c) you can't tell from this information.
16. As you walk, do you make use of friction with the ground in both the forward and the backward directions? (*Hint:* Consider what happens when you are walking along and then you stop.)
17. Faced with getting a shopping cart over a curb in a parking lot without tilting it, would you push it or pull it? Explain.
18. When you get up from a sitting position on the floor, your feet push down with a force equal to (a) your weight (b) more than your weight (c) less than your weight.
19. If you are on a scale in an elevator that is going upward at a constant speed, will the scale's indicator show (a) more than your weight (b) less than your weight (c) exactly your weight?
20. If you find it takes a push of 3 lb to slide your dictionary at a constant speed across a table, then what is the value of the kinetic frictional force retarding its motion?
21. A man (m = 65 kg) in an elevator accelerates upward at 1.2 m/s^2. What is the net force on him?
22. Standing on a slope, your friend (m = 50 kg) begins to slip. If the component of your friend's weight that points down the slope is 100 N, and the kinetic friction on your friend's shoes is 40 N, find your friend's acceleration along the slope.
23. When a weightlifter cleans and jerks a 250-lb barbell (lifts it from the ground and pushes it overhead), the force exerted on the barbell must be (a) less than (b) equal to (c) greater than 250 lb. Why?
24. When a solid object moves through air, does its mass affect the amount of drag on it?
25. When a falling object reaches its terminal speed through the air, is it still accelerating? Is it still moving?

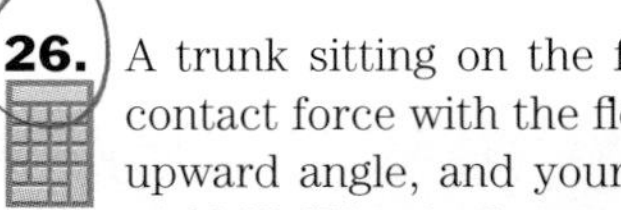

26. A trunk sitting on the floor weighs 250 N. What is its contact force with the floor? You pull on the trunk at an upward angle, and your pull has a vertical component of 60 N. What is the contact force of the trunk with the floor while you are pulling on it?
27. What reason can you think of for the bottom of an elephant's foot to be so large and flat? Why should the small foot (hoof) of a deer be so hard?
28. The practice of putting horseshoes on horses originated in Europe during the Middle Ages. Explain how the shoes protect the horse's hoofs from damage when they carry heavy loads—such as knights in metal armor during the Crusades.
29. If there is a wind blowing, an eagle will land on its nest into (or against) the wind. Why? (It's the same reason airplanes take off and land into the wind.)
30. The bones in your fingers are flat on the gripping side. What's the advantage of flattened finger bones compared with round finger bones when you grip something firmly or lift something heavy?
31. Give *two* reasons why it is harder to stop a car quickly when it is going downhill.
32. On sea ice too thin to support a walking person, a polar bear lies on its belly and glides, pushing itself along with gentle strokes from its arms and legs. How can it get away with that?
33. If you walk on your heels at the beach, why do you sink deeper into the sand? To avoid sinking in snow, how does it help to wear snowshoes or skis?

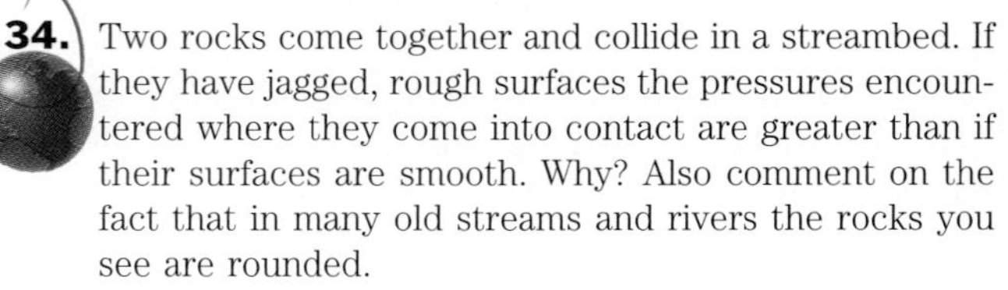

34. Two rocks come together and collide in a streambed. If they have jagged, rough surfaces the pressures encountered where they come into contact are greater than if their surfaces are smooth. Why? Also comment on the fact that in many old streams and rivers the rocks you see are rounded.

35. When your car travels with constant speed on level ground, the forward push your tires get from the road balances the air's drag on the car. True or false? Why?

36. Why is a faulty trailer hitch more likely to break if a car is going uphill at a constant speed than on a level road at that same constant speed?

37. Give the effects on the force needed to turn a car if (a) the car's mass (m) is doubled, (b) the car's speed (v) is doubled, (c) the radius (r) of the car's path is doubled; (d) m, v, and r are all doubled.

38. A lineman on a football team often tries to block by getting his body under his opponent's and pushing upward. What does this do to the frictional force on the opposing lineman's feet? How does that affect the opposing lineman's acceleration?

39. If a speeding car suddenly goes over a crest in the road, do the passengers feel heavier or lighter? If the speeding car is going down a hill and then the road suddenly rises, do they feel heavier or lighter? Explain why, using diagrams in each case showing the forces.

40. When someone you are riding with takes a curve too fast and you slide up against the car door, what force moved you over? What force turned you after you pressed against the door? What force acted as a centripetal force?

41. You've been traveling on a four-lane interstate highway just after a rain, and you see a curve ahead. You want to maintain your speed, minimizing the risk of skidding. Which lane should you take and why?

42. What is your motion like if you let go from the edge of a child's merry-go-round that is spinning fast?

43. A certain car can accelerate along a road at a maximum rate of 2 m/s^2 without spinning its tires. If it is hooked up to a trailer with twice its own mass, how fast could it accelerate without burning rubber?

44. Aristotle believed that the earth must not rotate, or else he'd be flung off! Discuss his conjecture.

45. Which way do you lean if you hold a basket of wet clothes or a heavy bag of groceries in front of you? Why? Which way do you lean if you strap a heavy pack on your back? Why?

46. Which of the three ways depicted in Figure 4-14 is the hardest way to do sit-ups? Should taller people have a harder time doing sit-ups?

47. A bowling ball that is released with no spin slides along the lane and immediately begins to roll. Why?

48. Why is it easier for you to walk over a log across a creek, or on one of the iron rails of a railroad track, if you hold your hands out to the side? Why does it help even more if you hold one of your shoes out in each hand as you walk?

49. Maybe you've noticed that large fan blades take longer to stop rotating than small ones do after the current has been turned off. Why should that be?

50. Comment on why redwood trees, the tallest on earth, grow so vertically. (Trees are stable only because their root systems can summon torques from the ground to counter the torques exerted by wind.)

51. Why does the front end of your car rise a little when you accelerate rapidly?

52. Why do tightrope walkers carry those long, long poles? Discuss.

53. Explain why a cyclist on a 10-speed bike will slide as far back as the seat will allow when forced to make an emergency stop.

54. Connie wishes to replace her model airplane attached to a string. To keep her sport relatively easy on her hands, should she replace her model with one that weighs the same but goes twice as fast, or one that goes just as fast but weighs twice as much?

55. The world's fastest elevator is in a building in Japan and reaches a speed of 2475 ft/min. If the passengers are not to be subjected to an apparent weight increase of more than 0.1 g, how long would it take that elevator to reach that speed?

56. A fully loaded 747 airplane weighs 600,000 pounds. How much lift must this plane have from its wings when it is cruising with zero vertical motion? Its wings have about 7000 square feet of area contributing to the lift, so what is the net average upward pressure?

57. When you push open a heavy door, how much harder would you have to push at the center than you would at the edge far from the hinges just to get the door moving?

58. The camera operator and camera on the lightweight boom in Figure 4-15 weigh 500 lb. How heavy must the counterweight be to balance the boom? (Ignore the weight of the boom.)

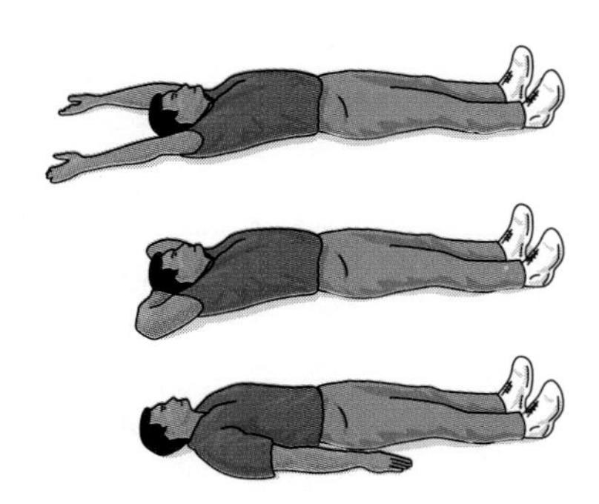

FIGURE 4-14

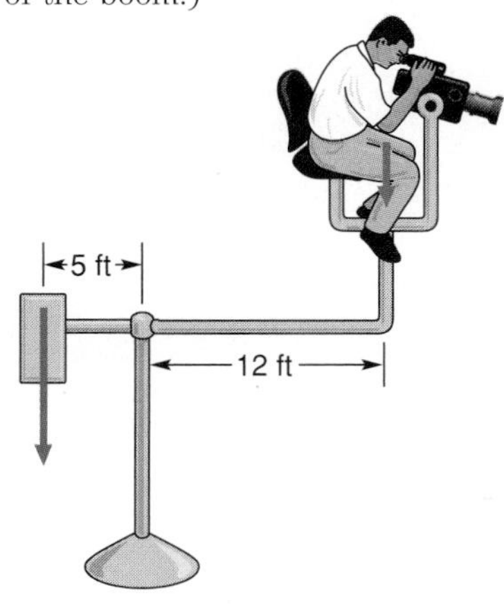

FIGURE 4-15

Demonstrations

1. Make missiles of slippery watermelon seeds. Place one between your thumb and forefinger, pinch down hard, and the seed pops out at a high rate of speed. The wet surface of the seed is almost frictionless and its sides are never perfectly parallel, so when you press against them the opposing forces from your two fingers don't quite cancel. As shown in Figure 4-16, a net force pushes the seed along between the fingers.

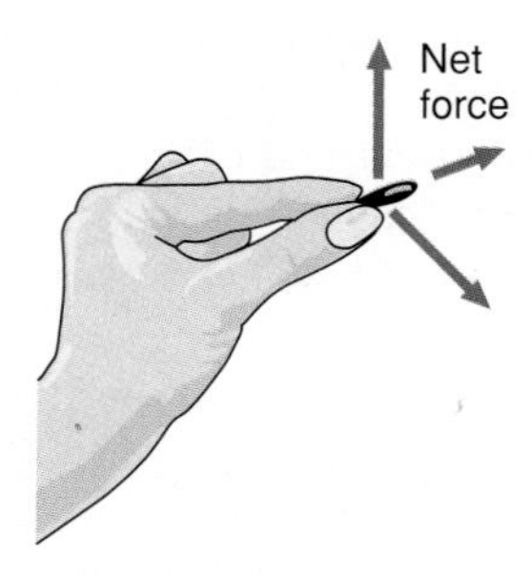

FIGURE 4-16

2. Pick up a book with one hand and a small sheet of paper with the other. Hold them horizontally and drop them side by side. The sheet of paper descends slowly while the book zips down, accelerating at the rate of g. Next, hold the book horizontally and then lay the sheet of paper on the book's cover. When you drop the book again, there'll be no air drag on the paper. Now crumple the paper into a small wad and drop the paper and the book side by side again. Although the mass of the paper is the same as before, the change in the size and shape of its surface area obviously affects the paper's acceleration.
3. Hold a penny or a dime horizontally between thumb and forefinger. Slowly decrease your fingers' pressure until the coin falls. It's almost impossible to keep that coin from rotating a little. Why? The chances of both fingers releasing it simultaneously are almost zero. The last finger touching the coin acts for a brief interval as an axis of rotation for the coin. Gravity acting at the coin's center of mass exerts a torque that makes it slowly rotate about this axis as it falls free of that last finger.

FIVE

Momentum

Airborne after ascending a large mogul, these motorcross riders swing their legs outward to counter the rotations of their dirt bikes. This chapter is about conservation laws of motion, both straight-line and rotational.

As we saw in the last two chapters, we can predict something's motion once we know the forces acting on it. Consider what happens when a kicker's foot strikes a football, though. The force begins to grow at the instant of contact and continues to grow as the foot compresses the football and accelerates it. The force then drops in value as the compressed football springs forward, finally going to zero when the ball loses contact with the foot and goes on its way. In such a situation it is nearly impossible to know the rapidly changing forces very well, but the physics discussed in this chapter can help us to understand events like these more clearly. This chapter takes up the subject of *momentum*, and its properties give us another way to deal with motion, especially when it is impossible or impractical to use $\vec{F}_{net} = m\vec{a}$.

IMPULSE AND MOMENTUM

To stop a car, you can slam on the brakes to stop quickly or you can apply the brakes gently for a longer time. In other words, a large force exerted for a short time results in the same change in speed as a small force exerted for a long time. For motion along a straight line, we can rearrange Newton's second law to see exactly how time and force change something's speed. Since $F_{net} = ma$, we can say net force = mass × change in speed/time. Multiplying both sides by time so that force and time appear together, we have

PUTTING IT IN NUMBERS—IMPULSE AND MOMENTUM

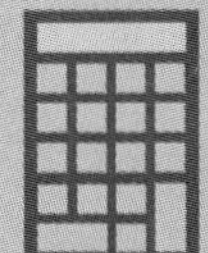

net force × time = mass × change in speed

$$F_{net}\, t = m(v_f - v_i) = mv_f - mv_i$$

The quantity net force × time is called **impulse**, while the product of mass and speed, *mv*, is called **momentum**. The equation above tells us that **impulse = change in momentum**. The units of impulse are newton-seconds (N·s) in the SI system and pound-seconds (lb·s) in the British system. The units of momentum most often used are kilogram·meter/second, or (kg·m/s). **Example**: This formula tells us that if 800 newtons of force can bring a car to rest in a time of 3 seconds, then 200 newtons of force can do the same job in 12 seconds. That is, (800 N × 3 s = **2400 N·s**) or (200 N × 12 s = **2400 N·s**) causes the same change in speed. Any product of force and time equal to this value would bring your car to rest.

FIGURE 5-1 THIS FIRE ENGINE IS ABOUT 13 TIMES AS MASSIVE AS THIS CAR. AT THE SAME SPEED, THE FIRE ENGINE WOULD HAVE 13 TIMES AS MUCH MOMENTUM AND REQUIRE 13 TIMES AS MUCH IMPULSE TO STOP AS THE CAR WOULD.

The right-hand side of the impulse–momentum formula shows that a car that's twice as massive as another needs twice as much impulse to stop from the same speed (Figure 5-1). That makes sense. If you've ever pulled a trailer or driven a car overloaded with passengers, you know that extra mass makes your car more difficult to stop. Also, you're aware that the faster you go, the harder it is to stop. If your car goes twice as fast, the right-hand side of the formula tells you that again you need twice as much impulse to stop it. So speed and mass have equal effects on how much impulse is needed. Both impulse and momentum are vector quantities, with the direction of the impulse being the direction of the net force and the direction of the momentum being the direction of the velocity. However, we will focus on what happens with these quantities when the motion is along a straight line, in one dimension.

Keep in mind that it is the impulse of the *net* force on an object that equals the object's change in momentum. You can push against a wall for ten minutes and calculate $F \times t$, but the wall's momentum doesn't change at all! That's because your push on the wall is *not the net force* on it. The reaction forces the wall gets from the building and the ground balance your force so that the impulse from the net force is zero.

If you jump down from a small stool, you automatically flex your legs at the knees. Why? Because that brings the upper part of your body to a stop more slowly. If you hit the ground stiff-legged, your body stops suddenly and the force on your feet, which is passed on through your bones, is enough to jar your teeth. So you bend your legs and they act as shock absorbers, reducing the *force* your body feels by increasing the *time* of impact (Figure 5-2). Notice that the final change in momentum is the same no matter how you land, because when you have stopped your momentum is zero! Bending your knee when you walk down stairs has the same effect, spreading the impact of each step over a longer time to ease the shock of decelerating the rest of your body. You intuitively arrange the impulses on your body to make life easier and safer.

Modern cars are designed to "give" and crumple if they suffer a front-end collision, which brings the rest of the car to a stop over a more extended time. If the passengers are not fastened down in shoulder harnesses and seatbelts, they leave the car seat as it slows beneath them (Newton's first law) and continue moving until the dashboard or steering wheel stops their motion. There they decelerate very quickly and feel a much larger force than if they had remained in the seat. Passengers wearing seatbelts and shoulder harnesses come to rest more slowly, with the rest of the vehicle, and they feel a smaller force for a relatively longer time interval.

THIS BUILDING, WITH A HUGE MASS, HAS LESS MOMENTUM RELATIVE TO THE GROUND THAN ANY CAR THAT PASSES IT.

Momentum is a familiar concept. We all learn early just how hard it is to stop a moving object and that it's not just speed that matters. Most of us would have no qualms about trying to catch a softball moving at 20 mph (a slow pitch) even without a glove, but we know better than to try the same stunt with a firetruck even if it's only moving at 5 mph. Because of its huge mass, the firetruck's momentum is much greater than the softball's, even though the ball moves much faster. So if you pushed as hard as you could (F), it would take a much longer time (t) to bring the firetruck to rest. $F \times t =$ change in mv.

High speed or great mass can make momentum large. That's why the linemen of professional football teams should be massive. The opposing linemen are positioned close together in a game and can't gather much speed before making contact. Nevertheless, their large masses give each one substantial momentum. Yet momentum is also why lighter-weight running backs traveling at full speed can be so hard to stop.

FIGURE 5-2 RIDERS OF DIRT BIKES OR MOUNTAIN BIKES ON ROUGH TERRAIN STAND ON THE FOOTPEGS SO THEIR KNEES CAN FLEX, LETTING THEIR BODIES ABSORB THE SHOCKS FROM BUMPS OVER LONGER TIME INTERVALS.

SIMPLE COLLISIONS

Collisions were an active topic of physics research in Isaac Newton's day. Soon after the Royal Society of London was formed in 1662 (see page 33), its members began to collect facts about collisions and set up experiments at their meetings. A *simple* collision is one where two objects collide and only their contact forces push them apart. That is, there are no attached springs or explosives that can be set off to push them apart faster than they came together. Experiments with simple, straight-line collisions show there are natural limits to what takes place. For example, suppose two freight cars on a railroad track bump into each other and couple together, as in Figure 5-3. Because they stick together, their relative speed after the collision is zero, even if they move off together. Such a collision is called a **perfectly inelastic** collision.

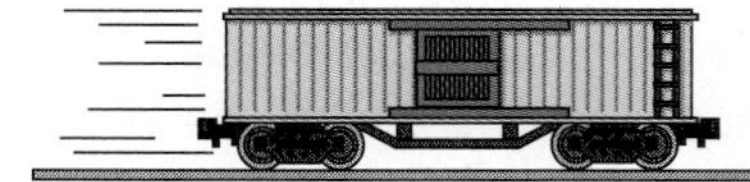

(a)

(b)

FIGURE 5-3 TWO FREIGHT CARS THAT COLLIDE AND STICK TOGETHER HAVE A PERFECTLY INELASTIC COLLISION.

CRAWLING WITH MOMENTUM

To get from the launch pad into its initial orbit, the space shuttle soars some 60 miles up and 800 miles out over the Atlantic in only ten minutes. But to get to the launch pad from the assembly building at Cape Kennedy, a distance of about 3.5 miles, the shuttle spends eight hours in the slow lane. The crawler beneath it is the size of a major league baseball diamond and weighs 12 million pounds. As twin 2750-horsepower engines ease this monstrous moving platform along, 16 hydraulic jacks constantly tilt it in small amounts to keep the shuttle upright—even a slight sea breeze exerts a dangerous force on a spaceship that large. With a maximum speed of only 1 mph, the crawler must deliver the shuttle to within two inches of a center point on the launch pad.

At such a leisurely pace, you'd think the drivers could take it easy as well as slow, but that's not the case. There are two turns on the way to the pad. Though they look gentle by most standards, they look like Grand Prix corners in the driver's mind. The enormous mass of the crawler and its load gives it a huge momentum, making turns difficult and quick stops or adjustments impossible. The driver must start easing into each turn well before the crawler gets there, getting help only from a crew of spotters. Controlling that much momentum, even at a snail's pace, has to be done smoothly to avoid disaster.

A different kind of simple collision is often seen on a pool table. An incoming ball can slam into a target ball at rest. The moving ball stops dead, and the target ball scoots across the table with the same direction and (almost) the same speed the incoming ball had. That is, even though the individual speeds for the balls change, the *relative* speed after the collision is almost the same as the relative speed before. If *no* relative speed is lost, the collision is called **perfectly elastic,** and experiments show that's the upper limit for relative speeds after simple collisions. Most collisions fall somewhere between perfectly inelastic and perfectly elastic, so that some relative speed is lost, but not all. These are called **inelastic** collisions. (We'll see in Chapter 6 that bending, breaking, or friction during a collision reduces the relative speed afterward.)

Two tennis examples illustrate these natural limits. Suppose you use a tennis racket to serve a ball made of very soft and sticky clay. It would stick to the racket, so after the collision its relative speed with respect to the racket would be zero. That's a perfectly inelastic collision. Next, consider what happens when someone serves a real tennis ball (Figure 5-4). The ball's speed is almost zero just before the racket hits it, so the relative speed between the ball and the racket is just the speed of the oncoming racket, which we can call v_r. Because the racket and the player's moving arm are so massive compared to the ball, after the collision the racket's speed is still close to its original value, v_r. What would the ball's speed be? If the collision were perfectly elastic, the served ball would move at a little less than $2v_r$! Why? After a perfectly elastic collision, the ball would move *away from the racket* with a speed of v_r, the same relative speed it had before the collision. (Remember, that is as fast as it can move away after a simple collision.) Since the racket is also still moving

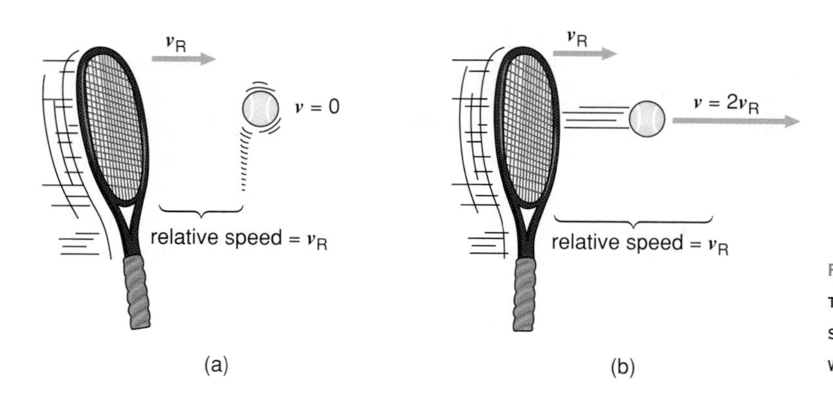

FIGURE 5-4 IF THE COLLISION BETWEEN A TENNIS RACKET AND THE BALL AS IT IS SERVED COULD BE PERFECTLY ELASTIC, THIS WOULD BE THE RESULT.

at almost v_r after the collision, the ball must move at v_r *plus* the racket's speed after the collision, or almost $2v_r$. So if your racket moves at 40 mph, the tennis ball you serve can approach 80 mph.

Most collisions, as we've said, fall between the extremes of perfectly elastic and perfectly inelastic. If a new baseball is struck with a bat, the relative speed afterward is only about 55 percent of the relative speed before. Even golf balls do not keep all of their relative speed before the collision with the club, as you can see in Figure 5-5. Once again, we will return to these points in Chapter 6.

FIGURE 5-5 A STROBE PHOTO OF A GOLF BALL BEING HIT. YOU COULD MEASURE THE RELATIVE SPEED BEFORE AND AFTER THE COLLISION IN THIS PHOTO AND DETERMINE HOW MUCH OF THE RELATIVE SPEED BETWEEN THE BALL AND THE CLUB WAS LOST DURING THE COLLISION.

CONSERVATION OF MOMENTUM

When Newton tallied up the values of the momentum for the colliding objects in his experiments, he found a remarkable fact. Whenever two bodies collide, the momentum of each will change, but whatever momentum is lost by one body is gained by the other so that the *sum* of the momenta of the bodies is constant—*so long as there is no net external force acting on either body.* Newton found this to be true even when the collisions took place in two or three dimensions, and even when there were more than two bodies colliding. In the absence of net external forces, the total vector momentum *before* a collision is the same as the total vector momentum *after* the collision.

Let's see what this means. If a truck collides with a car, as in Figure 5-6,

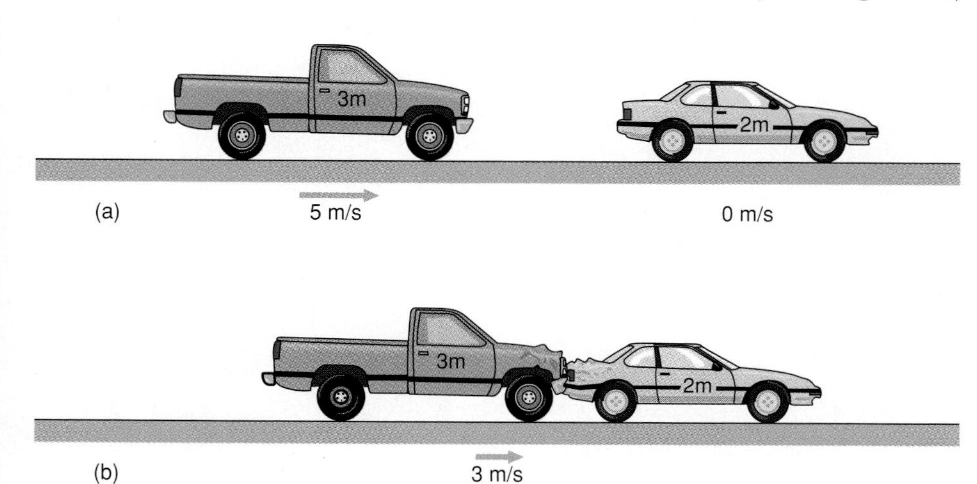

FIGURE 5-6 A COLLISION THAT SHOWS CONSERVATION OF MOMENTUM. (A) THE INITIAL MOMENTUM IS ALL BECAUSE OF THE TRUCK'S MOTION AND EQUALS ($3M \times 5$ M/S). (B) THE FINAL MOMENTUM HAS TO BE EQUAL TO THAT SAME VALUE, AND SINCE BOTH CARS MOVE OFF WITH THE SAME SPEED, ($5M \times V_{FINAL}$) = ($3M \times 5$ M/S), SO $V_{FINAL} = 3$ M/S.

each gets an impulse that is equal to the change in its momentum. So if the momentum lost by the truck is gained by the car, they must receive equal but opposite impulses. That is, the force exerted by the truck on the car must be equal to but opposite the force exerted by the car on the truck, since they exert forces on each other for the same amount of time—the time of contact during the collision. Equal but opposite forces is Newton's third law. This is how Newton discovered the third law, through his experiments with collisions.

Newton's discovery of the constancy of momentum goes beyond simple collisions: Even if interacting objects have internal springs that release to push them apart faster than they came together, or explosions take place to hurl the matter in every direction, no matter what sort of interaction, the net momentum of the group of participants stays the same. It's been true in any interaction that has ever been checked, so it has the status of a physical law. This far-reaching property of motion is called **conservation of momentum.** *The net momentum never changes during a collision, so long as there are no net external forces acting.* Momentum, which Newton called the "quantity of motion," is an extraordinary property of matter!

When using the law of conservation of momentum, you must take every interaction into account. Throw a soccer ball and you've changed its momentum, but you took part in this; you're part of the interaction. Your momentum and the ball's momentum were both zero initially. When you gave the ball momentum, conservation of momentum says you should receive an opposite amount of momentum so that the sum would still be zero. You don't seem to move, however, and so it looks as if you have created momentum! Remember,

THE IMPORTANCE OF A CONSERVATION LAW

Because it holds true *no matter what forces occur,* the law of conservation of momentum lets us calculate what happens in interactions even if the forces are not known. The details of the forces involved in a two-car collision are both complicated and impossible to know exactly, yet the total momentum of the cars just before the wreck will be equal to their total momentum just afterward. This conservation law lets us bypass using forces and still make predictions about what happens when two or more objects interact with forces—anywhere, anytime. A conservation law is a powerful realization about nature!

PHYSICS IN 3D/A COLLISION IN 3 DIMENSIONS

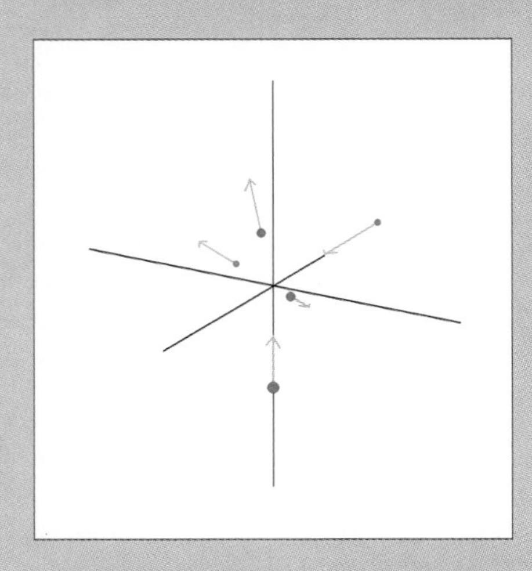

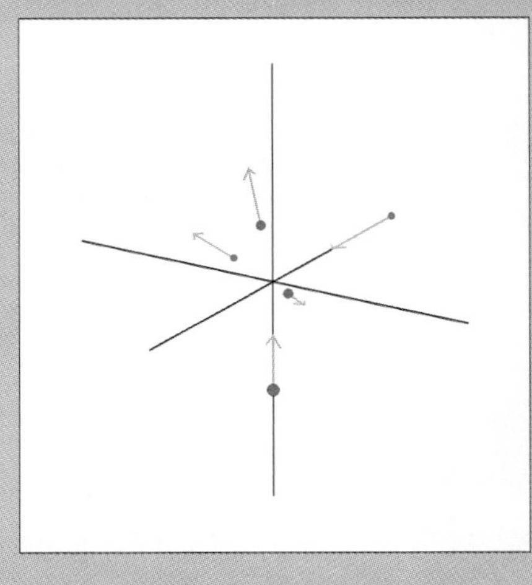

Two objects collide at the origin of the axes, and one breaks into two pieces afterward. The total vector momentum before such a collision will always equal the total vector momentum afterward, even though more objects carry away the momentum.

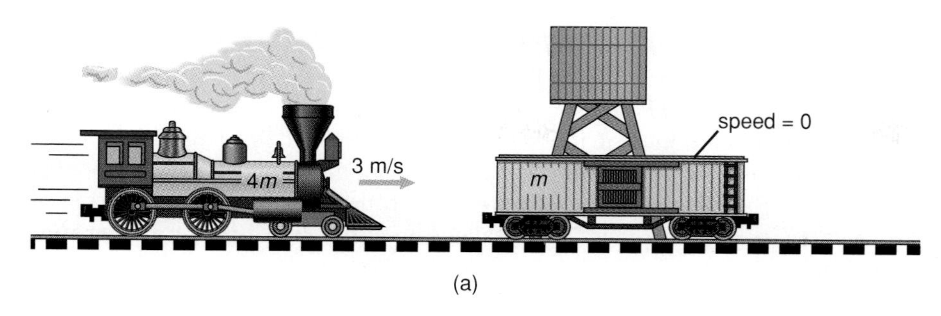

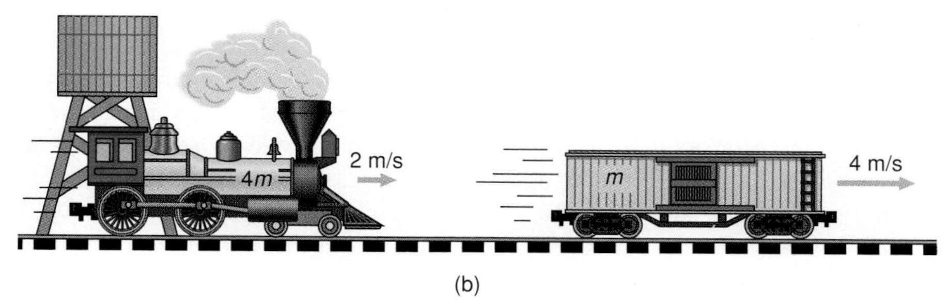

FIGURE 5-7 IS MOMENTUM CONSERVED IN THIS COLLISION? THE INITIAL MOMENTUM IS DUE TO THE MOTION OF THE ENGINE, $4M \times 3$ M/S. THE FINAL MOMENTUM IS DUE TO THE MOTIONS OF BOTH. IS THE INITIAL MOMENTUM EQUAL TO THE FINAL MOMENTUM?

however, that you interact with the earth's surface while you interact with the soccer ball. The earth is part of the system in this interaction. Friction stops you from moving backward, so *you and the earth move backward together* (an unobservable amount, to be sure) while the ball is traveling through the air. Then, when the ball hits the ground, friction brings it to rest. The ball's forward momentum is transferred back to earth and to you, stopping your ever-so-tiny backward motion. Momentum is conserved after all. For a numerical example of this law, see Figure 5-7.

ANGULAR MOMENTUM

There is momentum associated with spinning motions as well as translational motions, and the "spinning" type of momentum has its own conservation law. The following examples point this out. Picture an Olympic ice skater spinning with her arms outstretched. As she brings them in close to her body, her spin rate soars until she is just a blur. In the last chapter we saw that only a torque from an external force could cause a change in the spin rate of a *rigid* body. But the skater changes her shape, causing her spin rate to change without torques from external forces. You can get the same effect with a playground merry-go-round. Pack some friends around the edge, start the rotation, and then have everyone step to the center. The rotational speed of merry-go-round plus friends picks up greatly.

The skater's spin rate changed because she brought the mass of her arms closer to her axis of rotation. This shift of mass caused her moment of inertia to decrease, and in turn her rate of rotation increased. The same thing took place when the people on the merry-go-round moved to the center. The moment of inertia of the system (of merry-go-round and riders) became less and the rate of rotation rose. Measurements of such systems where there is no net external torque show that the moment of inertia and the angular speed change in such a way that their product remains constant. This product is called the **angular momentum.**

angular momentum = moment of inertia × angular speed

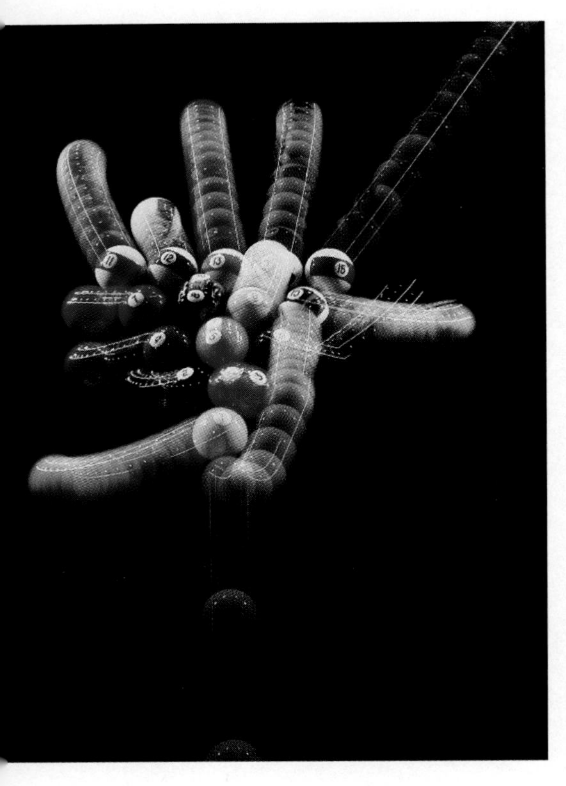

FIGURE 5-8 COLLISIONS ON POOL TABLES SHOW THE CONSERVATION OF LINEAR AND ANGULAR MOMENTUM. HERE THE CUE BALL COMES FROM THE BOTTOM OF THE PHOTO AND GIVES THE OTHER BALLS THAT WERE AT REST SOME OF BOTH TYPES OF ITS MOMENTUM, AS YOU CAN SEE FROM THEIR SUBSEQUENT MOTIONS.

CONSERVATION OF ANGULAR MOMENTUM

The spin rate of a rigid body is constant so long as there is no external torque. This means the product of its moment of inertia and its angular speed is constant, just as for the ice skater who changes her moment of inertia by moving her arms. In the absence of a net external torque, the angular momentum of *any* system is constant. This fact is called the **conservation of angular momentum.**

A visit to the campus pool hall lets you experiment with the conservation of linear and angular momentum. Figure 5-8 shows the collision of a fast-spinning cue ball (from the bottom of the photo) with a collection of balls at rest. The cue ball carries both linear momentum and angular momentum (due to its spin), and after the collision some of the other balls leave spinning, with angular momentum taken from the cue ball. (You can see those balls turning as they travel, because they spin against the felt of the table rather than simply rolling, summoning kinetic friction which turns them.) And while the balls scatter in many directions, you can see that the net linear momentum of the balls after the collision is toward the top of the photo, as was the cue ball's linear momentum before the collision. Since the balls react with the table, their linear and angular momentum does not remain at a constant level, however. You know they will slow and stop and also stop spinning, and we will discuss why in the next chapter.

A change in linear momentum takes place if an object's direction of travel changes, and a change in angular momentum takes place if the direction of an object's spin axis changes. A child's top demonstrates what can happen when a torque acts to change the direction of an object's spin axis.

Gravity's pull on the top in Figure 5-9 acts to topple it over. Figure 5-10 shows what happens when the top is spinning rapidly. Now the top does not fall over. Instead, gravity's torque and the torque from the contact force from the table cause the axis of rotation to swing around, to "circle" a line in the vertical direction, maintaining the same angle. This fascinating motion is called **precession.** The greater the angular momentum of a spinning object, the harder it is to change the direction of its axis of rotation and the smaller the rate of precession is for a given torque.

The stability of an object's spin direction is why quarterbacks throw spiral passes. The spinning football keeps traveling through the air in its most streamlined position, making its airborne motion predictable. Even if the axis is not perfectly aligned when the ball is thrown, a slight torque from the air will cause only precession of the ball's spin axis. Then it will wobble, but it won't

FIGURE 5-9 WHEN RELEASED AT AN ANGLE TO THE VERTICAL, A TOP THAT ISN'T SPINNING SIMPLY FALLS OVER.

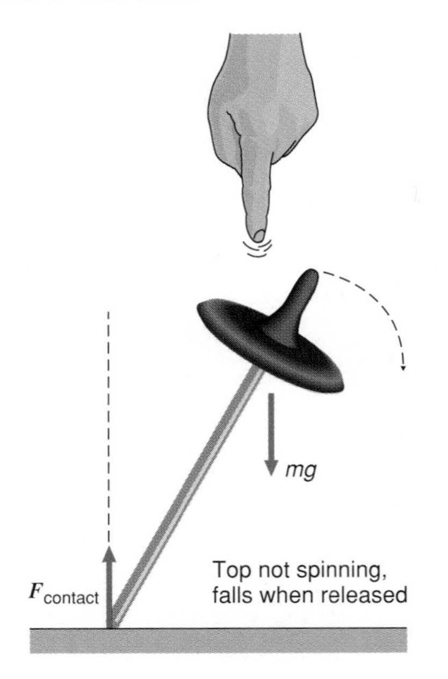

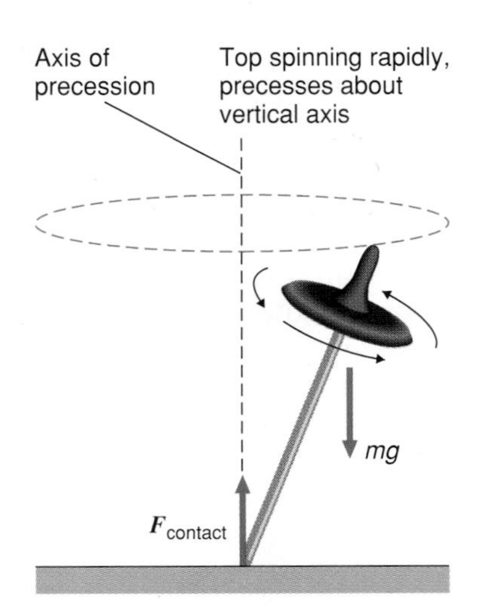

FIGURE 5-10 IF THE TOP IS SPINNING RAPIDLY, ITS ANGULAR MOMENTUM CAUSES IT TO MOVE AROUND THE VERTICAL LINE (PRECESS), KEEPING ITS ORIGINAL ANGLE WITH RESPECT TO THE VERTICAL DIRECTION.

turn end over end and increase its air resistance as a nonrotating football might. This is also why a gun barrel contains grooves that set the bullet spinning. The angular momentum of the spinning bullet helps prevent tumbling.

Angular momentum can be large because of a large moment of inertia, as with the earth, or because of a large angular speed, as is the case with a gyroscope. A modern gyroscope is a very rapidly spinning toplike object mounted on nearly frictionless swivels and bearings. The mount is built to turn in any direction without exerting a torque on the gyroscope. As its mount changes directions, the whirling gyroscope stays pointed almost perfectly in the same direction. A gyroscope keeps a predetermined reference direction in spacecraft, airplanes, submarines, and guided missiles, even through the most complicated motions.

CHANGING THE SPIN RATE OF EARTH; EL NIÑO

EARTH'S ROTATION

In 1976 the U.S. National Meteorological Center in Washington, DC, began keeping records of wind speeds worldwide. These records let researchers estimate the total angular momentum of the earth's atmosphere. In January 1983, when El Niño (a huge warming of the Pacific Ocean) was at its peak, winds moving from west to east raised the angular momentum of the atmosphere by nearly 10 percent. Scientists involved in measuring our planet's rotation rate noticed a drop in earth's angular speed during that same period, and the loss of angular momentum by the solid earth matched the gain of angular momentum of the atmosphere.

Summary and Key Terms

In a straight line, ***impulse*** is the product of net force and time, and ***momentum*** is the product of mass and speed. Impulse and momentum are vector quantities: Impulse points in the direction of the net force, and momentum points in the direction of the velocity.

impulse–66
momentum–66

We often use the *impulse = change in momentum* relation by instinct. You bend your knees as you jump down from a chair in order to increase your time of impact with the floor. Since your change in momentum is the same no matter how long it takes you to stop, increasing time serves to decrease the force on you, making the impact safer for your body. Either a large mass or a large speed increases the impulse necessary to stop an object in motion.

There are natural limits to what can take place in a *simple collision,* one in which only the contact forces push the colliding objects apart. In a ***perfectly inelastic*** collision, two objects stick together after impact with a relative speed of zero. In a ***perfectly elastic*** collision, two objects collide such that the relative speed after the collision is the same as before. Most simple collisions fall between these extremes such that some relative speed is lost but not all; these are called ***inelastic*** collisions.

perfectly inelastic–67
perfectly elastic–68
inelastic–68

Newton found that, *in the absence of a net external force, the total momentum never changes during a collision.* Each object's momentum might change, but the *sum* of their momenta is constant; the net momentum *before* and *after* the collision is the same. This law, called the ***conservation of momentum***, is true even in two or three dimensions, even when more than two bodies collide, and even in collisions that are not simple. Momentum is conserved no matter what forces act, so it allows us to predict motion even when it is impractical to use $\vec{F} = m\vec{a}$.

conservation of momentum–70

An object's ***angular momentum*** is the product of its moment of inertia and angular speed. *In the absence of a net external torque, an object's angular momentum is constant,* according to the ***law of conservation of angular momentum***. An ice skater's spin rate increases when she brings her arms closer to her axis of rotation because of the decrease in her moment of inertia. An external torque can cause a ***precession***, a wobbling in the direction of a spinning object's axis of rotation.

EXERCISES

Concept Checks: Answer true or false

1. An impulse of a net force on an object is equal to the momentum change of the object.
2. If an object moves only along a straight line its momentum is its mass times its speed, mv.
3. The momentum of an object (in three dimensions) is equal to its mass times its velocity.
4. In one dimension, $F_{net} \times$ time = mass $\times$ change in speed.
5. If the relative speed of two objects immediately after a collision is zero, the collision is perfectly elastic.
6. Isaac Newton referred to a "quantity of motion." We call the same thing *momentum* today.
7. Momentum is a vector, but impulse is not a vector.
8. In any interaction where momentum is exchanged, the total vector momentum before and after is the same. That means the momentum of every one of the interacting bodies is the same.
9. The angular momentum of a rotating object is a product of its moment of inertia and its rate of spin.
10. When there are no external torques, the angular momentum of a rotating system of bodies is constant.

Applying the Concepts

11. A jet liner flies level with constant momentum. Which statements are true? (a) No impulse is being delivered to the plane. (b) No forces are acting on the plane. (c) The net force on the plane is zero.
12. Walking with a certain speed, you have a certain momentum. Would your momentum be different if you were walking at that speed on the moon? (Remember that you would weigh much less on the moon, $mg/6$.)
13. Which of these collisions is perfectly inelastic? A bullet smacking into a tree, a football caught by a tight end, a drummer's sticks impacting on a drum, a tennis ball hit by a racket, an egg thrown against a wall.
14. Why do high-impact aerobics classes often use very thick floor mats?
15. A student leans with one hand against a brick wall, exerting a force. Is the student providing an impulse to the wall? If so, then why does the wall not have a change in momentum?
16. What impulse will it take to stop a 2000-kg sports car that is moving at 25 m/s?
17. Which has the greater momentum, a freight train at rest or a drop of water falling from a leaky faucet?
18. To ease the pain, should you catch a fast-moving softball while your hand is moving toward the ball, while it is at rest, or while it is moving in the oncoming ball's direction? Why?
19. How can an airbag that expands from the steering column during a crash help the driver? Explain in terms of impulse and momentum. Does the area of impact figure in for the driver?
20. Calculate the momentum of a bicycle rider (55 kg) and her mountain bike (12 kg) if they are moving at 6 m/s. Compare that to her momentum if she were running at 8 m/s or traveling in a car at 27 m/s (60 mph), or skydiving at a terminal speed of 54 m/s.
21. Dumbo the elephant shoots a peanut from his snout while standing on a frictionless surface. The same force is applied to the peanut and the elephant for the same fraction of a second. Which has the greater momentum change?
22. If you drop your keys, they gain momentum constantly as they fall toward the ground. Doesn't this violate conservation of momentum? Why or why not?
23. A .22 long rifle bullet fired in a pistol exits at 320 m/s, while if fired in a rifle it exits at about 400 m/s. Discuss the difference in terms of the impulse the bullet receives.
24. Using impulse = change in momentum, calculate the force needed to stop a fastball with a catcher's mitt in 1/10 of a second. (m_{ball} = 0.149 kg; $v_{fastball}$ = 40 m/s, or about 90 miles per hour.)
25. A truck takes a curve with a constant speed. Does its momentum change? If so, what force is providing the impulse?
26. Observe motorcross racers as they go over the huge bumps on the course. Exactly why do they stand on the foot pegs rather than sit on the motorcycle seats?
27. Compare the momentum of a 50-kg soccer player moving at 7 m/s with that of a 30-06 bullet (mass about 0.1 kg) moving at 850 m/s. Next, suppose such a slug hits a villain in a western flick. If his mass is 60 kg and he is initially standing at rest, about what would his final velocity be if he absorbs the full momentum of the bullet?

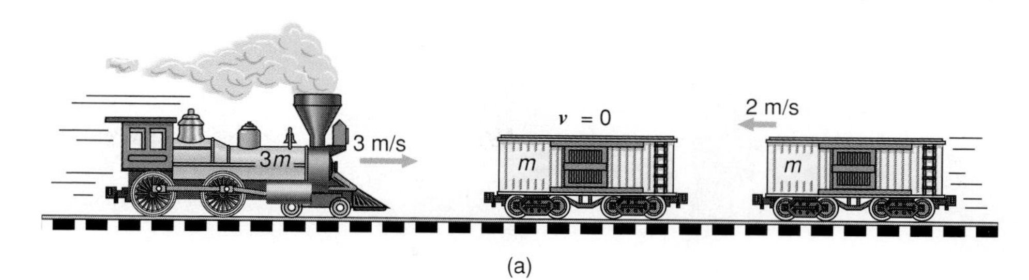

(a)

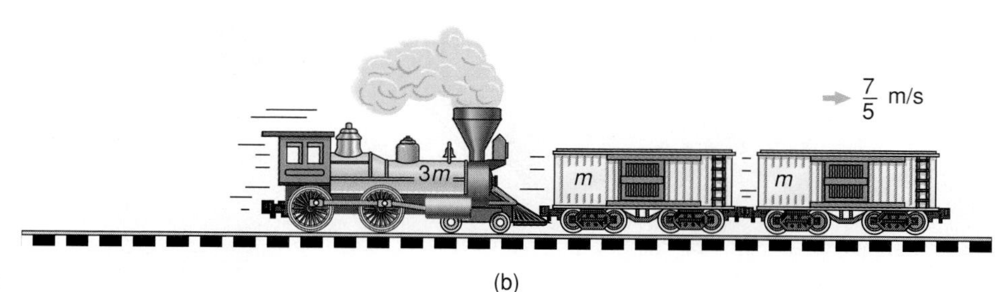

(b)

FIGURE 5-11

28. If you either double the force or double the time it is applied, you double the impulse. True or false?

29. If momentum is conserved during a tennis serve, the momentum gained by the ball was lost by the racket. So how could we claim that the racket's speed doesn't change very much?

30. A car coasts down a hill, slowly gaining momentum. What force is delivering most of the total impulse?

31. Explain why it's easier for a batter to hit a home run from an incoming fastball than from one that the batter tosses up in practice.

32. Farmers and outdoor adventurers who must sometimes tow mired vehicles from muddy fields prefer ropes to towchains. Why?

33. Is an object's momentum in the same direction as that of the net impulse that acts on the object? (Careful!)

34. Imagine that Congress passed a law requiring cars to be perfectly elastic in order to lower the damage done to cars in collisions. What effect would that have on the cars and the drivers?

35. Two ice skaters talk on an ice rink. One's mass is 90 kg; the other's is half as much. After an argument they shove each other. Using the conservation of momentum, discuss their final speeds.

36. Figure 5-11 shows the results of an inelastic collision between a railroad engine and two railroad cars. Check the numbers to see if momentum was conserved. (If not, what does this tell you about the numbers on the drawing?) Then find the relative speeds before and after the collision and tell whether the collision was elastic or inelastic.

37. As it prepares to land but while still high in the air, the space shuttle turns in giant S-curves to the left and right. How does this motion increase the impulse delivered to the shuttle to lower its speed?

38. During World War II baseballs were made from inferior-quality materials. A 1943 National Bureau of Standards research paper showed that the balls rebounded with only 41 percent of an 85-ft/s incoming speed as opposed to 54.6 percent with today's baseballs. What effect could this lower "rebound" speed have had on the game?

39. Why should a hammer be made of iron rather than plastic, since a plastic hammer would be lighter and easier to swing? (*Hint:* Consider the point of view of a nail that's in need of an impulse.)

40. If you stand in the front of a canoe and walk to the rear, what happens to the canoe? What does conservation of momentum tell you about the canoe's momentum?

41. Why does it help to have tennis shoes on if you jump down from a stepladder?

42. A diesel train engine weighs 3 times as much as a fully loaded flatcar. If the diesel coasts into the car at 2 m/s and they couple and coast off together, what is their speed as they move off together?

43. As an avalanche progresses down a mountainside, it may gather more snow and ice. What does this additional mass do to its momentum?

44. Why does a prizefighter move toward his opponent while throwing a punch rather than just stand still?

45. Someone throws a medicine ball to you while you are standing on a skateboard. You catch it and roll backward with the skateboard. Is your momentum change greater if (a) you catch the ball and hold it, (b) you catch the ball then throw it back, (c) your outstretched hands don't yield to the ball; they stop it and it falls straight down to the ground as you roll backward?

46. A waterfall has a nearly constant downward momentum since about the same amount of water is falling every second. Does this mean the earth is constantly gaining upward momentum? Discuss.

47. In Figure 5-12, which person has the greater impulse exerted on his shield? Which likely receives the greater force? (You can answer this by considering the momentum changes of the arrows.)

48. Give two reasons a parachutist doesn't land stiff-legged but instead rolls and turns his or her body to one side so one leg, hip, and the side of the chest strike the ground.

FIGURE 5-12

49. A fastball comes across the plate at 90 mi/h. The bat travels forward at 70 mi/h. (a) What is their relative speed before they collide? (b) Suppose the batter knocks a line drive back to the pitcher. If the relative speed after the ball is struck is 50 percent of the relative speed before, what is the speed of the ball as it comes at the pitcher?

50. Two hailstones of the same mass and size fall at the same speed. One hits the ground in a cornfield and has an inelastic collision, while the other hits pavement and has an elastic collision. Which undergoes a larger momentum change?

51. Why is it better to set bullets spinning before they leave the gun barrels? Why are most passes in football thrown with a spiraling motion?

52. The next time you are at a large party, try to get the partygoers to go outside and run toward the east. What might this do to the length of the earth's day? (If it's a big party?)

53. Perhaps you (or one of your friends) can balance a basketball on the tip of a finger by spinning it with the other hand. What would happen if you tried to balance it when it was not spinning? (Try it and see!)

54. In earth's core there are liquid regions. Some geophysicists think iron (with relatively high density) is sinking in this region even today. If true, what effect would this have on the earth's rotation rate?

55. Suppose the ice on the continents near the polar regions all melted. Earth's ocean level would rise substantially (several hundred feet), even at the equator. What effect would this have on the earth's spin rate?

56. In 1992 at the U.S. Open, the fastest men's serve clocked was 129 mph. If the tennis ball and racket could have had a perfectly elastic collision, how fast would the racket have been moving when it struck the ball? In reality, was the racket moving faster or slower than that speed?

57. In 1992 at the U.S. Open, the fastest women's serve clocked was 108 mph. Assume the tennis ball had 60 percent of the relative speed after the collision. About how fast was the racket moving when it struck the ball?

58. Coasting toward the moon, an Apollo spacecraft's velocity had to be fine-tuned, and this was done by a small rocket engine with a thrust of 70 lb. How long would this engine burn in order to change the 150,000-lb spacecraft's speed by 2 ft/s? (Ignore the tiny loss of mass as the rocket used fuel during the burn.)

59. A pool ball moves parallel to the length of the table with 5 units of momentum. It hits a ball at rest, giving that ball 2 units of momentum parallel to the length of the table and 2 units perpendicular to the length. What is the new momentum of the initial ball?

60. A fully loaded dumptruck collides with a car at rest. Stuck together, they move off at 1/3 the dumptruck's original speed. What is the mass of the car in terms of the dumptruck's mass?

Demonstrations

1. Blow up a rubber balloon and let it go. In effect, it's a small rocket! The air rushes out as the stretched balloon contracts, and the change in momentum for the outrushing air is equal to the "rocket" impulse on the balloon. In rockets, chemical reactions produce large volumes of hot gases under high pressure, which exit through a nozzle. The faster the flow, the more momentum the gas molecules carry, and the greater the impulse given to the rocket.
2. The children's game "Catch the Egg" demonstrates the effect of impulse on momentum. Two players face each other and toss an egg back and forth. Each time a player catches the egg, they both take a step backward, increasing the distance the egg must be thrown. As they throw the egg higher it's easier to break during a catch. The trick is to increase the time you use to stop the egg. As you catch the egg, swing your arm and hand backward (in the egg's direction of motion), yielding with the egg as it slows.
3. Watch the earth rotate! Indirectly, of course. On a bright, sunny day notice the tip of the shadow of a tall flagpole. Pay close attention, and you'll see the shadow or sunlight move slowly as the earth turns.
4. Take a fresh egg from your refrigerator. Spin it, really get it going, then stop it dead with your fingers, and quickly let the egg go. The yolk inside will still be spinning. It rubs against the shell, and the entire egg is soon spinning again.

SIX

Energy

The leopard frog exploding from the pond's surface in this photograph accelerates much faster than we can, in part because of its smaller inertia and the strength of its leg muscles. Nevertheless, we use our muscles to control our actions much like the frog does.

About 200 muscles come into play in our bodies whenever we walk down a street. What gives muscles the ability to exert forces? Food you ate previously acts as fuel. Eat a hot fudge sundae, and you may soon feel hyperactive—you might even do something to "work off" that dessert. If you exercise enough to get tired, then you might say, "I'm out of energy." *Energy* is what your muscles use when they contract. This chapter is about energy and the related concepts of work and power.

WORK

In everyday language, "work" usually means the same as "job." In physics, however, work is related to a physical action. Work is a measure of *how productive* an applied force is. Push as hard as you can on the wall of a building, and the wall won't budge. Use the same force on a stalled car, and it moves. Those identical forces perform differently. You might say your push on the car did something, while the one on the wall did not—it was unproductive.

When a force pushes or pulls something *through a distance,* we say the force does work. Work is the *force in the direction of motion multiplied by the distance the force acts through.* In this definition, the direction of the force is important. If you push a grocery cart through a supermarket, your horizontal push does work on the cart, moving it through a distance (Figure 6-1). The weight of the groceries also acts on the cart, but that force does no work on the cart since it has no component in the direction of the cart's motion. *If there is no distance covered in the direction of a force, that force does no work.* Often an applied force isn't entirely in the direction of the object's displacement, as in Figure 6-2, so the component of force parallel (∥) to the direction of motion does work while the component perpendicular (⊥) to the direction of motion does none. In general, then,

PUTTING IT IN NUMBERS—DEFINING WORK

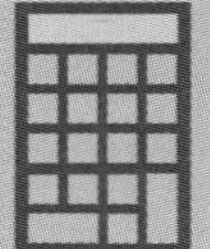

work = force parallel to displacement × displacement

$$W = F_{\parallel} \times d$$

The SI unit of work is a newton-meter, also called a joule (J). The British unit of work is a foot-pound (ft·lb). Although work is defined in terms of a force component, work is *not* a vector quantity.

Example: A delivery person slides a refrigerator through a distance of $\frac{1}{2}$ m into its place in the kitchen with a 400 N horizontal push. What work did his force do? $W = F_{\parallel} \times d =$ 400 N × 0.5 m = **200 J**

FIGURE 6-1 THIS HORIZONTAL FORCE PUSHES WHILE THE CART HAS A HORIZONTAL DISPLACEMENT, AND THE WORK DONE IS $W = F \times D$.

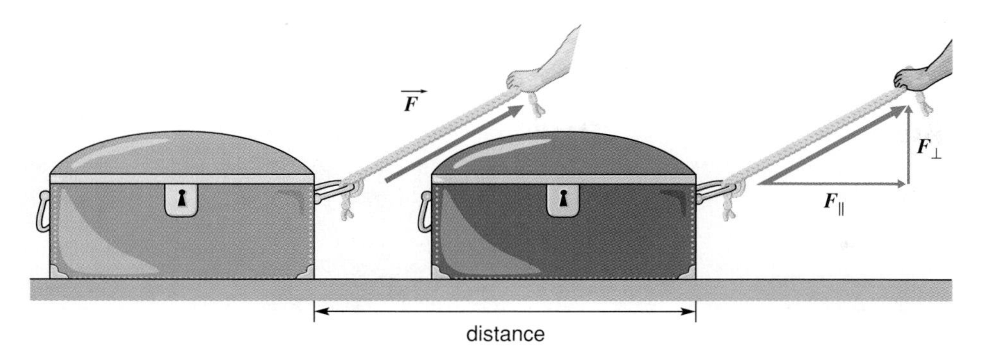

FIGURE 6-2 THE FORCE APPLIED TO THE TRUNK HAS TWO COMPONENTS, ONE PARALLEL TO THE DISPLACEMENT ($F_\parallel$) AND ONE PERPENDICULAR TO IT ($F_\perp$). ONLY THE COMPONENT PARALLEL TO THE DISPLACEMENT DOES WORK, SINCE THE TRUNK DOES NOT MOVE IN THE DIRECTION OF THE PERPENDICULAR FORCE.

WHAT IS A JOULE?

A JOULE, WHICH IS ONE NEWTON-METER, IS THE WORK YOU WOULD DO TO LIFT A MEDIUM-SIZED APPLE (WEIGHT = 1 NEWTON) STRAIGHT UP FOR A DISTANCE OF ONE METER.

WHAT IS A FOOT-POUND?

A FOOT-POUND, WHICH IS EQUAL TO ABOUT 1.36 JOULES, IS THE WORK YOU WOULD DO TO LIFT A PINT CARTON OF MILK (WEIGHT ABOUT 1 POUND) STRAIGHT UP FOR ONE FOOT, OR TO LIFT THIS BOOK (WEIGHT ABOUT 3 POUNDS) STRAIGHT UP FOR A DISTANCE OF 4 INCHES.

MAKING JOBS EASIER: SIMPLE MACHINES

Sometimes there are easy ways and hard ways to do the same amount of work, and the key to making work easier is in the definition $W = F_\parallel \times d$. Any time you can decrease the force while increasing the distance, you can do the same amount of work more easily. A **simple machine** is a device that lets you do work with less force, and the lever is perhaps the simplest machine of all.

Figure 6-3 shows the use of a simple lever. If the miner applies a force far from the pivot, a distance l_{applied} away, and the mining car on the other side

USING A SCREWDRIVER AS A LEVER TO PRY OPEN A PAINT CAN. THE OUTSIDE EDGE OF THE CAN IS THE PIVOT. NOTICE HOW FAR THE LEFT END OF THE SCREWDRIVER MUST MOVE TO RAISE THE LID LESS THAN A CENTIMETER.

PUTTING IT IN NUMBERS—MULTIPLYING FORCE WITH LEVERS

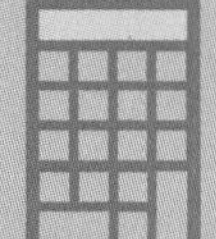

measurements show that the force you apply times the lever arm (which is the distance to the pivot) is equal to the force the object gets times its lever arm (or its distance to the pivot)

$$F_{\text{applied}} \times l_{\text{applied}} = F_{\text{on object}} \times l_{\text{object}}$$

Example: A screwdriver is used to pry open the lid of a paint can. The edge of the can is the pivot, and the tip of the screwdriver that pushes up on the lid is 3/8 inch from the pivot. If you push down on the handle 7 inches from the pivot with a force of 4 pounds, what force acts to pry the lid up? From the equation above,
4 lb × 7 in. = $F_{\text{on lid}}$ × 3/8 in., so $F_{\text{on lid}}$ = **75 pounds.**

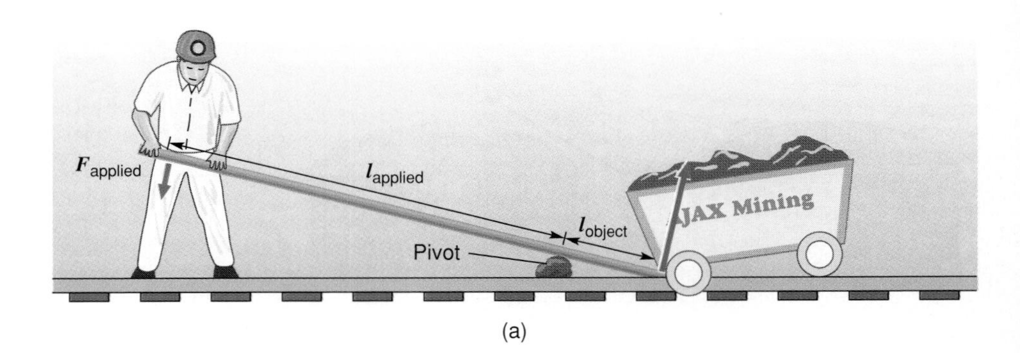

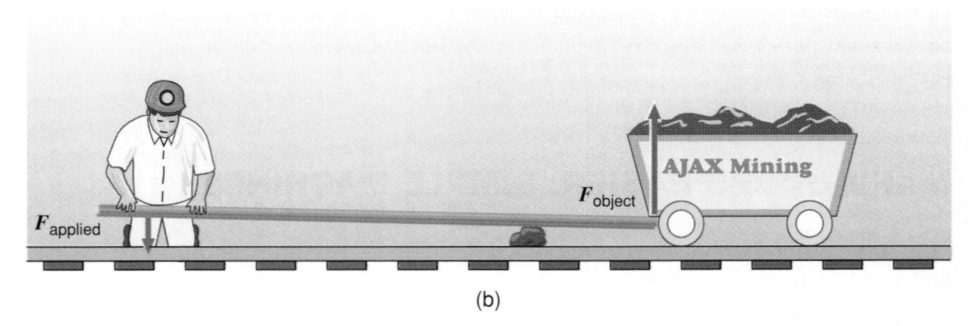

FIGURE 6-3 USING A STEEL BAR TO PUT A MINING CAR BACK ON ITS TRACKS. THE FORCES TIMES THEIR RESPECTIVE DISTANCES TO THE PIVOT ARE EQUAL.

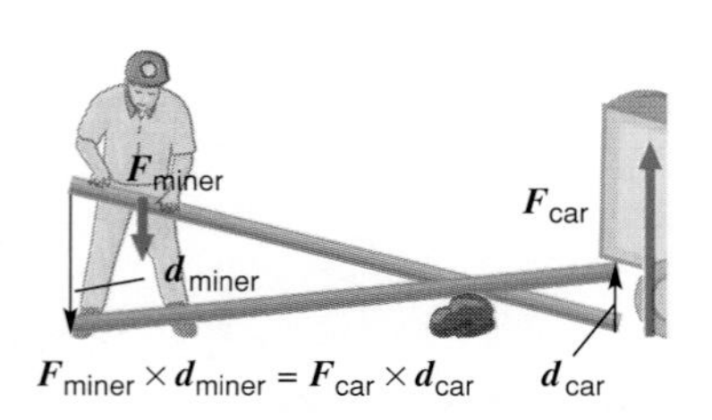

FIGURE 6-4 FINDING THE WORK THE MINER DID ON THE CAR IN FIGURE 6-3. $F_{MINER} \times D_{MINER} = F_{CAR} \times D_{CAR}$, SO THE WORK DONE BY THE MINER IS TRANSFERRED TO THE OBJECT THROUGH THE LEVER.

is closer to the pivot as shown, the force the car gets is greater than the force the miner applies. Notice exactly what the lever in Figure 6-3 does. A smaller force is multiplied by the lever to exert a greater force. We can also see that the work done *by* the miner in Figure 6-3 equals the work done *on* the mining car. Look at Figure 6-4. The miner's push moves through the distance d_{miner}, so $W_{miner} = F_{miner} \times d_{miner}$. The car moves up a distance d_{car}, so $W_{car} = F_{car} \times d_{car}$. In the absence of friction or flexing of the bar, as we'll see later, these two amounts of work are equal.

The lever, or any other simple machine, trades extra distance for a smaller force, making it easier (you use less force) to do the work. Pliers (Figure 6-5), scissors, and bottle openers trade distance for force and make it easy for you to apply a much larger force to squeeze, cut, or pry than you could otherwise.

FIGURE 6-5 PLIERS ARE LIKE TWO LEVERS COUPLED TOGETHER. LIKE LEVERS, PLIERS TRADE FORCE FOR DISTANCE.

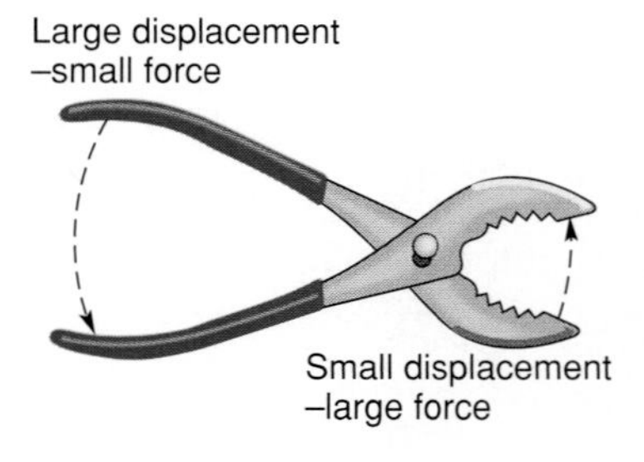

ANOTHER SIMPLE MACHINE: THE INCLINED PLANE

Figure 6-6 shows two ways to place a stone into a wall. To lift the stone straight up from the ground requires a force slightly greater than the weight of the stone, mg. (A force equal to mg balances the stone's weight; a little extra force then accelerates the stone upward.) If you push it on rollers (to reduce friction) up the ramp, or *inclined plane,* you need a force only slightly greater than the component of mg parallel to the plane to push the stone up, and that force is much less than the stone's weight. However, as you can see, you have to push the stone much farther than if it is simply lifted to get it to the top.

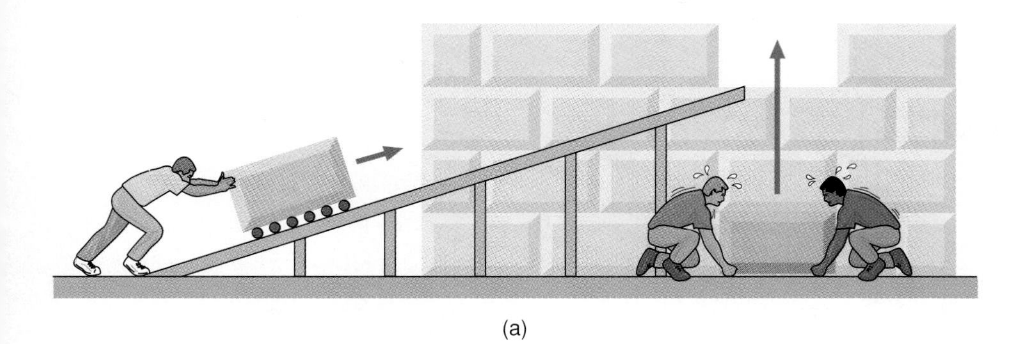
(a)

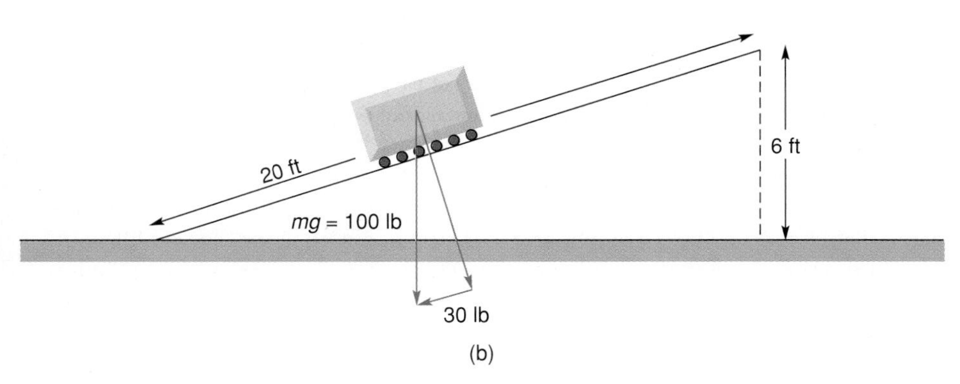

(b)

FIGURE 6-6 WITH AN INCLINED PLANE, A LITTLE FORCE THROUGH A LONGER DISTANCE DOES THE SAME AMOUNT OF WORK AS A LARGER FORCE THROUGH A SHORTER DISTANCE.

The inclined plane, like the lever, trades extra distance for a smaller force. Also like a lever, an inclined plane doesn't save you any work. It only helps you do the work while exerting less force. Suppose the stone in Figure 6-6 weighs 100 lb and is 6 feet below the empty space in the wall. Lifting it straight up requires 100 lb × 6 ft = 600 ft·lb of work. Pushing it up the ramp requires the same amount of work, if friction is negligible. If the ramp's length is 20 ft, then you have to apply 30 pounds of force to get the stone to the top (30 lb × 20 ft = 600 ft·lb). This simple machine works wonders!

POWER

A tree root grows under a section of cement sidewalk and over a ten-year period lifts that heavy slab up. Two workers with crowbars could lift an identical section of sidewalk to the same height in a second or so. The same work would be done in both cases, but it would be done at very different rates. The rate of doing work is called **power.**

PUTTING IT IN NUMBERS—DEFINING POWER

Power is the work done divided by the time it takes to do the work.

$$P = \frac{W}{t}$$

The units of power are joule/second and foot-pound/second. One joule/second is called a watt (W). Other common power units are the kilowatt (1000 watts, or 1 kW) and the horsepower (hp), which is 550 ft·lb/s or 745.7 W. **Example:** You lift a heavy suitcase (weight = 60 lb) a distance of 3 feet off the floor to carry it. If you do this in 0.3 second, what power did you use? W = 60 lb × 3 ft = 180 ft·lb, so $P = W/t$ = 180 ft·lb/(0.3 s) = 600 ft·lb/s = 1.1 hp or **810 W**.

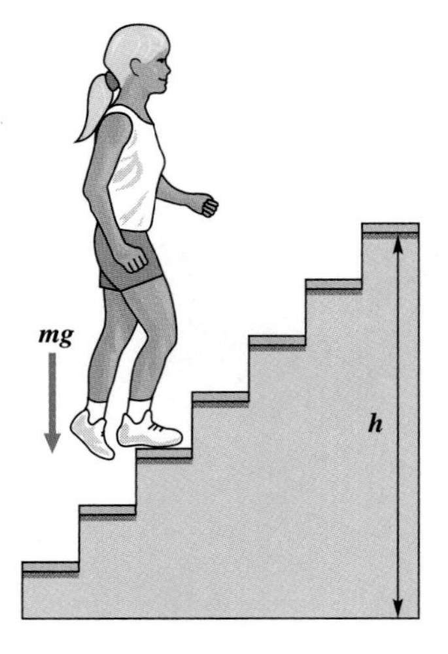

FIGURE 6-7 IF YOU WALK UP STAIRS, IT IS MUCH AS IF YOU WERE THE BLOCK ON THE INCLINED PLANE IN FIGURE 6-6. THE WORK YOU DO IS EQUAL TO THE FULL FORCE OF GRAVITY *MG* TIMES *H*, THE VERTICAL HEIGHT YOU GAIN. THE POWER YOU USE DEPENDS ON HOW LONG YOUR ASCENT TAKES.

WHAT'S A WATT?

A SINGLE WATT IS A SMALL AMOUNT OF POWER. FOR INSTANCE, THIS BOOK'S MASS IS A LITTLE OVER 1 KILOGRAM, SO ITS WEIGHT *MG* IS A BIT OVER 10 NEWTONS. IF YOU LIFT THIS BOOK UP 1/10 METER (ABOUT 4 INCHES) IN A SECOND, YOU'VE "SPENT" A WATT OF POWER. NOW HOLD TWO BOOKS LIKE THIS ONE, ONE IN EACH HAND, AND LIFT THEM ONE METER, *TWICE* EACH SECOND. YOUR POWER OUTPUT THEN IS 40 WATTS, WHICH IS THE POWER A 40-WATT LIGHT BULB USES. SEE IF YOU CAN CONTINUE THIS RAPID LIFTING FOR A MINUTE. TO SUSTAIN A POWER OUTPUT OF 40 WATTS IS IMPRESSIVE!

It takes more power for you to run up a flight of stairs than to walk up. Either way, you do the same amount of work—lifting yourself through the same distance. The stairs are just an inclined plane with steps, so you lift your weight mg through a vertical distance equal to the height h of the stairs (Figure 6-7). The faster you go up, however, the more power you use.

NET WORK AND KINETIC ENERGY

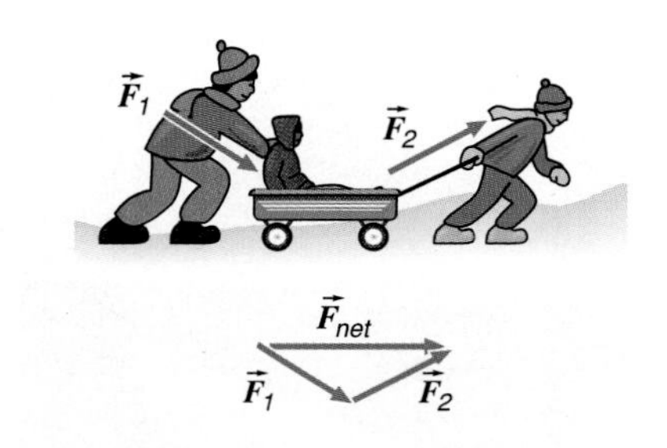

FIGURE 6-8 IF THERE IS NO FRICTION ON THE SLED'S RUNNERS, THE NET FORCE IS THE VECTOR SUM OF $\vec{F}_1$ AND $\vec{F}_2$. THAT FORCE TIMES THE DISTANCE THE SLED MOVES WHILE THE NET FORCE IS APPLIED GIVES THE NET WORK DONE ON THE SLED.

Almost anything that moves has more than one force acting on it, and any force with a component parallel to the object's displacement does work on the object. When more than one force does work at the same time, as in Figure 6-8, you can add the forces to find the net force. Then the component of the net force that is parallel to the displacement times the displacement is called the **net work.** The *net work* is a measure of the productivity of all the forces acting.

$$W_{\text{net}} = F_{\text{net}\,\parallel} \times d$$

Hit a volleyball and you do work on it—you push it through a distance. The volleyball gains a great deal of speed from your work, and it flies off. It's not hard to show the connection between the *net work* done on an object and *how its speed changes* (this is derived in Appendix 2b).

$$W_{\text{net}} = \tfrac{1}{2}mv_{\text{f}}^2 - \tfrac{1}{2}mv_{\text{i}}^2$$

This formula tells us the net work on something brings about a change in the quantity $\frac{1}{2}mv^2$. The expression $\frac{1}{2}mv^2$ is called the **kinetic energy** (or ***KE***) of the moving object. In words, the formula says:

PUTTING IT IN NUMBERS—RELATING NET WORK AND KE

net work on object = change in its kinetic energy

$$W_{\text{net}} = KE_{\text{final}} - KE_{\text{initial}}$$

The kinetic energy $\frac{1}{2}mv^2$ has the same units as work, joules or foot-pounds. **Example:** A sprinter, m = 50 kg, goes from zero to 8 m/s in 1.6 s. What was the net work she did? $W_{\text{net}} = KE_{\text{f}} - KE_{\text{i}} = \frac{1}{2}(50\text{ kg})(64\text{ m}^2/\text{s}^2 - 0)$ = **1600 J.** What power did she use to do that? $P = W/t$ = 1600 J/1.6 s = **1000 W.** (See Figure 6-9.)

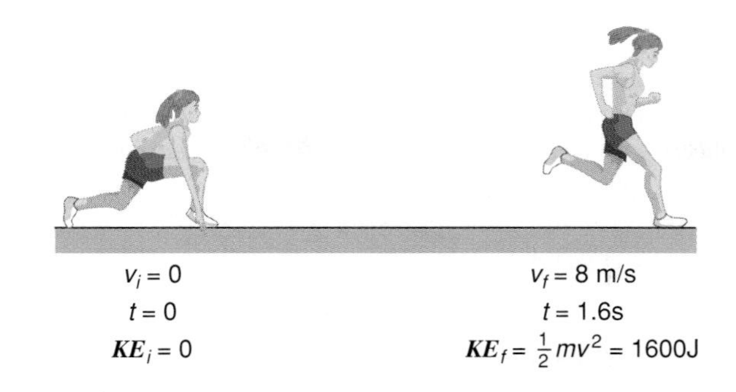

FIGURE 6-9 CALCULATING THE CHANGE IN A SPRINTER'S KINETIC ENERGY.

Notice that kinetic energy, like work, is not a vector quantity. No matter what direction the sprinter runs her kinetic energy is just $\frac{1}{2}$ her mass times her speed squared. Also notice what happens if the work being done slows an object, decreasing its speed. Then $KE_{\text{final}} - KE_{\text{initial}}$ is a negative number, so work that slows an object is always *negative* work. A *force that is directed opposite to the direction of motion always does negative work.*

To see what kinetic energy means, consider a hammer that pushes a nail into a wall and stops. The hammer has lost all its kinetic energy. The contact force from the nail acted on the hammer, doing negative work on the hammer and stopping it. At the same time, however, the nail had the same amount of positive work done on it by the hammer. (Action equals reaction, and both the nail and hammer moved the same distance while exerting forces on each other.) That means the kinetic energy of the hammer went into doing work on the nail, so **the kinetic energy of a moving object is a measure of that object's ability to do work.** (In a later section we'll see that some of the work done on the nail and hammer appears as heat, another form of energy.)

KINETIC ENERGY IS OFTEN CALLED "ENERGY OF MOTION"

EXAMPLES WITH NET WORK AND KINETIC ENERGY

Imagine giving a push to a friend who's wearing in-line skates. If there's no friction involved, the net work on your friend is the work you do, and this work is your push (the average force) times the distance you push through. Assuming your friend began at rest, after your push your friend's kinetic energy, $\frac{1}{2}mv^2$, is equal to the work you did, and it remains at that value until other forces, such as friction, do work on your friend to change it.

If the *KE* of an object doesn't change, the net work on it must be zero. For example, if you gently slide this book several feet across a table, you'll do work on the book. A friction force from the table opposes the book's motion, however, doing negative work on the book while you are doing positive work on it. When you stop pushing, the book stops moving. So the change in the book's kinetic energy from start to finish is *zero*. That means the net work from the two forces is zero, or

$$W_{\text{net}} = W_{\text{your hand}} + W_{\text{friction}} = 0$$

The work done on the book by friction ($-F_{\text{friction}} \times d$) is the negative of the work you did ($F_{\text{hand}} \times d$). The frictional force is negative since it points opposite the direction of the displacement. That means the *average* force from your hand was exactly equal to the value of the average frictional force.

When a cheerleader gets a lift high into the air, her partner's push moves her up through a distance, doing work on her (Figure 6-10). Gravity also acts on her as she rises. Her partner's force is greater than her weight, *mg,* to

FIGURE 6-10 AT REST ON THE GROUND, THEN AT REST OVERHEAD, A CHEERLEADER'S CHANGE IN *KE* IS ZERO, SO NO *NET* WORK HAS BEEN DONE ON HER. GRAVITY'S NEGATIVE WORK CANCELED HER PARTNER'S POSITIVE WORK.

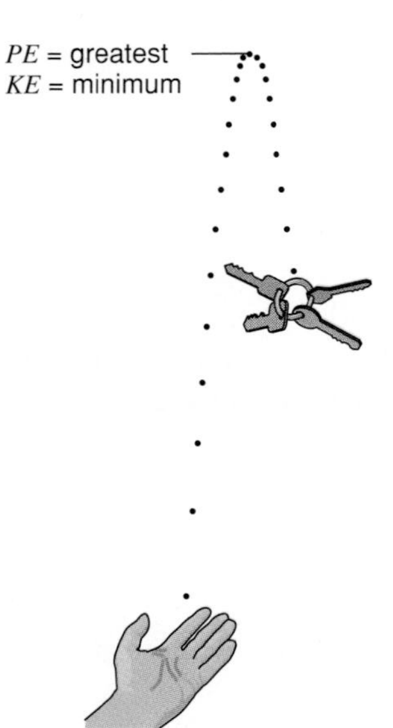

TRACING THE MOTION OF TOSSED KEYS AS WITH A STROBE. AS THE KEYS RISE, GRAVITY OPPOSES THEIR MOTION, DOING NEGATIVE WORK AND DECREASING THE KEYS' KINETIC ENERGY, $\frac{1}{2}mv^2$. WHEN THEIR KINETIC ENERGY IS FINALLY DEPLETED, THE KEYS BEGIN TO FALL AND GRAVITY DOES POSITIVE WORK ON THEM, INCREASING THEIR KINETIC ENERGY AGAIN.

begin with, so she gains speed (and kinetic energy) and moves up. But then he pushes up a little less than gravity pulls down and she slows. She comes to rest at the height of the lift far above the ground, so the change in her kinetic energy was zero and the net work done on her was zero. Overall, gravity's (negative) work on her canceled her partner's (positive) work on her, and the average force he exerted was exactly equal to her weight.

POTENTIAL ENERGY

Consider what happens if you toss your keys straight up over your head. You give the keys kinetic energy, $\frac{1}{2}mv^2$, and as they rise, the force of gravity pulls them back toward the ground, doing negative work that reduces their kinetic energy. At the top of the path, their speed is zero instantaneously, and the work done by gravity has brought their *KE* to zero. Then they fall downward, and the force of gravity does work again. For each centimeter they fall, gravity does the same amount of work as it did when the keys moved upward through that distance. Only now the work is positive because gravity pulls in the direction of their motion. So as they fall, the keys *gain* kinetic energy step-for-step in the identical amounts they *lost* earlier. Just as they reach your hand, the keys have the same amount of kinetic energy as when they left your hand. (The work that air drag does on the keys is negligible because their slow speed means the drag will be very small compared to mg.)

The pull of gravity first took away the keys' kinetic energy and then gave it back. In effect, the keys' kinetic energy is momentarily traded for their gain in height, and then vice versa. It's as if the keys' kinetic energy is *stored* by the action of gravity as they rise and then *returned* as they descend. This idea of stored energy is very important. *Such energy exists wherever an object has a force ready to do work on it.* Keys or anything else above the

FIGURE 6-11 FEW PEOPLE WOULD PICNIC BENEATH THIS BOULDER IN UTAH ON A WINDY DAY. BECAUSE OF ITS ELEVATION, IT HAS THE POTENTIAL TO DO A LOT OF WORK ON ANYTHING BELOW. THAT IS, THE BOULDER HAS A LARGE POTENTIAL ENERGY.

ground have the potential to have gravity do work on them, so we say they have **potential energy** (or ***PE***). The potential energy due to gravity (also called the *gravitational* potential energy) for an object that is a height h above the ground is just mgh, which is the work gravity will do on the object if it descends. While kinetic energy is energy of *motion,* potential energy is energy of *position.* $PE_{\text{gravity}} = mgh$

Returning to the falling keys, they lost potential energy as their height h decreased, but they gained an equal amount of kinetic energy. Then they did work on your hand as they stopped. That means potential energy is **a measure of an object's potential to do work.** Figure 6-11 shows an object that has enough potential energy to do a lot more work than your car keys did.

An arrow held in an archer's hand has a small amount of potential energy due to gravity. If dropped, it falls the few feet to the ground. In the archer's drawn bow, however, the arrow has much more potential energy (Figure 6-12). It's in a position to have a lot of work done on it as a result of the elasticity of that bent bow. We say the bow stores *elastic* potential energy. A flick of the fingers, and the string and bow spring back to their original positions, releasing stored energy to the arrow, doing work on it. The arrow then leaves with a great deal of kinetic energy.

FIGURE 6-12 AN ARCHER GIVES THE ARROW A GREAT DEAL OF POTENTIAL ENERGY BY PLACING THE ARROW AGAINST THE STRETCHED BOW'S STRING.

As another example of elastic potential energy, compress a metal spring under a jack-in-the-box. So long as the top of the box is closed, the coiled spring remains compressed, ready to do work the instant the lid is released. The spring stores elastic potential energy. That brings us to an interesting point.

Remember from the discussion of elasticity in Chapter 3 that atoms and molecules in solids act as if they are connected by springlike interatomic forces. These bonding forces can either store potential energy or give it up, just as springs do. During *chemical reactions,* molecules are put together or taken apart, and interatomic bonds form or break. This is the kind of chemical potential energy (also called *chemical* energy) that comes from gasoline as it burns in car engines. It's also the kind of energy we get from food. In each process the chemical potential energy is transformed into work by the cars or our bodies. Not all the work goes into kinetic energy, however, as we'll see in the next section.

THE EXPLOSION OF FIREWORKS IS DUE TO THE RAPID RELEASE OF CHEMICAL ENERGY.

MOVING OUT WITH POTENTIAL ENERGY

Most animals on the run store some of their energy of motion, their kinetic energy, between strides. Elastic tendons connect muscles to the feet and leg bones, and with each step these tendons are stretched. While under tension, they retain some of the energy it took to stretch them, and during the next stride, they snap back, helping the foot to push off the ground. Like springs, the tendons store some kinetic energy as potential energy and give it back again.

Kangaroos, however, have a running technique that sets them apart. Treadmill tests show that when a red kangaroo (the largest of them all) triples its hopping speed, for example from 8 mph to 24 mph, it uses up to 10 percent *less* energy per minute. It uses less energy if it goes faster! A cheetah or a horse must use over 100 percent *more* energy per minute for a similar gain in speed. At cruising speeds up to 40 mph, the red kangaroo bends far over forward as it lands and straightens out as it pushes off for the next hop. It uses its entire body like a spring.

Scientists think this method of running helps the kangaroos survive in the Australian desert. Thunderstorms are rare and miles apart, but they bring green grass out quickly. To get to this food before the desert sun parches it, often in just a few hours, the kangaroos must move fast. But they must move efficiently, too, or else they might expend more energy getting to that faraway dinner table than they can consume once they arrive.

FRICTION AND INTERNAL ENERGY: HEAT

Throw your keys once more, but this time toss them along the floor. They'll slide along for some distance, slowing as friction does negative work on them, reducing their kinetic energy (Figure 6-13). When the keys come to rest, the kinetic energy you gave them is gone. But this time they can't turn around as they did when you pitched them upward, gather speed from the same force that acted to slow them, and jump back into your hand with the same kinetic energy. Once the motion stops, the friction stops. There is no force to restore the kinetic energy of the keys as gravity did. *Frictional forces cannot give objects potential energy.* But the kinetic energy that you gave the keys doesn't just disappear.

The molecules of any solid object vibrate in all directions even though they are held in place by bonds with the molecules around them. When molecules of the keys and the floor strike each other, they both get pushed through a tiny

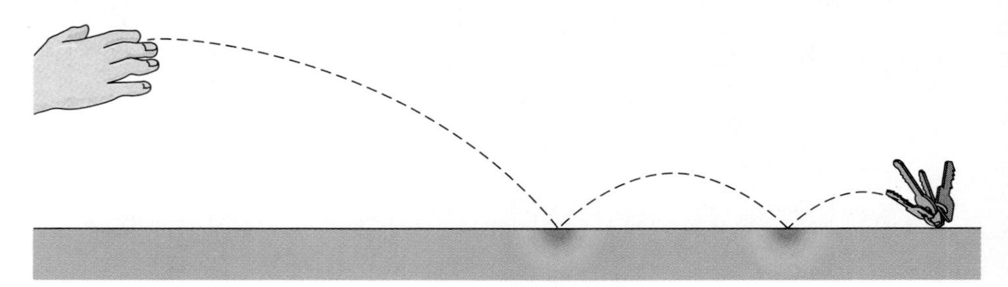

FIGURE 6-13 THE WORK DONE BY FRICTION SLOWS THE KEYS. THE KEYS CANNOT RECOVER THE LOST KINETIC ENERGY THIS TIME, AS THEY COULD WHEN GRAVITY ACTED TO TAKE THEM BACK OVER A SIMILAR DISPLACEMENT. THE FRICTIONAL FORCE DOES NOT GIVE THE KEYS POTENTIAL ENERGY.

distance. This work gives them extra *KE* and they vibrate even faster. They slam into all their nearest neighbors so that the neighbors move a bit faster. The kinetic energy you gave the keys disappears as they stop, and most of it scatters aimlessly among the molecules of the keys and the floor in this fashion. When their molecules move faster, the keys and the floor become warmer—their temperature increases. (In Chapter 12 we'll discuss details of the connection between temperature and the kinetic energy of molecules.) That chaotic, random kinetic energy of vibration has then become part of what we call the **internal energy** of the keys and the floor, which is the sum of the kinetic and potential energies of all of their molecules. Whenever frictional forces do work, some of that work goes into increasing the internal energy of matter, and this energy spreads out, making it less effective for producing work. **Heat** refers to the part of an object's internal energy that moves because of a temperature difference, as we'll see in later chapters.

USING ENERGY TO DO WORK: EFFICIENCY

EFFICIENCY OF MACHINES

When gasoline burns in the cylinders of a car's engine, the result is a mixture of very hot gases under high pressure. The gases push the pistons through a distance, but the work done on the pistons is only about 25 percent of the chemical potential energy released from the gasoline's chemical bonds. What about the rest of the energy? Some energy goes into raising the temperature of the cylinder walls, and that heat spreads outward, doing no useful work. Some leaves with the still-hot exhaust gases through the tailpipe, once again doing no useful work.

The **efficiency** of a machine, animal, or any other energy processor is defined as **work done/total energy expended in doing the work.** The efficiency of a car's engine is 0.25 if 25 percent of the energy released by the fuel goes into work that moves the car (Figure 6-14). Because of friction between the moving parts and the loss of energy through the cylinder walls and the exhaust gases, an engine's efficiency can never be equal to 1 (or 100 percent). The same is true for us, where a large percentage of energy released in metabolism goes to keep our internal energy high (in other words, to keep us warm). And if levers have friction at their pivots or if friction occurs with the use of an inclined plane, even those simple machines have efficiencies less than one.

The links between work done by friction and internal energy and heat were not fully understood until the mid-1800s. Once that connection was made, it quickly led to an important insight about energy, as we will see next.

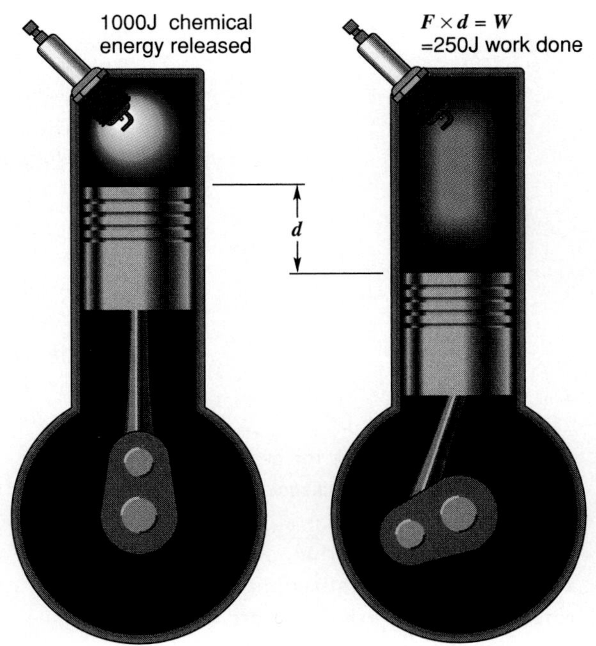

FIGURE 6-14 AN ENGINE HAS AN EFFICIENCY OF 25% IF FOR EVERY 1000 J OF CHEMICAL ENERGY RELEASED IT CONVERTS 250 J TO MECHANICAL ENERGY. 250 J/1000 J = 0.25.

CONSERVATION OF ENERGY

If you throw your keys along the floor, you can hear them hit and jingle. Sound travels from the keys to push your eardrum back and forth, which means the sound does work. Therefore sound carries energy away from the sliding keys.

ANOTHER TYPE OF ENERGY: ROTATIONAL KINETIC ENERGY

A yo-yo sent downward by a flick of the wrist descends to the end of the string. Set spinning as the string unwound, it stays there at the end of the string and spins until another flick of the wrist lets it catch the string and rise back to the hand. *The yo-yo's spinning motion contains energy that lets it rise against gravity.*

The particles of the yo-yo that circle its central axis all had kinetic energy, $\frac{1}{2}mv^2$, in their circular paths. The sum of all their kinetic energies is called the yo-yo's *rotational kinetic energy.* The yo-yo's spinning motion slows noticeably as it rises because its rotational kinetic energy is traded for gravitational potential energy each step of the way.

Furthermore, any scratches made on the floor or on the keys mean that molecules have moved, their bonds broken. A displacement of matter (or a deformation) takes work, and to do work takes energy. And heat in both the keys and the floor caused by the impact and by friction will spread out into the surroundings.

Many different experiments involving every sort of interaction have verified that energy may change forms but its total value in all its forms is constant, or *conserved.* This fact is expressed as a law called the **conservation of energy:** *In every interaction of any kind within a closed system—the collection of interacting material objects and their energy in all its forms—the total energy after the interaction is always the same as the total energy before the interaction.* The kinetic energy you gave to the keys changed into heat, sound, and other forms, but all of the initial kinetic energy is accounted for when we include these various forms in the closed system where the measurements are made.

For an example, consider a child swinging in a playground swing. The kinetic energy of the child–swing system is a maximum at the bottom of the path and is zero at the top, where the energy is entirely potential energy. Energy of motion turns into energy of position, and the cycle repeats. If no air were there to resist the motion, and if no frictional forces acted at the swing's pivot at the top, the motion would repeat indefinitely. Those frictional forces decrease the kinetic energy, however, and the swing's motion finally stops. The energy didn't vanish because the air around the swing is a little warmer and so are the connections between the swing and the supporting bar above it. This increase in temperature came at the expense of the kinetic energy and potential energy of the child and the swing. The energy of the child–swing system does not stay constant, but if we expand the system to include swing's support, the air surrounding the swing, and so on, energy is conserved. The closed system must be large enough to contain the initial energy in all its subsequent forms. (See Figure 6.15.)

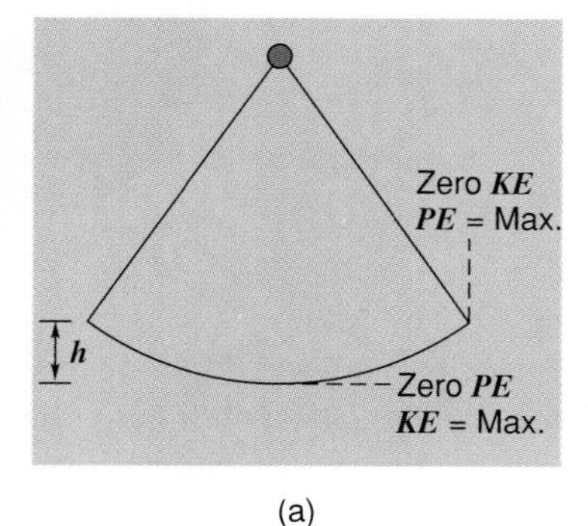

(a)

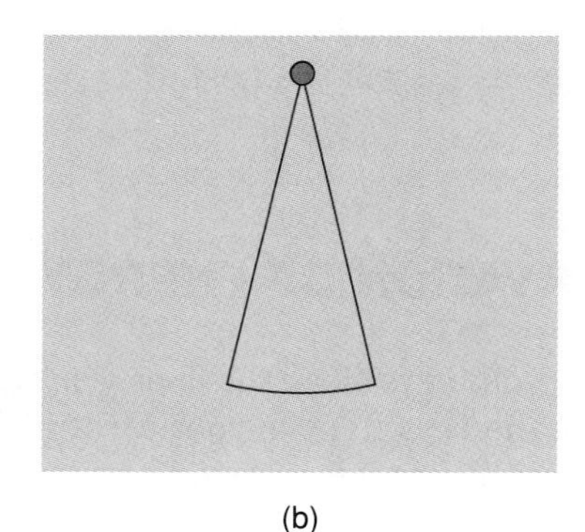

(b)

FIGURE 6-15 (A) AT THE TOP OF THE SWING'S ARC, THE KINETIC ENERGY IS ZERO BUT THE POTENTIAL ENERGY IS AT ITS GREATEST, *MGH*. AT THE BOTTOM, THE KINETIC ENERGY IS AT ITS GREATEST AND THE POTENTIAL ENERGY IS ZERO—BECAUSE THE SWING CAN GO NO FARTHER DOWN. (B) AS THE CHILD SWINGS, FRICTION TAKES THE *KE* (AND *PE*) AWAY. THE ENERGY DISPERSES, PARTLY INTO THE AIR AND THE SWING'S PIVOT, SO THE SYSTEM IN WHICH ENERGY IS CONSERVED HAS TO INCLUDE THE SWING'S SURROUNDINGS.

ENERGY AND COLLISIONS

In simple collisions, as discussed first in Chapter 5, there are natural limits on what can happen. Understanding that energy can change forms gives us an understanding of why those limits occur.

Perfectly inelastic collision (after the impact the objects stick together, their relative speed is zero): There is a *maximum* loss of kinetic energy when this happens. The lost kinetic energy goes into internal energy or does work deforming the bodies, as when a bullet penetrates a block of wood and stays there.

Other inelastic collisions (the relative speed after the collision is less than before but not zero): If two cars with a small relative speed run together in a parking lot, the bumpers may bend like springs, stop the cars, and push them gently apart. Even if there is no damage, there is always some molecular friction when solids stretch or bend. The friction turns some of the cars' kinetic energy into internal energy, and the cars will separate with less relative speed than when they came together. When the impact is hard so that bumpers and fenders crumple, even more kinetic energy disappears as work is done to change the shapes. *The more work done to deform material and the more work done by friction, the less elastic the collision.*

Perfectly elastic collision (the relative speeds before and after the collision are the same): In this situation *no* kinetic energy is lost. The kinetic energy of the approaching objects changes into potential energy during the collision, and all that potential energy turns into kinetic energy again as they part. Each object might take away a different amount of kinetic energy than it brought in, but their sum remains the same.

COLLISIONS BETWEEN CARS ARE EXAMPLES OF INELASTIC COLLISIONS.

THE IMPORTANCE OF ENERGY

From this chapter on, you'll meet the word energy time and time again. You'll see that sunlight carries energy that warms the ground, the oceans, and the air (Figure 6-16). Growing plants use the energy of sunlight much as our bodies use chemical energy from food. And near the end of this text you'll read about Albert Einstein's discovery in 1905 that mass is a form of energy. Stars emit light by converting mass to energy deep in their interiors.

Energy is used to describe the exercise of joggers, the production and output of thunderstorms, the destructive power of earthquakes, and the output and input in chemical reactions. It is the *conservation* of energy that makes the concept of energy so valuable in describing changes in nature. If only part of the energy can be measured in a situation, the rest can be deduced from the application of this law. Energy is important to all of the natural sciences because it is so useful in understanding *any* physical process.

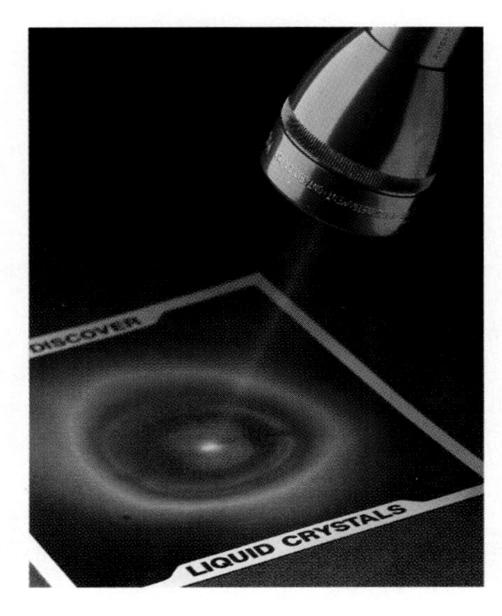

FIGURE 6-16 HEAT-SENSITIVE LIQUID CRYSTALS SHOW THAT LIGHT FROM THE FLASHLIGHT CARRIES ENERGY THAT WARMS THEM.

Summary and Key Terms

- work–78
- simple machines–79
- power–81
- net work–82
- kinetic energy (*KE*)–82
- potential energy (*PE*)–85
- internal energy–87
- heat–87
- efficiency–87
- conservation of energy–88

In physics, ***work*** is a measure of *how productive* an applied force is. Work equals the force in the object's direction of motion multiplied by the distance the force acts through, or $W = F_{\parallel} \times d$. Work is not a vector quantity. ***Simple machines*** trade force for distance to help you do work more easily. *Levers* and *inclined planes* are simple machines that allow you to lift objects with less force by moving them through a greater distance. ***Power*** is the rate of doing work. $P = W/t$.

Net work is equal to the component of the *net* force parallel to the object's motion multiplied by the distance it acts through, $W_{\text{net}} = F_{\text{net}\parallel} \times d$. When net work is done on an object, its ***kinetic energy***, $\frac{1}{2}mv^2$, is changed. $W_{\text{net}} = \frac{1}{2}mv_f^2 - \frac{1}{2}mv_i^2$. An object's kinetic energy, or ***KE,*** is the energy it has as a result of its motion. Kinetic energy, like work, is not a vector quantity.

While kinetic energy is energy of motion, ***potential energy*** (or ***PE***) is energy of *position,* or a measure of an object's potential to do work. *Wherever an object has a force ready to do work on it, the object has potential energy.* Any object above the ground has gravitational potential energy, *mgh,* the potential for gravity to do work on it. Bows and springs have *elastic* potential energy, and molecules, through their bonds, have *chemical* potential energy.

Kinetic frictional forces convert an object's kinetic energy into ***internal energy***, making the energy less effective at producing work. ***Heat*** refers to the part of an object's internal energy that moves because of a difference in temperature. The ***efficiency*** of a machine, animal, or any other energy processor is defined as work done divided by the energy expended to get the work done. In doing work, some energy always goes to increasing a system's internal energy, so no machine can ever be 100 percent efficient.

The ***conservation of energy*** takes into account all of energy's forms: *In every interaction in a closed system, the total energy afterward is always the same as the total energy at the beginning.* Energy may change forms, as it does in any inelastic collision, for example, but the total energy in a closed system is constant.

EXERCISES

Concept Checks: Answer true or false

1. Work is done on an object only if an applied force has a component along the direction of the object's motion.
2. If an object moves while a force acts on it, the force always does work.
3. A simple machine can multiply your force, which saves you work.
4. Power is a rate, work divided by time. If the same amount of work is done in less time, the power has a larger value.
5. Work can be done on something by more than one force at a time.
6. The change in an object's kinetic energy is equal to the net work done on it, so the units of *KE* and *W* are the same.
7. If net work is done on an object, it can be at rest afterward.
8. The kinetic energy of a bowling ball depends upon its mass and its speed, but not its position.
9. Every force can store potential energy.
10. For an object to have potential energy there must be a force ready to do work on it.
11. If you climb a flight of stairs, you increase your potential energy but you do no work.

12. A squirrel 10 meters up in a tree climbs until it is 20 m from the ground. It has doubled its potential energy.
13. Chemical energy is another name for the potential energy stored in the bonds of molecules.
14. Work can cause a change in something's speed, its position, or its temperature.
15. The law of conservation of energy tells us that if all its forms are accounted for, the total energy in any closed system is a constant.

Applying the Concepts

16. Is it possible to do work on an object that remains at rest? Explain.
17. Can net work be done on something without changing its speed?
18. When a pool ball rolls across a level billiard table, does gravity do work on the ball?
19. Does gravity do any work on you while you are lying in bed? When you are sitting in a car that is moving along a level road?
20. Why, in terms of work, is it important to have level floors in buildings?
21. When you walk, does the frictional force on your feet from the ground do any work on you? Discuss.
22. Does carrying a heavy backpack along a level trail require that you do any work on the backpack? Does carrying it up a hill require that you do any work on it?
23. Calculate your kinetic energy (in joules) when you stroll along at 1 m/s. (Don't know your mass in kg? 1 kg *weighs* 2.2 lb.)
24. A 15,000 lb (mass = 6800 kg) backhoe goes along the highway at 25 mph (11 m/s). What is its kinetic energy in joules?
25. When you walk down a flight of stairs, does gravity do work on you? When you walk up the stairs? Discuss.
26. Are roads that wind back and forth as they go to the tops of mountains like inclined planes? Discuss.
27. Is the person shown in Figure 6-17 doing any work on the boat while he is rowing? (*Hint:* Does the boat's *KE* change while he is rowing?) Discuss your answer.

FIGURE 6-17

28. Can something have potential energy and kinetic energy at the same time? Does the potential energy of an object depend upon its speed?
29. Is it correct to say that gravity supplies the energy for a downhill skier's run?
30. When you carry a heavy suitcase up a flight of stairs, do you do more work on it if you run rather than walk?
31. A woman (60 kg) climbs 3 flights of stairs (9 m). How much work does she have to do because of gravity?
32. Approaching a steep hill in a car, many drivers will accelerate to gain speed. How does this change things in terms of energy?
33. A cat climbs a tree. Describe its ascent in terms of its potential energy, the work the cat does, and the work gravity does on the cat.
34. A stick of dynamite is small and lightweight, and yet it contains all of the energy released when it explodes. Where does the energy come from?
35. Identify a major source of potential energy for the following things: (a) a squirrel in a tree, (b) a bullet in a gun, (c) a banana split, (d) a waterfall.
36. During a brake test, a Honda Accord stopped from a speed of 30 mph in 44 feet, and at 60 mph it required 162 feet to come to a stop. If the average braking force was the same in both cases, are these numbers consistent with the *net work = change in kinetic energy* formula?
37. When an avalanche comes to rest at the base of a mountain, where has its potential energy gone?
38. In Figure 6-18, how much more potential energy does ball B have than ball A? (*Hint*: use a ruler!)

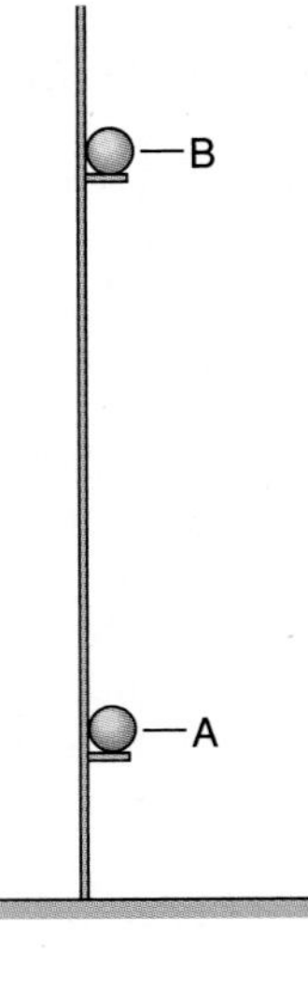

FIGURE 6-18

39. Books on diets and jogging often have tables that show the approximate energy (in Calories, where 1 Calorie is 4180 joules) used during various activities. In these tables you can read that walking down a flight of stairs requires more energy than walking at the same speed on level ground. Explain why this is true.

40. Bicycle manufacturers go to great lengths to make bike frames rigid. Explain how this helps a cyclist with regard to energy.

41. A child throws a ball of clay against a wall and it sticks. Has energy been conserved?

42. After exercise, you can probably see your pulse in the veins of your arms and feet. Each heartbeat forces blood through your arteries and veins, raising the pressure, and the elastic blood vessels expand. But what about the time between heartbeats—does the blood flow stop? No, and the answer has to do with potential energy. Can you guess why the blood keeps flowing?

43. It is ten times more efficient to get usable aluminum from recycled aluminum than from aluminum ore. Recycled glass requires 32% less energy to turn into usable glass than do its ingredients, sand, soda, and lime. Discuss recycling in the light of these facts.

44. A shot-putter pushes a 7-kg shot from rest to 12 m/s in 1.5 seconds. How much work did he do (ignoring any work he had to do against gravity)? What was the average power required?

45. A typical load rate for sanitation workers in New York City is to lift some 45 bags (total weight of 2925 lb) in 27 minutes. Calculate the average horsepower these workers exert while they work.

46. If a child pulls a wagon along a sidewalk at a constant speed, does the child's pull do any work? Why doesn't the *KE* increase?

47. A person who runs 10 miles in 60 minutes uses approximately 1200 Calories of food energy, but a walk for that distance uses only 200 Calories. (1 Calorie is 4180 J.) Why is there such a big difference in the amounts of energy used? (*Hint:* Muscles used in running give off up to 20 times the heat they do at rest.)

48. We saw in Chapter 4 that a centripetal force acts to turn something, but it doesn't speed the object up or slow it down. Why doesn't a centripetal force do work on the object it turns? (Remember that a centripetal force is always perpendicular to an object's velocity vector.)

49. From 5 to 8 percent of a marathon runner's energy output goes into pushing through the air. What strategy should reduce this amount? Would you expect a jogger who moves half as fast as a world-class marathon runner to use more or less energy pushing through the air? Why?

50. If you drop a tennis ball onto a concrete floor, it will bounce back farther than if you drop it on a rug. Where does the lost energy go when it strikes the rug?

51. A raccoon of mass 5 kg climbs 4 m up a tree. Calculate its potential energy. Find its kinetic energy just before it hits the ground if it slips, neglecting air resistance.

52. Does it take more work for you to run through soft deep sand than to run on an asphalt track? Explain.

53. Should you *really* worry about your car's efficiency? Consider that LA residents drive over 140 million miles *every day,* about the distance from the Sun to Mars. Discuss.

54. Use the net work-kinetic energy relation to find the speed of a diver as she hits the water if she falls from a 3-meter diving board. What is her speed if she first springs 1/2 m above the board?

55. A pickup truck (2000 kg) and a car (1000 kg) each going 90 km/hr collide head-on and stick together, moving off at 30 km/hr. How much kinetic energy was lost?

Demonstrations

1. Find a long board and prop one end up to make an inclined plane. Then attach one or more rubber bands to the front end of a child's toy, perhaps a heavy truck or tractor. Pull the rubber bands to tow the toy up the incline to the top. Next, place the toy on the floor beside the board and lift it by the rubber bands straight to the top of the incline. Notice how much farther the rubber bands stretch and how much more force you have to use when you lift the toy directly up. It's hard to believe the amount of work done by each method is the same, but they are. Simple machines help that much.

2. A screw is just a wrap-around inclined plane! Cut a triangle of any shape from a piece of paper—this is the inclined plane. Holding a pencil vertically, wrap the inclined plane around the length of the pencil to form the threads on a "screw." The more threads per centimeter the inclined plane makes on the screw, the more distance will be traded for force when the screw does work.

3. Investigate the potential energy in food. Skewer a shelled pecan on a straightened paper clip and use a match to set the nut on fire. Then hold a small aluminum cup of water over the burning pecan and watch the water boil from the heat produced. This principle is how laboratories determine the number of Calories (1 Calorie = 4180 joules) in food. They burn a quantity of beans or carrots or some other edible item and carefully measure the heat transferred from the food.

THE DISCOVERY

During the plague year of 1665, 23-year-old Isaac Newton retreated to his mother's farm in the English countryside, where he concentrated on natural philosophy for the next year and a half. One of the mysteries he turned his attention to was gravity. In the absence of air resistance, the attractive force exerted by the earth gave everything around him exactly the same acceleration, *g*. Why should earth pull on all objects, no matter what their mass, with precisely the right amount of force to accelerate them all at the same rate? With a great leap of imagination, he solved this puzzle. He later told a biographer he was inspired by a falling apple, so that legend is apparently true.

Perhaps Newton was the first to think of the moon as being attracted by earth's gravity, just like that apple that fell on his mother's farm. A force must act to keep the moon from gliding out of sight along a path tangent to its orbit (see the figure), and Newton figured out that this force on the moon is exerted by the earth.

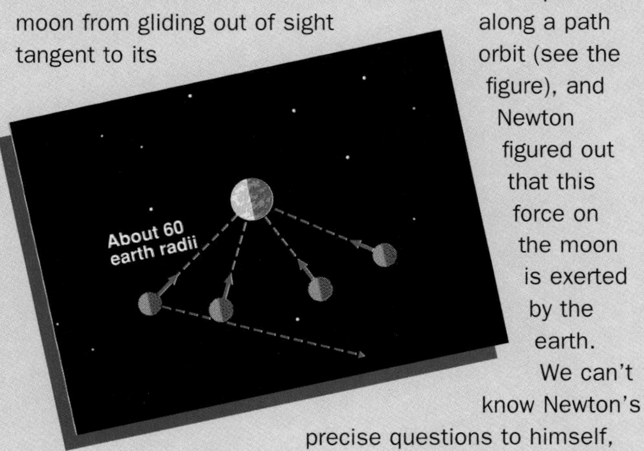

We can't know Newton's precise questions to himself, but we do know his answers. The moon's speed v is the distance around its circular orbit, $2\pi r$, divided by the time it takes to complete one orbit. With v^2/r, Newton calculated that the rate at which the earth accelerates the moon is 0.009 ft/s^2, as compared to 32 ft/s^2 for the falling apple. Newton knew that the earth–moon distance is about $60R_{\text{earth}}$. Comparing the acceleration of the moon with that of the apple, he could see that the acceleration of the moon toward the earth is about $g/3600$, or about $g/(60)^2$. The acceleration, and thus the force, decrease with distance d according to the factor $1/d^2$. That force depends on the mass of the attracted object, too, since something twice as massive weighs twice as much. Likewise, because action equals reaction, the mass of the attracting object must figure in as well. Putting the facts together, Newton deduced the formula for the force of gravity:

$$F_{\text{gravity}} = G\frac{m_1 m_2}{d^2}$$

where G is a constant found by experiments, m_1 and m_2 are the masses of the attracting objects, and d is the distance between them. To see if this idea was correct, Newton needed a test.

The scholars of Newton's time have a great legacy from earlier atronomers. Johannes Kepler (1571–1630) found in 1604 that the planets orbit the sun along elliptical paths. He also found relationships between their average distances from the sun and how fast they move (1609) and how long it takes them to complete their orbits (1619). These three "laws" showed how the planets move, but no one could say *why* they moved that way. Newton showed his gravitation law would cause the planets to move according to Kepler's laws. It was a magnificent discovery. Then Newton put his calculations away and left them, perhaps because he had satisfied himself. The discovery of the law of gravity would lie unannounced for almost 20 years.

In 1683 Edmund Halley (Halley's comet is named for him) and Christopher Wren (famous for his architecture) put a question to the English scientist Robert Hooke. They asked what force from the sun might cause the planets to move in elliptical paths. Hooke guessed that the force must decrease with distance according to the factor $1/d^2$. Perhaps Hooke drew an analogy between the sun's gravitational force and the sun's light; scientists knew that the intensity of a candle's light fades with increasing distance from the source according to $1/d^2$. Hooke tried to prove it. He did experiments on the tops of buildings in London to see if earth's gravity was less there, and he went down into wells to see if earth's gravity increased, but all to no avail. Hooke's remarkable guess went unproved.

The next year Halley went to Cambridge and asked Newton what path a planet would follow should the force from the sun be proportional to $1/d^2$, to which Newton replied, "Ellipsis." When asked how he knew, Newton said, "Why, I have calculated it." Halley insisted that Newton publish his discoveries. In 1687, with Halley's financial backing, Newton published, in Latin, his *Principia (The Mathematical Principles of Natural Philosophy).* In it were his three laws of motion, the law of gravity, and a number of applications of his laws, explaining things that had never before been understood. He derived Kepler's three laws of planetary motion, he explained why there are tides in the oceans, he showed why a rotating planet bulges at its equator, and much more. Experiments proved his laws to be true time after time, and their success soon affected thinking in such diverse areas as philosophy, psychology, politics, and law. Newton's accomplishments were a catalyst for the Age of Reason, when science first emerged as an important influence in human culture.

SEVEN

Gravity

The stars leave trails in this photograph because the earth rotated while the camera's shutter was open. The motions of the stars, the planets, the sun, and the moon played an enormous role in the history of science. We'll see here that the moon's motion was Isaac Newton's key to understanding gravity.

Long ago in the high desert plains of the American West, Papago Indians explained their surroundings in legends. Because year after year the stars rose in the same patterns, the Papagos believed the stars were somehow connected to the earth. They told their children that in the beginning spider people used their webs to sew the heavens to the desert. Spider webs, so nearly invisible and yet remarkably strong, seemed to be the perfect threads to whirl the glittering stars around. Today we know the Papagos were not far off. There is a connection between our planet and the heavens. Now we call this web gravity, and its threads really are invisible, and they really do stretch all the way to the stars. Like the Papagos, we also pass our explanations along, but as science rather than legend.

THE LAW OF GRAVITY

There is a force in nature that's everywhere—in all of space and acting on every particle that exists. It's a mutual attraction between all the particles in the universe, an attraction that acts over long distances, through empty space, and directly through any other particles in between. Every particle exerts this pull on every other particle and in return gets an equal and opposite pull from each of those particles. This remarkable property of matter is called gravity. This force is the same one you must overcome each morning when you get out of bed. The earth's mass attracts you because you, too, have mass. Your right arm has mass and therefore attracts your left arm, and so it goes throughout the universe.

Newton found that any two particles pull directly toward each other with forces that are equal and opposite. Newton's law of gravity states:

PUTTING IT IN NUMBERS—THE FORCE OF GRAVITY

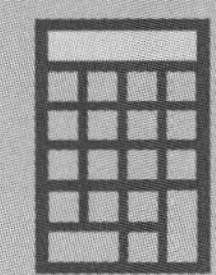

The force between two particles varies as the product of their masses, m_1 and m_2, divided by the square of the distance d between them:

$$F_{\text{gravity}} = \frac{G\,m_1 m_2}{d^2}$$

G is a constant which is apparently the same throughout the universe. Its numerical value depends on which units are used for force, mass, and distance. In SI units, G is $6.67 \times 10^{-11}\ \text{N}\cdot\text{m}^2/\text{kg}^2$. **Example:** Find the force of attraction between a 50-kg sports car enthusiast and a 1200-kg Ferrari located 100 meters away from her down the street in a showroom. $F_{\text{gravity}} = 6.67 \times 10^{-11}\text{N}\cdot\text{m}^2/\text{kg}^2 \times 50\ \text{kg} \times 1200\ \text{kg}/(100\ \text{m})^2 = \mathbf{4 \times 10^{-10} N}$. (That is less than a billionth of a newton, so she should be able to resist it.)

The farther apart the particles are, the smaller the gravitational attraction they have for each other. As d becomes larger, Gm_1m_2/d^2 becomes smaller, as Figure 7-1 shows. Nevertheless, the distance must become *infinitely large* before F_{gravity} vanishes completely. Here's a force that acts over a distance as great as you can imagine, straight through anything that's in its way!

However, applying this force to an object of any size can be tricky. Suppose you want to use the law of gravity to see how large the gravitational attraction

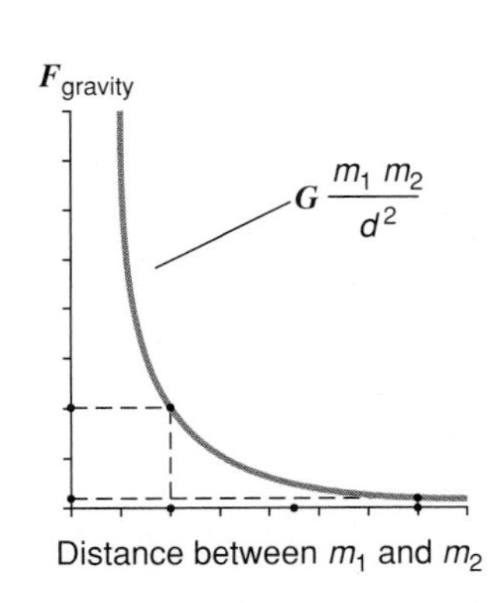

FIGURE 7-1 THE FORCE OF GRAVITY SHRINKS QUICKLY WITH INCREASED DISTANCE. IF TWO OBJECTS A DISTANCE D APART ARE MOVED 3 TIMES FARTHER APART, THE FORCE BETWEEN THEM IS ONLY 1/9 AS LARGE: COMPARE $1/D^2$ AND $1/(3D)^2 = 1/(9D^2)$

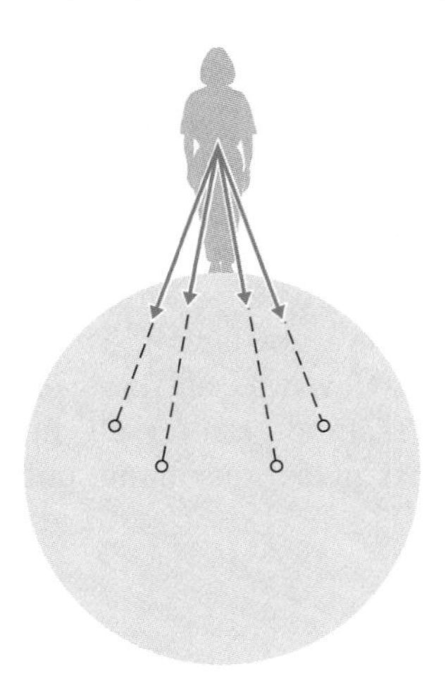

FIGURE 7-2 ALL OF EARTH'S MASS ATTRACTS YOU. THESE PARCELS OF MATTER AT DIFFERENT DISTANCES ATTRACT YOU WITH DIFFERENT AMOUNTS OF FORCE AND FROM DIFFERENT DIRECTIONS.

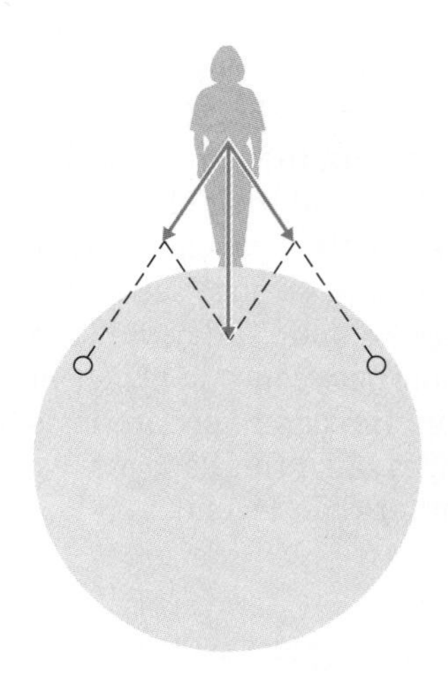

FIGURE 7-3 BECAUSE EARTH IS SYMMETRICAL, THE NET ATTRACTION FROM ALL ITS PARTS IS TOWARD ITS CENTER. AS YOU STAND ON THE SURFACE, THERE IS JUST AS MUCH MASS PULLING YOU DOWN AND TO THE RIGHT AS DOWN AND TO THE LEFT. HERE TWO PARTICLE'S ATTRACTIONS ARE SHOWN, GIVING A NET FORCE TOWARD EARTH'S CENTER.

is between you and the earth. The molecules on one side of earth are much closer to you than those on the other side, so their pulls are very different, and each of those tiny pulls is a vector, so all those vectors have to be added. Figure 7-2 illustrates the problem. However, despite the staggering number of attracting particles, there is only one *net* force, and you can see that the "average" attraction of all the particles in earth points approximately toward its center (Figure 7-3). Newton worked out how those small forces add together for a spherical planet, and he found that the net force exerted is *the same as if all the particles were located at the planet's center!* Earth is nearly spherical, so to find the force of earth's gravity on something at its surface, or even out in space, we can just assume that the total mass of the earth is at its geographical center. This point is called the **center of gravity.** The center of gravity of an object is also the point where the *net* force of gravity from an *external* object acts. Figure 7-4 shows how to find the center of gravity of small, nonspherical objects.

THE CENTER OF GRAVITY IS, FOR OUR PURPOSES, THE SAME AS THE BALANCE POINT OF AN OBJECT CALLED THE CENTER OF MASS, FROM CHAPTER 4.

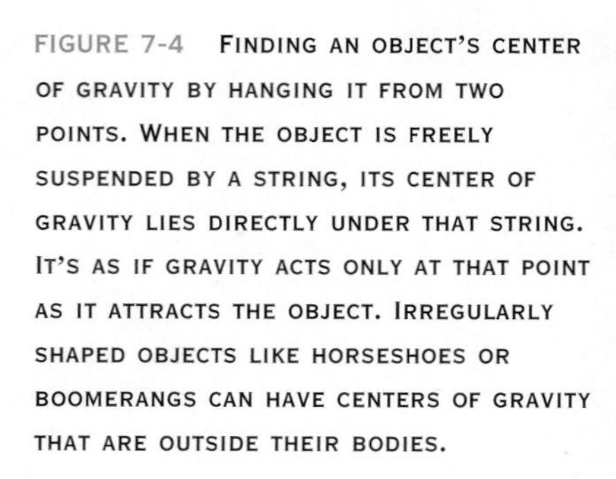

FIGURE 7-4 FINDING AN OBJECT'S CENTER OF GRAVITY BY HANGING IT FROM TWO POINTS. WHEN THE OBJECT IS FREELY SUSPENDED BY A STRING, ITS CENTER OF GRAVITY LIES DIRECTLY UNDER THAT STRING. IT'S AS IF GRAVITY ACTS ONLY AT THAT POINT AS IT ATTRACTS THE OBJECT. IRREGULARLY SHAPED OBJECTS LIKE HORSESHOES OR BOOMERANGS CAN HAVE CENTERS OF GRAVITY THAT ARE OUTSIDE THEIR BODIES.

VISUALIZING GRAVITY: THE GRAVITATIONAL "FIELD"

FIGURE 7-5 GRAVITATIONAL FIELD LINES LET US PICTURE THE FORCE OF GRAVITY. A PARTICLE WITH MASS PLACED ANYWHERE NEAR EARTH IS ATTRACTED IN THE DIRECTION OF THE FIELD LINES.

Lines sketched at equal intervals straight into a body such as earth, called **gravitational field lines,** show the direction of the body's gravitational pull on any matter in the space around it (Figure 7-5). Where the field lines are closer together, the value of the force is relatively larger. Farther from the mass, where the field lines are spread out, the force is smaller. Like the spider webs of the Papagos, these field lines are imaginary, but they give us a helpful picture of the force of gravity throughout space.

ORBITS CAUSED BY GRAVITY

An arrow leaves a bow extremely fast and then follows a curved path, attracted toward earth's center by gravity. At the circus a human cannonball explodes from the mouth of the cannon and arcs over into a net. Each of these is a *projectile*, which is any object that is given a velocity and released to the influence of gravity and air drag. The path of a projectile is called a *trajectory.*

Newton took trajectories a step beyond the human experience of his day. In an illustration he placed a cannon on an imaginary mountaintop above the earth's atmosphere. Using the law of gravity, he derived the paths a cannonball would take in this airless environment if it was launched horizontally at various speeds. Figure 7-6 shows paths much like those that Newton drew 300 years ago. Low-speed projectiles arc over and fall to earth. The greater the projectile's speed, the farther it goes around the earth before landing. Finally, at some very great horizontal speed, a cannonball might fall toward earth but never reach the ground. Such a trajectory is called an *orbit*.

To orbit the earth requires a very great speed if the projectile is not far away from earth's surface. At the relatively close distance of 100 to 200 miles up, for example, the necessary speed is about 17,000 mph. That sort of orbit was first achieved by Sputnik 1 in October 1957, and since then thousands of artificial satellites and spacecraft have been placed in orbits. A rocket lifts a satellite above the atmosphere, gives it the necessary speed to orbit, and releases it. Then, with almost no drag, the satellite glides along in a path dictated almost entirely by the earth's gravity.

Newton proved that the gravitational orbits of one body around another are *ellipses,* curves that resemble flattened circles. (See Demonstrations 1 and 2 for the details of ellipses. A circle is a special case of an ellipse with no flattening.) The paths Newton drew for cannonballs that did not have enough speed to orbit the earth are also elliptical as far as they go, that is, until the cannonballs strike the ground.

When the space shuttle or an artificial satellite orbits our planet in an elliptical orbit, the earth's center isn't at the center of the ellipse. It is at a

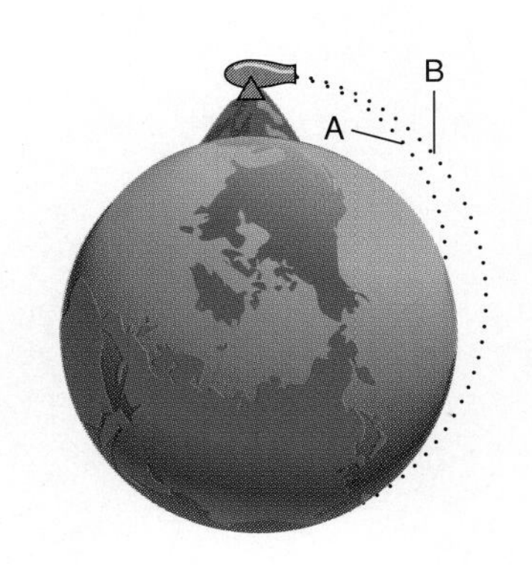

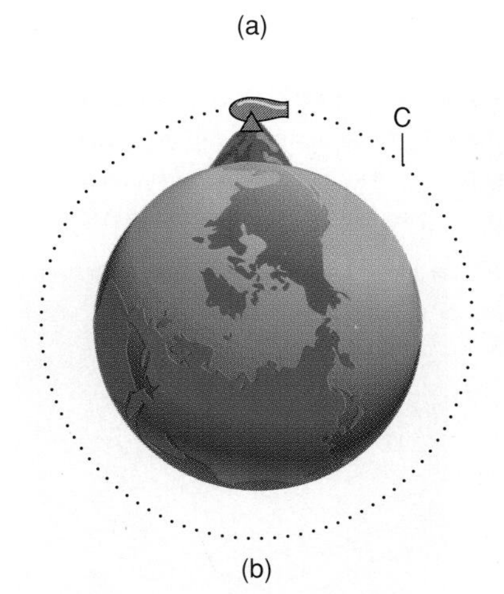

FIGURE 7-6 (A) SHOT FROM AN IMAGINARY MOUNTAINTOP FAR ABOVE THE ATMOSPHERE, A CANNONBALL WOULD FALL FAR "AROUND" THE EARTH BEFORE HITTING THE SURFACE. PATH B WOULD BE TAKEN BY A CANNONBALL WITH A GREATER INITIAL SPEED THAN ONE THAT WOULD FOLLOW PATH A. (B) AT ONE CERTAIN LAUNCH SPEED, THE RATE OF TRAVEL TO THE SIDE (THE HORIZONTAL SPEED OF THE BALL) WOULD BE GREAT ENOUGH SO THAT, AS THE BALL FELL, ITS DISTANCE TO THE EARTH'S SURFACE WOULD REMAIN THE SAME. THE BALL WOULD THEN FOLLOW PATH C, A CIRCULAR ORBIT.

PHYSICS IN 3D/THE SPACE SHUTTLE IN ORBIT

THE SPACE SHUTTLE IN ORBIT, FOLLOWING A TRAJECTORY WITHOUT AIR DRAG. PHOTOGRAPHY FROM THE SHUTTLE GIVES US A NEW PERSPECTIVE OF EARTH.

PHYSICS IN 3D/A STEREO VIEW OF AN ISLAND OFF GREECE

LOOKING DOWN FROM THE SPACE SHUTTLE, PHOTOGRAPHS TAKEN A FEW SECONDS APART GIVE US THIS STEREO VIEW OF AN ISLAND OFF GREECE. THE MOTION OF THE SHUTTLE IN THAT BRIEF TIME IS ENOUGH TO GIVE A "RIGHT" EYE VIEW AND A "LEFT" EYE VIEW THAT WE CAN USE TO SEE AN EXAGGERATED STEREO, OR HYPERSTEREO.

point toward one end called a *focus*. (See Figure 7-7 and Demonstration 1.) An elliptical orbit brings an orbiting body sometimes closer and sometimes farther away from its parent mass. The closest point in an earth orbit is called the *perigee;* the farthest, the *apogee*. Corresponding points for an elliptical orbit around the sun are *perihelion* and *aphelion,* and for such an orbit around the moon, *perilune* and *apolune*.

FIGURE 7-7 EARTH IS AT ONE FOCUS OF THE ELLIPTICAL ORBIT OF AN ARTIFICIAL SATELLITE. THE OTHER FOCUS, MARKED BY THE ×, IS EMPTY (SEE DEMONSTRATION 1).

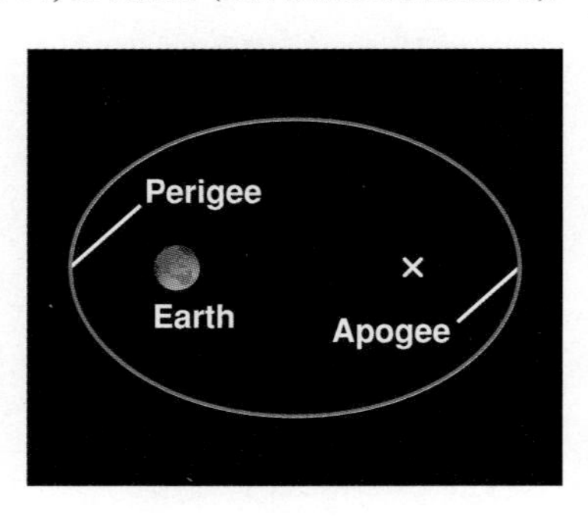

ESCAPE VELOCITY

The path of a projectile shot with an even greater speed than those drawn by Newton is shown in Figure 7-8. A launch speed of about 25,000 mph (40,000 km/h) would catapult Newton's cannonball out of a potential earth orbit. That speed is called the **escape velocity** from earth. A projectile can never get

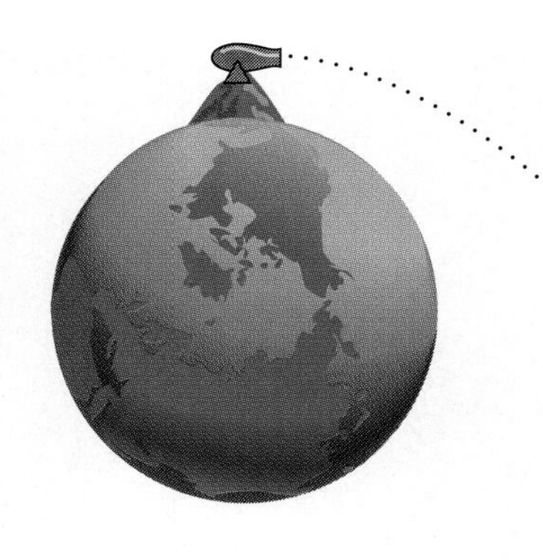

FIGURE 7-8 AN OBJECT TRAVELING AT A SPEED EQUAL TO EARTH'S ESCAPE VELOCITY CAN LEAVE EARTH PERMANENTLY. THOUGH IT CONTINUES TO SLOW AS IT MOVES OFF, IT IS NEVER STOPPED AND TURNED AROUND BY EARTH'S GRAVITY.

"beyond" earth's gravity, but it can move fast enough so that earth's gravity cannot pull it back. That doesn't mean the object could leave the solar system. Although a speed of 25,000 mph lets a spacecraft pull away from earth, it would need, from the vicinity of earth, a speed of almost 100,000 mph to escape from orbiting the sun.

The escape velocity from a planet determines how much atmosphere the planet can have. The earth, for instance, continuously loses fast-traveling molecules that escape from the upper edges of its atmosphere. A molecule

EARTH SATELLITES: MONITORS OF THE WORLD

Today satellites placed in earth-orbits by rockets or gently launched from an orbiting space shuttle provide platforms for monitoring our weather, the condition of the ozone layer, gathering data on the oceans, and much more. Recently satellites with radar have measured the flow rates of glaciers and the uplift and subsidence of ground areas after an earthquake in California—which amounted to only several inches! The figure below shows the track of a typical satellite for a 10-day period, so data can be collected over most of earth in a very short while. Figure 7-9 shows the distribution of the production of phytoplankton in the oceans and in lakes on land over a three year period. (Reds and yellow show very high production, while blue shows moderately high production. See the bands in the oceans at the equator.) Such information would be impossible to acquire in such detail by any other method.

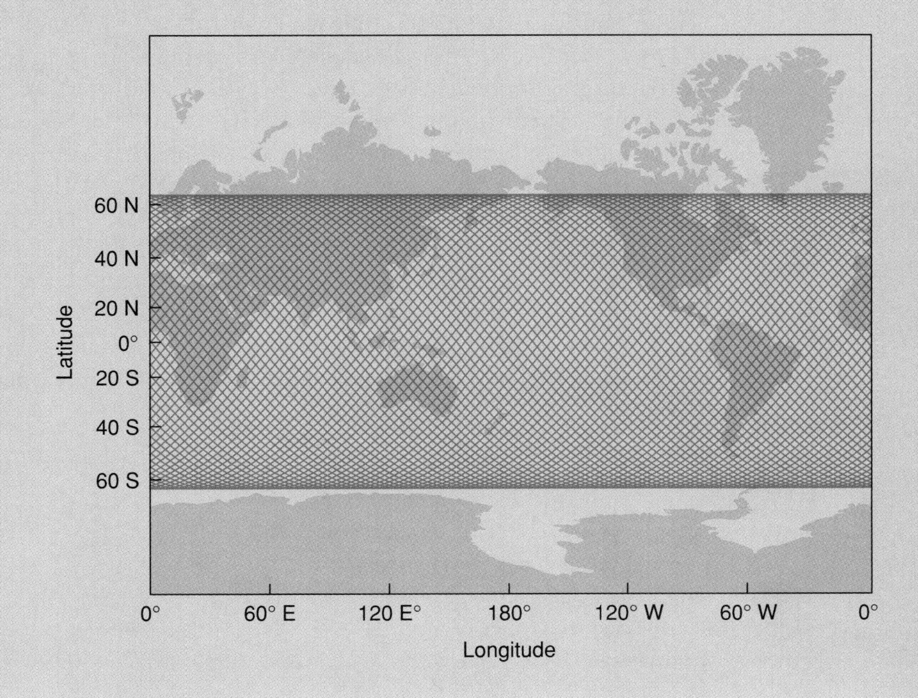

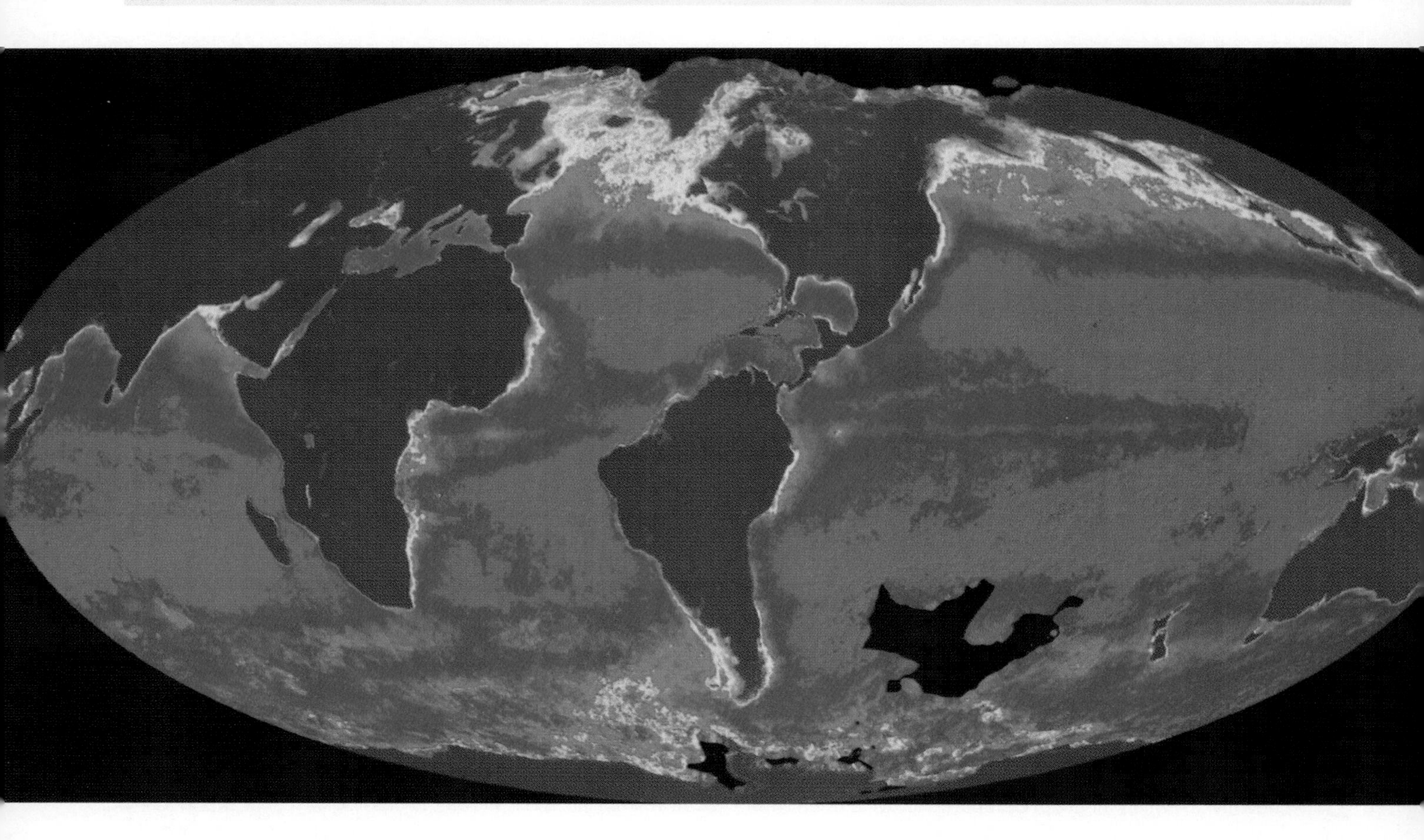

FIGURE 7-9 AREAS OF PHYTOPLANKTON PRODUCTION OVER A THREE-YEAR PERIOD AS MONITORED FROM A SATELLITE IN EARTH ORBIT.

heading outward at 25,000 mph won't return unless it hits another particle and gets turned around. Because Mars is only about 1/10 as massive as earth, its gravitational force is much weaker than earth's. Mars can't hold onto any outward-bound molecules that travel at speeds greater than 11,200 mph. Consequently, the Martian atmosphere is extremely thin, exerting less than 1% of earth's atmospheric pressure at its surface. Because of the moon's low escape velocity, about 5100 mph, it can't hold gas molecules at all—the moon has no atmosphere (Figure 7-10). Sunlight heats the moon's surface and would quickly raise the temperature of any gases there, increasing the kinetic energy of the molecules enough to let them escape. Even the exhaust gases left by the liftoffs of the lunar landers are long gone. Unfortunately, the gases from automobiles and fossil-fueled power plants on earth do not escape!

FIGURE 7-10 THE MOON HAS NO ATMOSPHERE BECAUSE OF ITS SMALL ESCAPE VELOCITY, WHICH IS DUE TO ITS SMALL MASS. NOTICE THE BLACK "SKY" ABOVE THE MOON'S HORIZON DUE TO ITS LACK OF AN ATMOSPHERE.

GRAVITY AND YOU

Single-cell organisms that float about in water apparently don't respond to gravity in any way. Larger land dwellers, however, are built to perform in earth's gravitational pull. Each time you stand up, pressures adjust and muscles do work against the force of gravity. Unlike the one-cell floaters, you can fall, and at considerable peril. The danger from falling increases with size and mass. American bison wouldn't cross railroad tracks for fear of tripping and falling, so the first transcontinental railroad in the 1800s permanently divided the buffalo into a northern herd and a southern herd. People capture elephants in India by enticing them into an area laced with sugar cane and nearly surrounded by a shallow ditch—just deep enough so the elephant's trunk won't touch the bottom. With its poor eyesight, an 11,000-lb elephant won't risk a fall to escape.

Astronauts have shown how much gravity has to do with our physical condition. The first Skylab crew (in 1973) was "weightless" for 29 days. In spite of vigorous exercise, the members flunked their cardiovascular exams after two weeks in orbit. In free fall, blood is not pulled downward into the lower limbs as it normally is, and the astronauts' faces puffed up with body fluids. Each crew member lost up to 25 percent of his red blood cells and significant amounts of calcium from the bones of his legs. After touchdown, they had very erratic blood pressure and each had trouble walking. Soviet cosmonauts who were in orbit over 200 days were taken in wheelchairs from their spacecraft and still had problems with walking a month later. Inactivity on earth, as occurs with bed-ridden patients, has similar effects. The heart does less work pumping blood through a horizontal circulatory system, and skeletal bones experience much less stress and strain. Each time you climb out of bed in the morning, you begin the exercising vital to the performance of a healthy body.

CATTLE GUARDS, SO FAMILIAR TO WESTERNERS, KEEP COWS FROM CROSSING FOR FEAR OF CATCHING A HOOF AND FALLING. BECAUSE OF THEIR MASS (AND HENCE WEIGHT) A FALL IS A SERIOUS MATTER. MODERN CATTLE GUARDS ARE OFTEN JUST WHITE STRIPES PAINTED ACROSS THE ROAD, AS SHOWN HERE. THE CATTLE APPARENTLY CAN'T TELL THE DARK AREAS BETWEEN THE STRIPES FROM HOLES, SO THEY WON'T CROSS THEM.

GRAVITY AND THE OCEANS: TIDES

Since the dawn of navigation, mariners have marked deep channels in ports and rivers to avoid sandbars and mud flats at low tide. For centuries port cities have kept records of the times and levels of high and low tides. Table 7-1 lists the high and low tides at the Battery in New York City for several days. It reveals a general pattern: There are usually two high tides and two low tides each day; the high tides are about 12 hours and 25 minutes apart, as are the low tides. This pattern was the key to discovering the cause of the tides. While 24 hours is the period of the earth's rotation with respect to the sun (sunrise to sunrise, say), 24 hours and 50 minutes is the earth's rotation period with respect to the moon (or moonrise to moonrise). This is twice the tidal period of 12 hours and 25 minutes, and it leads to the fact that *the moon is the principal cause of the ocean's tides.*

Table 7·1 Tides at the Battery, New York, 1982

October 26	high	3:11 A.M.
	low	9:36 A.M.
	high	3:22 P.M.
	low	10:12 P.M.
October 27	high	4:10 A.M.
	low	10:28 A.M.
	high	4:23 P.M.
	low	10:57 P.M.
October 28	high	5:02 A.M.
	low	11:16 A.M.

The earth and moon pull on each other, according to Newton's law of gravity, with exactly the same amount of force. So as the moon glides along on its large orbit, the earth, with about 81 times the mass of the moon, performs

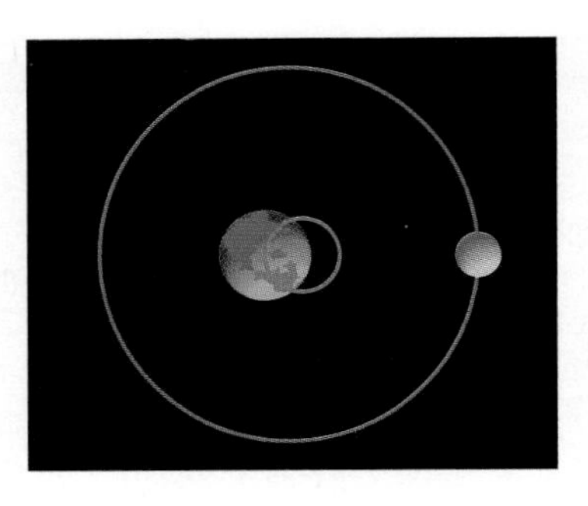

FIGURE 7-11 AS THE MOON ORBITS EARTH, EARTH TRAVELS ALONG A SMALLER ORBIT OF THE SAME SHAPE. ACTION EQUALS REACTION WITH GRAVITY AS WITH ANY OTHER FORCE. EARTH'S LARGER MASS MEANS ITS ACCELERATION, AND HENCE ITS ORBIT, ARE MUCH SMALLER THAN THE MOON'S. $M_E A_E = M_M A_M$ (THE MOON'S ORBIT IN THIS DRAWING IS NOT TO SCALE.)

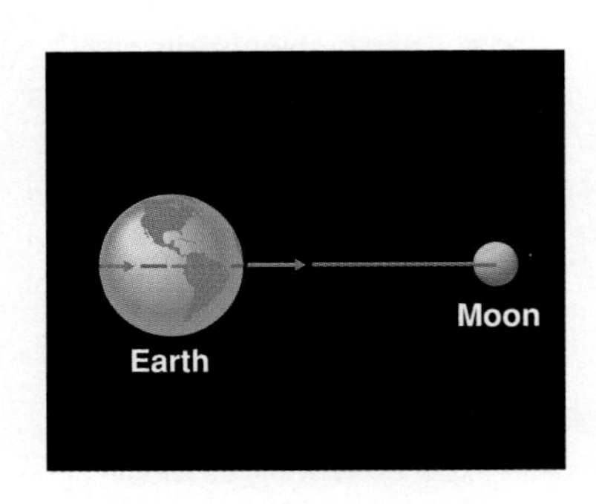

FIGURE 7-12 A COMPARISON OF THE RELATIVE STRENGTHS OF THE MOON'S GRAVITATIONAL PULL AT THREE POINTS, THE CLOSEST AND FARTHEST POINTS TO THE MOON FROM EARTH'S SURFACE AND AT THE CENTER OF THE EARTH. (THE MOON IS ACTUALLY MUCH FARTHER AWAY FROM THE EARTH THAN SHOWN HERE, AND THE FORCES ARE MORE NEARLY THE SAME AT THE THREE POINTS.)

a miniature orbit of its own (Figure 7-11). Both are in free fall around each other! At the same time, each body is immersed in the other's gravitational field, and the strength of these fields varies somewhat over their surfaces and throughout their bodies. Let's assume for now that the earth is perfectly rigid and accelerates toward the moon at a rate determined by the *average* pull on it. The oceans, which are not rigid, cover about 70% of earth's surface and are all connected. Because they are connected, they act much like a bowl of jello does when a spoon touches it; if the surface of the water goes down somewhere, it has to rise somewhere else.

Figure 7-12 shows some details of the moon's attraction. On the side of earth nearest the moon, the moon's pull on the ocean waters is slightly stronger than its pull on the rigid earth below. On the side of earth farthest from the moon, the moon's pull on the ocean waters is weaker than its pull on earth. The result is that ocean water nearest the moon actually weighs less by a small percentage because of the moon's gravity. If earth's gravity weren't holding it to the surface, that part of the ocean would "outrun" the earth toward the moon. The water on the far side of the earth is lighter as well because if earth's gravity didn't hold it against the surface, earth would "outrun" those waters toward the moon, leaving them behind. You can feel this effect in an elevator that's accelerating downward; you feel lighter. You still stay on the elevator floor, of course, and the ocean still stays on the surface of the earth, but the "reduced weight" effect is real in both cases.

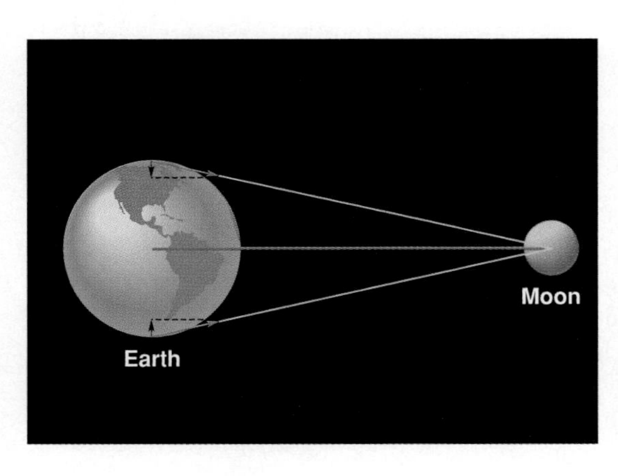

FIGURE 7-13 AT POINTS ON EARTH'S SURFACE THAT ARE PERPENDICULAR TO THE EARTH–MOON LINE, THE MOON'S GRAVITATIONAL PULL HAS A SMALL COMPONENT THAT PRESSES MATTER TOWARD EARTH'S CENTER, MAKING THINGS SLIGHTLY HEAVIER THERE. (DRAWING NOT TO SCALE.)

At the same time, ocean water that is about the same distance from the moon as earth's center is slightly heavier due to the moon's gravity. That's because the moon's pull in that area isn't horizontal; it has a downward component that presses the ocean water directly against the surface, adding to its weight (Figure 7-13).

The effect of all these forces is to stretch the oceans out a tiny bit along the straight line between the centers of the earth and moon, and to squeeze them in between. As a result, the ocean's surface becomes, to a very small degree, football-shaped. To understand this we can look at the gravitational force at points on the surface relative to the *average* gravitational force from the moon on the earth. If the average force for the rigid earth is subtracted from all the force vectors in its surface, Figure 7-14 is the result. This force is called the moon's **tidal force** at the surface of earth.

This tidal force is very small, or else you would feel heavier or lighter as the earth's rotation took you through the regions of Figure 7-14. But the oceans respond to these tiny forces over their immense areas, and tiny bulges and dips are the result—no drop of water moves very far, but the oceans' surfaces rise and fall as earth's rotation carries them through the areas of different tidal force. The bulges in the oceans cause high tides along the shores of continents and islands as earth rotates. The depressions, encountered about six hours after each high tide, cause the low tides (Figure 7-15).

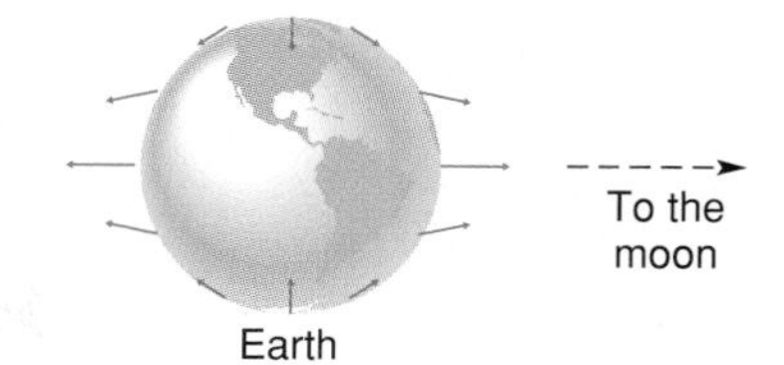

FIGURE 7-14 IF THE VECTOR FORCES AT EARTH'S SURFACE DUE TO THE MOON'S GRAVITY ARE DRAWN AND THE PULL OF THE MOON'S GRAVITY AT EARTH'S CENTER IS SUBTRACTED FROM EACH, THE RESULT IS THE ACTION OF THE MOON'S GRAVITY WITH RESPECT TO THE EARTH'S CENTER. THAT FORCE IS CALLED THE TIDAL FORCE. IT CAUSES A BULGE (HIGH TIDE) IN THE OCEANS UNDER THE MOON AND ON THE OPPOSITE SIDE FROM THE MOON, AND LOW-TIDE REGIONS IN BETWEEN.

FIGURE 7-15 THE (EXAGGERATED) SHAPE OF EARTH'S OCEANS BECAUSE OF THE MOON'S TIDAL FORCE AS IN FIGURE 7-14. (THE DISTORTIONS CAUSED BY EARTH'S ROTATION ARE NOT SHOWN.)

TIDES FROM THE SUN

Ocean tides at mid-ocean due to the moon amount to only $1\frac{1}{2}$ feet from high to low. The sun, too, pulls harder on the ocean waters nearest it and less on those farthest away; its high-to-low tides amount to about 1 foot at mid-ocean. Its tides are smaller than the moon's even though the sun's net gravitational pull on earth is far stronger! This is because the *percentage difference* in the sun's force of gravity from one side of the earth to the other is less than the moon's, and that is what causes the tides. When the sun and moon tides align, at full moon and new moon, as in Figure 7-16 a and b, the tides swing $2\frac{1}{2}$ feet and are called *spring tides*. At first quarter moon or third quarter moon (Figure 7-16 c), their tides subtract; that is, the bulges due to the sun occur in the depressions due to the moon and vice versa, and the tides are only about $\frac{1}{2}$ foot. These are called *neap tides*. Coastline indentations, such as bays and rivers, can funnel the rising tidal waters and bring about a large horizontal current and a higher rise in the local water level. A noteworthy example is the Bay of Fundy in Nova Scotia, where during spring tides the water sometimes rises 50 feet from low tide to high tide! These enormous volumes of water can be temporarily trapped and sent through turbines which drive electrical generators. In France one such plant has been in operation since 1966, generating the electrical needs of Brittany. Another smaller plant is operated in the former Soviet Union, and studies are underway for a tidal plant along the coast of Maine. Unfortunately, only 30 locations world-wide have tides large enough to generate significant power, so this alternative source of energy is very limited.

There is a tidal effect wherever a gravitational field changes appreciably in size and/or direction across the dimensions of a body. Tidal forces are everywhere, but for the most part they are too tiny to matter. There's no obvious tidal effect from the moon on a pond or on a person's body because every point in a pond or a person gets almost exactly the same gravitational pull from the moon. Without much difference in the force from point to point, there is little of that squeezing or stretching effect.

Sun Moon Earth

(a)

(b)

(c)

FIGURE 7-16 (A) NEW MOON AND (B) FULL MOON CAUSE TIDES THAT ALIGN WITH THE SUN'S TIDES, GIVING THE VERY LARGE "SPRING" TIDES. (C) NEAP TIDES, VERY LOW TIDES, OCCUR WHEN THE EARTH–SUN LINE MAKES A RIGHT ANGLE WITH THE EARTH–MOON LINE. THAT HAPPENS WHEN THE MOON IS IN ITS FIRST OR THIRD QUARTER PHASE. (DRAWING NOT TO SCALE.)

TIDAL STRENGTHS AND TIDES IN SOLID BODIES

The strongest tides any object can exert are found in the regions closest to the object. Figure 7-17 plots gravity's strength versus distance, and you can see that the closer something is to the attracting body, the greater the *difference* in gravity's pull across its surface, which causes tidal effects. Also, the larger the

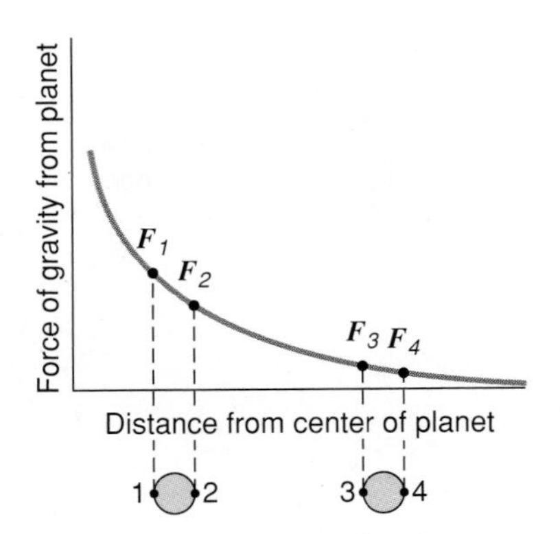

FIGURE 7-17 A SMALLER BODY AT TWO POSITIONS IN THE GRAVITATIONAL FIELD OF A MORE MASSIVE BODY. THE FARTHER A BODY IS FROM AN ATTRACTING MASS, THE SMALLER THE PERCENTAGE DIFFERENCE IN THE GRAVITATIONAL FORCE FROM ONE SIDE OF THE BODY TO THE OTHER. AS A RESULT, THE TIDAL FORCE IS LESS IF THE BODY IS FARTHER AWAY AND INCREASES IF THE BODY COMES CLOSER.

mass, the larger its tidal force at any given distance. Figure 7-18 compares the gravitational field strengths of earth and Mars, which has only a little more than 10% of earth's mass. The force of gravity of a larger mass exerts greater tidal forces than a smaller mass at the same distance.

Though the planet earth is rigid compared to water, the moon's tidal action stretches and squeezes the earth throughout its volume. Even solid rock is slightly elastic, so it yields. There is a tide in the solid earth due to the moon that at the surface amounts to about one foot, high to low. Its strength is impressive: Twice a day the earth tide lifts Mt. Everest up a foot and lets it down again. We don't see an up-and-down motion of the ground because we all rise and fall with the buildings, lakes, and everything else around us. But petroleum engineers monitor pressure in large underground reservoirs of oil, and they see an effect of these solid earth tides. The liquid-filled cavity in the rock below is stretched and squeezed as the tides deform the earth, and the pressure rises and falls on their gauges twice each 24 hours and 50 minutes.

The moon is squeezed and stretched by the tidal force from the earth's gravity. Because the same face of the moon points toward earth all the time, however, there are no daily highs and lows at points on the moon's surface as there would be if the moon rotated with respect to the earth. Yet the moon does respond to the earth's strong tidal grip: changes in earth's tides in the solid moon set off moonquakes. The moon's orbit around earth is an ellipse that brings it as near as 226,000 miles (its perigee point) and sends it as far away as 252,000 miles (its apogee). The earth's tidal force on the moon is strongest at perigee, and seismographs left on the moon by astronauts detected swarms of small tremors each time the moon reached perigee. At apogee those moonquakes are more rare.

AP OR APO = AWAY FROM, PERI = NEAR, AND LUNA = MOON.

Earth's tidal force on the moon is so strong that if the moon ever had rotated with respect to the earth, this tidal force would have stopped the rotation. Calculations show that the tides in the moon's solid surface would be 25 feet (7.6 meters) or more from high to low. The tremendous tidal stretching and compressing would cause great internal grinding and friction.

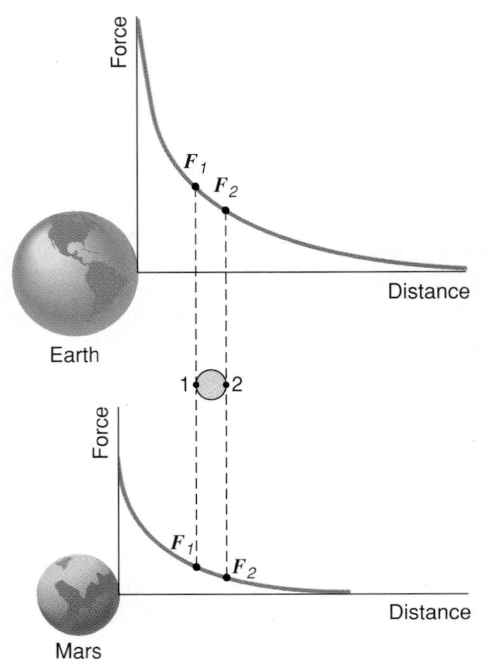

FIGURE 7-18 A COMPARISON OF THE TIDAL FORCES DUE TO EARTH AND MARS ON AN OBJECT THAT IS THE SAME DISTANCE FROM EACH. THE CURVED LINES INDICATE THE FORCE OF GRAVITY FROM EACH PLANET. THE FORCE ON THE NEAR SIDE, F_1, DIFFERS FROM THE FORCE ON THE FAR SIDE, F_2, FOR BOTH PLANETS. SO BOTH PLANETS EXERT A TIDAL FORCE ON THE OBJECT. BUT F_1 IS GREATER THAN F_2 BY A LARGER PERCENTAGE IF THE BODY IS NEAR THE EARTH, MEANING THE EARTH EXERTS A STRONGER TIDAL FORCE AT THE SAME DISTANCE THAN MARS (WITH ABOUT 1/10 OF EARTH'S MASS) DOES.

EINSTEIN'S REVOLUTIONARY THEORY

From Newton's time through the 19th century, scholars debated the nature of gravity. Newton's law of gravity worked extremely well, both with calculations and with predictions. The physical action Newton's law of gravity describes goes like this: One body exerts a force on another over a far distance with absolutely nothing in between them. How could any body such as earth reach out through nothing and exert a force on another body such as the moon? It seemed illogical. How can the one body know the other is there? No one really understood this *action at a distance*, as it came to be called. It was a great puzzle.

Albert Einstein revolutionized the understanding of gravity with his **theory of general relativity.** This theory did away with the concept of action at a distance and *even with the concept of the force of gravity.* In this complex mathematical theory, an object with mass distorts both the space and the time around itself. Space is curved in the vicinity of a massive object, and the greater the mass, the greater the curvature. In Einstein's theory of gravity, objects aren't pulled by a force at all; instead, they follow lines called *geodesics*, which trace out the local curvature of space and time. We cannot visualize in 4 dimensions (3 of space and 1 of time), but the warping of space near a large mass is something like what happens when a large, thin rubber sheet is stretched out horizontally and an object is placed on it. An object of small mass causes only a barely noticeable dimple in the sheet, but a large mass causes a major depression. Now think of a marble shot from someone's hand and passing far from the large mass's position. It would go straight, unaffected by the curvature. But shot along a path closer to the mass it would dip and swerve toward the depression before climbing back out and continuing. Newton's theory would say the marble swerved because the mass exerted a pull on the marble. Einstein's theory says the marble curved *because of the curved surface on which it moved.*

This turned out not to be just two ways of saying the same thing. Einstein's theory predicted that even rays of light, which have no mass in the everyday sense (you cannot stop light to weigh it; if stopped, it simply vanishes), would follow his geodesics. In a famous experiment performed during an eclipse of the sun in 1919, physicists observed that light from a star that passed near the edge of the sun swerved a tiny amount toward the sun just as Einstein's theory predicted. Since that time, his theory has correctly predicted the results of many carefully done experiments where Newton's theory could not. But Newton's law of gravity fails only when the masses are *very* large or the speeds *very* high or the times *very* long. Today scientists working in areas where those conditions do not exist still use Newton's law of gravity for calculations, such as those done for the trajectories that sent the Apollo astronauts to the moon and back.

PHYSICS IN 3D/THE CURVED SPACE OF RELATIVITY

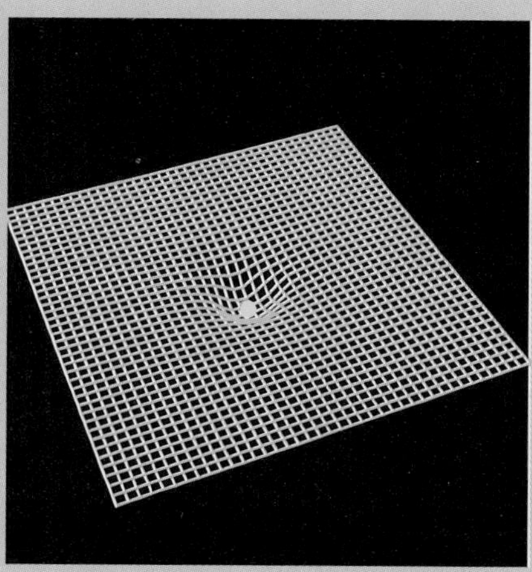

THE RUBBER-SHEET ANALOGY FOR THE CURVED SPACE AROUND A MASSIVE OBJECT. THE OBJECT'S MASS DISTORTS SPACE AND TIME AROUND IT.

The heat generated would eventually escape into space. As this happened, the moon's rotation would slow because the energy lost due to friction during the lifting and squeezing would come from the rotational kinetic energy of the moon. Conservation of energy tells us that, as the moon heated, the rate of spin would decrease until that smaller body faced the earth as it does today.

Summary and Key Terms

gravitational force–95

center of gravity–96

gravitational field lines–97

escape velocity–98

tidal force–102

theory of general relativity–105

The ***gravitational force*** is a mutual attraction between all objects with mass. Newton found that for two objects a distance d apart, $F_{\text{gravity}} = Gm_1m_2/d^2$. Newton also found that the value of the net gravitational force exerted by the earth on another object is *the same as if all of earth's particles were located at its center.* Every object has such a ***center of gravity***, a point from which its net gravitational force acts on other objects *and* where the net gravitational force from other objects acts on it. ***Gravitational field lines*** let us visualize the gravitational field of an object's mass.

An artificial satellite in earth orbit continuously falls around earth in an elliptical trajectory. Spacecraft can be given the ***escape velocity***, the minimum speed they need to escape from an orbit around earth. No object can really move beyond another body's gravitational pull, however; objects can only move fast enough to keep the attracting body from pulling them back.

A body's gravitational force exerts a stretching and squeezing force called a ***tidal force*** on other objects. A tidal effect occurs wherever a gravitational field changes appreciably in size and/or direction across the dimensions of the object it acts upon. The closer an object is to an attractive body, the greater the *difference* in gravity's pull across its surface and the larger the tides. The moon's gravity causes tides in earth's oceans and in the solid earth itself. The sun, too, produces tides in earth and its oceans, but these tides are smaller because the percentage difference in the sun's force of gravity from one side of the earth to the other is less than that of the moon. Such tidal forces are everywhere, but for the most part they are too tiny to be noticed.

Einstein's ***theory of general relativity*** revolutionized our understanding of gravity. This theory, in which both space and time are warped in the vicinity of any mass, does away with the concepts of the force of gravity and action at a distance.

EXERCISES

Concept Checks: Answer true or false

1. Gravity is the attraction of mass for all other mass. It acts over long distances, through empty space, and through matter. The force of gravity pulls any two masses directly toward each other.
2. Anything on earth's surface is attracted toward its center just as if all earth's mass were concentrated there. For any object, such a point is called the *center of gravity*.
3. Because you have mass and this book has mass, you are attracted to it because of gravity and it is attracted to you *with exactly the same amount of force.*
4. The gravitational orbit of a smaller body (with very little mass) around a much larger body (with a great deal more mass) has the shape of an ellipse with the larger body at the center of the ellipse.
5. If the moon's gravitational force did not change with distance, it would not create tides in earth's oceans.
6. In mid-ocean, over a 24-hour-and-50-minute period, there are usually two high tides about 12 hours and 25 minutes apart and 2 low tides about 12 hours and 25 minutes apart.
7. Spring tides occur only in the spring, and neap tides occur only when the moon is nearest the earth in its orbit (at perigee).
8. The strength of the tidal force depends on the distance and mass and diameter of the attracting body.
9. In Einstein's theory of general relativity, the force of gravity is replaced by a curvature of space and time near any object with mass.

Applying the Concepts

10. Does your weight depend upon earth's mass? On your distance from the center of the earth? On the universal constant G? On anything else?

11. Do you think you would weigh more or weigh less on the top of Mt. Everest (elevation 29,028 feet, about 8850 meters) than you do at home? Why?

12. Earth and the moon exert equal but opposite gravitational pulls on each other. True or false?

13. Discuss this verse in light of the law of gravity: ". . . thou canst not stir a flower without troubling of a star." (Francis Thompson)

14. Is there some point between the moon and the earth where the net gravitational force on a spacecraft drifting there would be zero? Discuss.

15. The acceleration due to gravity on the surface of Mars is about 1/3 of earth's. What would your weight be on Mars? What would your mass be?

16. Why are the planets, the sun, and the moon all more or less spherical in shape?

17. If you know the acceleration due to gravity at the surface of another planet, you'd only need to multiply that acceleration by your mass to find your weight on that planet. True or false?

18. How much gravitational force does your body exert on the earth?

19. A barrel of oil weighs less than half as much as a barrel of ordinary rocks. What effect might a huge underground oil reservoir have on the value of g at the surface above it? Explain why.

20. An object suspended by a cord doesn't usually point exactly toward the center of the earth. For example, near the base of the Himalayas, a suspended object is measurably deflected a small amount toward the mountains. Why?

21. A 747 airplane travels almost horizontally at 35,000 feet (8900 meters), while above it the space shuttle travels almost horizontally at an elevation of 125 miles. Why do the passengers on the airplane feel their normal weight, yet the shuttle's passengers are "weightless"?

22. In the summer of 1985 an astronaut on a shuttle repair mission "dropped" a power screwdriver. Discuss its probable fate.

23. Set $F = ma = GmM_{\text{earth}}/d^2$ for an object of mass m in free fall and show how this tells us that g is the acceleration for *all* bodies in free fall near earth's surface *no matter what their masses are!*

24. When a satellite is in a circular orbit about a larger mass, does gravity do any work on the satellite? Discuss.

25. Explain what would happen to the earth if something took away its orbital speed around the sun.

26. Calculate the gravitational force the sun exerts on your body and compare it to your weight. $M_{\text{sun}} = 2 \times 10^{30}$ kg; and the average distance between the sun and earth is about 1.5×10^{11} m.

27. With respect to work done by gravity, discuss the relative speeds an earth satellite should have at apogee and perigee.

28. A simple first aid procedure to use on someone who has fainted is to lower the persons' head to the level of their heart. Why should this help?

29. The weather pictures you can see daily on the TV news come from a GOES spacecraft (Geostationary Operational Environment Satellites) in a stationary position 22,300 miles above earth's equator. Exactly how can it stay in one position over the equator day after day? What would happen if it were closer in? Farther out? (It helps to draw a sketch of this.)

30. Calculate the force of gravity between two people who weigh 160 lb (mass = 72.7 kg) and 120 lb (mass = 54.5 kg) if they are standing 10 ft (3.05 m) apart.

31. If a large asteroid collides with the moon, its mass is added to the moon's. Does earth attract the moon more after that happens?

32. Why couldn't a satellite have an orbit only 6 miles above our planet? (It could over the moon.)

33. Figure 7-19 is a graph of altitude, for the first 14 hours, of the Soyuz and Apollo spacecrafts that docked together in orbit on July 17, 1975. (a) About how many revolutions of the earth did the Soyuz make before the Apollo launch? (b) Describe the orbits these crafts had during the times shown on the graph. (c) At the arrow a rocket changed the Soyuz speed by 3.1 m/s. Was the speed

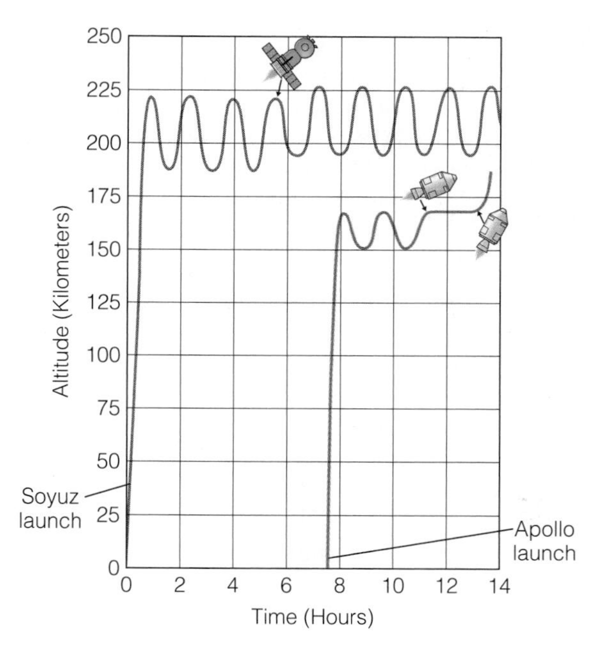

FIGURE 7-19

increased or decreased? Was its speed in its new orbit increased or decreased (compared to the old orbit)?

34. Earth is closest to the sun in January and farthest from the sun in July. In which month is the earth traveling fastest along its orbit?

35. Could an astronaut drop something to earth from an orbiting space shuttle? Discuss.

36. Set $m_{earth}v^2/r$ equal to the sun's gravitational force on earth and find how fast the earth speeds along as it travels around the sun. (The distance to the sun is 1.5×10^{11} m, and $M_{sun} = 2 \times 10^{30}$ kg.)

37. About how many neap tides are there each month? How many spring tides?

38. Find the average time between the consecutive high tides in Table 7.1. Then find the average time between consecutive low tides in that table. Do these correspond to an average time of 12 hours and 25 minutes?

39. Which positions of the moon relative to earth and sun bring the highest ocean tides? (Refer to Figure 7-16.)

40. If a lifeguard at the beach tells you high tide is 9 A.M., when should you expect the next high tide? The next low tide? When will high tide be the next morning?

41. Does the strongest tidal force on our bodies come from (a) the moon, (b) the sun, or (c) the earth?

42. If the solid earth had its present mass but was larger in diameter, would the ocean tides due to the moon be larger or smaller? Discuss.

43. If the moon had its present mass but was somewhat larger or smaller in diameter, would the moon's tidal force on earth change? Discuss.

44. Earth's atmosphere, like earth's oceans and the solid earth itself, responds to the lunar tidal force. Would you expect these tides to be detectable?

45. Describe what the ocean tides would be like if the moon did not exist.

46. When scientists discovered that moonquakes occurred most frequently at the moon's perigee, they decided to check the dates of recorded earthquakes to see if the moon's tidal force on earth could be involved in setting off quakes here. They found no correlation between the times of earthquakes and times when the moon was at perigee. Discuss (with respect to earthquake prediction).

47. The sun is closest to earth in January, farthest in July. When are the sun's tidal forces on earth's oceans largest?

48. Would you expect the sun to be at rest in the center of our solar system? Why or why not?

49. Do the speeds of the planets around the sun depend on their masses? Their distances from the sun? The sun's mass? To see the answer, set the centripetal force equal to the gravitational force.

50. If you were in orbit one earth radius above earth's surface, what would be the force of earth's gravity on you, expressed as a percentage of mg? If you were 2 earth radii up? If you were in orbit in the space shuttle some 200 miles up? $M_{earth} = 5.98 \times 10^{24}$ kg, and $r_{earth} = 6.38 \times 10^6$ m.

Demonstrations

1. An ellipse is a curve such that every point on the curve is at the same total distance from two points. Each of these points is called a focus of the ellipse. An ellipse is easy to draw with a pencil, a piece of string, and two thumbtacks (see Figure 7-20). Try it! When a satellite follows an elliptical orbit caused by earth's gravity, the center of the earth is at one focus of the ellipse.

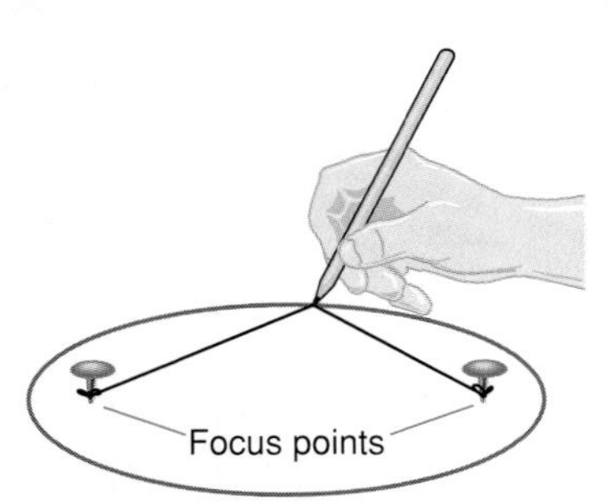

FIGURE 7-20

2. Turn on a single bright light (preferably without a shade) in an otherwise dark room, step to the far wall, and hold a phonograph record or a compact disc so that its shadow is projected there. The outline of the shadow of the disc is an ellipse. If the light comes in perpendicular to the wall and the disc is held vertically, the shadow's edge is circular. Turn the disc at an angle, and the circle becomes flattened in an ellipse. If viewed from an angle, any two-dimensional circular shape, even a basketball hoop, has the shape of an ellipse.

3. For a visualization of a $1/d^2$ field, all you need is a vacuum cleaner and some talcum powder. If sunlight comes through a window into an otherwise dark room, you can sometimes see small bits of dust floating in the air. A puff of talcum powder puts many such floating particles there. Hold the end of a vacuum cleaner hose into the cloud, without an attachment, and turn it on. The nearest particles zip in, and particles twice as far away move toward the hose at $\frac{1}{4}$ that speed.

EIGHT

Atoms and Molecules

Solids, liquids, and gases, the most common states of matter on earth, are all seen here. All of matter, including life, is made of atoms and molecules, the subject of this chapter.

Over 2500 years ago, the philosophers of Greece made guesses about the structure of matter. Aristotle thought that nature was built from four basic entities that could be divided without limit: fire, earth, air, and water. But another philosopher, Democritus (460–370 B.C.), speculated that every material substance was made of tiny invisible units that could not be broken down farther. He called these tiny units **atoms,** from the Greek *atomos* ("indivisible" or "uncuttable"). Today we know that that idea of Democritus was remarkably on target. About 100 years ago, investigators developed ways to probe matter at the submicroscopic level. Years of experiments and theories followed, and today physics has an accurate description and understanding of the nature of atoms and how they combine.

THE SIZE OF THE SMALLEST STRUCTURES

When astronauts in the space shuttle look at earth from 180 miles up, they see few signs of life (Figure 8-1). From that distance, the naked eye cannot see the teeming cities and crowded highways. We stand over the world of the atom in much the same way, but from an even more distant overlook. What, for instance, can we tell about the structure of water from looking at a single drop? Water flows, which means its smallest units can slip past each other, and left alone, it evaporates in quantities too small for you to see (Figure 8-2). However, water does have a striking property that gives us a clue about its smallest units. If you fill a glass bottle to the top with water, insert a snug-fitting cork and push really hard, the volume of water doesn't get noticeably smaller. Water's invisible structure is tough, as shown by this near incompressibility. Other liquids and solids share this incompressibility because it is a property of every kind of atom.

FIGURE 8-1 FROM 180 MILES OVER FLORIDA, ASTRONAUTS IN THE SPACE SHUTTLE HAVE VIEWS LIKE THIS. FEW HUMAN ACTIVITIES ARE RECOGNIZABLE FROM THAT DISTANCE.

Today, sophisticated experiments measure the sizes of atoms and give us pictures of them. The smallest atoms are a little more than 1×10^{-8} cm across, and even the largest have diameters only three times that size. Let's use the atoms on the surface of your fingernail to see what this small diameter means. The number of atoms lying side by side on 1 cm of your fingernail is about $1\ \text{cm}/(1 \times 10^{-8}\ \text{cm/atom}) = 10^8$ atoms. That's one hundred million, and it's not easy to relate to a number that large. So imagine that you are the right size to stroll from one atom to the next on your fingernail, using them as stepping stones. If each stepping stone were 1 meter across, you'd find them continuing for about 100,000 km (about 60,000 mi)—over two times around the world! All that just to get across 1 cm of fingernail, or across 1 cm of this page, since all atoms are roughly the same size.

ELEMENTS, COMPOUNDS, AND MOLECULES

Early chemists (Figure 8-3) mixed precise quantities of materials that would react with each other, sometimes heating them to bring about changes, and then they carefully weighed the products that appeared. With this procedure, they gained insight into the unseen world of the atom. Underlying all material

(a)

(b)

FIGURE 8-2 (A) WATER FLOWS, WHICH MEANS ITS SMALLEST UNITS MOVE PAST EACH OTHER EASILY, AND IT EVAPORATES WITH UNITS SO SMALL THEY ARE INVISIBLE. (B) YET WHEN THE SMALLEST UNITS OF WATER ARE PRESSED AGAINST EACH OTHER, THEY PROVE TO BE ALMOST INCOMPRESSIBLE.

FIGURE 8-3 A RENAISSANCE PAINTING, "L'ALCHEMISTA," OF EARLY CHEMISTS EXPERIMENTING. FROM BEGINNINGS LIKE THESE, CHEMISTS SLOWLY LEARNED TO IMPROVE MEASURING TECHNIQUES, LEADING TO THE PRINCIPLES OF MODERN CHEMISTRY.

things—the air you breathe, the foods you eat, the fragrances you smell—is a striking simplicity. As of 1994 there were only 109 different types of atoms known, and these are called the **chemical elements.** Ninety of these are found naturally occurring on earth, and the others have been made in laboratories.

Fewer than a dozen of the elements were known in antiquity because most are normally found stuck together in combinations with other elements, called compounds, making the different types of atoms hard to separate and

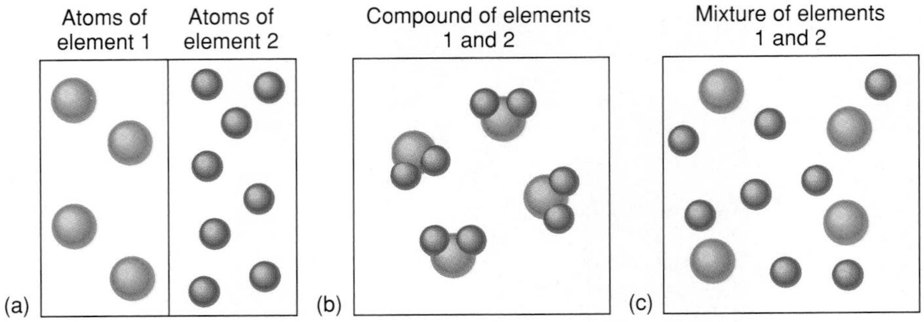

FIGURE 8-4 (A) ATOMS OF TWO TYPES, WHICH CAN BE MIXED. (B) IF THEY REACT CHEMICALLY, THEY FORM A COMPOUND. (C) IF THEY DO NOT REACT, THEY FORM A MIXTURE.

identify. **Compounds** are *combinations of atoms of two or more elements, tightly bound by chemical bonds*. Chemical reactions can break down compounds into their individual elements, whereas elements are substances that will not break down into simpler substances by chemical means. Compounds, then, are completely different from *mixtures,* where the ingredients do not react chemically and can be added in any proportions (Figure 8-4). Sometimes compounds and elements exist in molecular forms. A **molecule** is a "package" of two or more atoms bound together that reacts and behaves as a single, distinct object—much like the atoms of the elements do. Forming millions of compounds, the atoms of the 90 natural chemical elements on earth make up everything around you, and your body as well.

In the late 1700s Antoine Lavoisier made the most accurate measurements of chemical reactions to that point in time. He discovered that the mass of the reacting substances was always equal to the mass of the end products, and he is often called the father of modern chemistry for his work. Today we know that his discovery means "*during a chemical reaction, atoms are neither created nor destroyed.*" Compounds can be rearranged or added to or broken apart, but the number of each type of atom present after any chemical reaction is the same as the number present before the reaction.

SUBATOMIC PARTICLES

Today scientists investigate atoms and their combinations in a variety of ways. They break atoms up (by means we'll see in later chapters) and use the pieces as projectiles to pierce other atoms, watch what comes out, and work backward to find out what must have happened. They illuminate matter with X-rays or other forms of radiation and investigate how the matter and radiation interact. Such experiments have given us an amazing and detailed picture of what atoms are like.

An atom is made up of three kinds of particles: **electrons, protons,** and **neutrons.** All atoms occupy a roughly spherical volume, but they do not have a sharp outer boundary the way a billiard ball does. That's because an atom is not a solid ball of matter but instead is almost entirely empty space! The sphere is just a volume in which the electrons move. Early in this century, electrons were thought to be planetlike particles whirling at superhigh speeds about a massive speck at the atom's center called the **nucleus,** which was compared to the sun (Figure 8-5). (In Chapter 25, we will see that the behavior of atoms is quite different from that of ordinary objects.) The nucleus, which is about 10^{-13} cm across, is the home of the atom's tightly packed protons and neutrons. A proton has about 1836 times as much mass as an electron, and a

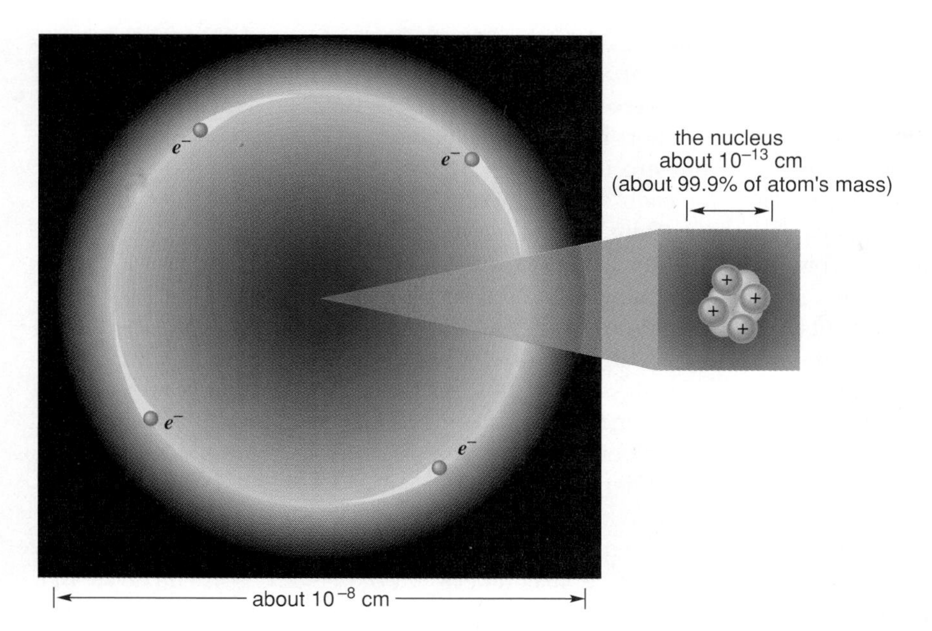

FIGURE 8-5 A VISUALIZATION OF AN ATOM AND ITS NUCLEUS. REMIND YOURSELF THAT THIS MODEL IS ONLY A WAY OF THINKING ABOUT ATOMS AND NOT THE WAY THEY ARE. WHEN A FOOTBALL COACH DRAWS OOO AND XXX TO ILLUSTRATE A PLAY, IT'S A SIMPLE REPRESENTATION OF THE PLAY. JUST AS A FOOTBALL PLAYER ISN'T REALLY A LITTLE X WITH AN ARROW AND LINE FOR MOTION, NEITHER ARE ELECTRONS LITTLE PLANETS ORBITING A MINIATURE SUN.

neutron is about 0.1 percent more massive than a proton (Figure 8-6). However, calculations showed an electron would circle the nucleus about 10^{15} times every second. A visualization of such motion cannot be realistic, but such whirlwind orbiting gives an impression that one might see a blurry cloud.

FIGURE 8-6 THE MASS OF A QUART OF MILK COMPARES WITH THIS CAR'S MASS AS AN ELECTRON'S MASS COMPARES WITH A PROTON'S MASS. THE ELECTRON IS THE TRUE "FEATHERWEIGHT" OF THE ATOM.

HOW ATOMS STAY TOGETHER

Electrons and protons stay together in atoms because of a powerful force of attraction between them. This **electric force** acts between any proton and electron, and it is ~10^{39} times as strong as the force of gravity between them. Like gravity, however, the electric force loses strength rapidly with increased distance between the particles. The electric force works in two ways: Protons and electrons attract each other, but a proton repels any other proton, and any two electrons repel each other as well. To describe these interactions, we say the proton has a positive charge and the electron has a negative charge of exactly the same size, and that like charges repel each other, while unlike charges attract each other (Figure 8-7). The third type of subatomic particle, neutrons, have no electric charge and don't respond to an electric force; hence they are called *neutral*. (We will discuss electric charge and the electric force in more detail in Chapter 17.)

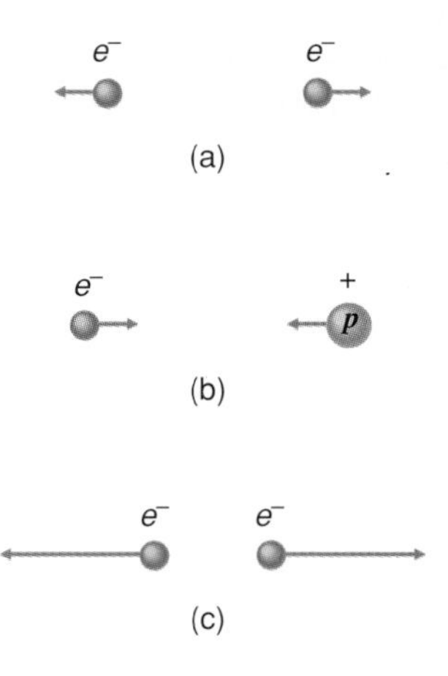

FIGURE 8-7 (A) ANY TWO ELECTRONS REPEL EACH OTHER. (B) IF A PROTON AND AN ELECTRON ARE THE SAME DISTANCE APART AS THE PAIR OF ELECTRONS WERE, THE PROTON AND ELECTRON ATTRACT EACH OTHER WITH THE SAME AMOUNT OF FORCE AS THE REPULSIVE FORCE BETWEEN THE TWO ELECTRONS. (C) IF THE DISTANCE BETWEEN TWO ELECTRONS (OR BETWEEN TWO PROTONS) IS HALVED, THE REPULSION INCREASES BY A FACTOR OF 4. (THE ELECTRIC FORCE IS DISCUSSED IN DETAIL IN CHAPTER 17.)

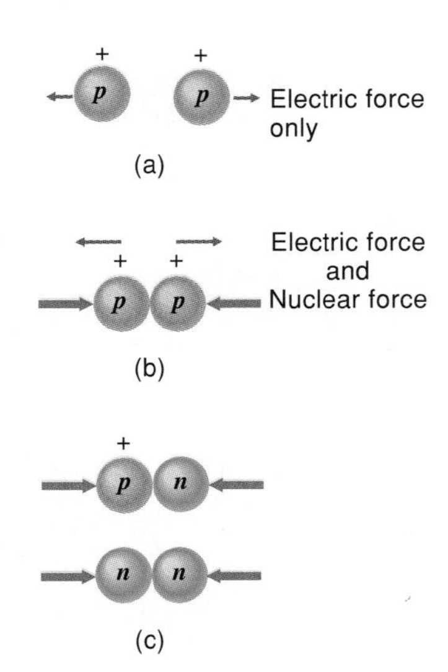

FIGURE 8-8 THE NUCLEAR FORCE BETWEEN TWO NUCLEONS ACTS ONLY AT CLOSE RANGE. (A) WHEN TWO PROTONS ARE A FEW NUCLEON DIAMETERS AWAY FROM EACH OTHER, ONLY THE ELECTRIC FORCE ACTS BETWEEN THEM. (B) WHEN THE PROTONS "TOUCH" WITHIN A NUCLEUS, THE NUCLEAR ATTRACTION IS OVER 100 TIMES AS STRONG AS THE ELECTRIC REPULSION BETWEEN THEM. (C) WITHIN A NUCLEUS, PROTONS AND NEUTRONS ATTRACT EACH OTHER, AS DO NEUTRONS AND NEUTRONS.

Another force, the strongest yet detected in nature, acts only between the nuclear particles, or *nucleons,* another name for protons and neutrons. This **nuclear force,** or **strong force,** is an attraction between nucleons (Figure 8-8). It has a very short range, acting only when the protons and neutrons "touch," and electrons are immune to it. This force overcomes the protons' electric repulsion for each other, since at the tiny distances found between the particles in the nucleus it is about 100 times stronger than the electric force. It binds the protons and neutrons together in that small, dense ball at the center of the atom called the nucleus.

Because protons and neutrons contain most of the mass of the atom and because the nucleus is so tiny, the matter in the nucleus weighs about a billion tons per cubic inch. Needless to say, nothing in our everyday world approaches this density. The surrounding electrons keep the nuclei in the atoms of ordinary matter far apart from each other. How? The electrons of the atoms make them nearly incompressible. When you clap your hands or chomp down on a piece of celery, the electric force keeps the colliding atoms from going right through each other. When two atoms "touch," the edges of their electron-regions come close together. The electric repulsion between the electrons in one atom and those in the other atom becomes enormous as the distance between the atoms decreases (Figure 8-9).

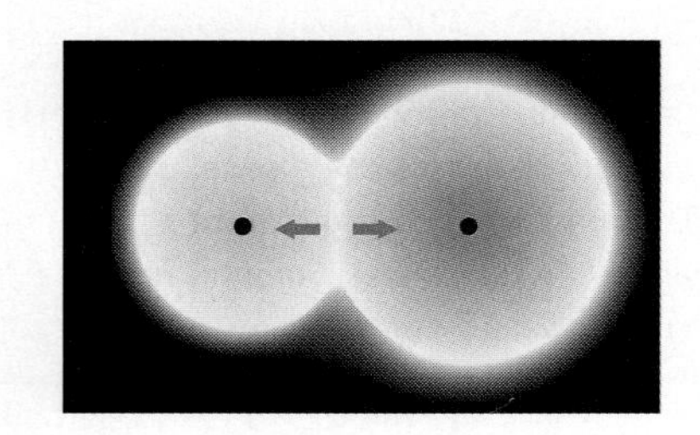

FIGURE 8-9 WHEN ATOMS COLLIDE, THEIR ELECTRON REGIONS REPEL ONE ANOTHER AS SOON AS THEY "TOUCH," AND THIS REPULSION KEEPS THE NUCLEI OF THE ATOMS VERY FAR APART. MATTER AS WE KNOW IT IS ALMOST ALL EMPTY SPACE.

MOST MATTER IS NEUTRAL

Since the electric force is so much more powerful than the gravitational force, why don't we see more evidence of it around us? It's because most atoms and most combinations of atoms are electrically neutral. They don't exert a net electric force on anything else because they have equal amounts of positive and negative charge. Suppose another charged particle, say an electron, happens to pass by a neutral atom. The electrons of that atom repel the passing electron with just about as much force as the protons in that atom's nucleus attract it (Figure 8-10). Those electric forces essentially cancel; it's as if there were no charge there at all, and that's what the word neutral means.

Apparently there are exactly as many electrons as protons in the universe. Because their electrical attraction is so strong, these charged particles tend

FIGURE 8-10 AN ATOM OR MOLECULE THAT HAS EQUAL NUMBERS OF ELECTRONS AND PROTONS APPEARS NEUTRAL TO A PASSING CHARGE BECAUSE THE NET ELECTRIC FORCE ON THAT PASSING CHARGED PARTICLE IS APPROXIMATELY ZERO. THAT IS, NO MATTER WHETHER THE PASSING PARTICLE IS POSITIVE OR NEGATIVE, IT RECEIVES (ABOUT) EQUAL AND OPPOSITE ELECTRICAL FORCES FROM THE PROTONS AND ELECTRONS, WHICH CANCEL. (SEE THE FORCE VECTORS ON THE CHARGE AT THE RIGHT.)

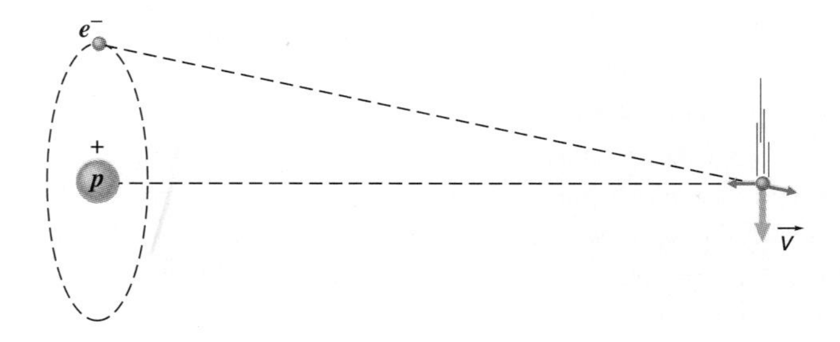

to pull themselves together and form the neutral units we call atoms. Even in stars, where the temperatures are so high that the collisions between atoms knock off some or all of their electrons, there cannot be a large-scale imbalance of electric charge. In any fistful of star-stuff, there is always the same amount of positive and negative charge. Were it otherwise, the powerful electric force would quickly pull the separated positive and negative charges to each other, making the matter neutral again though it would still be far too hot for atoms to form.

THE MASSES OF ATOMS

Hydrogen has the smallest mass of any element. An ordinary hydrogen atom has only one electron circling a solitary proton, and it has a total mass of 1.67×10^{-24} grams. Atoms of any other element have masses very nearly equal to that of some whole number of hydrogen atoms. The mass of a typical carbon atom, for example, is about 12 times the mass of a hydrogen atom, and an oxygen atom is about 16 times more massive than a hydrogen atom. Thus the hydrogen mass of 1.67×10^{-24} g, which is 99.9 percent due to its proton, is a convenient unit to use for comparing the masses of atoms. It is called the **atomic mass unit,** and the abbreviation for it is the letter "u." (The modern definition of u is 1/12 of the mass of a carbon atom, since carbon is easier to work with than hydrogen in laboratories.)

As we've mentioned, the mass of the electrons of any atom is almost nothing compared with the mass of its nucleus. To find the mass of an atom, first count the protons and neutrons in its nucleus. That number is the **mass number** of the atom. Multiplying that number by 1 u gives a good estimate of the mass of the atom, since the mass of an electron is only about (1/1836) u = 0.0005 u. The hydrogen atom, with a lone proton in its nucleus, has a mass number of 1 (Figure 8-11). The next element in the periodic table is helium, with two protons and two neutrons in its nucleus. Helium's mass number is thus 4, and the mass of a helium atom is about 4 u (or $4 \times 1.7 \times 10^{-24}$ g = 6.8×10^{-24} g). The helium atom has two electrons, whose negative charges balance the positive charges of the two protons in the nucleus.

The number of protons in an atom's nucleus is called its **atomic number.** Helium's atomic number is thus 2 while hydrogen's is 1. The atomic number of a nucleus determines which element it is because the number of protons determines how many electrons are in the neutral atom, and those electrons

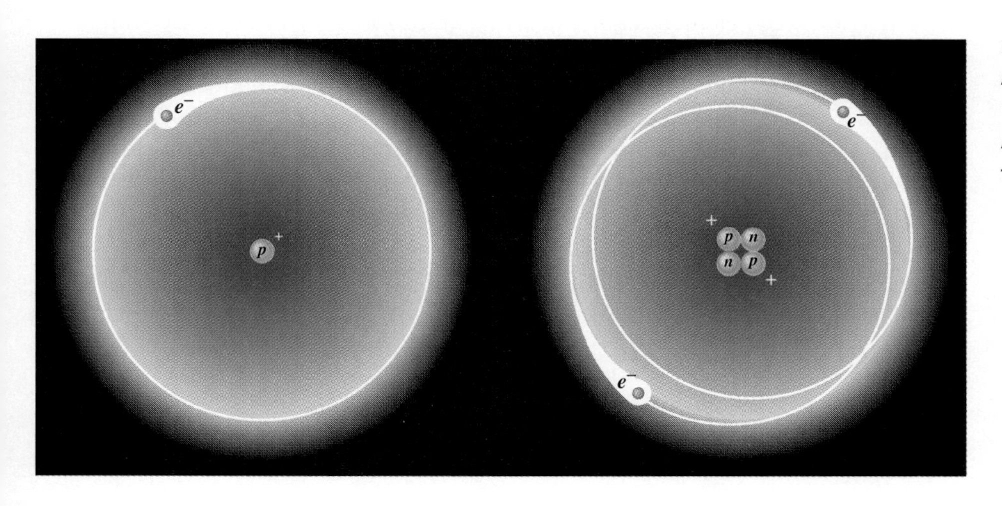

FIGURE 8-11 VISUALIZATIONS OF HYDROGEN AND HELIUM, WITH THE NUCLEI DRAWN ABOUT 10,000 TIMES TOO LARGE RELATIVE TO THE ATOMS. HELIUM'S NUCLEUS HAS ABOUT 4 TIMES THE MASS OF HYDROGEN'S NUCLEUS.

FIGURE 8-12 THE NUCLEI OF DEUTERIUM (MASS NUMBER = 2) AND TRITIUM (MASS NUMBER = 3). BOTH HAVE ATOMIC NUMBERS OF 1, BEING ISOTOPES OF HYDROGEN, SO BOTH HAVE ONLY 1 ELECTRON IN THEIR NEUTRAL ATOMS.

determine how that atom reacts chemically. So any nucleus with one proton in its nucleus is hydrogen, and any nucleus with two protons is helium, *regardless* of the mass of the nuclei, which may vary. For example, there is a form of hydrogen atom that acts normally, entering into chemical reactions just as every other hydrogen atom does, except that its nucleus is twice as massive; its mass number is 2 (Figure 8-12). About 15 of these turn up in every 10,000 hydrogen atoms in nature. The reason for the extra mass is a neutron that is bound to the hydrogen's proton by the nuclear force. Sometimes a hydrogen atom has two neutrons attached to the proton, making it three times as massive as ordinary hydrogen, but only a tiny trace of these nuclei exists. So there are three varieties, or **isotopes,** of the chemical element hydrogen. The double-mass atom is called *deuterium,* and the triple-mass hydrogen is called *tritium.*

All the other elements have isotopes, too. For example, the most common carbon atom has a mass number of 12, with six protons and six neutrons in its nucleus, but about one percent of all carbon nuclei have one more neutron, giving them a mass number of 13. Both forms of carbon have six electrons that balance the charge of the six protons, and both sets of electrons act the same in chemical reactions.

PATTERNS OF CHEMICAL REACTIONS

In the 1800s, chemists discovered remarkable similarities in how some of the chemical elements combine with others. For example, the element sodium (chemical symbol Na) combines with nitrogen (chemical symbol N) to make a compound called sodium nitride. In this compound three sodium atoms unite with one nitrogen atom, so its *chemical formula*—which shows its exact composition in symbols—is Na_3N. Potassium (K), another soft, silvery-white metal like sodium, combines with nitrogen in the same ratio to make potassium nitride, K_3N. Whenever sodium or potassium combines with other atoms or groups of atoms, they combine with them in the same ratios. Though sodium's atomic number is 11 and potassium's is 19, they have similar chemical behavior.

In 1869 a Russian chemist, Dmitri Mendeleev (1834–1907), arranged the elements known at the time in a table that's now called the **periodic table.** (See the inside back cover of this text.) The columns in the table contain groups of elements with similar chemical behaviors. The elements are arranged by their atomic numbers, which increase from left to right and from top to bottom.

The reason the elements have these patterns of behavior was discovered in the early 1900s. The electrons in atoms can be thought of as occupying layers, or shells, around the nucleus, as shown in Figure 8-13 and that picture helped to explain a great deal. Physicists found that each shell-like region could have only a specific number of electrons (we'll read more about this in Chapter 25). Most atoms have their innermost shells filled to capacity. If the outermost shell of an atom is filled as well, the atom tends to be inert, that is, normally it doesn't react at all with other atoms to make molecules or compounds. Such outer shells are called *closed* shells. If the element has a *partially* filled outer shell, however, it will be chemically active, meaning it can form **chemical bonds** when brought into contact with certain other elements. To form chemical bonds, the atom must either *give up* one or more electrons or *accept* one or more electrons from the outer shell of another atom (Figure 8-14), or else it must *share* electrons with one or more atoms.

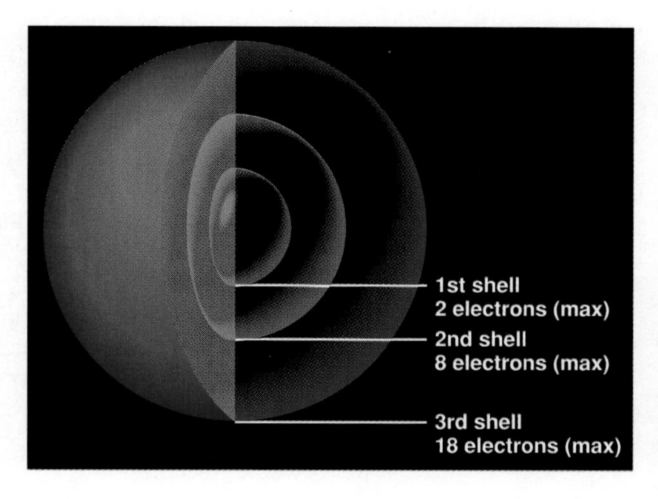

FIGURE 8-13 VISUALIZATION OF ELECTRON SHELLS IN AN ATOM, NOT TO SCALE. THE FIRST SHELL (FROM THE NUCLEUS) CAN ACCOMMODATE ONLY 2 ELECTRONS, THE SECOND, 8, AND THE THIRD, 18. ALL SHELLS BUT THE FIRST HAVE SUBSHELLS.

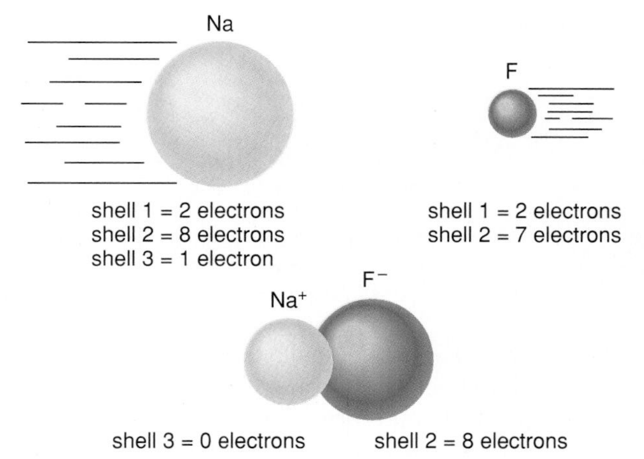

FIGURE 8-14 A SODIUM ATOM AND A FLUORINE ATOM UNITE WHEN THE SODIUM ATOM, HAVING ONLY ONE ELECTRON IN ITS OUTER SHELL (THAT COULD HOLD 18), LOSES THAT OUTER ELECTRON TO FLUORINE, WHICH HAS ONLY ONE VACANCY IN ITS OUTER SHELL. THE SODIUM IS LEFT WITH A NET POSITIVE CHARGE OF ONE PROTON'S CHARGE, WHILE THE FLUORINE HAS A NET NEGATIVE CHARGE OF ONE ELECTRON'S CHARGE. THE POSITIVE SODIUM ION AND THE NEGATIVE FLUORINE ION THUS ATTRACT EACH OTHER; THIS IS AN EXAMPLE OF AN IONIC BOND.

The rearrangement of electrons leaves the reacting atoms with filled outer shells. (Each electron shell except the one closest to the nucleus is further divided into subshells, and a chemical reaction may leave a subshell filled rather than an entire shell. A closed outer subshell is also a relatively stable arrangement.)

When chemical bonding takes place, the atoms generally lose energy as the electrons rearrange themselves. The combination of atoms then has a lower total energy and is more stable. That is, energy would have to be added to break the combination apart. Most of the atoms on earth are found in compounds for this reason, explaining why only a handful of chemical elements were known in antiquity.

As you look along the rows of the periodic table from left to right, each element has one more proton in the nucleus of its atom than the preceding one and is usually more massive as well. Since any neutral atom has one electron for every proton, heavier elements also have more electrons. It follows that the progression of atoms fills up one electron shell after another across the periodic table. The reason sodium (Na) and potassium (K) form many compounds in the same way is that they both have similar outermost shells. Each has only one electron in its outermost shell. It doesn't matter much in chemical reactions that potassium has an extra shell full of electrons beneath

1 H																	2 He
3 Li	4 Be											5 B	6 C	7 N	8 O	9 F	10 Ne
11 Na	12 Mg											13 Al	14 Si	15 P	16 S	17 Cl	18 Ar
19 K	20 Ca	21 Sc	22 Ti	23 V	24 Cr	25 Mn	26 Fe	27 Co	28 Ni	29 Cu	30 Zn	31 Ga	32 Ge	33 As	34 Se	35 Br	36 Kr
37 Rb	38 Sr	39 Y	40 Zr	41 Nb	42 Mo	43 Tc	44 Ru	45 Rh	46 Pd	47 Ag	48 Cd	49 In	50 Sn	51 Sb	52 Te	53 I	54 Xe
55 Cs	56 Ba	57 La	72 Hf	73 Ta	74 W	75 Re	76 Os	77 Ir	78 Pt	79 Au	80 Hg	81 Tl	82 Pb	83 Bi	84 Po	85 At	86 Rn
87 Fr	88 Ra	89 Ac	104 Rf	105 Ha	[106] Sg	[107]	[108]	[109]									
				58 Ce	59 Pr	60 Nd	61 Pm	62 Sm	63 Eu	64 Gd	65 Tb	66 Dy	67 Ho	68 Er	69 Tm	70 Yb	71 Lu
				90 Th	91 Pa	92 U	93 Np	94 Pu	95 Am	96 Cm	97 Bk	98 Cf	99 Es	100 Fm	101 Md	102 No	103 Lr

Metals
Semimetals
Nonmetals

FIGURE 8-15 THE MODERN PERIODIC TABLE SHOWING THE ELEMENTS' CHEMICAL SYMBOLS AND THEIR ATOMIC NUMBERS. (THE FULL PERIODIC TABLE WITH THE AVERAGE ATOMIC MASSES IN U IS INSIDE THE BACK COVER.)

its chemically reactive shell. Chlorine (Cl), fluorine (F), and bromine (Br) each lack only one electron to fill their outermost shells, so they too behave alike chemically and form similar compounds.

Elements that have outer shells that are less than half full typically lose those electrons in chemical reactions, leaving them with a closed outer shell. These are called **metals,** and they occupy the far left (except for hydrogen) and center of the periodic table. Elements that have near-full outer shells tend to gain electrons during chemical reactions, and they are called **nonmetals.** A few elements have properties that fall between those of metals and nonmetals, and they are called *semimetals*. Figure 8-15 shows these three groupings of the elements on the periodic table.

Uranium (U), with atomic number 92, is the heaviest element found naturally on earth, with a mass of about 238 hydrogen atoms. The elements listed on the table that have more than 92 protons have been formed in nuclear physics labs, and all are radioactive. That is, they are unstable and their nuclei change into other, lighter elements. Elements with atomic numbers 84–92 are also radioactive, but they still exist naturally on earth. They will be discussed in Chapter 27. Two lighter elements, technetium (atomic number 43) and promethium (atomic number 61) are also highly unstable and are not found on earth, and this leaves our planet with only 90 naturally occurring chemical elements.

TYPES OF CHEMICAL BONDS

The chemical bonds between atoms come in several varieties, with shades in between. Sometimes an electron from one atom transfers almost entirely to an atom of another type, as we saw in Figure 8-14. This is called an **ionic**

bond, from the word *ion,* which means charged particle. (Any atom with an imbalance of charge is an ion, for example.) Sodium and chlorine bind together this way in ordinary table salt, NaCl (Figure 8-16). The sodium atom gives up an electron and becomes a positive ion, while the chlorine atom takes that electron into its outer shell and becomes a negative ion. The ionic bond is very strong since the positive and negative ions attract each other directly with the electric force. Ionic bonding is the principal way metals from the left of the periodic table react with nonmetals from the far right (not including the inert gases in the last column). Such compounds are usually crystalline solids, where the ions are arranged in regular patterns (as in NaCl) due to the strong electric forces between them. Ionic compounds are usually brittle and have high melting points.

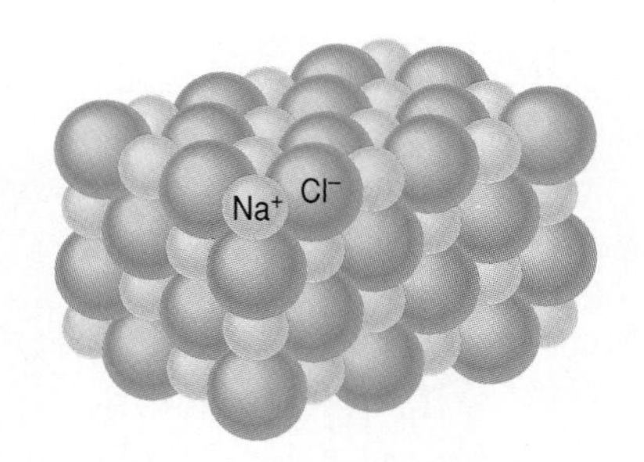

FIGURE 8-16 THE ARRANGEMENT OF CHARGED ATOMS (IONS) IN TABLE SALT, SODIUM CHLORIDE. A SALT CRYSTAL IS BRITTLE, A CHARACTERISTIC OF IONIC BONDING.

A **covalent bond** is an attraction between two atoms that *share* electrons in the regions between them to obtain closed outer shells (Figure 8-17). These bonds are the way nonmetals react with other nonmetals; covalent bonds are the bonds that keep molecules together. Molecular substances have a phenomenal range of properties. Water, alcohol, peaches (indeed, all foods), plastic bags, and gasoline are examples of molecular substances. The simplest molecules are diatomic molecules such as nitrogen (N_2) and oxygen (O_2), the major components of air, and hydrogen (H_2). A large molecule such as the DNA found in living cells may consist of over a million atoms bonded together (Figure 8-18).

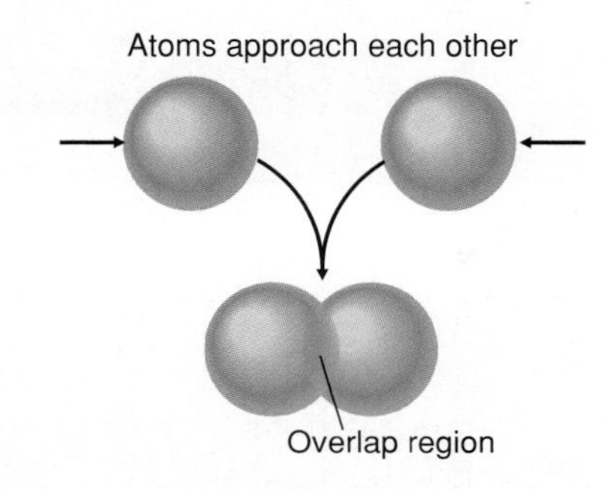

FIGURE 8-17 A HYDROGEN MOLECULE, H_2, WHICH CONSISTS OF 2 HYDROGEN ATOMS UNITED WITH A COVALENT BOND, IS THE SIMPLEST TYPE OF MOLECULE. NITROGEN AND OXYGEN, THE TWO MAJOR COMPONENTS OF AIR, ARE ALSO FOUND IN DIATOMIC MOLECULES, N_2 AND O_2.

In a single covalent bond, two electrons share the space between the two atoms. Double covalent bonds contain two pairs of electrons, and triple bonds have three pairs. When atoms form covalent bonds in a molecule, the bonds regulate the positions of the atoms according to the electron concentrations all around them. In molecules, then, every atom is attached at a specific angle; covalent bonds are called *directional* bonds.

Carbon reacts chemically by forming covalent bonds. A single atom of carbon can bond with up to four other atoms at once (Figure 8-19). More important, it bonds with other carbon atoms in chains and even in rings. That helps to account for the many complex *organic* molecules found on earth. (Organic refers to a chemical compound containing carbon.) The great majority of the approximately 10 million substances known today are compounds formed with covalent bonding. On the average, covalent bonds are almost as strong as ionic bonds. Much of modern molecular biology depends on 3D imaging of complex molecules to understand the way the molecules interact. The stereo figures on the next page show examples.

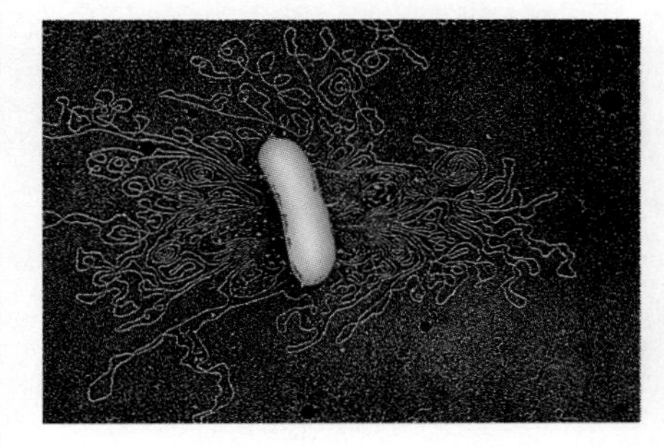

FIGURE 8-18 A MOLECULE OF DNA HAS SPILLED FROM THE BROKEN CELL WALL OF THIS BACTERIUM. DNA MOLECULES TYPICALLY HAVE MOLECULAR MASSES OF 6–16 MILLION U.

In metallic elements such as copper, silver, gold, and aluminum, the atoms bond by sharing outer electrons in the most general way possible. In **metallic bonding,** the outermost electrons of the atoms can wander from one atom to another, shared not just between two atoms but by all the atoms on the metal object. It's this mobility of the electrons that makes many metals good conductors of electric current and also of heat. Metallic bonds are generally

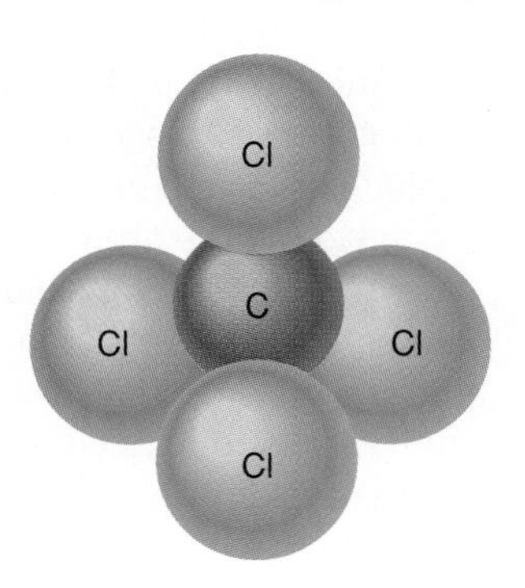

FIGURE 8-19 A MOLECULE OF THE CLEANING FLUID COMPOUND, CARBON TETRACHLORIDE. HERE CARBON HAS BONDED COVALENTLY WITH 4 CHLORINE ATOMS.

PHYSICS IN 3D / MOLECULES ENTERING A BACTERIUM

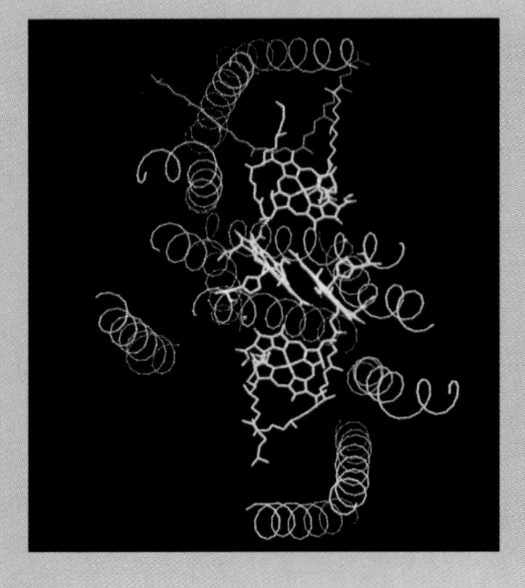
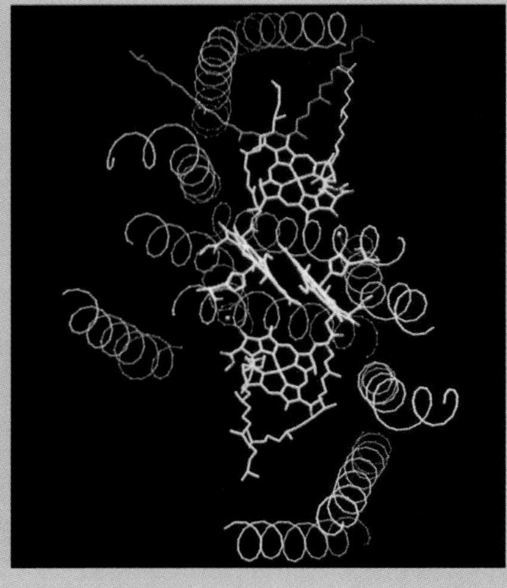

HELICES IN BLUE, PURPLE, AND BROWN GOING THROUGH THE SURFACE OF A LARGE PROTEIN MOLECULE THAT SPANS THE MEMBRANE OF A BACTERIUM. THE FIRST MEMBRANE-SPANNING PROTEIN MOLECULE WHOSE 3-D STRUCTURE WAS SOLVED, IT EARNED A NOBEL PRIZE FOR JOHANN DEISENHOFER, ROBERT HUBER, AND HATMUT MICHEL.

PHYSICS IN 3D / DRUG BINDING TO A COLD VIRUS

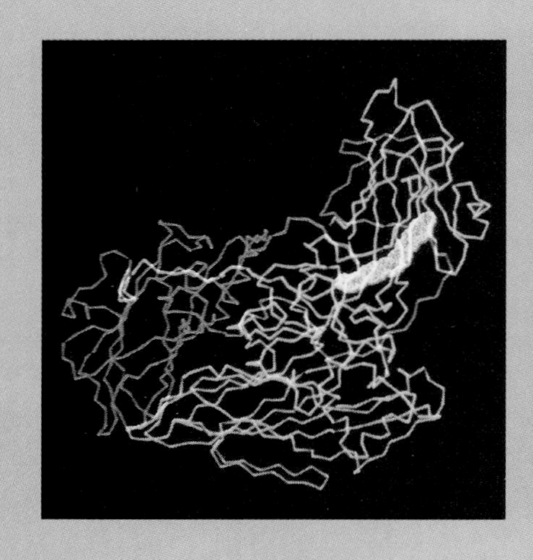
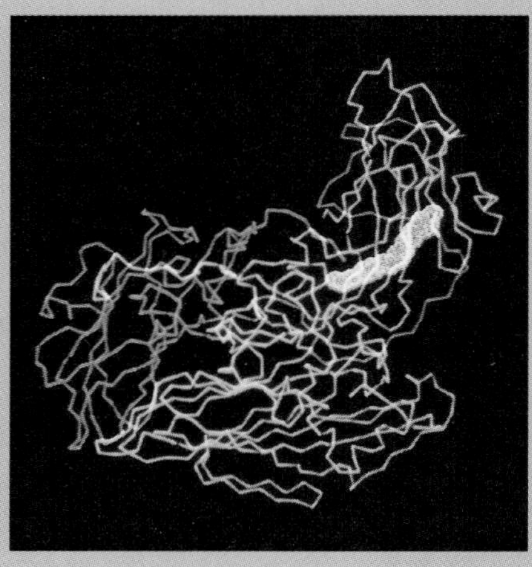

AN ANTIVIRAL DRUG (YELLOW) FITS INSIDE A POCKET OF A COMMON COLD VIRUS THAT NEVER CHANGES, THOUGH OTHER PARTS OF THE VIRUS FREQUENTLY MUTATE. THE DRUG STIFFENS THE VIRUS'S STRUCTURE SO THAT IT CAN'T RELEASE ITS GENETIC MATERIAL INSIDE AN INFECTED CELL.

much weaker than covalent or ionic bonds. One result of this is that metals are usually *malleable* (they can be pounded into thin sheets) and *ductile* (they can be drawn into wire).

Another type of bond occurs between any two neutral atoms or molecules that get close together. Because of the motions of electrons in atoms or molecules, at any instant there might be more on one side than on the other. Then we say the atom (or molecule) is *polarized,* having a negatively charged side and a positively charged side. Suppose a polarized atom has its negative side facing a neutral atom. The electrons on that neighboring atom are repelled and move somewhat to the far side of their nucleus, leaving the side toward the initially polarized atom positive. The two atoms then have oppositely charged regions facing each other, and they attract. That is the **van der Waals**

bond; it comes from the natural fluctuations of the charge throughout the atom due to the electrons' motions, and it is the weakest bond of all.

The kaleidoscopic forms and functions of the matter in our world come from the properties of compounds. Nearly all of the science of chemistry and most of the physical properties of matter around us have to do only with the electrons in the shell farthest from the nucleus. Those electrons make mercury flow, make lead soft, and make diamonds hard. The outermost electrons' behavior governs the chemical reactions of all substances.

GOLD (LOWER LEFT) CAN BE POUNDED INTO VERY THIN SHEETS, EXHIBITING THE PROPERTY OF MALLEABILITY DUE TO METALLIC BONDING. COPPER (UPPER RIGHT) MAY BE DRAWN INTO WIRES, DEMONSTRATING THE PROPERTY OF DUCTILITY DUE TO METALLIC BONDING.

THE ENERGY OF CHEMICAL REACTIONS

The strength of a chemical bond is measured in terms of the energy the bond represents. Every chemical reaction either releases or absorbs energy. If energy is released, the chemical reaction is **exothermic;** and if energy is absorbed, the reaction is **endothermic.** Because of the strong electric attraction of the nucleus, the outer electron shells of atoms have a great deal of potential energy, and any rearrangement results in a greater or smaller total potential energy for the atoms. That change in potential energy is where energy comes from (or goes to) during chemical reactions.

Many important chemical reactions do not occur spontaneously; they need an energy nudge before they proceed. Gasoline, vaporized and mingling with air, is a good example. At ordinary temperatures, the molecules of oxygen and gasoline collide and bounce off without reacting. A simple spark, however, provides enough extra energy, the *activation* energy, to some of these molecules to make them react, and the energy they release provides the activation energy to the surrounding molecules of gasoline and oxygen. An analogy to such a reaction is a car that is parked next to a curb at the top of a cliff. If someone pushes it up over the curb (adds the activation energy), gravity takes over, and the change of potential energy into kinetic energy (an exothermic reaction) is automatic.

Summary and Key Terms

Everything in our world is made of ***atoms***, which we can visualize as negatively charged ***electrons*** orbiting rapidly around a ***nucleus***, which is composed of positively charged ***protons*** and neutral ***neutrons***. The different types of atoms are called the ***chemical elements***, substances that do not break down into simpler substances by chemical means. There are 90 elements found naturally on earth, but there are 109 presently on the ***periodic table*** of the elements.

atoms–110
electrons–112
nucleus–112
protons–112
neutrons–112
chemical elements–111
periodic table–116

Protons and neutrons have far more mass than electrons, so we find an atom's total mass using only the masses of its nuclear particles. Hydrogen has a mass of 1.67×10^{-24} grams, about one ***atomic mass unit*** (u). To find any atom's mass, count the protons and neutrons in its nucleus to find its ***mass number***, then multiply that number by 1 u. All elements have ***isotopes***, forms with different numbers of neutrons in the nucleus. The number of protons in a nucleus, its ***atomic number***, determines the number of electrons held by the nucleus in a neutral atom and so determines its chemistry and thus the chemical element to which it belongs.

atomic mass unit–115
mass number–115
isotopes–116
atomic number–115

Two forces hold atoms together. The ***electric force***, a force vastly stronger

electric force–113

than gravity, holds an atom's electrons in the space around its nucleus. The ***nuclear force*** holds an atom's protons and neutrons together in its nucleus. It is the strongest known force in nature, but it is effective only over a short range.

nuclear force–114

The chemical properties of an atom depend mainly on the electrons in its outer *shell* or *subshell*. If this shell is *closed,* the atom tends to be *inert*. If it is *partially* filled, an atom can enter a chemical reaction and form ***chemical bonds***. Chemical bonds bind different elements tightly together in ***compounds***. Both compounds and elements can exist as ***molecules***, the smallest units of combined atoms with their own chemical identity. There are four principal types of bonds between atoms. ***Metals***, elements whose outer shells are less than half full, react with ***nonmetals***, elements whose outer shells are nearly full, through ***ionic bonds***. In ionic bonds, electrons from metal atoms transfer almost entirely to nonmetal atoms. Nonmetals react with nonmetals through ***covalent bonds***, in which two atoms *share* electrons in the region between them. Metal atoms stick together with ***metallic bonds***, in which the outermost electrons of the atoms wander freely, shared by all the connected metal atoms. The ***van der Waals bond*** occurs when the motions of an atom's electrons *polarize* it, creating positively and negatively charged regions that attract any other neutral atoms in close proximity. If energy is released when bonds are formed or broken, the process is ***exothermic***. Chemical reactions that absorb energy, on the other hand, are ***endothermic*** reactions.

chemical bonds–116
compounds–112
molecules–112
metals–118
nonmetals–118
ionic bonds–118
covalent bonds–119
metallic bonds–119
van der Waals bond–120
exothermic–121
endothermic–121

EXERCISES

Concept Checks: Answer true or false

1. The 90 elements found naturally on earth are different types of atoms that are distinguished by their chemical activity.
2. A compound is a substance that's formed when two or more species of atoms unite with chemical bonds.
3. In a molecule, two or more atoms are united. They have to be different types.
4. Electrons, protons, and neutrons are bound together by two forces in neutral atoms. One force acts only between charged particles while the other acts only between nucleons.
5. An electron repels any other electron, but an electron is attracted to all the particles in the nucleus.
6. The atomic mass unit (u) is a convenient unit for comparing masses of atoms. Its value is about equal to the mass of a hydrogen atom.
7. The atomic mass unit multiplied by the mass number gives a good estimate of the mass of an atom.
8. The mass number is the sum of the numbers of protons and electrons in the atom.
9. The different isotopes of an element contain different numbers of protons and neutrons in their nuclei.
10. The mass number distinguishes one chemical element from another.
11. The periodic table groups the elements in columns according to their chemical behavior.
12. An exothermic chemical reaction is one where energy is released in the form of heat.

Applying the Concepts

13. In Shakespeare's play *A Midsummer Night's Dream*, he spoke of a "small chariot" driven by a "team of atomies." This was before the days of modern physics or chemistry, so is that surprising? Comment.
14. At least ten of the chemical elements were known in antiquity. Do you think they would be the most reactive elements or the least reactive? Guess a few of these ten.
15. Defend this statement: The stability of chemical compounds is why your teeth don't stick to a carrot that you eat.
16. The volume of matter comes more from (a) the electrons (b) the protons (c) the nuclei (d) the neutrons.
17. If you know only the atomic number of a certain atom, could you find its mass from the periodic table? Why or why not?
18. What is the mass of 1 atom of gold (Au)?

19. Does a molecule have to be a compound? If not, give an example. Does a compound have to have a molecular form? If not, give an example.
20. Explain the difference between an atom's mass and the same atom's mass number.

21. The small compressibility of matter comes more from (a) the electrons (b) the protons (c) the nuclei (d) the neutrons.

22. The mass of a molecule is to a very high accuracy the sum of the mass of its individual atoms. Find the molecular mass in u of the following molecules: (a) CH_4 (methane), (b) H_2O (water), (c) CO_2 (carbon dioxide), (d) N_2 (nitrogen gas), (e) O_2 (oxygen gas).

23. Compute the molecular masses of the following molecules in u: (a) sugar, $C_{12}H_{22}O_{11}$, (b) carbon tetrachloride (cleaning fluid) CCl_4, (c) KCl (a salt substitute), (d) $CaCO_3$ (chalk, calcium carbonate).

24. Sodium chloride, NaCl, is common table salt. Look at the periodic table inside the back cover of this book and find at least three other elements that should combine with sodium in the same way as chlorine does.

25. In 1970 chemical publications listed about 3 million pure substances. Over 2.5 million of those were organic, containing carbon atoms. Comment on this fact. By 1992, approximately 10 million chemical compounds were known. Comment on this.

26. Classify each of the following as (a) mixture (b) compound (c) element: ink, milk, ice, gold, apple pie, copper, bread, air, silver, sugar, a soft drink.

27. The nucleus of nitrogen has seven protons and seven neutrons. What is its mass number? Its atomic number?

28. Researchers have discovered that if deuterium atoms are substituted for hydrogen atoms in oral polio vaccine, stronger chemical bonds result, and the vaccine is stable against heat for an extra 2 or 3 days. This is an enormous benefit to inoculation programs in countries without widespread refrigeration or air conditioning, and would not have been possible except for the field of nuclear physics. Do you think the general public is much aware of such positive connections with medicine and nuclear physics? Comment on whether your friends would be likely to contribute donations for research in nuclear physics through a foundation, as compared to medical research.

29. You can bite into an apple with ease, but you cannot bite into a rock. Chemically speaking, what are some of the differences between an apple and a rock?

30. The contact forces between objects come more from (a) the electrons (b) the protons (c) the nuclei (d) the neutrons.

31. Discuss the illustration in Figure 8-20. What property of atoms and molecules limits the number of people that can squeeze into a car or a telephone booth?

FIGURE 8-20

32. Consult the periodic table to see if calcium (Ca) should form compounds more as potassium (K) does or more as magnesium (Mg) does.

33. When you buy a pound of table salt (NaCl), what percentage of that pound is sodium?

34. Some 0.2 percent of all oxygen atoms have masses of 18 u rather than 16 u as most do. What is the difference in their structures?

35. Think of how flexible the insulated copper wires on the extension cords of electrical appliances are. Metals are often malleable (easy to shape) and ductile (easy to draw into wires). What does this say about the strength of the metallic bonds?

36. Copper (Cu) and silver (Ag) are both good electrical conductors. Would you expect gold (Au) to be a good electrical conductor? Why or why not?

37. The periodic table tells you that a gold (Au) nucleus has 1 more proton than a platinum (Pt) nucleus. Yet the gold nucleus on the average has 3 u more mass than a platinum nucleus. Explain why.

38. Since a hydrogen atom has a mass of about 1 u, show that there are about 6×10^{23} hydrogen atoms in one gram of atomic hydrogen. (That number, 6.02×10^{23}, is called *Avogadro's number.*)

39. Referring to Exercise 38, do you think you will ever see 10^{23} separate objects in your lifetime? Assume 80 years and show that in each and every second of your life you'd have to inspect between 10^{13} and 10^{14} objects.

40. A molecule of nitroglycerine bursts apart into smaller molecules—water (H_2O) and carbon dioxide (CO_2) among others—with the slightest of shocks, in a highly exothermic reaction. When nitroglycerine is made in a laboratory, would you expect the chemical reaction to be exothermic or endothermic? Why?

41. Which has the most protons in its nucleus? (a) gold (Au) (b) silver (Ag) (c) copper (Cu).

42. A red blood cell (erythrocyte) in your body is about 7×10^{-4} cm in diameter. If you could count the atoms in a line across that cell, about how many would you find?

43. The mass of matter comes more from (a) the electrons, (b) the protons, (c) the nuclei, (d) the neutrons. Explain.

44. The most accurate balances detect differences in mass of about one-millionth of a gram (10^{-6} g). What does this imply about finding an atom's mass with a balance?

45. Why should the distance from the nucleus affect how well electrons are held by an atom?

46. Show that if the atoms in 1 gram of atomic hydrogen were side-by-side in a straight line, the line would be about 200 times as long as the diameter of the earth's orbit around the sun. (The diameter of the earth's orbit is 3×10^8 km, and the length of that line would be about 6×10^{23} atoms $\times\ 1 \times 10^{-8}$ cm/atom.)

47. In the U.S. automobile tires are used up at the rate of 5 million per week! A Ford Escort goes about 2 meters down the road for each revolution of its tires, which, when new, have treads about 1 centimeter thick. If it goes 80,000 km (50,000 mi) before its tread is gone, find the average "thickness" of tire lost during each revolution. Compare this thickness to the diameter of the smallest atoms, and discuss.

48. The planets have potential energy because of the force of the sun's gravity, which can do work on them. What force gives electrons in atoms potential energy?

49. Methanol is an alcohol used as a solvent and cleaner, and it is poisonous to humans in even very small dosages. Since the early 1980s methanol has been added to some gasolines sold for cars. Death could result if you siphoned this gas from a car and ingested some accidentally. Starting September 1, 1993, all new cars had anti-siphoning necks to protect people from such accidents. Discuss the merits and disadvantages of that approach as compared with simply putting a warning label on the necks of all the gas tanks.

50. Figure 8-21 shows the chemical reaction when methane (natural gas) burns, or combines with oxygen. The coefficients in front of the chemical symbols show how many molecules of each type enter the reaction. There must be exactly the same number of each type of atom (not molecule!) on each side of the equation. Such an equation is said to be balanced. Are the following equations balanced?
(a) $2H_2 + O_2 \rightarrow 2H_2O$
(b) $Al + 6HCl \rightarrow 2AlCl_3 + 3H_2$
(c) $2Na + 2H_2O \rightarrow 2NaOH + H_2$

51. Gasoline is about 84.9 percent carbon, 14.76 percent hydrogen, and 0.08 percent sulfur. A gallon of gas weighs about 5.67 pounds. Estimate about how many pounds of sulfur, mostly as SO_2, sulfur dioxide, an average car emits into the atmosphere each year.

52. Suppose you were terribly unlucky and all the carbon atoms in your body, which make up about 20 percent of your mass, had only ^{13}C atoms rather than the usual ^{12}C atoms. How much more would you weigh as a result?

53. An average coal-burning power plant ejects about 1000 tons of sulfur dioxide into the air each day. That SO_2, then undergoes the following reactions.
$SO_2 + O_2 \rightarrow 2SO_3$
$SO_3 + H_2O \rightarrow H_2SO_4$ (sulfuric acid)
Are these two equations balanced? (This is the principal cause of acid rain.)

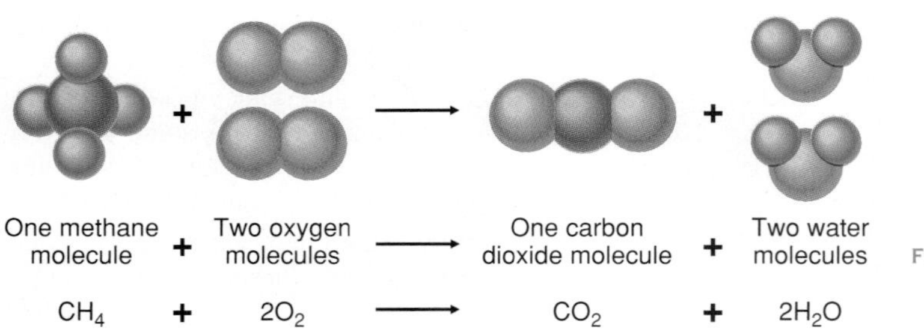

FIGURE 8-21

Demonstrations

1. Sprinkle some pepper over the surface of a bowl of water. Drop a small drop of cooking oil into the middle of the bowl. As the oil races outward over the surface of the water, the pepper particles are pushed to the side very rapidly by the spreading oil film. The oil film spreads to the edges of any container you use, but if you drop a very small quantity of oil on a very large surface of water, such as a small pond, the oil stops spreading when the layer of oil is extremely thin. That thin layer is one molecule thick. Benjamin Franklin and others realized that if the initial volume of the oil were known, it would equal the area of the slick times the diameter of the oil molecules, providing an early way to estimate the sizes of atoms.

2. One very familiar exothermic chemical reaction is rapid oxidation, or burning, which we associate with flames. Most fuels, such as oil, gasoline, alcohol, natural gas (see Exercise 50), coal, kerosene, and candle wax, are organic hydrocarbons, containing carbon and hydrogen. When they burn completely, the carbon combines with oxygen from the air to form carbon dioxide, CO_2, and the hydrogen combines with oxygen to form water, H_2O. These two compounds are always among the end products of burning hydrocarbons.

Let a candle burn for a moment or two and blow it out. A stream of smoke will rise from the snuffed candlewick. Immediately hold a lighted match an inch or more above the candle in that stream of smoke. The flame from the match will race downward along the unburned fuel in the smoke to relight the candle.

NINE

Solids

Technical rockclimbers depend on the strength of the solid rock as they ascend. If the climber's weight is placed on small footholds or handholds, the pressure on the rock can be tremendous.

We live on *terra firma,* the 29 percent of the earth's solid crust that lies above sea level. We live in solid houses, drive cars made of solids, and eat solid foods. If we are asked to define a solid, however, we may run into difficulty because a solid is one of those familiar things that are hard to define. One description of a solid is that it tends to keep its shape when left alone, but that doesn't mean a solid is rigid. Shoes, rubber bands, books, clothes—these flexible materials maintain their shapes only to some degree. They aren't rigid, but they are solid. The strength and rigidity of a solid depend only partially on the strength of the bonds between its atoms. Diamond, the hardest natural substance, and graphite, so soft and slippery it is used to lubricate door locks, are both made up of carbon atoms held together with covalent bonds. We shall see that the difference that makes one so hard and the other so soft is the structural arrangements of the atoms.

THE STRUCTURE OF SOLIDS

When there is a repeating pattern in the location of the atoms in a solid, the solid is called **crystalline.** Examples of crystalline solids are table salt, diamonds, quartz, and ice. If the atoms have no particular arrangement, fitting together in a random way, the solid is called **amorphous.** Plastics, glass, and cement are examples of amorphous solids. Many solids are a mixture of crystalline and amorphous parts; granite, for instance, is an amorphous *composite* of small crystals of different chemical compositions.

THE SAME PROCESS THAT DETERMINES CRYSTAL GROWTH IN ROCKS AFFECTS THE QUALITY OF ICE CREAM. TO GET THE SMOOTH CONSISTENCY PRIZED IN TOP-QUALITY ICE CREAM, COMMERCIAL PRODUCERS CONTROL THE CRYSTALLIZATION PROCESS. THEY MUST TAKE THE ICE CREAM MIXTURE TO −40°F AS QUICKLY AS POSSIBLE. ICE CREAM THAT IS FROZEN TOO SLOWLY IS VERY GRANULAR BECAUSE OF THE LARGE CRYSTALS; RAPID FREEZING OF THE MIXTURE PRODUCES ONLY MICROSCOPIC CRYSTALS.

Whether a solid is crystalline or amorphous can depend on how it is formed. For example, melted rock that vents from a volcano at earth's surface cools and solidifies very fast. The molecules have no time to get arranged into a crystalline pattern. The molten rock hardens into an amorphous solid; sometimes it even looks like glass. When molten rock cools underground, it cools more slowly and under pressure. This pressure tends to push the atoms into an orderly crystalline arrangement, and the slower cooling gives the atoms time to move into place in the pattern. The resulting rock has grains of crystals in it, giving it a rough texture. The slower the cooling, the larger the crystals, as seen in Figure 9-2. Other solids, such as table salt, form only into crystalline arrangements due to their strong ionic bonding. A piece of salt, no matter what the size, is always crystalline.

(a)

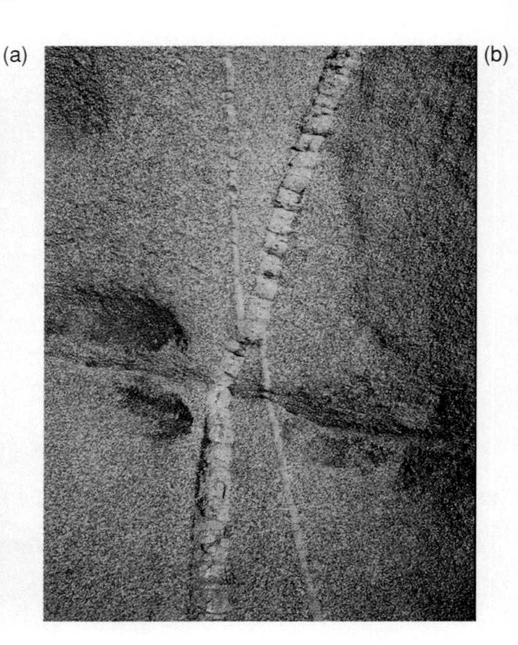
(b)

FIGURE 9-1 (A) THIS DIKE (ARROW) WAS FORMED WHEN MOLTEN ROCK INTRUDED INTO A CRACK IN THE LARGER ROCK STRUCTURE. (B) THE MATERIAL IN THE DIKE, INSULATED BY THE ROCK IT ENTERED, COOLED MUCH MORE SLOWLY, AND THE CRYSTAL GRAINS ARE VERY LARGE THERE.

THE STRENGTH OF SOLIDS

Diamond and graphite owe their great differences in strength to different crystalline structures. Diamonds form slowly at high temperature and under great pressure. The atoms in the diamond structure are closely packed, held rigid by covalent bonds pointing at various angles in all three dimensions (Figure 9-2). In graphite, formed at low temperature and pressure, covalent bonds bind the atoms together in flat sheets. These sheets are held in place only by weak attractions between neighboring layers. Thus, the individual sheets of atoms can easily slip past each other.

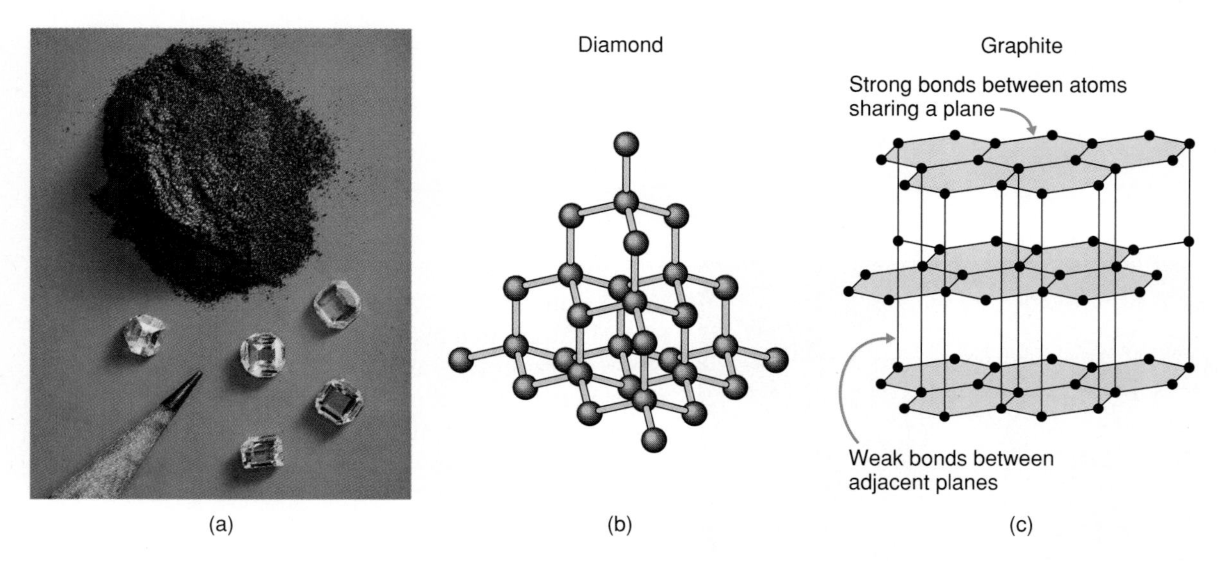

FIGURE 9-2 (A) ARTIFICIAL DIAMONDS ARE MANUFACTURED FROM GRAINS OF CARBON (IN THE BACKGROUND), THE SAME MATERIAL A PENCIL "LEAD" IS MADE OF. (B) THE CRYSTALLINE STRUCTURE OF DIAMOND, WITH BONDS EXTENDING IN MANY DIRECTIONS. (C) THE CRYSTALLINE STRUCTURE OF GRAPHITE, WHICH IS STRONGLY BONDED ONLY IN PLANES, WITH WEAK ATTACHMENTS BETWEEN ADJACENT PLANES.

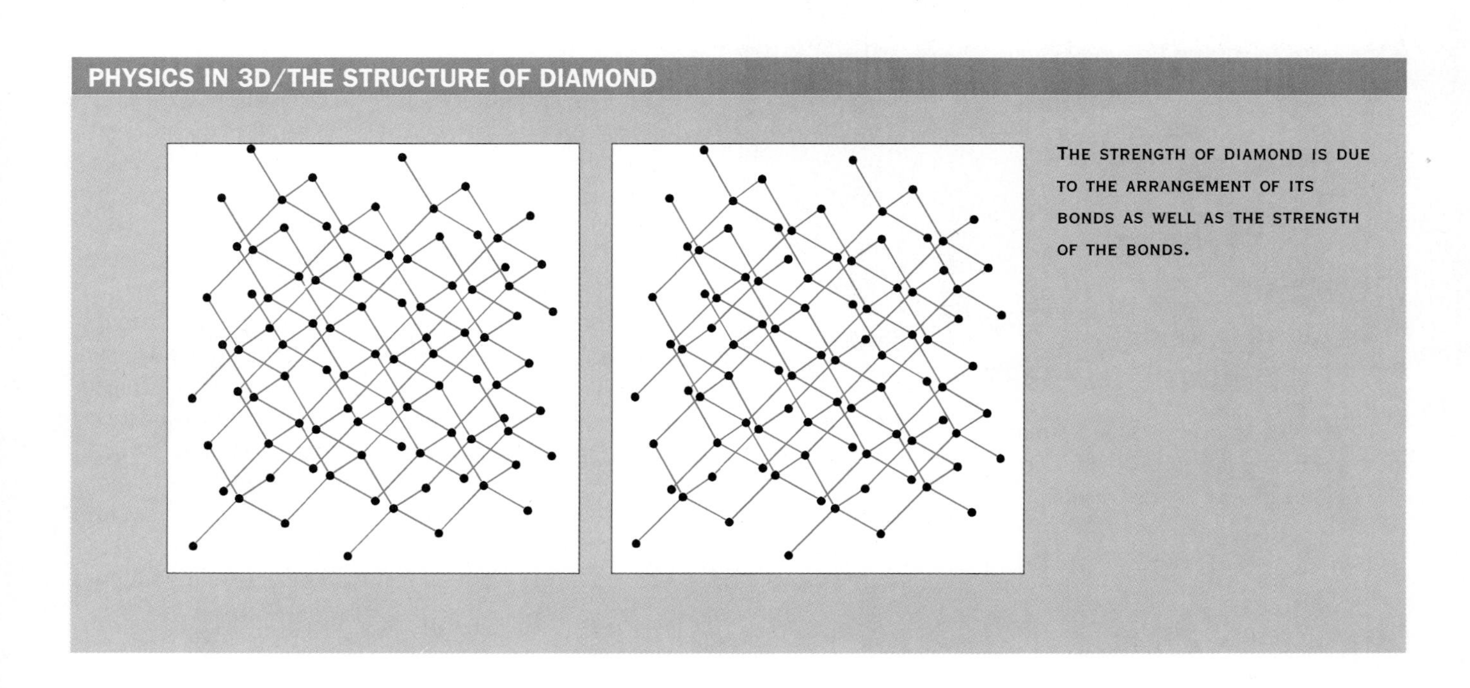

THE STRENGTH OF DIAMOND IS DUE TO THE ARRANGEMENT OF ITS BONDS AS WELL AS THE STRENGTH OF THE BONDS.

PHYSICS IN 3D/THE STRUCTURE OF GRAPHITE

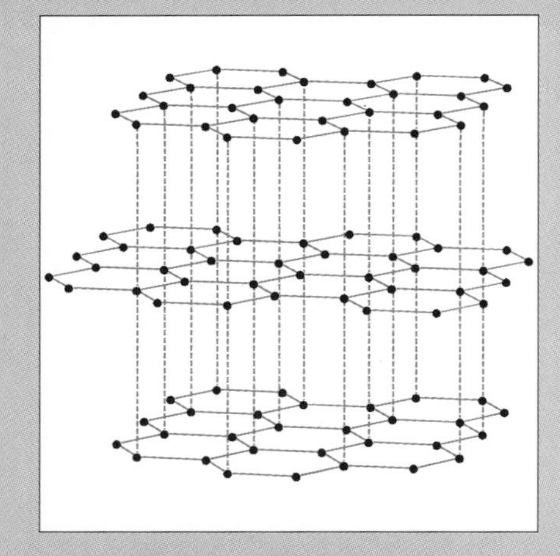

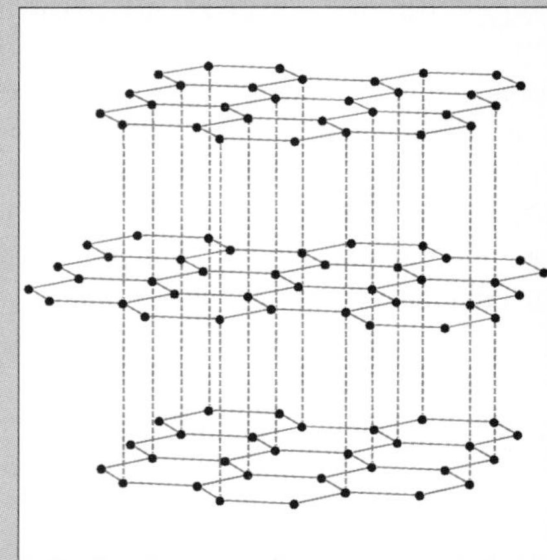

GRAPHITE YIELDS TO PRESSURE BECAUSE THE CARBON ATOMS ARE STRONGLY BOUND ONLY IN PLANES, WITH THE PLANES BEING WEAKLY ATTACHED TO EACH OTHER.

FIGURE 9-3 A GEM CUTTER MUST AIM THE BLOW EXACTLY ALONG A NATURAL PLANE OF ATOMS IN THE CRYSTAL OR RISK FRACTURING THE GEM.

Large crystals often have surfaces that consist of intersecting planes. These planes are called *faces* of the crystal. Atoms in a plane are usually bonded strongly to other atoms in the same plane but weakly to atoms in other planes. Because of these weaker bonds, crystals can be *cleaved,* or split, with a relatively small force along directions parallel to crystal faces. With a single tap of a gem cutter's mallet, a diamond, the world's hardest substance, breaks cleanly along such planes (Figure 9-3).

For an amorphous substance such as glass, there are no planes of symmetry and it breaks or shatters rather than cleaves. A break like this is called a *fracture*. Crystalline solids can fracture, too, if a large force acts in a direction not parallel to a crystal face. The great care a gem cutter takes in cleaving a diamond is partly to ensure that the gem won't crack across its internal planes.

Often a crystal can yield, or slip, along crystal planes without cleaving into two pieces. Such slippage is called a *dislocation* (Figure 9-4). It is possible to make a crystalline solid stronger by using small, internal dislocations to block the easy directions of yield in its structure. When dislocations occur throughout the crystal, they become twisted and tangled and stop each other's potential motion.

Two other methods can strengthen crystalline materials by making it more difficult for dislocations to form. One method is to mix some small amount of an impurity atom or molecule into the material. Often such impurities can block dislocations and give greater strength to the material. Another method is to make the solid grainy, composed of smaller crystals locked together in all directions. This is done by heating the solid close to its melting point, a process called *annealing,* and cooling it quickly, a process called *quenching*. Then the scrambled crystalline grains, if bonded tightly, make formation of dislocations more difficult and cleavages impossible. For example, a blacksmith heats a horseshoe to soften it. The extra thermal motion lets the metal flow enough to reduce stress and dislocations in the iron. Once the metal is "plastic," it can be shaped with hammer and anvil. Then, quenching the horseshoe forms tiny crystals that harden the metal again.

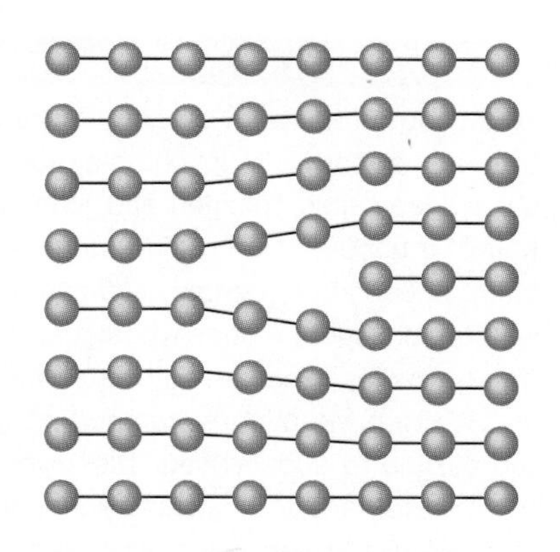

FIGURE 9-4 ONE TYPE OF DISLOCATION IN A CRYSTAL. MANY DISLOCATIONS HELP TO BLOCK A MAJOR SLIPPAGE, THEREBY HARDENING THE MATERIAL. (SOMETIMES IMPURITIES IN THE CRYSTAL LATTICE CAN SERVE THE SAME PURPOSE.)

ALLOYS

Like gold and silver, the metal copper (Cu) is sometimes found in the pure state in nature, and archaeologists have found bits of worked copper that date back to about 8000 B.C. Being durable and yet easy to hammer into various shapes, copper was no doubt highly prized by early peoples. Later, copper ores were melted in kilns as a coloring agent on pottery. Another element occasionally found as a free metal in nature is tin (Sn), a soft, bluish-silver-white element. Around 4000 B.C., someone melted copper and tin and mixed them. That metalworker invented bronze, a copper–tin **alloy,** and civilization was soon in the Bronze Age. Bronze is far stronger than copper and holds a cutting edge. Razors, sickles, chisels, even safety pins appeared. Bronze rims made wooden wheels durable for the first time. The Bronze Age lasted until after 1000 B.C., when iron ores were melted in furnaces. By 600 B.C. the stronger alloys of iron had replaced bronze, and civilization entered the Iron Age.

Small amounts of carbon mixed with molten iron (Fe) provide alloys with a variety of properties. When iron contains 1 percent carbon or less, it is called *steel.* Structural beams for buildings and bridges are made of steel with less than 0.2 percent carbon. One-percent carbon steel is good for low-speed cutting tools, such as drill bits and saw blades. If even more carbon is added to molten iron, it begins to flow as freely as water. Iron with up to 4 percent carbon is *cast iron,* which is poured into molds and cast into shapes such as frying pans and engine blocks.

As early as 500 years ago, gold (Au) was used to fill cavities in teeth because it is malleable (easy to work into a different shape) and also inert (doesn't easily react chemically to form compounds). Much earlier, gold was worked into crude bridges for missing teeth (Figure 9-5). To increase its durability in coins, gold was alloyed with copper (90 percent Au, 10 percent Cu). The 14-karat gold of wedding rings is 58 percent gold alloyed with various percentages of copper and silver (Ag). (On the karat scale of purity, pure gold is 24 karat; so 14-karat gold means $14/24 = 0.58$ or 58 percent gold.) Platinum gold, one form of white gold, is 60 percent gold and 40 percent platinum (Pt).

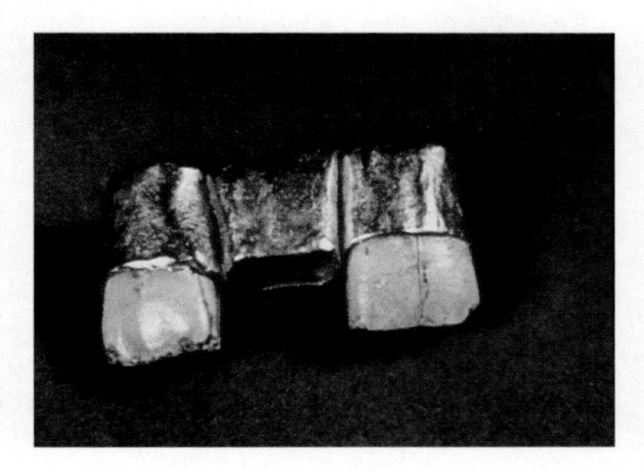

FIGURE 9-5 AN ETRUSCAN GOLD BRIDGE, FROM 600 B.C. THE CENTER GAP HELD THE FALSE TOOTH, PROBABLY MADE OF WOOD OR IVORY. THESE BRIDGES WERE MADE NOT FOR IMPROVED CHEWING BUT FOR COSMETIC REASONS, AND THEY APPARENTLY WENT OUT OF FASHION BY THE TIME OF THE ROMAN EMPIRE.

Many hundreds of alloys are used today for all sorts of special purposes. Some, such as tungsten steel and stainless steel, are made for strength, durability, and resistance to rust; others, such as spring steel, have special elastic properties. Aircraft alloys of aluminum, magnesium, and titanium provide strength with low weight. There are dozens of kinds of brass (copper and zinc); there is pewter (tin and copper with traces of other metals), and solders that are useful because of their low melting points.

THE ELASTICITY OF SOLIDS

Stretch or compress a steel spring, and it snaps right back as soon as you let go. As explained in Chapter 3, a spring behaves in this way because the bonds between the atoms act like springs themselves. If the stretching or compressing isn't too great, the atoms return to their original positions as soon as you stop pushing or pulling.

THE STRENGTH OF NATURAL AND FABRICATED MATERIALS

Your bones are an alloy of sorts; they are made of two materials that are strong only in combination. One is a very elastic substance called collagen and amounts to a little more than half of a bone's volume. The other is a brittle mineral of calcium, phosphorus, oxygen, and hydrogen that is interspersed with the collagen. A bone whose calcium–phosphorus mineral has been dissolved can be twisted into a circle without breaking. Without the mineral component, your skeleton would be too flexible to support you. Dissolve the collagen instead of the mineral, and the calcium–phosphorus compound, though rigid, breaks under only a little pressure. Without the elastic collagen to fill out its porous structures, your bones would disintegrate, fracturing immediately under your weight.

A blade of grass or a piece of bamboo splits easily along its length, but either material stubbornly resists tearing across its width. Each has long, cellular fibers aligned along its principal direction of growth. It's relatively easy to cleave the fibers but not to cut across them. Wood, too, is easiest to split with the grain. A sheet of paper is strong because the wood fibers (in some papers, cotton fibers) twist and overlap in every direction, forming a flexible mat.

A woven rope or even a steel cable gains strength because the component fibers or wires act together to resist tension. The bending and twisting around each other bring the contact force into play between the fibers or the wires, and friction between the strands lets all of them share any tension that is applied. You can easily fold your clothes because cloth is woven to be flexible. In a simple vertical and horizontal weave, such as in handkerchiefs and tablecloths, the material does not stretch much in either the horizontal or the vertical direction, but knitted materials allow some flexibility in all directions, as in the fabric of a t-shirt.

THE FILAMENTS OF A CLIMBER'S ROPE, WHICH ARE LEFT STRAIGHT SO THE ROPE WILL BE FLEXIBLE. IF LARGER BRAIDED STRANDS OF FILAMENT ARE TWISTED TOGETHER, AS IN STEEL SUPPORT CABLES FOR BRIDGES, MUCH FLEXIBILITY IS LOST.

AN ENLARGED VIEW OF A THREADED NEEDLE. MUCH OF THE THREAD'S STRENGTH COMES FROM THE FIBERS SHARING ANY TENSION PLACED ON THE THREAD. SINCE THE FIBERS ARE TWISTED AROUND EACH OTHER, FRICTION BETWEEN THEM PROMOTES THE SHARING OF THE TENSION.

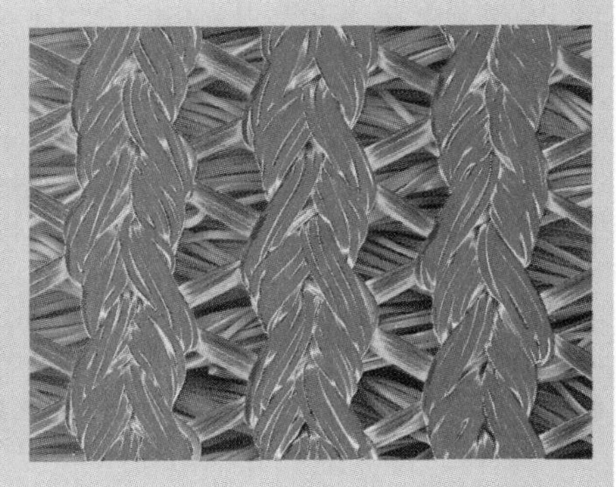

THE WOVEN FABRIC OF LYCRA TIGHTS, MAGNIFIED 18 TIMES. THE WEAVE OF THIS MATERIAL LETS IT STRETCH IN EVERY DIRECTION WITHOUT TEARING, WHICH MAKES THE FABRIC A NATURAL CHOICE FOR GYMNASTS AND OTHER ATHLETES.

Even the most rigid solid objects show some degree of elasticity, but every substance also has a point where bonds break and the deformation becomes permanent. Sometimes solids that don't seem to be elastic are. You may think of a building as a perfectly rigid structure, but in a strong wind, tall buildings sway. Watch the wingtips of an airplane move up and down when you take a flight. Your body, too, is elastic, even if you aren't a gymnast. As you walk around during the day, the weight of your body gradually compresses the elastic connections between the vertebrae in your spinal column. By nighttime, you may be as much as half an inch shorter. Then, during sleep, you stretch out again.

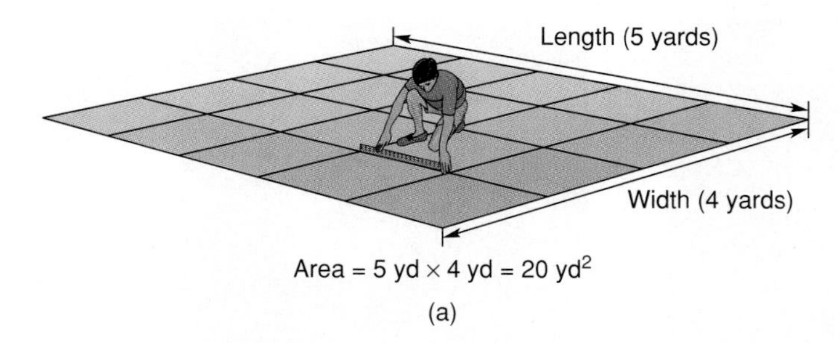

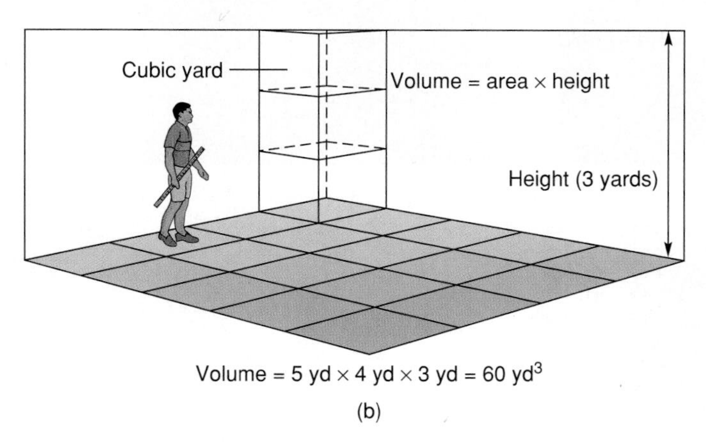

FIGURE 9-9 (A) TO MEASURE THE AREA OF A FLOOR OR ANY OTHER RECTANGULAR FLAT SURFACE, YOU NEED TO MULTIPLY THE LENGTH TIMES THE WIDTH, $A = L \times W$. (B) TO FIND THE VOLUME OF A ROOM, YOU NEED TO MULTIPLY ITS LENGTH TIMES ITS WIDTH TIMES ITS HEIGHT. IN OTHER WORDS, YOU CAN TAKE THE AREA OF THE FLOOR AND MULTIPLY BY THE HEIGHT OF THE ROOM TO GET ITS VOLUME. $V = L \times W \times H$ OR $V = A \times H$.

VOLUME, SURFACE AREA, AND DENSITY

The amount of space any flat object occupies is called its **surface area** (or sometimes merely its **area**), and the amount of space occupied by any three-dimensional object is called its **volume** (Figure 9-6).

The volume of a body does not change merely because its shape changes. For example, the volume does not change when a small ball of clay is flattened. What changes in this case is the surface area of the clay, not the volume occupied by its atoms. The volume of anything, regardless of its shape, is always just the number of cubic centimeters (or other units of volume) enclosed by its surface. Several ounces of gold may be pressed into a thin foil to cover the dome of a statehouse; although its surface area becomes enormous, its volume remains the same.

If something has an irregular shape, you can use an indirect way of measuring its volume. If you totally immerse a nonabsorbent solid of any shape in water, it pushes aside, or *displaces,* a volume of water equal to its own volume (Figure 9-7). The volume of water displaced is exactly equal to the volume of the submerged solid.

FIGURE 9-7 THE NUMBER OF CUBIC CENTIMETERS THE WATER LEVEL RISES IS EQUAL TO THE VOLUME DISPLACED BY THE SUBMERGED OBJECT, SO LONG AS THE OBJECT DOESN'T ABSORB ANY WATER.

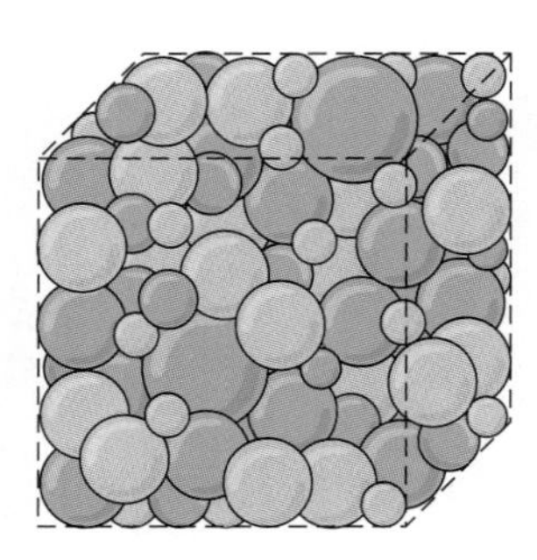

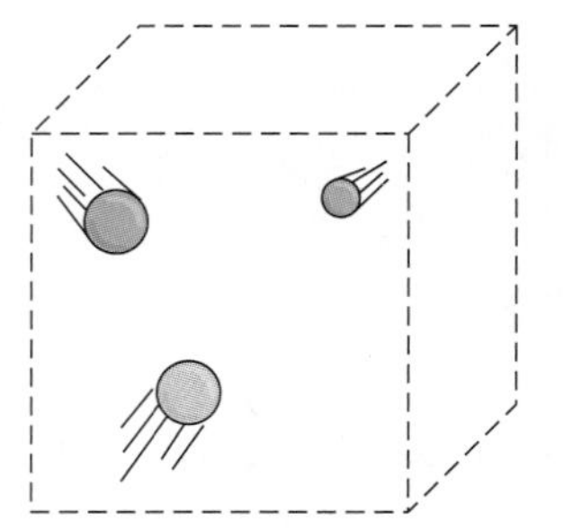

FIGURE 9-8 THOUGH THE VOLUME OF THESE TWO CUBES IS THE SAME, THE NUMBER OF ATOMS THEY CONTAIN IS QUITE DIFFERENT. EVEN IF THE ATOMS ARE OF THE SAME TYPE, THE DENSITY OF MATTER IS VERY DIFFERENT IN THE TWO CUBES. DENSITY = MASS/VOLUME.

Sometimes different amounts of mass occupy the same volume (Figure 9-8). The quantity that shows the relationship between mass and the space it fills is **density,** the ratio of mass to volume:

PUTTING IT IN NUMBERS—DEFINING DENSITY

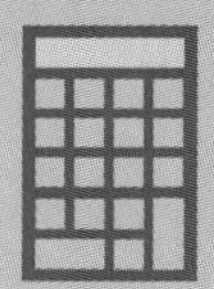

average density = mass/volume

Example: A lead fishing weight is dipped into a graduated cylinder (as in Figure 9-10), where it displaces 2 cm^3 of water. On a scale the lead weight's mass is found to be 22 grams. What is its density? density = 22 g/2 cm^3 = **11 g/cm^3**.

Whenever matter is spread evenly throughout a volume, its density is uniform. Density is often measured in grams per cubic centimeter. For instance, the average density of granite is 2.7 g/cm^3; gold's density is about 19.3 g/cm^3, and the density of water is 1 g/cm^3 (Tables 9-1 and 9-2).

Table 9·1 Approximate Densities (In Grams Per Cubic Centimeter)

Element	Density
Calcium	1.5
Carbon: graphite	2.25
diamond	3.5
Aluminum	2.7
Iron	7.8
Copper	8.9
Silver	10.5
Lead	11.3
Mercury	13.6
Gold	19.3

THE RELATIONSHIP BETWEEN SURFACE AREA AND VOLUME

If an object maintains its shape as it expands, or is scaled up in size, its volume increases more rapidly than its surface area. We can see the relationship between surface area and volume easily with the cubes shown in Figure 9-9, where one cube is twice as long, twice as wide, and twice as tall as the other. Think of the larger cube as a collection of 8 of the smaller cubes. The volume

THESE THREE ELEPHANTS ARE ROUGHLY THE SAME SHAPE BUT QUITE DIFFERENT IN SIZE. SO THE SMALLEST ELEPHANT HAS MORE SURFACE AREA PER UNIT OF VOLUME THAN THE ONE IN THE MIDDLE, AND THE LARGEST HAS THE LEAST SURFACE AREA PER UNIT OF VOLUME OF THE THREE.

Table 9·2 Approximate Densities

Substance	Density (g/cm^3)
Cork	0.25
Paper	0.7–1.15
Butter	0.86
Ice	0.917
Beeswax	0.96
Human body	1.0
Fat	0.9
Muscle	1.1
Nylon	1.09–1.14
Sugar	1.6
Bone	1.7–2.0
Brick	2
Clay	2
Granite	2.7
Cement (hardened)	3

of the larger cube is 8 times the volume of the smaller cube. Each face of the larger cube has only 4 times the surface area of each face of the smaller cube, however, and so the surface area of the larger cube is only 4 times the surface area of the smaller cube. The larger cube, then, has more volume per unit of surface area than the smaller cube. That is, the ratio V/A is greater for the larger cube than for the smaller cube. *This result holds for any shape.* For any given shape, a larger object has more volume per unit of surface area and, conversely, a smaller object has more surface area per unit of volume.

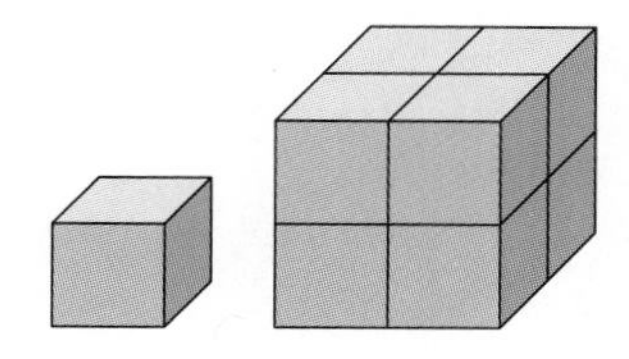

FIGURE 9-9 IF ONE CUBE IS TWICE AS WIDE AS THE OTHER, IT HAS EIGHT TIMES THE VOLUME AND FOUR TIMES THE SURFACE AREA OF THE SMALLER ONE. COUNT THE FACES AND THE CUBES TO SEE THIS FOR YOURSELF. THIS MEANS THE PRESSURE UNDER THE LARGER CUBE IS ONLY TWICE AS MUCH AS THE PRESSURE UNDER THE SMALLER CUBE.

The relationship between surface area and volume is important to such phenomena as chemical reactions, heat transfer, and drag from motion through fluids. For example, when a lollipop dissolves in your mouth, it turns to liquid, but only at its surface. If it is in the shape of a sphere, it has the least possible surface area per unit of volume (see Demonstration 4). If you bite down and crack the lollipop into several pieces, those smaller pieces have more surface area per unit of volume and dissolve faster than the original shape would. Another example is found in raindrops. Every drop is pulled downward by the force of gravity, mg, and this force is the drop's weight. How much the drop weighs, and thus how much mass it has, is determined by how much water is in it—in other words, by its volume. Because they have more surface area *per unit of volume* than larger drops, smaller drops experience more air drag per unit of volume, or per unit of mass, than larger drops. Because air drag is larger for them *per unit of mass,* smaller drops decelerate quicker due to air drag and have much lower terminal speeds as they fall.

PRESSURE AND SOLIDS

If the two cubes shown in Figure 9-9 are made of the same material, which exerts more pressure on its base? The larger cube weighs 8 times as much as the smaller. Yet that doesn't mean the pressure it exerts is 8 times more than the pressure exerted by the smaller cube. As we saw in Chapter 4, pressure is force/area, and so we have to take into account the larger area the big cube

FIGURE 9-10 IN COLLISIONS, THE SURFACE AREA PER UNIT OF VOLUME CAN BE VERY IMPORTANT. THE SMALLEST CREATURE, WITH A GREATER SURFACE AREA PER UNIT OF VOLUME (AND HENCE MORE SURFACE AREA PER UNIT OF MASS), RECEIVES LESS PRESSURE WHEN IT COLLIDES AND HENCE HAS AN EASIER TIME OF IT. THIS IS ONE REASON A TODDLER CAN TRIP AND TAKE A SPILL WITH LESS RISK OF BREAKING A BONE THAN AN ADULT WOULD HAVE.

rests on. Each face of the larger cube has 4 times the surface area of each face of the smaller cube, so the larger cube rests on 4 times as much surface area as the smaller cube does. With 8 times the weight (F) supported by 4 times the area (A), the larger cube exerts only *twice* as much pressure on its base as the smaller cube does.

The relationship between volume, surface area, and pressure has important consequences. Imagine this highly improbable but perilous scene: You are traveling in a boxcar with a mouse, a cat, and a hippopotamus when the train suddenly stops. Innocently following Newton's first law, all four of you will slam against the front wall of the boxcar (Figure 9-10). Just as with the cubes, the pressure is least on the smallest creature because it has less mass per unit of surface area of contact. If all of you hit the wall in the same manner, the mouse will be better off than the cat, and the cat will be better off than you. The hippo will have the worst time (unless you are caught between the hippo and the wall!). This is why a small pet might endure a car crash better than its owner, and why an insect along for the ride is in even less danger.

PRESSURE AND THE SOLID EARTH

At Carleton, South Africa, workers mine for diamonds almost 4 km (2.4 mi) beneath the earth's surface. Each 3.8-m (12.5 ft) depth of rock adds about 1 atmosphere of pressure (1.01×10^5 Pa or 14.7 lb/in^2) on the rock below. Because they withstand such high pressure, we can surmise that the bonds that keep the rock together around that shaft are *very* strong.

The porous rock where oil and natural gas are found is not that strong. When an oil well is drilled, the hole made by the drill is subject to almost certain cave-ins. To prevent this, the crews start with a large-diameter drill. At a certain level, the drill is pulled out and a hollow pipe is inserted into the drill hole. The area between the sides of the hole and the outside of the pipe is then pumped full of cement. This artificial rock, stronger than the porous rock around it, preserves the expensive well. Then small explosive charges are lowered to the proper depth and set off. The explosion perforates the cement-lined well to let any oil or gas through to be pumped to the surface.

Deep in any well or mine, the temperature rises, the result of heat released from the natural radioactivity of some of the minerals in the earth's interior. At the lowest level of the Carleton mine mentioned above, the temperature is 55°C (131°F). One natural-gas well drilled in Oklahoma reached a depth of almost 10 km (6 mi). The temperature there was 246°C (475°F). Even deeper into the earth, the pressure grows relentlessly, squeezing the atoms more tightly to make the rock more dense. At depths from about 8 to 40 km the rock can flow wherever there is an imbalance of pressure.

Another unseen happening takes place at depths of about 160 km (100 miles) and temperatures of about 1500°C (2700°F): The crystals we call diamonds slowly form in the molten rock. Diamonds are brought near the surface only when volcanic action occurs. Mines like those in South Africa tunnel into ancient volcanic vents, or "pipes," to reach those precious stones.

Understandably, pressure is greatest at the center of the earth. The atoms there support all the overlying weight of earth and the atmosphere. They squeeze close together, increasing the density of the core material to nearly 20 g/cm^3 for combinations of atoms that might have densities of only 7 to 9 g/cm^3 at the surface.

FIGURE 9-11 THE ALASKAN PIPELINE, ELEVATED TO AVOID MELTING THE SUBSURFACE ICE.

Of course, any solid will melt or vaporize if its temperature gets high enough. At earth's surface, this happens at much lower temperatures than in earth's interior, where the solids are under so much pressure. For one example, at earth's surface, ice turns into water at 0°C (32°F). One place this had to be taken into account was in the construction of the Alaskan pipeline (Fig. 9-11). Heat from the warm oil would creep down the pipeline's supports and melt the ice just beneath the soil if it were not for the heat-radiating fins connected to each support.

HEAT AND PIPELINES

CEMENT AND CONCRETE

In the earliest civilizations, building stones were chiseled or carved to fit. Later, however, the Romans perfected a stonelike cement to anchor building stones together. Their natural cement was a calcium carbonate (limestone) paste that hardened when mixed with volcanic ash and water. This cement was used in the construction of the Colosseum, the Pantheon, the Appian Way, and the aqueducts. Still standing after 2000 years, these structures prove how well the Romans mastered that art. With the fall of the Roman Empire, however, the secret of cement making was lost for many centuries.

In 1796 in England, James Parker burned impure limestone, added water, and watched the mixture slowly harden into a material he named Roman cement. In 1824 Joseph Aspdin, an English bricklayer, patented a cement of superior strength. He burned limestone and clay together, the clay melted and fused with the lime, and he cooled the resulting substance and ground it into a powder. When water was added, this powder slowly combined chemically with the water in an exothermic reaction. The paste hardened into a stonelike material resembling limestone quarried from the region called Portland on the coast of England. Today all over the world, this Portland cement is used to make concrete in those thousands of cement-mixer trucks that roll out of cement plants to building sites.

Concrete is a mixture of cement and an aggregate such as sand or gravel. As the cement hardens, it tightly locks the aggregate particles into place. The largest concrete structure in the world is the Grand Coulee Dam on the Columbia River in the state of Washington. The total amount of heat released as the poured concrete cured in this dam was enormous. Left alone to cool, the concrete dam would have taken decades to come to the temperature of its surroundings. To speed up the cooling process, over 2000 miles of water pipes were laid throughout the concrete as it was poured, and cold water was pumped through the pipes. This great manufactured stone, precision-fit to a river channel, backs up water in a lake that stretches 250 kilometers upriver.

Summary and Key Terms

crystalline solid–126
amorphous solid–126

The rigidity of a solid depends on the strength and arrangement of the bonds between its atoms. In a ***crystalline solid***, such as table salt or diamond, the molecules or atoms have a repeating pattern. In an ***amorphous solid***, such as glass or cement, the molecules or atoms fit randomly together. Whether a solid is crystalline or amorphous can depend on how it was formed. When melted rock cools quickly, the molecules have no time to move into a crystalline pattern, so an amorphous solid forms. When molten rock cools underground, however, it cools slowly and under pressure, and the resulting rock has crystals in it. *Dislocations,* small slippages along crystal planes, can help to stop any potential motion (or *cleavage*) along a crystal's planes. A metal can sometimes be strengthened by adding small amounts of other metals to form an ***alloy***.

alloy–129

Stretch or compress a steel spring, and it will snap back when you let go because the bonds between its atoms act like springs themselves, attracting a bit more when the atoms are pulled apart and resisting when they are squeezed together. Every solid shows some degree of elasticity, but every solid also has a point where bonds break and deformations become permanent.

volume–131
density–132
surface area–131

Every object has ***volume***, the amount of space enclosed by its surfaces. The relationship between an object's mass and the space it fills is its ***density***: Average density = mass/volume. For any given shape, a larger size has more volume per unit of ***surface area***, *V/A,* than a smaller size. Conversely, a smaller object has more surface area per unit of volume, *A/V,* than a larger object of the same shape. A sphere is the shape that has the smallest surface area per unit volume.

EXERCISES

Concept Checks: Answer true or false

1. Both the arrangement of atoms in a solid and the bonds between them help determine how strong the solid is.

2. There is no such thing as a perfectly rigid solid.

3. An alloy can have very different physical properties if its components are varied by a tiny percentage.

4. A broken bone and a crushed aluminum can share this description: Both have exceeded their limit of elasticity.

5. The volume of space enclosed by the surface of a solid is described by units of length cubed, such as cm^3 or m^3.

6. If you change the shape of a piece of gold or modeling clay, you change its surface area but not its density or volume.

7. Drops of mist fall more slowly through the air than raindrops because the mist drops have more surface area per unit of mass.

8. Far beneath the earth's surface, the weight of the overlying material produces great pressure. This pressure increases the density of matter there and decreases the volume of matter there.

Applying the Concepts

9. How do we know without seeing them that the atoms in a solid generally keep their places with respect to each other?

10. Small grains of zircon found in Australia are the oldest pieces of solid material known on earth. These grains are estimated to have formed 4.1 to 4.2 billion years ago, very soon after the earth's crust formed. What would be your bet about zircon's chemical reactivity, solubility, and reactions to high and low temperatures?

11. The Incas did not use cement to build the walls shown in Figure 9-12, yet the walls have withstood the test of time. Comment on the reasons for their stability.

12. Several species of frogs breathe through their skin. One

FIGURE 9-12

species has pendulous folds of skin, and another has many tiny protrusions on its hindquarters. How do these folds and protrusions affect the skin's ability to breathe?

13. Both diamond and graphite are pure crystals of carbon atoms. How can you explain their different densities (Table 9-1)?

14. An auto manufacturer claims to use magnesium in the wheels because it is "lighter" than steel. How should the company modify this statement to make it correct?

15. The Mayans melted calcified rocks in fires to make lime, but first they broke up the rocks into pieces the size of oranges. Exactly how did that help?

16. Why are sugar and salt made into such fine grains?

17. A sterling silver spoon is 92.5 percent silver and 7.5 percent copper. What function does the copper serve?

18. Suppose you have a cube of modeling clay. If you slice it in a direction parallel to any side, you will add two extra faces of surface area without changing the total volume of the clay. How many such slices should you make to expose twice the surface area of the original cube?

19. Contemporary Egyptian goldsmiths have hammered gold into leafs so thin that 367,000 of them made a pile an inch high. About how many gold atoms "thick" is one such leaf? (The diameter of a gold atom is about 3×10^{-8}cm.)

20. A heavy metal object resting on a metal surface at only a few points can stick to that surface, a phenomenon known as "cold welding." Explain.

21. To cool a drink quickly, should you use shaved ice or ice cubes? Why?

22. Up until about 900 A.D., horses were prone to foot disease and hoof breakage when they carried heavy loads. Then the metal horseshoe, shaped to the horses' hooves, was invented; without them the heavily armored knights could not have ridden off to the crusades. Explain exactly how the metal horseshoe helps in terms of pressure.

23. Give an argument for keeping coffee beans whole and grinding them just before making coffee.

24. Pure uranium reacts with oxygen in the air, as do many other pure metals. If a small amount of uranium is machined on a lathe or drilled, any small particles of uranium released will burst into flame, whereas a larger chunk will not. Discuss.

25. Why does it help to use small twigs rather than large limbs and branches to start a campfire? Once a fire is going and you want it to burn for a long time, should you use thick logs or thin logs? Why?

26. Why will deli-sliced turkey spoil faster than a whole turkey in the refrigerator?

27. Why does it help your digestive system if you chew food well before swallowing?

28. A solid gold nugget weighs 1.4 pounds (so its mass is 0.63 kilograms). What is its volume in cubic centimeters? (Use Table 9-1 and the formula for density.)

29. Explain why snowflakes fall more slowly than hailstones.

30. The surface area of the skin of an average adult male is about 2.1 square meters, while females average 1.4 square meters. Which of the two would you expect to have the greater surface area per unit of volume?

31. The inside of an average pair of human lungs has enough surface area to cover a 30-foot $\times$ 30-foot floor. How does having so much surface area help the lungs to exchange oxygen and carbon dioxide with the blood?

32. When the winds have been right, extremely tiny particles of sand from the Sahara have drifted thousands of miles through the air, even passing over Florida, where they were visible as a white haze in the sky. Discuss in terms of surface area and mass (or weight) how this is possible.

33. A milkshake froths with bubbles when air is beaten into the milk and ice cream mixture. Does the density of the milkshake change? Does its surface area per unit of volume increase, decrease, or remain the same?

34. A lead brick is twenty-five cm long, ten cm wide, and six cm thick. What are its mass and weight? (Use Table 9-1 and $V = l \times h \times w$ as in Figure 9-6.)

35. A 1-lb rock of basalt displaces 758 cm^3 of water. What is the rock's average density in g/cm^3? (A 1-kg mass weighs 2.2 lb.)

36. The next time you see a rotary lawn sprinkler, watch it carefully. Why doesn't the sprinkler wet only a circle at a certain distance rather than the entire area within the circle?

37. Why does the Federal Food and Drug Administration demand that a box of crackers display the weight rather than the volume of the box?

38. Bush pilots use this trick to deliver fresh eggs to expeditions in wilderness areas in Alaska. They wrap each egg in a crumpled sheet of newspaper, put them in a cardboard box, and drop the box from the plane as they fly over the adventurers. Discuss the factors that contribute to the soft landing of the eggs.

39. Modern steel was first made in 1815, when a researcher, looking no doubt for a hard alloy of iron, melted iron wires together with diamond dust. Steel isn't made that way today. Why not?

40. The density of this book is about 1 g/cm^3. Measure its length, width, and thickness in centimeters and find its mass.

41. A child puts a penny on a railroad track and picks it up after a train passes over it. (a) Has its mass increased, decreased, or remained the same? (b) Has its surface area increased, decreased, or remained the same? (c) Has the ratio of surface area to volume increased, decreased, or remained the same? (d) Has its density increased, decreased, or remained the same?

42. A single hydrogen atom has a mass of 1 u and a diameter of about 1×10^{-8} cm. One of the largest molecules known has a molecular mass of about 10 million u, yet its diameter is only about 1×10^{-5} cm. Explain how something that is only a thousand times bigger than something else might have a mass 10 million times greater.

43. The rate at which a lollipop dissolves in your mouth depends on how much surface area the lollipop has. True or false?

44. Two lollipops weigh the same, but one is spherical and the other is a flat disk. Which would dissolve at the faster rate initially?

45. When you are eating a lollipop, there is more surface area per unit of volume for the last cubic centimeter that dissolves in your mouth than for a new lollipop. True or false?

46. A dedicated student taking this course decides to wallpaper a wall of his room with the pages of this book to make it easier to study. The pages are about 8 inches by 11 inches, and his wall is eight feet high and twelve feet long. How many pages are needed to cover the wall? (Unless he used pages from 2 copies of the book, he'd be able to study only every other page!) 1 ft^2 = (12 in)^2 = 144 in^2.

47. Which of these factors influence the rigidity of a solid? (a) temperature, (b) pressure, (c) crystalline structure, (d) chemical bonding.

48. What do you think would happen if the force of gravity suddenly ceased to exist throughout the earth?

49. If 24-karat gold is 100 percent gold and 14-karat gold is 58 percent gold, what percentage is 18-karat gold?

50. A gallon of a certain paint will cover 400 square feet. About how many gallons will cover a 9-foot high wall that is fifty feet long and thirty-five feet wide?

51. When you pop popcorn, does its mass change significantly? Its surface area? Its volume? Its density?

52. The mineral that makes up most of your bones is *hydroxyapatite*, and it is found in a pure, nonporous form in the enamel of your teeth. Recently scientists have synthesized a form of this mineral that is stronger than the form your body makes and, when grafted, is accepted by bones just as if it were the natural substance. They can produce it in a porous form or in a form similar to enamel. Speculate on some of the applications it could have.

53. In 1990 over 195 million tons of garbage was produced in the United States, and only a little more than 10 percent was recycled. To produce less garbage, you can pay attention to the size of the containers you buy. Should you buy the jumbo economy (very large) size of laundry detergent or the smaller size? (Which will save some plastic?)

54. A 50-lb bag of fertilizer covers 2000 ft^2. How many bags will cover a rectangular lot that is 150 ft by 90 ft if a 1800 ft^2 house is on the lot?

55. A certain telephone pole (treated with creosote) has a density of 1.3 grams per cubic centimeter and is 12 meters high. What pressure does it exert on its base when it is placed erect in the ground? Find the answer in N/m^2 and convert that to atmospheres.

56. Suppose the telephone pole in Exercise 55 were made of solid aluminum. What pressure would it exert on its base? (Use Table 9-1.)

57. The metal feet on the rollers in Figure 9-13 help compact ground that will be paved over as a road is built. How do they work differently from a smooth steel drum?

FIGURE 9-13

Demonstrations

1. Adding a little at a time, stir salt or sugar into a cup of warm water until no more will dissolve. Warm the solution on the stove—don't boil it—and add more salt or sugar while stirring. Pour the solution (which should not contain any undissolved salt or sugar) into a clean glass and suspend a thread. Make certain the thread almost touches the bottom of the glass. Let the solution and thread stand undisturbed for a few days. Crystals will form on the string. If you grow a really big crystal of salt, use a single-edge razor blade and a small hammer to cleave it. If you grow crystals of sugar instead, that's rock candy. Help yourself!

To make an amorphous solid from sugar, heat a few spoonfuls in a small pan and slowly melt it. (If you heat it too fast, you'll get the broken sugar molecules called caramel—have some apples ready!) Place the hot pan into a larger container of cold water. The liquid sugar cools quickly into the clear, glossy (and amorphous) candy familiar in cough drops, candy mints, and lollipops.

2. Using a pair of pliers and protecting your hand with an oven mitt and your eyes with safety glasses, hold one end of an iron nail in a candle flame. The nail may get red hot but not much else happens. Then use the pliers to hold a piece of steel wool (which is essentially the same stuff) in the flame. It will burn brightly. The increased surface area is responsible for the fireworks. As oxidation proceeds over the surface of the fine strands of steel, their temperature is raised enough to increase the rate of oxidation to the point we know as burning.

3. Work a large lump of modeling clay with your hands. No matter how you shape it, you don't change its density or its volume. Each cubic centimeter of clay still contains the same number of molecules (so its mass is the same), and the lump still displaces the same volume when submerged in water. You change only its surface area. To make the lump as compact as possible so that it has the least surface area for its volume, you want to round it into a ball. A sphere has less surface area per unit of volume than any other shape. Or, from another viewpoint, a sphere encloses more volume per unit of surface area than any other shape. Use your fingers to press a large elephant ear out from each side of the ball of clay and pull a nose out. You've obviously increased the surface area significantly without changing the volume of the clay. When you smooth them back into a round ball, the surface area for the same volume decreases.

TEN

Liquids

The solids of the preceding chapter and the fluids of this chapter and the next are seen in this photograph. Solids, liquids, and gases are the three most common phases of matter on earth.

To contrast the properties of solids and liquids, look at a waterfall over a rock face. The solid rock stays upright because its strong bonds and structural arrangements hold its atoms and molecules tightly in place, while the water *flows.* It is fluid because the weaker attractions holding its molecules together let them move past each other. A quantity of liquid, then, flows unless it is in a container. It will fall down as far as it can because of gravity, and it flows downhill if there is any slope where it falls. The waters of the oceans fill most of the lowest points on earth, and rains keep rivers and streams flowing over land.

The attractive forces between molecules differ in strength from one liquid to another, which leads to a certain fuzziness when labeling some substances. Gasoline molecules have smaller attractions than water molecules, for example, while the molecules of toothpaste flow easily under a little pressure but much more slowly without that applied pressure. Even though modeling clay also yields under pressure, it holds its shape much better than toothpaste does and is considered a solid. It is easier to contrast a liquid and a gas, since in a gas the molecules are on the average separated by large distances. Yet gases flow, much as liquids do, and for that reason these two states of matter are grouped together under the name of *fluids.* Liquids are the subject of this chapter, while the next chapter discusses gases.

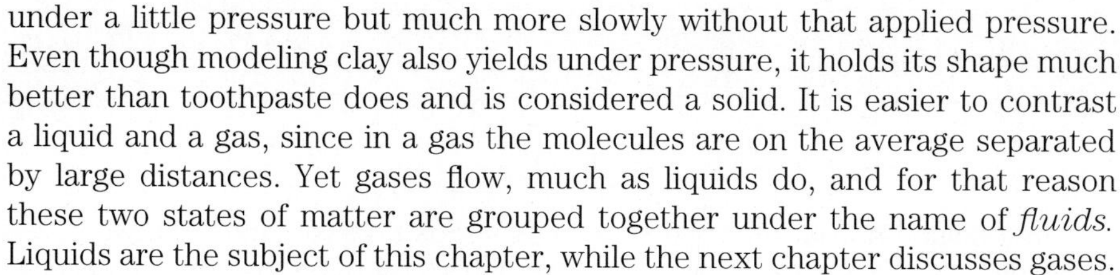

MOLECULAR MOTION AND PRESSURE

If you look through a microscope at a sliver of ice, chances are you can find a tiny particle of dust trapped inside. Within the ice, this particle has no observable motion. When the ice melts, however, the dust particle suspended in the water becomes animated, jerking frantically like a puppet on a string. In the liquid state, the rapidly moving water molecules continuously strike this larger particle from all directions. You can't see the water molecules through the microscope, but you can see the dust particle dancing in the sea of invisible torpedoes. Microscopic dust particles in air do a similar jig as air molecules punch them from all directions.

The three-dimensional lurching of a small particle in a fluid is called **Brownian motion** (Figure 10-1), and it gives us a clue to what is happening on the molecular level. In a glass of water from your kitchen, the typical water molecule has a speed of about a thousand miles per hour. Yet the molecules get nowhere fast because they are packed tightly, and, in a single second, each molecule undergoes several *billion* collisions. Like that jiggling dust particle of Figure 10-1, the molecules move in a "drunkard's walk," changing directions continuously.

The nonstop activity of the water molecules means they beat constantly on the sides of the glass. So many molecules slamming into each small area of the glass every second produces a *steady* pressure on the glass surface. What's more, the water molecules themselves feel pressure because of the constant pounding from their neighbors; because these molecules all move in three dimensions, *at any one point in a liquid the pressure is transmitted equally in all directions.*

FIGURE 10-1 A PARTICLE UNDERGOING BROWNIAN MOTION. BOMBARDED BY LIQUID MOLECULES COMING FROM EVERY DIRECTION, PARTICLES SUSPENDED IN LIQUIDS MOVE A LOT BUT GET NOWHERE FAST. FOR THIS REASON, A DROP OF INK CAREFULLY PLACED IN A STILL GLASS OF WATER TAKES A LONG TIME TO COLOR ALL THE WATER.

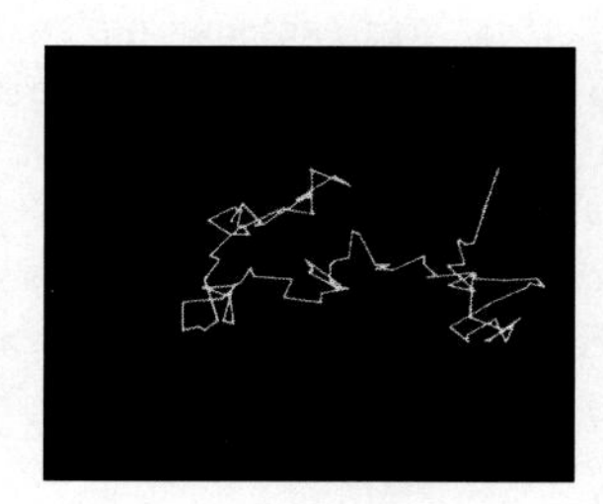

PHYSICS IN 3D/BROWNIAN MOTION

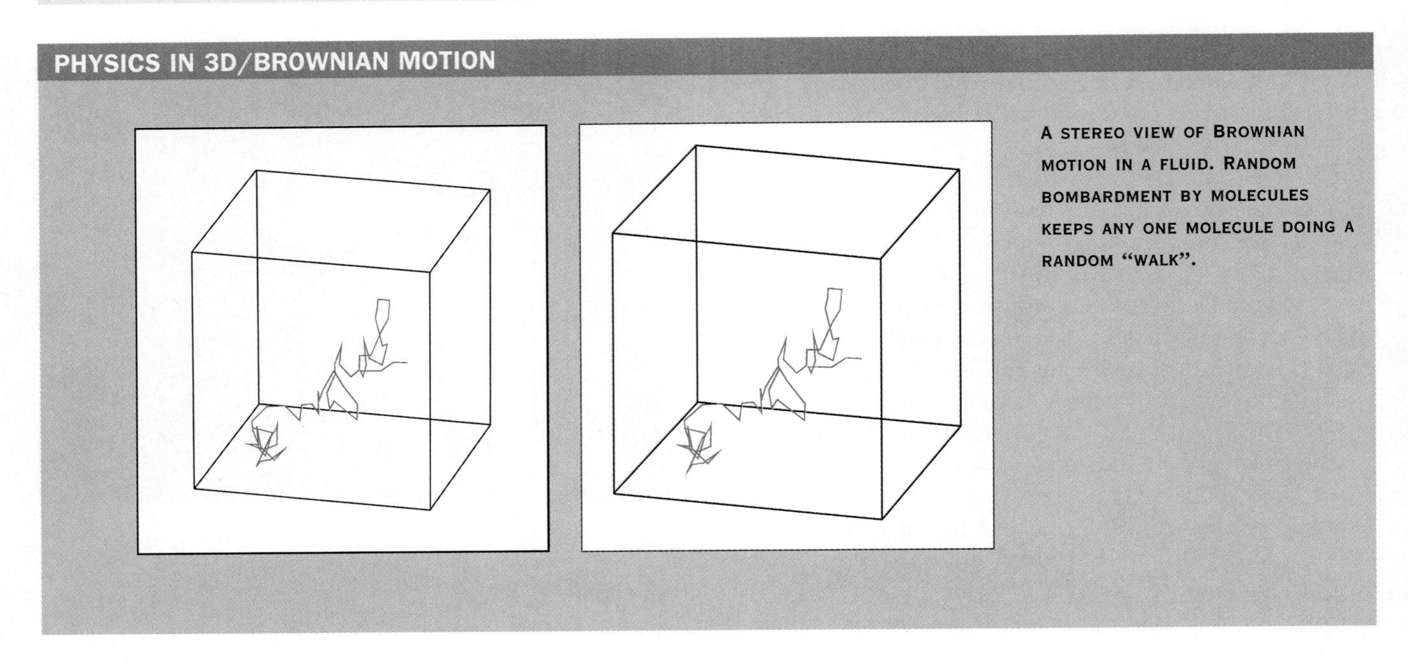

A STEREO VIEW OF BROWNIAN MOTION IN A FLUID. RANDOM BOMBARDMENT BY MOLECULES KEEPS ANY ONE MOLECULE DOING A RANDOM "WALK".

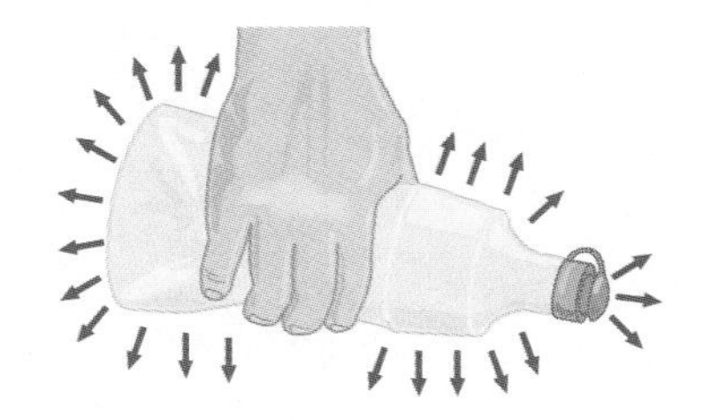

FIGURE 10-2 WHEN YOU SQUEEZE A PLASTIC SHAMPOO BOTTLE, THE PRESSURE YOU APPLY IS TRANSMITTED IN EVERY DIRECTION INSIDE THE BOTTLE.

In a plastic bottle full of liquid soap, there is no extra space between the molecules. If you squeeze the bottle with its cap on tight, you can't compress the liquid. The pressure you apply only shoves one soap molecule into the next throughout the bottle. This demonstrates an important property of fluids: *Pressure applied to a confined liquid (in a container) is transmitted uniformly throughout the fluid.* This is called **Pascal's principle,** after Blaise Pascal (1623–1662), a natural philosopher and theologian. The metric unit of pressure, the pascal (1 Pa = 1 N/m^2), was named for him. If you uncap the bottle and squeeze on the sides, the soap comes out of the open end even though the force you apply is not in that direction. The molecules, no longer completely confined, flow because of the unbalanced pressure. *A fluid will flow from a region of greater pressure toward a region of lesser pressure.*

USING LIQUID PRESSURE

Figure 10-3 shows a liquid form of a lever (Chapter 6). A chamber has two cylindrical holes in its ends. The holes have different diameters, and each is plugged with a tight-fitting, movable piston. Liquid, usually a lightweight oil, fills the chamber. Suppose the smaller piston has an area of 1 square centimeter and someone pushes on it from the outside with a force of 10 newtons. The piston presses on the liquid in the chamber, exerting a pressure $P = F/A$ of 10 newtons per square centimeter. The pressure *everywhere in the liquid goes up by 10 newtons per square centimeter and that pressure is felt all over the face of the large piston. If that larger piston has 5 square centime-*

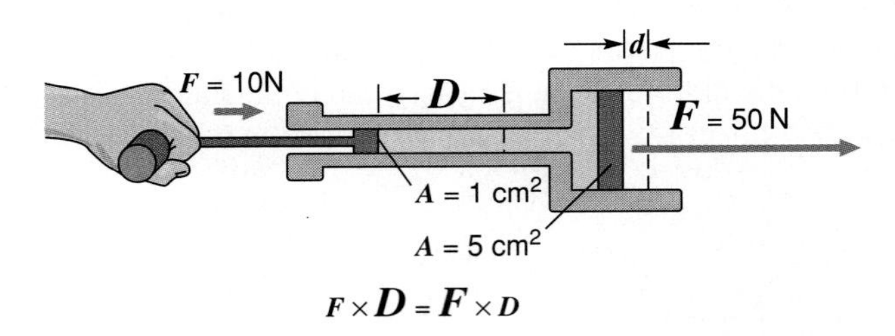

FIGURE 10-3 THE PRESSURE THE SMALLER PISTON EXERTS IS TRANSMITTED THROUGHOUT THE OIL IN THE CHAMBER. BECAUSE THE LARGER PISTON HAS 5 TIMES AS MUCH SURFACE AREA, THE FORCE EXERTED ON IT IS 5 TIMES GREATER THAN THE FORCE EXERTED BY THE SMALLER PISTON. THIS DEVICE IS A LIQUID "LEVER," INCREASING FORCE BY TRADING FORCE FOR DISTANCE. THE SMALLER PISTON HAS TO MOVE 5 CM FOR EVERY 1 CM THE LARGER PISTON MOVES.

FIGURE 10-4 INDUSTRIAL MACHINERY AND FARMING EQUIPMENT OFTEN USE LIQUID (OR HYDRAULIC) "LEVERS" IN ORDER TO APPLY LARGE FORCES AT VARIOUS ANGLES, WHICH WOULD BE DIFFICULT TO ACHIEVE WITH MECHANICAL LEVERS. SINCE PRESSURE IN A FLUID IS TRANSMITTED THROUGHOUT THE FLUID'S CONTAINER, SMALL HOSES CARRYING LIQUID LET FORCE BE TRANSFERRED TO THE PLACE WHERE IT IS NEEDED FOR APPLICATION.

ters of area, the force on the piston is $F = P \times A = 10\ \text{N/cm}^2 \times 5\ \text{cm}^2 =$ 50 newtons. So the larger piston presses outward with a force that is 5 times as great as the force the person applied to the smaller piston.

Though this simple machine multiplies force, you still can't get more work out of it than you put into it—conservation of energy guarantees that. Remember from Chapter 6 that levers and pliers multiply force, too, and in each case there is a trade-off between force and distance (since work = force × distance). In this cylinder the larger piston moves at the most only $\frac{1}{5}$ as far as the smaller piston does. Such liquid, or *hydraulic,* "levers" push cars up on racks at service stations and are the basis for hydraulic brakes on automobiles. They lift drawbridges, move the compressing doors of garbage trucks, push and lift the shovels of bulldozers and backhoes, and lift the long arms of cranes (Figure 10-4).

THE PRESSURE FROM THE WEIGHT OF LIQUIDS

At any depth, the moving water molecules in a glass of water or in the ocean support the weight of the water above them. In a glass full of water, the bottom of the glass supports the force of the water's weight, mg (Figure 10-5). The *pressure* on the bottom of the glass (of area A) is $P = F/A = mg/A$. The mass of the water is its volume V, times its density, m/V, and its volume is its depth d times the area of the bottom. The pressure due to the liquid's weight is:

FIGURE 10-5 THE BOTTOM OF THE GLASS, WHICH HAS AN AREA *A*, SUPPORTS THE WEIGHT OF THE WATER, *MG*. THE VOLUME OF WATER IS EQUAL TO THE DEPTH, *D*, TIMES THE AREA *A*. KNOWING MASS = DENSITY × VOLUME, THE WEIGHT OF THE WATER, *MG*, IS EQUAL TO ITS DENSITY TIMES ITS VOLUME TIMES *G*.

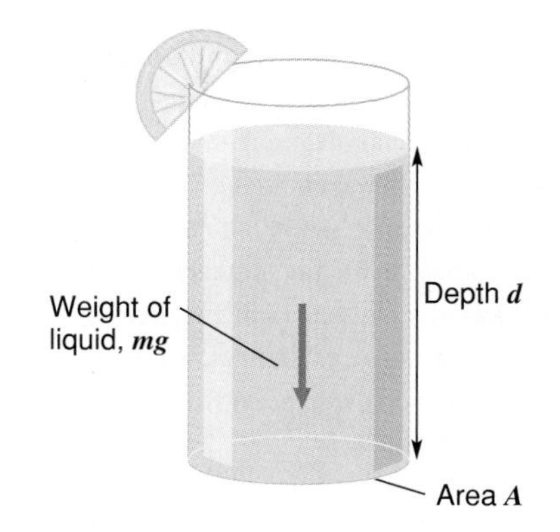

PUTTING IT IN NUMBERS—PRESSURE FROM THE WEIGHT OF A FLUID

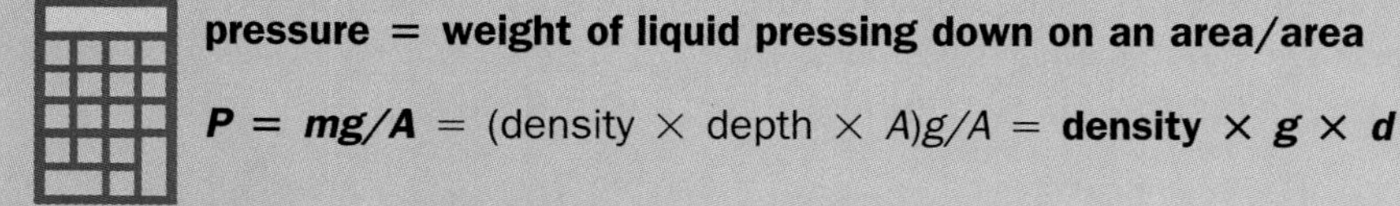

pressure = weight of liquid pressing down on an area/area

$P = mg/A$ = (density × depth × $A)g/A$ = **density × g × d**

Example: Water's density is 1000 kg/m³. Find the water pressure on a swimmer's foot that is 1.5 meters beneath the surface in a swimming pool, as in Figure 10-6. P = density × $g \times d = 1000\ \text{kg/m}^3 \times 9.8\ \text{m/s}^2 \times 1.5\ \text{m} =$ **14,700 N/m² (2.1 lb/in²)**

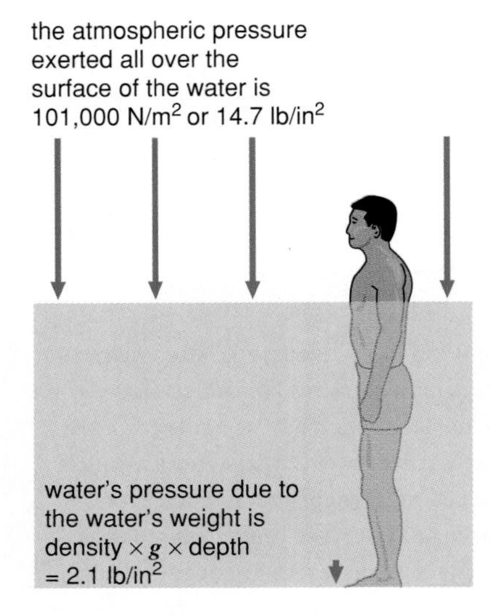

FIGURE 10-6 THE PRESSURE DUE TO THE WEIGHT OF THE WATER ABOVE THIS PERSON'S FOOT IS ONLY 2.1 LB/IN^2, AS CALCULATED IN THE TEXT. HOWEVER, THE AIR'S PRESSURE, WHICH IS EXERTED ALL OVER THE SURFACE OF THE WATER, IS 14.7 LB/IN^2 (AT SEA LEVEL). THAT PRESSURE IS TRANSMITTED THROUGHOUT THE WATER AS WELL, SO THE TOTAL PRESSURE ON HIS TOES IS 16.8 LB/IN^2.

This formula is correct *no matter what the shape of the container.* Note that pressure is directly proportional to depth. For instance, if the depth of a liquid is tripled, the pressure at the bottom will be tripled. For another example, the pressure halfway down a glass full of water is half the pressure at the bottom, *regardless* of the shape of the glass.

The pressure calculated in the example above is only the pressure due to the liquid's weight. The atmosphere above a swimming pool or a glass of water also exerts pressure on the water's surface, and that pressure is also transmitted throughout the liquid, though it does not appear in the formula above. The atmosphere's pressure is relentless above any surface of water, keeping the molecules under great pressure. (We'll see later that if it were not for this air pressure, the oceans would evaporate!) The *total* pressure on the foot in Figure 10-6 is about 2.1 lb/in^2 + 14.7 lb/in^2 if the pool is at sea level. Since the air's pressure is everywhere around us we often ignore this extra pressure if it doesn't influence the effect we are studying.

Some towns use elevated water towers to supply water to homes and businesses (Figure 10-7). Because pressure = density × g × depth, the water pressure at a home faucet depends on the vertical distance (the depth) between the faucet and the surface of the water in the tower. Larger city systems use pumps at points along the water lines to supply extra pressure.

FIGURE 10-7 WATER ELEVATED IN A TOWER SUPPLIES WATER UNDER PRESSURE TO NEARBY HOMES AND BUSINESSES. THE GREATER THE HEIGHT OF THE WATER ABOVE THE OUTLET, THE GREATER THE PRESSURE. WATER PUMPS ARE USED TO PUSH THE WATER UP INTO THE STORAGE TOWER, AND THEN THE WEIGHT OF THE WATER SUPPLIES THE PRESSURE TO DELIVER IT TO LOWER ELEVATIONS.

BUOYANCY: ARCHIMEDES' PRINCIPLE

If you put your arm under water, the water pressure from above pushes it downward while water pressure from beneath pushes it upward (Figure 10-8). The water below your arm, however, is farther below the surface than the water above your arm. That means the upward pressure exerted by the water on the bottom of your arm is greater than the downward pressure on the top of your arm. Consequently, *your arm gets a net push from the water in the upward direction.* That net push from the water's pressure is the **buoyant force,** F_b. If you get into the water up to your shoulders, you can stand on the very tips of your toes with no trouble at all—a feat few of us can manage on dry land. The buoyant force helps support you that much! It is easy to hold a friend up in a swimming pool, too, because F_b counteracts much of your friend's weight.

Simple experiments show exactly how large the buoyant force is. Carefully place a toy boat on the surface of a container filled to the brim with water and catch the overflow (Figure 10-9). Since the floating boat neither rises nor falls, the net force on it is zero. The buoyant force, then, is equal to the boat's weight and points in the opposite direction. Weigh the overflow you collected, and you'll find it is equal to the boat's weight. *The buoyant force on a floating object is equal to the weight of the displaced fluid.*

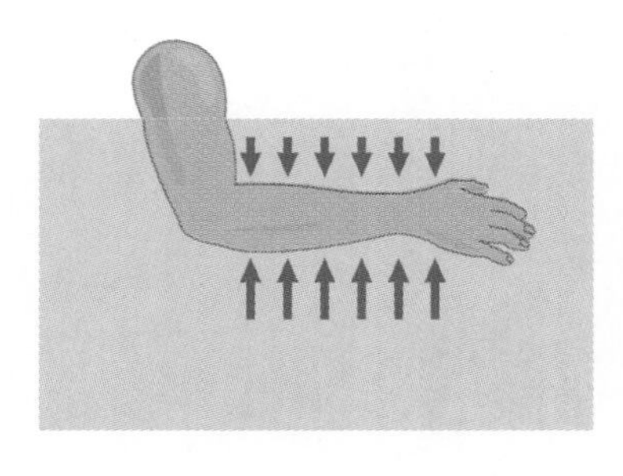

FIGURE 10-8 THE UPWARD-POINTING PRESSURE ON THE BOTTOM OF THE ARM IS GREATER THAN THE DOWNWARD-POINTING PRESSURE ON THE TOP OF THE ARM. THIS MEANS THE WATER EXERTS A NET UPWARD FORCE, THE BUOYANT FORCE, ON THE SUBMERGED ARM.

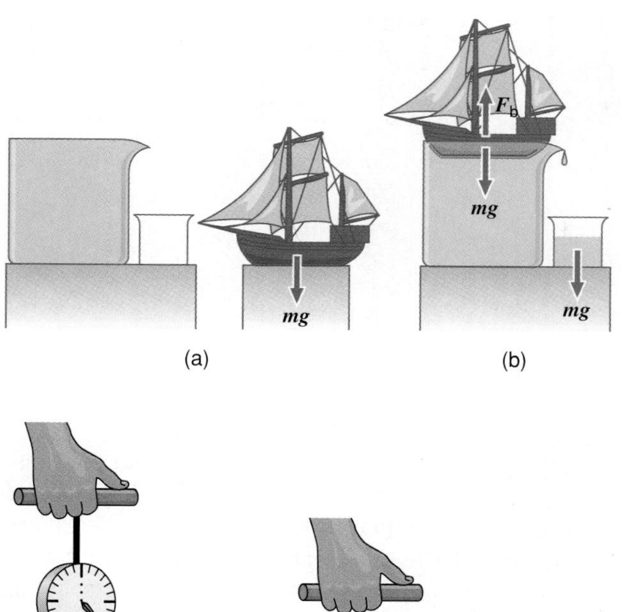

FIGURE 10-9 AN EXPERIMENT TO DEMONSTRATE THAT THE WEIGHT OF THE WATER DISPLACED BY A FLOATING OBJECT IS EQUAL TO THE BUOYANT FORCE OF THE WATER ON THE OBJECT. IN (A) THE LARGE BEAKER IS FULL OF WATER. IN (B) THE BOAT HAS BEEN CAREFULLY PLACED IN THE WATER SO THAT ALL OF THE OVERFLOW IS COLLECTED IN THE SMALLER BEAKER. THIS OVERFLOW IS THE VOLUME OF WATER DISPLACED BY THE BOAT, WHICH IS THE VOLUME OF THE BOAT THAT IS BELOW THE SURFACE. THE WEIGHT OF THIS DISPLACED WATER IS EQUAL TO THE BUOYANT FORCE ON THE BOAT.

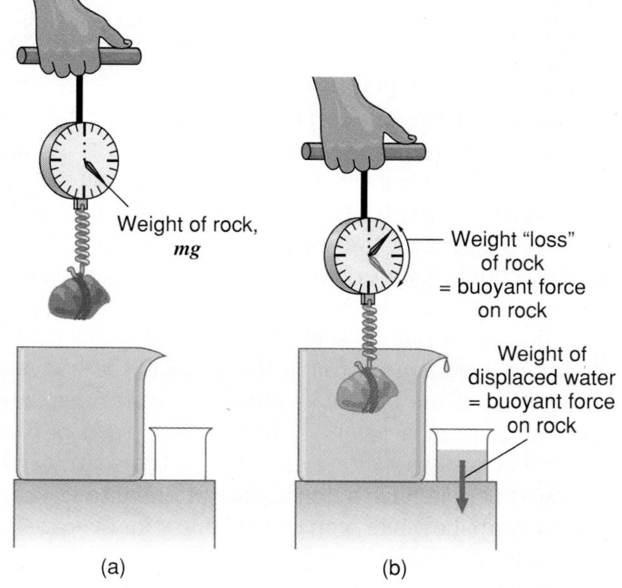

FIGURE 10-10 AN EXPERIMENT TO DEMONSTRATE THAT THE WEIGHT OF THE WATER DISPLACED BY A SUBMERGED OBJECT IS EQUAL TO THE BUOYANT FORCE OF THE WATER ON THE OBJECT. THE WEIGHT OF THE ROCK BEFORE IT IS IMMERSED MINUS THE BUOYANT FORCE (THE WEIGHT OF THE DISPLACED WATER CAUGHT BY THE SMALLER BEAKER) GIVES THE NET FORCE ON THE ROCK WHEN IT IS UNDER WATER AS MEASURED BY THE SPRING SCALE.

What is the buoyant force on a totally submerged object? Immerse a rock suspended from a spring scale; its pull on the scale is less than its weight mg (Figure 10-10). The difference between its weight and the pull it exerts on the scale when it is under the water is the buoyant force on the rock. The volume of displaced water this time is equal to the submerged rock's volume, and the weight of that displaced water is once again equal to the buoyant force. Just as for the floating boat, *the buoyant force on a submerged object equals the weight of the displaced fluid.* This is called **Archimedes' principle.**

FLOATING, SINKING, AND DIVING

If you hold a bar of soap under water and then let it go, two forces act on it, its weight and the buoyant force. They push in opposite directions, and their difference is equal to the net force on the soap (Figure 10-11). If the buoyant force is exactly equal to the soap's weight, the soap will hover wherever you leave it under the surface. If the buoyant force is less than the soap's weight, the soap will sink. If the buoyant force is greater than the weight of the soap, the bar will rise to the surface and float.

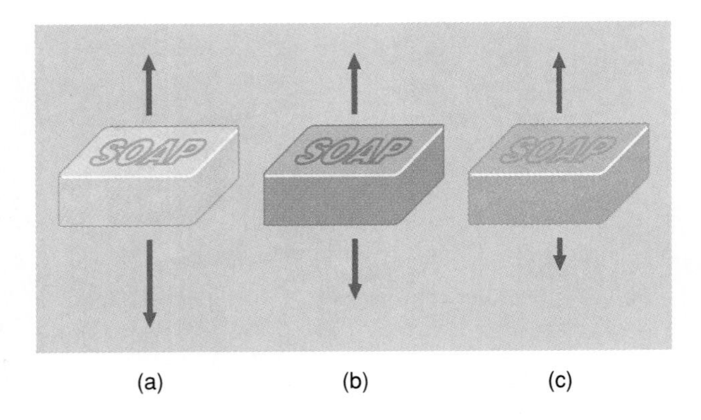

FIGURE 10-11 THREE BARS OF SOAP WITH IDENTICAL VOLUMES BUT DIFFERENT DENSITIES ARE ALL UNDER WATER. SINCE THEY ALL DISPLACE THE SAME VOLUME OF WATER, THE BUOYANT FORCE IS THE SAME ON ALL THREE. THEIR DIFFERENT DENSITIES, HOWEVER, MEAN THEIR WEIGHTS ARE DIFFERENT. (A) WITH A DENSITY GREATER THAN WATER'S DENSITY, THIS BAR HAS A NET FORCE POINTING DOWN AND WILL SINK. (B) THIS BAR'S DENSITY IS EQUAL TO WATER'S DENSITY, SO IT HOVERS, NEITHER SINKING NOR RISING; THE NET FORCE ON IT IS ZERO. (C) BECAUSE ITS DENSITY IS SMALLER THAN WATER'S, THIS BAR HAS A NET FORCE UPWARD AND WILL RISE TO FLOAT AT THE SURFACE.

What determines whether the buoyant force is greater or less than the weight? To answer this question, recall that the submerged soap displaces a volume of water equal to its own volume. Therefore any difference between the weight of the displaced water and the weight of the soap *is due to a difference in density.* If the density of the soap is greater than the density of the water, the bar weighs more than the water it displaces. That means the bar's weight is greater than the buoyant force, and the bar sinks. If the soap's density is less than the water's, the soap rises. A hovering bar of soap has a density equal to water's density. Even the smallest difference in density will cause something to rise or sink.

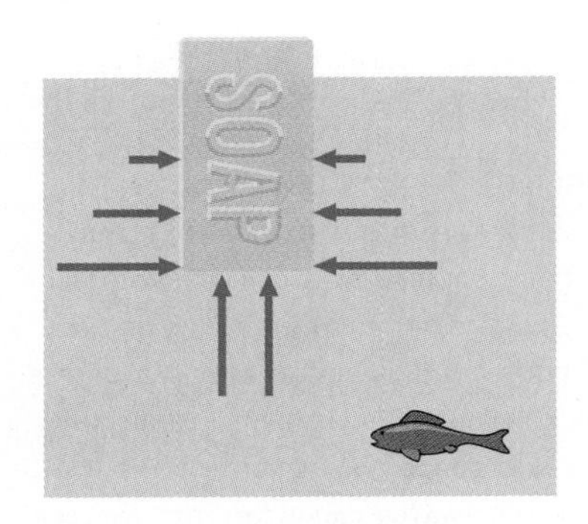

FIGURE 10-12 THE WATER PRESSURE ON THE BOTTOM OF A FLOATING OBJECT SUPPORTS IT. NOTICE THERE IS NO NET FORCE IN THE HORIZONTAL DIRECTION. (NOT SHOWN IS THE WEIGHT OF THE SOAP.)

Suppose a bar of soap rises to the surface of the water, as in Figure 10-12. As it pops up through the surface, the buoyant force changes because the portion of the bar that's above the water displaces no water. The bar will settle to float at a level where its submerged portion displaces a volume of water that weighs exactly as much as the soap weighs.

Most people have a density about equal to water's density, 1 g/cm^3. Unlike a bar of soap, however, you can change your average density by a small amount. In a bathtub almost full of water, breathe in deeply and you will float higher in the water. Then exhale as far as possible and you'll sink. Expanded lungs add volume to your body, and since a breath of air has little mass, your average density, m/V, decreases. Likewise, deflated lungs mean your body's volume is less, so your average density increases. The hippopotamus in Figure 10-13 uses this principle to forage on the bottom of a riverbed.

SCUBA IS AN ACRONYM FOR SELF-CONTAINED UNDERWATER BREATHING APPARATUS.

When a scuba diver goes under water, the water pressure on the diver's body increases as the diver descends. At a depth of about 34 feet in fresh

Table 10·1 Densities of Some Liquids

Liquid	Density (g/cm^3)[a]
Gasoline	0.66–0.69
Ethyl alcohol	0.791
Kerosene	0.82
Turpentine	0.87
Olive oil	0.918
Castor oil	0.969
Water (at 4°C)	1.0
Sea water	1.025
Milk	1.028–1.035
Mercury	13.6

[a] To convert to kg/m^3, multiply by 1000.

water, the pressure from the water's weight equals an atmosphere of pressure, almost 15 pounds per square inch. Yet divers go to depths of 150 feet or more, adding 4 or 5 atmospheres of water pressure all over their bodies.

Except for our lungs, our bodies, being mostly liquid, are almost incompressible. To keep the water's great pressure from compressing the lungs, the pressure of the compressed air a diver breathes (through a "regulator") must match the water pressure at the diver's depth. Otherwise, the extra water pressure transmitted through the diver's body would crush the lungs.

IF YOU CAN FLOAT, YOUR DENSITY IS SLIGHTLY LESS THAN THE DENSITY OF THE WATER YOU ARE FLOATING IN. FLOATING IN THE OCEAN IS A BIT EASIER THAN FLOATING IN A LAKE BECAUSE SEA WATER'S DENSITY IS SLIGHTLY GREATER THAN FRESH WATER'S DENSITY. THE DRINK BALANCED ON PAULA'S STOMACH AFFECTS HOW HIGH SHE FLOATS IN THE WATER. HER BODY MUST SINK A TINY AMOUNT TO DISPLACE AN EXTRA AMOUNT OF WATER THAT WEIGHS AS MUCH AS THE DRINK DOES.

LIQUIDS IN MOTION: WHAT HAPPENS AT A BOUNDARY

Along the banks of a river, flowing water meets solid earth, and the result is friction. Any water molecules pressed firmly against the bank won't move along, any more than you can move your hand sideways if you press it hard against a sheet of sandpaper. In other words, the speed of the layer of liquid molecules next to a solid surface is *zero.* Those immobile liquid molecules in the *boundary layer* slow the movement of the liquid molecules next to them, which in turn slow their neighbors, and so on. Farther from the boundary, the water flows more freely.

Because of this boundary effect on a flowing fluid, when a film of oil is used to lubricate the moving parts in a car engine or an electric motor, it must be at least a few molecular layers thick to work effectively. The oil molecules can't move freely and provide lubrication unless they are several layers away from the two passing metal surfaces (Figure 10-14).

FIGURE 10-13 A HIPPOPOTAMUS IS BUOYANT AND HAS TO EXHALE VIOLENTLY BEFORE IT CAN DESCEND TO EAT VEGETATION ON THE BOTTOM OF A RIVER. TO SURFACE AGAIN, IT MUST PUSH UPWARD AND SWIM OR TAKE AN UNDERWATER WALK TO A SHALLOW AREA OF THE RIVER, PERHAPS AT THE RIVER'S EDGE.

VISCOSITY AND TURBULENCE

Friction with boundaries isn't the only thing that slows a liquid. Within the liquid, passing molecules—or groups of molecules—collide with each other, interfering with the flow. This internal friction is called **viscosity.** It is greater for liquids like ketchup and syrup and less for those like water and gasoline. The thick liquid we know as toothpaste stays on the top of the bristles of a toothbrush because of its very high viscosity. Viscosity depends greatly on temperature because the faster molecular motion at higher temperatures

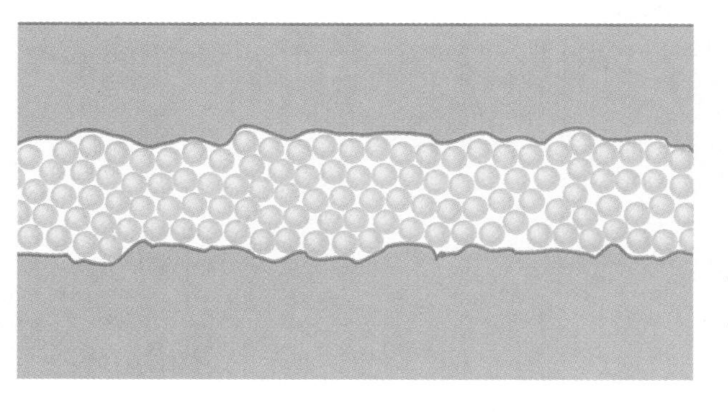

FIGURE 10-14 BOUNDARY EFFECTS LIMIT LUBRICATION UNLESS THERE ARE AT LEAST SEVERAL LAYERS OF LUBRICANT MOLECULES BETWEEN THE TWO SOLID SURFACES. AS ILLUSTRATED HERE, THE LAYER OF MOLECULES NEXT TO THE SOLID IS IMMOBILIZED DUE TO FRICTION, AND THEY SLOW THE PASSAGE OF THE NEXT LAYER OF LUBRICANT MOLECULES, AND SO ON.

FIGURE 10-15 VERY VISCOUS AT 80°F, AS SEEN WHILE IT IS BEING MIXED ON THIS MACHINE, SALT WATER TAFFY JUST FROM THE REFRIGERATOR IS ROCK HARD. THE VISCOSITY OF HONEY ALSO INCREASES DRAMATICALLY WHEN IT IS COLD, ANOTHER FAMILIAR EXAMPLE OF THIS PROPERTY OF LIQUIDS.

FIGURE 10-16 WHEN WATER IS PUSHED ENERGETICALLY WITH A PADDLE, AN EDDIE APPEARS. THIS SIGNIFIES TURBULENT MOTION, AS OPPOSED TO SMOOTH, OR LAMINAR, FLOW OF THE WATER.

makes it easier for the molecules to slide past one another. For instance, honey that's initially very slow to move can be made to flow freely by placing the jar in warm water. The taffy of Figure 10-15 provides another example of how temperature affects viscosity. Straight from the refrigerator, a piece of taffy is as hard as a rock, but once in your mouth it becomes soft and chewable.

The relative speeds between different portions of a moving stream can promote internal friction. This can cause a *smooth* (or *laminar*) flow to become **turbulent,** which means that small whirlpools, or eddies, appear (Figure 10-16). Eddies interfere with the stream's passage, sending some of the liquid sideways and some even backwards, retarding the flow. In a pipe or tube of small diameter, there is little distance between any edge (where the stream's speed is zero) and the center (where the flow is fastest), so turbulence occurs for even small rates of flow. Once turbulence starts, extra pressure of the liquid at one end of the pipe does not cause an increase in the rate of flow. Any extra work (or energy) given to the stream just goes into making more eddies.

When a liquid flows smoothly through a pipe, the boundary effect and viscosity influence the rate of flow. If the diameter of the pipe is large, the flow at the center isn't influenced as much by the friction at the sides, and the average flow is therefore much faster.

SURFACE TENSION OF A LIQUID

Inside a drop of dew (or of any other liquid), a molecule is attracted almost symmetrically by all its neighbors, as Figure 10-17 shows. For molecules at the surface of the drop, however, the attractive forces from neighboring molecules are unbalanced. There is a net force directed toward the inward side,

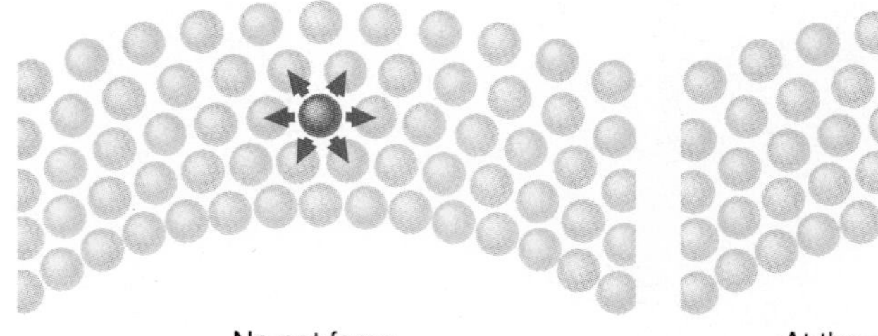

FIGURE 10-17 (A) INSIDE A VOLUME OF WATER, A SINGLE MOLECULE FEELS ATTRACTIONS FROM ALL SIDES BY ITS NEAREST NEIGHBORS. THE FORCES ADD TO ESSENTIALLY ZERO ON THE AVERAGE. (B) A MOLECULE AT THE SURFACE FEELS UNBALANCED FORCES FROM ITS NEIGHBORS, AND THE NET FORCE (RED) POINTS BACK INTO THE WATER. THIS IS THE CAUSE OF SURFACE TENSION.

CHECKING YOUR BLOOD PRESSURE WITH TURBULENCE

The flow of blood through your arteries and veins is ordinarily smooth. However, a constriction in a major artery causes turbulent flow as the blood squeezes through, a fact used to measure blood pressure. Your blood pressure consists of two numbers—"120 over 80," for instance. Let's see what these two values refer to. To measure your blood pressure, a cuff with a pressure gauge circles the arm. Pressure is increased until the flow of blood through the main artery in the arm is cut off. With no flow, the person listening with the stethoscope hears nothing. Then the pressure exerted by the cuff is slowly reduced, and when the maximum blood pressure caused by the beat of the heart is great enough to squeeze the blood through the compressed artery, the turbulent flow is heard through the stethoscope. The pressure at that point is the larger number (in millimeters of mercury, where 1 mm of mercury is equivalent to 133 Pa) of the blood pressure reading, and it is called the *systolic* pressure. As the cuff pressure continues to decrease, the artery gradually assumes its normal diameter, the flow becomes smooth again, the turbulence and noise vanish, and the person listening again hears nothing. The pressure at this point is the *diastolic* pressure, which is the blood's pressure in your arm between heartbeats. This is the lower number in the reading.

into the dewdrop. It is as if there is an invisible skin at the surface, making it difficult for molecules there to escape or even for a jostled molecule from below to squeeze its way into the surface layer. So the surface of a liquid is kept under tension, much like the tension in an inflated balloon; we call it **surface tension.** Because of this effect, a liquid resists any change in shape that would add surface area.

A small dewdrop on a leaf is close to spherical (Figure 10-18); its surface tension pulls it into that shape because a sphere has less surface area per unit of volume than any other shape. A larger drop lying on the same leaf is much flatter because the drop's weight presses it downward, overcoming the smaller force of surface tension. In a volume as large as a glass of water, the surface of the water is almost flat. In space flights, however, where everything is in free fall, very large drops of water pull themselves into spheres as they float in the air of the cabin.

The strength of surface tension varies greatly from liquid to liquid because of the differences in the attractive force among the molecules. Water has two to four times as much surface tension as gasoline, ether, or alcohol. If you've ever broken a mercury thermometer, you've probably noticed how persistently the small drops of mercury stay together as they roll around; that's because mercury has about seven times as much surface tension as water.

Oil and water don't mix because water molecules attract one another so much more than they are attracted to oil molecules. A duck floats high on the

FIGURE 10-18 (A) DEWDROPS ON A LEAF, SHAPED BY SURFACE TENSION AND GRAVITY. FOR THE SMALLER DROPS, SURFACE TENSION IS THE LARGER INFLUENCE. FOR THE LARGER DROPS, GRAVITY'S INFLUENCE APPEARS, FLATTENING THE DROPS CONSIDERABLY. (B) SURFACE TENSION PULLS LARGE DROPS OF WATER INTO SPHERES IN THE SPACE SHUTTLE.

(a)

(b)

FIGURE 10-19 ANHINGAS DRY THEIR WINGS AFTER FISHING UNDER WATER. LACKING OIL, THEIR FEATHERS ARE SATURATED WITH WATER EACH TIME THEY FEED BENEATH THE SURFACE.

water because the air trapped in its feathers makes the duck buoyant. Oil on the feathers repels the water, and a duck preens regularly to distribute the oil from its oil glands. If a baby duck tries to swim in a bathtub of soapy water, the oil dissolves, the feathers become saturated with water, and the baby duck sinks! Another waterbird, the anhinga (Figure 10-19), lacks oil glands to provide water repellency. Each time it dives to feed on fish, water seeps into its feathers, and they get soaked. When the bird surfaces, it struggles out of the water and flies heavily away. Then it spreads its wings and dries them in the sun before its next fishing trip.

COHESION, ADHESION, AND CAPILLARY ACTION

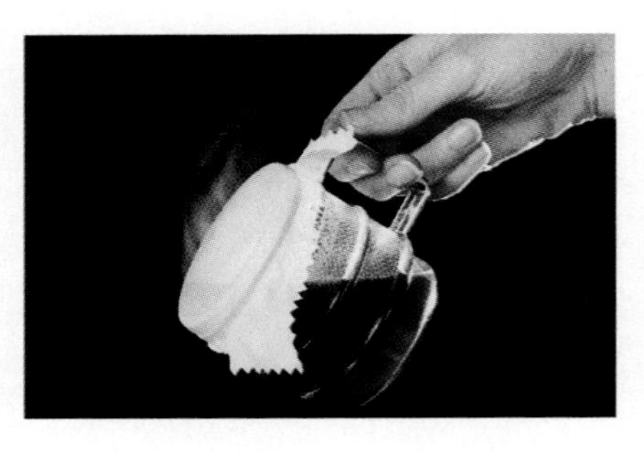

THE HOLES IN THE FILM OF GORE-TEX ® ARE SO SMALL THAT LIQUID WATER (OR HERE, COFFEE) CANNOT PASS THROUGH THEM DUE TO WATER'S COHESION. WATER VAPOR, HOWEVER, DOES PASS THROUGH EASILY, MAKING SUCH FILMS GOOD FOR RAIN GEAR.

Dip the corner of a paper towel into a spot of water on a countertop and watch the water climb. The forces that make this happen are the attractive forces between molecules. Molecules of all kinds attract each other. The attraction between molecules of the *same* kind is called **cohesion,** while the attraction between *different* kinds of molecules is called **adhesion.**

When a drop of water is put onto a horizontal plate of glass, the water curves outward at the edges of the drop where the water meets the glass. That's because the water molecules are attracted more to the molecules of glass than they are to each other. We say the water *wets* the glass. If water's surface tension were less, the drop would creep over the plate's surface to cover it with a very thin film, as a drop of oil or gasoline does. When a drop of mercury is put on the same glass, the mercury turns inward where it touches the glass. In mercury, the force of cohesion is greater than the force of adhesion with glass, and the mercury is therefore *nonwetting.*

Strong adhesive forces let wetting liquids rise against the pull of gravity into small vertical spaces. For example, if a small, hollow glass tube is held vertically so that its lower end dips into water, water molecules at the edge of the inside surface of the tube adhere to the glass and rise into the tube.

THE LIQUID INSIDE YOU

About once a second, your heart muscle contracts, squeezing any blood it contains and pushing it into the aorta, the large artery connected to the heart. Like any other liquid, blood is nearly incompressible. As a result, when your heart muscle pushes out the blood in your heart, the blood throughout your body moves.

When you are standing, the pressure of the blood in your feet is greater than in your brain because pressure from the weight of any fluid depends upon depth. Lie on the floor for a minute and then stand up quickly. You might feel a little lightheaded until your heart increases the pressure to pump blood up to your head. You can get the same feeling at the fair if a ride accelerates you upward. Along with everything else in your body, your blood becomes "heavier" with an upward acceleration, and then it's harder for the heart to pump it up to your head. If the acceleration of the ride is so great that your heart isn't strong enough to pump the heavier blood to your head, you will have a blackout. For most of us this would happen at over 3 *g*'s if standing and at a somewhat greater acceleration if sitting. Fighter pilots with special suits can withstand 7*g*'s for short periods of time.

Your heart works all the time to pump blood, against the force of gravity, from your feet up to your scalp. To keep the rising blood from draining back down between heartbeats, small bands of muscles, called shunts or cuffs, clamp down around the ends of the smaller vessels, ensuring that blood flows in only one direction. When one of these one-way valves fails for some reason, the blood pools, and the result is a varicose vein.

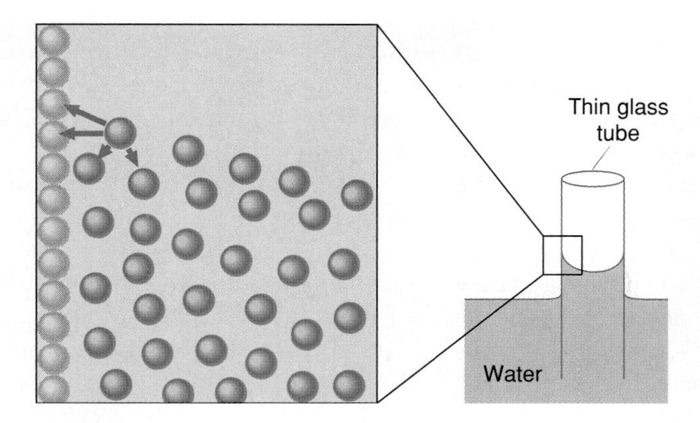

FIGURE 10-20 LIQUID RISES IN A CAPILLARY TUBE WHEN THE FORCE OF ADHESION TO THE WALLS IS GREATER THAN THE FORCE OF COHESION AMONG THE LIQUID MOLECULES. THE STRONGER ATTRACTION FOR THE GLASS LIFTS WATER MOLECULES AT THE EDGE, AND THEY DRAG OTHERS UP BEHIND THEM.

As a result, the water surface inside the tube is curved. Surface tension at the upward-curved edges pulls up on the rest of the surface layer. The edges of the water surface move upward by adhesion, and as they move, cohesion causes them to pull a column of water along behind themselves. The column rises until the forces of adhesion are balanced by the weight of the fluid that is pulled up in the tube. (See Figure 10-20.) This process is called **capillary action.**

When you light a candle, the heat of the flame melts the wax at the top of the candle. Then capillary action pulls the melted wax up through the wick, where it burns in the flame. In a kerosene lantern, the kerosene rises up a braided cloth wick. Certain fabrics in clothes can also "wick" perspiration away from the body through capillary action.

THE MIXING OF GASES AND LIQUIDS

Whenever a liquid and a gas meet, as at the surface of a lake, some of the bombarding gas molecules penetrate the liquid's surface, or **dissolve** in the liquid. At the same time, some of the liquid molecules **evaporate** into the gas. As you might expect, more gas dissolves in the liquid if the gas is under a lot of pressure. At any given temperature and pressure, there is a maximum quantity of gas that will remain dissolved in a liquid. When that level is reached, the solution is **saturated** with that gas. Then if either the pressure is reduced or the temperature is increased, gas will come out of the liquid. This happens when you uncap a carbonated drink, which reduces the pressure on its surface. Dissolved CO_2 comes out in bubbles that you both see and hear.

Fish absorb dissolved oxygen through their gills. Much of this oxygen comes from the air above the lakes, ponds, and oceans, and some is released by aquatic plant life. Tumbling mountain streams are especially rich in dissolved oxygen, since air is bubbled through the falling waters. In a home aquarium, an air pump keeps enough oxygen dissolved in the water to support the fish. In some springs that come from deep underground, the water has very little dissolved oxygen. No fish can live in these oxygen-poor springs. As the water flows away, though, air dissolves into its surface, and a short distance down the stream fish can thrive. On the other side of the coin, dissolved oxygen is deadly to most forms of an anaerobic bacteria, the kind that abounds in sewage. Treatment plants aerate water as part of their process.

THE TUMBLING OF WATER IN A MOUNTAIN STREAM INCREASES THE AMOUNT OF AIR DISSOLVED IN THE WATER. FISH SUCH AS TROUT, WHICH NEED MORE OXYGEN THAN MANY OTHER SPECIES, THRIVE IN THE OXYGENATED WATER.

DEADLY LAKES

In the mountains of Cameroon in equatorial Africa, a series of deep lakes lie in the large central vents (called calderas) of old volcanoes. In the 1980s disaster struck at two of these, Lake Manoun and Lake Nios. A landslide at the shoreline plunged to the bottom of Lake Manoun, mixing the water. The churning of the deep layers released large amounts of dissolved carbon dioxide gas, absorbed from springs at the lake's bottom. Surfacing, this heavier-than-air gas hugged the ground as it spread over the nearby area, pushing the oxygen-bearing air away. The people and animals of a lakeside village suffocated before winds could disperse it.

At Lake Nios there was no landslide; investigators think CO_2 had reached the saturation level in the lake and that the gas was released suddenly when the water could absorb no more, erupting something like champagne does if the cork is popped when the bottle is too warm. The water churned so violently as the CO_2 surfaced that waves swept 60 feet above the lake shore. In a matter of minutes, the gas covered a ten square-mile region. Over 1700 people died in a village 2 miles from the lake, along with uncounted livestock. In the future, scientists hope to artificially induce mixing to keep the CO_2 levels of those Cameroon lakes under control.

DEADLY GASES IN SOLUTION

Summary and Key Terms

Brownian motion–141
Pascal's principle–142

Molecules of water have billions of collisions per second. These collisions occur in all directions, as seen in ***Brownian motion***, the random lurching of a microscopic particle suspended in a liquid. ***Pascal's principle*** says that pressure applied to a confined fluid is transferred to every point in the fluid. The pressure at a depth beneath the surface of a liquid due to its weight is $P =$ **density** $\times$ ***g*** $\times$ **depth**.

buoyant force–144
Archimedes' principle–145

The ***buoyant force*** is the upward force that a fluid exerts on an object either totally or partially immersed in it. ***Archimedes' principle*** states that *the buoyant force on a floating or submerged object is equal to the weight of the fluid it displaces.* If the object's density differs from that of the fluid, there is a difference in weight between the object and the fluid it displaces, and the object rises or sinks.

viscosity–147
turbulence–148

Friction retards the speed of the layer of liquid molecules next to a solid surface. These immobile molecules slow the molecules next to them, and so on to zero. Liquid flows also slow due to ***viscosity***, internal friction arising between the liquid's own molecules. Viscosity decreases as temperature rises because faster molecules slide past one another more easily. ***Turbulence*** occurs because of relative motion between different parts of a stream, initiating small eddies in the liquid.

surface tension–149
cohesion–150
adhesion–150
capillary action–151

Liquids have ***surface tension*** because the net molecular attractions on surface molecules pull them into the liquid. ***Cohesion*** is the attraction between molecules of the same type, while ***adhesion*** is the attraction between molecules of different types. Strong adhesion can cause ***capillary action***, the rising of a wetting liquid into a vertical space.

dissolve–151
evaporate–151
saturated–151

Whenever a liquid meets a gas, some of the bombarding gas molecules ***dissolve*** into the liquid and some of the liquid molecules ***evaporate*** into the gas. At any given temperature and pressure, there is a level at which a liquid becomes ***saturated***, which means that the maximum quantity of a gas is dissolved in that liquid. Earth's lakes and oceans serve as storehouses for various gases, including oxygen, which fish and other aquatic life need to survive.

EXERCISES

Concept Checks: Answer true or false

1. Because of its weight, a liquid exerts pressure only in the downward direction.
2. An applied pressure on an enclosed liquid is passed along to every molecule in the liquid, no matter what shape the liquid's container has.
3. At any point in a confined liquid, pressure is transmitted in every direction, even upward.
4. The pressure in a glass of liquid changes with depth. Besides that, the value of the pressure depends on the density of the liquid and on the value of g.
5. When a solid object is immersed in a liquid, the buoyant force on it is exactly equal to the weight of the liquid that the object displaces.
6. An object that's held under water can rise to float only if the weight of the water it displaces is greater than the object's weight.
7. Because of friction between molecules of the liquid and the solid, the speed of a liquid at a stationary boundary is zero.
8. Viscosity is a measure of the internal friction in a liquid, while turbulence decreases the flow rate of a liquid by causing eddies to form.
9. Surface tension gives small liquid drops their spherical shape and resists any increase in the surface area of the drops.
10. The adhesive force is between molecules of the same type, while the cohesive force is between molecules of different types.
11. The ability of some liquids to rise against the pull of gravity into small vertical spaces is called capillary action. In this action, adhesion is greater than cohesion.

Applying the Concepts

12. Do people weigh less when they are swimming? Discuss.
13. Ice floats in water. What does this tell you about the density of ice? (See Tables 9-2 and 10-1.) Would ice float in kerosene? Gasoline?
14. If a fish is motionless under water, is the buoyant force on it equal to its weight?
15. When an ice cube melts in a glass of water, does the water level rise?
16. Explain why a steel battleship floats in water even though the density of steel is much greater than that of water.
17. Explain why toothpaste comes out of the end of an open tube even though you squeeze the tube from the side.
18. Have you ever stood up quickly after lying down for a while and felt faint? What causes this? Why does your heart do less work when you lie down?
19. Giraffes have systolic blood pressures of 350 mm and hearts that weigh about 25 pounds. What feature of the giraffe makes that necessary?
20. When you pour ketchup, at what part of the container's opening does it move fastest? Why?
21. If ketchup pours too slowly, what can you do to decrease its viscosity?
22. People who snorkel know that after a while the act of breathing becomes tiring. What causes this to happen?
23. If fresh water weighs 62.4 pounds per cubic foot, how much does a cubic foot of seawater weigh? See Table 10-1.
24. If you go twice as deep in liquid, do you double the pressure due to the liquid's weight? If g were twice as large, what would happen to the pressure in a liquid? If one liquid is twice as dense as another, how do their pressures compare at the same depth?
25. Why are towels dried on a clothesline stiff and scratchy while those dried in a clothes dryer are soft and fluffy?
26. Why does a warm soft drink emit a larger "whoosh" when it is opened than an identical but colder drink?
27. Traveling downstream in a canoe, where should you keep the canoe in the stream to take advantage of the current? Why? Traveling upstream, where should you put the canoe? Why?
28. Why is hot soapy water better for cleaning than cold soapy water?
29. What holds contact lenses on the eyes?
30. Ships are charged according to weight when they pass through the lock systems of some canals. The canal tenders read from markers on the side of a lock how many feet or meters of water the ship displaces in the lock, and this reading can be converted to the volume of water the ship displaces. If a freighter displaces 1750 cubic feet of ocean water, how much does it weigh?
31. It is easier to pick up an egg from its carton if your fingers are damp. Why?
32. A tailor sometimes licks the end of a thread before pushing it through the eye of the needle. Why? What effect does this action use?
33. A compound that you can apply to a car's windshield causes rain to collect in much larger drops so you can see better while driving. What must happen at the molecular level to do that?

34. Once you have drunk all the liquid, why does the last ice cube in a glass stick so well to the bottom even when you tip the glass upside down?

35. Why are varicose veins found more in legs than in arms?

36. Why should some patients exercise in water after an operation or a long illness?

37. The average depth of the Pacific Ocean is 4300 meters. What is the pressure of the water at this depth? (Ignore the air pressure on the ocean's surface.)

38. What is the total water pressure on the bottom of a swimming pool that is 12 feet deep, including the transmitted pressure of the air above the pool?

39. After swimming in a pool or in the ocean, you will dry off by evaporation after a while even if you don't use a towel. The next time you do that, notice the hairs on your arms. Why are they matted together in little clumps?

40. Does capillary action help as you are drying off with a towel?

41. A 1000 ton ship floats in the ocean. What weight of seawater has it displaced?

42. It is easier for a stream to move pebbles and gravels than to pick up a particle of clay from a clay bed over which it moves. Since the smaller particle has more surface area, it would seem drag should affect it more. Can you explain the seeming contradiction?

43. If you jump into a swimming pool and float at the surface, do you increase the pressure on the bottom of the pool?

44. Which should be built stronger, a 10-foot dam to hold back a small pond or a 10-foot dike in Holland to hold back the ocean? (Ignore ocean waves when you answer this question.)

45. To turn the page in a new book, you might first moisten a finger. Explain why.

46. The Dead Sea is much saltier than the Atlantic Ocean. From Demonstration 2, do you think that you would float higher or lower in the Dead Sea than in the Atlantic Ocean?

47. If you save leftovers, notice that if you wet the rim of a bowl before you stretch a plastic film over it, the plastic sticks better and seals more tightly. Why?

48. A family of four uses, on the average, an acre-foot of water each year (about 325,000 gallons). If the average home has a floor area of 1200 square feet, how deep would one year's supply of water fill that house? (1 acre = 43,560 square feet.)

49. Liquid copper at 1131°C has a surface tension 15 times greater than that of water at room temperature. How should this very high surface tension affect impurities that fall onto the surface of copper? (Anyone who has used a soldering iron has probably noticed dirt or dust floating high on the melted solder.)

50. Before swimming across a lake, an armadillo gulps in a good deal of air, enough to inflate its stomach and intestines. What does this tell you about its normal density?

51. Why are thicker motor oils sometimes recommended for summer months, when a car's engine runs warmer?

52. Why will a scoop of sour cream sometimes float on cold soup but sink in the same soup if the soup is hot? When we say cold soup is "thicker" than hot soup, do we mean it is more dense?

53. Some desert plants protect their internal liquid from evaporation by having minimal surface areas, such as the tiny plants in Figure 10-21. Why is this feature more important for smaller plants?

FIGURE 10-21

54. The Hoover Dam is 221 meters high. In June 1983 the runoff from heavy snows during the winter raised the level of Lake Mead, behind the dam, to the top, cresting the dam for the first time since an early test when the dam was new. What was the water pressure at the base of the dam?

55. A raft built from solid spruce (density 0.6 g/cm^3) is 1/3 m thick, 4 m long, and 2.5 m wide. What weight in newtons and in pounds will it support without submerging completely in fresh water? (1N = 0.225 lb)

Demonstrations

1. Iron is about eight times as dense as water, but it doesn't always sink in water. Lay a sewing needle on a small piece of tissue paper and then lay the combination gently on the surface of some water in a bowl. In less than a minute, the paper becomes saturated and falls to the bottom of the bowl. But not the needle! It is supported by surface tension. Find the reflection of an overhead light on the surface of the water, move your head until the light's reflection approaches the needle, and watch carefully. The needle depresses the water, exactly as if the water had a stretchable skin over it.

Next remove the needle and stir a few drops of liquid soap into the water. Repeat your steps with the needle and another piece of tissue. This time the needle sinks with the paper. Soap reduces the water's surface tension so that the water can no longer support the needle. This property of soapy water lets it penetrate the small spaces within fabrics to get clothes clean.

2. Prove that salt water is denser than fresh water. Put a raw egg into a glass of water. The egg will sink (unless it is old enough to have formed some gas bubbles inside). Add a few tablespoons of table salt, stir, and watch the egg rise. The denser salt water exerts a larger buoyant force on the egg than the tap water does.

3. Pressure increases with depth, which is why when you are standing the blood pressure in your feet is higher than in your head. Take off a shoe and look at the veins in the top of your foot. You can probably see those that stand out because of blood pressure. Then sit down and slowly raise that foot (prop it on a desk). As your foot gets to the level of your heart or slightly higher, the veins no longer bulge outward because the blood pressure in them is now less. This also works with the veins on the back of the hand. Hang your arm by your side and clench your fist. The veins pop out. Then slowly raise your hand above shoulder level and watch the veins disappear.

4. If you plunge your arm into a sink filled with water, the pressure is so uniform that you don't really feel a change in pressure. However, try wrapping your arm in a waterproof plastic bag and then slowly submerging it up to your elbow. As the pressure grows, the wrinkles in the bag become taut and pull unevenly on your skin, making you aware of the change in pressure. It is dramatic!

ELEVEN

Gases

Astronaut Robert L. Steward maneuvers outside the space shuttle. High-speed bursts of nitrogen gas provide the small forces he needs to accelerate in this almost frictionless environment 240 kilometers above the curve of the earth's surface. At this distance there are only traces of air, or such spacewalks (he is moving at about 30,000 kilometers per hour) would be impossible due to friction.

The most common gases on earth are the gases mixed together in its atmosphere. Clean, dry air is so nearly invisible that on a clear day in the western United States, you can easily see a mountain range 200 kilometers (125 miles) away. Yet at that same distance straight up, where space walks take place, there is only a scant trace of air. If its gases cooled and liquified, the liquid air would cover earth only to a depth of about the height of a telephone pole.

Gases are **fluids,** sharing many properties with liquids. Like a body of water, the air presses down everywhere on earth's surface. And yet, even though we live and breathe in this blanket of air, it is easy to be oblivious to its properties since it can be so transparent.

FIGURE 11-2 AIR, LIKE LIQUIDS, IS A FLUID. WHEN AN EAGLE (A) OR OTHER BIRD GLIDES THROUGH AIR AT MODERATE SPEEDS, THE AIR PARTS SMOOTHLY AROUND IT (B). THIS IS CALLED *LAMINAR* FLOW. WHEN THE BIRD PUSHES ITS WINGS HARD AGAINST THE AIR TO GAIN LIFT OR TO DECELERATE, THE FLOW AROUND THE WINGS BECOMES *TURBULENT* (C AND D). THIS TRANSMITS A GREAT DEAL OF ENERGY TO THE AIR. THE LINES SHOWN HERE ARE CALLED *STREAMLINES;* WHERE THEY ARE CLOSER TOGETHER THE FLUID FLOWS FASTER.

(a)

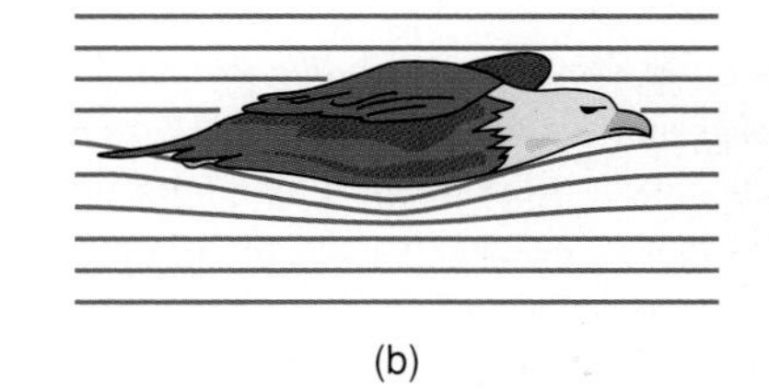

(b)

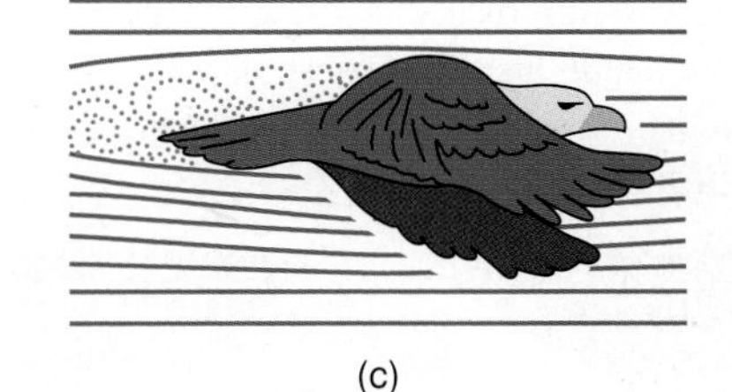

(c)

(d)

PROPERTIES OF GASES

Gases behave much like liquids. For example, if two parcels of air at the surface of the earth are at different pressures, the air at higher pressure will move into the region of lower pressure—which is what is happening whenever you feel a breeze outside. A flow of air past a solid surface has the same boundary effect due to friction that a liquid has: The speed of the air at the solid's surface is zero. Like liquids, gases are viscous and subject to turbulence. Insects and birds push against the air with their wings to fly (Figures 11-1 and 11-2).

You can demonstrate some of the properties of gases with very little effort in your kitchen. Take a balloon, inflate it with a few breaths, and tie its end. You've forced the mixture of gases called air into the balloon, expanding it. Figure 11-3 shows what is happening inside the balloon. The molecules inside collide with the balloon's inner surface, causing the balloon to stretch outward. Said another way, the air inside exerts pressure, F/A, on the balloon's inner surface. Next, inflate a second balloon to the same size and place it in a freezer for a few minutes. It becomes noticeably smaller. When a parcel of gas cools, its molecules move more slowly and exert less force when they hit. With less pressure from within, the balloon shrinks—its volume decreases. Since the

FIGURE 11-1 MOTHS, BIRDS, BUTTERFLIES, AND MANY INSECTS PUSH AGAINST THE AIR IN ORDER TO FLY. HERE A LACEWING PUSHES OFF INTO THE AIR, PERFORMING A HALF-LOOP AS IT DOES. SOME MOTHS MUST BEAT THEIR WINGS SO HARD AT TAKEOFF THAT THEY LOSE SCALES FROM THEIR WINGS EACH TIME.

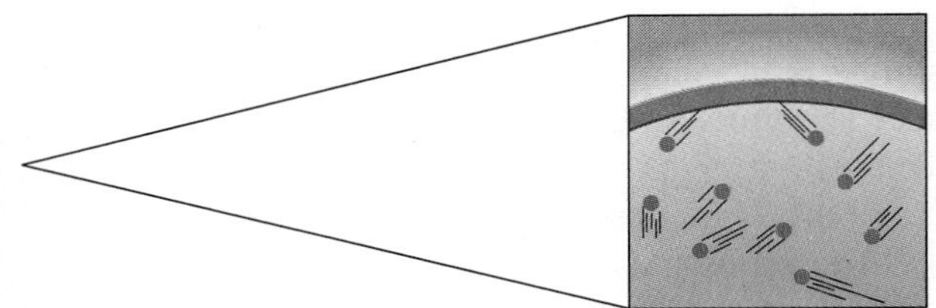

FIGURE 11-3 DETAILS OF WHAT IS HAPPENING IN AN INFLATED BALLOON. THE AIR TRAPPED INSIDE EXERTS PRESSURE AGAINST THE INNER SURFACE OF THE BALLOON THROUGH COLLISIONS WITH IT, AND THIS PRESSURE STRETCHES THE BALLOON OUT. THE GREATER THE PRESSURE INSIDE, THE GREATER THE BALLOON'S VOLUME—UNLESS ITS LIMIT OF ELASTICITY IS EXCEEDED AND IT POPS.

THE COLLISIONS OF THE AIR MOLECULES, WHICH TRANSMIT THE PRESSURE, TAKE PLACE IN ALL DIRECTIONS—UP, DOWN, AND SIDEWAYS—JUST AS IN A LIQUID. AT ANY POINT IN A FLUID, PRESSURE IS TRANSMITTED EQUALLY IN EVERY DIRECTION.

same number of molecules now occupy less volume in the balloon, the density, m/V, of that air increased as it cooled down. Likewise, if the balloon is warmed, it will expand and the density of the air inside will decrease. To see this, just leave the cold balloon on the countertop; it will return to its original size as it warms.

A car tire is basically a balloon with a valve to insert or remove extra air. A tire pressure gauge can measure the pressure inside, as shown in Figure 11-4, where the pressure is recorded in both SI units (220 kPa) and British units (32 lb/in^2). However, a tire gauge does not measure the *total* pressure. It measures the "gauge pressure," which is only the pressure in excess of atmospheric pressure. The atmosphere's pressure is considerable, as we'll see next.

MEASURING THE PRESSURE OF THE AIR

A straight line between your nose and your toes intersects about a billion molecules at any instant. The molecules of air are invisible, and when you walk you push them aside easily. You can *feel* air if you run through it or even if you just take a breath and blow across the hairs on your arm. These molecules, moving at the speed of bullets, hammer the surface of any solid in the air as well as the surface of any liquid exposed to them, exerting a fairly steady pressure.

To show that air exerts pressure, fill your kitchen sink about halfway with water. Lower a drinking glass in sideways, and fill it completely as in Figure 11-5a. Without letting the water run out, turn the glass upside down and ease it up out of water until the rim is just below the surface, as in Figure 11-5b. The water in the glass doesn't flow out even when most of the glass is out of the water. The water in the glass has weight, but something *counteracts* that weight and keeps the water up in the glass. The water in the sink just below that column of water in the glass *exerts pressure that won't let the water flow out.* That pressure comes from the air, which presses down on the surface of the water in the sink. That air pressure is transmitted to every point in the water (Pascal's Principle), in every direction, and it supports the weight of the water in the glass. In fact, experiments show that the water would not empty from any glass at sea level with a height of up to about 10.3 meters (33.8 feet), as Figure 11-6 shows. This tells us what **atmospheric pressure** is. That tall column of water exerts about 101 kPa, or 14.7 lb/in^2, at its bottom, which is equal to the pressure from the air that keeps the column up.

FIGURE 11-4 THE AUTHOR'S MOTHER CHECKING THE PRESSURE IN A TIRE WITH A TIRE GAUGE SHOWING BOTH SI AND BRITISH UNITS. THE READING, CALLED THE *GAUGE* PRESSURE, DOES NOT TELL YOU THE TOTAL PRESSURE IN THE TIRE. THE TOTAL PRESSURE INCLUDES THE ATMOSPHERIC PRESSURE OUTSIDE THE TIRE, WHICH IS TRANSFERRED TO THE INSIDE JUST AS HAPPENS IN A BODY OF WATER, AS WE SAW IN THE PRECEDING CHAPTER.

(a)

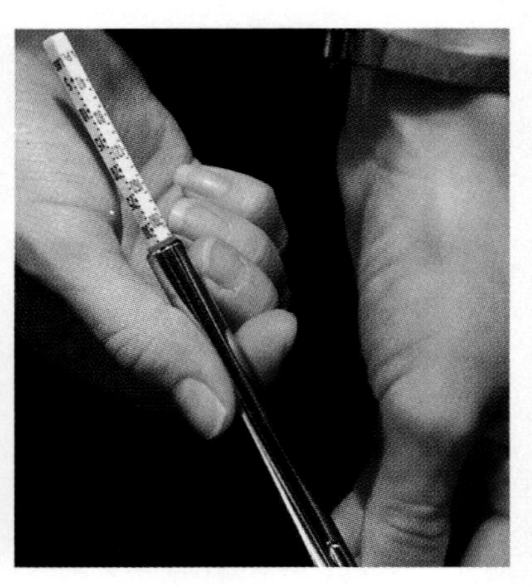

(b)

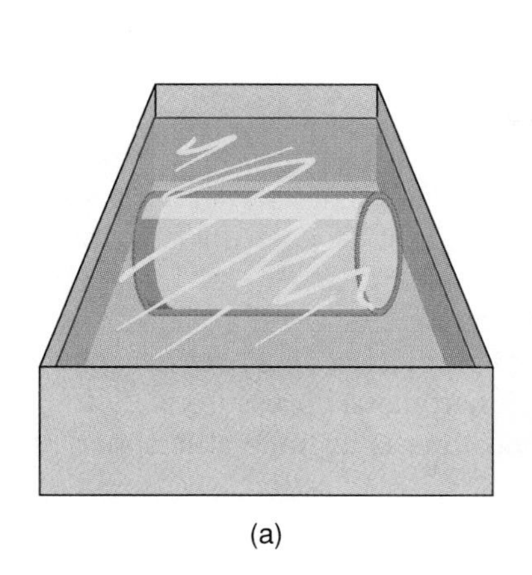

(a)

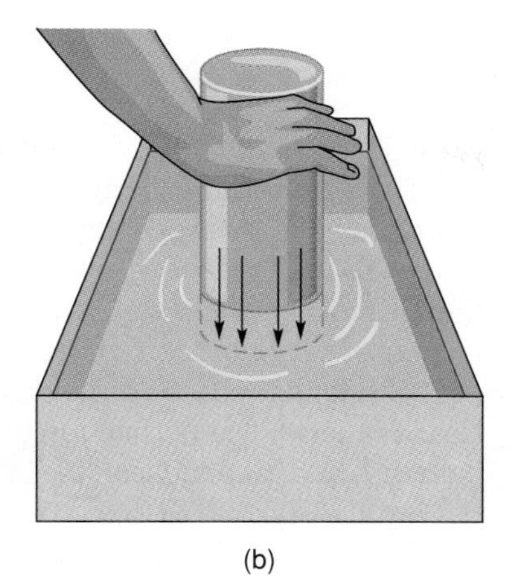

(b)

FIGURE 11-5 (A) FILLING A GLASS WITH WATER TO SHOW THE EXISTENCE OF AIR PRESSURE. (B) WHEN THE WATER-FILLED GLASS IS POSITIONED SO THAT ITS RIM IS JUST BELOW THE SURFACE OF THE WATER IN THE SINK, THE WATER IN THE GLASS DOES NOT FLOW DOWNWARD INTO THE SINK. WHAT SUPPORTS ITS WEIGHT? IF THE WATER IN THE GLASS FLOWED OUT, THE WATER LEVEL IN THE SINK WOULD RISE. THAT DOESN'T HAPPEN BECAUSE THE AIR'S PRESSURE ON THE WATER IN THE SINK IS FAR GREATER THAN THE PRESSURE EXERTED BY THE WATER IN THE GLASS.

(a)

(b)

FIGURE 11-6 (A) THIS OLD WOODCUT ILLUSTRATION SHOWS THE CONSTRUCTION OF A WATER BAROMETER. THE AIR'S PRESSURE ON THE SURFACE OF THE WATER IN THE TUB AT THE BOTTOM (AFTER THE SPIGOT AT THE TOP OF THE TUBE WAS CLOSED) COULD NOT SUPPORT A COLUMN OF WATER HIGHER THAN 10.3 METERS. EXPERIMENTS LIKE THIS SHOWED THE PRESSURE THROUGHOUT THE AIR AT SEA LEVEL TO BE ABOUT 14.7 LB/IN^2, OR 101 KPA. (B) IN WATER PUMPS SUCH AS THIS ONE, THE PUMPING ACTION PARTIALLY EVACUATES THE AIR IN A PIPE THAT REACHES THE GROUNDWATER BELOW. ANOTHER AIR PASSAGE EXPOSES THE UNDERGROUND WATER TO THE ATMOSPHERE'S PRESSURE, WHICH THEN PUSHES THE WATER UP THE EVACUATED PIPE—BUT NO HIGHER THAN 10.3 METERS AT SEA LEVEL, AND LESS IN THE MOUNTAINS WHERE THE AIR PRESSURE IS LESS.

Why does earth's atmosphere produce such a great pressure? Just like the pressure exerted by water underneath the ocean's surface, the air's pressure is due to its weight. We are at the bottom of what has been called "an ocean of air," and that phrase isn't too misleading. At every elevation in our atmosphere, the air molecules must support the weight of those above. No matter that there is a lot of space between the molecules. They are some 10 molecular diameters apart on the average at sea level. Their high speeds cause them to continually collide with one another—at sea level an air molecule has several billion collisions a second. Through these collisions, the total weight of the atmosphere is transmitted through the air to earth's surface.

In 1643 Evangelista Torricelli, a student of Galileo, thought of using mercury in a closed tube to determine air pressure (Figure 11-7). Since mercury's density is 13.6 times that of water, the column of mercury that balances the atmospheric pressure is 1/13.6 as tall as that 10.3-meter column of water, or about 760 millimeters (30 inches). The device shown in Figure 11-7 is called a **barometer.** Today many other types of barometers are in use, and air pressure is often called barometric pressure.

THE PRESSURE ON YOUR FEET DUE TO YOUR WEIGHT WHEN YOU STAND BAREFOOT IS PROBABLY AROUND 5 LB/IN^2. TO GET ONE ATMOSPHERE OF PRESSURE (ABOUT 15 LB/IN^2) THERE, YOU WOULD NEED TO HOLD ON YOUR SHOULDERS TWO PEOPLE WHO EACH WEIGH THE SAME AS YOU. YOU CAN IMAGINE HOW THAT WOULD CAUSE YOUR BARE FEET TO FEEL! ONE ATMOSPHERE IS A *LOT* OF PRESSURE.

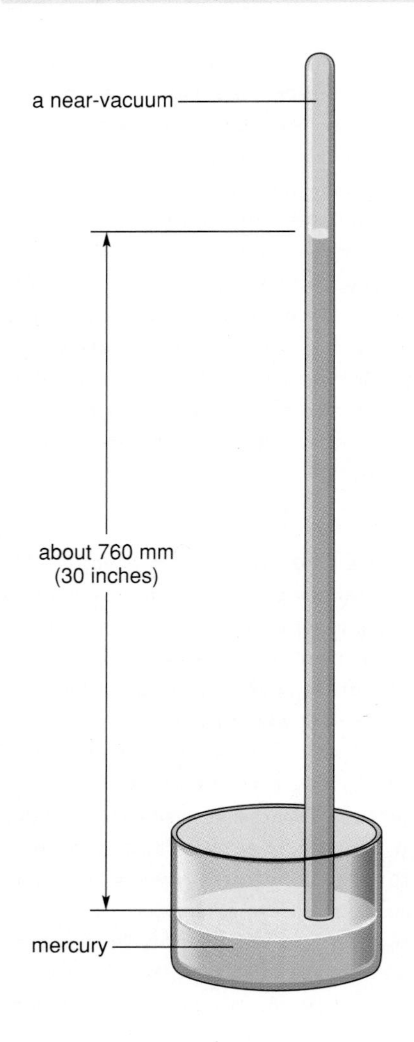

FIGURE 11-7 BECAUSE MERCURY HAS A HIGHER DENSITY THAN WATER, A MERCURY BAROMETER IS MORE COMPACT AND EASIER TO WORK WITH THAN A WATER BAROMETER.

The air pressure at any location varies slightly from day to night and from day to day. We saw with the balloons earlier that warmer air is less dense than colder air. So when the air over some location gets warmer, the air pressure there drops a bit. Because it is less dense, warm air weighs less than an equal volume of cold air. Likewise, the barometer climbs (which means the air pressure increases) if a large parcel of relatively cold air sweeps over an area. Near sea level, the normal range of barometric pressure is 29 to 31 inches of mercury, and 29.9 inches (14.7 lb/in^2) is the average atmospheric pressure at sea level.

Since air is a mixture of gases (Table 11-1), each gas contributes a certain percentage to the total pressure air exerts. The pressure exerted by any one gas in a mixture is called the **partial pressure** of that gas. The atmosphere's net pressure is the sum of the partial pressures exerted by the various gases that make up air. Nitrogen molecules account for almost 4/5 of the pressure, while oxygen comes in second, contributing about 1/5 of the total. In later chapters, we shall see that the partial pressure of water vapor in the air has a great deal to do with our weather.

Table 11·1 Gases in the Atmosphere

	Percentage of Molecules in Dry Air
Nitrogen (N_2)	78.09
Oxygen (O_2)	20.95
Water vapor (H_2O)	0–4
Argon (Ar)	0.93
Carbon dioxide (CO_2)	0.03
Neon (Ne)	0.0018
Helium (He)	0.000524
Krypton (Kr)	0.0001
Radon (Rn)	6×10^{-18}

FLOATING IN AIR

Another property of air is that it provides buoyancy. As we saw in Chapter 10, liquids exert a buoyant force because the pressure in the liquid increases with depth. Likewise, at points successively closer to earth's surface the air's pressure increases since there is more air overhead to support. A parcel of heated air is less dense than the cooler air around it and thus weighs less than the same volume of cool air. So the buoyant force on that parcel of hot air is greater than the parcel's weight, and the hot air rises. When hot air fills a hot air balloon and the balloon rises, it is because the combined weight of the balloon, its cargo, and the hot air inside is less than the weight of the cool air they displace.

Remember that the force of buoyancy is equal to the weight of the displaced fluid. Your body displaces air, so air exerts a buoyant force on you! Your total volume is only several cubic feet, however, and several cubic feet of air only weighs several ounces. Nevertheless, when you step on a scale it reads several ounces *less* than your weight, mg, due to air's buoyant (upward) force on you.

GASES ARE COMPRESSIBLE FLUIDS: BOYLE'S LAW

Push on any trapped gas and it compresses; the molecules can crowd a little closer together since there is so much empty space between them. Robert Boyle (1627–1691), a contemporary of Newton's, discovered a rule about how gases compress. If you increase the pressure on a confined gas without allowing its temperature to change, its volume decreases by the same factor. Likewise, if the pressure decreases by any amount while the temperature remains constant, the volume increases by the same factor. In other words, the *product* of pressure P and volume V doesn't change:

PUTTING IT IN NUMBERS—BOYLE'S LAW

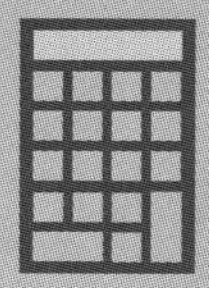

initial pressure × initial volume = final pressure × final volume

$$P_i V_i = P_f V_f \text{ (Boyle's law)}$$

Example: If 60 cubic inches of air (V_i) at 15 pounds per square inch (P_i) is squeezed into a space of 30 cubic inches (V_f) by a bicycle pump, the new pressure (P_f) will be 30 pounds per square inch (psi), provided the temperature of the gas is kept constant. $P_f = P_i \times (V_i/V_f) = 15 \text{ psi} \times 60 \text{ in}^2/30 \text{ in}^2 =$ **30 psi**

HOT-AIR BALLOONS

In the summer of 1782, Spain had the British garrison at Gibraltar under siege by land and sea. Joseph Montgolfier, a Frenchman who sympathized with the English, considered a novel escape route for the English: through the air. He and his brother soon launched a paper balloon almost 16 meters in diameter. The hot air came from a fire in an iron brazier hung from chains beneath the balloon. During its short flight, the two passengers were kept busy dousing fires that sprung up in the paper balloon. The dispute at Gibraltar was settled without balloons, which is probably just as well considering the primitive state of these early balloons.

Scientists of that time knew that hydrogen, the lightest gas, is much more buoyant than hot air, but no one knew how to make a lightweight material that hydrogen would not escape through. Then the French physicist Jacques Charles used turpentine to dissolve natural rubber and then washed silk cloth in the solution. When it dried, the rubberized silk was impervious to hydrogen. On December 1, 1783, Charles rose over Paris in a hydrogen balloon 9 meters in diameter. The sun had just set, but as the balloon rose, Charles saw the sun again in the west—no doubt the first person ever to see two sunsets in one day.

(A) HEATING THE AIR IN THE BALLOON AS IT INFLATES. (B) UP AND AWAY. . .

The pressure in Boyle's law is the total pressure, or **absolute pressure;** that is, it is the amount of pressure a gas exerts as compared to a perfect vacuum, an empty space where there is no pressure. When a tire gauge registers 32 lb/in^2, for instance, the *absolute* pressure is 32 lb/in^2 *plus the atmospheric pressure* wherever that tire is.

The molecular picture helps us understand Boyle's law. Gas pressure is the result of collisions between gas molecules and the walls of the container holding the gas. All the molecules that hit a square centimeter of the wall each second contribute to the pressure there. When you compress the molecules of a gas into half as much space, there are twice as many molecules per cubic centimeter (Figure 11-8) and therefore twice the number of collisions per second. That means the pressure doubles—provided the molecules' average speeds are the same before and after the compression, which means the temperature must be the same. (We'll see the connection between temperature and speed in the next chapter.)

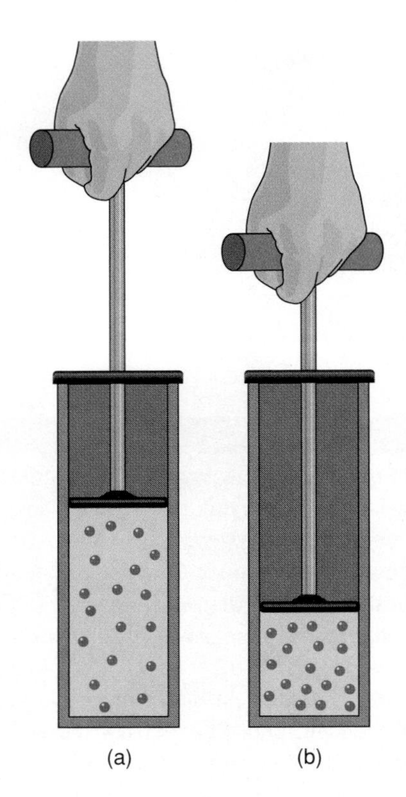

FIGURE 11-8 IF THE VOLUME OF AIR IN A BICYCLE PUMP IS SLOWLY COMPRESSED TO $\frac{1}{2}$ OF ITS INITIAL VOLUME, THE PRESSURE GOES UP BY A FACTOR OF 2 (PROVIDED THE TEMPERATURE REMAINS THE SAME). WE CAN UNDERSTAND THIS VOLUME–PRESSURE RELATIONSHIP BY NOTING THAT THE DENSITY IS 2 TIMES GREATER IN THE SMALLER VOLUME, SO COLLISIONS WITH THE WALL ARE 2 TIMES MORE FREQUENT.

MORE ABOUT EXPANSION AND COMPRESSION

Open your mouth wide and breathe on your palm. Then do it again with your lips nearly pressed together. This time the air expands a bit as it escapes. Feel the difference? Air that expands cools. Warm air that expands does work in pushing the cooler air out of the way. The collisions that push the cooler air back slow the warmer molecules that do the pushing, and so the temperature drops as the initially warm gas expands. In other words, the expanding gas loses energy as it does work pushing the cooler air back.

The reverse takes place when a gas is compressed. You do work as you operate a bicycle pump, compressing the air. The work you do gives the air molecules extra speed. The plunger pushes the air molecules head-on, and they pick up speed, like a tennis ball that's hit by a racket. Compressing a gas increases its temperature. In Figure 11-8, for example, the compressed gas must cool before Boyle's law holds exactly.

PRESSURE IN A MOVING FLUID: THE BERNOULLI EFFECT

The next time you take a road trip, look for convertibles moving along the highway. At highway speeds, the fabric top of a convertible balloons up (Figure 11-9a, b). This tells us that the air pressure above the top is less than the air pressure from below, inside the car. Figure 11-9c shows what happens. The air that passes over the windshield has to travel farther than air higher up, and to do so its speed must increase; it accelerates. The bulging convertible top indicates that *the pressure in the air decreases if the speed of the air increases.* This is called the **Bernoulli effect,** after Daniel Bernoulli (1700–1782), and it is displayed by any fluid. Note that it is the pressure within the flow itself that decreases; don't confuse the fluid's internal pressure with the pressure the fluid will exert if it impacts a surface head-on, as when you put a cupped hand out of the window to catch the air.

A straw in a glass of tea can demonstrate this effect. Blow sharply across the top of the straw. While the air rushes past, the pressure at the top of the straw drops, and the liquid rises in the straw. It moves into the area of lower pressure—since the liquid's surface outside the straw is still exposed to the atmosphere's full pressure. If the stream of air over the straw is very fast, the liquid can rise all the way up and you'll get a stream of spray. Perfume atomizers, aerosol cans, and airbrushes use this principle for their operation.

On a larger scale, a strong gust of wind rushing over the roof of a house makes the pressure on top of the roof less than the pressure inside the house.

(a)

(b)

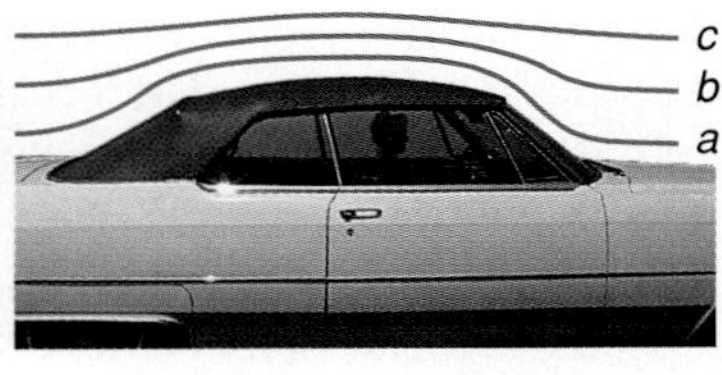

(c)

FIGURE 11-9 (A) AN ANTIQUE CONVERTIBLE AT REST. (B) THE SAME CONVERTIBLE AT 55 MPH. (C) LINES *A*, *B*, AND *C* SHOW PATHS TAKEN BY PARCELS OF AIR. AIR MOVING ALONG PATH *A* HAS TO TRAVEL FARTHER THAN AIR MOVING ALONG PATH *B* AND MUCH FARTHER THAN AIR MOVING ALONG PATH *C*. THE AIR THAT MOVES ALONG *A* MUST ACCELERATE TO GO FASTER AND COVER ITS LONGER DISTANCE IN ABOUT THE SAME AMOUNT OF TIME AS USED BY AIR MOVING ALONG *B* OR *C*.

If the wind speed is very high, the roof can be lifted, not by the wind above but by the air pressure from below. In most houses, metal strips connect the roof structure to the wall frames to keep this from happening.

THE DENSITY OF AIR IN THE ATMOSPHERE

The pressure caused by the weight of the air above compresses the air at lower elevations, pushing more molecules into each unit of volume. For this reason, the density of air closer to earth is greater than the density of air at higher elevations. To discover how air's density changes, you could take some jars with airtight lids from sea level up to the summit of a high mountain. At various altitudes, open a jar, swish it around, and turn it upside down to pour out the thicker air from sea level. (No kidding! This actually happens.) Then seal it again. Carefully weigh those jars on your return, and you'll find that the higher the sampling site, the lighter the jar. Then, by subtracting the weight of a jar pumped entirely free of air, you could show that 1 cubic foot of air at sea level weighs 0.08 pounds and the same volume of air at 18,000 feet, or 5500 meters, weighs only half as much. So the density of air at 18,000 feet is half the density of air at sea level.

The density of the air around us also changes with temperature. During the day, the ground absorbs sunlight and becomes warm, and it then warms the layer of air on top of it. The warmer the air, the faster its molecules move and the more they push back the cooler air around them; in short, the warmed air becomes less dense. At night, the ground cools (we'll see why in Chapter 13), and the air above it loses energy to the cooler ground. That parcel of cooled air shrinks under the pressure of the atmosphere. The denser air hugs the ground and flows just like a liquid down any slope into the lowest area available (Figure 11-10).

FIGURE 11-10 AT FIRST GLANCE, THIS LOOKS LIKE A WATERFALL ILLUMINATED BY MOONLIGHT (SEE THE STAR TRAILS). INSTEAD, IT IS AN AIRFALL! A SMOKE BOMB IN A DRY CREEKBED ABOVE SHOWS HOW THE COOLED AIR NEXT TO THE GROUND IS MORE DENSE AND FLOWS DOWNHILL ALONG THE STREAMBED AND OVER THE SIDE, JUST AS WATER WOULD.

BREATHING

When you inhale, muscles expand your chest wall and pull your diaphragm down, allowing the elastic lungs (which have no muscles of their own) to expand. This expansion causes the pressure inside your lungs to drop to 1 or 2 percent below atmospheric pressure, and the higher-pressure air from outside enters your lungs. When your chest muscles relax, the stretched lungs relax back to their original volume and the air in them is pushed out—you exhale.

With each breath of air, oxygen comes into your lungs. It dissolves into the thin, moist walls of the alveoli, the hundreds of millions of bubblelike globules on the inner surface of your lungs. These alveoli have forty times as much surface area as your entire skin! At the same time, carbon dioxide comes out of the blood there and escapes when you breathe out.

THE ATMOSPHERE

Earth's atmosphere has distinct regions of different properties as outlined below.

The **ionosphere:** At the upper edge of the atmosphere, intense sunlight and fast-moving particles from the sun (the solar wind) and from interstellar space (cosmic rays) hammer the widely-scattered molecules. Many of these molecules lose electrons, becoming ionized (thus the name of this layer). We shall see later that this layer affects radio communications and that the colorful auroras take place there. Sparse as the molecules may be, the air in the ionosphere is dense enough to offer substantial friction to the incoming interplanetary debris we know as meteors (or popularly, "shooting stars"). By the time they reach an altitude of 100 kilometers (65 miles), these tiny bits of rock glow white-hot.

The **mesosphere,** literally "in the middle": The region from 100 to 60 kilometers (60 to 35 miles) above earth's surface. Even large, bright meteors usually don't get through the mesosphere, due to increased density of the air there.

The **stratosphere:** Immediately below the mesosphere, 60 kilometers (35 miles) over our heads, the stratosphere begins. If you were there looking down, you could just make out things as large as football stadiums. The powerful sunlight there separates some of the O_2 molecules into two individual atoms which can combine with undisturbed O_2 molecules to form O_3, which is *ozone,* a molecule that absorbs the harmful ultraviolet radiation in the sun's light. Without this invisible guardian, ultraviolet rays would destroy life on earth. The ozone is most dense at an altitude of about 24 kilometers (15 miles) above the earth's surface. For comparison, the highest-flying commercial jets cruise at about 14 kilometers (9 miles) above sea level, just above the stratosphere's lower edge.

The **troposphere:** Beneath the cold, thin stratosphere lies the troposphere, a region 11 kilometers (7 miles) deep that harbors most of the molecules in our atmosphere. Our weather is produced here by the combination of sunlight, oceans, and the land below. It is normally a region of rapid mixing, where water vapor rises, cools, and turns first into clouds and finally into rain or snow.

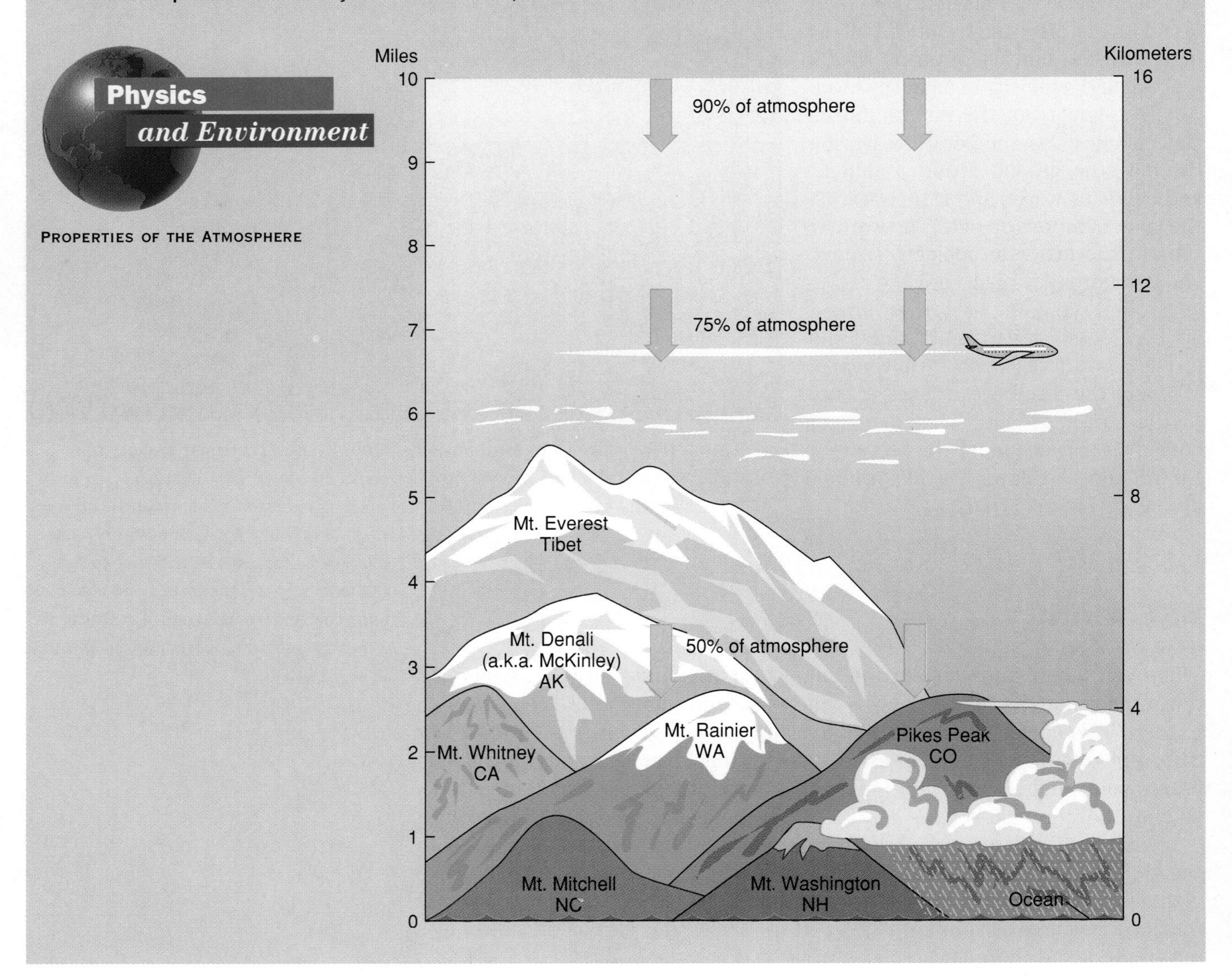

PROPERTIES OF THE ATMOSPHERE

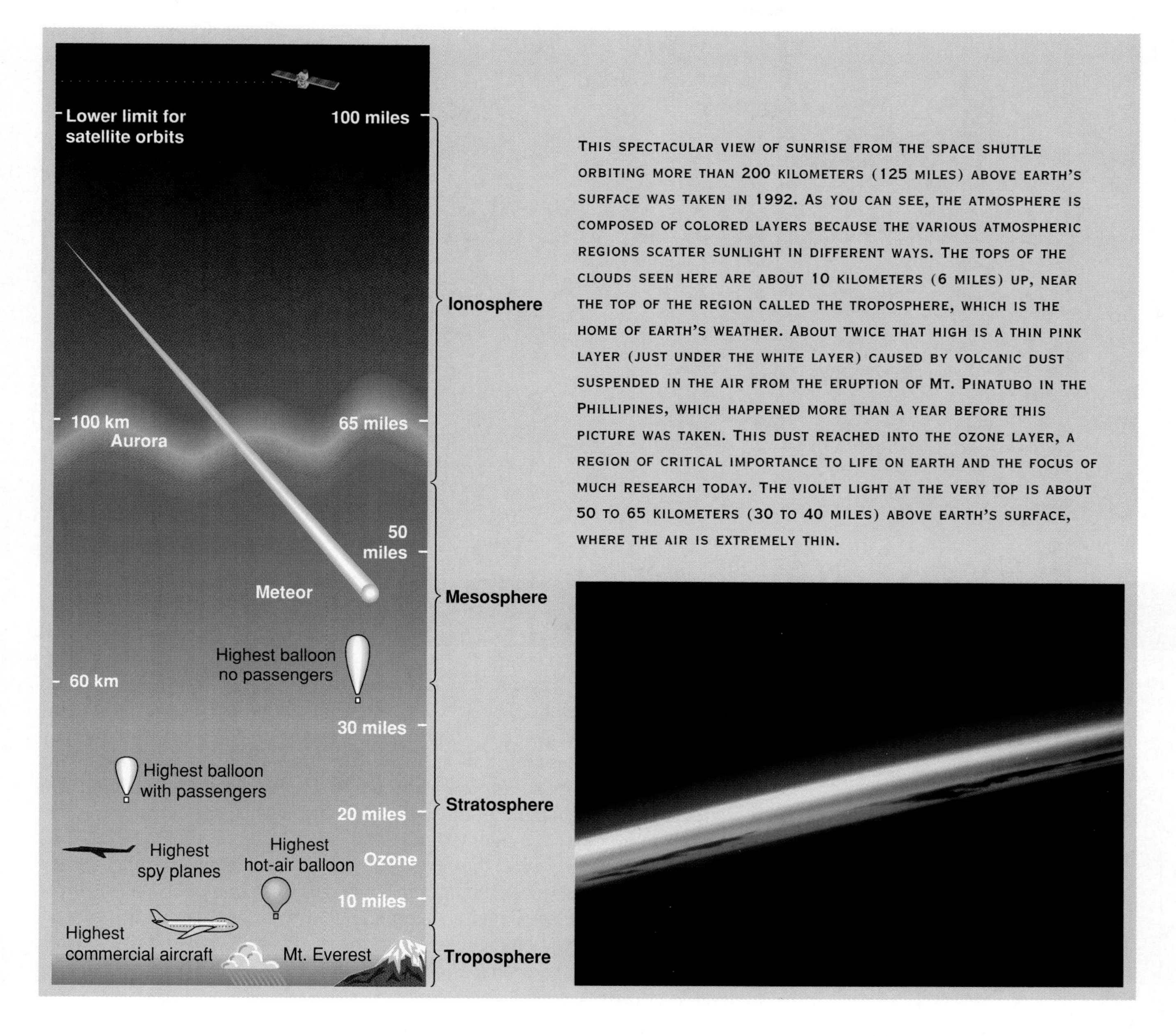

THIS SPECTACULAR VIEW OF SUNRISE FROM THE SPACE SHUTTLE ORBITING MORE THAN 200 KILOMETERS (125 MILES) ABOVE EARTH'S SURFACE WAS TAKEN IN 1992. AS YOU CAN SEE, THE ATMOSPHERE IS COMPOSED OF COLORED LAYERS BECAUSE THE VARIOUS ATMOSPHERIC REGIONS SCATTER SUNLIGHT IN DIFFERENT WAYS. THE TOPS OF THE CLOUDS SEEN HERE ARE ABOUT 10 KILOMETERS (6 MILES) UP, NEAR THE TOP OF THE REGION CALLED THE TROPOSPHERE, WHICH IS THE HOME OF EARTH'S WEATHER. ABOUT TWICE THAT HIGH IS A THIN PINK LAYER (JUST UNDER THE WHITE LAYER) CAUSED BY VOLCANIC DUST SUSPENDED IN THE AIR FROM THE ERUPTION OF MT. PINATUBO IN THE PHILLIPINES, WHICH HAPPENED MORE THAN A YEAR BEFORE THIS PICTURE WAS TAKEN. THIS DUST REACHED INTO THE OZONE LAYER, A REGION OF CRITICAL IMPORTANCE TO LIFE ON EARTH AND THE FOCUS OF MUCH RESEARCH TODAY. THE VIOLET LIGHT AT THE VERY TOP IS ABOUT 50 TO 65 KILOMETERS (30 TO 40 MILES) ABOVE EARTH'S SURFACE, WHERE THE AIR IS EXTREMELY THIN.

PROBLEMS NEEDING SOLUTIONS

Air also brings into your lungs suspended particles and chemical compounds that are called air *pollutants* when their concentrations are high enough to cause harm. Although some comes from natural sources, about 50 percent of the human-caused portion is due to motor vehicles and over 30 percent is due to the burning of fossil fuels by power plants and other industrial plants. Emissions of the principal air pollutants are all regulated to some extent in the United States, Canada, and other industrialized countries. Automobiles today emit 95 percent less pollutants than they did in 1975, when all new cars sold in the United States were required by law to have catalytic converters.

The major pollutants are carbon monoxide (most comes from natural sources), sulfur dioxide and nitrogen oxides (which cause acid rain), and volatile organic compounds (VOCs) from engine and power plant emissions which can react in the air to cause smog. Also very fine pollution particles that aren't filtered by hairs or mucous in your airways are drawn deep into the lungs where they can remain and cause tissue damage such as found in emphysema. We'll return to these topics in later chapters.

PHYSICS IN 3D/CLOUDS FORM AT DIFFERENT LEVELS IN THE TROPOSHPERE

LOOKING DOWN ON CLOUD LAYERS IN THE TROPOSPHERE OVER BOLIVIA FROM THE SPACE SHUTTLE. NOTICE THE DIFFERENT ELEVATIONS OF THE CLOUD LAYERS.

Summary and Key Terms

atmospheric pressure–158

barometer–159

partial pressure–160

Boyle's law–161

absolute pressure–161

Bernoulli effect–162

Gases exert pressure through random molecular collisions, and gases, unlike liquids, are compressible. The weight of the air above earth's surface compresses the air around us. The ***atmospheric pressure*** at sea level is about 101 kPa, or 14.7 lb/in^2. This pressure supports a column of liquid mercury about 760 mm high in a mercury ***barometer***. Each type of molecule or atom in the air contributes a certain percentage to the total air pressure: its ***partial pressure***.

Air buoys up objects immersed in it; the buoyant force equals the weight of the displaced air. By day, the air warms up, so its molecules move faster and its density decreases. At night, the air cools, so its molecules slow down and its density increases. When air expands, it uses energy to push back the cooler air around it, so its molecules slow down and *the expanding air cools off.* The molecules of a gas that is compressed gain speed as work is done on them, so *a compressed gas heats up.*

Boyle's law is a rule about how gases compress. When you compress a gas into half as much space, keeping temperature constant, the pressure doubles. $P_iV_i = P_fV_f$. The pressure here is ***absolute pressure***, the amount of pressure a gas exerts relative to a perfect vacuum, where the pressure is zero.

All fluids display the ***Bernoulli effect***: the pressure in a fluid drops whenever the speed of the fluid increases.

EXERCISES

Concept Checks: Answer true or false

1. Air, like liquids, flows when there is unbalanced pressure.
2. Pressure in a gas is transmitted by molecular collisions.
3. Collisions between the molecules of a gas cause them to change speeds and directions continually.
4. The weight of the atmosphere exerts a pressure of about 101 kPa on everything at sea level.
5. Unlike a liquid, a gas is compressible because of the space between its molecules.
6. The breeze you might feel on your face when you go outside comes from air that flows from regions of lower pressure toward those of higher pressure.
7. Anything immersed in air is supported by a buoyant force equal to the volume of air it displaces.

8. The Bernoulli effect states that the pressure in a fluid rises if its speed increases.
9. Boyle's law states that the pressure times the volume of a gas is constant so long as the temperature doesn't change.
10. In a perfect vacuum there are no molecules to exert pressure.

Applying the Concepts

11. At sea level there is about 14.7 lb/in^2 of air pressure pressing on your body. Why aren't you pushed around by it?
12. Should a propeller-driven plane fly more easily on a cold day or a warm day? (Think of a propeller as slicing the air and throwing it backwards.)
13. Gases, it's sometimes said, fill all the space available to them. So why doesn't the earth's atmosphere move out into space?
14. We live at a boundary of our atmosphere where gases meet solids. How does this help us with respect to winds?
15. Explain how a suction cup holds on to a surface.
16. If someone went very quickly from sea level to an elevation of 5500 meters (18,000 feet), what could happen to air spaces trapped in his or her body?
17. Jet aircraft must be moving noticeably faster when they land at the mile-high city of Denver than when they land at sea level in San Francisco. Why?
18. Show that 15 lb/in^2 is more than 2000 lb/ft^2.
19. A man has a surface area of 20 ft^2. Show that the total force ($F = PA$) of the air's pressure on his body is over 20 tons. (1 ton = 2000 lb.) Notice this is *not* the *net* force.
20. If you let air out of a tire, you can feel the tire valve cool off. Why does it do this?
21. A bubble of air in the blood can be fatal. Physicians call that an air embolism. Explain.
22. Analyze what happens when you drink from a glass with a straw.
23. A box sealed airtight and sitting on a very sensitive scale weighs different amounts during the week. Why?
24. If you cough to dislodge a foreign particle in your trachea (or *windpipe*), the air pushing through the trachea can cause a partial collapse. Why? The smaller airway then helps increase the speed of the air that follows. How does this help eject the foreign matter?
25. Air weighs 0.08 lb/ft^3. What volume of air weighs the same as you do?
26. Calculate the weight of sea-level air in a 10-ft by 12-ft bedroom with an 8-ft ceiling. (Air weighs 0.08 lb/ft^3.)
27. In a smoothly flowing stream, floating twigs or people in innertubes tend to drift toward the center, where the current is fastest. Explain.
28. If you hold a very deep breath while standing on a scale, do you weigh more, less, or the same when you exhale?
29. Relatively large particles, with diameters of 10 microns or more (1 micron = 1 μm = 10^{-6} m), settle out within a day or two due to gravity when they are emitted into the air. But finer particles, with diameters of less than a micron, can remain in the troposphere for weeks, and if they are carried to the stratosphere they can stay aloft for up to five years! Why is there such a difference in their fates?
30. Prairie dogs maintain large burrows, with at least two entrances that are piled up to very different elevations (Figure 11-11). Explain how this difference in height helps with ventilation in a burrow. (The air in a large burrow is exchanged about every ten minutes.)

FIGURE 11-11 AN ENTRANCE TO A PRAIRIE DOG BURROW. (NOT SEEN IS THE SECOND ENTRANCE AT A DIFFERENT ELEVATION.)

31. Calculate the net force on the front door (6.5 ft × 3 ft) of a closed house if the outside pressure suddenly dropped 15 percent, as it could if a tornado swept past.
32. Show that a 1-inch drop of mercury in the barometric pressure is equal to a drop of about 70.6 lb/ft^2. (You can also show that a drop of that much over an area of the ocean can cause the ocean's level to rise 13 in!)
33. At a gas station, you put 25 lb/in^2 of air pressure into the car's tires. Does it make any difference whether you are in Denver or Miami at the time? Discuss. (Presume the temperature is the same in both cities.)
34. Speculate on what eventually happens to a child's helium-filled balloon as it rises.
35. Would water in car tires work as well as air does? Would water-filled tires work better than solid rubber tires? Which tires would probably produce the best gasoline mileage? Why?
36. How can dust stay on a car traveling at 55 mph?

37. Inspect the blades on a house fan that's in use. Invariably there are small dust particles that stay on even while the blades are moving. Why?

38. Some scuba divers use double tanks that hold 142 ft^3 of sea-level air. What is the weight of the air in a filled double tank? (Air weighs 0.08 lb/ft^3.)

39. A standard scuba tank is filled with 71.2 ft^3 of sea-level air. The pressure gauge on the tank then registers 2250 lb/in^2. What is the volume inside that tank?

40. Can you hold your breath for a little over a minute? That's about the time it takes your blood to circulate through your body. Do you think those two acts are related? Discuss.

41. If the acceleration of gravity near the earth's surface suddenly became 0.5*g*, give two ways this would affect the flight of birds.

42. People who use fireplaces know that a slight breeze outside usually makes the fireplace "draw" better, meaning there is less likelihood of smoke entering the room. Explain.

43. Suppose the atmosphere of earth suddenly vanished. Would you weigh (a) more, (b) less, or (c) the same if you were standing on a scale at the time? What would happen to a hovering hot-air balloon?

44. Empedocles (500–440 B.C.), a Sicilian, held his finger firmly on the top of a straw and placed the straw into a glass of water. What, he asked, kept the water out of the submerged end of the straw? Can you answer that?

45. The Goodyear blimp uses about 5660 m^3 of helium. If air has a density of 1.2929 kg/m^3, and helium at atmospheric pressure has a density of 0.1785 kg/m^3, find the buoyant force on the blimp due to that volume of helium.

46. The particles in an aerosol spray slow down quickly as they leave the nozzle of the can. Why? What would happen to the stream from a spray can on the moon?

47. The space shuttle's main engine generates 1.67 million newtons of thrust at sea level but 2.09 million newtons in space. Why should there be a difference?

48. Mt. Denali in Alaska (Figure 11-12) has been seen from 370 kilometers (230 miles) away by people on the ground, while the minimum elevation for a space shuttle orbit is only 185 kilometers (115 miles). What does this difference imply about our atmosphere?

49. If you want to throw your fastest fastball, should you try it in New York City or in Denver? On a hot day or a cold day?

50. Why do hot-air balloonists prefer either early-morning or early-evening flights?

51. Why should a gasoline can have a small hole on the top in addition to the spout? Why is it easier to pour chocolate syrup if you punch two holes on opposite sides of the top of the can?

52. Why should the blades of a windmill be placed high above the ground on a tower?

53. Where would you be safer during a hailstorm, in Leadville, Colorado, elevation 3000 m (10,000 ft), or Boston harbor? (Assume each area has hailstones that are the same size.)

54. A small bubble of air trapped in the brake lines of a car lessens the pressure the brakes apply on the brake shoes or pads. Why? (Car mechanics "bleed" the brake lines after working on brakes to eliminate any air bubbles.)

55. A skydiver who leaps from a plane at 4200 meters (14,000 ft) reaches terminal speed in 10 to 12 seconds. What happens to this terminal speed as the skydiver approaches earth?

56. How can you account for the fact that workers in a mine shaft 1600 meters beneath the earth's surface need no regulators to breathe, while scuba divers only 10 meters beneath the water's surface could not breathe air that was only at sea level pressure?

57. In a record effort, a swimmer once held his breath for 3 minutes and 40 seconds while he descended 100 meters. What was the ratio of his lung volume at that depth to his lung volume when he returned to the surface?

58. The air pressure in the tires of a Boeing 747 is 200 lb/in^2. How many square inches of tire must support a fully loaded 747 that weighs 600,000 lb?

FIGURE 11-12 MOUNT DENALI (FORMERLY MT. MCKINLEY) AS SEEN FROM ABOUT 50 KILOMETERS (30 MILES) AWAY. IT HAS BEEN SEEN FROM 370 KILOMETERS (230 MILES) AWAY ON EXCEPTIONALLY CLEAR DAYS.

Demonstrations

1. Place five new playing cards side by side on a cloth surface. Moisten your thumb, press on the center of the first one, and it comes up with your hand. Put the first one squarely on the next one and press again; they'll both come up—and so on down the line. Air pressure keeps the cards together.

2. Place a dime or a penny near the edge of a table. With a deep breath, blow sharply across the top of the coin. Normal air pressure from beneath the coin pushes it upward into the region of lower pressure. (Remember, the stream of air needs a high speed to reduce the pressure over the surface.) With only a little practice, you can make the coin land in a tilted cup (or someone's hand) a few inches away.

3. With the very tip of a spoon's handle between your thumb and forefinger, hang the spoon parallel to a moderate stream of water from a faucet and ease the back of the spoon over into the flow. Although it seems as if the spoon should be knocked away by the water's pressure, it isn't. Instead, it moves into the stream. That's Bernoulli's principle at work. Now increase the flow from the faucet and watch the effect increase. (Watch the stream of water below the spoon. Does it show action–reaction?)

4. Use a quart jar to measure the volume of a breath of air from your lungs. Run some water into a sink, fill the jar completely, and invert it, keeping the rim under water. Then use a plastic straw to exhale under the jar so that the air rises to displace water in the jar. Mark the point of the new water level on the jar. Take the jar out of the water, empty it, and fill it with water to that mark. Then measure this volume with a graduated cylinder or a measuring cup.

TWELVE

Temperature and Matter

In prehistory people learned to use fire for cooking food, one of the first applications of controlling the temperature of matter. The energy stored internally in matter determines whether it is a solid, liquid, gas, or plasma. Here chef Bruce is flambéing a dessert, igniting the vaporized alcohol above the pan.

Two of the earliest words of children are "hot" and "cold" because they easily sense differences in temperatures. They quickly learn to avoid touching hot stoves or getting too close to fires, and to dress up against the bite of winter cold. We increase the temperatures of foods to cook them and preserve foods in refrigerators that keep the temperature low, and we routinely adjust air temperatures in homes, classrooms, and offices for comfort. Everyone knows what the word temperature refers to, but if you are asked what it means, it might be difficult for you to explain. This chapter is about what temperature measures and about temperature's effects on matter.

FIGURE 12-1 TO COOL A CAR'S ENGINE, LIQUID IS CIRCULATED THROUGH THE HOT ENGINE AND THEN THROUGH THE RADIATOR, WHERE IT IS COOLED BEFORE ITS NEXT PASS THROUGH THE ENGINE. BUT THE LIQUID'S VOLUME INCREASES WHEN THE TEMPERATURE OF THE LIQUID FIRST RISES, AND THE OVERFLOW TANK (ARROW) OF THE COOLING SYSTEM CATCHES ANY COOLANT FORCED OUT BY EXPANSION.

TEMPERATURE AND THE VOLUME OF MATTER

Matter—whether solid, liquid, or gas—usually expands if its temperature is increased. (Water near freezing is an exception, as we'll see shortly.) When you warm a liquid or a solid, the molecules move faster and stretch farther against their bonds, causing the substance's volume to increase. Because the bonds are stronger in solids, they don't expand as much as liquids do. Gasoline stored in an underground tank at a service station is cool. Pumped into a car's gas tank on a hot summer's day, it will expand as it warms and will soon overflow if you've filled the tank to the brim. Likewise the coolant in a completely filled automobile radiator will expand and spill over into its overflow tank as soon as the hot engine warms it (Figure 12-1).

When the heating system comes on in a cold house, the ducts and vents expand. They pop and snap as they expand unevenly and as they stretch against their anchors. Likewise, a house sometimes creaks and groans in the quiet of night as its parts shrink unevenly from the day's expansion. Different solids expand or contract by different amounts for the same temperature change due to differences in their bond strengths and atomic structures. Masonry, for example, expands only a third as much as steel for the same change in temperature. If heating or cooling is too uneven, a solid may crack, as Figure 12-2 shows. A chilled glass may fracture as you pour hot liquid into it, and a glass hot from the dishwasher may crack if you fill it with ice water. That happens because the inside of the glass expands or contracts too fast for the outside, causing a great stress in the material.

IN MANY CALIFORNIA OIL WELLS, THE UNDERGROUND OIL IS TOO THICK AND VISCOUS TO BE PUMPED UP TO THE SURFACE. HOT WATER IS FORCED DOWN UNDER PRESSURE TO CAUSE THE OIL TO FLOW. THESE LOOPS ARE EXPANSION JOINTS THAT TAKE UP THE SLACK (THE EXTRA LENGTH) AS THE PIPES FOR SUCH A WELL EXPAND GREATLY FROM THE GREAT TEMPERATURE CHANGES.

A practical use for expansion and contraction is found in home thermostats, which turn heaters and air conditioners on and off depending on the temperature. Thin strips of brass and steel (or two other dissimilar metals) are sandwiched together into a springlike coil (Figure 12-3). Because the two metals expand or contract by different amounts for the same temperature change, the coil winds or unwinds according to the temperature and, in doing so, triggers an on–off switch.

Different liquids also expand by different amounts for the same temperature change, but gases do not. Unless the gas is very dense, the molecules are too far apart on the average for their intermolecular attractions to influence the expansion or contraction of the gas. If two balloons are filled with different gases, such as oxygen and nitrogen, but each contains the same number of molecules, the gases in those two balloons (and hence the balloons themselves) will expand or contract by equal amounts for the same change in temperature.

FIGURE 12-2 A CERAMIC DISH PLACED IN A HOT OVEN TO KEEP FOOD WARM CAN DEVELOP CRACKS IN ITS GLAZED COATING. THIS OCCURS FOR TWO REASONS. THE SURFACE OF THE DISH GETS HOT BEFORE THE INTERIOR DOES, PUTTING A STRESS ON THE COATING, AND THE COATING AND THE CERAMIC BENEATH IT MAY EXPAND BY DIFFERENT AMOUNTS FOR THE SAME TEMPERATURE CHANGE.

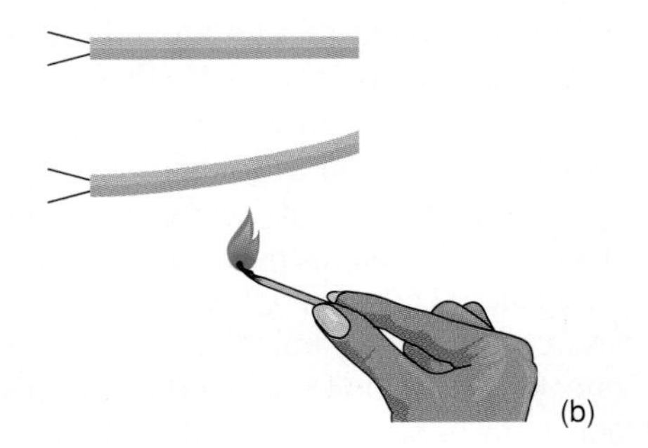

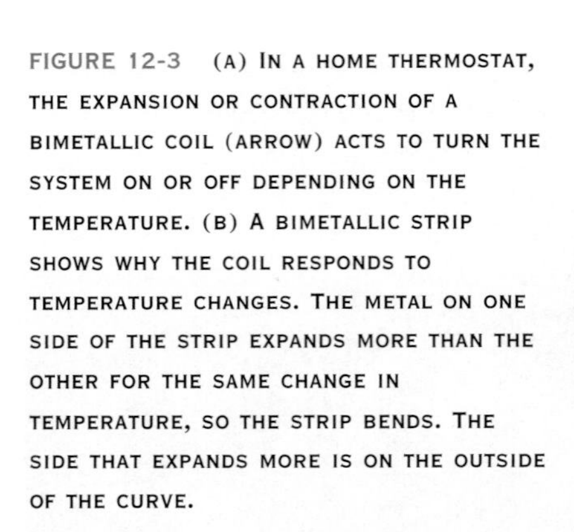

FIGURE 12-3 (A) IN A HOME THERMOSTAT, THE EXPANSION OR CONTRACTION OF A BIMETALLIC COIL (ARROW) ACTS TO TURN THE SYSTEM ON OR OFF DEPENDING ON THE TEMPERATURE. (B) A BIMETALLIC STRIP SHOWS WHY THE COIL RESPONDS TO TEMPERATURE CHANGES. THE METAL ON ONE SIDE OF THE STRIP EXPANDS MORE THAN THE OTHER FOR THE SAME CHANGE IN TEMPERATURE, SO THE STRIP BENDS. THE SIDE THAT EXPANDS MORE IS ON THE OUTSIDE OF THE CURVE.

MEASURING TEMPERATURE

Rough estimates of temperatures are all we need for many purposes. A new dad tests the temperature of a baby's formula by sprinkling a few drops on the underside of his wrist, for example, and a cook judges the temperature of the cooking oil by whether a drop of water sizzles in it. But for more exact estimates, we use thermometers. In a household thermometer, a bit of liquid enclosed in a glass bulb expands if the air warms it and contracts if the air cools it. The rest of the thermometer is a thick glass tube with a very slender cavity. As the pool of mercury or red-colored alcohol in the bulb expands, it has nowhere to go but into that skinny tube. (The glass expands and contracts a smaller amount.) A calibrated scale along the tube shows the temperature of the thermometer's bulb and hence the temperature of its surroundings.

There are three scales in use today for measuring temperature (Figure 12-4). The Celsius scale reads zero at the position of the liquid when the bulb

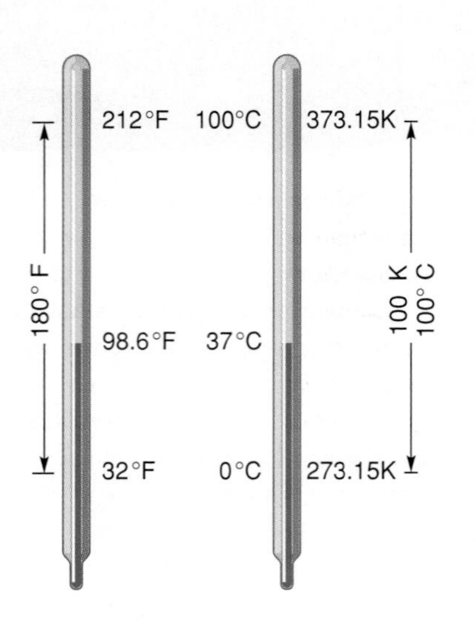

FIGURE 12-4 THESE TWO THERMOMETERS ARE IDENTICAL EXCEPT FOR BEING MARKED WITH DIFFERENT SCALES, CELSIUS AND FAHRENHEIT. ALTHOUGH WATER FREEZES AT THE SAME "TEMPERATURE," THAT POINT IS 0° ON THE CELSIUS SCALE AND 32° ON THE FAHRENHEIT SCALE.

is in freezing water and 100 when the bulb is in boiling water at sea level. The liquid in the thermometer expands evenly between these two points, so the distance between them along the glass tube is divided into 100 equal parts, or Celsius degrees (C°). The Fahrenheit scale reads 32 when the bulb is in freezing water and 212 when the bulb is in boiling water at sea level, so 180 Fahrenheit degrees (F°) separate the two points. The Kelvin scale has intervals that match the Celsius degrees. The difference in these scales is their "starting" temperatures. The Kelvin scale has its zero point at absolute zero, −273.15°C on the Celsius scale, a temperature we discuss in a later section. As a result, the Kelvin temperature is equal to the Celsius temperature plus 273.15 degrees. There is no degree symbol for a Kelvin temperature; 300 "degrees" on the Kelvin scale is written as 300 K, and we say this as 300 kelvins (*not* 300 degrees Kelvin).

Temperature changes also affect the way metals and other materials conduct electricity. Modern thermometers used by hospitals and physicians use that property to measure the body's internal temperature.

PUTTING IT IN NUMBERS—CONVERTING TEMPERATURES

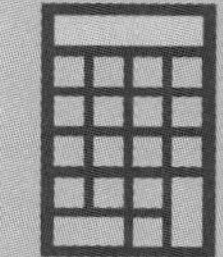

A simple formula converts temperatures from Fahrenheit to Celsius, or vice versa. If T_C stands for the Celsius temperature and T_F for the Fahrenheit temperature, then

$$T_F = \frac{9}{5}T_C + 32 \text{ or } T_C = \frac{5}{9}(T_F - 32)$$

Example: Find the average internal temperature of humans on the Celsius scale. That temperature is 98.6°F, so by the second formula,

$$T_C = \frac{5}{9}(98.6 - 32) = \mathbf{37.0°C}$$

WHAT TEMPERATURE ACTUALLY MEASURES

James Joule was an English brewer who studied heat and temperature in the 1840s because they are important in the brewing process. He placed a paddle wheel in a tub of water and measured the small temperature changes caused when the paddles stirred the water. He let a falling weight attached to a cord turn the wheel (Figure 12-5). The slowly falling weight did work on the wheel, and the paddles did work on the water, increasing its internal energy (the sum of the molecules' kinetic and potential energies). Joule found that the temperature of the water rose a precise fraction of a degree for every unit of work the wheel delivered to the water. This discovery—that equal amounts of work, or energy, cause equal increases in temperature—was a clue about what temperature means.

FIGURE 12-5 A SCHEMATIC OF JOULE'S EXPERIMENT. THE WEIGHT FELL, DOING WORK (*MGH*) AS IT TURNED THE PADDLES. THE PADDLES STIRRED THE WATER, RAISING ITS TEMPERATURE. JOULE FOUND THAT EQUAL AMOUNTS OF WORK RAISED THE TEMPERATURE OF THE WATER BY EQUAL AMOUNTS, SHOWING HEAT TO BE A FORM OF ENERGY.

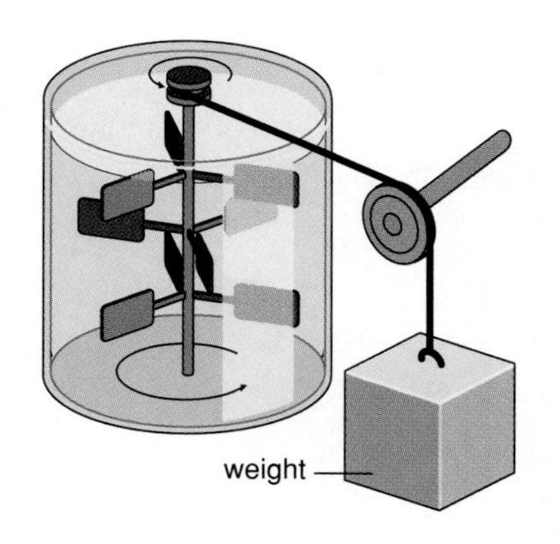

If Joule had stirred alcohol or molasses in his tub instead of water, the same amount of work would not have produced the same temperature change. For example, if 1000 joules of energy from the falling weight raised the temperature of water in his vat by 0.01C°, that same amount of energy would raise the temperature of an equal mass of alcohol by 0.02C°. As we'll see in the next sections, different substances store their internal energy differently.

The molecular picture of matter eventually explained this. This model showed that temperature is not a measure of matter's total internal energy; it measures only the portion of the energy that is in the form of random translational kinetic energy, $\frac{1}{2}mv^2$. A molecule or atom in a solid moves for short distances around its place in the solid, vibrating chaotically as it is pounded by its neighbors, jiggling like a leaf in a hurricane. Because of this random motion, that atom has an average translational kinetic energy, $\frac{1}{2}mv^2$, and its value grows higher if the solid's temperature climbs. The model shows that *the Kelvin temperature of matter varies as the* ***average kinetic energy*** *of its molecules.* In other words, it is the kinetic energy portion of the matter's total energy that the Kelvin temperature indicates.

UNITS FOR HEAT TRANSFER AND INTERNAL ENERGY

As we saw in Chapter 6, *heat is energy moving from one place to another because of a temperature difference.* Heat always moves *from* a warmer object *to* a cooler object when the two are placed in contact. If you are holding a glass of iced tea, you might say, "Cold things cool warmer ones." "Cold" doesn't move into your hand, however; heat flows from your hand to the glass because your hand is initially warmer. The internal energy of your hand decreases as its molecules transfer some energy to the glass, and as a result the internal energy of the glass increases.

The transfer of heat and the corresponding changes in internal energy are often measured in **calories** (cal), a term coined in the 1700s when scholars thought a fluid called "caloric" was passed from one body to another whenever a temperature change took place. By definition, **one calorie is the amount of heat energy that raises the temperature of 1 gram of water by 1 Celsius degree.** One thousand calories (1 kcal) equals 1 Calorie, with a capital C. On diet and nutrition charts, Calories are the units used for food energy. The relationship between calories and joules is

REMEMBER FROM CHAPTER 6 THAT A JOULE IS THE ENERGY USED TO LIFT A MEDIUM-SIZED APPLE (WEIGHT = 1 NEWTON) STRAIGHT UP FOR 1M. THE BRITISH THERMAL UNIT (BTU) IS ANOTHER MEASURE OF HEAT, STILL USED TO RATE AIR CONDITIONERS AND FURNACES IN THE UNITED STATES. ONE BTU (= 1054 J) IS THE HEAT NEEDED TO RAISE THE TEMPERATURE OF 1 POUND OF WATER BY 1 F°.

$$1 \text{ calorie} = 4.18 \text{ joules}$$
$$1 \text{ Calorie} = 1 \text{ kcal} = 4180 \text{ joules}$$

So if you consume 2400 Calories in a day, that's about 10 million joules of energy that your body turns into heat energy (mostly) and physical work by your muscular exertions.

HOW MATTER STORES INTERNAL ENERGY

Different types of matter, as we've said, store added energy differently. For example, 1 calorie of heat will raise the temperature of 1 gram of water by 1C°. That same calorie of heat would raise the temperature of 1 gram of copper almost 11 C°. So water stores more internal energy than the same mass of copper for each degree of temperature increase. The differences are indicated by a quantity called specific heat. The **specific heat** of a substance is *the quantity of heat that will raise the temperature of 1 gram by 1 Celsius degree.* By definition, 1 calorie of heat raises the temperature of 1 gram of water by 1C°, so the specific heat of water is 1 cal/g·C°. To raise the temperature of 1 gram of copper by 1C° requires only 0.093 calories of heat, however, so copper's specific heat is 0.093 cal/g·C°. Matter with a higher specific heat stores more energy for each degree of temperature change.

The temperature rise of an object depends not just on its specific heat but also on the amount of matter. If 100 J of heat is added to a gallon of water, the temperature rise is twice as great as if the heat were added to 2 gallons of water, since there are only half as many molecules to share the energy in one gallon. The way that a quantity of heat added to or subtracted from an object changes its temperature is summarized in this relationship:

PUTTING IT IN NUMBERS—CALCULATING TEMPERATURE CHANGES

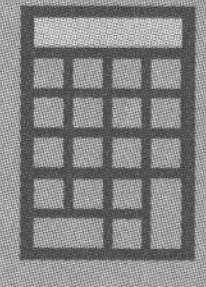

Heat added (or subtracted) = Specific heat × mass × change in temperature

Example: How much heat would it take to warm a liter (about a quart) of water from 15°C to 90°C? From Table 12-1 the specific heat of water is 1.0 cal/g·C°, and one liter is 1000 cm³, which has a mass of 1000 g. The heat you'd need to add to the water is

(1.0 cal/g·C°) × (1000g) × (90°C − 15°C) = **75,000 calories**, or 75 food Calories.

The simplest kind of gas, such as a gas made up of single atoms of helium or argon or neon, also has the simplest manner of storing internal energy (Figure 12-6a). When heat is added to such a gas, all the added energy goes to increase the average kinetic energy $\frac{1}{2}mv^2$ of the atoms and hence its Kelvin temperature, T. For that reason, the temperatures of single-atom gases increase the most per gram per calorie added, so they have the smallest specific heats. Other substances store part of any added heat in other forms than kinetic energy, meaning more energy is needed to change their temperatures by a given amount—in other words, they have larger specific heats. For one example, the molecules or atoms in solids or liquids stretch apart when they gain kinetic energy, storing energy as potential energy in their spring-like bonds. That energy which is stored does not increase the molecules' kinetic energy and therefore does not increase the temperature (Figure 12-6b). As you can see from Table 12-1, the specific heat of water is very high compared

Table 12·1 ***Specific Heats (calories per gram · degree C)***

Gold	0.03
Platinum	0.03
Lead	0.03
Mercury	0.033
Silver	0.06
Copper	0.093
Iron	0.11–0.12
Air	0.2
Clay	0.2
Granite	0.2
Aluminum	0.21
Wood	0.4
Ice	0.55
Steam	0.5
Alcohol	0.58
Paraffin	0.7
Water	1.0
Apples	0.92
Beef	0.75
Chicken	0.80
Grapes	0.92
Pork	0.50
Potatoes	0.80

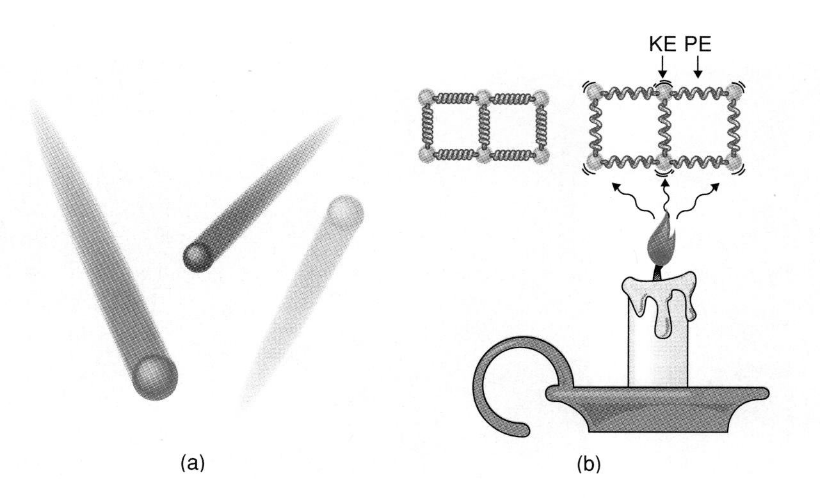

FIGURE 12-6 (A) HEAT TRANSFERRED TO A GAS OF SINGLE ATOMS IS "STORED" IN THE FORM OF THE KINETIC ENERGY OF ITS MOLECULES, $\frac{1}{2}MV^2$. (B) ATOMS OR MOLECULES OF A SOLID HAVE KINETIC ENERGY BECAUSE THEY VIBRATE; THEY ALSO HAVE POTENTIAL ENERGY BECAUSE OF THE BONDS THAT HOLD THE SOLID TOGETHER. IF THE ATOMS VIBRATE FASTER, THEY PULL OUT FARTHER AGAINST THOSE BONDS.

to that of most other substances, which is one of the reasons water is so useful in fighting fires, and why a water-rich slice of apple from a hot pie will burn your tongue when the crust of the pie won't.

CHANGING STATES

The ice cubes in your refrigerator's freezer are probably at −25°C (−13°F) or even colder. Drop those cubes into a glass of water, and they begin to warm from contact with the warmer water. The cubes' surfaces quickly warm to 0°C, and then they continue to absorb heat—but now their temperature doesn't climb any higher! Instead, the ice *melts,* absorbing a lot of heat without a change in temperature. The added energy doesn't disappear; it breaks the bonds of the crystallized ice, which raises the potential energy of the molecules—they would have to lose energy to become a solid again. That energy changes the *state* of the water from solid to liquid. It is called the *latent* (hidden) **heat of fusion.** For 1 gram of ice at 0°C to melt and become water at 0°C takes 80 calories of energy—so water's latent heat of fusion is 80 calories per gram (Figure 12-7). That's a lot of energy—for example, 10 grams of water from a faucet would have to cool by 8C° to give up enough energy just to melt 1 gram of ice. That's why only a few ice cubes will cool a large drink.

To freeze a gram of water at 0°C, 80 calories of heat must be taken from it. Any melting solid *absorbs its heat of fusion from its surroundings while*

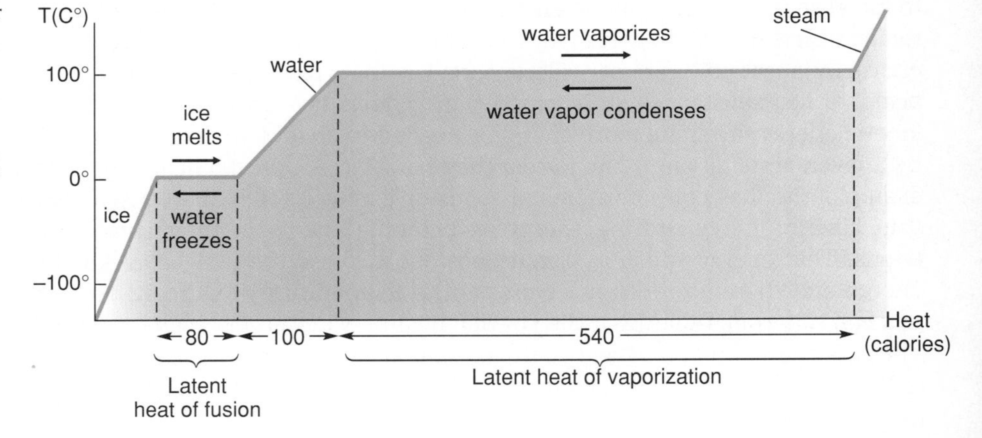

FIGURE 12-7 A TEMPERATURE VERSUS HEAT CURVE FOR ONE GRAM OF WATER. MOVING TO THE RIGHT ON THIS GRAPH, HEAT IS BEING ADDED; MOVING TO THE LEFT, HEAT IS BEING SUBTRACTED. NOTICE THAT WHEN THE WATER IS CHANGING STATE, THE LATENT HEATS BEING ADDED OR SUBTRACTED DO NOT CHANGE THE WATER'S TEMPERATURE.

TIM IS USING AN ACETYLENE TORCH TO CHANGE THE STATE OF A SMALL AMOUNT OF METAL TEMPORARILY AS HE WELDS.

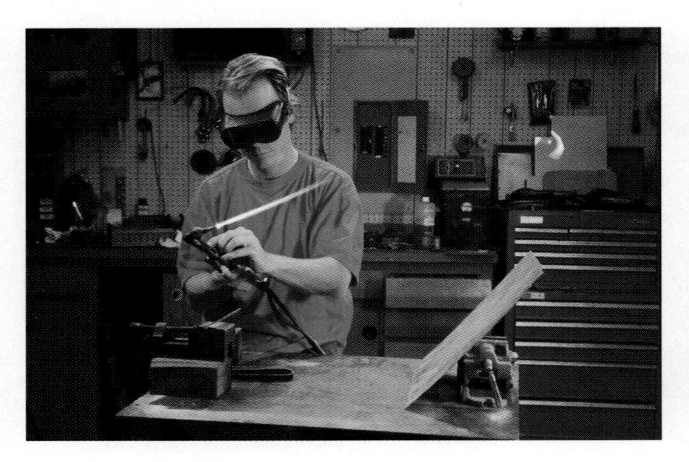

its own temperature doesn't change. Likewise, any freezing liquid *releases the heat of fusion to its surroundings while its own temperature doesn't change.*

Heat from a stove can quickly turn the water in a pot into steam, which is just water vapor, a gas of water molecules. Water with 1 atmosphere of air pressure on its surface can get no hotter than 100°C, at which point any extra heat transferred goes into *vaporizing* the water. The energy that goes to break the liquid's molecular bonds is called the *latent* **heat of vaporization** (Figure 12-7). At sea level, it takes a whopping 540 calories of heat energy to vaporize a gram of water that is at 100°C. In the reverse process, sea-level water vapor at 100°C can *condense,* turning into liquid water at 100°C, but only if the vapor loses 540 calories per gram of water vapor. Any vaporizing liquid *absorbs its heat of vaporization* from its warmer surroundings while its own temperature doesn't change. Likewise, any condensing gas *releases the heat of vaporization* to its cooler surroundings while its own temperature doesn't change. If a solid *sublimes,* changing directly into a gas (or vice-versa), the **heat of sublimation** is absorbed (or released).

As seen in Figure 12-7, once water has frozen, the ice can be taken to lower and lower temperatures. Once water has vaporized, the water vapor can be taken to higher and higher temperatures. It is only during the change of state that the temperature does not change as heat is added or subtracted.

FREEZING WATER

Water at room temperature (about 20°C) shrinks in volume if it begins to cool, just as other liquids do (see Demonstration 1). Its molecules slow, and the intermolecular bonds bring them closer together. At 4°C, however, the shrinking stops; the water is as dense as it ever gets. If cooled further, it expands (Figure 12-8). When water freezes at 0°C, the molecules bond together in a definite pattern. The structure of ice leaves extra empty space between the molecules; that is, they don't fit together as closely as they do in the liquid state (Figure 12-9). As liquid water at 0°C freezes, it expands and takes up about 10 percent more volume, which is why ice floats.

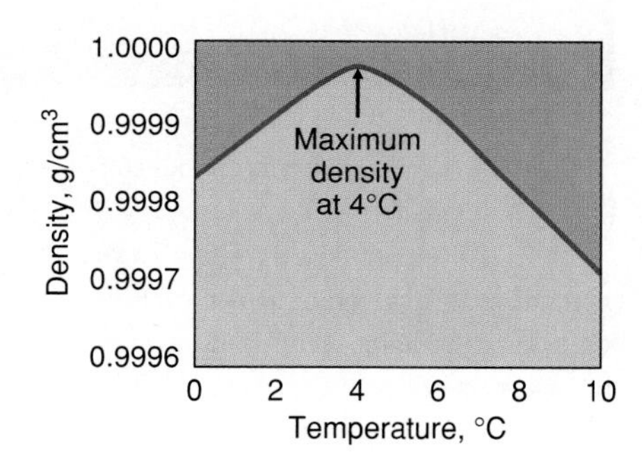

FIGURE 12-8 THE DENSITY OF WATER IS GREATEST AT 4°C. THE CHANGES IN DENSITY ARE VERY SMALL, AS SEEN FROM THE LEFT SIDE OF THIS GRAPH, BUT SINCE WATER IS A FLUID, THE SLIGHTEST DIFFERENCES IN DENSITY CAUSE IT TO RISE OR SINK BECAUSE OF BUOYANCY.

The fact that water is most dense at 4°C rather than 0°C affects how northern lakes and ponds freeze when the air temperature drops below 0°C. The water in a pond loses heat to the cold air at its surface. That cooled water is heavier than the water below it, so it sinks to the bottom, pushing the deeper, warmer water to the top. Then this water cools and sinks. Eventually the water at the top will be at 4°C, and it is as dense as water ever gets. When this water sinks to the bottom, it stays there. At that point, each drop on the

SEASONAL MIXING OF LAKES

Zaca Lake, a spring-fed lake in a remote part of central California, is a beautiful coral blue in summer. In winter, however, the surface waters cool, become denser than the waters below, and sink. The churning waters bring sediments up from the lake bottom 60 feet below, and the lake changes colors as it turns "upside down." It first becomes milky gray, then shades of purple, magenta, and pink. The Chumash Indians who once lived by the lake thought spirits were beneath it, causing the whirlpools, currents, and color changes.

Most lakes undergo seasonal mixing like Zaca Lake does, turning upside down when the temperature changes. In regions with little temperature change, however, deep lakes aren't stirred this way, and separate layers can develop very different properties. The volcanic lakes of Cameroon in equatorial Africa mentioned in Chapter 10 are such lakes, and the lack of natural circulation has had disastrous consequences.

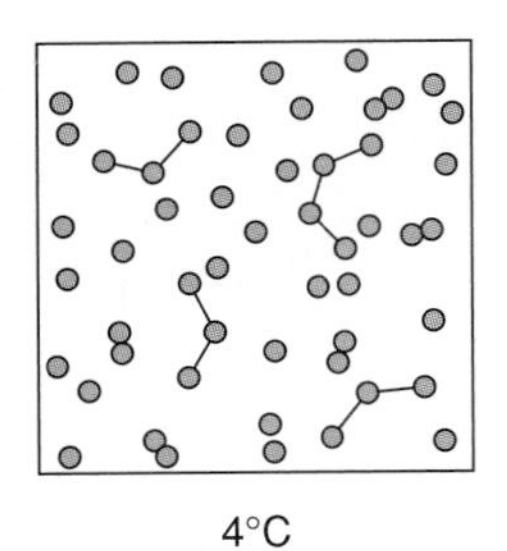

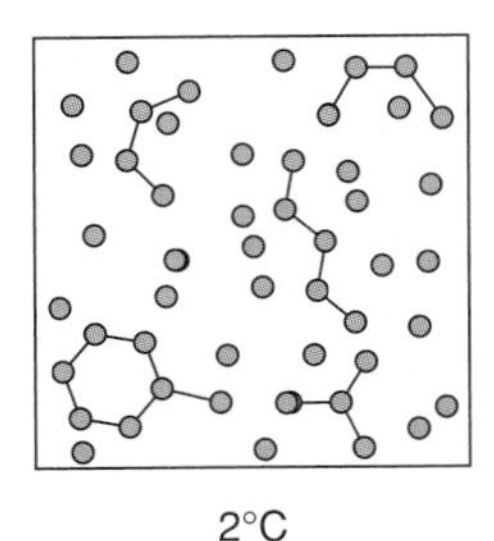

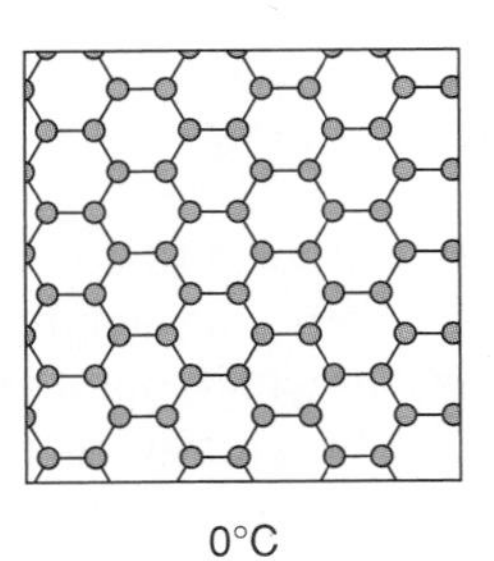

FIGURE 12-9 WATER EXPANDS WHEN ITS TEMPERATURE DROPS FROM 4°C TO 3°C, OR FROM 3°C TO 2°C, AND SO ON DOWN TO ZERO. THOUGH ICE CAN'T EXIST AT THOSE TEMPERATURES, SOME OF THE MOLECULES OF WATER ARE MOVING SLOWLY ENOUGH TO BOND TOGETHER BRIEFLY. BUFFETING BY OTHER WATER MOLECULES QUICKLY BREAKS THEM APART. NEVERTHELESS, AT ANY INSTANT, ENOUGH OF THESE SMALL "CRYSTALS" EXIST IN WATER THAT IS AT, SAY, 2°C TO LOWER THE WATER'S DENSITY. THIS HAPPENS BECAUSE THE MOLECULAR STRUCTURE OF ICE IS MORE OPEN THAN IS THE STRUCTURE OF WATER IN THE LIQUID STATE. ONCE AT 0°C, WATER WILL FREEZE SOLIDLY IF MORE HEAT IS REMOVED TO SLOW THE FASTER-MOVING MOLECULES TO THE POINT WHERE THEY CAN JOIN THE SOLID STRUCTURE.

SINCE EVAPORATION PROCEEDS FASTER WITH INCREASED TEMPERATURE, WATERING LAWNS AND CROPS IS BEST DONE IN THE EVENINGS OR EARLY MORNINGS. SPRINKLING SYSTEMS DISPERSE DROPS OF WATER INTO THE AIR, WHERE EVAPORATION PROCEEDS QUICKLY.

surface that cools to 4°C sinks, adding to the layer on the bottom *until the entire lake is at 4°C.* Then as water at the surface cools from 4°C to 0°C, it becomes less dense as the temperature drops and stays at the top, where it freezes as heat moves from it into the colder air. For each 80 calories of heat removed, a gram of ice forms. Once a layer of ice forms, it is more difficult for the water below it to freeze because heat lost by that water must travel through the ice before it can escape to the air. The thicker the ice, the more time that takes. This is why ponds and lakes that are more than a few feet deep do not freeze solid in winter, which would kill fish and other aquatic life.

EVAPORATION

Liquids don't have to be heated to the boiling point in order for molecules to leave their surfaces. It happens all the time in the process called **evaporation,** as when rain puddles on a sidewalk slowly disappear. At the surface of a liquid, the molecules have a wide range of speeds and travel in every possible direction. Attractions from nearby molecules within the liquid keep most of the molecules at the surface from leaving. However, when the most energetic molecules head almost straight out from the surface, they do escape. The rate of evaporation increases if the liquid is warmed because then more of its molecules have the

IF SOIL IS SATURATED WITH WATER, SOME WATER WILL BE HELD AT THE SURFACE IN CAPILLARY-SIZED (CHAPTER 10) HOLES. IF A HARD FREEZE OCCURS, THE WATER EXPANDS AND PUSHES OUT THROUGH AN OPENING TO THE SURFACE AS IT GOES FROM 4°C TO 0°C. WATER BELOW KEEPS RISING INTO THE SPACE BY CAPILLARY ACTION AND FREEZING AS IT EXITS, FORMING THESE THIN TUBES OF ICE THAT SOMETIMES LIFT UP CLODS OF SOIL.

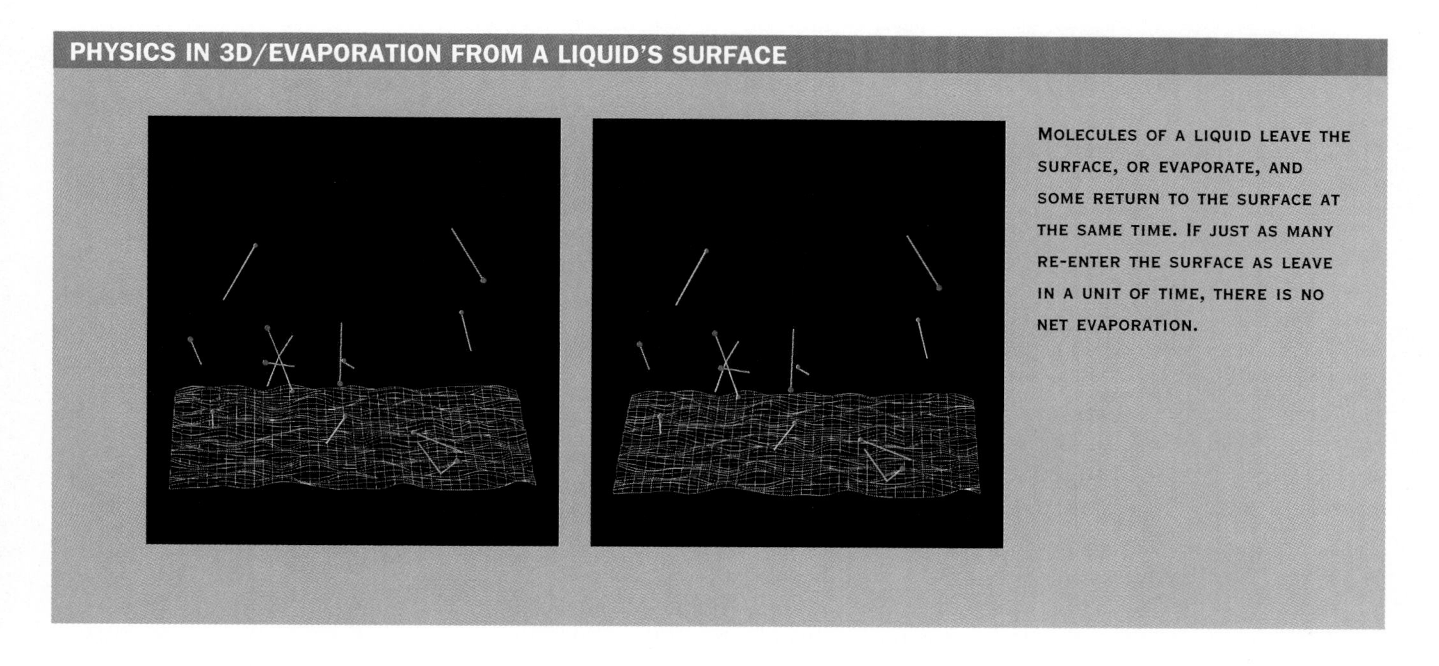

PHYSICS IN 3D/EVAPORATION FROM A LIQUID'S SURFACE

MOLECULES OF A LIQUID LEAVE THE SURFACE, OR EVAPORATE, AND SOME RETURN TO THE SURFACE AT THE SAME TIME. IF JUST AS MANY RE-ENTER THE SURFACE AS LEAVE IN A UNIT OF TIME, THERE IS NO NET EVAPORATION.

high kinetic energy needed to escape. No matter what its temperature, however, an evaporating liquid always absorbs the latent heat of vaporization from its surroundings. Even as molecules evaporate, some of the escaped molecules of liquid (or others previously in the vapor above) reenter the liquid's surface due to their random collisions with other molecules. So it is the *net* vaporization that is the rate of turning the liquid into a gas.

BOILING

A pot of water boils at 100°C at sea level, and no matter how high you turn the heat after the boiling begins, the temperature of the water doesn't go any higher. Small regions of water close to the heat source absorb enough energy to vaporize *even though they are under water.* The liquid molecules break away from each other, bursting into bubbles of water vapor by pushing nearby water aside, and the lightweight bubbles rise to the top. If the temperature beneath the pot is increased, more bubbles will form per second, increasing the boiling action, but the temperature of the water won't rise. So boiling is just, in a sense, very fast evaporation.

The temperature at which boiling takes place depends on the air pressure above the liquid. As we discussed in detail in Chapter 10, that pressure is transmitted throughout the liquid since the molecules of water are in contact with each other. At higher elevations, water boils at lower temperatures because the air pressure on its surface is less, making it easier for vapor bubbles to push back the water around them as they form near the pot's bottom. Many first-time backpackers in the mountains find this out when boiling water doesn't cook their food in the usual amount of time. Because the water is boiling, the camper thinks the usual cooking process is going on, but it is the water's temperature of 100°C that cooks the food (at sea level) at a certain rate, not the boiling. In the mile-high city of Denver, water begins to boil at about 94°C. If you could go near the top of the atmosphere where there's almost no air pressure, water would boil at the temperature of your room—and potatoes or rice boiled there would *never* get cooked.

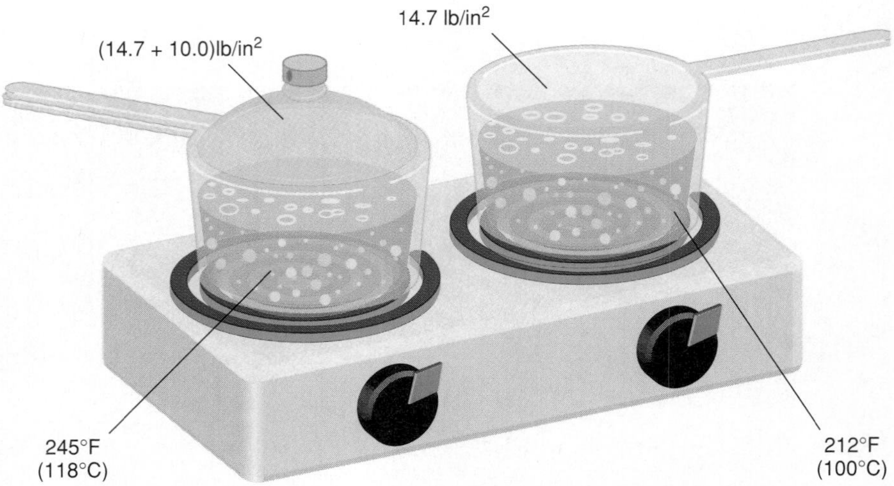

FIGURE 12-10 THE BOILING POINT OF WATER (AND OTHER LIQUIDS) DEPENDS ON THE AIR PRESSURE AT THE SURFACE OF THE LIQUID, SINCE THE BUBBLES OF VAPOR HAVE TO PUSH THE LIQUID BACK AS THEY FORM. PRESSURE COOKERS USE THIS FACT TO COOK WITH BOILING WATER AT TEMPERATURES HIGHER THAN 100°C.

In a pressure cooker, however, the story is different. There the water boils at temperatures higher than 100°C because the water in the pot is under a pressure greater than one atmosphere (Figure 12-10).

TURNING KINETIC ENERGY INTO HEAT

Each year in June, the racing world comes to Le Mans, France. There, over 8 miles of tight corners and straightaways have the full attention of some of the world's best racing drivers. They drive almost lying down, only inches from the road. In the straights they move at 220 mph, and everything looks blurred except the roadway in the distance. The longest straightaway is the one before the sharp turn known as the Mulsanne corner. The drivers see the turn coming, and then they have to wait. To maintain the highest average speed, a driver waits until the last possible instant to apply the brakes. Hard. To stay on the road, a car must go around that corner no faster than 35 mph. Spectators get a brief glimpse of the metal discs of the brakes glowing at 800°C (1500°F)—a bright cherry red—and the brake fluid boils in the lines next to that red-hot metal. At the corner, the car's brakes turn more than 95 percent of the car's kinetic energy into heat, on every lap, for 24 hours.

THE COLDEST TEMPERATURE: ABSOLUTE ZERO

As we mentioned in Chapter 11, an inflated balloon placed in a freezer shrinks noticeably in just a few minutes. As the air in the balloon cools, the molecules have a lower average kinetic energy, so they move more slowly and push less on the inside of the balloon. Any water vapor inside the balloon soon becomes frost on its inner surface. If the freezer could cool to −78.5°C, even the carbon dioxide molecules in the air inside the freezer would become a frost of dry ice. At much lower temperatures, the nitrogen and oxygen would become liquids and then solids. The balloon, of course, would be flat by then.

Is it possible to cool the air in the balloon—or the balloon or anything else—to a point where it can transfer no more heat? There is in fact a temperature at which the molecules of any substance have no more kinetic energy they can give up. At the temperature of −273.15°C (−459.7°F), matter is as cold as it can ever be. This coldest point is called **absolute zero,** and it is the zero on the Kelvin scale of temperature. Using sophisticated cooling apparatus, physicists today can produce temperatures of one millionth of a kelvin.

COMING HOME IN THE SPACE SHUTTLE

At an elevation of 150 miles, the space shuttle has a lot of potential energy. Moving at more than 17,000 mph, it has a lot of kinetic energy, too. Before it lands, the shuttle must lose all that energy. The first step is to slow the shuttle a bit so that it will angle down and catch the edge of the atmosphere. Turning the craft so the rocket engine points forward, the astronauts fire the engine briefly. When this retrofire has slowed the shuttle's speed by 200 mph, the engine is cut off, and the crew turns the shuttle back around. Then they tilt its nose upward and wait.

They feel the onset of deceleration about 90 miles up, as the shuttle's bottom becomes the world's largest brake shoe. At 60 miles up (where most meteors glow brightly and vaporize), the nose of the fast-moving craft ionizes the air upon impact. If the shuttle is on the dark side of earth, the astronauts see an eerie reddish-pink glow from that plasma. Because the ionized air reflects radio waves, the crew is out of touch with the ground for 16 (long) minutes. When they are still 10 miles up, they begin to steer the shuttle into sweeping S-shaped turns and wait for air resistance to lower the speed enough for a safe landing. From retrofire to touchdown, the trip takes half an hour—not a bad commuting time from work to home, especially considering the mileage.

HIGH TEMPERATURES: PLASMAS

Most solid things remain solid even if you dip them into boiling water and raise their temperatures to 100°C (373 K). An iron frying pan and clay bricks can survive the 1000 K temperature of a full-blown fire that consumes a burning house. Sand melts in a glassblower's oven at about 1400 K, however, and the temperature of molten steel flowing from a blast furnace might be over 1700 K. By the use of electric welding arcs, iron can be made to boil and vaporize at 2700 K. At the sun's surface, at 6000 K, matter exists only as a gas of atoms. Within a vapor such as that, any molecule suffers collisions so violent that it splits apart, separating into component atoms. When a gas is at a temperature of tens of thousands of degrees Kelvin, any colliding atoms can lose electrons. At this point, the gas becomes a **plasma,** which is an ionized gas, because the electrons and the partially stripped atoms have net electric charges. At the center of the sun the plasma is at a temperature of about 15,000,000 K, and at the heart of exploding stars, the temperature briefly reaches billions of kelvins.

When a stroke of lightning passes between the earth and a cloud, a plasma exists along that jagged, superheated path through the air. In modern street lamps, a tiny bit of plasma glows brightly and lights city streets, a kind of artificial lightning that has its temperature controlled electrically. Even the fluorescent lights in libraries and classrooms make use of thin, lower-temperature plasmas, although the light from these lamps comes from a fluorescent coating on the inside of the tubes rather than the plasmas themselves. When you strike a match, there are charged particles in the hot gases in the flame.

Summary and Key Terms

As a liquid or solid grows warmer, it generally expands. Liquids and solids expand when warmed because their molecules move faster, stretching outward against their bonds.

Temperature is expressed in degrees Celsius or Fahrenheit, or in Kelvins. At sea level, water freezes at 0°C (32°F) and boils at 100°C (212°F). The Kelvin scale has divisions that match the Celsius degrees, but the Kelvin scale

absolute zero–180

starts at ***absolute zero***, −273.15°C. The Kelvin temperature of matter is directly related to the average random kinetic energy, $\frac{1}{2}mv^2$, of its molecules. Absolute zero is the temperature at which the molecules of any substance have no more kinetic energy they can give up.

Heat is energy that moves due to a temperature difference, and heat flows from an object of higher temperature into an object at a lower temperature in contact with it. Heat is measured in joules or calories. By definition, one ***calorie*** is the amount of heat energy that raises the temperature of 1 gram of water by 1 Celsius degree. Since ***specific heat*** is the energy needed to raise a unit of mass of a substance by 1 Celsius degree, water's specific heat is 1 cal/g·C°. A higher specific heat means the substance can store more energy for each degree of temperature change: *Heat added (or subtracted) = specific heat × mass × change in temperature.*

calorie–174
specific heat–174

The energy that goes into breaking a solid's bonds to turn it into a liquid is called the latent ***heat of fusion***. Any melting solid absorbs the heat of fusion from its surroundings, and any freezing liquid releases the heat of fusion to its surroundings, while its own temperature doesn't change. The energy that goes into breaking a liquid's molecular bonds to turn it into a gas is called the latent ***heat of vaporization***. Any vaporizing liquid absorbs the heat of vaporization from its surroundings, and any condensing gas releases the heat of vaporization to its surroundings, while its own temperature doesn't change. If a solid sublimes, or changes directly from a solid to a gas, the ***heat of sublimation*** is absorbed. In ***evaporation***, the molecules of a liquid leave its surface even when no extra heat is added.

heat of fusion–176
heat of vaporization–177
heat of sublimation–177
evaporation–178

At very high temperatures, the colliding atoms of a gas can ionize, forming a ***plasma***. Absolute zero is the lowest possible temperature for any atom or molecule, where it will have no kinetic energy that it can transfer as heat.

plasma–181

EXERCISES

Concept Checks: Answer true or false

1. Liquids and solids generally expand when heated. Solids generally expand the most.

2. There are three temperature scales in general use, and the numerical values for the freezing point and boiling point for water at sea level are different on each scale.

3. Temperature is a measure of the average kinetic energy of the molecules of a substance.

4. Heat refers to energy that moves from one place to another because of a difference in temperature.

5. The temperature of a substance is proportional to the average total energy of its molecules.

6. The heat energy in an object depends on its mass, its specific heat, and its Kelvin temperature.

7. The specific heat of a substance indicates the heat needed to raise its temperature 1°C per unit of mass.

8. The latent heat of fusion of a substance indicates the heat needed to melt a solid, per unit of mass.

9. When matter changes from the solid to the liquid state, energy is transferred without a change in the temperature of the substance. Heat leaves the solid during this process.

10. Water at 3°C is less dense than water at 4°C.

11. The latent heat of vaporization indicates how much energy the substance loses per unit of mass if it condenses.

12. Absolute zero occurs at the same point on the Celsius and Fahrenheit scales.

Applying the Concepts

13. If you ever want to freeze some apple cider in gallon jugs, you'd be well advised to drink a cup from each full jug first. Why?

14. Discuss what would happen in a thermometer if mercury didn't expand at a very uniform rate between 0°C and 100°C. What would the scale have to look like?

15. Discuss this statement: Almost all the energy from a car's engine ultimately goes to pushing air around and heating the brakes.

16. Exactly how do materials expand when they are warmed? Do all materials expand when heated? Discuss.

17. A heated gas expands and cools if it can push back the matter that surrounds it. But in the kitchen device known as a pressure cooker, any gas inside is trapped in a space of constant volume. What happens as the gas is heated?

FIGURE 12-11

18. Why do ice cubes snap and crackle when put into a glass of warm tea?

19. How could the height of the Jefferson Memorial Arch in St. Louis (Figure 12-11) be used as a seasonal indicator?

20. Ancient miners (before the day of the iron pick) heated exposed copper ore inside the ground and then doused it with water. How could this help extract copper from the mine?

21. Don't leave your cassette tapes on the dashboard in the sun. Discuss what will happen if you do and why it happens.

22. James Joule found that a temperature increase of a substance was directly proportional to the energy added to it. True or false?

23. If you have a fever and find that your temperature is 39°C, what is that temperature on the Fahrenheit scale?

24. The highest recorded temperature for a location in the United States is 134°F. What is that on the Celsius scale?

25. Skiers sometimes have to melt wax on their skis and then iron out the drips to smoothly coat the skis' surface. What is the energy called that they supplied to do that?

26. A wise cook does not let a tightly fitted pot lid remain on a hot pot of stew that's left to cool. Why not?

27. For every 10°F temperature drop, a car's tires lose about 1 lb/in^2 of air pressure, and if the temperature rises the air pressure in the tires increases the same way. So when should you check the air pressure in your tires, before a trip or during the trip?

28. Once the water begins to boil in a pot of vegetables (or eggs to be hard-boiled), why should you adjust the stove element?

29. Exactly why does a drop of water thrown into a hot frying pan jump around?

30. Can two objects of the same mass but different composition have the same total internal energy if their temperatures are not the same? Discuss.

31. Discuss this statement. Without an atmosphere, earth's oceans would boil away.

32. Compare what would happen if you boiled one egg at sea level and another egg on a high mountain.

33. If you lick a silver spoon that's been in a very hot cup of coffee, it probably won't burn your tongue. But a spoonful of that coffee dropped on your tongue could leave a blister. Why?

34. How does water's high specific heat make it a good coolant for car engines? How does water's expansion upon freezing make it a poor coolant during subfreezing weather?

35. Pyrex glass, used to make bowls and dishes that can go directly into ovens, expands about 1/3 as much as ordinary glass does. How does that help prevent cracking when a casserole dish of pyrex goes into the oven?

36. When a teapot containing water nears the boiling point, you can usually hear distinct popping noises. Speculate why, and then be sure to read the answer.

37. A car's coolant system is closed by a tightly fitting radiator cap that maintains pressure within the system. Why is this done?

38. Show that an 1100-kg racing car slowing from 220 mph (98 m/s) to 35 mph (15.6 m/s) dissipates more than 5 million joules of kinetic energy. (You consume twice that much energy in a 2400-Calorie day.)

39. Someone consumes an amount of food that contains 2400 kilocalories in a day; calculate how much water (at 100°C) that much energy could turn into vapor.

40. Which could give you a more severe burn, 2 g of water at 100°C or 2 g of steam at 100°C?

41. Why can a tub of water placed in a food cellar on a farm keep vegetables from freezing on a cold night? (*Hint:* Fresh vegetables and fruit typically freeze at about −1°C, or 30°F.)

42. It helps to run hot water over the metal top of a jar of food when it's on too tight. Metals expand slightly more than glass for the same temperature change, but metals typically have much smaller specific heats than glass does. Explain how that could make the difference.

43. The specific heat takes into account all the ways in which matter stores energy, not just the kinetic energy of its molecules. True or false?

44. If you heat water to 100°C, does it necessarily boil?

45. You could know when a cooling liquid begins to freeze (or a warming liquid has reached its boiling temperature) by observing a thermometer in it. How would you know?

46. Can a thermometer placed in an ice tray in a freezer measure the latent heat of fusion? Can a thermometer placed in a pot of boiling water measure the latent heat of vaporization?

47. A spent lead bullet dug from a piece of wood it had penetrated is partially melted. Where did the heat to do that come from?

48. Why is cooking oil better for cooking at high temperatures than water is?

49. Why is water so useful in hot water bottles?

50. Oftentimes when you park a car and get out, you can hear small "ticking" noises under the hood. Why?

51. If a car's engine overheats, you should add water to the radiator slowly and keep the engine running as you do. Why?

52. Submarines can travel to the North Pole by remaining under the polar icecap, which never gets thicker than 4 meters even in the darkness of the polar winter. What keeps the ice from growing thicker there?

53. Mixtures of two different liquids have different freezing and boiling points from either of the pure substances. Pure antifreeze freezes at about 8°F, while a mixture of half antifreeze and half water freezes at −34°F. What other consideration might apply to the mixture you should use in a car's radiator?

54. Bumblebees sometimes need to warm up the larvae and pupae in their nests. When they do, they first decouple the flight muscles from their wings. Tell how this can help.

55. During the construction of a large dam, pipes are laid a few feet apart throughout the concrete, and water is run through them while the concrete hardens. Why is this necessary?

56. A 4.0 gram lead bullet traveling at 250 m/s enters a wooden block and stops. If 50 percent of its initial kinetic energy changes into thermal energy in the bullet, by how much does its temperature increase?

57. If you wipe a puddle of water over a countertop with a wet dishcloth, the small droplets of water the cloth leaves dry up quickly by evaporation, whereas the larger puddle would take longer to disappear. Why?

58. Have you ever seen tennis players sweep the water puddles on a court after a rain? How does this help?

59. Internal energy refers not only to translational kinetic energy of molecules but also to types of energy such as rotational energy, vibrational energy, and potential energy. True or false?

60. Why is the heating element in hot water heaters located at the bottom of the tank?

Demonstrations

1. Fill a 16-oz glass soft-drink bottle to the brim with water from your cold water faucet. Make certain the water is level with the opening. Then run some hot water into the sink and place the bottle in it. In no time you can see the water overflow the top of the bottle as the water warming in the bottle expands. The glass bottle expands too, but water and other liquids expand more than glass does. The water in the bottle stops expanding when it heats to (nearly) the temperature of the water in the sink. Then if you leave and come back in a few minutes, you'll find the level of the water has dropped below the rim. Warm water evaporates fast!

Next, rinse the bottle with water from the cold water faucet to reduce its temperature and fill it to the brim with cold water again. This time place it in your refrigerator and check it in 15 minutes. The decrease in volume is very noticeable. Except for its behavior between 4°C and 0°C, water that is being cooled behaves just as you'd expect it to; it shrinks. Some of the drop in level is due to evaporation; but not much, as you can show by placing the cold bottle of water into a sink of water at room temperature. The expansion brings it nearly back to its original level.

2. Air, a gas, expands even more than water for the same increase in temperature. Stretch a balloon flat over the rim of the same empty glass bottle you used in Demonstration 1 to prove it. Tie it tightly to the lip of the bottle so no air will escape, but without stretching the balloon very tightly across the opening. Then immerse the bottle in hot water or just run hot water over it, and place the bottle in the refrigerator. The expansion and contraction of air is almost 20 times that of water, as you will see.

For a variation, put the empty bottle into the refrigerator and get it cold. Then take it out and immediately place a wet nickel over the opening. Wrap your hands around the bottle to increase its temperature, and the expanding air inside makes the nickel clank against the bottle.

3. Sneakers too tight? Don't throw them away without trying this. Shove a plastic bag deep into each shoe, fill the bags with water and close the tops tightly (wrap them with packing tape), making sure there are no air bubbles trapped inside. Put the shoes in the freezer of your refrigerator for a day. The shoes will fit a foot that's 10 percent larger in volume!

THIRTEEN

Heat: Energy in Motion

When light travels through air of different temperatures, as when it passes through convection currents, its direction changes. The wavering convection currents distort the images of objects seen through them. This painting, taken from a photograph made with a telephoto lens, shows this effect dramatically.

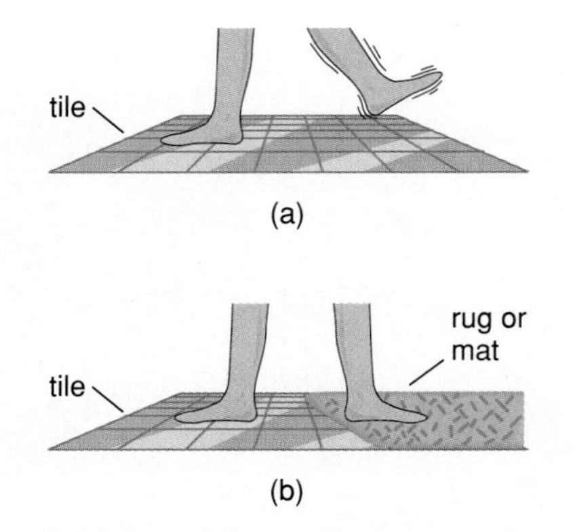

FIGURE 13-1 (A) A FOOT IN CONTACT WITH A COLD TILE FLOOR LOSES HEAT MORE RAPIDLY THAN A FOOT HELD UP IN THE AIR, AND SO THE FOOT ON THE FLOOR GETS COLD MORE QUICKLY. THE TILE IS A MUCH BETTER CONDUCTOR OF HEAT THAN THE AIR. (B) A RUG IS ALSO A POOR CONDUCTOR OF HEAT COMPARED TO TILE, SO HEAT FLOWS MORE SLOWLY FROM THE FOOT ON THE RUG THAN FROM THE FOOT ON THE TILE.

Heat, as we saw in Chapters 6 and 12, is *energy that moves from one place to another because of a temperature difference*. When a hamburger is cooked on a grill, energy flows into the hamburger. If a warm can of beer is placed in the refrigerator, energy flows out of the beer. Heat flow is sometimes called a **thermal** process, and the internal energy of an object that can be transferred to a cooler object is sometimes called **thermal energy.** So we can also say that heat is *thermal energy* in motion because of a temperature difference.

This chapter is about the ways thermal energy moves around and the ways we can use that energy to do work. We begin by looking at three ways heat moves: conduction, convection, and radiation.

CONDUCTION

When you throw back the covers on a cold morning and roll out of bed, the floor feels cold because heat flows from your feet into the floor as soon as the two make contact (Figure 13-1). As soon as foot meets floor, the molecules on the cooler surface of the floor are punched around by the faster-moving molecules of your skin and take energy from your foot. As these floor molecules in contact with your foot vibrate faster, they set the floor molecules under them and around them into faster motion, and the energy spreads, or *flows*, outward. When thermal energy passes this way from one object to another because the two are in physical contact, we say that heat is transferred by **conduction** (Figure 13-2).

Some materials conduct heat better than others. For example, a tile floor feels colder than a wooden floor that's at the same temperature because tile is a better thermal conductor than wood is. Materials that are very poor thermal conductors are called thermal *insulators*. Concrete blocks, plaster, and bricks, for instance, conduct heat relatively quickly, so these building materials are lined with insulating materials in homes.

Metals conduct heat far more rapidly than other solids do because many of the outer electrons of the atoms making up the metal leave the atoms to wander throughout the solid. Although the atoms can only vibrate in place and therefore pass heat relatively slowly from molecule to molecule, the free electrons move fast and travel a long way between collisions with atoms or

FIGURE 13-2 HEAT TRANSFER BY CONDUCTION. (A) TWO SOLIDS AT DIFFERENT TEMPERATURES. THE HIGHER THE TEMPERATURE, THE GREATER THE AVERAGE KINETIC ENERGY OF THE ATOMS MAKING UP A SOLID. (B) IF THE TWO SOLIDS TOUCH, HEAT MOVES FROM THE HOTTER ONE TO THE COOLER ONE BY CONDUCTION. (C) A COOL GAS IN CONTACT WITH A HOT SOLID. EACH GAS MOLECULE IS IN CONTACT WITH THE RAPIDLY VIBRATING SURFACE MOLECULES OF THE SOLID ONLY DURING COLLISIONS, AND LITTLE HEAT TRANSFERS BY CONDUCTION DURING THOSE BRIEF ENCOUNTERS. THEREFORE GASES ARE RELATIVELY POOR CONDUCTORS OF HEAT.

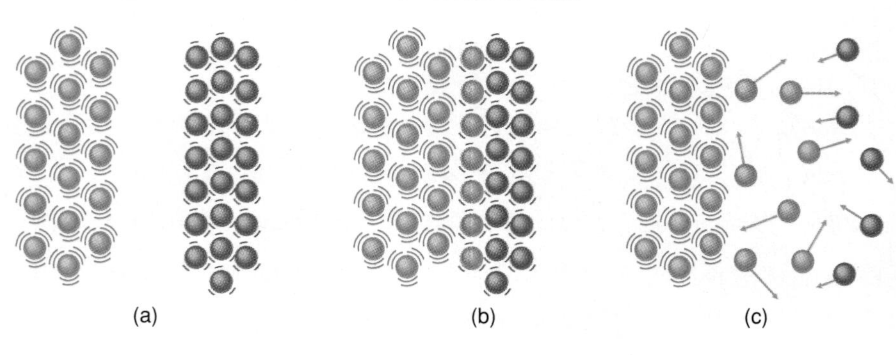

other electrons. In this way, heat moves quickly through the metal.

If you've ever grabbed a metal spoon that has been sitting in hot soup, you know that the heat transfer to your hand is very rapid. The *speed* of heat transfer by conduction is greater when the temperature difference is greater. Perhaps you've jumped into cold water in a swimming pool or a lake. The water feels cold because heat is conducted rapidly from your skin. Later, as your skin cools to a temperature closer to the water's temperature, the *rate of heat flow* is much less—so the water doesn't feel as cold as it did before. Because conduction is greater when the temperature difference is larger, you should always turn the heater or air conditioner down when you leave home for awhile. That makes the drop or rise in temperature between your home and the outside less, and you'll save energy.

AS SNOW MELTS NEXT TO THE WARM ROOF OF THE BUILDING (HEAT IS CONDUCTED INTO THE SNOW), THE WATER RUNS DOWN TO THE OVERHANG, WHERE IT IS EXPOSED TO COLDER AIR AND REFREEZES (HEAT IS CONDUCTED OUT OF THE WATER) INTO ICICLES.

The rate of flow of heat between two objects with different temperatures in contact with each other also depends upon the surface area in contact. Campers in sleeping bags on cold winter nights often leave only their noses exposed to the cold air. And the chameleon of Figure 13-3 is coping with a hot surface in the same way, by minimizing the surface area in contact.

CONVECTION

Gases or liquids can carry energy from one place to another very effectively, as they do in home heating systems that use fans to push air along. When *matter moves and carries thermal energy*, we say the energy moves by **convection.** For fluids, convection is often the fastest method of transporting thermal energy. For example, air conducts heat very poorly since there aren't many molecules per unit of volume to transfer the energy, but a *flow* of air moves (convects) heat quickly.

Sunlight coming through a window warms whatever it touches—a curtain, a carpet, a waxed floor. When air touches those warmed spots, it warms and expands. The warm air then rises as the cooler, denser (thus heavier) air around it flows in from the sides and forces it upward (Figure 13-4). The rising

FIGURE 13-3 A CHAMELEON CROSSING HOT ASPHALT GOES ON TIPTOES TO AVOID LARGE-SCALE CONDUCTION OF HEAT INTO ITS FEET. IT USES ONLY THE CLAW-LIKE NAILS OF ITS FEET BECAUSE DOING SO BOTH MINIMIZES THE AREA IN CONTACT AND USES GOOD THERMAL INSULATORS. THE GRASSHOPPER SEEMS TO HAVE DISCOVERED AN EASIER WAY TO CROSS THE ROAD.

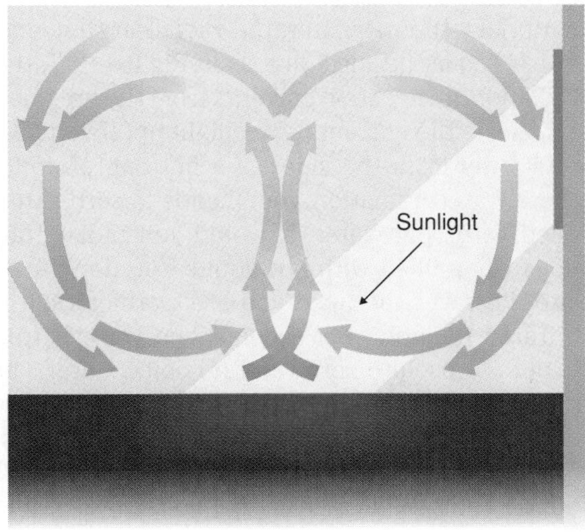

FIGURE 13-4 HEAT TRANSFER BY CONVECTION. SUNLIGHT IS ABSORBED BY THE RUG, AND THE RUG WARMS. THE AIR JUST ABOVE THE RUG IS WARMED BY CONDUCTION AND RISES SINCE ITS DENSITY IS REDUCED BY WARMING, BEING REPLACED BY COOLER, DENSER AIR FLOWING IN FROM THE SIDES. THE RISING CURRENT OF AIR IS A CONVECTION CURRENT, CARRYING THERMAL ENERGY UPWARD TO EVENTUALLY MIX WITH THE REST OF THE AIR IN THE ROOM.

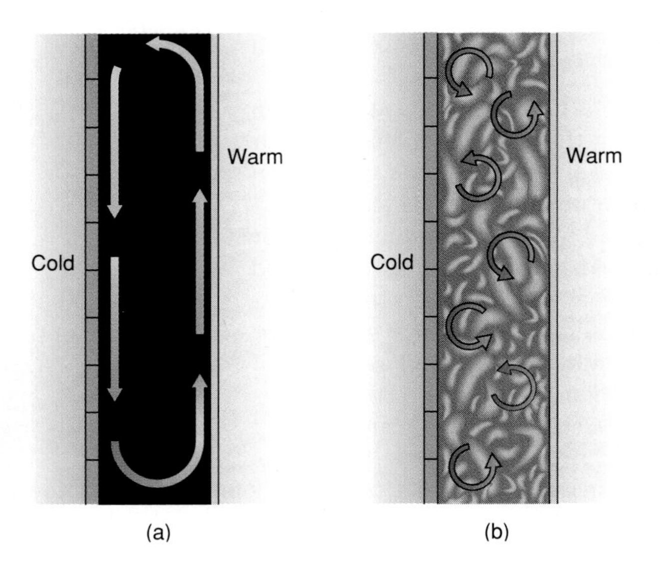

FIGURE 13-5 THE SPACE BETWEEN THE WALLS IN (A) IS NOT INSULATED, AND LARGE CONVECTION CURRENTS QUICKLY CARRY HEAT FROM THE WARM SURFACE TO THE COLD ONE. IN (B) INSULATION PREVENTS LARGE CONVECTION CURRENTS, A DESIGN THAT SLOWS THE RATE OF TRANSFER OF HEAT.

RETARDING HEAT FLOW

current of air is a *convection current*. Outside, large-scale convection currents mix the air in the atmosphere.

Home insulation slows the transfer of heat by convection, as Figure 13-5 shows. It breaks up large air spaces into much smaller spaces, preventing large-scale convection currents. Heat still moves through the insulating material by conduction and by smaller convection currents, but at a much slower rate.

Your own body is the source of convection currents when your skin is warmer than the air's temperature. Figure 13-6 shows a photograph that reveals these currents, which affect the way light travels through them.

FIGURE 13-6 AIR RISING IN CONVECTION CURRENTS PRODUCED BY THE HEAT FROM A MAN'S BODY. BY WEARING CLOTHES, WE AVOID SUCH LARGE-SCALE CONVECTIVE LOSSES OF HEAT FROM OUR BODIES.

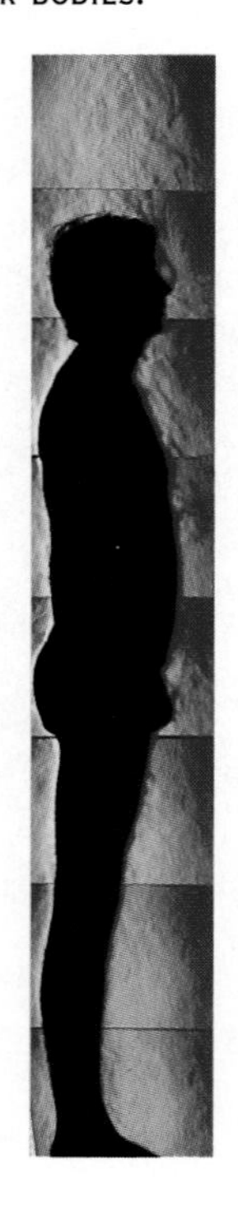

RADIATION

In the third form of heat transfer, matter does not carry the energy. Instead, the energy moves in the form of radiation called **heat rays,** or **infrared radiation** (IR). These waves are an invisible form of light and therefore can move through a vacuum, as sunlight does on its way to earth. We shall see what light rays and infrared rays are in later chapters, but here we can say that the electrons in vibrating or colliding atoms and molecules produce these rays. The faster the atoms vibrate—that is, the higher the temperature of the material—the more infrared radiation they emit.

If you sit beside a fire, as in Figure 13-7, the heat you feel is from infrared radiation. Convection currents from the fire don't warm you because the heated air rises—the currents go straight up. If you place your hands several centimeters away from the sides of a hot cup of coffee, you can feel the heat from the infrared radiation your hands absorb. Amazingly, if that same cup were filled with cold water, it would still radiate about half as much energy as it does when filled with hot liquid. You don't feel those heat rays from the cold cup, however, because you are warmer than the cold water and emit more infrared radiation per second than you get from the water. When any object is not at absolute zero (and no objects are!), its molecules are in motion and collide, producing infrared radiation. Therefore any surface emits heat rays and also absorbs heat rays given off by the matter around it. When more infrared radiation leaves an object's surface each second than is absorbed, the object cools—unless heat comes in by conduction or convection. Likewise,

FIGURE 13-7 HEAT TRANSFER BY RADIATION. HEAT FROM THE FIRE IS CARRIED UPWARD IN A CONVECTION CURRENT, AS ARE THE SPARKS AND SMOKE, WHILE RADIATION FROM THE FIRE WARMS THE PEOPLE AROUND IT. IN HOMES, THE CONVECTION CURRENTS FROM THE TOPS OF WOOD STOVES HELP HEAT THE AIR IN THE ROOM, SO THE ENERGY IS NOT LOST TO THE ENVIRONMENT AS QUICKLY AS IT IS HERE, WITH AN OPEN FIRE.

if more infrared radiation is absorbed each second than leaves, the object grows hotter.

Infrared radiation travels through empty space at the speed of light, 300,000 kilometers (186,000 miles) per second. It also moves through matter but more slowly since matter first absorbs the radiation and then reemits it, passing it along molecule to molecule much as in conduction. Another way to slow the passage of IR is to *reflect* some of the heat rays back to the matter that emits them. A smooth coating of aluminum (or gold or silver) reflects infrared rays (as well as visible light) very well (Figure 13-8), and that is why the backing of house insulation often has a shiny coat of aluminum. That's also why those reflective space blankets used by outdoor sports fans can keep them warm. In a thermos bottle (Figure 13-9), coffee stays hot and lemonade stays cold because the walls of the vacuum bottle have a shiny metallic coating to slow radiative heat transfer (either in or out). Meanwhile, convection and conduction are nearly zero because of the partial vacuum between the walls.

FIGURE 13-8 THE FACEPLATES OF ASTRONAUTS ARE COATED WITH A THIN LAYER OF GOLD TO REFLECT MOST OF THE INTENSE SUNLIGHT, INCLUDING BOTH INFRARED AND ULTRAVIOLET RAYS. THE SUITS ARE WHITE BECAUSE VISIBLE LIGHT IS ABSORBED MORE BY DARKER COLORS AND WOULD WARM THE ASTRONAUTS.

THERMAL EQUILIBRIUM

A glass of iced tea left on a table slowly warms. Air at the surface of the glass cools and falls (convection), to be replaced by warmer air, which conducts more heat into the glass (Figure 13-10). The tea and glass also absorb infrared radiation from their warmer surroundings, and heat is conducted from the warm table to the cool tea. (Often water vapor in the air condenses on the outside of the glass, giving up its latent heat of vaporization, and that also helps warm the glass.) Once the ice melts and the tea reaches the temperature of its surroundings, it is at **thermal equilibrium.** At thermal equilibrium, radiation energy in is equal to radiation energy out and convection and conduction are nonexistent because there is no temperature difference.

If the air in a closed room and the walls of the room were at the same

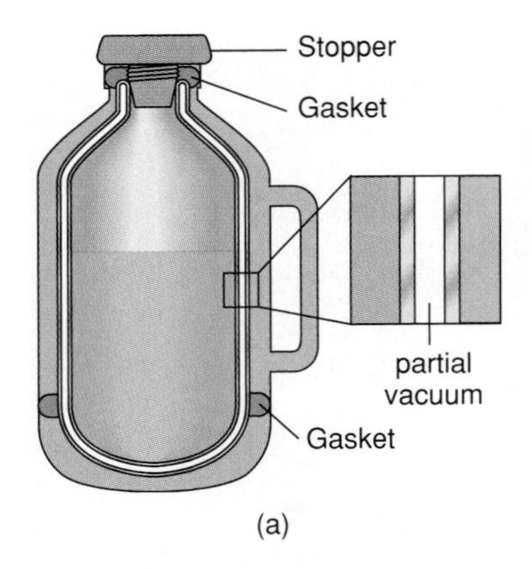

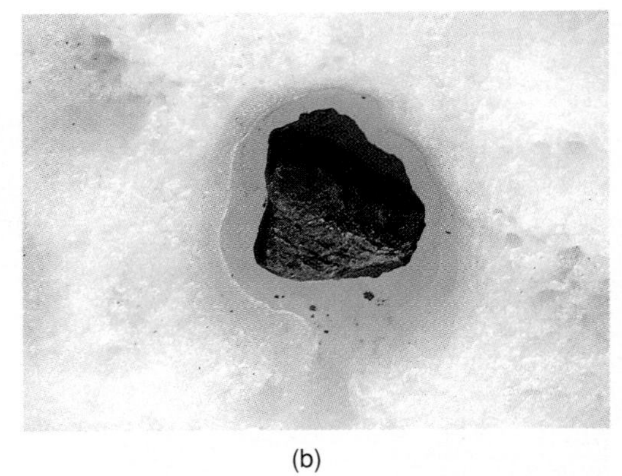

FIGURE 13-9 (A) DETAILS OF A THERMOS BOTTLE, WHICH KEEPS HEAT FLOW AT A MINIMUM. THE GASKETS ARE POOR THERMAL CONDUCTORS, BUT THE REAL INSULATORS ARE THE PARTIAL VACUUM AND THE REFLECTIVE SURFACES. THE VACUUM ENSURES THERE IS NO LOSS FROM CONVECTION OR CONDUCTION IN THE SPACE BETWEEN THE WALLS. (B) WHILE SHINY, METALLIC SURFACES REFLECT IR WELL, MOST SURFACES ABSORB BOTH IR AND VISIBLE LIGHT WELL. DARK SURFACES ABSORB MUCH MORE VISIBLE LIGHT THAN LIGHTER-COLORED SURFACES, AS YOU'VE NO DOUBT NOTICED IF YOU'VE TOUCHED WHITE AND BLACK CARS IN THE SUN ON A SUMMER'S DAY. HERE A DARK ROCK, FALLEN FROM A NEARBY CLIFF ONTO A GLACIER, HAS ABSORBED SOLAR ENERGY AND MELTED THE ICE AROUND IT.

temperature, there would be no convection currents. Thermal equilibrium would be impossible if you were in the room, however, because your breathing and uneven skin temperatures would cause convection currents, and you would stir the air as you moved around. The air molecules in a room typically have speeds well over 300 meters per second, so it might seem that a lot of mixing should take place. However, each molecule has some 10^{10} collisions every second. These collisions constantly change a molecule's direction, and no molecule gets anywhere fast. Without convection currents, for instance, an oxygen molecule in the room would need 8 or 9 hours to wander only 1 meter from its starting place, and heavier molecules would take even longer to go that far! Of course the air, walls, and surroundings are not in thermal equilibrium, and therefore convection currents abound. That is the reason you can smell food cooking in a kitchen even when you are in another room.

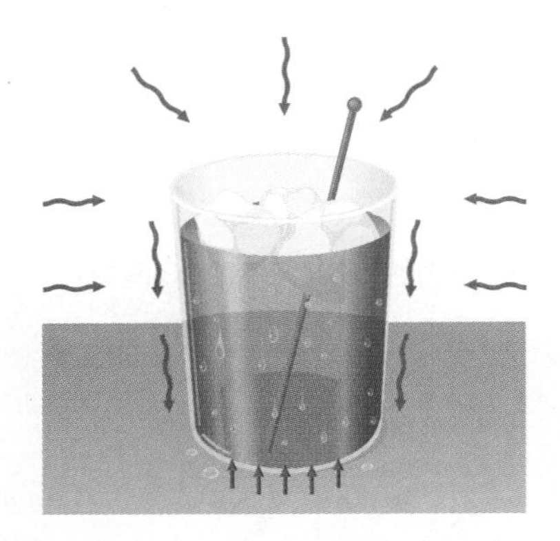

FIGURE 13-10 THE WAY A GLASS OF ICED TEA WARMS ON A WARM DAY. INFRARED RADIATION FROM THE ROOM COMES IN FROM ALL DIRECTIONS. CONDUCTION OF HEAT OCCURS WHERE THE GLASS TOUCHES THE WARMER TABLE. CONVECTION CURRENTS KEEP WARM AIR CLOSE TO THE GLASS'S SIDES AS AIR COOLED FROM CONTACT WITH THE GLASS FALLS ALONG THE SIDES. THE WARMER AIR CONDUCTS HEAT INTO THE SIDES FASTER THAN THE COOL BOUNDARY LAYER WOULD. FINALLY, MOISTURE FROM THE AIR CONDENSING ON THE COOL GLASS GIVES UP ITS LATENT HEAT OF VAPORIZATION, MUCH OF WHICH GOES TO WARM THE GLASS.

COOLING THINGS OFF: MOVING ENERGY BY EVAPORATION

As we learned in Chapter 12, evaporating molecules absorb a lot of energy from their surroundings without a change in their temperature. By hanging wet mats in doorways and windows, the ancient Egyptians, Greeks, and Romans cooled the summertime air coming into their homes. As the water evaporated, it took energy from the air that drifted through the moist mats.

Modern air conditioners and home refrigerators cool by evaporation, too, but the liquid doesn't get away in the process. A quick-to-evaporate (or *volatile*) liquid is cycled through a closed system consisting of a low-pressure region and a high-pressure region (Figures 13-11 and 13-12). The liquid enters the low-pressure side through an expansion valve. The drop in pressure lowers the liquid's boiling point, and it evaporates while traveling through the evaporator coils. As it evaporates, the liquid absorbs energy from the air (in the case of an air conditioner) or from the food (in the case of a refrigerator). Then the gas enters a compressor, which compresses it, raising its temperature. The hot, dense gas goes from the compressor into the high-pressure side of the system and passes into a condenser, where metal fins conduct and radiate heat away from it. A fan moves air over the condenser fins to speed the cooling, and the gas inside liquifies, giving up its latent heat of vaporization to the air in the kitchen in the case of a refrigerator and to the air outside the building in the case of an air conditioner. Once again a liquid, it is ready for another cycle through the system.

Freon was the liquid/gas used in refrigerators and air conditioners until both it and its chemical relatives, called chlorofluorocarbons or CFCs, were found to destroy ozone when released into the atmosphere. Today other volatile liquids are being put into use, since ozone blocks harmful UV from the sun.

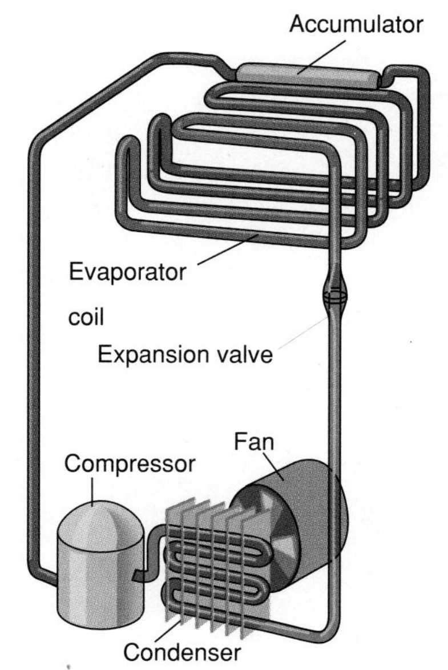

FIGURE 13-12 THE COOLING SYSTEM OF A REFRIGERATOR. THE EXPANSION VALVE LETS THE PRESSURIZED LIQUID PASS SLOWLY INTO THE LOW-PRESSURE SIDE OF THE SYSTEM, WHERE IT EVAPORATES IN THE EVAPORATOR COIL, ABSORBING A GREAT DEAL OF ENERGY. THE ACCUMULATOR CATCHES ANY UNEVAPORATED LIQUID (WHICH IS INCOMPRESSIBLE) AND PREVENTS IT FROM ENTERING THE COMPRESSOR AND CAUSING DAMAGE. RED HERE INDICATES HIGH PRESSURE, AND BLUE REPRESENTS LOW PRESSURE. HEAT FROM THE FOOD IS REMOVED TO THE ROOM IN THIS PROCESS.

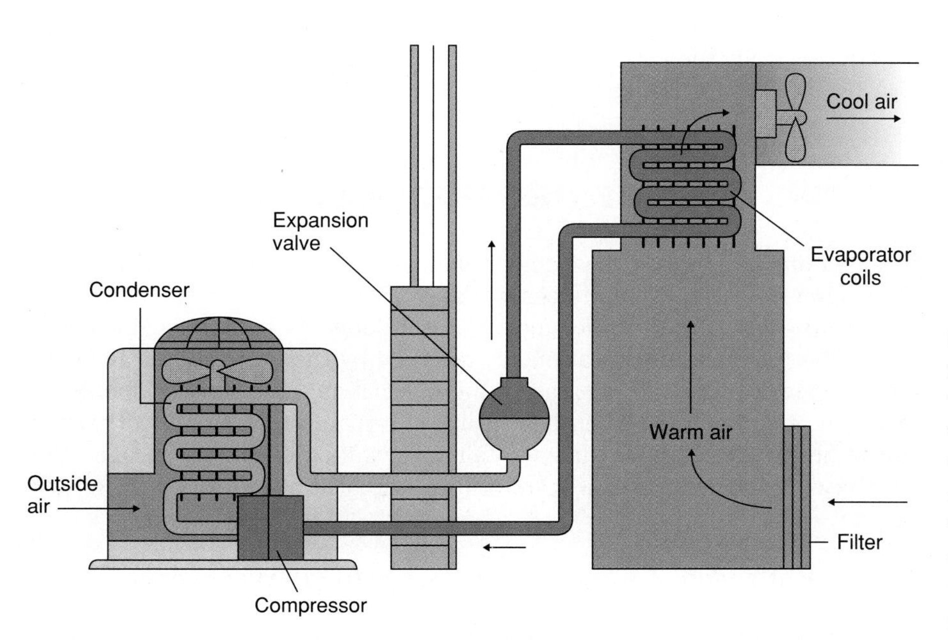

FIGURE 13-11 A CENTRAL AIR-CONDITIONING SYSTEM FOR A BUILDING. IT IS A LARGE VERSION OF A REFRIGERATOR, COOLING FILTERED AIR BY PASSING IT OVER THE EVAPORATOR COIL. THE CONDENSER, WHICH SHEDS THE HEAT GATHERED BY THE EVAPORATOR COIL, IS LOCATED OUTSIDE THE BUILDING. SO HEAT IS TAKEN FROM THE INSIDE AIR AND REMOVED TO THE OUTSIDE.

YOUR SENSE OF SMELL

With each breath you take, air comes up through your nose, curves over in your nasal cavities and goes down into your lungs. The incoming air carries molecules that have evaporated or otherwise escaped from nearby matter, and some of these molecules will strike the postage-stamp-size pieces of skin at the top of your nasal cavities. At these *olfactory surfaces,* the molecules come in contact with cells that transmit nerve signals to the brain. In order for you to smell something, airborne molecules from that something must stimulate those olfactory cells. A steel pot has no smell because steel does not evaporate at room temperature. Musk and vanilla evaporate with ease, however, and consequently are two of the most powerful odors known. When you breath in and out normally, only about 2 percent of the air you inhale strikes the olfactory surface. When you "sniff" the air, inhaling *quickly* and then pausing, turbulence creates eddies that hold the air next to the sensitive area for better exposure to any molecular scents.

We humans have a comparatively poor sense of smell. This sense in dogs, for instance, is over 70 times more sensitive than in humans. That is why a German Shepard that sniffs for explosives or drugs can detect scents that elude the nearby Customs Inspector.

Because heating increases evaporation, food being cooked is a factory for smells. If you are hungry, those odors can cause you to snap to attention. If those molecules were in absolutely still air, they would take hours to travel only a few meters from the stove. However, the mouth-watering scent of bread baking or a pot of stew simmering speeds to your nose by convection currents as the heated air rises and spreads out to mix with the air in the rest of the house.

STEAM TURBINES: USING HEAT TO DO WORK

Commercial power plants use steam turbines to generate electrical energy. First, heat from burning oil or coal or from a nuclear reactor turns water into steam in a boiler. The steam's temperature is then boosted in a superheater, which is made of coils that pass through the heat source under the boiler. This high-pressure steam is directed to strike sets of blades fixed around a movable axle, which is the working part of the turbine. The rushing steam spins the turbine blades around, providing the power to run other machines. Large steam turbines produce commercial electricity by turning electrical generators, while the axles of smaller steam turbines aboard large ships go to gear boxes that turn the propellers.

THE FIRST LAW OF THERMODYNAMICS

When air and fuel burn in the cylinder of an automobile, the hot gas produced pushes down on a piston and does work—work that is passed on to the tires to move the car. However, the amount of work done by such an engine never equals the full amount of heat energy released by the burning fuel. Some of the heat always goes into increasing the internal energy of the engine itself, raising its temperature. When heat transfer is involved in doing work, as is almost always the case, the law we call conservation of energy is called the **first law of thermodynamics.** In terms of the heat added to a system, this law states:

heat added = work done + change in internal energy

The portion of the added heat energy that goes into the internal energy of a system has important properties of its own, as we'll see next.

THE SECOND LAW OF THERMODYNAMICS

We all know this from experience: Something hotter than its surroundings always loses internal energy and cools off, and something cooler than its surroundings takes on internal energy from them and warms up. For example, let's suppose a cup of hot coffee on a table were to somehow gather energy from the cooler room to become even hotter. Even though energy could still be conserved, this process of a hot object absorbing heat from cooler surroundings never happens. Heat always flows from hot objects to cold objects. Thermal energy, when transferred, spreads outward, becoming more dilute. It never concentrates of its own accord from a more dilute or randomized form.

This principle is called the **second law of thermodynamics.** In simple terms, *there is a net heat flow from matter at higher temperatures to matter at lower temperatures.* This law is a summary of which processes occur in nature and which do not. For example, if a rock falls from a high cliff, its kinetic energy $\frac{1}{2}mv^2$ goes into other forms of energy as it stops. Both the rock and the ground below are a bit warmer after impact, but that thermal energy can never concentrate itself and kick the rock upward—it can only spread out into the ground and the air, becoming more and more dilute. Figure 13-13 shows another example of the second law in action. Once energy is in the form of thermal energy, it always disperses.

FIGURE 13-13 (A) IN FLIGHT, THE BULLET'S PARTICLES HAVE HIGHLY ORGANIZED KINETIC ENERGY IN ADDITION TO THEIR RANDOM THERMAL ENERGY. (B) STOPPED IN THE BLOCK OF WOOD, THE BULLET'S FORMERLY ORGANIZED KINETIC ENERGY IS RANDOMIZED. THE FIRST LAW OF THERMODYNAMICS STATES THAT THE ENERGY BROUGHT IN BY THE BULLET DOESN'T DISAPPEAR. THE SECOND LAW STATES THAT THE ENERGY SPREADS OUTWARD, BECOMING LESS ORGANIZED AS TIME GOES ON.

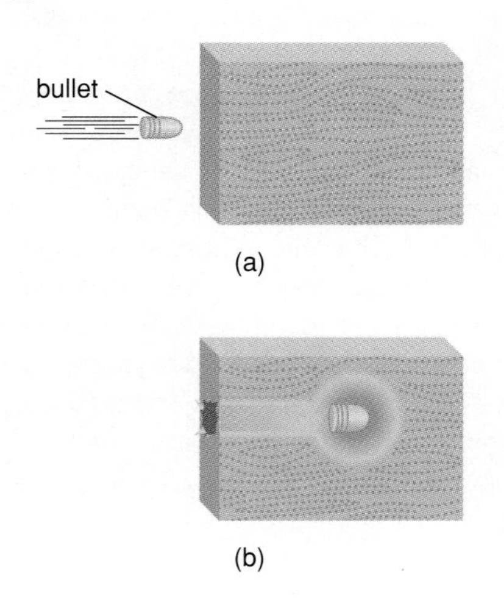

ANOTHER STATEMENT OF THE SECOND LAW: ENTROPY

The second law of thermodynamics can be stated in other terms: *Natural processes tend to move toward a state of greater disorder.* Notice the words "tend to" in this statement. They tell us that the second law, unlike the first law, is about *probabilities*. To illustrate, suppose you have a dozen red marbles and a dozen white marbles in a box, separated so that you have all red at one end and all white at the other. If you then close the box and shake it, the marbles become thoroughly mixed. Common sense tells you that no amount of shaking is likely to separate the two types of marbles again. The ordered set of marbles became disordered through the randomizing process of shaking. The system—all the marbles—spontaneously went from a state of order to a state of disorder as energy was added.

The same thing happens if one end of a spoon is heated. The added energy spreads out along the spoon, being shared by more and more molecules, and you would never see this spread-out (randomized) energy gathering together in only one end of that spoon again.

Now it could happen that the red and white marbles become ordered again after much shaking—especially if there were only, say, four marbles. Attaining order isn't as likely if there are a few dozen marbles, though, and is even less probable if there are more. In physical systems such as an automobile engine or a coffee cup, the number of molecules that take part in the spreading out of internal energy is enormous. This fact makes the statement "heat flows from hot to cold and never backward" a virtual certainty.

A quantity called **entropy** measures the disorder of a system. The greater the disorder, the greater the entropy. So another statement of the second law is that *entropy (or disorder) always tends to increase in systems that are isolated from their environments.* If the system isn't isolated, energy from outside could come in and reorganize things. For example, you could reach in and reorder those marbles.

A new deck of cards has low entropy, since the cards are ordered in suits and in numerical sequences. Shuffling randomizes their order, increasing their

Physics *and Environment*

LIVING WITH THE LAWS OF THERMODYNAMICS

Of course the laws of thermodynamics apply to living organisms. The first law says that you can't use more energy than you take in. The second says that you can't put all the energy you get from food to work for you; much is turned into heat and disperses into your surroundings.

Your body extracts about 95 percent of the potential energy available in food, but the ways you use that energy are much less efficient. On the average, humans and other animals convert only 10 percent of the energy they extract from food into building molecules of protein and fat and doing physical work. The other 90 percent is transferred as heat to the environment. In terms of efficiency, then, humans have a rating of about 10 percent. (Remember from Chapter 6 that efficiency = work done/energy used.) Even with such a low efficiency rate, however, we are among the most efficient organisms on earth. Plants convert only 1 percent or less of the energy they get from sunlight into the potential energy that is stored in the chemical bonds of their molecules.

The laws of thermodynamics even affect population sizes. Higher organisms in the animal kingdom feed on lower forms, which in turn feed on lower forms in a progression that ends with plant life. This progression is called a *food chain.* Since each organism is inefficient in using energy, the lowest levels of a food chain must be the most abundant. To make a pound of tuna that you might use in a recipe, a tuna has to eat about 10 pounds of herring. In turn, the 10 pounds of herring required some 100 pounds of smaller fish that grew from consuming about 1000 pounds of microscopic plants. Although plants and animals use potential energy from food to construct complex, high-energy molecules from simpler ones, *overall* the energy flow is toward disorganization, in agreement with the second law of thermodynamics.

entropy. Imagine the surprise if during a game a well-shuffled deck became perfectly ordered again. In a randomizing process, such as the shuffling of cards or the spreading of heat in matter, entropy tends to increase.

Every time friction or some other process turns organized kinetic or potential energy into chaotic internal energy, the energy spreads out and becomes more dilute. *The total entropy of the universe increases in each such interaction.* For example, the light and heat from the stars go out into space, becoming less concentrated. Thus we see that the energy of the universe is thinning out, winding down into less concentrated amounts that are less able to do organized work. Called by philosophers the "heat death" of the universe, this model lends itself to much speculation.

There is one familiar process in which entropy decreases, however. Plants and animals are more concentrated forms of energy than the simple ingredients they come from. Does this imply that living organisms violate the second law of thermodynamics? Only at first glance, as you will see in the box on page 193.

KEEPING COOL AND KEEPING WARM

PHYSICS OF TEMPERATURE REGULATION

Your internal temperature must remain constant so the many chemical reactions necessary to sustain life can proceed normally. Therefore your body constantly adjusts the heat flow to keep its core temperature at approximately 37.0°C (98.6°F). When you do physical work, your muscles produce extra heat, raising the core temperature slightly. This temperature increase triggers the hypothalamus, a region in your brain that acts as your body's thermostat. It signals the blood vessels nearest the skin to *dilate* (become larger), letting larger volumes of the hot blood come nearer to the surface and radiate the extra heat away. This extra blood close to the skin can give you a flushed appearance.

In the summer, when the air temperature isn't very different from the skin's, it's not easy to lose heat by radiating, convecting, and conducting. To help you keep cool in a warm environment, the hypothalamus causes sweat glands in your skin to release perspiration. This liquid evaporates, drawing much of its latent heat of evaporation from your skin, so you become cooler.

Your body loses heat most rapidly when you are naked. About 50 percent of this heat leaves as radiation, and most of the rest leaves by convection currents next to your skin. Clothing reduces the net radiative heat lost to only a few percent, and loose-fitting clothes reduce the convective loss since they break up the air space close to the body and eliminate large-scale convection currents. A sleeping bag made of goose down works the same way. Down compresses into a small volume for convenient carrying, but it expands when you shake it, trapping a large volume of air inside with the small feathers. The many feathers prevent circulation of this air, and air that can't circulate is a good insulator. The feathers also have an enormous amount of surface area, which absorbs and reemits infrared radiation, slowing its passage. Hop into one of these bags, fluff it up around your body, and you get as warm as toast in only a few seconds. The source? Your own body heat. You may have noticed that birds fluff up their feathers on cold days as in Figure 13-14, trapping more air and providing more efficient insulation.

FIGURE 13-14 BIRDS VARY THEIR INSULATION BY FLUFFING UP THEIR FEATHERS, ADDING AIR POCKETS ON COLD DAYS TO SLOW CONDUCTION.

If your skin temperature falls, tiny thermoreceptors (organs that can sense hot and cold) in the skin relay the message to the hypothalamus, and the blood vessels just under your skin *constrict*. When people get cold, they may shiver. Then their muscles vibrate, contracting and relaxing at a fast rate, doing work and generating heat to maintain their normal internal temperature.

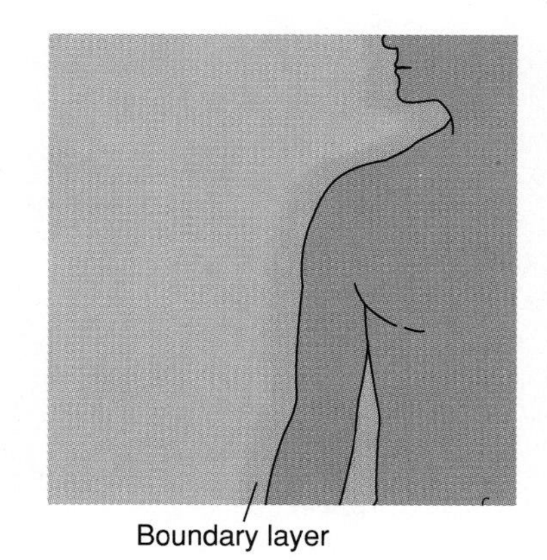

FIGURE 13-15 IN STILL COLD AIR, A BOUNDARY LAYER NEXT TO YOUR SKIN STAYS WARM, AND THIS WARM-AIR LAYER SLOWS THE TRANSFER OF HEAT FROM YOUR SKIN. EVEN THOUGH CONVECTION CURRENTS MOVE THE WARM AIR SLOWLY DOWNWARD, THE BOUNDARY LAYER, SLOWED BY FRICTION, IS EFFECTIVE INSULATION. WHEN A WIND STRIKES YOUR BODY, THE WARM BOUNDARY LAYER IS SWEPT AWAY AND YOU LOSE HEAT VERY RAPIDLY BY COMPARISON.

During strenuous exercise, too, the body's larger muscles release up to 20 times more heat than resting muscles do.

Since your body loses heat from its surface, your rate of heat loss drops if you reduce the exposed surface area. For example, you stay warmer if you cross your arms and put your hands under your arms. Likewise, your nose, toes, fingers and ears all get cold quickly in cold air since they have so much surface area per unit of mass (see the discussion on surface area per unit volume or mass in Chapter 9).

COOLING AND HEATING EFFECTS OF THE WIND

Cool air warms when it comes in contact with your skin. Once warm, the layer of air touching you slows any further heat loss from the body, since the speed of heat transfer is large only for great differences in temperature (Figure 13-15). If there is a wind, though, it sweeps away this warm layer of air and replaces it with fresh cold air, and you chill more quickly. The wind can't take the skin's temperature below the temperature of the air—remember, no net heat flows when there's no temperature difference—but by keeping the cold air next to you, the wind does help remove heat more rapidly. That effect is measured by the *wind chill factor*. For example, when it is $-15°C$ (5°F) in Fairbanks, Alaska, and the wind chill factor is $-29°C$ ($-20°F$), the moving air at $-15°C$ cools as rapidly as quiet air would at $-29°C$. However, the air inside an unheated cabin there would not drop below $-15°C$. The wind chill factor is only a gauge of the *rate of cooling* of the moving air. A wind chill chart is shown in Figure 13-16.

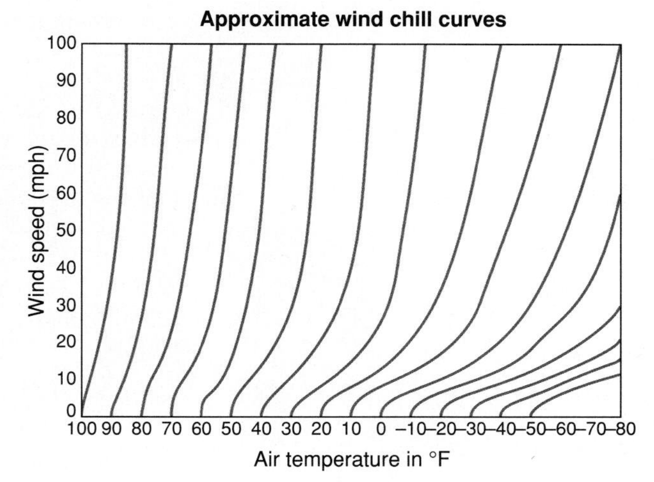

FIGURE 13-16 A WIND-CHILL FACTOR CHART. SUPPOSE THE AIR TEMPERATURE IN FAIRBANKS, ALASKA, IS $-10°F$ AND THE WIND IS 40 MPH. YOU DETERMINE THE WIND CHILL FACTOR BY FOLLOWING THE CURVED BLUE LINE UP FROM $-10°F$ AT THE BOTTOM OF THE CHART. AT THE POINT WHERE THIS BLUE LINE CROSSES THE 40-MPH LINE COMING FROM THE LEFT, DROP BACK DOWN TO THE TEMPERATURE SCALE AT THE BOTTOM AND YOU'LL SEE THE WIND CHILL FACTOR IS ABOUT $-70°F$.

The same effect that causes wind chill to lower temperature can work in reverse. A convection oven is designed so that a fan blows heated air directly onto the food. This moving air eliminates insulation by a boundary layer of still air at the food's surface and transfers heat faster, reducing the cooking time (Figure 13-17). If it were not for the countering effects of evaporation, air hotter than your body blown against you would heat you up, too.

THERMODYNAMICS AND EFFICIENCY

The first law of thermodynamics tells us something about the operation of a living organism, a steam turbine, an air conditioner, or the motor in a food blender: *energy in = work done + change in internal energy*. In the case

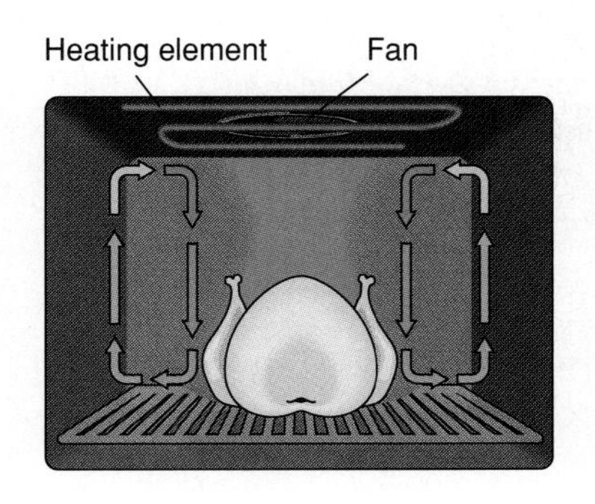

FIGURE 13-17 IN A CONVECTION OVEN, AN ARTIFICIAL WIND SWEEPS AWAY THE BOUNDARY LAYER AROUND THE FOOD, WHICH PUTS VERY HOT AIR NEXT TO THE FOOD AT ALL TIMES, SPEEDING THE RATE OF HEAT TRANSFER AND THUS THE RATE OF COOKING.

of a steam turbine, this statement becomes heat in = work done + heat out. As we've seen, the efficiency of a turbine (or an engine or motor) is how much work it does compared to the energy that's put into it, or

efficiency = work done/energy used in doing that work

The second law of thermodynamics says even more. No machine can convert into work all of the energy that goes into the machine. Some energy always leaves as heat into the surroundings. In a heat-driven engine like the steam turbine, efficiency is limited by the temperatures of the heat that goes in and the heat that comes out. The maximum efficiency, expressed in terms of the final and initial Kelvin temperatures of the heat-carrying medium, is

PUTTING IT IN NUMBERS—CALCULATING EFFICIENCY

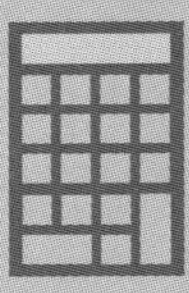

maximum efficiency = $(T_{in} - T_{out})/T_{in}$ or $(T_{hot} - T_{cold})/T_{hot}$

This is called the Carnot efficiency, after Sadi Carnot (1796–1832), a French engineer who investigated the efficiency of heat-using engines. **Example:** Steam comes out of a ship's boiler at 490°F (528 K), pushes through the steam turbine, and exits into a condenser that is kept at 77°F (298 K) by circulating seawater. The turbine's maximum efficiency is

(528 K − 298 K)/528 K = **0.44,** or 44 percent

Friction and heat lost through conduction, convection, and radiation always prevents achievement of this maximum efficiency.

This equation shows that it is important to have the highest temperature possible for the steam used to drive a steam turbine. An engine could have 100 percent efficiency only if the exit temperature (T_{cold}) were absolute zero. Then all the heat transferred into it would have been converted to work. Such total conversion is impossible, however, since heat always leaves an engine, as in a car's exhaust gases or the spent steam from a turbine. There is no sidestepping the second law of thermodynamics; entropy always increases when heat is either generated or used to do work.

Summary and Key Terms

heat–186
thermal energy–186
conduction–186
convection–187
infrared radiation (IR)–188

The workings of steam turbines, air conditioners, and even living organisms all rely on the motion of thermal energy. Thermal energy is the part of matter's internal energy that can move. ***Heat*** is ***thermal energy*** *that moves because of a temperature difference.* ***Conduction*** is heat transfer between objects that are in direct contact with each other. Conduction is faster in solids than in fluids, and it is fastest in metals, where free electrons move rapidly over long distances to transmit heat quickly. ***Convection*** occurs when *moving matter carries thermal energy,* usually in convection currents. This is often the fastest way for energy to move in fluids. All objects above absolute zero also radiate heat, ***IR***, as their molecules have collisions. This energy is a form of light, so it moves through a vacuum at the speed of light. It also moves through matter, but more slowly since molecules absorb and then reemit it.

An object at the same temperature as its surroundings is said to be in ***thermal equilibrium*** with its environment. Radiation into the object equals the radiation out and there is no convection and no conduction.

thermal equilibrium–189

The ***first law of thermodynamics*** says that the total energy of a system, in all of its forms, is constant, such that *heat added* = *work done* + *change in internal energy*.

first law of thermodynamics–192

The ***second law of thermodynamics*** says that *thermal energy spreads outward and becomes more dilute, never concentrating of its own accord from a more random form*. So in a system isolated from its environment, any object hotter than its surroundings loses energy and cools off, and any object cooler than its surroundings absorbs energy and warms up. The second law of thermodynamics is about *probabilities*. Natural processes that are isolated from their environments *tend* to move toward a state of greater disorder, or ***entropy***. In any randomizing process, such as the spreading of heat through matter, *entropy tends to increase*.

second law of thermodynamics–192

entropy–193

Efficiency is the ratio of the work a system does to the energy it uses, and the maximum efficiency of any heat-using engine or turbine is equal to $(T_{in} - T_{out})/T_{in}$, where T is the Kelvin temperature. No engine can be 100 percent efficient, converting all added heat into work.

efficiency–196

EXERCISES

Concept Checks: Answer true or false

1. Convection occurs in fluids but not in solids, while conduction is equally important in solids and fluids.

2. A metal is a good conductor because of the spatial arrangement of its atoms.

3. Insulation in the walls of a house slows heat transfer by conduction, convection, and radiation.

4. Heat cannot be transferred through a vacuum.

5. Heat flows spontaneously from an object of higher temperature to one of lower temperature.

6. The rate of heat conduction between two objects depends on their temperature difference and on the amount of surface area in contact.

7. An object is at thermal equilibrium when it quits emitting or absorbing heat.

8. When coolant is cycled through a refrigerator or an air conditioner, it absorbs heat as it evaporates, and the vapor loses energy when it condenses.

9. The first law of thermodynamics is the law of conservation of energy expanded to include thermal energy and heat transfer.

10. The second law of thermodynamics states that disorder occurs spontaneously but order usually doesn't.

11. Entropy describes the order or disorder of matter. A higher entropy means a greater disorder.

12. The wind chill factor is a gauge of the rate of cooling of moving air, but not a gauge of its temperature.

Applying the Concepts

13. Have you ever heard the expression, "the cold creeps in"? Explain why it's inaccurate.

14. How does it help to put long aluminum nails through potatoes to be baked?

15. Have you ever licked an aluminum ice cream scoop just after it's been used? What happened? Would the tip of your tongue stick to a wooden spoon that was at the same temperature?

16. Pick up a pair of scissors and a wooden pencil. Which feels colder even though they are both at room temperature? Why?

17. The air in an attic on a summer's day might be 17 C° (30 F°) hotter at the ceiling than at the floor. Explain why heat coming in through the roof doesn't get transferred easily to the floor.

18. Lighted candles won't burn well in the orbiting space shuttle. Explain why and suggest a solution.

19. How does a fan help cool you on a hot day? (Incidentally, Leonardo da Vinci made a water-powered fan about 1500 A.D., perhaps the first practical powered fan.)

20. If you jump into cold water in a swimming pool or a lake, before long it won't seem as cold and you might say you have gotten used to the water. What has happened?

21. The average person uses about 90 Calories per hour (90 kilocalories/hour) while lying down and resting. Show that this is equal to about 100 watts. (The energy usage while at rest is called the basal metabolic rate.)

22. How does turning down the thermostat in a house save fuel in winter?

23. On a hot day, an elephant keeps cool by throwing wet mud onto its back. How does this help?

24. A fire in a fireplace gives out more heat after it has been burning for some time. Why?

25. On a cool evening, a dog or a cat might curl up to sleep, but a litter of puppies or kittens not only curl up but also huddle together. Explain why all this helps keep the animals warm.

26. Would leaving the refrigerator door open on a hot day cool the kitchen? Explain.

27. During the summer in Death Valley, some of the few residents turn off their home water heaters. Then they use the hot water faucets for cold water and the cold water faucets for hot water. Explain how this works.

28. A naked person emits infrared rays effectively from only about 85 percent of the skin's surface area. Why?

29. Suppose you need 2400 Calories per day just to stay at your present weight. If you went on a strict diet of 2000 Calories/day, how long would it take you to lose 5 pounds? (1 pound of fat = about 4100 Calories.)

30. Calculate how long you could live on just your body's supply of fat. (On the average, 15 percent of an adult male's weight and 22 percent of an adult female's weight are in the form of fat.) If you use 90 Calories/hour while resting, how many days would it take to get rid of your fat if you were on a starvation diet? (1 pound of fat = about 4100 Calories.)

31. In the late 1700s, Dr. Charles Blagden took some friends, a dog, and a raw beefsteak into a room heated to 260°F. They all emerged unharmed three-quarters of an hour later; the steak, however, was cooked. Why weren't the people and dog cooked as well?

32. Travelers of long ago discovered that a damp cloth cover on a canteen helps cool the liquid inside. Explain how.

33. Explain why you could stick your hand into an oven at 260°C (500°F) for 15 seconds without harm, whereas you would never put your hand into boiling water at 100°C (212°F) for even a second.

34. Faced with a bowl of thick, steaming hot vegetable soup, most people would stir it to make it cool faster. How would that help?

35. In a convection oven, should it take a large turkey (a) more time, (b) less time, or (c) the same amount of time to cook than a small turkey at the same temperature? (Cooks beware!) Explain your answer.

36. The odor of onions and garlic won't spread through a kitchen if they are stored in a dark corner or a cupboard. Why not? What happens if you put them in a sunny window for an hour?

37. The tiny pygmy shrew, a mouselike mammal, has the misfortune to be warm-blooded. Explain why its small size means it loses heat energy relatively rapidly. (It must eat 2.5 times its own weight each day to survive.)

38. Heavy-bodied insects shiver awhile before their first takeoffs of the day. Why?

39. Do you think a jogger on a hot day would prefer to run with a slight summer breeze blowing in her direction or to run directly into the same breeze? (If you jog, chances are excellent that you know the answer.)

40. When you cook rice or potatoes, the pot you fill too full of water will always overflow when boiling. Why?

41. Using the example in the box "Living with the Laws of Thermodynamics," estimate how many joules of sunlight began the process that put a 1-pound can of tuna on the shelf at the grocery store. (1 ounce of tuna contains about 30 Calories.)

42. The rate of heat flow from an object depends on its surface area. What would be the best shape for a house in order to minimize heat loss in winter?

43. Why do bridges freeze over faster than roadways in a winter cold snap (Figure 13-18)?

FIGURE 13-18

44. A trainee in a cold-weather survival program is told to pack 8 inches of snow onto her lean-to shelter but no more, or she'll waste her own energy. Discuss.

45. Explain why a glass of iced tea takes longer to warm up than a cold can of beer, which needs only a few minutes to become lukewarm.

46. Use Figure 13-16 to estimate the wind chill of 45°F air moving at 15 mph.

47. Why does the air in a deep underground cave have a constant temperature all year?

48. "Cold-blooded" is a misnomer. On a sunny day, a lizard or a snake can have a blood temperature greater than a human's. Explain how, and suggest ways reptiles can regulate their temperature on sunny days.

49. Spheres have the smallest ratio of surface area to volume of any shape, and the larger the sphere, the less surface area per unit of volume. Does this fact benefit round-shaped penguins in cold climates or elephants in hot ones?

50. In a convection oven, the cooking temperature often can be up to 25° C (about 45° F) less than in a conventional oven being used for the same cooking task. Why?

51. A traveler driving farther and farther north notices that the various species of wild rabbits and foxes have smaller and smaller ears (Figure 13-19). Why should they?

FIGURE 13-19

52. Ground squirrels, woodchucks, and big brown bats hibernate in winter, their metabolic rates dropping to less than 1/50 of normal and their core temperatures falling to only 1 to 2° F above the temperature of their surroundings. What does this temperature drop accomplish?

53. A rabbit feels warm to you because its body temperature is about 39.4° C (103° F). If you look at its big ears, you'll notice many small blood vessels near the surface. What purpose do they serve?

54. In the summer you might be very comfortable in 21° C (70° F) air in a movie theater. If you jump into a spring whose water temperature is 21° C, however, it feels like ice water. What is the difference?

55. Why do wine glasses have long stems?

56. The engine of a small- to medium-sized motorcycle uses about 1 gallon of gas in an hour's operation (for instance, it goes 50 miles on a gallon at 50 mph). A gallon of gasoline yields 32×10^6 calories of energy. If the motorcycle is 22 percent efficient, how much work does the engine do in 1 hour? How much energy is lost to the surroundings?

57. At a commercial electrical power plant, superheated steam enters the turbine at 1000° F (811 K) and is condensed at about 110° F (316 K). What is the theoretical maximum efficiency of the turbine?

58. There are experimental projects under way to generate electricity from heat engines that use the temperature difference between the surface of the ocean (at about 25° C) and the ocean's depths (about 4° C). What is the maximum efficiency of such an engine? Why is this idea interesting despite the low efficiency?

Demonstrations

1. Use a candle to explore the convection currents in a house. The slightest motion of air bends a flame. If you hold a lighted candle first near the top and then near the bottom of a partially opened door, you can see which of the two rooms the door separates is warmer and which is cooler. Warmer air rises, spilling through the doorway at the top, and cooler air flows in the opposite direction near the floor to replace it. The flame points in the moving air's direction.

2. The glass or plastic lenses of an ordinary pair of eyeglasses let visible rays pass through them but block infrared rays. Sit with your face close to the fire in a fireplace and close your eyes gently. Your eyelids are very sensitive to the heat brought by the infrared rays. Now slip a pair of glasses on while your eyes are still closed and feel instant relief. Your eyelids cool immediately because the infrared radiation is stopped by the glass or plastic. (The heat you feel is IR, since all the air heated in a fireplace goes up the chimney. Only 10% of the energy released travels out into the room itself as infrared radiation.)

3. Here's a way to feel cooling by evaporation and wind chill. Splash some rubbing alcohol on your forearm and feel the coolness as the alcohol evaporates. Now blow over it, and the cooling effect is even greater. The wind speeds the rate of evaporation, and heat leaves the skin much faster.

4. To prove that a refrigerator puts out lots of heat, the next time you have a wet pair of shoes, just place them in front of the refrigerator. Warm air exiting from the condenser will dry them overnight.

5. You can feel the equivalent of wind chill in water. The next time you take a dip in a cold spring, pond, or even the ocean (out past the breakers), stand still so you don't stir the water around you. Your body soon warms the layer of water next to your skin. Then move around a bit. Heat begins to leave your skin in a hurry. The layer of water that had been warmed by your body is swept away and replaced by a colder layer, and you feel the chill immediately.

FOURTEEN

Heat and the Earth

Earth seen by an Apollo crew as they orbited the moon. Blue light reflects from the oceans, white light from clouds, and browns, greens, and reds from the continents. The atmosphere, the oceans, and the land masses all absorb some of the incoming sunlight. The light's energy becomes thermal energy and the oceans and atmosphere redistribute that energy and cause the climates we have.

(a)

(b)

FIGURE 14-1 (A) LIQUID ROCK COMES THROUGH THE CRUST OF THE EARTH WHEN VOLCANOES ERUPT. THE EARLY EARTH WAS ENTIRELY MOLTEN, AND TODAY THE INTERIOR STILL HAS MOLTEN ZONES KEPT HOT BY THE DECAY OF RADIOACTIVE ELEMENTS. (B) LAVA FLOWS INTO THE OCEAN FROM HAWAII. THE CLOUD IT CREATES AS IT VAPORIZES WATER CONTAINS LARGE AMOUNTS OF SO_2. (THE WHITE STREAK IS STARLIGHT REFLECTING FROM THE BREAKING WAVES ON THE BEACH.)

Scientists today have a good idea of earth's history, gathered from overlapping lines of evidence from physics, chemistry, and geology. Earth was born about 4.5 billion years ago as smaller bodies orbiting the sun were brought together by gravity. At first the young earth was largely molten rock from which dissolved gases slowly leaked. Then the outer surface cooled and solidified into the crust, and gases dissolved in the molten rock below continued to surface, or *outgas* through the crust. Even today, such gases continued to leak through the crust from active volcanoes (Figures 14-1 and 14-2).

Volcanic gases contain almost 60 percent water vapor, more than 20 percent carbon dioxide, about 12 percent sulfur compounds, 5 percent nitrogen, and no oxygen. So today's atmosphere of 78 percent nitrogen, 21 percent oxygen, and only traces of carbon dioxide and sulfur compounds is nothing like the early atmosphere on our planet. The water vapor condensed to form the oceans, and the carbon dioxide dissolved in the new oceans and reacted to form calcium carbonate, which marine animals use to build their shells. These shells accumulated on the ocean floor, compressing under their own weight to form the rock we call limestone. The sulfur compounds came down with rains—the first acid rains. Because nitrogen is relatively inert, most of the nitrogen that escaped from earth remains in the atmosphere today. Oxygen appeared in the air much later, a byproduct of *photosynthesis,* which plants use to build carbohydrate molecules. During photosynthesis, plants remove carbon dioxide from the air and release oxygen.

Today the atmosphere, the oceans, and the land surfaces share the heat they acquire from sunlight, and their mutual thermal interactions make the climates that determine which plants can grow where, which in turn determines where most species of animals live. These properties and interactions are the subject of this chapter.

FIGURE 14-2 MT. ST. HELENS ERUPTING IN WASHINGTON STATE, TWO MONTHS AFTER ITS EXPLOSIVE ERUPTION OF MAY, 1980.

SUNLIGHT ON EARTH

When sunlight strikes the oceans and the land, some is reflected but most is absorbed. The absorbed light warms the matter, and its molecules move faster, collide harder, and radiate more heat in all directions, including outward into the atmosphere and eventually into space. These warmed areas continue to radiate heat even during the night, so night is a time of cooling.

When the land and the oceans absorb sunlight, they don't warm by the same amounts. The ocean's temperature changes more slowly than the land's for several reasons. One is that water's specific heat (Chapter 12) is about 5 times as large as the specific heat of soil and sand. For the same amount of absorbed sunlight, then, the temperature of a given mass of land increases up to five times as much as the temperature of the same mass of water. Likewise, if the land were to lose the same amount of heat at night as water does, its temperature would fall five times as much. Another reason for the ocean's slow temperature change is that sunlight penetrates more deeply into water, so that its energy is shared by a larger mass than on land, where sunlight is usually stopped within a centimeter or less of the surface.

The orientation of the spinning earth as it orbits the sun is the major factor determining how hot or cold a given place is. The earth's rotation axis tilts $23\frac{1}{2}°$ from the perpendicular to the plane of its orbit around the sun. Figures 14-3 and 14-4 show how the tilt causes varying amounts of day and night during a year at every location except the equator. In our summertime, earth's tilt points the Northern Hemisphere more toward the sun's direction, and each day locations north of the equator get more hours of daylight and fewer hours of dark. Meanwhile, the Southern Hemisphere is pointed more away from the sun, keeping it in the dark longer than 12 hours and in the sunlight less than 12 hours each day.

Having more hours of daylight than of night in any 24-hour period is one reason summers are warmer than winters. A second reason has to do with the angle of the incoming sunlight at a given location. The sun's daily path in the sky as seen from the ground is noticeably lower in the winter than in the summer (Figure 14-5). That means in winter sunlight strikes the earth from

FIGURE 14-3 THE TILT OF EARTH'S AXIS WITH RESPECT TO THE SUN CAUSES THE SEASONS. HERE THE EARTH IS SHOWN AT THREE POSITIONS IN ITS ORBIT. THE EXPANDED VIEWS (1), (2), (3) SHOW THE EARTH AT THOSE DATES AND TIMES AS SEEN FROM THE SUN. IN VIEW (1), JUNE 22, THE NORTHERN HEMISPHERE IS TILTED MORE TOWARD THE SUN THAN AT ANY OTHER TIME DURING THE YEAR. AS YOU CAN SEE FROM VIEW (1), AT THAT TIME THE NORTHERN HEMISPHERE FACES THE SUN MORE DIRECTLY THAN THE SOUTHERN HEMISPHERE DOES. THAT'S SUMMER FOR NORTH AMERICA, OF COURSE, AND FOR EUROPE AND ASIA—ALL OF THE LAND MASSES IN THE NORTHERN HEMISPHERE. AT THE SAME TIME, IT IS WINTER FOR THE ENTIRE SOUTHERN HEMISPHERE. THE SITUATION IS REVERSED IN VIEW (2), DECEMBER 22. IN VIEW (3), ABOUT MARCH 21 (AND SEPTEMBER 23, NOT SHOWN), THE HEMISPHERES FACE THE SUN AT EQUAL ANGLES. FOR MORE DETAILS ON THE CAUSE OF THE SEASONS, SEE FIGURES 14-4 AND 14-5.

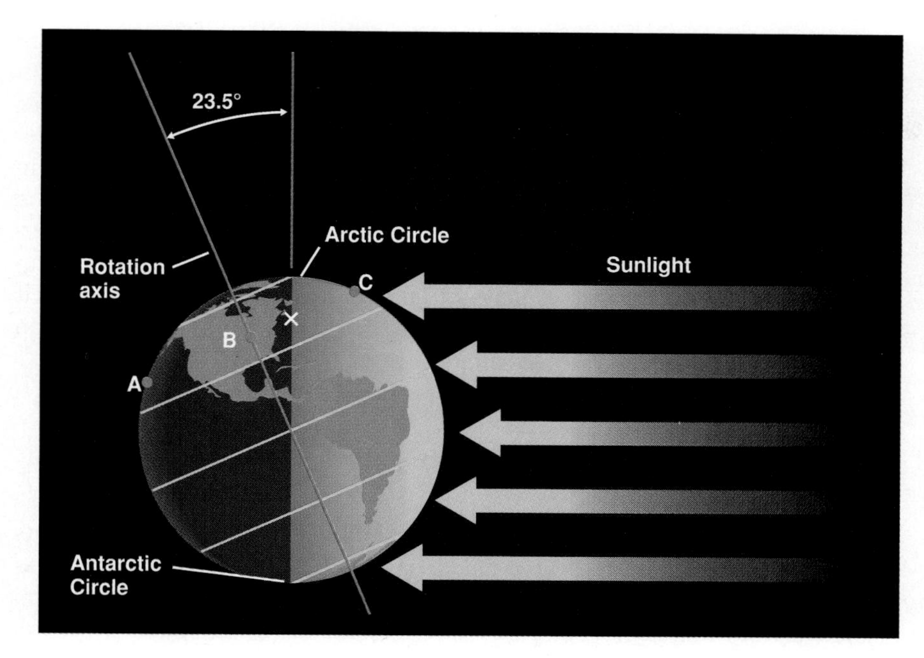

FIGURE 14-4 ANOTHER VIEW OF EARTH ON DECEMBER 22. THE LINE BETWEEN DARKNESS AND DAYLIGHT (CALLED THE *TERMINATOR*) IS VERTICAL SINCE THE SUN IS DIRECTLY TO THE RIGHT. A, B, AND C SHOW WHERE ST. LOUIS, MISSOURI, IS AT MIDNIGHT (A), 6 AM (B), AND NOON (C), WHICH IS HALF OF ITS 24-HOUR DAY. NOTICE THAT SUNRISE, WHEN ST. LOUIS COMES OUT OF EARTH'S SHADOW AT THE POINT MARKED WITH AN X, OCCURS MUCH LATER THAN 6 AM. THEREFORE ST. LOUIS HAS MORE THAN 12 HOURS OF DARKNESS AND FEWER THAN 12 HOURS OF DAYLIGHT ON THIS DATE, AS DOES ALL OF NORTH AMERICA. POINTS IN SOUTH AMERICA THAT ARE SOUTH OF THE EQUATOR, HAVE MORE THAN 12 HOURS OF DAYLIGHT, AND LESS THAN 12 HOURS OF DARKNESS, WHILE POINTS ON THE EQUATOR HAVE 12 HOURS OF DAYLIGHT AND 12 HOURS OF DARKNESS ALL YEAR LONG. NOTICE THAT POINTS BELOW THE ANTARCTIC CIRCLE ARE IN DAYLIGHT FOR 24 HOURS ON THIS DATE, WHILE POINTS ABOVE THE ARCTIC CIRCLE ARE IN DARKNESS FOR 24

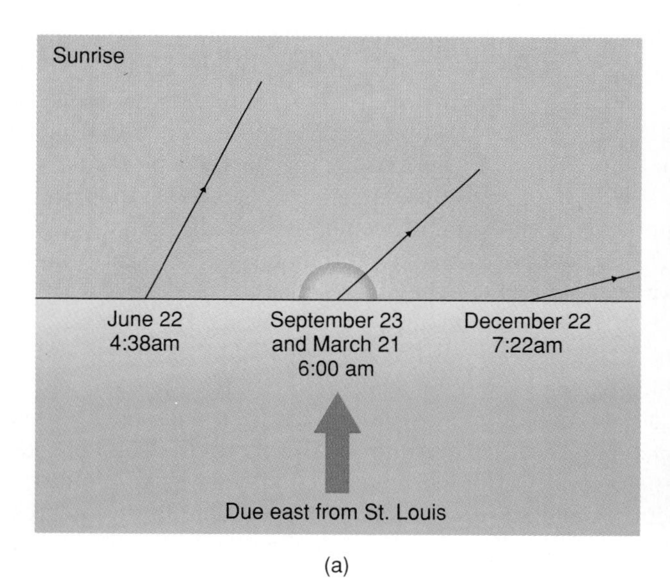

(a)

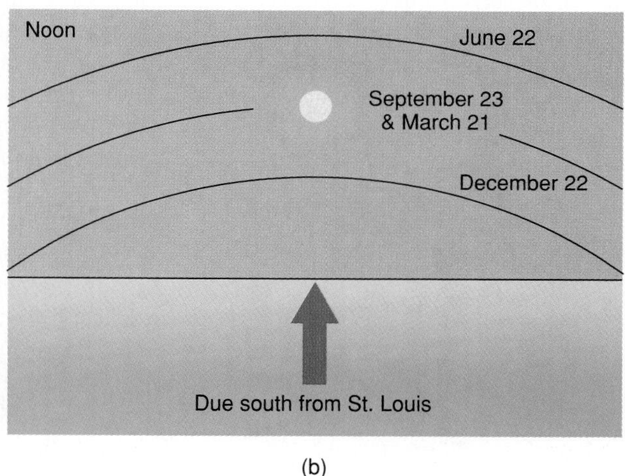

(b)

FIGURE 14-5 VIEWS FROM A POINT ON EARTH'S SURFACE IN THE CENTRAL UNITED STATES, SHOWING THE DAILY PATH OF THE SUN AT DIFFERENT TIMES OF THE YEAR. THE SUN RISES DIRECTLY IN THE EAST AND SETS DIRECTLY IN THE WEST ONLY ON ABOUT MARCH 21 AND SEPTEMBER 23.

a lower angle, and a given amount of sunlight spreads over a much larger area (Figure 14-6). The land and water get less energy per second per unit of area in the wintertime, and this difference, along with shorter days, causes lower temperatures in winter.

During any given day, summer or winter, the energy per second falling on each square meter from sunlight is greatest when the sun is highest in the sky—about 12 noon, local time. Yet the hottest time during a clear day is a few hours after 12 noon. Why? For several hours *after* the peak input of heat energy, the sun still adds more heat energy per second than the ground is losing by radiation, convection, conduction, and evaporation. So the temperature continues to rise into the early afternoon hours.

The seasons show a similar lag in temperature change. From Figure 14-5 you can see that the sun reaches its highest angle in the sky on June 22 for points in the Northern Hemisphere above $23\frac{1}{2}°$ latitude. Summer's hottest months, however, are usually July and August. Though not at its highest angle

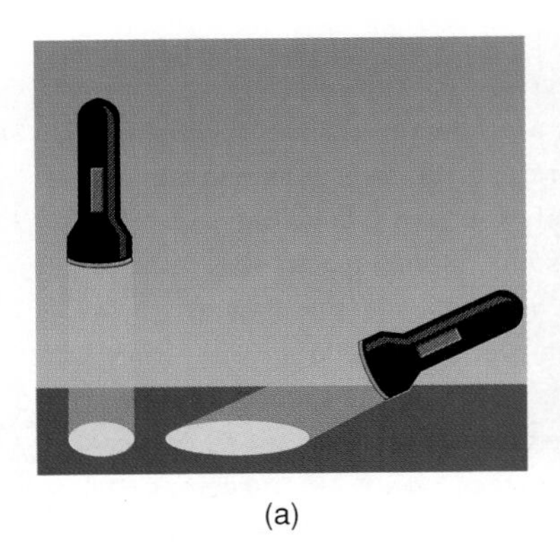

(a)

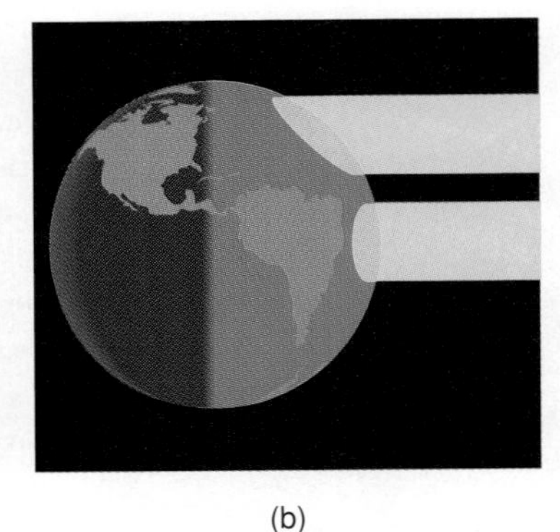

(b)

FIGURE 14-6 (A) LIGHT THAT STRIKES A SURFACE AT AN ANGLE IS SPREAD OVER MORE AREA AND SO DELIVERS LESS ENERGY PER UNIT OF AREA. (B) IN THE SAME WAY, SUNLIGHT THAT STRIKES THE EARTH AT A GREATER ANGLE SPREADS ITS ENERGY OVER A GREATER AREA.

in those months, the sun continues to add more heat energy per day than the hemisphere radiates away during 24 hours, so the Northern Hemisphere continues to warm. Similarly, December 22 is the lowest point for the sun in the skies of the Northern Hemisphere, but the coldest months are usually January and February. During those months, the Northern Hemisphere still radiates away more energy than it receives each day, making them colder than December.

CONVECTION CURRENTS IN THE ATMOSPHERE

Because of solar heating, earth's atmosphere abounds with convection currents of all sizes. Warmed by sun-heated fields, parking lots, roofs, or roads, parcels of air rise, buoyed upward by the cooler, denser air that hugs the ground and pushes in from the sides (Figure 14-7). Winds occur when higher pressure, relatively dense air moves into areas of lower pressure where the air has warmed and expanded.

Within less than a centimeter above the ground, convection currents can't move the air effectively because of friction. In this boundary layer, molecules must get around by moving randomly because of their collisions rather than being swept along in a current of molecules. Fortunately for us, above the first centimeter and all the way to the upper atmosphere, winds thoroughly mix and stir the air. It's difficult to appreciate this fact enough. Air is a poor conductor of heat. Without convection currents to carry heat away, the temperature of the air next to the ground would rise relentlessly as the ground absorbed sunlight and got hot each day. Automobile exhaust gases and industrial emissions would stay where they were created, next to earth's surface. So without convective mixing, the air we breath would be deadly.

Convection currents keep earth's air well mixed. Some of the carbon dioxide molecules exhaled today by someone in North America will turn up in Europe in just a couple of months. And in another year some of them will be in the air a penguin breathes in Antarctica. Daily updrafts of air take some molecules from sea level in the morning to 10 kilometers (6.2 miles) up by noon. In a violent thunderstorm, molecules can rise that far in minutes.

Occasionally, most often in natural basins or areas partially circled by hills or mountains, temperature "inversions" occur. A layer of warm air will lie over cooler air that is next to the ground so that for a time convection currents will be subdued. If the immobilized air is over a city or an industrial site, the concentration of pollutants will rise rapidly. The resulting smog, from chemical reactions with the products of automobile exhausts as well as industrial emissions, then poses a serious health problem. Several decades ago London had such a severe episode of smog that some 4000 people died as a result.

FIGURE 14-7 DIAGRAMS OF (A) CONVECTION CURRENTS THAT CAUSE STIRRING OF THE AIR LOCALLY AND (B) LARGE-SCALE HORIZONTAL MIXING DUE TO A MASS OF COLDER AIR PUSHING INTO AND UNDER A MASS OF WARMER, LESS DENSE AIR.

local mixing

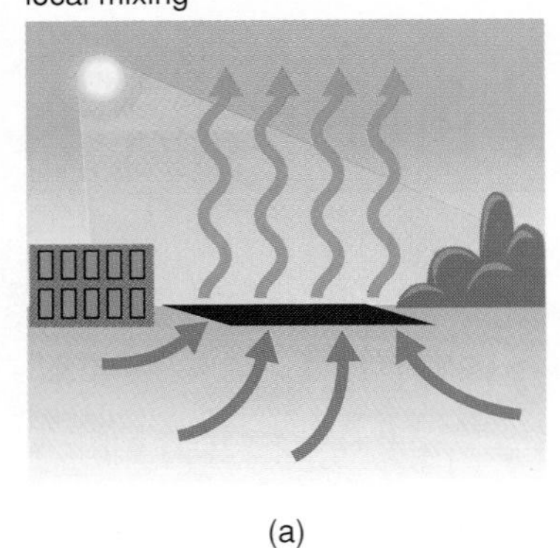

(a)

large-scale horizontal mixing

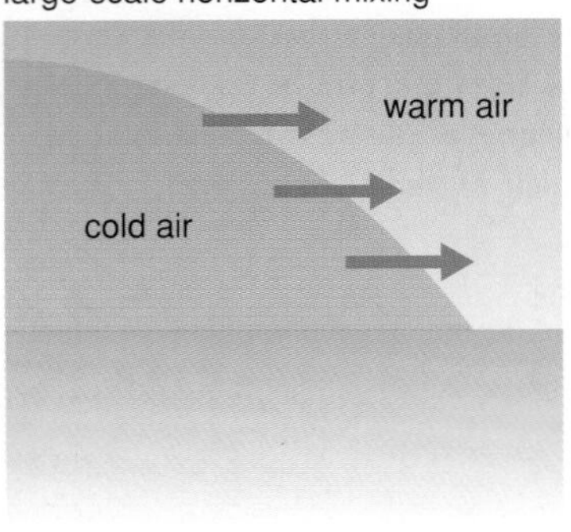

(b)

PREVAILING WINDS

Although local updrafts do a lot to mix the air vertically, the long-range horizontal mixing in our atmosphere comes from **prevailing winds,** wind patterns with definite directions on a global scale. These winds come from solar heating not over just a small area but over the entire earth. We've seen that the

GLOBAL WARMING

In the atmosphere certain molecules play a large role in "trapping" infrared radiation that would otherwise escape from earth's surface. That is, they absorb the IR's energy, increasing the temperature of the air. Since greenhouses trap thermal energy generated by absorbed sunlight, thes gases—water vapor, carbon dioxide, and other trace gases—are called "greenhouse gases." Some of these are increasing due to human activities, and there is great concern that global warming will result. One model predicts that a doubling of greenhouse gases would warm the earth from 1°C to 5°C, changing climates drastically and raising the ocean's level significantly.

Carbon dioxide levels are now 25 percent higher than they were before the industrial revolution, and methane, also a greenhouse gas, has doubled in that time. CFCs (Chapter 13), which were nonexistent 100 years ago, absorb 10,000 times as much IR per molecule as CO_2 does. In only a decade the increase in CFCs and other pollutant gases will equal the effect of CO_2 in trapping IR.

The problems in predicting what is happening are immense. For one example, particulates suspended in the upper atmosphere due to human activities and natural causes such as volcanic eruptions reflect incident sunlight, so a fraction less will reach earth to be turned into thermal energy. Deforestation is proceeding in the rainforests at an alarming rate, and since trees absorb CO_2 this will have some effect. The oceans absorb about half the CO_2 emitted each year from burning fossil fuels, but even that process is not completely understood. For instance, when Mount Pinatubo in the Philippines exploded in 1991, it pushed ash and SO_2 high into the stratosphere. (The mass of SO_2 exceeded the SO_2 emissions produced by human activities in an entire year.) Global temperatures dropped slightly and the CO_2 level in the atmosphere did not rise as fast in 1992 and 1993 as it had been rising in earlier years. Scientists are still debating why that happened. Clearly, not enough is understood about all the causes and effects at this time. But much of the scientific community has the opinion that understanding and acting on global warming will be the major environmental problem of the twenty-first century.

PROBLEMS NEEDING SOLUTIONS

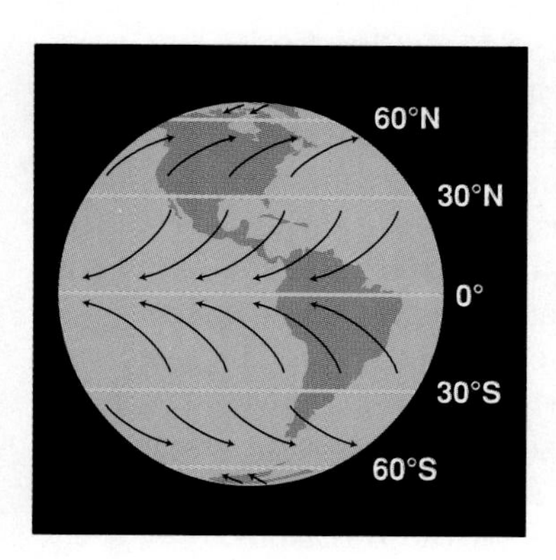

FIGURE 14-8 THE ZONES, OR BELTS, OF THE PREVAILING WINDS ON EARTH'S SURFACE. THE WINDS BETWEEN THE EQUATOR AND 30° N LATITUDE ARE CALLED THE NORTHEAST TRADE WINDS. FROM 0° TO 30° S, THE WINDS ARE CALLED THE SOUTHEAST TRADE WINDS. FROM 30° N TO 60° N IS THE ZONE OF THE PREVAILING WESTERLIES, WHICH GOVERN THE CLIMATE TO A LARGE EXTENT IN THE CONTINENTAL UNITED STATES. THE WINDS BETWEEN 30° S AND 60° S ARE ALSO CALLED WESTERLIES.

equatorial regions get almost direct sunlight all year and the polar regions get sunlight only at a low angle. Therefore the energy per second per unit of area brought in by sunlight is a great deal less at the poles than at the equator. Warmed tropical air rises, cools, and spreads outward at a high elevation, while cold polar air flows outward close to the ground. This suggests a picture of ground winds traveling toward the equator from the poles and high-altitude winds moving from the equator to replace the colder air. Earth's rotation turns the moving air, however, as we shall see later in this chapter, breaking the circulation pattern in each hemisphere into the three distinct zones shown in Figure 14-8. The prevailing winds follow those patterns, though local winds may come from any direction. Prevailing winds cause the principal features of the ocean's surface currents, too; compare Figure 14-8 to Figure 14-9 to see this connection.

Prevailing winds are often a major factor in climate, especially where land meets ocean. In winter the oceans are slower to cool off than the land. The pleasant winter climate in San Francisco is due to the westerly winds (this means "from the west") off the Pacific Ocean, which bring in relatively warm air from over the water (Figures 14-10 and 14-11). The ocean current called

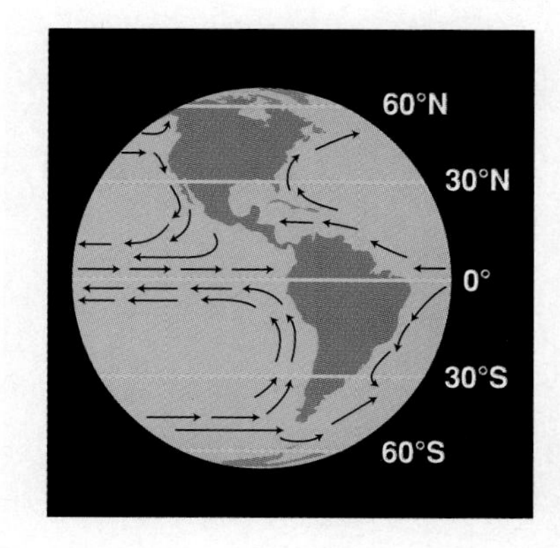

FIGURE 14-9 THE MAJOR SURFACE CURRENTS OF THE OCEANS. EXCEPT FOR A COUNTERCURRENT DIRECTLY ON THE EQUATOR AND THE INTERFERENCE OF LAND MASSES, THE SURFACE CURRENTS FOLLOW THE PATTERN OF THE PREVAILING WINDS SHOWN IN FIGURE 14-8.

FIGURE 14-10 THESE TREES AT POINT REYES, CALIFORNIA, JUST NORTH OF SAN FRANCISCO, SHOW THE EFFECT OF THE PREVAILING WESTERLY WINDS ON THIS SLOPE.

FIGURE 14-11 SATELLITE MAP OF SURFACE WATER TEMPERATURES OFF THE WESTERN COAST OF THE UNITED STATES. BLUES AND PURPLES SHOW COLD WATER UPWELLING FROM BELOW AS CURRENTS HIT INCLINES. THE ORANGES AND REDS SHOW WARMER WATER NEAR THE SHORES.

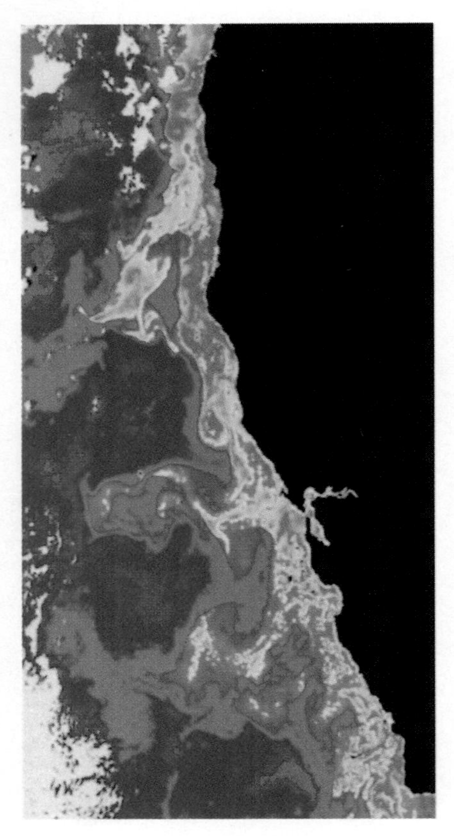

the Gulf Stream carries warm water from the Caribbean area along the East Coast of the United States all winter long. The air warmed by this current doesn't usually get to Washington, D.C., or New York City, however, because the prevailing winds bring chilled air from the interior of the continent instead. On the peninsula of Florida, the westerly winds come from the Gulf of Mexico, giving the state its warm winter weather. The Gulf Stream continues north, then east, then south to warm the British Isles, which are at the same latitude as Hudson Bay (Figure 14-12). Great Britain and western Europe would have far colder climates without the Gulf Stream and the prevailing winds.

When summer arrives, the ocean is cool compared to the land, and once again that temperature difference works to San Francisco's advantage. In the center of a large continent, however, because land temperature changes so fast with solar heating or the lack of it, the summers are relatively hot and the winters relatively cold. Siberia is infamous for those reasons.

WATER IN THE ATMOSPHERE

Water vapor is never a major component of earth's air. Even the muggiest summer air is only about 4 percent water vapor by weight. Water gets into the air by evaporation from the oceans (about 85 percent) and land masses (about 15 percent). Over 100,000 cubic miles of water rise into the air each year to fall back as rain, snow, hail, and dew. Since water vapor absorbs infrared radiation, a cloud cover keeps a night warmer than a crystal clear night when there is little water vapor overhead.

The portion of air pressure that is due to the water molecules in air is called the partial pressure of the water vapor or the **vapor pressure** of water. If the vapor pressure of water in the air is not changing, any liquid water or ice located below the air is in equilibrium with the water vapor above. That is, as water molecules evaporate from the water or sublime from the ice, other water molecules condense onto those surfaces from the air at the same rate. Most often, there is a net flow of molecules in one direction or the other, as the following example illustrates.

Suppose a dehumidifier dries the air in a closed kitchen, removing all the

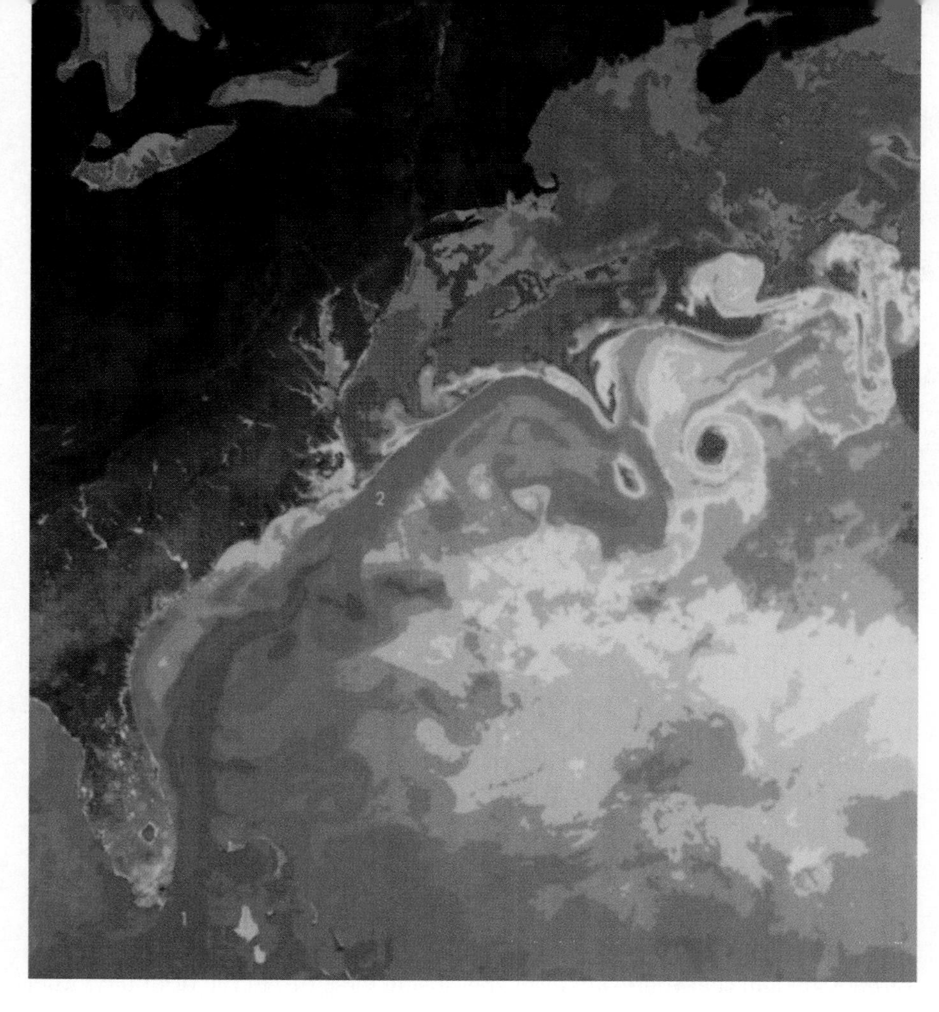

FIGURE 14-12 SATELLITE MAP OF SURFACE WATER TEMPERATURES OFF THE EASTERN COAST OF THE UNITED STATES. THE REDS AND ORANGES SHOW THE WARMEST WATER, AND THE GULF STREAM, FLOWING NORTH PAST THE COAST, IS CLEARLY VISIBLE. THOUGH ITS TEMPERATURE DROPS AS IT GOES EAST AND COMES DOWN PAST THE BRITISH ISLES, WATER IN THE GULF STREAM IS STILL WARMER THAN THE COLD WATERS FARTHER NORTH, SHOWN IN BLUES AND PINKS.

water vapor. Next, suppose the sink is filled with water. Suppose further that the temperature of the room is 26°C and the dry air registers a pressure of 760 millimeters of mercury. As water evaporates from the sink, the barometer gradually rises because of the extra pressure from the water vapor. In a day or so, the air pressure in that kitchen will be 760 + 25 = 785 millimeters of mercury and, at the temperature of 26°C, will rise no higher. We say the air is now **saturated** with water vapor, meaning that the air now contains all the water vapor it can hold at 26°. The water level in the sink will drop no more because now the number of water molecules evaporating from the water's surface is the same as the number reentering from the saturated air. The saturated vapor pressure of water at 26°C is 25 millimeters of mercury. If the temperature increased, though, more water could evaporate because saturated vapor pressure values depend on temperature (Table 14-1).

The saturated vapor pressure doesn't depend on the presence of air. If there had been no air in that kitchen, the barometer would have risen only from 0 to 25 millimeters of mercury. Without air's molecules to beat some of the would-be escapees back into the water, the barometer would have risen much faster, however. In other words, the *rate* of evaporation *does* depend on air pressure. And there is a net evaporation from the surface of water so long as the air above it is not saturated.

Table 14·1 Equilibrium Vapor Pressure of Water at Various Temperatures

Temperature (°C)	Vapor Pressure (mm of Hg)
0	4.6
5	6.5
10	9.2
15	12.8
20	17.5
21	18.6
22	19.8
23	21.1
24	22.4
25	23.8
26	25.0
30	31.8
35	42.2
40	55.3

RELATIVE HUMIDITY

Most air contains some water vapor, but air is rarely saturated, especially near the ground. To describe the amount of moisture in the air at a given temperature, we compare its vapor pressure to the *saturated* vapor pressure at that temperature. The ratio of these quantities is called the **relative humidity.**

PUTTING IT IN NUMBERS—RELATIVE HUMIDITY

relative humidity = vapor pressure/saturated vapor pressure

Example: If the water vapor pressure in a room at 25°C is 17 millimeters of mercury and the saturated vapor pressure at that temperature is 23.8 millimeters of mercury (from Table 14-1), the relative humidity is 17/23.8 = 0.71, or **71 percent**. (Usually the relative humidity fraction is multiplied by 100 to express it as a percent.)

Relative humidity depends on air temperature as well as moisture content, as you can see from Table 14-1. Air at 30°C, for instance, can hold more moisture than air at 25°C. That is, the saturated vapor pressure value is higher for 30° air than for 25° air. If air at 25°C is heated to 30°C while its moisture content remains the same, its relative humidity drops. Likewise, should the temperature of that same volume of air drop below 25°C, its saturated vapor pressure would be less and its relative humidity would rise.

Due to the prevailing westerly winds, the relative humidity in the western United States is usually very low. Moisture in the air from the Pacific Ocean is left in the Sierras as rain or snow as the air rises to pass over them. In the eastern United States, however, moisture from the Gulf of Mexico and from the north often pushes in despite the prevailing westerlies, and the relative humidity is much higher. People from the east who travel in the western states often complain that their skin dries out, losing moisture in the drier air. Other substances, like wood, also lose moisture if the relative humidity plummets, and this loss of moisture causes the material to shrink. Pioneers going west in the 1800s saw their wooden wagon wheels shrink and begin to wobble as soon as they entered the area west of the 100th meridian, which gets too little moisture to grow most crops without irrigation.

If the temperature drops sufficiently, any volume of air that contains water vapor will become saturated and its relative humidity will become 100 percent. The temperature at which this occurs is called the air's **dew point** and is the point at which the water vapor begins to condense.

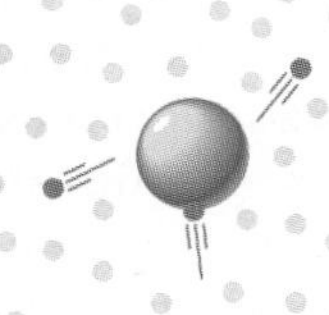

(a)

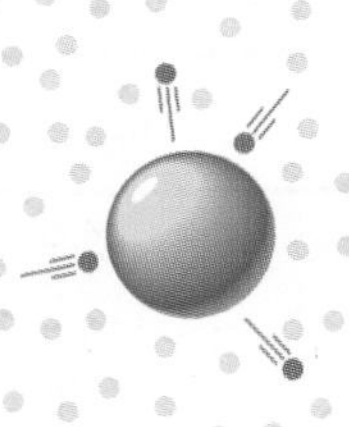

(b)

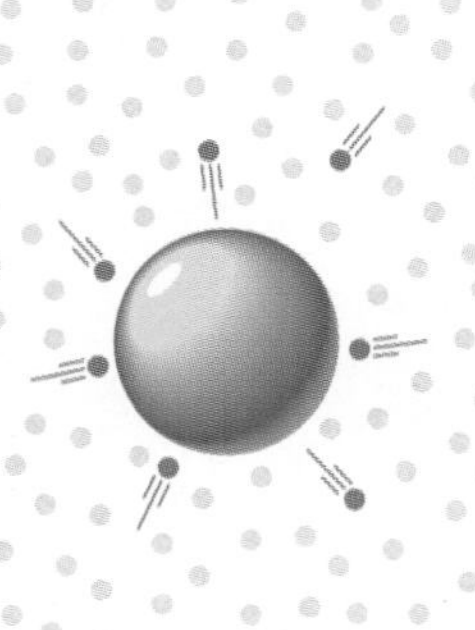

(c)

CLOUDS AND RAIN

Picture a droplet of water floating in the air. Water molecules evaporate from its surface to the surrounding air, and water molecules from the air hit its surface and are absorbed (Figure 14-13). One of three conditions applies.

1. If the relative humidity of the surrounding air is less than 100 percent, the drop loses more molecules to the air than it gains each second and therefore shrinks.
2. If the relative humidity is 100 percent (the air is saturated), just as many

FIGURE 14-13 (A) WHEN THE RELATIVE HUMIDITY IS BELOW 100 PERCENT, THE NUMBER OF WATER MOLECULES LEAVING THE RAINDROP IS GREATER THAN THE NUMBER ENTERING FROM THE AIR AND THE SIZE OF THE DROP DECREASES. (B) WHEN THE RELATIVE HUMIDITY IS 100 PERCENT, THE NUMBER OF WATER MOLECULES LEAVING THE DROP IS EQUAL TO THE NUMBER ENTERING THE DROP AND ITS SIZE REMAINS UNCHANGED. (C) WHEN THE RELATIVE HUMIDITY IS GREATER THAN 100 PERCENT, THE NUMBER OF WATER MOLECULES LEAVING THE DROP IS LESS THAN THE NUMBER BECOMING PART OF IT, AND ITS SIZE INCREASES.

molecules go back into the surface as escape each second. The drop's size stays the same.

3. If the relative humidity is greater than 100 percent, the number of water molecules leaving the air and entering the drop's surface each second is greater than the number leaving the drop. The drop grows. The air is said to be **supersaturated;** it contains more water than it can maintain indefinitely at that temperature.

It is supersaturated air that causes rain. When warm, moist air rises in the atmosphere, it cools. As it cools, the saturated vapor pressure value drops and therefore its relative humidity rises. If a volume of moist air in an updraft reaches 100 percent relative humidity *and continues to rise and cool,* condensation is certain to occur.

Condensation mostly begins on small particles in the air that provide surfaces where water molecules can stick and collect. Called **condensation nuclei,** these air-borne particles have numerous sources. For example, many water drops are launched upward in the spray from the top of an ocean wave. The smallest droplets evaporate and leave behind microscopic salt particles that the wind carries aloft. Over land, breezes pick up particles from forest fires, geysers, volcanoes, farming, and modern industrial activities.

When growing water drops in air become large enough to reflect light well, we see a cloud. The clouds we see are rarely more than 8 kilometers (5 miles) above sea level, where the air is cold enough to cause the moisture to condense. Cloud droplets are tiny and have *very* low terminal speeds and are dragged along by the slightest breeze. If these droplets grow, however, they have more mass per unit of surface area, and the air affects their motion less (Figure 14-15). Then they fall from the clouds as rain.

FIGURE 14-14 THE TEMPERATURE OF THE AIR GENERALLY DECREASES AT HIGHER ELEVATIONS THROUGH THE TROPOSPHERE. THE AIR'S LOWER DENSITY MEANS IT CANNOT HOLD AS MUCH HEAT, AND ITS TRANSPARENCY TO BOTH VISIBLE AND IR MEANS IT CANNOT ABSORB MUCH HEAT. HERE THE DECREASE IN TEMPERATURE WITH ELEVATION IN THE TETONS OF WYOMING IS CLEAR. THERE ARE GREEN MEADOWS BELOW, BUT NOTICE THE THINNING OF THE TREES UP THE SIDES OF THE MOUNTAINS UP TO THE TREE LINE, WHERE EXTREME TEMPERATURE AND MOISTURE CONDITIONS DO NOT ALLOW THE GROWTH OF TREES AT ALL.

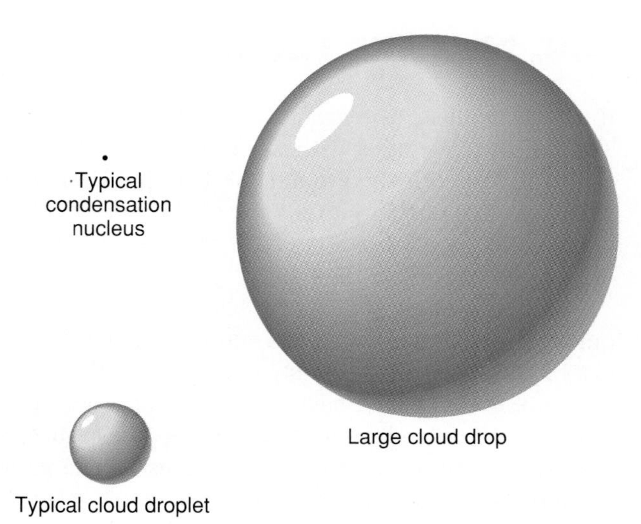

FIGURE 14-15 COMPARING THE DIAMETER OF A CONDENSATION NUCLEUS AND THE EVENTUAL RAINDROP THAT GROWS AROUND IT. THE CLOUD DROPLET SHOWN HERE IS 0.075 MILLIMETER IN DIAMETER.

HUMIDITY AND COMFORT

Relative humidity plays a part in how comfortable we are, especially on warm days. In hot air, we perspire to keep our internal temperature steady, and the perspiration must evaporate, absorbing its heat of vaporization from our skin, in order to cool us. If the relative humidity is high, our perspiration will not evaporate quickly.

If evaporation is retarded, we can be uncomfortable or even be in danger of overheating, with the result being serious illness or even death. In the United States more people die each year from heat stress than from tornadoes or hurricanes. The National Weather Service uses a heat stress index (Table 14-2) to indicate the *apparent*, or *perceived*, air temperature depending on the humidity. As an example of the table's use, if the air temperature is 85°F and the relative humidity is 90 percent, the air feels like 102°F air.

Table 14.2 Heat Stress Index

Temperature	Relative Humidity									
	10%	20%	30%	40%	50%	60%	70%	80%	90%	100%
80° F	75	77	78	79	81	82	85	86	88	91
85° F	80	82	84	86	88	90	93	97	102	108
90° F	85	87	90	93	96	100	106	113	122	
95° F	90	93	96	101	107	114	124	136		
100° F	95	99	104	110	120	132	144			
105° F	100	105	113	123	135	149				
110° F	105	112	123	137	150					
115° F	111	120	135	151						

Danger Category
- No discomfort
- I. Caution
- II. Extreme caution
- III. Danger
- IV. Extreme danger
- Not observed

Data from National Weather Service.

Physics and Environment

TEMPERATURE AND RELATIVE HUMIDITY

A RESEARCHER IN A CLOUD FOREST IN COSTA RICA COLLECTS AND MEASURES DEW THAT FORMS ON TEFLON ELEMENTS. IN SUCH FORESTS, AND IN SOME MUCH DRIER AREAS ALONG THE MEDITERRANEAN SEA, DEW PLAYS A MAJOR ROLE IN DELIVERING WATER TO THE PLANT LIFE.

FOG, DEW, AND FROST

If a night is clear, the ground cools rapidly by radiating its heat into space. Then the air in contact with this ground is cooled. If that air is humid, a cloud can form just over the ground. We call this kind of cloud **fog.** Fog also forms when warm, moist air moves over cold water or cold land and cools. In Alaska a winter fog is often made of ice crystals rather than water droplets. Called *ice fog,* it sparkles in the light.

Moisture often appears on blades of grass and other surfaces that cool quickly by radiation when the sun goes down. The air temperature is lowest just before sunrise, so its relative humidity is highest then. Cooling by radiation, leaves of grass (which have a lot of surface area per unit of mass) may well reach lower temperatures than the air does, lower even than the cool air's dew point. Water vapor then condenses on the grass. This is called **dew.** A concrete sidewalk bordering the grass won't get wet, however, because as it radiates heat away, more heat is conducted up from the ground beneath it, keeping it warmer than the grass. Leave a cold soft drink can or a glass of iced tea resting on a table in the summertime, and water vapor from the air condenses on the cold surface and runs down the sides. That, too, is dew.

FIGURE 14-16 THE SURFACES OF PLANTS COOL BY RADIATION AT NIGHT UNTIL THEIR TEMPERATURE IS BELOW THE AIR TEMPERATURE. AIR NEXT TO THE PLANTS BECOMES SUPERSATURATED AS IT IS COOLED BY CONTACT WITH THE PLANT, AND DEW OR FROST FORMS ON THE PLANT'S SURFACE. A LIGHT FROST FORMED ON THE EDGES OF THIS LEAF, WHICH HAS MORE EXPOSED SURFACE AREA PER UNIT OF MASS THAN THE CENTER. THEREFORE THE EDGES COOL OFF FASTEST, AND THAT IS WHERE THE FROST FORMED.

On nights below freezing, leaves of grass can radiate heat so fast they drop below freezing before the air does, and **frost** rather than dew forms as water vapor turns directly into ice on the leaves (Figure 14-16).

RAINSTORMS AND HURRICANES

IN MARCH 1991, A HUGE HAILSTORM STRUCK THE AREA AROUND AND INCLUDING THE AUTHOR'S UNIVERSITY. MANY THOUSANDS OF CARS WERE DENTED IN A MATTER OF A MINUTE OR TWO. LUCKILY THESE OCCURRENCES ARE RARE; OVER 200,000 CLAIMS WERE PROCESSED OVER THE NEXT SEVERAL MONTHS FOR AUTOMOBILES AND HOUSE ROOFS IN A 3-COUNTY AREA.

A typical summer raincloud forms when a buoyant plume of moist, warm air rises quickly, cools, and produces rain and pellets of ice called **hail.** Though hail is probably present in all thunderstorms, it usually melts or sublimes before it reaches the ground. Surprisingly, most raindrops don't make it to the ground either. Only about 20 percent of the rain in a typical summer thunderstorm falls to earth, the rest evaporates or disperses into less dense clouds.

Hurricanes are gigantic storms born over the warm tropical oceans in late summer. These violent storms may be several hundred kilometers across, with winds of at least 74 mph by definition (119 km/h) and sometimes up to about 200 mph (320 km/h). The energy to drive the winds comes from the warm water beneath the storm. The center of the storm is an area of low pressure (called the eye), and winds come toward it from all directions. As they do, earth's rotation turns them (Figure 14-17). As a result, the eye remains calm but there is a wall of rapidly ascending air around it fed by warm, moist winds sweeping in over the choppy surface of the ocean. This ascending column of wet air soars straight up to cooler elevations. As it does, the moisture in it condenses into rain and releases its latent heat of vaporization. That heat

PHYSICS IN 3D/FOREST FIRE PLUME

WERE IT NOT FOR CONVECTION, SMOKE FROM FIRES SUCH AS THIS ONE IN THE MOUNTAINS OF PERU WOULD NOT RISE. (PHOTO FROM SPACE SHUTTLE.)

PHYSICS IN 3D/STORM CELLS FROM ABOVE

LARGE CONVECTION CELLS OVER THE OCEAN STRETCH OUT TO THE HORIZON, AS SEEN FROM THE SPACE SHUTTLE. THEY ARE CONFINED TO THE TROPOSPHERE (EXCEPT FOR THEIR VERY TOPS).

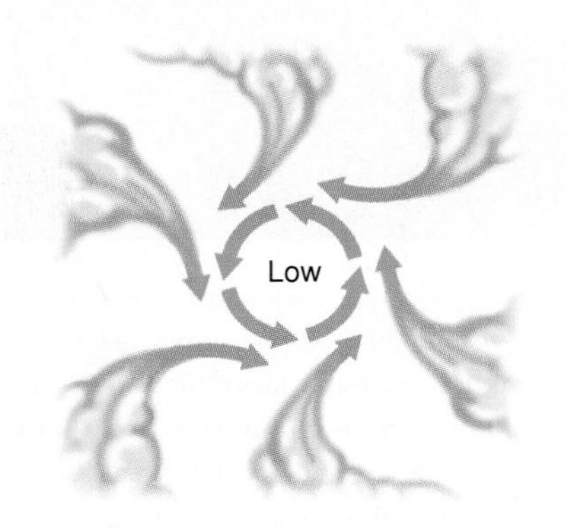

FIGURE 14-17 WINDS TRAVELING NORTHWARD IN THE NORTHERN HEMISPHERE CURVE EASTWARD BECAUSE THEY MOVE INTO A REGION WHERE THE EARTH'S SURFACE HAS A SLOWER EASTWARD SPEED THAN THEIR OWN. LIKEWISE, WINDS MOVING SOUTHWARD IN THE NORTHERN HEMISPHERE DEFLECT TO THE WEST BECAUSE THEY MOVE INTO REGIONS TRAVELING FASTER TOWARD THE EAST THAN THEY ARE TRAVELING. THESE SPEED DIFFERENCES GIVE A HURRICANE A COUNTERCLOCKWISE ROTATION IN THE NORTHERN HEMISPHERE AND A CLOCKWISE ROTATION IN THE SOUTHERN HEMISPHERE.

energy feeds the winds. If a hurricane travels onto land or over water that is cooler than 27°C (80°F), it loses its source of energy, slows, and disperses (Figure 14-18).

CLIMATE AND LIFE

The sun warms the earth unevenly from the poles to the tropics. Winds and ocean currents then distribute this heat energy around the globe. Ice caps and glaciers, steamy tropical jungles and swamps, grassy plains and barren deserts, regions with distinct seasons and islands where there are no real seasons—such is the sun-driven climate of earth, home to life.

Most living organisms exist only within a narrow temperature range. Unless they are in direct sunlight, plants take on the temperature of the air or water they live in; their chemical activities proceed when the air or water is warm enough and cease when it is too cold. One-celled animals, having so much surface area per unit mass for heat to move through, do the same. Lizards and snakes, though larger, have internal temperatures that are usually only a few degrees above the temperature of their surroundings. These "cold-blooded" animals hibernate when it gets cold and their body temperature drops. Humans, other mammals, and birds maintain very steady internal temperatures, however. Because we are "warm-blooded," we can function just as well at midnight as at noon.

Today we know the earth's climate has undergone tremendous changes in the past. There have been several periods when many species of life disappeared, unable to cope with climatic changes. Now there is evidence that our own activities on earth may well influence the climate in the future, and scientists are intensively studying areas of concern, such as the greenhouse effect and the depletion of ozone in the atmosphere, which we will discuss in Chapter 21. Understanding how our activities affect our environment is one of the most important tasks scientists face today.

FIGURE 14-18 A LARGE STORM OVER THE BRITISH ISLES IN 1987 SHOWS THE CIRCULATION CAUSED BY EARTH'S ROTATION OF STORMS AND HURRICANES IN THE NORTHERN HEMISPHERE.

(a)

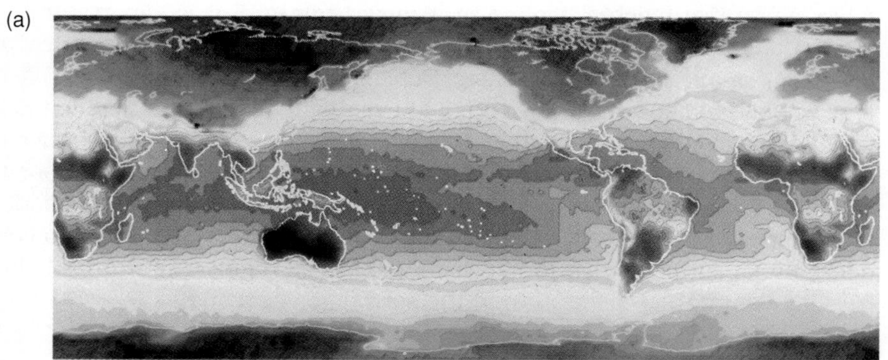

(b)

AVERAGE WATER AND LAND TEMPERATURES IN (A) JANUARY AND (B) JULY ON EARTH, TAKEN FROM WEATHER SATELLITES. TEMPERATURES BELOW FREEZING ARE GREEN AND BLUE, AND THE WARMEST TEMPERATURES ARE RED AND DARK BROWN. THIS IS A REMARKABLE LOOK AT HOW THE HEAT FROM THE SUN IS DISTRIBUTED OVER THE PLANET. THIS PERSPECTIVE OF OUR OCEANS AND LAND MASSES SYMBOLIZES WHAT WE HAVE LEARNED AND THE HOPE THAT WE CAN MAKE CAREFUL DECISIONS ON HOW OUR ACTIVITIES AFFECT THE EARTH'S SURFACE AND ECOSYSTEMS.

Summary and Key Terms

Each day, the oceans and land absorb most of the sunlight that strikes them, warming and radiating heat in all directions. This radiation continues through the night, so night is a time of cooling. The oceans warm and cool more slowly than the land because water's specific heat is larger than that of the land, and because they spread the absorbed energy over more mass.

Our seasons are due to the tilt in the earth's axis of rotation. In the winter, this tilt results in fewer hours of sunlight and more hours of darkness each day. Also, this tilt gives the winter sun a lower daily path, so a given amount of sunlight spreads over a much larger surface area. As earth's surface warms, the air above it forms convection currents that mix the air vertically. Long-range horizontal mixing in our atmosphere comes from global air currents called ***prevailing winds***.

prevailing winds–204

The portion of air pressure due to water molecules is water's ***vapor pressure***. Water's ***saturated vapor pressure*** is the highest vapor pressure possible before condensation begins. At higher temperatures, air holds more water molecules, so the saturated vapor pressure increases. The amount of moisture in the air is the air's ***relative humidity***, the ratio of its actual vapor pressure to its saturated vapor pressure. Relative humidity depends on temperature. If air is heated while its moisture content remains constant, its saturated vapor pressure increases and its relative humidity drops; if air is cooled while its moisture content remains constant, its saturated vapor pressure decreases and its relative humidity rises. At the air's ***dew point***, the air becomes saturated and its relative humidity reaches 100 percent.

vapor pressure–206
saturated vapor pressure–207
relative humidity–207
dew point–208

If the relative humidity rises above 100 percent, the air is ***supersaturated*** and condensation begins. A rainstorm occurs when solar heating causes warm, moist air to rise quickly, cooling as it moves up, until it is supersaturated. Then, water molecules collect and condense on small particles in the air, called ***condensation nuclei***. When the water droplets grow large enough, they reflect light enough to show up as a cloud, and eventually they fall as rain and hail, much of which never reaches the ground. ***Fog*** forms when the cool surface of the ground or ocean cools warm moist air above it, causing water vapor to condense as a low cloud. ***Dew*** forms when an object cools quickly by radiation to a temperature lower than the air's dew point, causing water vapor to condense on its surface. ***Frost*** forms when an object's temperature drops below freezing before the temperature of the surrounding air does.

supersaturated–209
condensation nuclei–209
fog–210
dew–210
frost–211

EXERCISES

Concept Checks: Answer true or false

1. The two most abundant gases in the earth's atmosphere are nitrogen and oxygen. The two most abundant gases in volcanic emissions are water vapor and carbon dioxide.
2. Sunlight that falls on water is absorbed by a larger amount of mass than sunlight that falls on land.
3. The portion of air pressure due to water molecules is what we call the vapor pressure of water.
4. A volume of air above a lake is saturated with water vapor when the number of water molecules moving from the air into the lake each second is the same as the number moving from the lake into the air. (Assume the water and air have the same temperature.)
5. Relative humidity is the ratio of the maximum water vapor pressure the air can have at that temperature divided by the actual water vapor pressure in the air.
6. Both the moisture content of the air and its temperature determine the relative humidity.
7. The net rate at which a liquid evaporates depends on the pressure of any gases over its surface and on its own vapor pressure at the surface.
8. The dew point for air is the maximum amount of water it can hold at a given temperature.
9. Dew is water vapor that condenses on cool surfaces.

Applying the Concepts

10. Oxygen is a highly reactive gas, combining readily with metals and other elements. Speculate about what would happen to the oxygen in our air if plant life ceased to exist today.

11. About 50 percent of earth's surface is covered with clouds in the daytime. What would happen to earth if that percentage were greater? If it were less?

12. Explain why the air temperature drops more on clear nights than on cloudy nights.

13. On a typical night, the average temperature in New York City, Washington, D.C., or Los Angeles drops about 11°C (20°F). Speculate on what would happen to the atmosphere if the sun didn't shine tomorrow morning, and the next, and the next, etc. (Breathe deeply! You're enjoying solar energy.)

14. Inspect Figures 14-3 and 14-4 to convince yourself that points on the equator get 12 hours of sunlight and 12 hours in earth's shadow (in other words, darkness) every day.

15. In the Arctic, the sun is never far above the horizon, so the sunlight comes from a very low angle and brings little energy per square meter per second. Yet the Arctic is where many species of whales go to feed in the summertime and store up thick layers of blubber. Why should there be an abundance of phytoplankton (plant life) in Arctic waters in the summer?

16. Explain why January and February are the coldest months in the Northern Hemisphere even though the sun is getting higher in the sky throughout these months.

17. When you exhale, the air coming from your warm, moist lungs is at 100 percent relative humidity. Explain why it turns into a white cloud on a cold day.

18. The relative humidity of air emerging from an air conditioner is close to 100 percent. Why doesn't it form a fog in the air as it emerges from a vent in a room?

19. The rate at which water evaporates from a lake depends on the air pressure on the lake surface. Does this mean a high mountain lake has a higher evaporation rate than a similar lake at sea level, if all other factors are equal?

20. In a thunderstorm, the large drops of falling rain are accompanied by a strong downdraft of air, sometimes called the rain draft. What causes this?

21. When the temperature of air is 30°C, its saturated vapor pressure is 32 mm of mercury. If the relative humidity in a room at that temperature is 28 percent, what is the vapor pressure of the water in that room?

22. Make a graph of the data of Table 14-1 with temperature on the horizontal axis, and estimate the saturated vapor pressure of water at 28°C and 37°C.

23. You leave a glass of water by your bedside at night, and in the morning the water level is unchanged. What can you say about the air in your bedroom?

24. Explain why dew forms on a cold can of soda pop.

25. In a closed house, what could you do to lower the relative humidity?

26. Look at Figure 14-8 and explain how every year some Monarch butterflies from the United States wind up in Europe. (It's about a 3-day trip for them.)

27. A closed classroom can become noticeably "stuffy" during a lecture. Speculate about what would happen to the air in the room if the room remained sealed for 24 hours with the students and the instructor inside.

28. Key West, Florida, is at the same latitude as Karachi, Pakistan, and Saudi Arabia, yet its climate is more moderate. Why?

29. To tell which way the wind is blowing, wet a finger all around and hold it up in the air. How does this old trick work?

30. In the southwestern states, places 1400 meters (4600 feet) above sea level or less get 25 centimeters of rain a year, while places about 2100 meters (6900 feet) above sea level get 50 centimeters or more. Discuss why this happens.

31. Speculate on what might happen to the climate in the United States if the prevailing westerly winds we now have become prevailing easterly winds. (Florida's climate wouldn't change too much.)

32. Breezes most often come up the slope of a mountain at noon and go down the slope after midnight. Why?

33. Using the graph of Exercise 22, estimate the relative humidity in a room at 28°C where the vapor pressure of water is 13 mm of mercury.

34. You've probably seen someone use moisture from a puff of breath to clean eyeglasses. How does that moisture condense on the lens? Does it make a difference if the person huffs or purses the lips to blow? Why or why not?

35. At the beach, where does the breeze most often come from in the daytime? Why? At night? Why?

36. Ninety percent of the ice on earth is in Antarctica, where the depth of ice over the continent averages 1800 meters (5900 feet) and gets as deep as 3700 meters (12,000 feet). If all this ice melted due to global warming, the oceans would rise about 61 meters (200 feet) or more. Find an atlas and see which major cities would be under water. (Almost all of Florida would be.)

37. In Chicago, cold winter air sweeping across warm Lake Michigan early in the winter dumps inches of snow in a mile-wide band along the lakefront but no farther. It is called "lake-effect" snow. Explain what happens.

38. Hailstones form when air carries raindrops quickly upward in a convection current to freeze at high altitude. If the hail is to reach the ground, it must descend in a hurry. How can it come down fast?

39. People cannot live in air that is perpetually at 35°C (95°F) and 100 percent relative humidity. Discuss how this combination would thwart the body's cooling mechanisms.

40. Much of the solar energy received at the equator is transported north and south. Approximately 20 percent of that moving thermal energy is carried by ocean currents. How must the rest be transported?

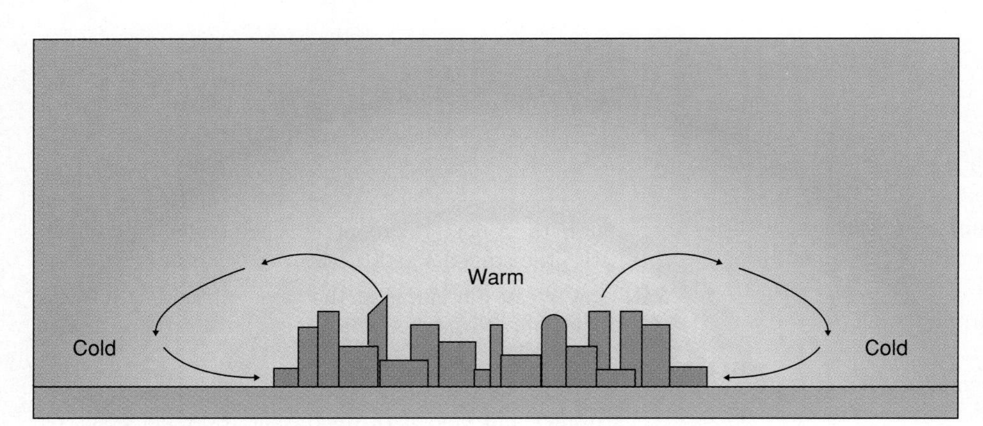

FIGURE 14-19

41. Inspect Figure 14-8 and discuss why the fastest bicycle times across the United States are much worse going east to west than going west to east.
42. Arctic willows grow parallel to the ground, and all Arctic flowers lie flat on the ground. How does this benefit them?
43. Figure 14-19 shows the idealized circulation around a city on a clear, calm night. The air coming in from the side is sometimes called a "country breeze." Explain why this circulation exists.
44. The rain falling from the clouds shown in Figure 14-20 falls through dry air and evaporates before it reaches the ground. Called *virga,* this is a common sight in the arid West. Discuss why the rain never reaches the ground.
45. In one minute an average adult male breathes in 0.009 kg of air. Calculate how much air he inhales in 24 hours, and compare this to the 2 kg of water and 1.2 kg of food he uses. This gives everyone a compelling reason to be concerned about air pollution.

FIGURE 14-20

Demonstrations

1. Air conditioning units always drip because the air that passes over and around the evaporator coil cools to its dew point and condensation occurs on the cold metal surfaces. The relative humidity of the cold air that comes from the air conditioner is close to 100 percent. Stand in this blast on a hot day, and there will soon be an invisible layer of moisture on your skin. Then step outside into warmer air, and you will shiver for a split second as that layer of moisture evaporates almost instantly and takes a lot of heat from your skin. This effect is especially noticeable if the relative humidity of the warmer air is low.
2. To simulate the formation of clouds and rain, boil water in a teapot with a spout. The steam (hot water vapor) leaving the spout is invisible, but as it expands, it cools to its dew point, where the relative humidity is 100 percent. It then condenses and becomes visible as a cloud. (Most people call that cloud "steam" also.) To produce rain from this cloud of water vapor, hold a large metal spoon in it. The water vapor in the cloud will condense on the metal surface (which conducts heat away fast). As the droplets condense and collect there, they grow large and fall as "rain."
3. Darken your kitchen, turn on a flashlight, and quickly open the freezer. Blow into it, using your flashlight to see the faint cloud of fog that forms as your breath cools. The warm air you breath out has a relative humidity of 100 percent from its close contact with your moist lungs. As it cools, the water vapor condenses.

FIFTEEN

Waves in Matter

Surfers catch rides along the wavefronts of large waves, which then "break." The physics of waves is the subject of this chapter.

THERE ARE TWO KINDS OF WAVES, THOSE THAT TRAVEL ONLY THROUGH MATTER AND THOSE THAT CAN TRAVEL THROUGH MATTER OR EVEN EMPTY SPACE. THIS CHAPTER DISCUSSES THE WAVES THAT TRAVEL ONLY THROUGH MATTER. IN CHAPTER 21 WE DISCUSS THE OTHER TYPE OF WAVE.

Surfers see a wave coming long before it reaches them. If they let the wave pass and wait for the next one, the water gives them a rolling motion that picks them up when the wave arrives and sets them down as it departs. The water itself does not travel along the surface as the wave does. The disturbance called a **wave** carries energy through matter without taking the matter along with it. The matter, be it solid, liquid, or gas, is called the **medium** of the wave. While you can *see* the waves in the water of a baby's bath, anyone standing close to a moving train can *feel* waves coming through the ground as the train passes. This chapter is about waves in matter.

SOME FACTS ABOUT WAVES

A long flexible spring can show you a lot about waves. Attach one end to a stationary object and stretch the spring out, which puts tension in it. Now gather some of the coils together at your end. Release them and they snap apart (because of the tension) and create a disturbance—a wave—that moves along the length of the spring. If you jerk the end of the spring back and forth quickly in the direction along its length, a series of disturbances travel along. As seen in Figure 15-1, such a wave consists of regions called **compressions,** where the coils are close together, alternating with regions called **expansions,** where the coils are stretched farther apart. A compressional wave is also called a **longitudinal** wave. In a longitudinal wave, the medium moves back and forth along the same line that the wave moves along.

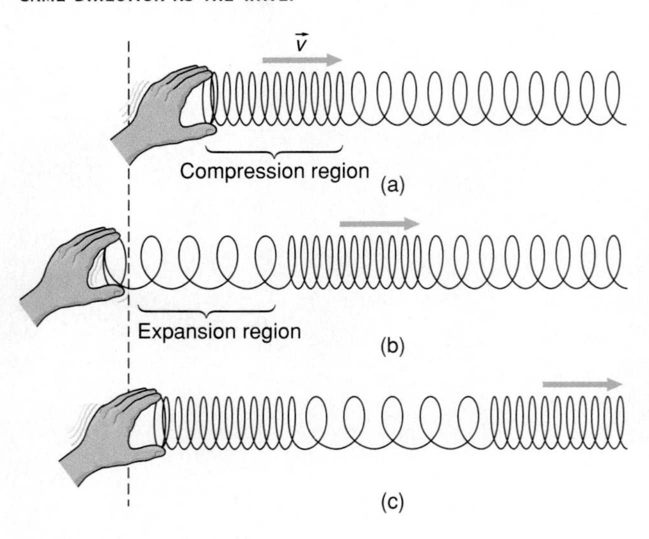

FIGURE 15-1 COMPRESSIONS AND EXPANSIONS OF A LONGITUDINAL WAVE ON A SPRING. AS THE COMPRESSIONS AND EXPANSIONS MOVE AWAY FROM THE HAND THAT CAUSED THEM, THE MOLECULES IN THE SPRING MOVE BACK AND FORTH SLIGHTLY. BECAUSE THIS WAVE IS LONGITUDINAL, THE MOLECULES OF THE MEDIUM MOVE IN THE SAME DIRECTION AS THE WAVE.

Next, move the end of the spring up and down with a regular motion that is perpendicular to the spring's length. Now a different wave pattern moves along, where the motion of the spring is perpendicular, or transverse, to the direction of the wave's motion (Figure 15-2). Such **transverse** waves occur only if there are forces that restore the molecules of the medium to their original positions after they have been pushed to the side by a wave. If the molecules of a liquid or gas move to one side, other molecules immediately move in behind them. Therefore there is no force to return them to their former places. Consequently, transverse waves cannot travel through fluids; they travel only through solids. Longitudinal waves, however, travel through matter in any of its states.

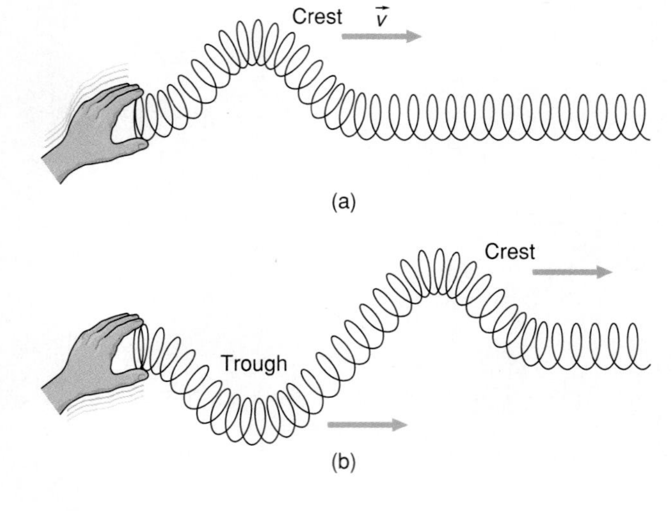

FIGURE 15-2 IN A TRANSVERSE WAVE, THE DIRECTION OF THE MEDIUM'S MOTION IS PERPENDICULAR TO THE DIRECTION OF WAVE MOTION. THE UPWARD PEAK SHOWN IN (A) IS A *CREST* OF THE WAVE, AND THE DOWNWARD DIP FOLLOWING THE CREST SHOWN IN (B) IS CALLED A *TROUGH.*

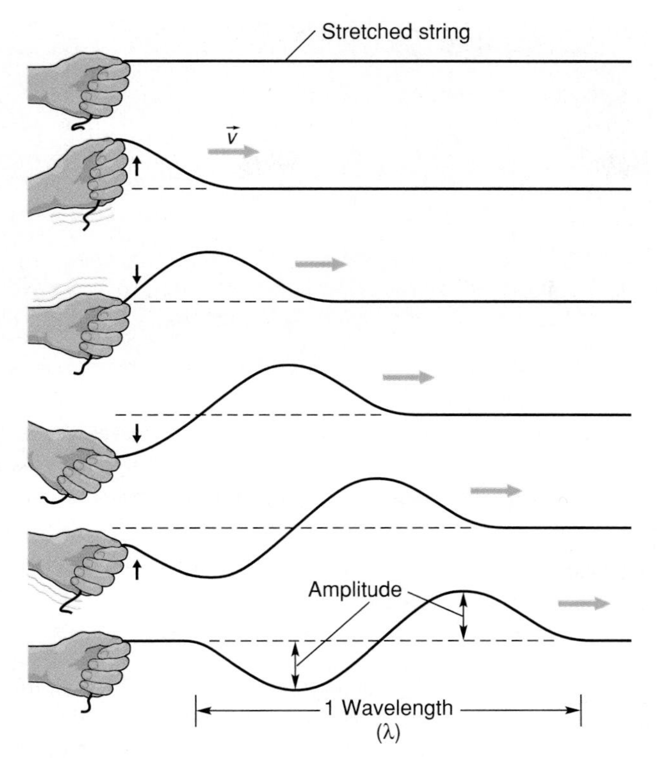

FIGURE 15-3 MAKING A WAVE ONE WAVELENGTH LONG ON A STRETCHED STRING. A TUG UP AND BACK TO CENTER PUTS A CREST ON THE LINE, AND A TUG DOWN AND BACK TO CENTER MAKES A TROUGH. ONE COMPLETE CENTER-UP-CENTER-DOWN-CENTER MOTION IS CALLED ONE CYCLE OR ONE OSCILLATION. THE DISTANCE FROM THE UNDISTURBED LINE TO A CREST OR TROUGH IS THE *AMPLITUDE* OF THE WAVE.

FREQUENCY, AMPLITUDE, AND WAVELENGTH

Next, let's look at waves on a stretched string, as in Figure 15-3. Flipping the string up and down with a regular motion sends a succession of crests and troughs along it. The motion that puts one crest and one trough on the string is called either a *cycle* or an **oscillation.** The vertical distance from the *undisturbed position* of the string to the crest or to the trough is called the **amplitude** of the wave.

By flipping your wrist up and down quickly, you can perform 3 or 4 oscillations in a second if you don't move the string very far (small amplitude). The number of oscillations per second is called the **frequency** of the wave, and its units are 1/s, which is called a **hertz** (Hz). The frequency is also the number of up-and-down trips a particle farther along the string makes in 1 second as the wave passes. A **wavelength** is the length of the wave made by one oscillation. It is also the distance over which the wave pattern repeats itself—for example, from one crest to the next or from one trough to the next. Wavelength and amplitude are measured in any convenient unit of length.

THE SPEED OF A WAVE

Suppose you make a 3-Hz wave on a stretched string. In 1 second, 3 wavelengths move down the string, one after the other, as in Figure 15-4. Since 3 wavelengths of the wave move along the string in 1 second, the wave's speed is 3 wavelengths/1 second. In other words, the frequency of oscillation f (3 Hz) multiplied by the wavelength λ (Greek letter lambda) gives the speed of the wave, v.

FIGURE 15-4 IF 3 WAVELENGTHS ARE PUT ON A STRING IN 1 SECOND, THE WAVE'S SPEED ALONG THE STRING IS 3 WAVELENGTHS/SECOND, OR 3λ/S.

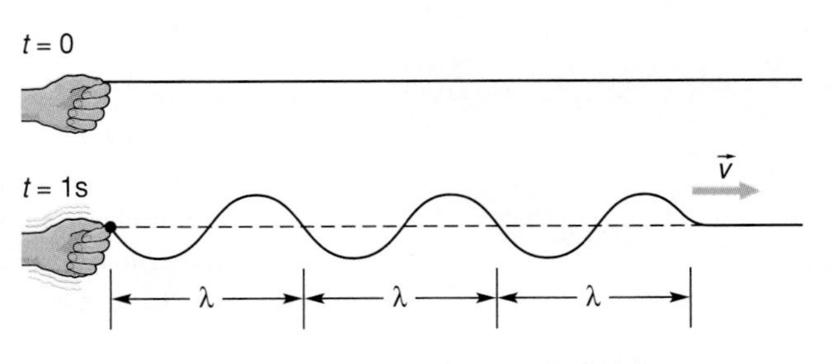

PUTTING IT IN NUMBERS—WAVE SPEED, FREQUENCY, AND WAVELENGTH

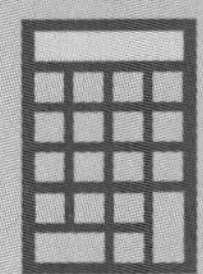

wave speed = frequency × wavelength

$v = f \times \lambda$

Example: If you oscillate the end of a string at 3 Hz and the speed of the wave is measured to be 9.6 meters/second, the wavelength is: wavelength = wave speed/frequency = v/f = (9.6 meters/~~second~~)/(3/~~second~~) = **3.2 meters**

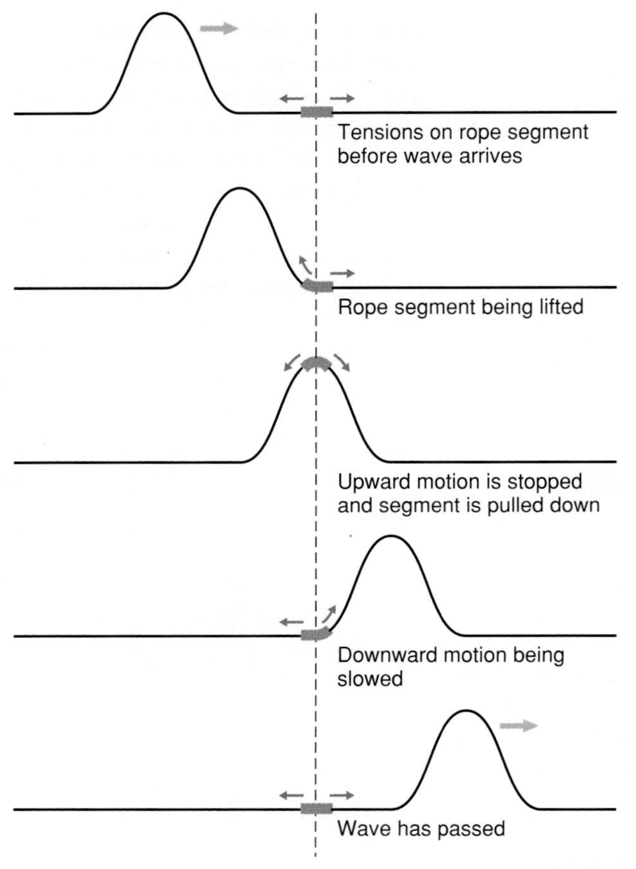

FIGURE 15-5 WHENEVER A TRANSVERSE WAVE TRAVELS ALONG A STRETCHED STRING, THE TENSION BOTH RAISES AND LOWERS THE STRING AS THE WAVE PASSES BY. IF THERE WERE NO TENSION IN THE STRING, THE WAVE COULDN'T MOVE ALONG IT.

For transverse waves on a string, the restoring force is the tension in the string (Figure 15-5). Stretch the string more tightly, and the waves travel faster. A greater restoring force increases the wave's speed, while a greater inertia of the medium decreases it. To see the effect of inertia, use a heavier string or even a rope and watch the wave move more slowly. If the tension in a string remains unchanged, experiments show that all waves moving on it have the same speed, *no matter what their frequencies*. This is not true for all types of waves in every medium, but in later chapters we shall see that it is an important property of sound waves and light waves.

Longitudinal waves move faster through solids than through liquids or gases because of the stronger bonds connecting the atoms or molecules of a solid, which provide a stronger restoring force. A 1000-Hz longitudinal wave travels through steel at 18,000 km/h (11,000 mph), through water at less than 1/3 that speed, and through air at less than 1/12 that speed. Transverse waves travel through solids at less than half the speed of compressional waves. That's because the restoring force exerted during an up-and-down medium displacement is less than the restoring force exerted by a back-and-forth displacement (Figure 15-6). Demonstration 3 lets you feel this effect.

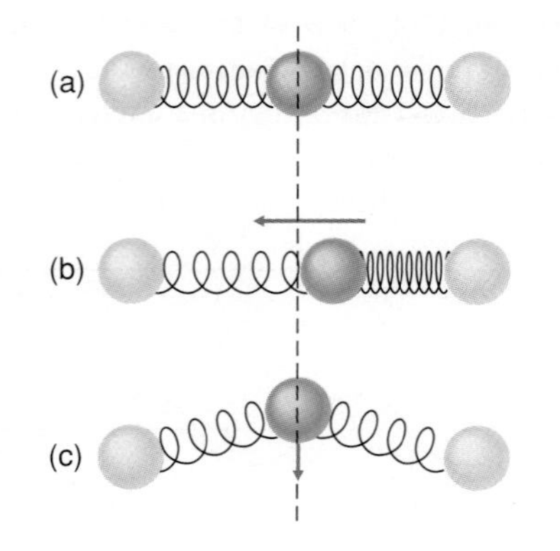

FIGURE 15-6 (A) THREE UNDISTURBED MOLECULES IN A SOLID. (B) A HORIZONTAL DISPLACEMENT OF THE CENTER MOLECULE TO THE RIGHT, AS WHEN A LONGITUDINAL WAVE COMES PAST. THE RESTORING FORCE IS STRONG BECAUSE THE SPRINGS, WHICH REPRESENT THE BONDS, ARE COMPRESSED ON THE RIGHT AND STRETCHED ON THE LEFT. (C) A VERTICAL DISPLACEMENT OF THE CENTER MOLECULE, AS WHEN A TRANSVERSE WAVE PASSES. THE STRETCHING OF THE BONDS IS NOT AS SEVERE AS IN (B), AND CONSEQUENTLY THE RESTORING FORCE IS SMALLER.

ENERGY AND MOMENTUM IN WAVES

A wave carries energy, not matter. The work you do as you shake a stretched string provides the energy the wave carries with it. Since work is force times distance, the amount of energy the wave carries depends on how far you pull the string from its resting position. That is, the energy in a wave depends on the wave's amplitude. If you shake the string faster, you put more energy per second into making that higher-frequency wave, and the wave delivers more energy per second to the particles of the string as it moves past (Figure 15-7). Not all the energy makes it to the far end, however. Some is left behind as heat due to friction within the medium as the wave moves through. Any wave through matter eventually dies out because of friction. Waves also carry momentum since the particles of the medium have mass and the passing wave gives them speed. Waves, like everything else, obey the laws of conservation of energy and momentum.

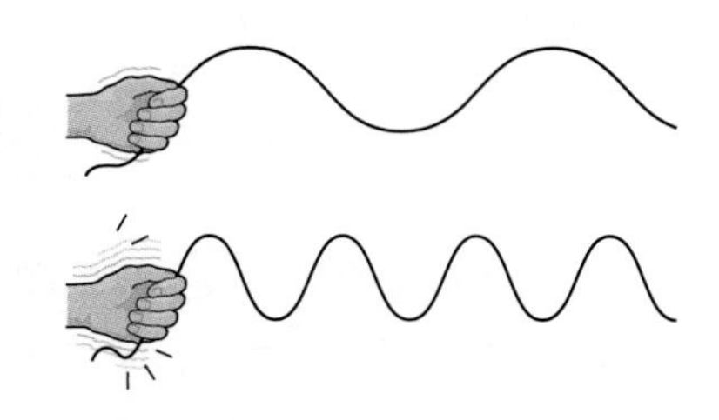

FIGURE 15-7 TO GENERATE A HIGH-FREQUENCY WAVE REQUIRES MORE ENERGY PER SECOND THAN TO GENERATE A LOW-FREQUENCY WAVE. THUS, A HIGH-FREQUENCY WAVE CARRIES MORE ENERGY THAN A LOW-FREQUENCY WAVE OF THE SAME AMPLITUDE.

REFLECTION, TRANSMISSION, AND INTERFERENCE

Figure 15-8 shows a "train" of several wavelengths moving along a stretched rope toward a tree. The wave reaches the tree, bounces off much like a ball thrown against a wall, and comes back along the rope. A wave **reflects** like this whenever it meets another medium (here, the tree) where the wave's speed changes. If the change in speed is not great, much of the wave's energy will be **transmitted,** passing on into the second medium, and only a little will be reflected. If the speed increases a significant amount as it does for this situation, much of the wave is reflected. The slow-moving wave on the rope almost totally reflects from the solid tree, where the tiny transmitted portion of the wave has a very high speed through the firmly-held molecules of the tree.

Suppose the end of the rope in Figure 15-8 is flipped back and forth continually, so that a train of wavelengths travels down the rope. As soon as the initial part of the wave reflects from the end tied to the tree, the rope plays host to two waves at once, the one coming from the left and the one returning from the reflecting end. Wherever the two waves merge along the rope, the rope gets a push from both waves at that point and moves according to the *net* force it receives. When two crests of the same amplitude, one

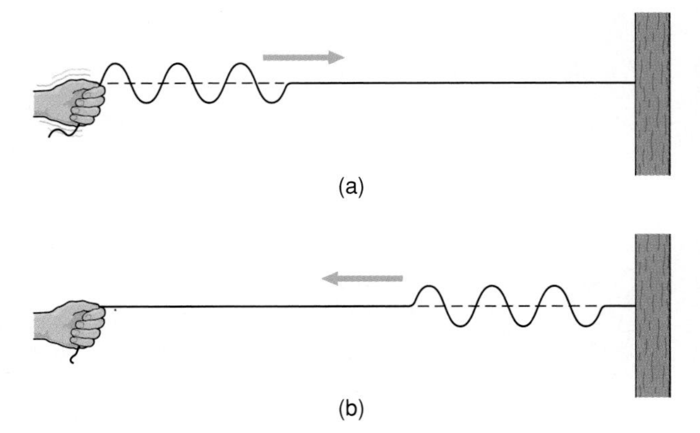

FIGURE 15-8 A WAVE REFLECTS FROM THE BOUNDARY OF A MEDIUM IN WHICH IT HAS TO TRAVEL AT A DIFFERENT SPEED. SOME OF THE WAVE IS TRANSMITTED INTO THE SECOND MEDIUM, AND SOME IS REFLECTED BACK INTO THE FIRST MEDIUM. HERE A WAVE REFLECTS FROM A VERY STIFF MEDIUM (THE TREE TRUNK), AND LITTLE OF THE WAVE'S ENERGY PASSES INTO THE TREE TRUNK.

coming from each direction, meet at some point, the rope at that point rises twice as high as it would have if only one wave were there; *the amplitudes of the crests add*. This is known as **constructive interference** and is shown in Figure 15-9(a). After the two crests pass, their forms are the same as before they met. When a crest moving in one direction meets a trough of the same amplitude coming from the opposite direction, the rope *doesn't move at all*. The upward pull from one wave is canceled by the downward pull of the other. That is called **destructive interference** and is shown in Figure 15-9(b). These waves also pass through each other, and afterward their forms are the same as before they met.

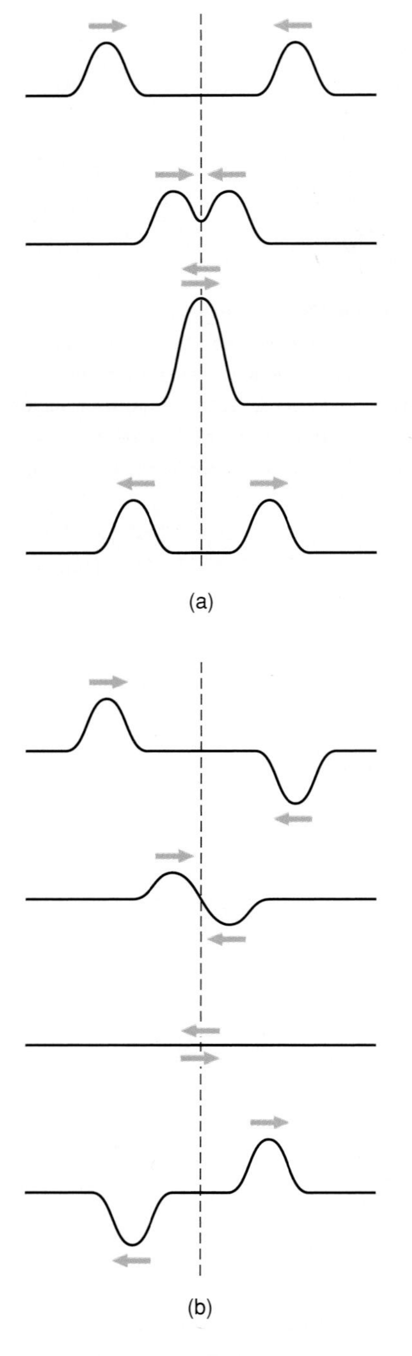

FIGURE 15-9 (A) TWO CRESTS ADD TOGETHER IN CONSTRUCTIVE INTERFERENCE WHEN THEY MEET AND THEN PASS ON UNALTERED. (B) A CREST AND A TROUGH OF EQUAL AMPLITUDE MERGE AND THEN DISAPPEAR FOR AN INSTANT IN AN ACT OF DESTRUCTIVE INTERFERENCE. FOR THAT INSTANT, AN OBSERVER CANNOT TELL WAVES ARE ON THE STRING. IMMEDIATELY AFTERWARD, THE INTERFERING WAVES PASS, AND EACH CONTINUES IN ITS ORIGINAL DIRECTION.

STANDING WAVES AND RESONANCE

If you fix one end of a rope to a wall and then oscillate the other end at different rates, you can find the right frequencies of oscillation to make patterns like those shown in Figure 15-10. For certain frequencies, the reflected wave

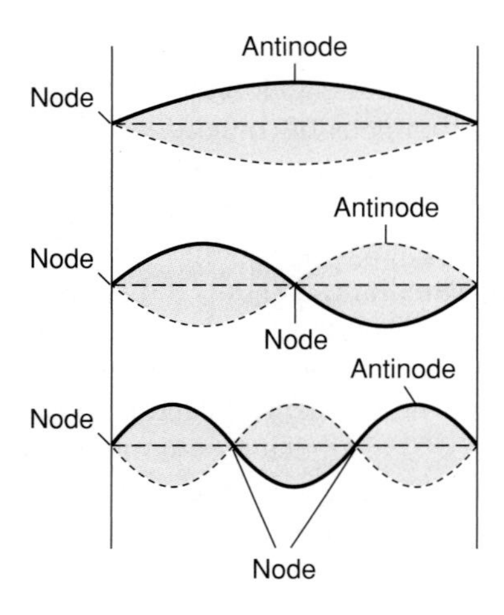

FIGURE 15-10 THE THREE LOWEST-FREQUENCY STANDING WAVES ON A STRETCHED STRING. EACH WAVE HAS A NODE AT EACH END. THE GREATER THE FREQUENCY OF THE STANDING WAVE, THE MORE NODES IT HAS. (NOTICE THAT FOR THE VERY LOWEST-ENERGY STANDING WAVE ONLY ONE-HALF OF A WAVELENGTH OCCUPIES THE ENTIRE STRING. THE NEXT STANDING WAVE HAS ONE WHOLE WAVELENGTH, AND SO ON.)

THERE IS NO MOTION OF THE MEDIUM AT A NODE OF A STANDING WAVE, SO THE BIRD AT THE CENTER OF THE WIRE IS UNAFFECTED BY THIS PARTICULAR STANDING WAVE. THE OTHER BIRD, HOWEVER, WAS LESS FORTUNATE. PERCHED ON THE PART OF THE STRING THAT BECAME AN ANTINODE, THAT BIRD IS FEELING THE FULL EFFECT OF THE STANDING WAVE.

interferes with the wave you are sending to cause a steady, or stationary, pattern on the rope. That is, the rope moves up and down *in a pattern that doesn't travel*—you can't see a shape moving in either direction. These are called **standing waves** because they move neither right nor left. Notice that there are points in Figure 15-10 where the rope *never* moves. At those points the two waves moving in opposite directions always cancel. Those points are called the *nodes* of the standing wave. The points where the rope flips up and down with maximum amplitude are the *antinodes*.

(a)

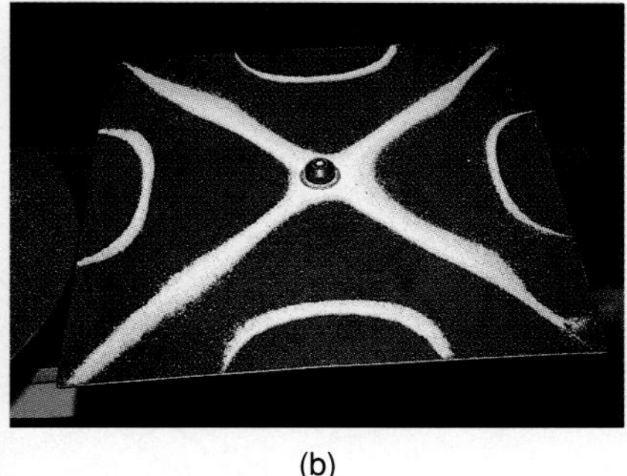

(b)

FIGURE 15-11 (A) WHEN A BOW IS DRAWN ACROSS THIS CIRCULAR IRON PLATE AT THE EXPLORATORIUM IN SAN FRANCISCO, SAND ON THE PLATE MOVES INTO THE NODES OF THE STANDING WAVE CREATED BY THE BOWING. THE CLEAR AREAS BETWEEN THE LINES OF SAND ARE THE ANTINODAL AREAS. (B) SIMILAR PATTERNS FOR A RECTANGULAR PLATE.

Standing waves aren't found just on ropes or strings in one dimension. Strike any elastic object and it vibrates with a standing wave throughout its whole volume. You can see patterns of vibrations on the surface of a plate of steel that is clamped down, as in Figure 15-11. When sand is sprinkled on it and a violin bow drawn over the edge to cause vibrations, a standing wave develops in the plate, sending the sand away from the antinodal areas on the surface, where the vibrational amplitude is largest, toward the nodal areas, where the sand isn't disturbed. You can make standing waves on the surface of tea or coffee in a cup sitting on a countertop. Tap the countertop with the heel of your hand, and circular standing waves appear on the liquid's surface. They are circular because all the reflections are at the circular edge of the cup.

The frequencies of oscillation that give rise to standing wave patterns are called the **natural frequencies** of the medium. Any object that supports waves—a stretched string on a guitar, air in a clarinet, a suspension bridge—has natural frequencies. Any time an object vibrates with a frequency equal to one of its natural frequencies, a standing wave is created. If the vibration continues, energy is added to the standing wave, and its amplitude grows. That process is called **resonance,** and for this reason natural frequencies are also called **resonant frequencies.** An object's resonant frequencies depend on its size, shape, and mass and on the details of its elastic properties. Musical instruments are crafted to vibrate with standing waves of certain frequencies. Strings, reeds, columns of air, bells, cymbals, and drumheads all are set into standing-wave vibrations by various means, and these vibrations cause the expansions and compressions in the adjacent air that we hear as sound. When the patterns of the sequences of frequencies created by musicians on these instruments please us, we call it music.

SURFACE WAVES ON WATER

Before the sun comes up in the morning, a lake's surface can be as smooth as glass. Later, sunlight heats the land and water, and breezes begin to stir. A gentle breeze causes ripples on a pond, but a persistent breeze in one direction over a large lake pushes up larger crests of water that move along.

Once such waves are in motion, gravity acts as a restoring force, much like the tension in a stretched cord does, tending to flatten any wave or ripple on the surface. Indeed, a cross section of ocean waves offshore looks just like the patterns that travel down a stretched rope (Figure 15-12a). A water wave isn't a transverse wave, however, nor is it longitudinal. As a wave train passes

PHYSICS IN 3D/A TRAVELING WAVEFRONT

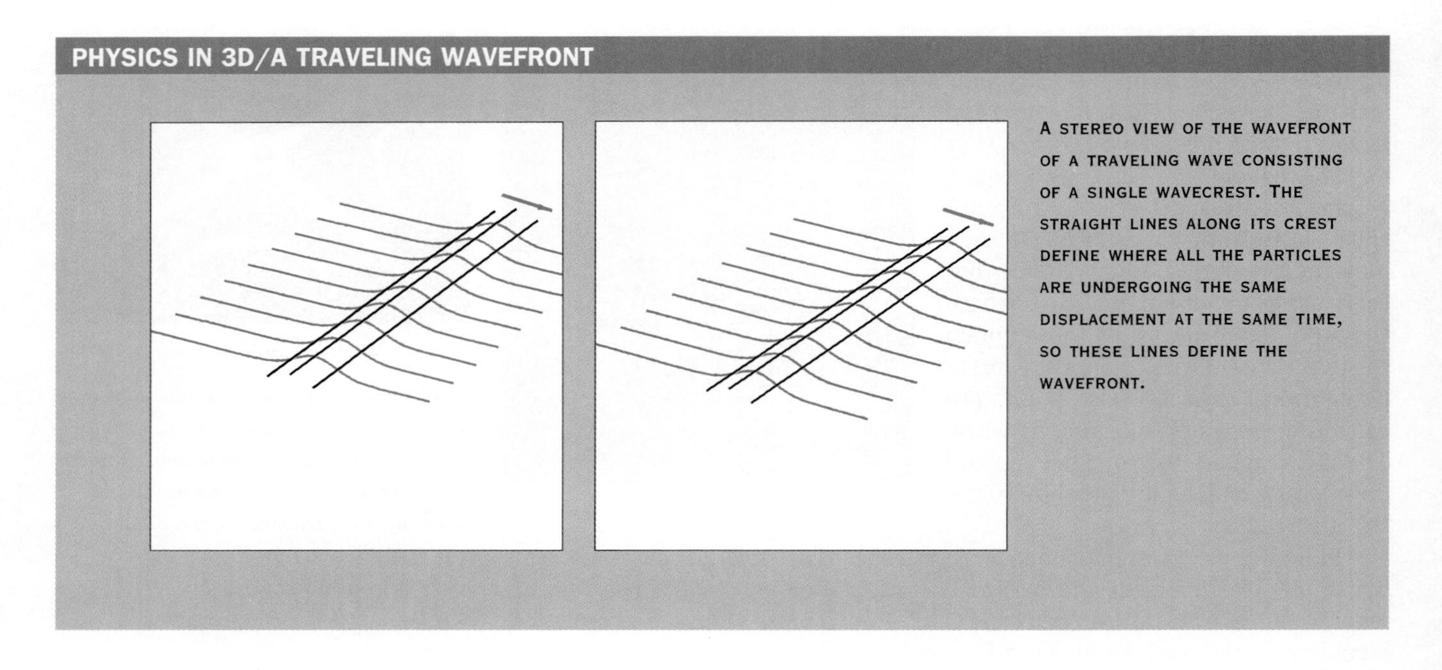

A STEREO VIEW OF THE WAVEFRONT OF A TRAVELING WAVE CONSISTING OF A SINGLE WAVECREST. THE STRAIGHT LINES ALONG ITS CREST DEFINE WHERE ALL THE PARTICLES ARE UNDERGOING THE SAME DISPLACEMENT AT THE SAME TIME, SO THESE LINES DEFINE THE WAVEFRONT.

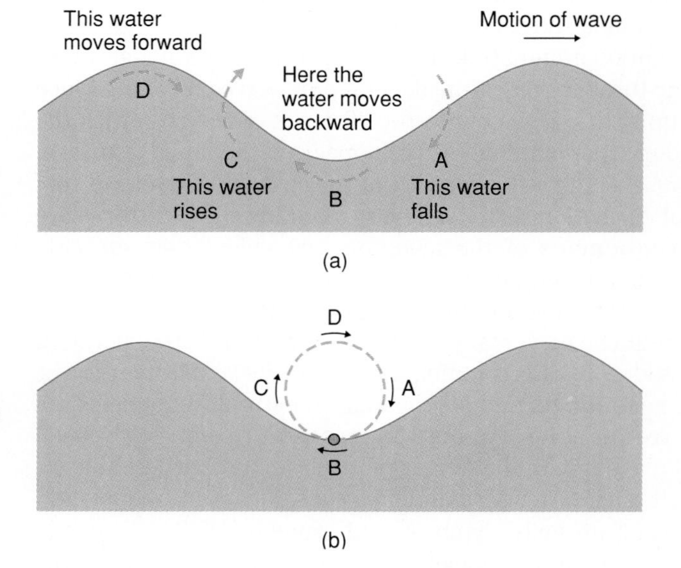

FIGURE 15-12 (A) THE MOLECULES AT THE SURFACE OF DEEP WATER MOVE IN A CIRCULAR PATTERN WHEN A SURFACE WAVE PASSES THEIR POSITIONS. (B) A SMALL FLOATING CORK MOVES WITH THE WATER AND SHOWS THE SAME PATTERN OF MOTION.

through deep water, the water at the surface moves in a circle. A cork floating on the surface shows the action of the water as a wave passes (Figure 15-12b). Swimming in the ocean, you can feel this back and forth motion if you float just beyond the point where the waves begin to break up near the shore.

When a surfer catches an ocean wave and scoots along the front side of the crest, as in the opening photograph of this chapter, he or she is riding a *wavefront*. A **wavefront** is a surface where a passing wave affects all the medium's particles in the same way at the same time, and it is a surface that is perpendicular to the wave's velocity at every point. Wavefronts can be seen if you toss a pebble into a quiet pond. Waves move out in all directions from the point of disturbance. The circular patterns made by the crests and troughs are the wavefronts (Figure 15-13). As the waves move away from the point of the disturbance, the wavefronts grow in circumference while the amplitude of the waves diminishes. That means the energy the waves carry becomes less concentrated as the waves spread out.

FIGURE 15-13 LAPPING WITH A STEADY FREQUENCY, THIS TIGER PRODUCES A SERIES OF WATER WAVES OF THE SAME WAVELENGTH. THE CIRCULAR PATTERNS IN THE WATER ARE THE WAVEFRONTS OF THESE WAVES.

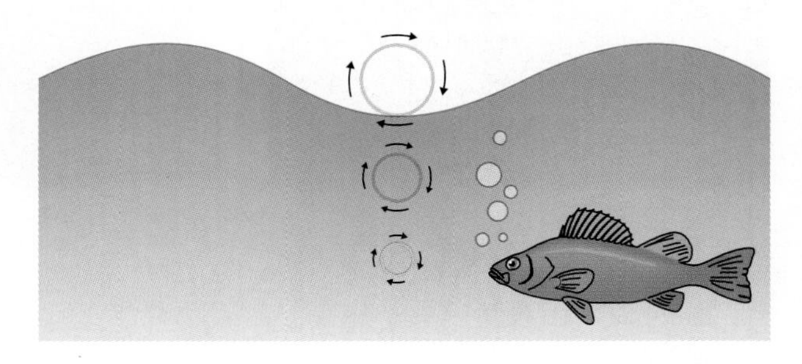

FIGURE 15-14 THOUGH MOLECULES OF WATER AT THE SURFACE SWEEP THROUGH LARGE CIRCULAR PATHS AS A WAVE PASSES, THOSE BENEATH THEM MOVE IN SMALLER CIRCLES. AT A DEPTH EQUAL TO ONE-HALF THE LENGTH OF THE PASSING WAVE, THE DISTURBANCE OF THE WATER IS ALREADY SMALL, AND 1 WAVELENGTH BELOW THE SURFACE, THE MOTION CAUSED BY THE WAVE IS EXTREMELY SMALL.

More than one wavelength below the surface, wave motion is negligible (Figure 15-14). Therefore in water that is many wavelengths deep, the solid lake bottom or ocean bottom has no effect on the wave at all. When a wave arrives at a beach, however, the water is shallow enough that friction with the shore's bottom soon interferes with the wave's motion. The crest may then break, which means it falls over in front of the portion of the wave below it, which has been slowed by friction with the bottom.

DIFFRACTION OF WAVES

Whenever water waves pass a boundary, such as a jetty that protects a harbor in bad weather, they spread out into the region behind the obstacle. This phenomenon is called **diffraction.** Figure 15-15 shows what happens when straight wavefronts come to an opening in a barrier. The incoming wave disturbs the water in the opening, and this disturbed water emits a circular wavefront into the region that the straight wavefront cannot reach. Diffraction is most pronounced when the barrier's opening is about the same size as the wavelength of the incoming wave. Nevertheless, when any kind of wave passes a sharp boundary, some diffraction occurs.

(a)

(b)

FIGURE 15-15 (A) DIFFRACTION IN A WAVE PASSING THROUGH AN OPENING IN A BARRIER ON A SMALL SCALE. (B) DIFFRACTION ON A LARGER SCALE, ALONG THE BIG SUR IN CALIFORNIA. (C) WAVES REFRACT LITTLE WHEN THE OPENING IS MUCH LARGER THAN THE WAVELENGTH. (D) WAVES REFRACT THE MOST WHEN THE OPENING IS ABOUT THE SAME SIZE AS THE WAVELENGTH.

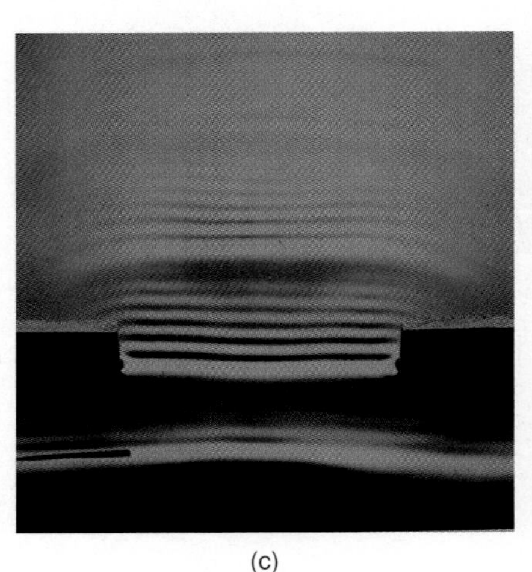

(c)

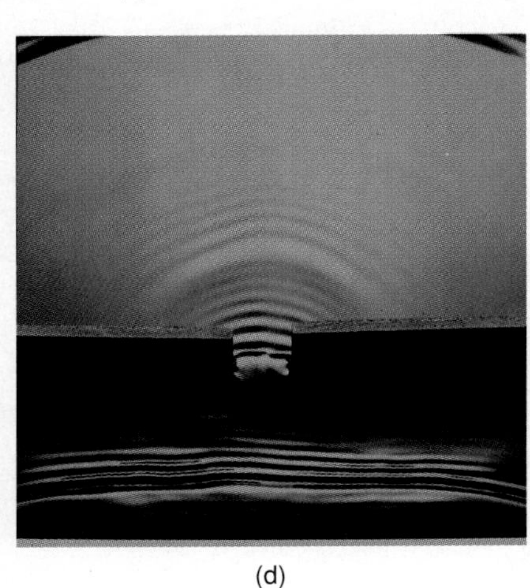

(d)

FIGURE 15-16 A CAR ANGLES INTO A SOFTER PORTION OF A ROAD THAT OFFERS MORE RESISTANCE. LIKEWISE, A WAVEFRONT ANGLES INTO A MEDIUM WHERE ITS SPEED IS SLOWED; THIS CHANGE IN DIRECTION OF A WAVE IS CALLED REFRACTION.

REFRACTION

When a wave comes to a region where its speed changes, such as when it changes mediums, its direction of travel may change. If it enters that region straight on, no change in direction takes place. If it enters at any other angle, the wave's direction of travel changes in an action called **refraction.** You can get an idea of how refraction works from experiences you may have had while driving a car. Picture this. A car is traveling along a straight and level road when its wheels on the right side slip off the pavement into some soft sand. They get extra resistance to their motion while the left wheels do not. This difference in resistance causes the car to swerve to the right (Figure 15-16). That is, the car angles, or turns, *into* the medium (the soft sand) where it travels more slowly.

Water waves do the same thing as they approach a beach. Ocean waves some distance from the shore often come toward a beach at a great angle. Then, as they meet the rising slope of the ocean bottom near the shore, their interaction with the bottom slows them, and they change direction. Their wavefronts turn until they are almost parallel to the shoreline (Figure 15-17). In Hawaii, however, large waves often come onto the beach at a considerable angle. The ocean bottom near those beaches falls off too fast to turn the waves parallel to the shore.

FIGURE 15-17 REFRACTION OF OCEAN WAVES TURNS THEM TOWARD A BEACH, WHERE THE SLOPING BOTTOM SLOWS THE INCOMING WAVES MORE AND MORE. THE WAVEFRONTS FINALLY COME ALMOST DIRECTLY INTO THE SHORE.

INTERACTING WITH WAVES

Floating on a raft out past the breakers in the ocean, a beachgoer bobs up and down and back and forth as waves pass, enjoying a smooth ride (Figure 15-18). However, a small boat whose length is about one wavelength of the passing waves falls from a crest into a trough and is pushed upward again before it stops moving downward, taking a pounding. An ocean liner is many wavelengths long, of course, so the up-and-down pushes of the waves along the length of the boat more or less cancel; the ship's large mass means it does not accelerate much anyway. In short, an object's size and its mass determine how much energy it can absorb from a wave of a certain wavelength. We'll encounter this fact again in later chapters.

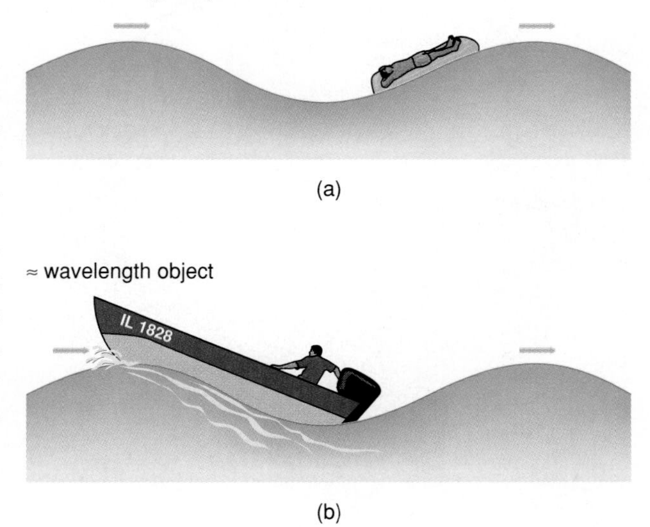

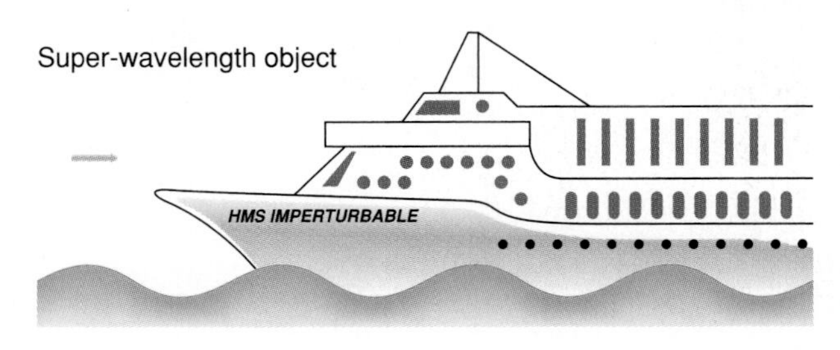

FIGURE 15-18 THREE OBJECTS INTERACTING WITH AN OCEAN WAVE. THE AMOUNT OF INTERACTION DEPENDS TO A LARGE EXTENT ON THE SIZE OF THE OBJECT. (A) AN OBJECT WHOSE LENGTH IS MUCH LESS THAN THE WAVELENGTH GAINS LITTLE ENERGY FROM THE WAVE AND DOESN'T DISTURB THE WAVE MUCH. (B) AN OBJECT WHOSE LENGTH IS ROUGHLY THE SAME AS THE WAVELENGTH CAN ABSORB A LOT OF ENERGY AND DISTURB THE WAVE. (C) AN OBJECT MUCH LONGER THAN THE WAVELENGTH ABSORBS LITTLE ENERGY SINCE THE UP AND DOWN PUSHES TEND TO CANCEL.

EARTHQUAKES AND TSUNAMI

Maybe you have felt the ground tremble as a very heavy truck passed close by. What you felt were small waves traveling along the surface of the ground. Waves through and on the ground also arise during earthquakes. Every day, tiny slippages along cracks in the earth's crust, called faults, cause waves, or *tremors*. More than half a dozen times a year, however, there is a major shift of underground rock somewhere on earth. Waves carry energy from that underground point, called the *focus* of the earthquake, outward in all directions. During a very large earthquake, the energy released in the form of waves may be as high as the combined energy of a hundred of the largest nuclear bombs. Because that energy is released in only a few seconds, the power of an earthquake is awesome. As the waves reach the surface, their shaking motions can level buildings, cause landslides, and more.

As an earthquake wave spreads outward from the focus, its energy does likewise. The waves of greatest amplitude at the earth's surface are at the *epicenter*, the point on the surface directly over the focus and closest to it. Interior earthquake waves are of both types, longitudinal, called **P** waves, and transverse, called **S** waves. (Geology students often remember P waves as "push-pull" waves and S waves as "side-to-side" waves.)

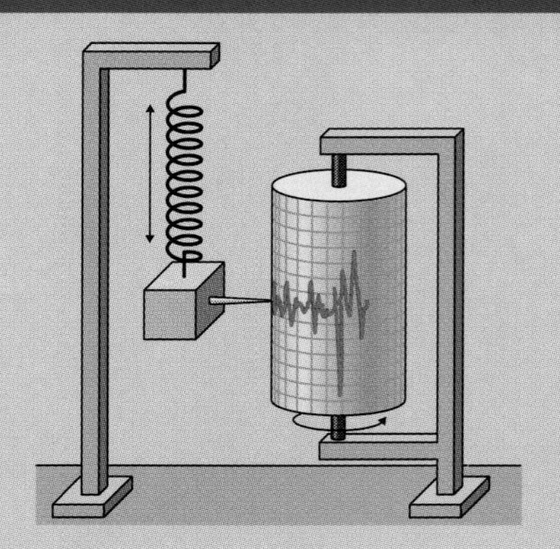

A SIMPLIFIED SEISMOGRAPH. BOTH PILLARS ARE ANCHORED FIRMLY TO THE GROUND, AND THE MASS TENDS TO REMAIN AT REST WHEN THE PILLARS SHAKE, RECORDING THE RELATIVE AMPLITUDE OF THE PASSING WAVES ON THE ROTATING DRUM.

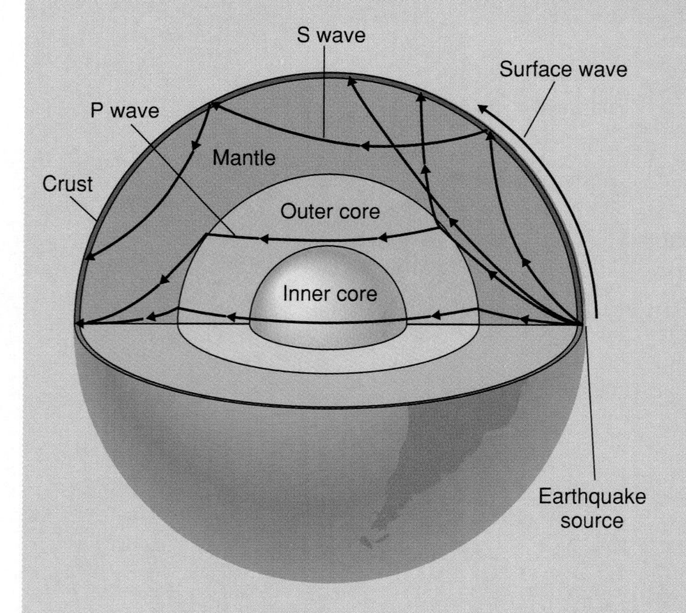

Instruments called *seismographs* detect these waves when they reach the surface. The working part of a seismograph is just a suspended mass that tends to remain at rest while the earth vibrates beneath it, and a pen connected to the mass, drawing a line on a moving strip of paper and recording the relative size of any waves. In addition, some earthquake waves travel along the surface after an earthquake, and these surface waves cause houses, cars, and people on the surface to move in a circular motion as they pass, much as an ocean wave moves a cork on the ocean's surface.

THE RICHTER SCALE

An earthquake rated as less than 3 on this scale isn't noticeable to us. Normally only six to twelve earthquakes a year rate 7.5 or larger, worldwide. The destructive power of an earthquake wave is related both to its amplitude and to its frequency. If the frequency matches a resonant frequency of a building, the building absorbs more energy and is more likely to be damaged. Each extra magnitude means about 32 times more energy in released.

Much of what we know about the earth's interior comes from the study of earthquake waves. A major earthquake sends S and P waves to most points on the earth's surface, as the accompanying figure shows. The waves reflect when there is

VALDEZ, ALASKA, AFTER THE MARCH 1964 TSUNAMI. ALMOST 200 METERS OF THE SHORELINE OF VALDEZ SLID INTO THE SEA. THE FISHING FLEET WAS TOTALLY LOST, AND THE WAVE PENETRATED ABOUT TWO BLOCKS FARTHER INTO TOWN,

an abrupt change in their speed due to a change in the properties of the rock they are passing through, much as a wave on a rope reflects from an end tied to a tree. The waves also curve upward because deeper rock is denser and under greater pressure than shallow rock and the waves travel faster there. Refraction turns them upward. The S waves stop abruptly in the earth's outer core, about 2900 kilometers (1800 miles) below our feet. Since S waves cannot travel through fluid matter, it is thought that the outer core is liquid.

If the epicenter of a large earthquake is located on the ocean floor, the motion there can lift and then drop an enormous volume of water. The ocean's surface over a large area may rise or fall as much as a meter, creating a surface wave with a relatively small amplitude but a very long wavelength. This wave, called a **tsunami** (pronounced "su-nah-mee"), can travel across the open ocean at speeds over 950 km/h (600 mph). When a tsunami hits shallow water somewhere, such as at the shore of an island or a continent, the shallow ocean floor slows its passage dramatically, allowing the back of the wave to catch up with the front. The wave's amplitude grows as the wave slows. When it hits a beach, the wave may be traveling at only 50 km/h (30 mph), but it can be as high as a 10-story building. Once called "tidal" waves, they have nothing to do with the tides, and the Japanese word for them is used today. Tsunami means "harbor wave" in Japanese, since

PHYSICS OF EARTHQUAKES AND TSUNAMI

if one strikes a harbor, it piles up even higher as it funnels in, causing even more damage.

The Richter Scale

Magnitude	Category	Description of Damage
8	Great earthquake	Total damage, surface waves visible
7–7.9	Major earthquake	Serious damage, buildings collapse
6–6.9	Destructive earthquake	Slight to large building damage
5–5.9	Damaging earthquake	Felt by everyone, slight building damage
4–4.9	Minor earthquake	Felt by many people
3–3.9		Smallest generally felt
2–2.9		Recorded only by seismographs

IN THE OCTOBER 1989 EARTHQUAKE IN NORTHERN CALIFORNIA, THIS HOUSE COLLAPSED ALONG WITH MANY OTHERS IN SAN FRANCISCO. WHAT YOU SEE HERE IS ONLY THE TOP FLOOR OF A PREVIOUSLY THREE-STORY HOUSE. THE BOTTOM FLOOR GAVE WAY AND THE SECOND STORY WAS CRUSHED UNDER THE TOP FLOOR WHEN THEY BOTH FELL.

Summary and Key Terms

wave–218
medium–218
longitudinal wave–218
compressions–218
expansions–218
transverse wave–218
oscillation–219
frequency–219
wavelength–219
amplitude–219
reflects–221
transmits–221
constructive interference–222
destructive interference–222
natural frequencies–223
standing wave–223
resonance–223
wavefront–224
diffraction–225
refraction–226

A ***wave*** moves through a ***medium*** of solid, liquid, or gas, carrying energy through matter without taking the matter along with it. A ***longitudinal wave*** moves the particles of a medium back and forth along its direction of travel in a series of ***compressions*** and ***expansions***. A ***transverse wave*** consists of a series of *crests* and *troughs* that move the particles of a medium back and forth perpendicular to the direction of wave travel. Transverse waves cannot travel through fluids.

If you create a transverse wave on a rope, each ***oscillation*** of your hand causes one crest and one trough. The number of oscillations per second is the wave's ***frequency***, f, and the distance from one crest (or trough) to the next is its ***wavelength***, λ. Any wave's speed, v, is equal to its frequency multiplied by its wavelength: $v = f \times \lambda$. In some media, waves of any frequency have the same speed. The maximum distance a particle in a medium moves from its undisturbed position as a wave passes is the wave's ***amplitude***, and the amount of energy a wave carries depends on its amplitude and frequency.

A wave ***reflects*** as well as ***transmits*** whenever it encounters a different medium, where its speed is different. Wherever two waves merge, or *interfere*, the medium responds to the *net* force it receives. In ***constructive interference***, two crests meet at a point and the particles there rise higher because the amplitudes of the crests add. In ***destructive interference***, a crest from one wave meets a trough from another and the particles at that point get opposing pulls. All objects have certain ***natural frequencies*** at which waves reflect within them and interfere to create a ***standing wave***. ***Resonance*** occurs when a disturbance at one of the natural frequencies excites a standing wave.

A ***wavefront*** is a surface where the passing wave affects all the particles of the medium in the same way at the same time. Wavefronts are perpendicular to the wave's velocity. When waves come to an opening in a barrier, they ***diffract***, emitting a circular wavefront that spreads out into the region behind the barrier. Diffraction is most pronounced when the barrier's opening is about the same size as the wavelength of the incoming wave. When a wave moves at an angle into a region where its speed changes, the wave ***refracts***, changing its direction, turning into the area where it moves more slowly.

EXERCISES

Concept Checks: Answer true or false

1. A wave is a disturbance that transports energy through a medium without permanently displacing the matter in the medium.
2. If the particles in a wave-carrying medium move back and forth along the direction of the wave's motion, it is a transverse wave.
3. Compression and expansion in a longitudinal wave take place perpendicular to the wave's velocity.
4. The distance between two successive peaks or two successive troughs of a wave is the wavelength.
5. The frequency of a traveling wave is the same as the number of oscillations per second of the particles as the wave passes.
6. The amplitude of a transverse wave is the maximum height of the wave measured from the level of the undisplaced medium.
7. A point that doesn't move in a standing wave is called an antinode.
8. The peaks on a standing wave move up and down but not right and left.
9. Diffraction refers to the spreading of a wave that passes a sharp boundary. Refraction refers to the turning of a wave whose speed changes when it enters a different medium at an angle, not straight-on.
10. Transverse waves cannot move through fluids, whereas longitudinal waves cannot move through solids.

Applying the Concepts

11. If you make a compressional wave on a stretched spring by moving one end back and forth with your hand, how does the frequency of your hand's motion correspond to the frequency of the wave on the spring?

12. If you are making a wave on a spring and you suddenly increase the frequency, what happens to the wavelength? Will the wave with the higher frequency overtake the one with the lower frequency?

13. With a hammer you can make both longitudinal and transverse waves in a long metal bar. Explain how.

14. When rain falls into a lake, each raindrop creates small waves. Why don't the numerous waves add to form waves with huge amplitudes?

15. When there are tall trees surrounding a small pond, the choppy water caused by the wind is located mostly in the middle of the pond. Explain.

16. When someone begins to pull on one end of a stretched rope, does the other end begin to move at the same instant? Is this also true for a steel cable? Discuss any differences in these situations.

17. Why can't transverse waves travel through fluids?

18. Visitors to Florida never see large alligators on the lakes when the wind is causing large-amplitude waves on the water (Figure 15-19). Why not?

FIGURE 15-19

19. A large bird takes off from a utility wire, leaving a wave that travels out along the wire in both directions. What happens to the energy of the wave?

20. Give some reasons why two tin cans with a string stretched between them could be a better way to communicate than merely shouting through the air.

21. The Waldorf-Astoria Hotel in New York City is built over a subway tunnel. The hotel's foundation rests on several centimeters of cork and lead. Comment on the function of that material.

22. Two identical clotheslines are stretched side by side in a yard. You pluck each of them and notice the disturbance travels faster on one than on the other. What does this tell you about the tensions in the clotheslines?

23. When a breeze blows into shore on a fresh-water lake, you can see an interesting phenomenon if there are weeds along the shore. The short-wavelength waves on the lake don't penetrate a weedbed, yet the long-wavelength waves do (Figure 15-20). Discuss.

FIGURE 15-20

24. Explorers who have crossed the ice packs off Antarctica have noticed that ice shifts occur up to 2 days before a storm arrives. What could be happening?

25. Everyone knows from experience how easy it is to excite standing waves on the surface of a cup of coffee or tea. Discuss the shape of that standing wave.

26. If an earthquake wave of 30 Hz travels at 6 km/s, how far apart are the crests?

27. What is the frequency of a water wave that has a wavelength of 15 cm and moves at 40 cm/s?

28. The snout of a mole is covered with hairs that connect to sensitive nerves. What purpose could this feature serve? Blind cave fish can evade a biologist's net because of nerve endings on their heads and sides. How could this feature help?

29. Discuss this statement: Cooks intuitively avoid resonances while beating eggs in a small bowl. (In other words, what would happen if the eggs were pushed back and forth at a resonant frequency?)

30. The sort of vibrations made in the ground by a rumbling train is familiar to most of us. Even in rural locations away from highways and railroads, however, earth's surface moves enough with another kind of wave action that disturbs the most sensitive seismograph. These devices must therefore be lowered into holes several hundred meters deep to escape the surface disturbances. What could be causing those waves at the surface?

31. A 1000-Hz compressional wave moves through aluminum 19 times faster than it moves through air. Explain why.

32. Explain why you can't stir sugar at the bottom of a full glass of iced tea by jostling the glass to make waves on the surface. (Try this; if you pour out some of the liquid a little at a time, the sugar will eventually move as the container is jostled.) What shape of glass would best lend itself to that action?

33. A solitary swimmer jumps into a swimming pool and climbs out. What becomes of the energy in the waves created by the swimmer?

34. Why do scuba divers 10 meters beneath the ocean's surface feel little motion from the 5-meter-wavelength waves passing overhead?

35. Sound in sea-level air at 25°C travels at about 1090 feet/second. How many seconds does it take to go 1 mile?

36. During a nighttime thunderstorm, the thunder arrives 8 seconds after you see the clouds light up. About how far away was the lightning strike? (See Exercise 35.)

37. In 1831 troops marching in step across a suspension bridge in England unwittingly set the bridge into an oscillation (they excited a standing wave) that grew until the bridge exceeded its elasticity and collapsed. Tell what must have happened. How could this situation have been prevented?

38. Is it possible to ride a pogo stick by jumping at any frequency you want? Would a child's frequency of jumping vary from an adult's?

39. In the earliest versions of the gigantic Saturn rockets that would eventually take astronauts to the moon, the fuel in the tanks would suddenly bounce up and down in "pogo-stick" vibrations, shaking the rocket too violently for human riders. What could cause such vibrations? Could you suggest a cure?

40. Seismic waves are caused by one-directional motion, as when two blocks of crustal material move past each other. They create mainly S waves. How do you think the waves formed by an underground explosion would differ from seismic waves? (This difference helps to distinguish an underground nuclear test in a foreign country from an earthquake that originates there.)

41. A wave in air made by opening a door quickly is a longitudinal wave (see Demonstration 2). Opening a door that swings into a room causes a compression that travels across the room. What would happen if you stood by that door and quickly shoved it back and forth? How could you test your assertion?

42. A couple on a small boat on a lake noticed that the waves lifted their boat about once every 2 seconds. Then they looked at the waves and estimated their wavelength to be 6 meters. What was the speed of the water waves? Draw a sketch of this and show the waves that will hit the boat in the next 4 seconds.

43. A surfer is lifted and then lowered as the crest and the trough of an ocean wave pass. At the top she was 4 meters higher than she was at the bottom. What was the amplitude of that wave?

44. Explain how you could stand at one end of a stretched rope and find the speed of transverse waves on it if you knew the rope length. What else would you have to measure?

45. Explain how you could use a fishing pier to find the speed of the wavefronts coming into a beach.

46. How many wavelengths are put onto a stretched string in 3 seconds if the wave has a frequency of 4 Hz?

47. Why are earthquakes most severe over the focus?

48. When the earth slips along a fault, the motion may be horizontal, vertical, or at an angle. Which type of dislocation do you think would be most likely to cause a tsunami?

49. A clothesline is 10 m long, and you find that, by shaking one end, you can put 3 wavelengths on the line in a second. At that time the front wavelength hits the far end. What is the speed of the wave?

Demonstrations

1. Fill a soft plastic rectangular ice tray with water until it overflows. Then tap the counter it's resting on and watch the standing waves that appear because of the straight boundaries. Both in the long direction and across the width of the tray, these standing waves intersect and make grid-like patterns. Identify the nodes and antinodes.

2. Close the windows and all but one door in a room, leaving that one door slightly open. Then quickly open a door on the opposite side of the room. In a split second, the door left ajar will move. (It opens or closes depending on how the doors are hinged.) A wave travels through the air in the room to move that door. Air molecules pushed by the door you moved push into other air molecules, which then bump into others, and some of the energy you put into the air when you moved the door travels across the room in a compressional wave.

3. Hook identical rubber bands onto both ends of a paper clip and have someone pull the rubber bands in opposite directions, exerting tension. The paper clip is like an atom in a chain of atoms in a solid, and the stretched rubber bands represent the "bonds" holding it to its nearest neighbors. Move the paper clip with your fingers, first along the direction of the rubber bands (the longitudinal direction) and next in the direction perpendicular to the rubber bands (the transverse direction). The difference in effort needed points out why longitudinal waves travel faster through solids than transverse waves do: The restoring force in the longitudinal direction is much larger.

SIXTEEN

Sound

Deep Purple in concert.

EARLY RESPONSES TO RHYTHMS

If you have normal hearing, you began to respond to sound while still in your mother's womb. A loud sound can make a fetus jump and its heartbeat quicken; expectant mothers have been forced to leave loud concerts because of their disturbed passenger's sharp kicks and punches. In the *Republic,* Plato advised mothers on the type of music they should listen to before their babies were born, to influence the baby's disposition. Modern research has shown this notion of Plato's to have some basis. Thirty-week-old fetuses were played the theme song from a popular soap opera; then, after birth, the babies became more alert when they heard that tune as compared to other tunes. In another study a group of pregnant women read Dr. Seuss's *The Cat in the Hat* several times a day to their abdomens for the last 45 days of their pregnancy. After birth, the infants would modify their frequencies of sucking to make a tape machine play the Dr. Seuss tale rather than another verse with a different rhythmic pattern.

For most of us sound is as much a part of everyday life as the downward tug of gravity. While the last chapter covered the description and properties of waves in general, now we focus on the waves that we hear.

THE NATURE OF SOUND

A tightly stretched string plucked at its center snaps against the air. It shoves the air molecules in its way closer together, creating the compression shown in Figure 16-1. Because the molecules in the compression run into molecules just outside that region, pushing *them* closer together, the compression moves away from the string. Meanwhile, when the string whips back to its equilibrium position, it leaves an expansion in its wake. Then part of the air in the original compression falls back into the partial vacuum of the expansion, leaving an expansion where the compression had been. Then the expansion moves away from the string, following the compression. As long as the string vibrates, compressions and expansions spread outward through the air. If the frequency of this longitudinal wave is between about 20 Hz and 20,000 Hz, the wave is audible to those of us with "perfect" hearing. It is called a **sound wave.** Compressional waves with frequencies higher than 20,000 Hz, inaudible to people, are called **ultrasonic** waves. Those with frequencies lower than 20 Hz are called **infrasonic** waves. We should keep in mind, however, that the ear (discussed in a later section) senses the vibrations in the air and triggers nerve impulses that signal the brain. Any sound you hear is only your brain's interpretation of the pressure changes in the air as the wave enters your ear. Your perception of the **loudness** of a sound depends on how much energy per second the wave brings to the ear. The amount of energy carried depends on the amplitude of the wave as well as the wave's frequency.

Sound can also move through liquids and solids, which is why we can hear sound under water or through a wall. Not only could the train robbers in early cowboy movies hear the approaching train through the rails, they could get the news faster that way than through the air. As we saw in Chapter 15, waves

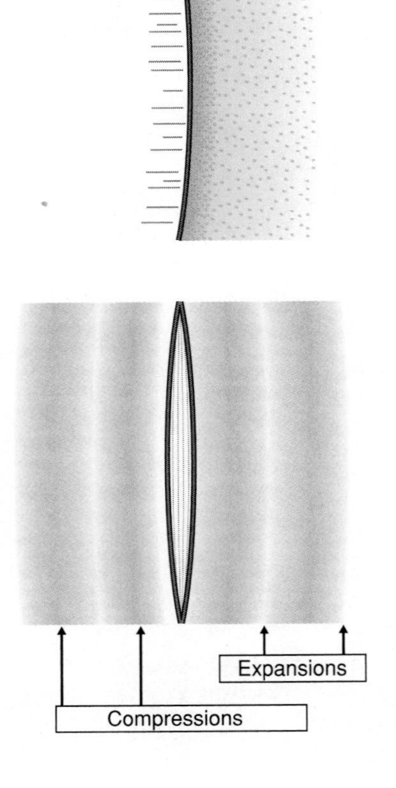

FIGURE 16-1 (A) THE RAPID MOTION OF A STRETCHED STRING THAT IS PLUCKED CREATES A COMPRESSION OF THE AIR ALONG MUCH OF ITS LENGTH. THIS COMPRESSION TRAVELS OUTWARD, WHILE THE STRING STOPS AND ACCELERATES IN THE OTHER DIRECTION, LEAVING AN AREA OF EXPANSION IN ITS WAKE. (B) AS THE STRING VIBRATES, AREAS OF COMPRESSION AND EXPANSION FOLLOW EACH OTHER OUTWARD, AND THIS TRAIN OF WAVES IS CALLED A SOUND WAVE.

travel more quickly through solids, where the molecules are bound by strong, springlike bonds, than through gases, where the wave travels only by random molecular collisions.

DETAILS OF SOUND WAVES

As with any other wave, a sound wave's frequency times its wavelength is equal to its speed. One complete oscillation of an air molecule takes place each time one wavelength of the wave passes, so if a molecule vibrates back and forth 300 times in a second, the wave's frequency is 300 Hz, and a total of 300 wavelengths moved past in that second. The measured speed of sound in sea-level air at 20°C is 343 m/s (767 mph). Let's find the wavelength of the lowest sound we can hear when the air's temperature is 20°C. The formula relating speed, frequency, and wavelength is

PUTTING IT IN NUMBERS—WAVE SPEED, FREQUENCY, AND WAVELENGTH

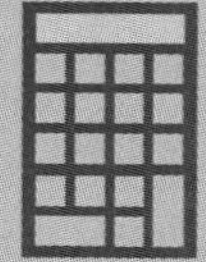

$$v = f\lambda$$

Example: The lowest frequency we can hear is 20 Hz. Find the wavelength associated with this frequency.

wavelength = speed/frequency = (343 m/s)/20 Hz = **17 m** (56 ft)

Likewise, you can calculate that the wavelength of the highest frequency we can hear, 20,000 Hz, or 20 kHz, is about 1.7 cm (2/3 in).

The speed of sound in air depends on the average speed of the air molecules, which means the sound speed depends on the temperature of the air (Table 16-1). However, the speed does *not* depend on the rate at which the pressure fluctuates as a wave passes, so *sound waves of all frequencies have the same speed*. Long-wavelength (low-frequency) sound waves take the same time to cross a room as short-wavelength (high-frequency) sound waves. Imagine the confusion if this weren't so! Each word you speak is a mixture of different frequencies of sound waves. If those waves moved at different speeds, the words (the patterns of the waves that your brain interprets as words) would become garbled as the sound moved across a room.

Table 16-1 Speed of Sound in Air

Temperature (Celsius)	Speed
0°	331 m/s
10°	337 m/s
20°	343 m/s
30°	349 m/s

Snap your fingers, and compressions and expansions ripple out through the air in all directions. As the waves spread, their amplitudes decrease and the energy they carry becomes less concentrated (Figure 16-2). Consequently, the farther the wave gets from its source, the smaller the distances the air molecules move back and forth as the wave passes. That finger snap could startle someone sitting next to you, but a person a hundred feet away might not hear it because of the smaller amplitude of the molecular oscillations.

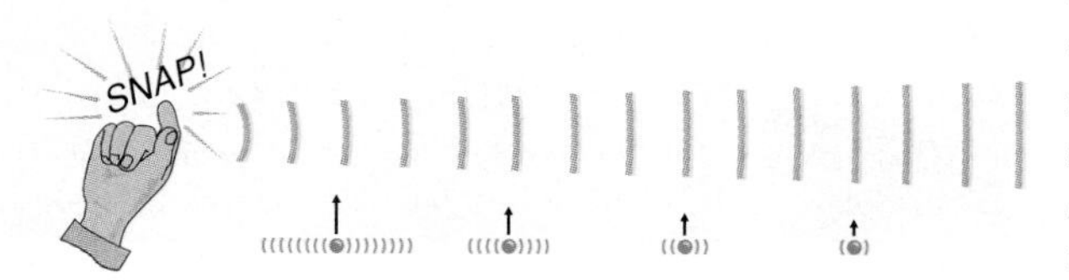

FIGURE 16-2 AS A SOUND WAVE SPREADS, SO DOES THE ENERGY IT IS CARRYING. THE MOLECULES OF AIR FARTHER FROM THE SOURCE VIBRATE BACK AND FORTH MUCH LESS AS THE SOUND WAVE PASSES THAN DO THE MOLECULES CLOSER TO THE SOURCE.

REFLECTION AND ABSORPTION OF SOUND WAVES

In an unfurnished house or apartment, your footsteps, your voice, and the "snick" of a closing door all sound hollow and harsh. This is because flat walls, floors, and ceilings make good reflectors of sound. As sound waves come in, they bounce off these surfaces with little distortion (Figure 16-3). That is, the waves that reflect keep the same shape, and the reflected sound, or echoes, resemble the original sound. We say these objects reflect sound in an orderly fashion, or **specularly.** On the other hand, the furniture, lamps, and pictures normally in a room scatter the sound waves in many different directions. We say they reflect sound **diffusely.**

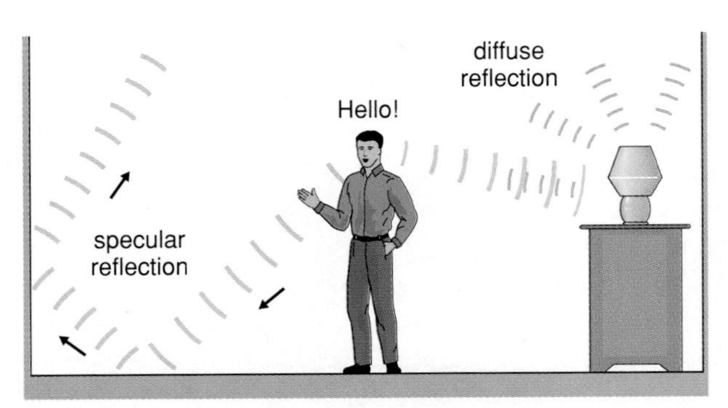

FIGURE 16-3 REFLECTIONS FROM FLAT, HARD SURFACES KEEP THE PATTERN OF THE ORIGINAL WAVE INTACT. THE REFLECTED SOUND WAVE SOUNDS MUCH LIKE THE ORIGINAL. THESE ARE CALLED *SPECULAR* REFLECTIONS. REFLECTIONS FROM IRREGULAR OBJECTS SCATTER THE SOUND IN ALL DIRECTIONS, IN AN ACTION CALLED *DIFFUSE* REFLECTION. THE ORIGINAL SOUND IS DIFFICULT OR IMPOSSIBLE TO HEAR IN THE DIFFUSELY REFLECTED WAVES.

Less rigid, porous materials, such as draperies and rugs, do more than just diffuse sound. They absorb a large percentage of the energy of the sound that strikes them, quieting echoes and softening noises. The energy of the sound is spread randomly among the molecules of the absorbing material.

The ceiling and walls of concert halls and lecture halls can be specially designed to reflect sound to the audience in the back. On the other hand, it is disturbing if echoes of the same sound come from different directions at different times. Acoustic technicians place panels of absorbing materials at key areas on the ceilings and walls to prevent such unwanted reflections (Figure 16-4). To tune an empty concert hall, they sometimes place sandbags in the chairs to simulate the absorption and diffusion of sound by the people that will sit there during a performance.

During a routine physical examination of a patient, a doctor gets a crude picture of the body's internal spaces by tapping on the patient's back and chest, a method known as *percussion*. The reflected sounds reveal the size of internal organs and the presence of fluids in cavities. For pictures of greater detail, a doctor may beam ultrasonic waves into the patient, which reflect from organs, bones, air, and so forth. The reflected sound waves are used to generate

FIGURE 16-4 ACOUSTIC PANELS USED BOTH TO ABSORB AND TO REFLECT SOUND TO THE AUDIENCE IN CASPARY AUDITORIUM, ROCKEFELLER CENTER, NEW YORK CITY.

USING SOUND TO MEASURE GLOBAL TEMPERATURE CHANGES

While air temperature measurements are easy to make, the fluctuations in the air's temperature are very rapid. To monitor global warming or cooling, the ocean's temperature should be a more accurate gauge. There, too, measurements taken at only one place or even along the line of a ship's path could show rapid variations due to currents. A scheme to measure the ocean's average surface temperature involves generating a sound near Heard Island in the southern Indian Ocean that is detected at points in all the major ocean basins of the world. By timing the arrival of the sound wave at those points, researchers can tell over the course of a decade any small temperature changes, since the speed of sound in water, as in air, depends on the temperature. The first measurements were done in 1991; a sound of 57 Hz was used to avoid confusion with lower-frequency sounds from ships and absorption of higher-frequency sounds by the water.

SOUNDING OUT THE OCEAN'S TEMPERATURE

pictures on a TV screen. Sonar gear on surface ships or submarines can detect other submarines through the water by emitting pulses of sound and listening for echoes. Commercial and sport fishing boats often use smaller models of sonar equipment to help locate schools of fish.

DIFFRACTION OF SOUND

If a door is closed, the sound of someone talking behind it is muffled. If the door is even slightly ajar, however, a clearer, louder sound is heard. Like other kinds of waves, sound waves spread, or diffract, somewhat as they come through an opening or past a corner. Diffraction of sound can let you hear someone who is standing around a corner. However, most of what you hear from around a corner or through an almost-closed door will be not diffracted sound but rather sound reflected from nearby surfaces (Figure 16-5). Outside, trees and bushes reflect sound and help us hear things better. Few people have been in places so isolated that reflected sounds are absent and diffraction alone lets them hear around corners. Rock climbers are an exception. High on a cliff away from other surfaces that could reflect sound, climbers not in direct sight of each other must shout mightily to be heard. If the angle is too large or the distance too great, climbers must use tugs on their ropes to signal each other. In a truly reflection-proofed room (even the floor must be soundproofed), people can hear the blood rushing through their ears, and a pistol discharged directly behind them sounds like a small "pop."

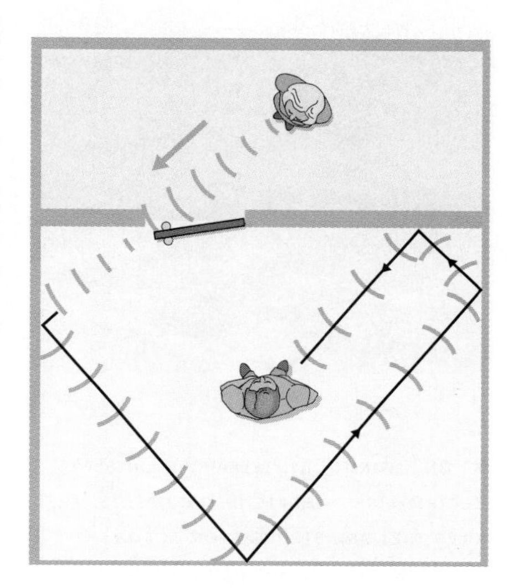

FIGURE 16-5 IF A WINDOW OR A DOOR IS BARELY OPEN, YOU HEAR SOUND FROM THE OTHER SIDE MUCH MORE CLEARLY THAN IF IT IS CLOSED. IF YOUR EARS ARE NOT IN A STRAIGHT LINE WITH THE SOUND SOURCE, SOME OF WHAT YOU HEAR IS DUE TO DIFFRACTION. BUT REFLECTIONS OF THE SOUND EMITTED INTO THE ROOM THROUGH THE OPENING, FROM THE REST OF YOUR SURROUNDINGS, ARE USUALLY RESPONSIBLE FOR THE LOUDER SOUND YOU HEAR.

INTERFERENCE OF SOUND

Sound waves that arrive at the same point at the same time interfere with each other. For instance, if a compression region of one sound wave occurs at some point when the expansion region of another sound wave is there, some destructive interference takes place. The molecules respond only to the net force they get when the two waves push them. Sound wave interference goes on around you all the time, but you rarely notice. That's because one sound is usually louder than the others, or there are different frequencies involved so the adding and canceling is not complete. On the other hand, if two stereo speakers emit the same sound (such as when you turn the amplifier to mono), you can sit in various locations in the room and find "dead spots" where the sound gets softer because of destructive interference and "live spots" where the sound is louder because of constructive interference.

SOUNDBOARDS AND RESONANCE

A stretched string that's plucked or struck sharply vibrates in the air and causes sound waves, but the sound produced is neither loud nor especially musical. The tones we hear from string musical instruments are primarily caused not by the strings or wires but by soundboards set into motion by the strings' vibrations. Guitar strings, for example, vibrate against a bridge that transmits the vibrations to the guitar's wooden body, which is its soundboard. This large surface area then vibrates in sympathy with the strings and gives rise to certain tones, or musical notes. (A tone is any single-frequency sound wave above about 100 Hz.)

(a)

(b)

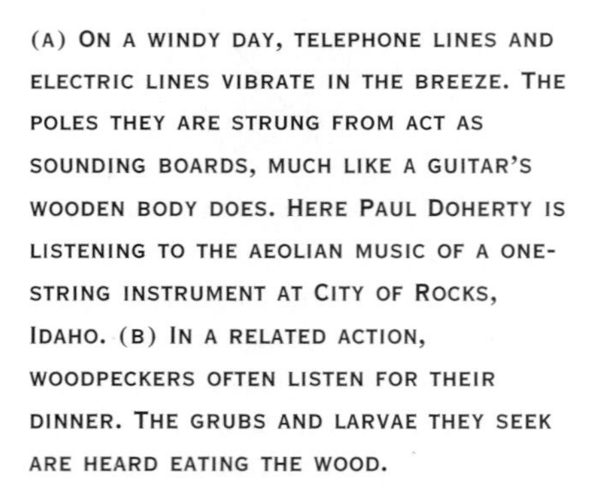

(A) ON A WINDY DAY, TELEPHONE LINES AND ELECTRIC LINES VIBRATE IN THE BREEZE. THE POLES THEY ARE STRUNG FROM ACT AS SOUNDING BOARDS, MUCH LIKE A GUITAR'S WOODEN BODY DOES. HERE PAUL DOHERTY IS LISTENING TO THE AEOLIAN MUSIC OF A ONE-STRING INSTRUMENT AT CITY OF ROCKS, IDAHO. (B) IN A RELATED ACTION, WOODPECKERS OFTEN LISTEN FOR THEIR DINNER. THE GRUBS AND LARVAE THEY SEEK ARE HEARD EATING THE WOOD.

In Chapter 15 we saw that any object has natural frequencies of vibration and that oscillations of the object at those frequencies set up, or *excite*, standing waves in the object. A vibrating guitar string excites one or more of the natural frequencies of the soundboard. The shape, thickness, and elastic qualities of the wood determine the natural frequencies and hence the musical quality of the sound it produces.

Alternating pushes and pulls by sound waves (or anything else) cause any object to vibrate to some extent. However, if a certain frequency of sound causes an object to absorb a lot of energy as it vibrates, the object is said to **resonate** with the disturbance. An empty cardboard shoe box with a lid resonates when low-frequency sound waves strike it. The air pressure on the box fluctuates as compressions and expansions arrive. Put the box in front of stereo speakers, turn up the volume, and you can feel the resonant vibrations in the box with your fingertips. On the other hand, a cloth packed into a bell deadens, or dampens, the oscillations of its surface if it is rung. In fact, most objects around us have enough natural damping to make resonance ineffective.

THE DOPPLER SHIFT

Pitch is the word we use that is most closely associated with frequency. What we characterize as high-pitch sound is high-frequency sound, and a low-pitch sound is low-frequency sound. If the horn of a parked car were stuck and you drove past in another car, you'd hear a change in pitch. As you move toward a source of sound, the pitch gets higher; as you recede from the sound source, the pitch gets lower. (Don't confuse the frequency or pitch of sound with the sound's loudness, which also changes as the car gets closer or goes farther away.) Any such change in frequency because of motion is called the **Doppler shift**. Here's why it happens.

Suppose a parked car's horn vibrates 400 times a second; that is, 400 compressions alternating with 400 expansions leave the horn each second. If

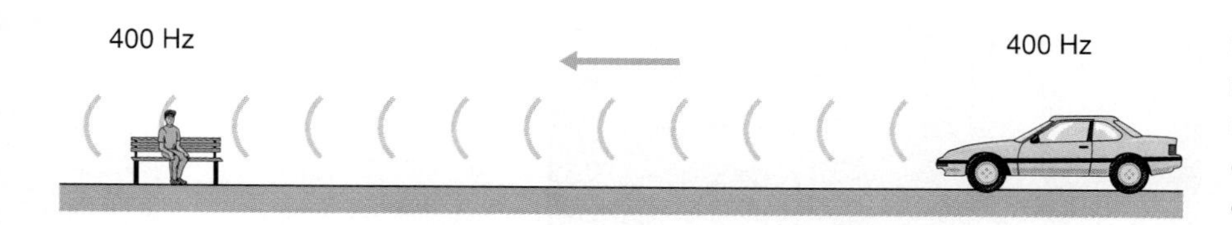

FIGURE 16-6 SITTING STILL, YOU WILL HEAR THE TRUE FREQUENCY OF A CAR'S HORN IF THE CAR IS ALSO AT REST. IF IT EMITS 400 OSCILLATIONS PER SECOND, THEN 400 OSCILLATIONS PER SECOND ENTER YOUR EARS.

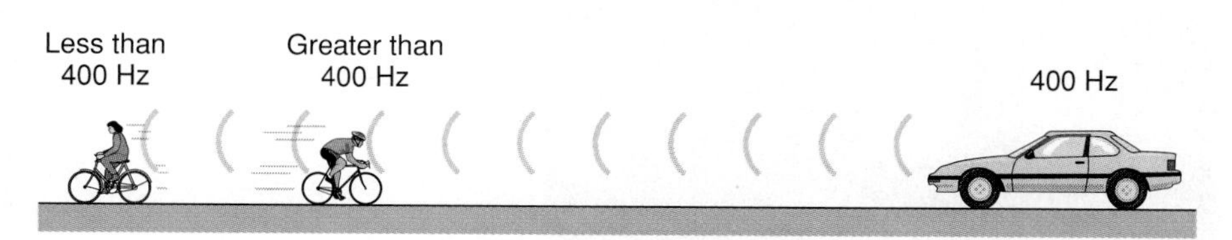

FIGURE 16-7 AN OBSERVER WHO IS MOVING TOWARD THE SOURCE OF A SOUND INTERCEPTS MORE WAVELENGTHS PER SECOND. AS A RESULT, THE FREQUENCY OF THE SOUND HEARD IS HIGHER THAN IF THE OBSERVER WERE STANDING STILL. FROM THE SAME HORN AS IN FIGURE 16-6, YOU WOULD HEAR A PITCH GREATER THAN 400 HZ AS YOU MOVED TOWARD THE SOURCE. LIKEWISE, IF YOU WERE MOVING AWAY FROM THE SOURCE, YOU WOULD INTERCEPT FEWER WAVELENGTHS PER SECOND AND HEAR A PITCH OF LESS THAN 400 HZ.

you were standing some distance away, 400 compressions and expansions would reach your ears each second and what you would hear is a 400-Hz sound (Figure 16-6). If you are moving toward that car, however, as in Figure 16-7, you run into *more* than 400 wavefronts per second and therefore hear a higher-pitched sound with a frequency greater than 400 Hz. Likewise, should you move away from the horn, fewer wavefronts would catch up to you each second. The frequency of the sound you'd hear would be less than 400 Hz.

Next, assume you stand still and the car is moving away from you. At the end of each oscillation, the horn would be farther from your position than it was when that oscillation began, as Figure 16-8 shows. The sound wave is stretched out by the car's motion, its wavelength is longer. Because fewer compressions and expansions reach you per second, the sound wave you hear has a frequency lower than 400 Hz. If the car were moving toward you, the wavelengths would be compressed instead, raising the frequency to over 400 Hz. So no matter whether you move, or the car moves, or both move, the pitch lowers if your relative motion increases the distance between you and the car, and the pitch increases if you and the car are approaching one another.

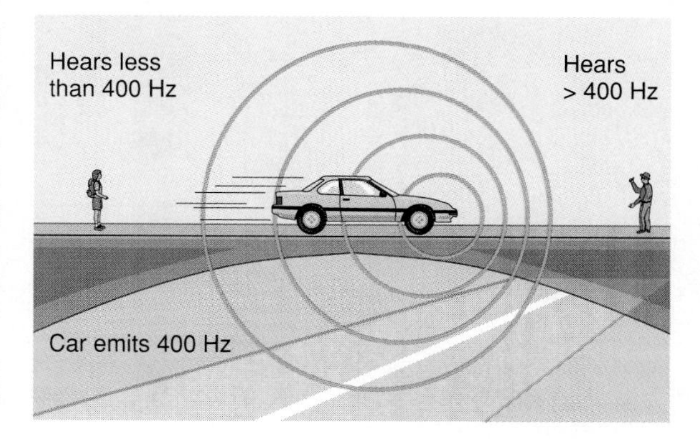

FIGURE 16-8 WHEN THE SOURCE OF A SOUND IS MOVING TOWARD AN OBSERVER, THE WAVEFRONTS ARE CLOSER TOGETHER THAN THEY WOULD BE IF THE SOURCE WERE STILL. THAT OBSERVER THEN HEARS A HIGHER-FREQUENCY SOUND. LIKEWISE, AN OBSERVER LEFT BEHIND BY A MOVING SOURCE RECEIVES SOUND WAVES OF LONGER WAVELENGTH (LOWER FREQUENCY).

SHOCK WAVES IN AIR: SONIC BOOMS

If a rifle bullet travels faster than the speed of sound, the sound waves it makes cannot travel in front of it. The sharp disturbance still travels outward at the speed of sound, but its wavefront forms a cone behind the bullet (Figure 16-9). We call this a **shock wave**. There are two shock waves for a bullet, a compression from the front end of the bullet and an expansion from the back end, but they are very close together because a bullet is a very short object. Therefore, when the compression and expansion arrive at someone's ear, the person hears only one distinct "crack." When a jet aircraft or the space shuttle flies faster than sound, however, people below its path hear two explosive sounds, the first from the compression of air by its nose and the other from the expansion in the wake of its tail (Figure 16-10). These sounds are called **sonic booms**. A shock wave from a plane contains a lot of energy. If a jet fighter breaks the speed of sound at an elevation as low as 2500 m (8000 ft), the energy in the waves can cause structural damage in buildings.

FIGURE 16-9 THE SHOCK WAVES FROM A SPEEDING BULLET, PHOTOGRAPHED WHEN THE BULLET BROKE THE FINE WIRE SEEN EXTENDING UPWARD FROM THE CANDLE. NOTICE THE COMPRESSIONAL WAVE COMING FROM THE LEADING EDGE OF THE BULLET AND THE EXPANSION WAVE FROM ITS BACK END, AS WELL AS THE TURBULENCE IN ITS WAKE.

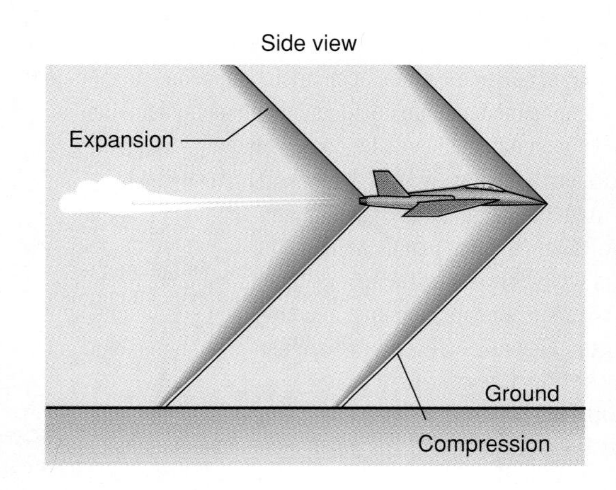

FIGURE 16-10 THE SHOCK WAVES FROM A PLANE TRAVELING FASTER THAN THE SPEED OF SOUND. THE COMPRESSION AT THE NOSE AND THE EXPANSION AT THE TAIL SPREAD OUT IN CONE SHAPES AND CAUSE EXPLOSIVE SOUNDS, OR SONIC BOOMS. (NOT SHOWN IS THE CURVING OF THE WAVEFRONTS DUE TO REFRACTION AS THEY TRAVEL THROUGH AIR OF CHANGING TEMPERATURES AND DENSITIES.)

THUNDER

Most of us have counted the seconds between seeing lightning and hearing the resulting clap of thunder. Sound travels about a mile in 5 seconds, so you can find the approximate distance to the place where the lightning struck.

The noise you hear as thunder is generated all along the lightning's path. The stroke superheats the air around it, raising the air temperature by thousands of degrees Celsius, and that heated air expands explosively, compressing the air around it. This compression creates a shock wave that moves outward and loses intensity as it spreads out. The result is a sharp thunderclap if you are very close or ordinary thunder if you are farther away. Sound waves leave all points along a stroke of lightning at essentially the same time. The rolling thunder you hear is the succession of sound that reaches you from higher and higher points along a stroke. If there are hills nearby, reflection also plays a part in the drum rolls of thunder.

No doubt you've seen lightning in a faraway cloud, but the thunder never arrived for you to hear it. That is *refraction* at work. Since the speed of sound decreases with temperature, sound refracts, or bends, into colder air when it enters it at an angle. Whenever you hear thunder, you can bet the lightning

THE SOUNDS FROM TREES AND BROOKS

Listen to the wind in a pine tree. The air's internal friction, or viscosity, causes turbulence as it encounters each needle and flows around to close in behind it. Eddies (tiny whirlpools) of air appear behind the needles, causing air pressure fluctuations, hence sound that travels outward. For pine needles, this soft sound begins at wind speeds of about 4 mph. If the breeze picks up speed, it causes more ripples per second, and the pitch of the sounds from the tree becomes higher.

In trees with broad leaves, the leaves flutter in the breeze, causing a rustling noise similar to the one made by flipping the pages of a book. These sounds are pitched lower than the sound from pine trees because pine needles, being smaller than leaves, vibrate faster and therefore emit a higher pitch. When the boughs of some trees lose their leaves in the autumn, the wind still makes a sound. The empty branches are larger in diameter and stiffer than the leaves and create fewer of the smaller eddies per second. The sound from the wind in the bare trees is lower-pitched and more somber than the sound of trees in full leaf.

In a slow brook where the water flows smoothly around stones or logs, practically no sound stirs the air. If you hear a quiet murmur, though, look closely at its source. That sound doesn't come from smooth waves or even from eddies in the water. Water flowing over or around an irregular object fast enough can trap air. Then bubbles burst at the surface or even under water, sending vibrations in all directions. In waterfalls, the water collides violently with the water or rocks below, trapping air and compressing it. These violent, large-scale compressions create the roar. When raindrops fall on bushes and trees, the leaves give with the impact and compressions are less violent, making a much softer sound.

SOUNDS IN NATURE

stroke was less than 24 km (15 mi) away. At that distance or greater, the sound waves moving almost parallel to the ground curve upward into the cooler air of the atmosphere and miss entering your ears.

SOUNDS FROM INSECTS AND OTHER ANIMALS

If you listen closely, you can hear a large butterfly fly. Most flap their wings at 12 cycles per second or less, below the frequencies we can hear. Nevertheless, if you are close, you can hear the air rippling from the edges of a butterfly's wings as they beat, much like the sound of wind going around a leaf. Big bumblebees move along with the help of 130 wingbeats per second, making a drone that's easily heard. The honeybee flaps its wings at 225 beats per second to make a higher-pitched hum. The mosquito owes its annoying high-pitched whine to 600 wingbeats per second. The sound from these insects comes from the pressure variations caused as their wings beat against the air.

FIGURE 16-11

Domestic dogs hear high-frequency squeaks up to 50,000 Hz. Those silent (to us) dog whistles put that ability to practical use. Cats hear even higher frequencies—up to 70,000 Hz. Bats sense ultrasound so well that they use their ears much like eyes (Figure 16-11). A bat sends out pulses at ultrasonic frequencies (50,000 to 120,000 Hz) and locates obstacles and insects by the echoes. These ultrahigh-frequency sound waves let the bats fly and hunt in total darkness. Porpoises send out ultrasonic sound and listen for reflections. They can easily sense schools of fish within a kilometer of their position. Killer

A RESEARCHER IN THE TOP OF A RAINFOREST CANOPY IN COSTA RICA RECORDS THE SOUNDS OF MICE WITH A SHOTGUN MIKE, WHICH FOCUSES SOUND INTO AN AMPLIFIER FOR MAGNIFICATION.

whales have been recorded talking to each other over distances of 30 kilometers. Other species of whales are thought to hear one another at a distance of 800 kilometers.

In fact, scientists are investigating whether the Heard Island experiment (in the box earlier in this chapter) could interfere with whale communications. Experiments should resolve that issue.

HOW WE TALK AND SING

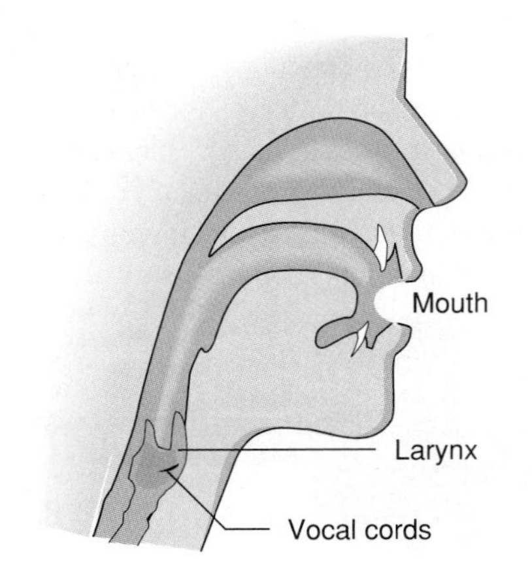

FIGURE 16-12

The human voice comes from the vocal cords in the larynx, with the mouth and nasal cavity working together as a resonant chamber (Figure 16-12). To speak, you must force air from your lungs through the narrow passage between your tightened vocal cords. This action causes the air in the acoustic cavity to resonate. Varying the shape and opening of the mouth and the placement of the tongue helps to give the amazing variations of sound in the human voice. Usually a person's voice in a tape recording sounds flat to that person and "not like myself." That's because the voice on the tape does not include the low-frequency resonances from the voice box that arrive at the speaker's ears via the bone and soft tissue surrounding the ear. So, in fact, others never really hear you the way you hear yourself.

During normal talking at low levels, the vocal cords vibrate over a range from 70 to 200 Hz for a man and 140 to 400 Hz for a woman. The gender difference is because a male's vocal cords are typically both longer and more massive than a female's, so the natural modes of vibration of a male's vocal cords are at lower frequencies. When you change the pitch of your voice, you are adjusting the tension in the muscles of your vocal cords. The tighter the cords, the higher the frequency of the sound you make. If you sing, you may increase the tension until your frequency range is more than doubled.

When you speak or sing, pitches much higher than those emitted by the vocal cords are present. Resonances in the vocal tract contribute, as well as the turbulence created as air is pushed through lips and teeth. High-pitched squeals and sustained whistles even enlist the Bernoulli effect. The increased speed of the fluid coming through a small constriction decreases the pressure and produces fast oscillations in the lips and vocal cords.

HOW WE HEAR SOUND: THE EAR

FIGURE 16-13 MECHANISMS OF THE HUMAN EAR. SOUND WAVES ENTER THE EAR CANAL AND VIBRATE THE EARDRUM, WHICH VIBRATES THE BONES OF THE MIDDLE EAR, WHICH VIBRATE THE COCHLEA, WHICH IS FILLED WITH FLUID. THE WAVES IN THE FLUID MOVE MEMBRANES WHICH REST ON CELLS THAT TRIGGER NERVES.

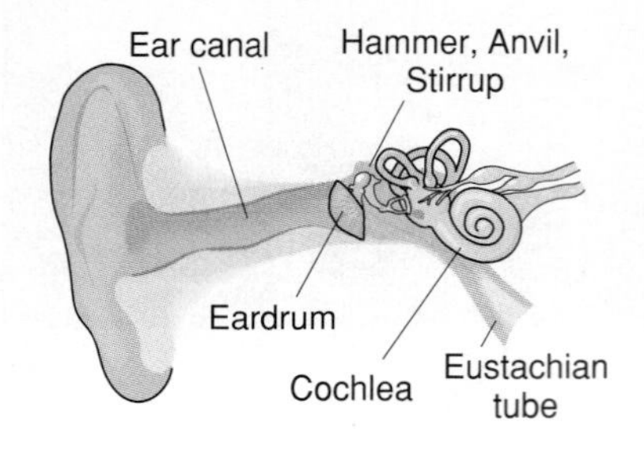

The sound waves that enter the ear canal are guided to the eardrum, a thin, conical, skin-covered membrane (Figure 16-13). Between the brain and the eardrum is an array of miniature equipment capable of detecting eardrum motion having an amplitude as small as one atomic diameter. Sound waves don't move the eardrum far. A movement of even 0.25 millimeter brings pain and/or damage. Sounds from close, violent explosions, for instance, cause total or partial deafness, as does repeated close exposure to the amplified sound from speakers at loud concerts or in cars.

Behind the eardrum is a region called the middle ear. Three tiny bones (the tiniest in your body) called the hammer, anvil, and stirrup occupy this chamber. As Figure 16-13 shows, the hammer has a handle that rests on the interior side of the eardrum. When that membrane vibrates, so does the hammer, and this vibration then moves the other two bones. The stirrup rests its footplace on another membrane, oval in shape. Behind this oval membrane is the cochlea, or inner ear, which is filled with fluid. The cochlea is a snail-shaped chamber built into the bone of the skull for protection. The waves that travel around its waterway move membranes that rest on cells that can trigger nerve impulses. Not an easy path for a sound wave—from gas to solid to liquid and finally to electrical impulses that the brain must interpret as sound. For most of us, however, it works very well.

SOUND INTENSITY AND LOUDNESS

The **intensity** of a sound wave is the energy it brings into a unit of area each second. The units of intensity are watts/meter^2. The human ear responds to intensities ranging from 10^{-12} W/m^2 to more than 1 W/m^2 (which is loud enough to be painful); for this reason, intensities are compared using powers of ten. The barely audible 10^{-12} W/m^2 is used as a reference intensity. A sound that has ten times that intensity (or 10^{-11} W/m^2) is 1 *bel* or 10 *decibels* more intense. Table 16-2 gives typical sounds and their intensities.

The ear senses some frequencies much better than others. For example, a sound wave with a frequency of 3500 Hz and an intensity of 80 decibels sounds about twice as loud to us as a sound of 125 Hz and 80 decibels because we sense the 3500-Hz sound much more. Therefore intensity by itself does not mean loudness. Loudness is how our ears detect and our brains perceive the intensity of sound waves. One reason that different frequencies are perceived differently has to do with resonance. The lowest-energy standing wave in our ear canals is excited with the least amount of energy. Depending on the shape of your ear canals, the frequency, or pitch, of that wave is between 2500 and 4000 Hz. Our ears become relatively insensitive at high frequencies, from 12,000 to 20,000 Hz.

Television commercials often seem louder than regular programs. A commercial's monitored intensity is equal to or below that of a sports program—believe it or not! Advertisers just concentrate the commercial's sound at frequencies where the human ear is more sensitive!

Table 16·2 The Decibel System for Sound Intensity

	Intensity (W/m^2)	dB
Threshold of unaided hearing	10^{-12}	0
Your own breathing	10^{-11}	10
A whisper, or rustling leaves, or standing on the rim of the Grand Canyon	10^{-10}	20
Room in a house when no one is talking (perhaps an air conditioner is on)	10^{-9}	30
Library's sound level	10^{-8}	40
A quiet office	10^{-7}	50
Someone talking normally 1 meter away	10^{-6}	60
Traffic, New York City (as heard on second floor of a tenement)	10^{-5}	70
Third-floor apartment next to Los Angeles freeway	10^{-4}	80
Very noisy street	10^{-3}	90
Inside a subway	10^{-2}	100
Intolerable sound over a long period	10^{-1}	110
Car horn at 1 meter	⅓	115
Threshold of pain (stick-fingers-into-ears-noise)	1	120
Jet taking off nearby	10^1	130
Shotgun blast	10^2	140
Standing *next* to a runway when jet takes off	10^3	150
Standing near the space shuttle at takeoff	10^6	180–up

Summary and Key Terms

sound wave–234
ultrasonic–234
infrasonic–234

As a string vibrates, compressions and expansions move outward in a ***sound wave***. We hear at best waves of frequencies between 20 and 20,000 Hz; waves above 20 kHz are called ***ultrasonic***, and those below 20 Hz, ***infrasonic***. Sound waves of all frequencies have the same speed as they move through air.

specularly–236
diffusely–236

We hear mostly reflected sounds. In an empty room, sound waves reflect ***specularly***, bouncing off the flat surfaces with little distortion, so echoes resemble the original sound. In a furnished room, sound waves scatter off objects, reflecting ***diffusely***, while objects like rugs absorb much of a sound wave's energy to soften noises. Sound waves also diffract around corners, and they refract, bending into air where they travel more slowly.

resonate–238

Soundboards create the tones of a guitar or violin. The vibrating strings excite one or more of the natural frequencies of the soundboard, inducing standing waves in it and causing it to ***resonate***. At certain resonant frequencies, an object can store a lot of energy in the standing wave, but most objects have natural damping that absorbs the energy instead.

Doppler shift–238
pitch–238

The ***Doppler shift*** is a shift in ***pitch***, or frequency, of a sound wave due to the relative motion of the source and the observer. If the distance between them is increasing, the pitch falls. If the distance between them is decreasing, the pitch rises.

shock wave–239
sonic boom–239

If an object travels faster than the speed of sound, the sound waves it makes cannot travel in front of it. The disturbance still travels outward at the speed of sound, but its wavefront forms a cone called a ***shock wave*** that spreads out behind the object. There are two shock waves for a jet plane traveling at a speed higher than the speed of sound: a *compression* from the front end and an *expansion* from the back. People below its path hear these waves as two loud booms, called ***sonic booms***.

intensity–243
loudness–234, 243

A sound wave carries energy away from its source, and the amount of energy it brings into a unit of area each second is its ***intensity***. The ear is more sensitive to certain frequencies than to others, so intensity is not loudness. ***Loudness*** is the way our brains *perceive* the intensity of sound waves.

EXERCISES

Concept Checks: Answer true or false

1. Sound waves are both compressional and transverse.
2. Audible waves in air have frequencies between 20 Hz and 20,000 Hz.
3. If sound waves reflect from smooth, hard surfaces, they keep the same shape when they are reflected. The reflected sound resembles the original sound.
4. A sound wave spreads out after passing an opening; in other words, it is diffracted.
5. If destructive interference occurs for sound waves in the middle of a room, you cannot hear those waves anywhere else in the room.
6. When an object absorbs significant energy from the sound waves striking it, it is said to resonate.
7. A refracting sound wave turns away from cooler air and into warmer air.
8. A Doppler shift in frequency occurs whenever there is any type of motion of either the sound source or the listener.
9. When a listener moves toward a stationary sound source, the pitch of the sound becomes lower.
10. A sonic boom is a shock wave.
11. The sound of thunder goes on and on because it is the sound from many lightning bolts.

Applying the Concepts

12. Researchers in Africa in 1993 reported watching giraffes stopping and waiting for other giraffes that were out of sight over a hill. The same researchers watched a young male giraffe try to break through a high brick wall to get to his mother, far away in a field on the other side. Yet the researchers heard nothing. What do these observations imply about any sounds giraffes might make?

13. Why does tightening a guitar string raise the pitch of the sound it makes? What effect would substituting a more massive string with the same tension have?

14. Does sound in a gas travel faster when the gas is warm or when the gas is cool?

15. Why must the volume of a stereo in a room with a wall-to-wall carpet be turned higher than in a room with a wooden floor?

16. What everyday fact tells us that the speed of sound doesn't vary significantly with the frequency of the waves?

17. Show that the wavelength of a 18,000-Hz sound wave is about 1.9 cm.

18. The air bladders of fish apparently enhance their hearing. How would that work?

19. Why should a bat, with tinier ear parts than our own, be able to sense higher frequencies than humans?

20. Comment on the fact that our ears are very sensitive in the frequency range of human speech.

21. When a bathtub faucet is only partly open, a smooth stream of water may flow with no noise. Open it full blast, though, and the gushing water is accompanied by much noise. Why?

22. Does sound travel faster or more slowly if the air pressure is lowered? Discuss.

23. Suppose the speed of sound in a certain room is 330 meters per second. If you wanted to adjust a sound generator to give waves that are 20 centimeters long, to which frequency should you set the dial?

24. Describe how you make and detect an infrasonic wave in a room with two doors just by moving one door back and forth in a regular motion.

25. The feathers under an owl's wing are much softer than those of a vulture or an eagle. What advantage does this give to the owl?

26. If builders could evacuate the air from the open spaces in the walls of apartment buildings, the rooms would be more effectively soundproofed, since no sound could move where there are no molecules to carry it. Why wouldn't this be practical? Discuss.

27. A sound wave striking a closed door pushes and pulls on it. However, little sound is heard on the other side. Why?

28. A small motorcycle engine revs up to 10,500 revolutions per minute. What insect does it sound like? (*Hint:* You need to convert to oscillations per second to make the comparison. Also see page 241.)

29. A sharp sound made at the edge of a pine forest returns an echo that has a higher pitch than the original sound, while woods containing trees with broad leaves return an echo more like the original sound. Explain.

30. Beer foam kills the sound if the glass mug is tapped with a spoon. Explain.

31. Show that sound traveling at 343 m/s takes about 5 seconds to travel 1 mile. This calculation is the basis for counting seconds after seeing a bolt of lightning and dividing them by 5 to determine how many miles away the strike occurred. What number would you need to divide by to find this distance in kilometers?

32. An arrow from a bow might travel as fast as 60 meters (200 feet) per second. Bow hunters know there is little chance to hit a deer at 100 meters even if their aim is accurate. Why?

33. A rock outcropping known as Stony Point on the Hudson River was removed in a construction project. As a result, riverboat captains found the river much more difficult to navigate in foggy conditions. Guess why.

34. Would you expect to hear sound for a greater distance in the desert or over a cool lake during the day? Discuss each situation.

35. If you submerge one ear into the water in a bathtub, you can often hear sound coming from other parts of the house or apartment building that you couldn't otherwise hear. Discuss.

36. If the speed of sound were much slower than it is, how would that change affect conversations? (Consider the Doppler effect while walking.)

37. The propellers of seagoing vessels push so hard against the water that the water separates as the propellers cut through it. This creates regions of partial vacuum called cavities. This cavitation makes noise that is easily picked up by submarine sonar devices. What causes the noise?

38. Discuss why the four strings on a violin produce sounds of different frequencies; after all, they all have about the same length!

39. A motorcycle rider in a box canyon approaches a distant wall with the horn of the cycle blowing. Discuss the frequency of the sounds a bystander near that wall hears and the frequencies the rider hears (Figure 16-14).

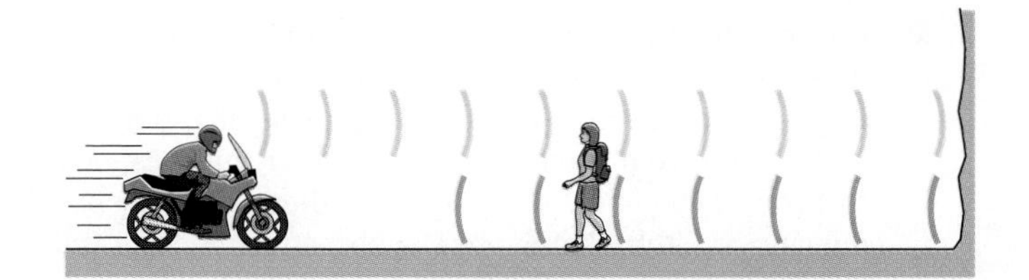

FIGURE 16-14

40. Sound waves reflect more from objects whose size is about one wavelength of the sound wave. Show that the 120,000-Hz wave emitted by a bat might well locate a flying insect that is about one-fourth of a centimeter in diameter.

41. Explain how the Doppler shift might be used with an ultrasound generator and detector to measure the speed of blood flowing through an artery or through the heart.

42. Why do you think cold winds howl but warmer winds do not?

43. Sometimes when you hear a plane that is far overhead, its sound isn't steady. Instead, the muffled sound rolls, fading out, then rising in loudness, over and over again. What's happening?

44. A bat called the horseshoe bat uses the Doppler shift to locate moving insects. It emits pulses at different frequencies so it can receive an echo of 83,000 hertz. How could reception of that single frequency be beneficial to the bat?

45. Why does an eavesdropper place a glass between ear and wall in order to hear what is said in the next room?

46. In order for a sound to be felt, it must be below 20 hertz in frequency and well above 100 decibels in intensity. Comment on this fact, considering Table 16-2.

47. The speed of sound at 20°C in air is 343 m/s, while in hydrogen gas (H_2) at the same temperature it is 1328 m/s. Why is it so much faster in hydrogen gas?

48. Just as different frequencies of equal intensity are perceived as different loudnesses, so are different intensities of equal pitch. If a soloist sings at a recital and is joined by a choir of 99, the 100 voices (each singing at the same intensity level) sound only about 4 times as loud as the soloist singing alone. Does this phenomenon expand or decrease the range of intensity of sound waves we can listen to comfortably?

49. A person talking normally emits at most 10^{-4} watts. If such energy could be used to power a light bulb, how many persons would have to talk to turn on a single 100-watt bulb?

50. Studies have shown that if a truck passing along a road at night makes a noise of intensity 75 dB in someone's bedroom near the road, 80 percent of the people tested would wake up. Advertisements for popular commercial smoke alarms for homes, which should wake up almost everyone in case of fire, state their noise level is at 85 dB. About how much more intensity do those waves carry than those from the noisy truck? Do you think that should be sufficient? Discuss.

51. A person going deaf can just barely hear someone talking normally. By how much would a hearing aid have to "magnify" sound to restore normal hearing for this person? (*Hint:* Use Table 16-2.)

52. The tone called middle C on a musical scale is a sound wave of 262 Hz. What is its wavelength in air at 10°C? (v = 337 m/s.)

53. If the sound 10 meters from the stage at a rock concert is at the 120-dB level, what is the intensity level 100 meters away? (*Hint:* The surface area of a sphere is $4\pi r^2$.)

54. On takeoff, a DC-10 airplane produces about 1500 watts of audible sound. What is the intensity of this sound at a distance of 10 meters? At 100 meters? At 1000 meters? (See exercise 53.)

Demonstrations

1. Excite some standing waves by tapping an empty drinking glass with a spoon. The sound it makes comes from oscillations of the glass, not from standing waves in the air column inside it. Prove this by putting a finger to the glass before you tap it again—there's not much sound then.

Using a soft drink bottle, you can prove it is the glass surface that makes the sound when you tap it and not the air column inside. To excite the air column, blow over the opening. As your breath oscillates in and out of the bottle, it excites the lowest-mode standing wave, and you hear a low hum. Then tap the bottle with a spoon at the same time to hear the two very different tones.

2. Listen to the Doppler effect on TV. During the broadcast of an automobile race or a motorcross, the trackside microphones record the sounds as the vehicles approach and recede from them, and the Doppler effect is easy to hear. Remember, however, it is the change in frequency, or pitch, that is the Doppler effect and not the change in loudness—which is also apparent.

3. A plastic straw is all you need to see how organ pipes work. Pinch the straw closed at one end and blow gently over the top of the other end until you hear a tone. You've set up a standing wave in the air column in the straw, a wave that draws energy from the stream of air from your lips. Now with the other hand, pinch the straw at the closed end again, but this time a little way up from the bottom. Then move your fingers upward while blowing across the top. The tone smoothly rises in pitch as the length of the air column decreases. (The wavelength of the standing wave becomes shorter, and a shorter wavelength means a higher frequency of oscillation.) Organ pipes of different lengths give different tones for the same reason.

4. High-speed motion pictures revealed in 1958 that the tip of a "cracked" bullwhip exceeds the speed of sound, which is why we hear the crack. In 1993 a video camera was used by high school students in North Carolina to show that the tip of a towel also exceeds the speed of sound when it is "snapped." Use a towel and show a friend the phenomena that causes the sonic boom when an airplane breaks the sound barrier!

SEVENTEEN

Electric Charge

Multiple lightning strikes over Tucson. The physics of such discharges is discussed in this chapter.

KNOWN TO THE ANCIENT GREEKS, THE NATURE OF ELECTRIC CHARGE FIRST CAME UNDER SERIOUS INVESTIGATION IN THE 1700S. THE CHARGED PARTICLES IN ORDINARY MATTER, ELECTRONS AND PROTONS, WERE DISCOVERED IN 1899 AND 1919, RESPECTIVELY.

Shred some tissue or newspaper into tiny pieces and then bring a plastic ballpoint pen close to them. Notice anything happening? Now rub one end of the pen vigorously against your clothing and bring the rubbed end close to the shredded paper. This time the pieces momentarily fly up to the pen, seeming to defy gravity. When you do this experiment, you join the company of the ancient Greeks who rubbed pieces of amber (yellow-brown fossilized resin used for jewelry) and wrote about its attraction for small bits of matter. Of course, those scraps of paper that jumped up to your pen don't really defy gravity. Another force, one stronger than gravity, is at work here: the electric force. This force depends not on the mass of the paper or that of the pen as gravity does but on a property of matter called electric *charge*. Usually you do not notice the electric force because the things around you are not charged; we say they are electrically *neutral*. The pen became charged only because the friction from rubbing somehow "charged" its surface. In Chapter 8 we saw that ordinary matter has equal numbers of electrons and protons, and their charges balance, or *cancel*, perfectly. For a review of the facts about atoms and charge, see Chapter 8.

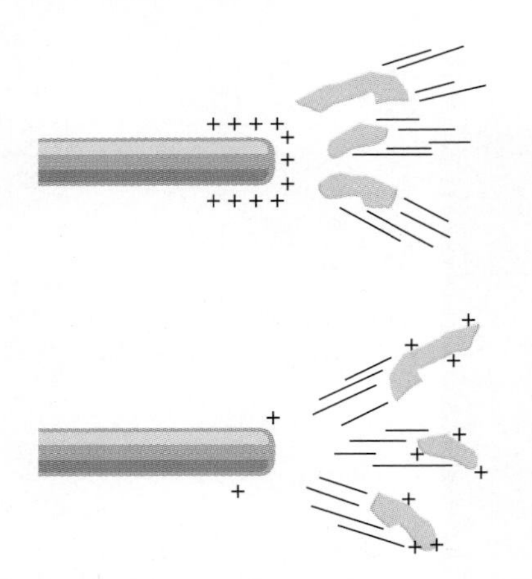

FIGURE 17-1 RUBBED WITH A SILK CLOTH, A GLASS ROD BECOMES CHARGED POSITIVELY. SMALL BITS OF PAPER ARE THEN ATTRACTED TO IT. WHEN THEY TOUCH THE GLASS, HOWEVER, THE BITS OF PAPER CAN GIVE UP SOME ELECTRONS TO THE POSITIVELY CHARGED GLASS, LEAVING THEMSELVES POSITIVELY CHARGED. THEY ARE THEN REPELLED BY THE POSITIVE CHARGE REMAINING ON THE GLASS.

THE TWO TYPES OF ELECTRIC CHARGES

Early investigators discovered that there are two types of electric charge. For example, a glass rod rubbed with a silk cloth becomes charged as in Figure 17-1, and if the rod then touches a tiny scrap of paper, that paper takes some of the charge from the rod. At the instant of taking on the charge, the paper flies away from the rod—because like charges repel each other. Likewise, a hard rubber rod becomes charged when stroked with cat's fur; if the rod is then touched to a scrap of paper, that paper too flies off. However, if those two charged scraps of paper come close to each other, they *attract* each other—because the two rods have opposite charges, and opposite charges attract each other.

In the late 1700s, Benjamin Franklin decided that these charges were an unseen, movable fluid that could be drawn from normal matter by friction. Any object with an excess of this fluid was charged one way, and any object with a deficiency was charged the other way. He decided to give the symbol + to the excess fluid and the symbol − to the deficiency; equal amounts of + charge and − charge canceled when all the fluid flowed back to the deficient material, leaving the matter once again in a neutral, uncharged state.

Franklin arbitrarily assigned the type of charge on rubbed glass to be positive and that on rubbed rubber to be negative. Today we know it is the rubber that has the excess "fluid" (it has extra electrons, transferred from the fur it was rubbed with) and the glass that has the deficiency (too few electrons to balance the charge of the protons in the atoms' nuclei in the glass rod; the silk took some electrons from the glass), as Figure 17-2 shows. Franklin's convention persists to this day, however, which is why we call electrons *negative* and protons *positive*.

FIGURE 17-2 WHENEVER POSITIVE CHARGE APPEARS ON MATTER, IT IS BECAUSE SOME OF THE MATTER'S ELECTRONS HAVE BEEN TAKEN AWAY. HERE SILK, RUBBED ON GLASS, GOES AWAY WITH SOME OF THE ELECTRONS OF THE GLASS, LEAVING THE GLASS POSITIVELY CHARGED.

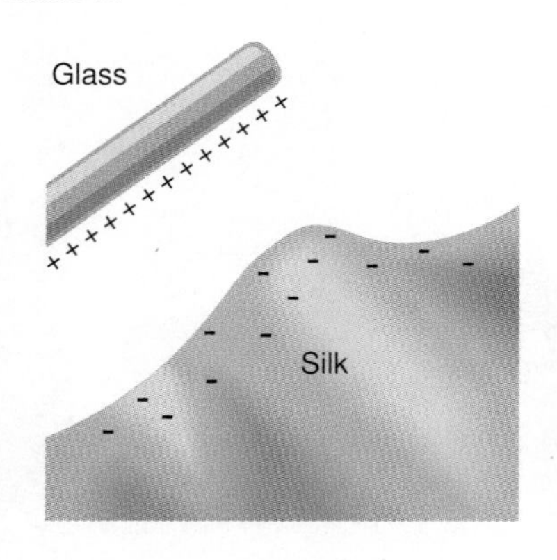

ELECTRON AFFINITIES

Amber rubbed with wool always gets a negative charge, and glass rubbed with silk always comes away positive. The reason is that each species of atom holds onto its outermost electrons with a different strength. We say atoms have different **electron affinities.** If an atom holds onto its outermost electrons tightly, it has a large electron affinity. If it loses its outermost electrons relatively easily, it has a small electron affinity.

When two materials come into contact, the one with the greater electron affinity pulls electrons from the one with the lesser affinity and therefore becomes negatively charged. The material that is left with fewer electrons is positively charged. Look at the list in Table 17-1 to see how some common materials become charged when rubbed together. Just pressing any two different materials together leaves each of them charged to some extent, but rubbing them together increases the amount of charge that is transferred from one material to the other.

The transfer of electrons is used to remove particulate matter from some power plants that burn coal. In the smokestacks of these plants, the dirty gases pass through negatively charged wires or go past a negatively charged central rod. The particles acquire a negative charge by contact, and are then repelled by those negative wires or rod and attracted to positively charged collecting plates or walls where they are deposited. These *electrostatic precipitators* typically remove 98% or more of the particulates.

Table 17·1*

(+) Positive Charge
Rabbit's fur
Glass
Wool
Quartz
Cat's fur
Silk
Your skin
Cotton
Wood
Amber
Rubber
Hard plastic (combs or ballpoint pens)
India rubber
(−) Negative Charge

*Rub any two of the substances in this list together and (usually) the upper will become positively charged and the lower will become negatively charged. In practice, the opposite result sometimes occurs; the results are very dependent on surface roughness, impurities on the surfaces, and the manner of rubbing.

CUTTING PARTICULATE POLLUTION

THE ELECTRIC FORCE: COULOMB'S LAW

In 1785, Charles Coulomb, a French scientist, found that the electric force between two charges obeys a law that looks a lot like the law of gravity (Chapter 7). This is **Coulomb's law:**

PUTTING IT IN NUMBERS—COULOMB'S LAW

The electric force between two charged particles varies as the product of the two charges and inversely as the square of the distance between them.

$$F_{\text{electric}} = k\frac{q_1 q_2}{d^2}$$

where q_1 is the electric charge on the first object, q_2 is the electric charge on the second, and d is the distance between them. The SI unit of charge is the **coulomb** (C), and it is equal to the charge of about 6×10^{18} electrons. The proton has a positive charge that balances the electron's negative charge exactly. In the SI system, k is about 9×10^9 N·m²/C². When a glass rod is rubbed with a silk cloth, about one ten millionth of a coulomb is transferred, which corresponds to about 600 billion electrons. The electron and the proton have the smallest amount of charge that has been isolated in nature.

The electric force is much, much stronger than the gravitational force. For example, the electric force between an electron and a proton is almost 10^{40} times as great as the gravitational force between them. Notice from the equation given above that, when charges are moved farther apart from each other, the electric force drops off rapidly (because $1/d^2$ becomes smaller very quickly when d increases). Conversely, when the charges are brought closer together, the force increases rapidly. Note that if either charge is doubled, the force itself only doubles.

THE TINY PARTICLES THAT CARRY THE CHARGE IN ATOMS—THE ELECTRONS AND PROTONS—ARE NEARLY INDESTRUCTIBLE. TODAY REACTIONS ARE KNOWN THAT CAN DESTROY ELECTRONS OR PROTONS—WHENEVER THESE PARTICLES MEET THEIR SO-CALLED ANTIPARTICLES, DISCUSSED IN THE FINAL CHAPTERS OF THIS BOOK. EVEN THEN, THE NET CHARGE REMAINS THE SAME, SO THERE IS ANOTHER CONSERVATION LAW IN NATURE, THIS ONE CONCERNING CHARGE. ELECTRIC CHARGE IS CONSERVED, JUST AS ENERGY IS, MEANING THE TOTAL AMOUNT OF CHARGE IS CONSTANT IN ANY INTERACTION.

Table 17·2

Conductors*	Insulators*
Silver	Amber
Copper	Hard rubber
Gold	Nylon
Aluminum	Porcelain
Brass	Beeswax
Iron	Glass and wood
Lead	Shellac
Mercury	Very pure water
Graphite	Air

In addition, water is a good-to-excellent conductor when it contains dissolved salts (or acids or bases).

*Each substance in the list is a slightly better conductor or insulator than the one below it.

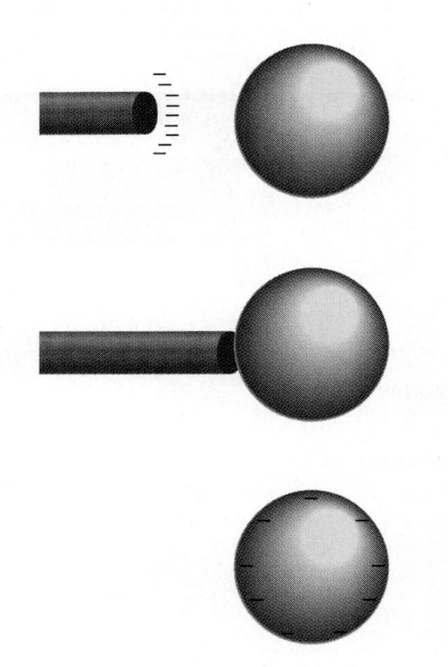

FIGURE 17-3 ELECTRONS ARE TRANSFERRED FROM THE NEGATIVELY CHARGED INSULATOR TO A METAL SPHERE AT THE POINT WHERE THE INSULATOR TOUCHES THE SPHERE. THESE ELECTRONS REPEL EACH OTHER AND WILL MOVE OVER THE SURFACE OF THE CONDUCTOR UNTIL THEY ARE AS FAR APART AS POSSIBLE.

ELECTRICAL INSULATORS AND CONDUCTORS

If you rub a plastic pen on one end, leaving it with a net electric charge, the other end won't pick up any scraps of paper. The charge imparted by the rubbing more or less stays at the place you stroked. It's the same for rubber and glass. On all these materials, the charge stays put unless something else touches the charged area. Because any charge on them remains immobile, glass, rubber, and plastic are called **electrical insulators** (or just *insulators*). Most solid materials except for metals are insulators. Since a charge on an insulator doesn't move around, we say it is a *static* charge.

Metals are **electrical conductors** (or just *conductors*). A conductor contains electrons that are free to move around both within it and on its surface. These movable electrons are called **conduction electrons.** Any *excess* charges on a metal also are moveable and they *remain on the surface* since they repel each other and therefore stay as far apart as possible, as Figure 17-3 shows. While this is easy to see with negative charges, which are carried by electrons, the motion of positive charge is not so easy to understand. The protons are bound in the nuclei of the atoms, and the atoms don't move while all this is going on. (If they did, the metal would flow as atoms left their locked-in positions in the object!) If a metal object receives a positive charge from a positively charged glass rod, we still say that the positive charge spreads out over the metal's surface. In reality, however, it is electrons that do the moving. When the positively charged rod touches the metal, as in Figure 17-4, the rod takes some free conduction electrons from that point on the

FIGURE 17-4 WHEN THE POSITIVELY CHARGED INSULATOR TOUCHES THE METAL SPHERE, ELECTRONS ARE TRANSFERRED FROM THE METAL TO THE INSULATOR. THIS LEAVES AN AREA ON THE SPHERE THAT IS DEFICIENT IN ELECTRONS, OR POSITIVELY CHARGED. MOBILE ELECTRONS IN THE METAL WILL MOVE INTO THIS REGION, ATTRACTED TO THE POSITIVE CHARGE, AND THAT MOVEMENT LEAVES THEIR FORMER AREAS WITH A POSITIVE CHARGE. ONLY WHEN THE POSITIVE CHARGE IS UNIFORM OVER THE SPHERE WILL THE ELECTRONS QUIT MOVING, FINDING NO NET ATTRACTION TO ANY AREA TO PUSH THEM ALONG. THE EFFECT IS THE SAME AS IF THE ROD TRANSFERRED POSITIVE MOBILE CHARGES TO THE SPHERE AND THESE CHARGES THEN REPELLED EACH OTHER (AS ELECTRONS WOULD) TO COVER THE SPHERE UNIFORMLY.

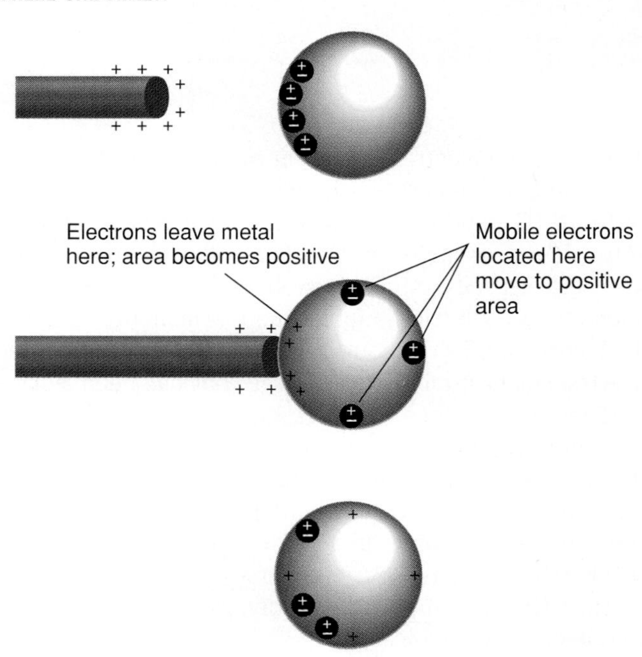

metal surface, leaving that area positive because it has a deficiency of electrons. Other conduction electrons in the metal are attracted to the positive area and so move in there. When they make this shift, they leave their previous locations with an electron deficiency and therefore slightly positive. The effect is the same as if free positive charges were present and repelled each other, spreading out to every point on the metal surface.

A charge detector called an electroscope, one of the first electrical devices built, demonstrates the mobility of charges on conductors (Figure 17-5). In this device, a conducting rod runs through an insulator. At the top of the rod is a conducting knob, and at the bottom hang two thin leaves of gold foil. Part of any charge transferred to the knob moves along the conducting rod to the foil. Since each leaf gets the same type of charge, the leaves repel one another and stand apart. The more charge the leaves receive, the greater the electrical repulsion between them and the farther they separate. The angle of separation is related to the amount of charge on the leaves.

Charge transferred to a metal sphere spreads out uniformly over the surface. If a charge is given to a metal object with a pointed area on its surface, however, a lot of the charge moves to that area. The forces drawn in Figure 17-6 show why. Later we'll see that lightning rods make use of that fact. If charge is transferred to a person, as in Figure 17-7,

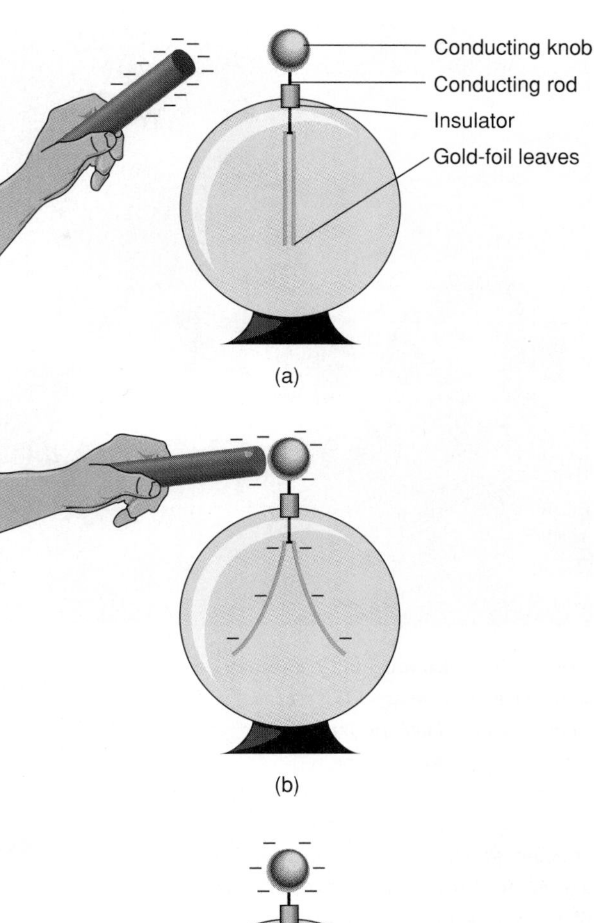

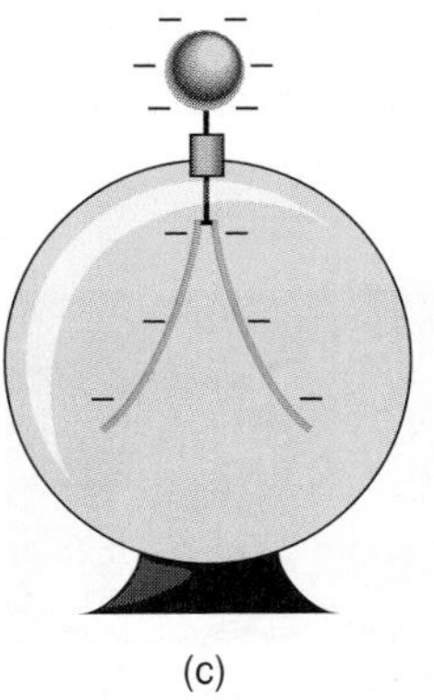

FIGURE 17-5 CHARGING AN ELECTROSCOPE WITH NEGATIVE CHARGE. (A) A NEGATIVELY CHARGED ROD IS BROUGHT TO THE ELECTROSCOPE, WHERE CONTACT IS MADE WITH THE CONDUCTING KNOB. (B) ELECTRONS TRANSFERRED TO THE KNOB TRAVEL THROUGH THE CONDUCTING ROD TO THE FOIL LEAVES, AND THE LIKE CHARGES ON THE THIN LEAVES PUSH THEM APART. (C) THE CHARGED ROD IS WITHDRAWN, AND THE ELECTROSCOPE IS LEFT WITH A NET NEGATIVE CHARGE.

FIGURE 17-6 A CHARGED METAL OBJECT WITH A POINTED TIP. CHARGES ACCUMULATE AT THE POINT. CONSIDER THE FORCES ON TWO ELECTRONS IF THE CHARGES WERE SPREAD *UNIFORMLY* OVER THE TWO AREAS SHOWN. THE NET FORCES ON THE TWO ELECTRONS FROM THE NEAREST FOUR ELECTRONS TO THEM SHOW WHY CHARGE MOVES TO THE POINTED AREA.

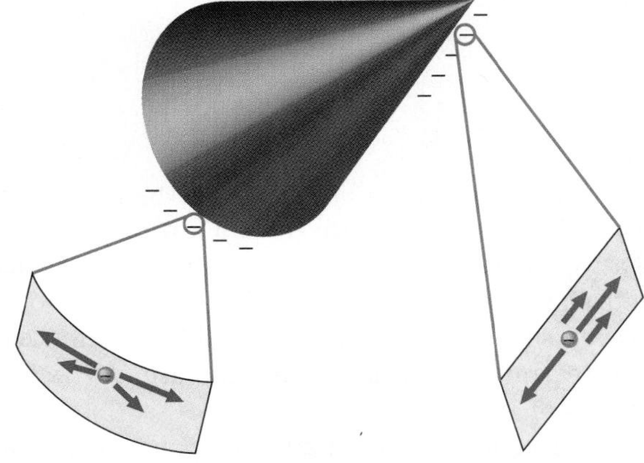

FIGURE 17-7 BY TOUCHING A CHARGED METAL SPHERE, SONESS GETS ELECTRIC CHARGE TRANSFERRED TO HER BODY. IT REPELS ITSELF, AND MUCH OF IT ACCUMULATES ON THE TIPS OF HER HAIR, WHICH CAUSES HER HAIR TO STAND OUT STRAIGHT BECAUSE THE CHARGES REPEL EACH OTHER. THE CHARGE WILL EVENTUALLY ESCAPE TO THE AIR OR TO THE GROUND.

it naturally flows to any pointed areas, such as the hair on the person's head. Since hairs are lightweight, even a small amount of charge on each hair causes them to spread as far apart as possible, showing that like charges repel each other.

SEMICONDUCTORS

Very pure crystals of the chemical elements silicon and germanium are electrical insulators. When atoms of certain other elements are introduced as impurities in these crystals, however, even in very small numbers, the crystals become *semiconductors*. A **semiconductor** is an electrical insulator that, with the addition of a small amount of energy (heat or light or electrical energy), becomes a conductor. The added energy prompts electrons from the impurity atoms to move as freely as electrons move in conductors. Tiny devices made with layers of semiconducting materials were the basis of an electronics revolution after the Second World War as transistors—which control the flow of electrons in electrical circuits by acting as electrical valves, switches, and more—replaced large, energy-inefficient vacuum tubes in electronic circuits.

ELECTRICAL CONDUCTION IN FLUIDS

Liquids and gases can be either electrical conductors or electrical insulators. To be a conductor, a fluid must have *ions* (charged atoms or molecules or even free electrons) to carry charge. Most water is a good conductor of electric charge because it contains dissolved minerals and salts that have broken up into ions. If you stir a cup of water with a finger, salt (NaCl) particles from your skin dissolve in the water, breaking into sodium ions (Na^+) and chlorine ions (Cl^-) and forming salt water, which is a good conductor. Pure water contains relatively few ions, however, making it a good electrical insulator. Likewise, dry air is normally an electrical insulator, since it also has very few ions compared with the number of neutral molecules.

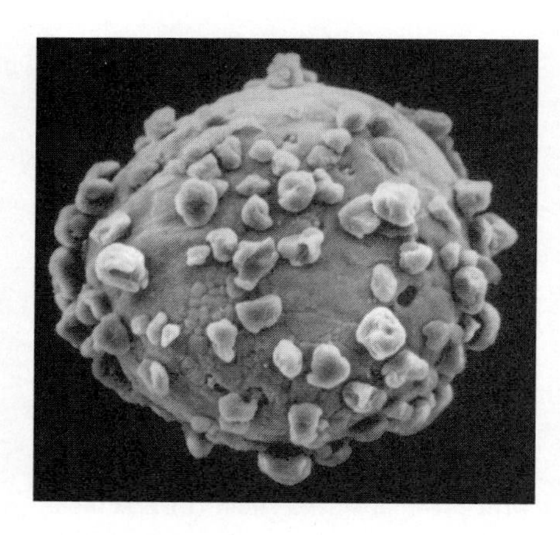

A CARRIER BEAD FROM A XEROX® PHOTOCOPIER. THE TONER PARTICLES (ATTACHED TO THE BEAD) AND THE BEAD CARRY OPPOSITE CHARGES.

INDUCTION

If a positively or negatively charged object is brought close to a conductor, some of the free electrons in the conductor move toward a positive external charge or away from a negative one (Figure 17-8). We say the external charge *induces* positive and negative areas on the conductor. The *net* charge on the conductor is still zero, however, since no charge has been added or subtracted. The charges have just been rearranged. **Induced charges** are *charges that become separated in the presence of an electric force*.

When a comb you've run through your hair or a plastic pen you've rubbed on your clothing picks up scraps of paper, it is because of induction. The negative charge on the comb exerts an electric force on the charges in the

SEMICONDUCTORS YOU USE IN THE LIBRARY

When you use a photocopy machine, you are using a semiconductor. Here's how it works. A large semiconducting drum in the machine is given a net charge. An image of the original to be copied is projected onto the semiconductor with the help of a very bright light. Wherever there is a white area on the original, the exposed semiconductor becomes a conductor, and the excess charge on the drum repels itself and darts away. (It is conducted to the ground; see the later section, Grounding.) Wherever there is a black area on the original, the semiconductor remains an insulator and the charge stays put.

Next, the semiconductor drum is bombarded with small charged beads covered with toner particles, which are specks of black plastic held to the beads by electrostatic attraction. As the beads strike the semiconductor surface, the toner sticks wherever the surface is charged (in other words, wherever there is a dark spot on the original). A sheet of paper is then brought over the drum, and its back side is charged so that the toner adheres to the paper. Heat is applied to melt the plastic toner particles onto the paper, producing a permanent copy of the original. When it exits from the machine, the paper may still may be warm from that last step.

paper. Paper normally picks up moisture from the air, making it a conductor. Negative charges run to the far side of a scrap (away from the comb), leaving the side nearest the comb positive. Because the positive side is closer to the comb, its attraction to the comb is stronger than the far side's repulsion (from Coulomb's law), and the paper flies to the comb. Once it touches the comb, however, some of the comb's negative charge can transfer to the paper. If this happens, the paper suddenly has a net negative charge and is pushed away from the comb.

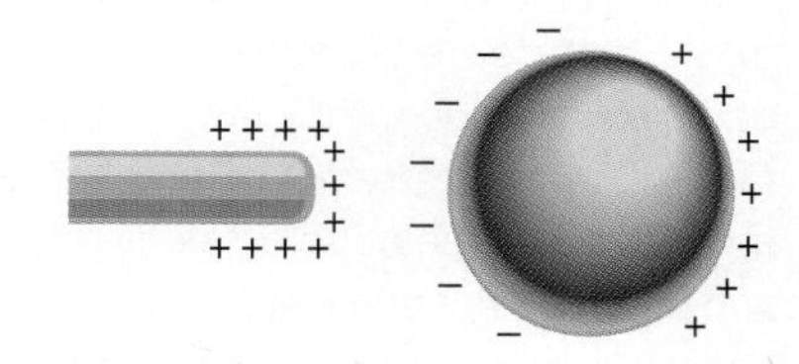

FIGURE 17-8 WHEN THE POSITIVE ROD IS BROUGHT NEAR THE METAL SPHERE, CONDUCTION ELECTRONS IN THE METAL ARE ATTRACTED AND MOVE NEARER TO THE ROD, WHICH LEAVES POSITIVE AREAS WHERE THEY WERE PREVIOUSLY. THIS SEPARATION OF CHARGE DUE TO THE ELECTRIC FORCE FROM THE EXTERNAL CHARGE IS CALLED INDUCTION.

Induction works for insulators as well as conductors. In the presence of an external charge, the molecules of insulators become positive on one side and negative on the other (Figure 17-9). That is, the external charge **polarizes** the molecules, and the insulating material itself becomes positive on one side and negative on the other.

CHARGING PAINT DROPLETS

Charges and induction are used in the electrostatic application of paint to many surfaces that are difficult to coat uniformly, such as automobile parts and bicycle frames. The particles of paint in the spray get a negative charge from a negatively charged tip. The object receiving the paint is grounded, so the droplets induce a positive charge on the surface and are attracted, even to the opposite side of the tubes of a bike frame. The resulting coat is more uniform, and less of the fine paint droplets escape into the air, reducing pollution and waste at the same time.

WHY STATIC CHARGES "DISAPPEAR"

When there is a lot of moisture in the air, any solid object exposed to the air usually has a thin coating of water molecules on its surface. That water dissolves ions from the solid surface and thus becomes a good conductor. A net charge on the solid surface can spread out and leak off to the ground or even into the air if the relative humidity is high. Water molecules are polar molecules,

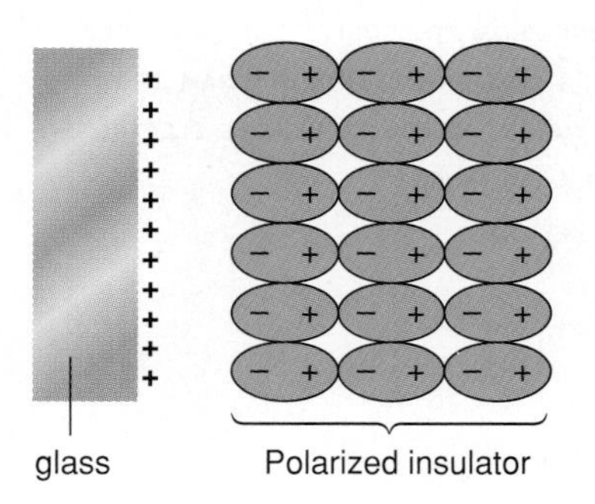

FIGURE 17-9 AN EXTERNAL POSITIVE CHARGE INDUCING CHARGE ON THE SURFACE OF AN INSULATOR. NONE OF THE ELECTRONS LEAVES ITS MOLECULE, BUT THE CHARGES ARE STRETCHED APART IN EACH MOLECULE, AND THIS STRETCHING LEAVES THE SURFACE NEAREST THE EXTERNAL CHARGE NEGATIVE AND THE FAR SURFACE POSITIVE.

CUTTING POLLUTION: ELECTROSTATIC PAINT GUNS

One component of human-caused atmospheric pollution is the release of volatile organic compounds (VOC). A very large source of this pollution is from industrial spray painting, where volatile organic compounds carry the paint pigments. These compounds evaporate directly into the air, both in the spray itself and as the painted surface dries. To comply with federal VOC emission standards, industries now use electrostatic devices which reduce the amount of VOC released in painting.

In an electrostatic paint spray gun, the spray is sent over a negatively charged tip and the paint droplets pick up electrons. The negative charges repel, breaking apart the larger droplets and ensuring other droplets don't coalesce, resulting in a very fine spray. But the real benefit comes as the paint droplets approach the surface to be painted. In this method those surfaces are grounded, and when the negative paint particles approach, they induce a positive charge on the surface of the bike frame or auto part or other object, which attracts all the paint particles directly to the object. The spray gun releases the paint with a very small velocity, so the object becomes surrounded in a fine mist of these mutually repelling charged droplets which by induction are self-attracted to the surfaces. This results in very uniform coatings of even hard-to-reach parts of the surfaces! Besides polluting much less, they save money. Now car manufacturers are experimenting with dry power "sprays" which release no VOCs. The charged dry particles are attracted by induction to the surfaces which then must be baked to "fuse" the paint particles to the surface. An experimental dry electrostatic paint gun is shown in the photo here.

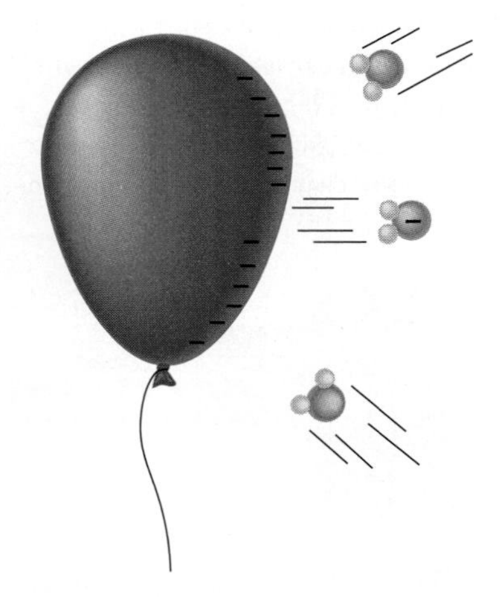

FIGURE 17-10 WATER MOLECULES, WHICH ARE POLAR AND THEREFORE HAVE THEIR CHARGE NATURALLY SEPARATED, CAN PICK UP CHARGE FROM THE SURFACE OF AN OBJECT DURING COLLISIONS. THIS DIAGRAM SHOWS HOW WATER MOLECULES REMOVE ELECTRONS FROM A SURFACE.

positive on one side and negative on the other because of the strong bonds between the oxygen and hydrogen atoms. Therefore water molecules in the air can quite easily take charges away from a surface that is exposed to the air (Figure 17-10).

GROUNDING

To take the charge from a charged electroscope, you could connect a wire from the conducting knob to a metal cabinet. The charge, separating as far as possible due to its repulsion, would flow out to cover the cabinet, leaving very little charge on the electroscope. Often a much larger quantity of charge needs to be disposed of, such as when a lightning strike brings charge to earth. That calls for a conductor larger than a cabinet, and all around us is the largest available conductor—the earth. A metal wire that leads underground or even a metal water pipe that runs into the ground can take away a very large amount of charge if it reaches damp soil, which is a good conductor. Furnishing such a conducting path to the ground is called **grounding.** In the next chapter we shall see why household appliances should be grounded.

THE ELECTRIC FIELD

The electric force exerted by a charge pushes any other like charge away and attracts any opposite charge even though nothing tangible bridges the gap between them. Let's suppose a charge Q is fixed in space at some point. If a

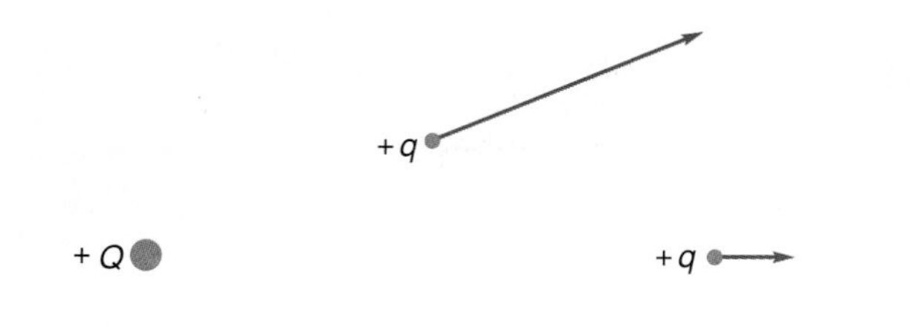

FIGURE 17-11 IF THE CHARGE $+Q$ IS FIXED AT THE POINT SHOWN, ANOTHER POSITIVE CHARGE AT *ANY* POINT NEARBY IN SPACE IS PUSHED STRAIGHT AWAY FROM $+Q$ BY THE ELECTRIC FORCE.

charge q of the same sign is placed at other points nearby, it is pushed straight away. A charge $2q$, twice as large, put at the same place is pushed twice as hard, and so on.

To express what happens to charges in the space around Q, we can define an **electric field,** $\vec{E}$, around Q that gives the force per unit charge exerted by Q on any other charge at any point. Every point in the space around Q has a unique value of $\vec{E}$ that acts on *any* quantity of charge q placed there:

PUTTING IT IN NUMBERS—DEFINING THE ELECTRIC FIELD

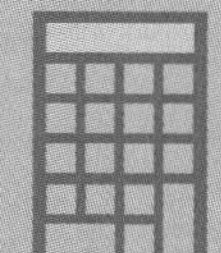

electric field = electric force per unit of charge

$$\vec{E} = \frac{\vec{F}}{q}$$

where E, the strength of the electric field without regard for its direction, is measured in newtons/coulomb and F is the magnitude of the electric force exerted by Q on the charge q, as given by Coulomb's law. If instead of one charge Q there are many charges, the net electric field at any point in space is just the vector sum of the fields due to those individual charges. **Example:** Several highly charged objects are close to a point P, and the net electric field at P has a strength of 25 N/C. What is the strength of the force on a 0.05-C charge placed at that point? $F = q \times E = 0.05\text{ C} \times 25\text{ N/C} = $ **1.25 newtons.**

PHYSICS IN 3D/THE ELECTRIC FIELD LINES AROUND A POSITIVE CHARGE

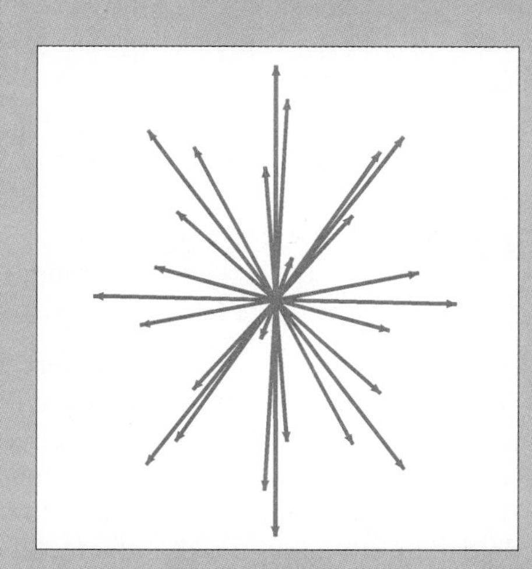

A STEREO VIEW OF THE ELECTRIC FIELD AROUND A POSITIVE CHARGE. THE FIELD AROUND A NEGATIVE CHARGE LOOKS THE SAME EXCEPT THE FIELD LINES WOULD POINT *INTO* THE NEGATIVE CHARGE.

PHYSICS IN 3D/THE ELECTRIC FIELD OF AN ELECTRIC DIPOLE

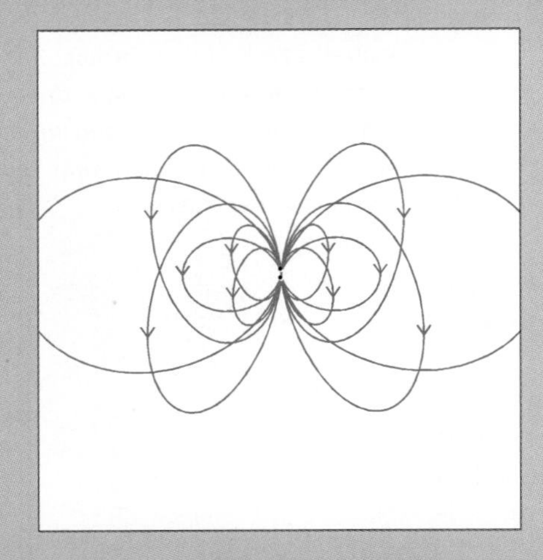

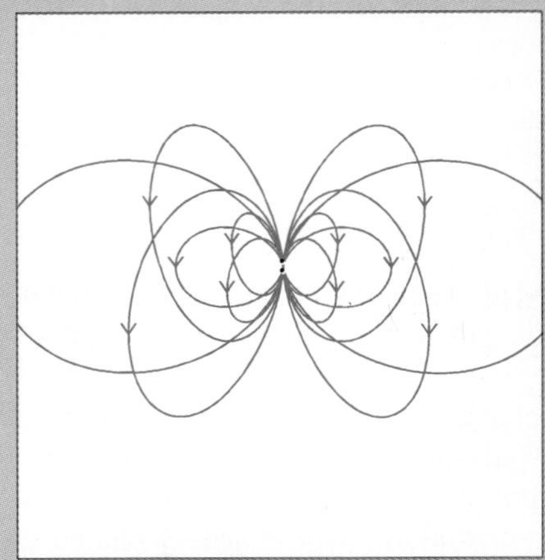

A STEREO VIEW OF THE ELECTRIC FIELD AROUND A DIPOLE ORIENTED IN THE VERTICAL DIRECTION. THE POSITIVE CHARGE IS AT THE TOP OF THE DIPOLE. THIS ELECTRIC FIELD IS THE VECTOR SUM OF THE FIELDS OF THE POSITIVE AND NEGATIVE CHARGES SEPARATED BY A SMALL DISTANCE.

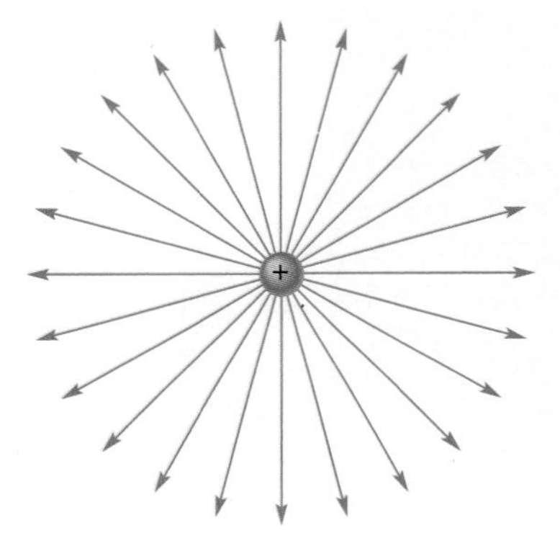

FIGURE 17-12 THE ELECTRIC FIELD OF A POSITIVE CHARGE IS REPRESENTED BY LINES RADIATING FROM THE CHARGE IN EVERY DIRECTION. THE FORCE EXERTED BY THE FIELD ON ANOTHER POSITIVE CHARGE NEARBY IS DIRECTED ALONG THE FIELD LINE THAT IS AT ITS POSITION. THE FORCE ON A NEGATIVE CHARGE NEARBY IS *OPPOSITE* THE DIRECTION OF THE FIELD LINE THAT IS AT ITS POSITION.

To visualize the electric field around a charge Q, we draw lines with direction arrows that show the direction of the force on a *positive* charge brought into the space nearby (Figure 17-12). These imaginary lines are called **electric field lines.** The field lines from an isolated positive charge Q point straight away from the charge, since another positive charge gets a push in that direction. The field lines around an isolated negative charge point straight *into* the negative charge, since that is the direction of the force felt by a nearby positive charge. Notice that electrons, because they have negative charge, experience a force in the other direction, *opposite* the direction indicated by the electric field lines. Where the field lines are close together (near the charge, for example), the force is strong. Where the lines are far apart, the force is weaker.

Figure 17-13 shows the field lines around positive and negative charges of the same magnitude that lie some fixed distance apart. A positive/negative charge arrangement like this is called a *dipole*. The electric field of a polar molecule, such as water, has much the same shape as the electric field around a dipole.

PROTECTION FROM ELECTRIC FORCES: SHIELDING

Interestingly, metallic objects can provide a shield from electric forces. *Any area surrounded by metal normally has no net electric fields within it.* The reason is that if a metal object has a net charge or is in a region where there is an electric field, the electrons that are free to move in the metal *will* move. These charges have electric fields of their own, and their fields add to the external field to give the *net* field at any point in or out of the metal object.

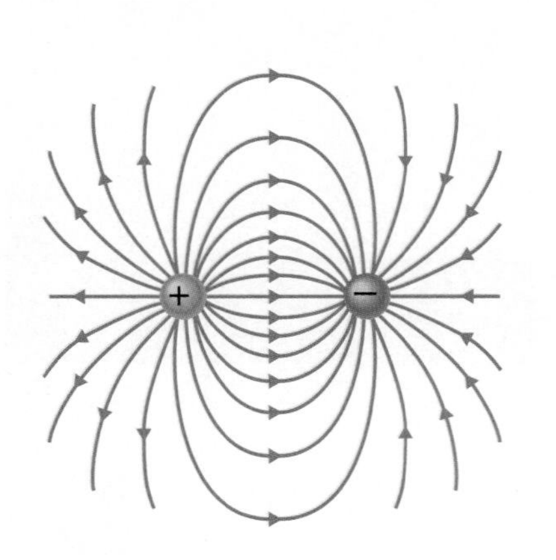

FIGURE 17-13 IF A NEGATIVE CHARGE IS BROUGHT CLOSE TO A POSITIVE CHARGE, THEIR FIELDS ADD LIKE VECTORS SINCE THE FORCES THEY EXERT ARE VECTORS. THE NET ELECTRIC FIELD AROUND EQUAL BUT OPPOSITE CHARGES IS PORTRAYED HERE. THIS ARRANGEMENT OF CHARGES IS CALLED A DIPOLE.

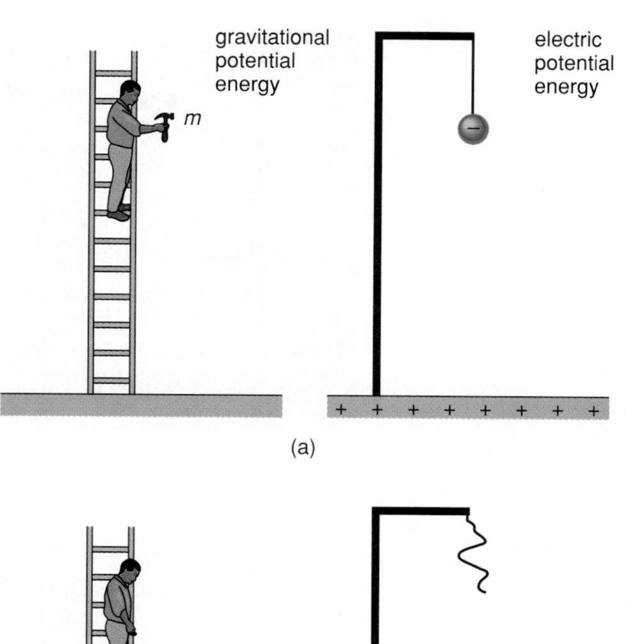

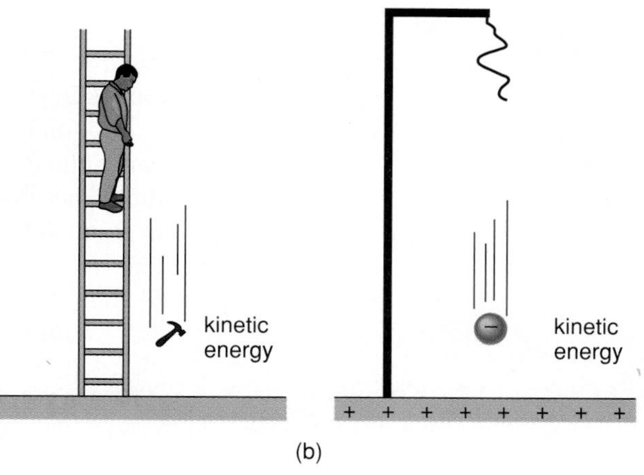

FIGURE 17-14 (A) THIS HAMMER HAS GRAVITATIONAL POTENTIAL ENERGY DUE TO ITS POSITION IN SPACE NEAR THE EARTH BELOW IT, AND THIS CHARGE HAS ELECTRIC POTENTIAL ENERGY DUE TO ITS POSITION NEAR THE CHARGE BELOW IT. (B) RELEASED, BOTH HAMMER AND CHARGE ACCELERATE BECAUSE OF THE NET FORCE ON THEM, AND THEIR POTENTIAL ENERGY IS CONVERTED TO KINETIC ENERGY AS THOSE FORCES DO WORK.

So those charges continue to move until all the electric forces on them are balanced—that is, until the net electric field *throughout* the interior is zero.

In effect, then, the safest place to be when there are large electric fields around is *inside* a conductor, which is why a car is a good place to be during an electrical storm. Maybe you have noticed metal boxes covering some of the parts in a stereo. Those metal covers serve as electrical shields to keep out unwanted electric fields that could interfere with performance. To guard against damage from lightning, some large computers are housed in rooms that are entirely lined with metal, with the metal shield grounded to the earth.

POTENTIAL ENERGY FROM THE ELECTRIC FORCE

A hammer held above the ground has gravitational potential energy (Figure 17-14). If it falls, it gains kinetic energy equal to its loss in potential energy and is able to do work on any object in its way. A charged particle in an electric field also has potential energy. The electric field exerts a force on the particle, and, if it is free to move, the particle does so and gains kinetic energy. A charged particle, then, has **electrical potential energy** whenever it is in an electric field.

Now look at Figure 17-15. The object with charge q has a larger potential energy at point A than at point B, and in moving from A to B it has a certain change in potential energy because the electric force does work on it. If that object had a charge of $2q$, it would have twice the change in potential energy since the electric force would be twice as strong at every point. Just as with the electric force earlier, it is very convenient to describe this effect of Q on nearby charges with a potential energy *per unit charge,* called the **electric potential,** V. Point A is at a certain electric potential whether or not a charge is there, and the same is true for B and all other points around Q. For a charge q at any point in an electric field, its electric potential energy is found by multiplying q times the electric potential at that point.

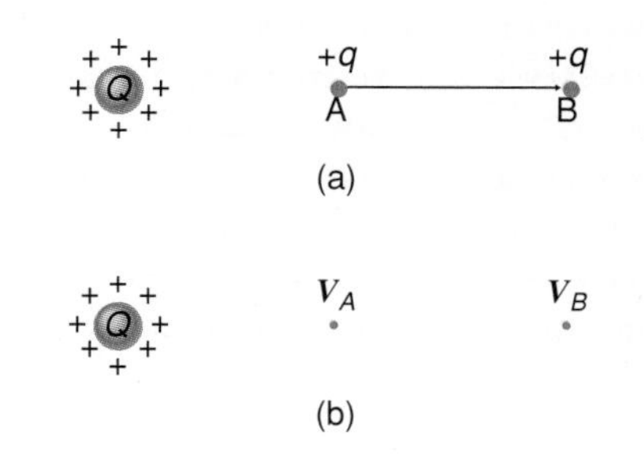

FIGURE 17-15 (A) A CHARGED OBJECT MOVING FROM A TO B LOSES ELECTRICAL POTENTIAL ENERGY AND GAINS KINETIC ENERGY; IF IT HAD TWICE THE CHARGE, IT WOULD GAIN TWICE AS MUCH KINETIC ENERGY. (B) DEFINING THE ELECTRIC POTENTIAL SIMPLIFIES THE UNDERSTANDING OF Q'S EFFECT ON CHARGES IN ITS VICINITY. EVERY POINT AROUND Q IS AT A CERTAIN ELECTRIC POTENTIAL. ANY CHARGE THAT MOVES BETWEEN THOSE TWO POINTS EXPERIENCES THE SAME POTENTIAL DIFFERENCE. TO FIND THE ENERGY GIVEN TO THE CHARGE, MULTIPLY THE POTENTIAL DIFFERENCE BY THE CHARGE.

PUTTING IT IN NUMBERS—DEFINING ELECTRIC POTENTIAL

electric potential = electric potential energy/charge

$$V = \frac{PE}{q}$$

The electric potential has units of joules/coulomb, and 1 J/C = 1 volt. Changes in the electric potential are called *potential differences.* **Example**: The potential difference between the two terminals on a car battery is 12 volts. If a charge of 3 coulombs moves between those two points, how much energy has the charge been given? Energy given to charge = change in *PE* of charge = $q \times V$ = 3 C × 12 V = **36 joules.**

Figure 17-16 shows the electric field lines and changing electric potential values in the space between two charged plates.

A comb takes electrons from your hair. The negative charge on the comb can induce charges on a scrap of paper, causing it to fly to the comb. The difference between the electric potential of the comb and that of the paper is about 5000 volts. And just as the force of gravity does work on a bowling ball that rolls off a shelf, the electric force does work on the paper as it moves to the comb.

Shuffle your feet across a nylon rug on a dry day, and you become positively charged. Touch a metal doorknob and you'll feel a slight shock as electrons flow from the doorknob to your hand to neutralize the excess positive charge on you. The difference between your electric potential and that of the doorknob is usually at least 5000 volts, but if a *spark* jumps between the doorknob and your hand, the potential difference must be 10,000 to 15,000 volts. Normally, dry air is a good electrical insulator. However, if a potential difference of 15,000 volts or more occurs between points only a centimeter or less apart, the few free electrons naturally in the air (from cosmic ray ionization and natural radioactivity) can smash into air molecules hard enough to ionize them. That ionized air offers a conducting path that large numbers of electrons will then follow due to the large potential difference. The spark you feel and hear is a miniature lightning bolt between your fingers and the doorknob. Even though the potential difference is large, the energy transferred is small since the charge q is small, and energy = $V \times q$.

THE POTENTIAL DIFFERENCE BETWEEN TWO POINTS IS MEASURED IN VOLTS, AND IN PRACTICAL USAGE THIS IS OFTEN CALLED THE "VOLTAGE" BETWEEN THE TWO POINTS.

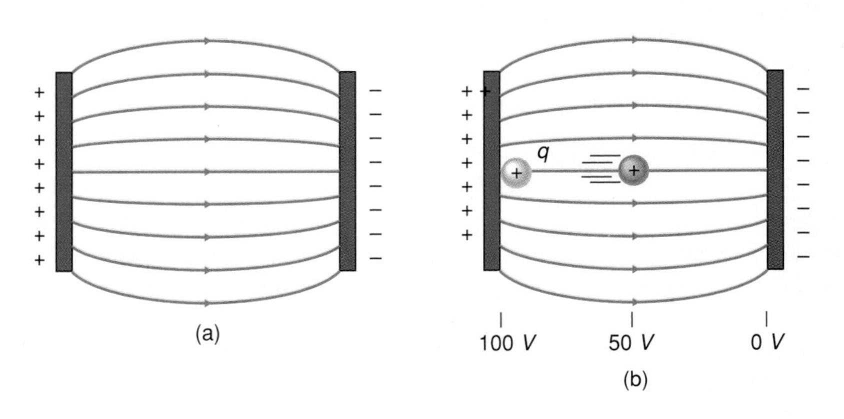

FIGURE 17-16 ONE EXAMPLE OF THE ELECTRIC FIELD AND ELECTRIC POTENTIAL FOR A LARGE NUMBER OF OPPOSITE CHARGES. (A) THE ELECTRIC FIELD BETWEEN TWO OPPOSITELY CHARGED METAL PLATES IS UNIFORM IN THE CENTER. (B) A POSITIVE CHARGE "FALLING" IN THE ELECTRIC FIELD BETWEEN THE PLATES (WITHOUT AIR RESISTANCE). THE ELECTRIC POTENTIAL AT POINTS ALONG THE WAY IS SHOWN. THE DIFFERENCE IN ELECTRIC POTENTIAL BETWEEN THE TWO POINTS TIMES THE CHARGE Q GIVES THE KINETIC ENERGY GAINED (AND THE POTENTIAL ENERGY LOST) BY THE CHARGE.

VERY LARGE POTENTIAL DIFFERENCES: LIGHTNING

Worldwide there are about 100 lightning bolts *every second* from the several thousand thunderstorms on the planet at any given moment. In a storm cloud, separated charges abound because of friction between rising and falling parcels of air, rain, and hail. When a lot of negative charge accumulates in one area and the corresponding positive charge collects in another area, a large electric field results, along with a great difference in electric potential. A bolt of lightning occurs when electrons jump from a negative area to a positive area. After the charge moves, the positive and negative areas are less charged than before. A lightning bolt is called an electric discharge for that reason.

For every lightning bolt that comes to earth, four others jump between clouds or between parts of the same cloud. In other words, only 20 percent of all lightning bolts come to the ground or, more rarely, jump from earth up to a cloud. A cloud usually becomes negatively charged near its base, and the negative charge induces a positive area on the ground and on all conducting objects beneath the cloud. Pointed objects below have the greatest induced charge, as we've seen, and thus the greatest attraction for the charges in the cloud. Just before the strike, the potential difference is typically 10 million to 100 million volts.

Any tall objects in effect bring the ground closer to the cloud's charge. Solitary trees are especially good targets for lightning. If you are caught out of doors in a thunderstorm, it is better to crouch in an open field than under a tall tree. Crouch, but don't lie down, because when lightning strikes, the charge can run along the ground. If such a charge travels from your head to your feet, it delivers more energy than if it travels between two feet close together. The farther the two points are apart, the greater the potential difference is likely to be.

In a house during a thunderstorm, stay away from any window with a tall tree nearby. Because charge moves most easily along conductors, stay away from metal appliances and out of the bathtub. Inside cars and airplanes is safe, and inside buildings with lightning rods is safe as well. Lightning rods, invented by Ben Franklin, should be a foot or more taller than any chimney or vent on the building, and they should be connected by a conducting cable to a conducting rod that goes at least 3 meters (10 feet) into the ground. A charged cloud above a lightning rod induces a larger charge at the tip of the brass or copper rod than at any other point on the roof, so the bolt is drawn to the metal point. Almost all of the charge goes down the cable and into the ground, causing no damage to the building.

PHYSICS IN 3D/CHARGED CLOUDS INDUCE LARGE CHARGES ON TALL CONDUCTORS

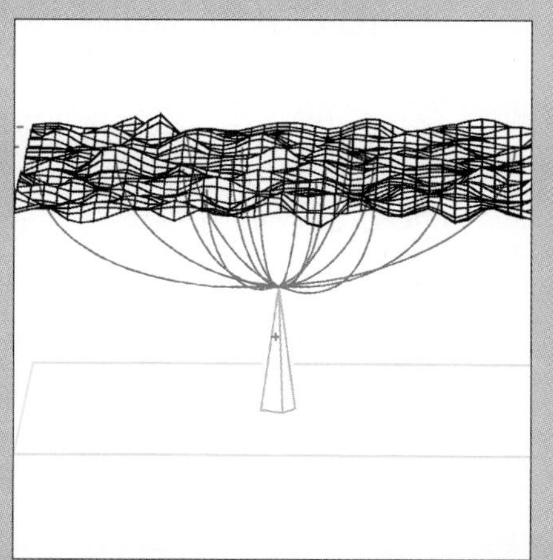

A STEREO ILLUSTRATION OF THE EFFECT OF A TALL CONDUCTING OBJECT BENEATH A CHARGED CLOUD ON THE ELECTRIC FIELD. THE OBJECT HAS AN INDUCED CHARGE OPPOSITE THAT OF THE BASE OF THE CLOUD ABOVE.

Summary and Key Terms

coulomb–249
Coulomb's law–249
electrical insulator–250
electrical conductor–250
conduction electrons–250
grounding–254
semiconductors–252
electron affinities–249
induced charge–252
electric field–255
electric field lines–256
electrical potential energy–257
electric potential–257

The *electric force* results from a property called *charge*. Matter carrying a net charge, measured in ***coulombs***, has either excess electrons, making it *negative*, or a deficiency of electrons, making it *positive*. Like charges repel one another, and unlike charges attract. Most matter is electrically neutral, having a balance of positive and negative charge. The electric force is given by ***Coulomb's law***: $F_{\text{electric}} = kq_1q_2/d^2$, where k is a constant. Like gravity, the electric force decreases rapidly as the charges move apart.

Most solids are ***electrical insulators***, materials on which electric charge is immobile, or static. Metals are ***electrical conductors***; they have ***conduction electrons*** free to move around within them. A net charge on a metal object with a pointed area on its surface concentrates on that area. Lightning rods supply a point where induced charges accumulate, and the strong electric field that results attracts any lightning strike in the immediate area. The lightning rod furnishes a conducting path to the ground, as does the third wire in large appliances. Such ***grounding*** lets charge flow easily to earth, the largest conductor available.

Semiconductors are electrical insulators containing impurity atoms. If energy is added, the insulator takes on the properties of a conductor.

Fluids can be conductors if they contain ions. Atoms of the different elements hold on to their outermost electrons with different strengths, which are referred to as ***electron affinities***.

If a negatively charged comb is held close to a neutral scrap of paper, the scrap's negative charges rush to its far side, away from the comb, and the side nearest the comb is left positive, so the scrap flies up to the comb. We say the comb causes ***induced charges*** in the scrap, charges that are separated in the presence of an electric force.

An ***electric field*** is equal to the electric force per unit of charge, or $\vec{E} = \vec{F}/q$. We visualize the electric field using ***electric field lines*** that show the direction of the force on a nearby *positive* charge: Field lines from a positive charge point *away from* the charge, and those from a negative charge point *into* the charge. The electric field is strongest where the field lines are closest together.

When a metal object has excess charge or is in an electric field, the electrons that are free to move *will* move until the net electric field *throughout* the object's interior is zero. So any area surrounded by metal is normally free from electric fields.

A charged particle in an electric field has ***electrical potential energy***, which depends on its position in the field and the field's strength. The potential energy per unit of charge due to an electric field is called ***electric potential***, and is measured in volts. When two points in space have different electric potentials and a charge moves from one point to the other, the energy given to the charge equals the potential difference (in volts) between the two points multiplied by the charge.

EXERCISES

Concept Checks: Answer true or false

1. Most matter on earth is electrically neutral; that is, it has no net charge.
2. Any material can become charged to some degree by bringing it into firm contact with some other material.
3. Like charges attract each other and unlike charges repel each other.
4. The two types of charge are fundamental properties of electrons and protons, and these particles carry the charge found in normal matter.
5. Charge residing on an electrical insulator cannot move around freely, whereas charge placed on a conductor can move around freely.
6. Induction works only for conductors and not for insulators.
7. If an object is grounded, it has a conducting path from it to the ground, which is itself a conductor.
8. The space around any charged object is said to contain an electric field, and this field gives the force per unit charge on any other charge brought to a point in that field.
9. The electric field is measured in newtons per coulomb.
10. Any net charge on a metal resides on its surface.
11. A charge in an electric field has electrical potential energy. Its potential energy divided by the quantity of the charge is called the electric potential.
12. The difference in electric potential between two points is measured in volts, which is the same thing as joules per coulomb.

Applying the Concepts

13. In a clothes drier, different fabrics are often tumbled together. What effect does that lead to when they are taken out?
14. The plastic wrap used in the kitchen is an insulating material designed chemically to become easily charged by friction. How does that help it to adhere to other things and to itself?
15. If you tear up small pieces of paper and try to stack them, you can't! Explain.
16. A bird's beak and its feathers are made of the same substance. When a bird preens its feathers, it closes up the spaces in them by zipping up hooklike structures along the edges. Why doesn't static electricity interfere with this frictional process?
17. While electrostatic precipitators remove 98% or more of the particulate matter from power plant emissions, the particles that get through are the smallest particles. In terms of the environment, is that good or bad?
18. To get rubbing experiments to work better on a humid day, you should first use a hairdryer or heat lamp to evaporate the unseen layers of wetness from the solid surfaces. Why?
19. In the 1700s a common practice during electrical storms was for people to go to the churches and pull the ropes to ring the bells in the steeples. It was thought that the noise would help break up the clouds and prevent lightning. There was a high mortality rate among the bellringers, however. Comment on the idea and the mortality rate. (The practice was finally outlawed for noise abatement reasons, not safety!)
20. When a ship moves through the ocean, it becomes slightly charged. Explain how.
21. Airplanes must be grounded after they land so that departing passengers won't get shocked. Why?
22. When a comb rubbed through your hair attracts a small scrap of paper, the paper sometimes touches the comb and then quickly jumps away. What happened?
23. A blouse or shirt taken straight from the clothes drier will cling to you. Explain.
24. Do you think the girl in Figure 17-7 is standing on an insulator or a conductor? Explain.
25. Sparks are extremely dangerous around grain storage bins, where there is a great deal of grain "dust" in the air. Explain. (For the same reason, medical personnel guard against sparks in operating rooms, where alcohol and ether vapors are in the air.)
26. If two like charges are taken twice as far apart, by how much does their repulsive electric force decrease?
27. Two charges repel each other with a force of 200 N. What will the force be if they are one-fourth as far apart?
28. Railroad tracks are often insulated from the earth by the wooden crossties. Explain what would happen to make them dangerous to be near during a thunderstorm.
29. Rubber tires get charged from friction with the road. What sign is their charge? Perhaps you have seen a gasoline truck trailing a metal chain beneath it. What purpose does the chain serve? (Modern tires for such trucks are made to be electrical conductors.)
30. Lightning sometimes occurs during large desert sandstorms. What mechanism is at work there?
31. Lightning bolts took a great toll on tall-masted wooden sailing vessels. As soon as Ben Franklin invented the lightning rod, sailors began to hoist copper chains from the water to the mast top when lightning threatened. On at least one occasion, sailors were killed when lightning

struck while they were lifting the chain. The metal ships that came onto the scene in the 1860s had no such problems. Why not?

32. Induced charges in a metal container shield the interior of the container against any strong electric fields present outside the container. Why can't we have a shield against the gravitational field in the same manner?

33. Usually a bolt of lightning consists of several quick strokes rather than just one discharge. If the wind is blowing, the second and third strokes do not follow the original path exactly. Explain.

34. Explain why it is difficult to keep your hair really clean if you brush it often.

35. Figure 17-17 is a facsimile of a sketch made by Michael Faraday in 1838. (You'll learn of his contributions to the physics of electricity and magnetism in the next few chapters.) This sketch showed Faraday's estimate of the electric field lines around an electric eel in a dish of water at the time of discharge. Faraday estimated the intensity of the field by dipping his bare hands into the dish at various points! Of what in this chapter does this field remind you?

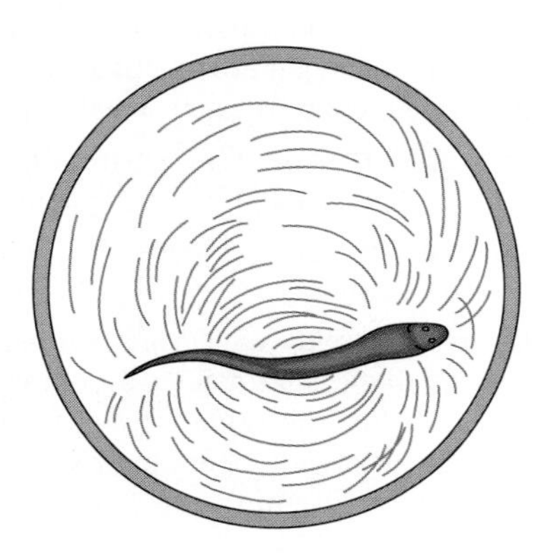

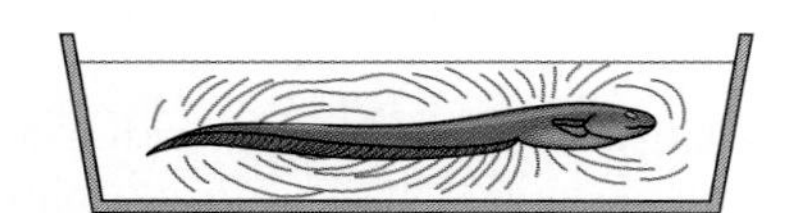

FIGURE 17-17 THE ELECTRIC FIELD OF AN ELECTRIC EEL AS SKETCHED BY MICHAEL FARADAY. HE USED HIS BARE HANDS TO ESTIMATE THE INTENSITIES.

36. Michael Faraday (see Exercise 35) built a metal cage and charged it until sparks flew to objects held nearby. Then he entered the cage and performed experiments to detect electric fields. He found none. Explain.

37. Explain how a neutral molecule or atom can be attracted to a charged particle.

38. If you live in a moist climate, explain why after a stroll across a rug you can summon sparks at a metal doorknob more easily in winter than in summer. If you live in a dry climate, explain why you can summon such sparks at a metal doorknob almost anytime.

39. A lightning bolt brings 25 coulombs of negative charge to earth through a potential difference of 100 million volts in only 0.005 second. What was the power expended?

40. If you have access to a black-and-white television, turn up the brightness and gently lay a piece of notebook paper on the screen. Why does the paper stick to the screen?

41. Calculate the force on a $+1$ C charge that is 10 meters from a -1 C charge. (One coulomb is a *lot* of charge!)

42. Two electrons are 1 meter apart. Calculate the electric force on each. (The charge on an electron is about -1.6×10^{-19} C.)

43. As far as experiments can tell, the negative charge of the electron and the positive charge of the proton balance each other perfectly. True or false?

44. Have you ever bitten down onto an aluminum foil wrapper of some candy or gum so that it made contact with a metal filling in a tooth? The small but sharp shock you felt was from a transfer of charge between two different metals. Discuss.

45. Explain why those movable electrons inside a copper penny that all repel one another don't rush out to the surface of the penny.

46. You should never attempt to free a kite that's snared in a power line by using a long pole. Why not?

47. Some historians of science have claimed that Coulomb's experiment was not accurate enough to prove the law now called Coulomb's law and that the fame should belong to others who verified it at later dates. What fact discussed in this chapter would lead you to believe this change probably will not happen?

48. If a high-voltage power line fell across your car while you were in the car, why should you *not* step out of the car?

49. Hundreds of years ago, sailors watched green, blue, or even violet "fire" crackle along the tall, pointed masts of their wooden ships at night. One of Columbus's sailors wrote of the "ghostly flame which danced among our sails." Explain what must happen to cause this phenomenon, known as "St. Elmo's fire."

50. A flashlight battery has a potential difference of 1.5 volts between its ends. If 0.001 coulomb of charge passes from one end to the other, how much energy does the battery give to those electrons?

51. Airplane pilots sometimes see St. Elmo's fire (see Exercise 49) glowing on the wings and even over the windshields of their aircraft at night. What do you think would be the source of the potential difference that causes this effect on airplanes?

52. A typical lightning bolt brings only 25 coulombs of charge to earth, an amount easily obtained from an ordinary car battery. So what is the source of the enormous amount of energy released by a lightning bolt?

53. A certain television has a potential difference of 30,000 volts to accelerate electrons toward the screen. Assuming an electron starts at rest, calculate its kinetic energy after passing through that potential difference and also its speed. ($q_{electron} = -1.6 \times 10^{-19}$ coulomb; $m_{electron} = 9.1 \times 10^{-31}$ kilogram.) Compare this speed with the speed of light.

Demonstrations

1. On a cold night, go into a dark room and quickly pull your wool or nylon sweater over your head. The snap, crackle, and pop and the flashes of light mean that potential differences of some 15,000 volts were generated by the friction.

2. Take a tray of ice cubes into a closet or a dark room. (Better still, sit in the dark for 5 or 10 minutes to get your night vision and have someone bring you the frozen tray.) Then twist the tray to free the cubes and you will see sparks as the ice breaks away. The bonds holding the solids together are electrical in nature, and when those bonds are broken, electrons are displaced in the solids. When the negative and positive charges get back together, it's like a miniature bolt of lightning—light is given off. Photographers who work in darkrooms see sparks if an ordinary piece of tape is ripped from a metal film canister. The explanation is the same. Crushing a wintergreen Lifesaver in the dark makes these bright flashes as well.

3. Charge a plastic pen or comb by rubbing it briskly against a dry garment on a day with low relative humidity. Then hold it close to a slow stream of water from a faucet. The water is attracted by the charged object. Yet if you tried this with some rubbing alcohol, nothing would happen. The polar molecules of water are responsible for this phenomenon. The plastic's negative charge repels the negative side of the water molecules and attracts the positive side. The tumbling water molecules respond and orient themselves with the negative side away from the plastic pen or comb. The closer, positive side of the molecules is tugged more strongly than the far side is repelled, and the stream of water moves toward the negatively charged object.

4. Take a wool sock and rub it for several minutes against a styrofoam plate. The plate will accumulate a significant charge. Then hold the plate vertically next to a friend's arm. Induced charges will make the hairs on the arm rise. If you slowly move the plate back and forth, the hairs turn to follow the plate, causing the sensation known as *formication*—the feeling of ants crawling on your skin.

EIGHTEEN

Moving Charges

When you turn on a light switch, a stereo, or a hairdrier, charges move and transfer energy. The energy is brought from power plants by high-voltage transmission lines such as these.

Most of the electric charge around us is bound in neutral atoms, and the electric attraction between the nuclei and the electrons is hard to overcome. In metals and in solutions that contain ions, however, some charges are free to move around, as we saw in Chapter 17. When free charges are exposed to an electric field, they move, as Figure 17-16 showed, and when they move, they carry energy. The energy can be substantial even though the mass of the particles is so small and even though they may not move far or fast! Today we use the motions of electrons to transport energy for lighting, heating and cooling, cooking, refrigeration—the list is very long. This chapter and the next two give details of how this is done.

ELECTRIC CURRENT

Acting under the force of gravity, water flows downstream; it naturally moves from points of high gravitational potential energy to points of lower gravitational potential energy (Figure 18-1). If there is a net electric field in a conductor, electrons move from places where they have a high electrical potential energy to places where their electrical potential energy is less. When that happens, we say electrons *flow.* Any flow of charged particles is called an **electric current,** or, more simply, a **current.** The current at some specific point is *the quantity of charge that passes by that point per second*—in other words, the *rate* at which charge moves past a given point. Current has the symbol i, so,

PUTTING IT IN NUMBERS—DEFINING ELECTRIC CURRENT

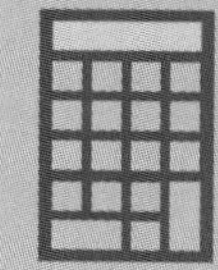

$$i = \frac{q}{t}$$

Current is measured in amperes, or amps (A), where one amp is one coulomb/second. (The current is positive if the flowing charges are positive and negative if the charges are negative, but for our purposes the sign of the current will rarely matter.) **Example:** If 2 coulombs of charge passes through a wire in 10 seconds, what is the current in the wire?

$$i = \frac{q}{t} = \frac{2\text{ C}}{10\text{ s}} = \mathbf{0.2\ amp,\ or\ 0.2\ A.}$$

FIGURE 18-1 THE WATER SPILLING OVER THIS DAM RESPONDS TO THE FORCE OF GRAVITY, GOING FROM POINTS OF HIGHER GRAVITATIONAL POTENTIAL ENERGY TO POINTS OF LOWER GRAVITATIONAL POTENTIAL ENERGY. ELECTRONS MOVING UNDER THE INFLUENCE OF AN ELECTRIC FORCE DO MUCH THE SAME THING, MOVING FROM POINTS OF HIGHER ELECTRICAL POTENTIAL ENERGY TO POINTS OF LOWER ELECTRICAL POTENTIAL ENERGY, WHETHER IN A VACUUM OR IN A CONDUCTING WIRE. (BECAUSE OF BEN FRANKLIN'S UNFORTUNATE CHOICE OF SIGNS, HOWEVER, ELECTRONS ARE PUSHED BACKWARD IN ELECTRIC FIELDS AND FLOW FROM POINTS OF LOWER ELECTRIC POTENTIAL TO POINTS OF HIGHER ELECTRIC POTENTIAL!)

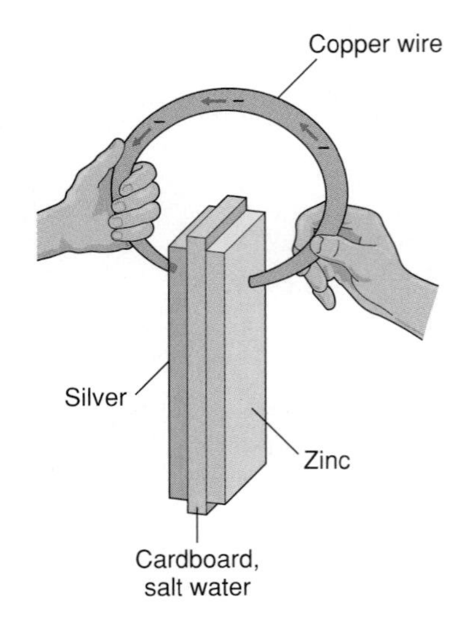

FIGURE 18-2 A SIMPLE ELECTRIC CELL, LIKE THE ONE MADE BY VOLTA IN 1800. WHEN A CONDUCTING WIRE IS CONNECTED TO THE UNLIKE METALS, A CHEMICAL REACTION PROCEEDS IN THE CELL, AND ELECTRONS MOVE THROUGH THE WIRE FROM THE ZINC TO THE SILVER. THE EFFECT OF THIS CURRENT IS JUST THE SAME AS IF POSITIVE CHARGES WERE MOVING FROM THE SILVER TO THE ZINC.

In many ways, electrons in a conducting wire (usually copper or aluminum) act like a liquid in a pipe. Pressure on the water in a pipe doesn't compress the water and change the number of water molecules in a given volume. Similarly, the electrons in a wire are incompressible when an electric field pushes them along. That is, there are just as many electrons in every cubic millimeter of the wire when it is exposed to an electric field as there are when no field is present. If electrons concentrated anywhere, making a portion of the wire negative, they would leave a region of the wire positive somewhere else, and that positive area would strongly attract those electrons, pulling them quickly back! This means that as electrons move through a wire, each cubic millimeter of the wire remains electrically neutral. So when electrons move through a conducting wire, just as many come in at one end as leave at the other—exactly as water flows through a pipe.

MAKING CHARGES MOVE

In 1800 Alessandro Volta sandwiched a strip of cardboard soaked with salt water between a plate of silver and a plate of zinc (Figure 18-2). Chemical reactions between the salt water and the metal plates made the silver plate positive (the chemical reaction at the silver plate *removed* electrons) and the zinc plate negative (the reaction there *released* electrons). When Volta then used a copper wire to connect the two plates, a current was created in the wire. That was the first **chemical electric cell** (or just *cell*). The cell does work on the electrons in the cell, taking them in at the positive plate and giving them up at the negative plate. These electrons then moved from the negative plate through the wire to the positive plate.

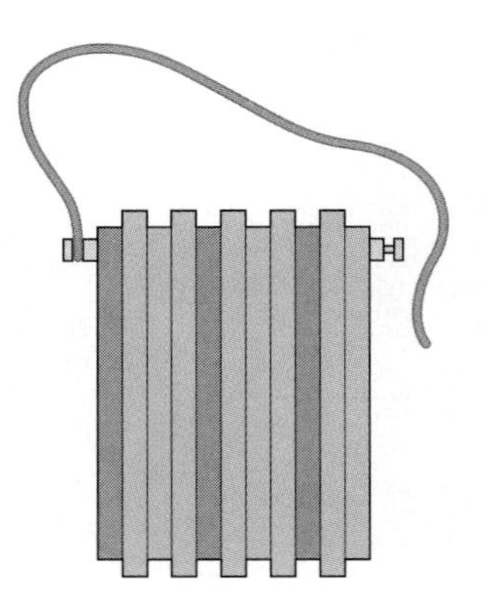

FIGURE 18-3 A SIMPLE BATTERY (A BATTERY OF CELLS). IF EACH CELL CONTRIBUTES 1.5 VOLTS OF ELECTRIC POTENTIAL, THIS BATTERY HAS A TOTAL ELECTRIC POTENTIAL OF 4.5 VOLTS ACROSS ITS ENDS, OR TERMINALS.

Volta then made the first **battery** of cells by stacking identical cells (Figure 18-3). When an electron moves through more than one cell, it has more work done on it. The total change in its electrical potential, V, is greater than in the single cell. So cells connected like those in Volta's battery have a larger potential difference between their outermost plates, or *terminals,* than a single cell does.

An ordinary 12-volt car battery, which contains 6 cells connected in a line, uses plates of lead and lead oxide in sulfuric acid. Each cell contributes about 2 volts to the total potential difference across the terminals of the battery. What most people call a flashlight battery is actually a single cell, the standard-size D cell. When new, it has a potential difference of about 1.5 volts between its terminals. The quantity of chemicals in a D cell is small compared with the quantity in a car battery, so the D cell delivers a smaller amount of electrical energy during its lifetime.

Although the negative electrons do the moving in a battery, the convention is that the current i flows from the positive terminal to the negative terminal, as if positive charges were moving. This is just two ways of saying the same thing.

THE IMPACT OF THE BATTERY

Volta's battery led to a burst of scientific activity with electricity. In 1802 Humphrey Davy, an English chemist, made a battery with 60 pairs of zinc and copper plates. He used the large electric potential to make currents that melted iron and platinum wires and to ignite gold, silver, and lead. By 1807 he had a battery of almost 300 plates, with which he was able to decompose chemical salts. The result was the discovery of potassium and sodium—new chemical elements!

By 1808 Davy had assembled 2000 pairs of plates. With this battery he created brilliant electric arcs and extracted the elements barium, strontium, calcium, and magnesium from compounds. His arc evaporated the hardest substances, even diamond and quartz. Electricity quickly took a front seat in the exploration of the nature of matter.

ARE BATTERIES "ON" ALL THE TIME?

In a fresh battery, two chemical reactions stand ready to proceed whenever they can. At the positive end of each cell, there's a reaction that will take place if electrons are available. At the negative end of each cell, there's a reaction that will occur if the reactants can shed electrons. However, neither reaction can proceed without a mechanism to shed or absorb those electrons. The instant a copper wire is connected between the terminals, a storm of chemical activity inside the battery releases electrons at the negative terminal and takes up electrons at the positive terminal. Electrons are pulled out of the end of the wire and into the positive terminal while other electrons are pushed into the wire at the negative terminal.

A large amount of charge can flow through a copper wire because there are about 10^{22} free electrons per cubic centimeter, about 1 per atom. If the wire between the terminals of a battery is disconnected, the chemical reactions stop—almost. So why do all batteries eventually go dead if left unused? One reason is that some charges flow even without a conducting wire between the terminals. Though air is a poor electrical conductor, each cubic centimeter of air contains a few ions, principally due to cosmic rays that ionize air molecules, and if positive ions strike the negative terminal and negative ions strike the positive terminal, the reactions in the battery creep along.

Under humid conditions, a thin layer of moisture bathes the battery's surface. Then charges can also flow over the surface. Pure water doesn't conduct well, but add a trace of impurities, even salt from fingers or fingerprints, and it's a different story. Ions are very mobile in water, and the battery can slowly discharge through water that has ions in it.

Then, too, the reactions in batteries are corrosive. In time, acids eat through most containers, ruining the cells. Until this happens or until the reactions have completely discharged, however, the battery gives a potential difference all the time, and charge will flow whenever there's a conducting path between its terminals.

RESISTANCE TO AN ELECTRIC CURRENT

Because of their thermal energy, the conduction electrons in a conductor have random motion even when there is no electric field to push them along. In fact, in a conductor at room temperature, the electrons move with an average speed that is much faster than a rifle bullet. Because the motion of the electrons is random, the average current past a point is zero. No more charges go in any one direction than in another. If you connect that conductor across the terminals of a battery, though, you supply an electric field that pushes the electrons from the negative terminal to the positive terminal.

EVEN THOUGH THE ELECTRONS INDIVIDUALLY CREEP ALONG AT A VERY LOW SPEED, THE CURRENT PAST A POINT CAN BE ENORMOUS SINCE EACH CUBIC MILLIMETER OF A COPPER WIRE CONTAINS ABOUT 10^{19} CONDUCTION ELECTRONS.

Confined to move within the wire, the electrons meet with friction: They often crash into vibrating atoms in the wire or even have close approaches to

PHYSICS IN 3D/ELECTRON MOTION IN THE CURRENT IN A CONDUCTING WIRE

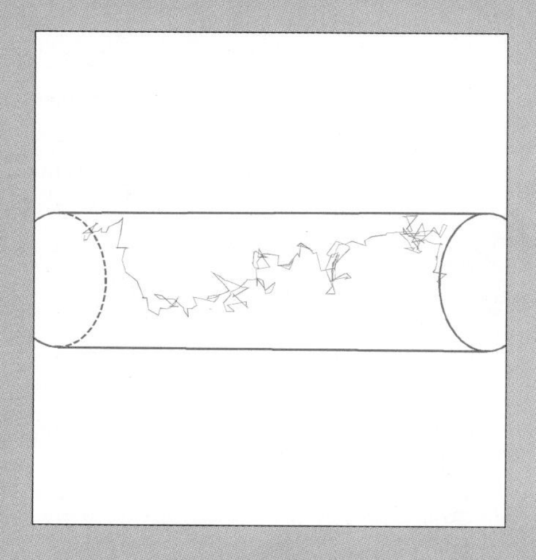

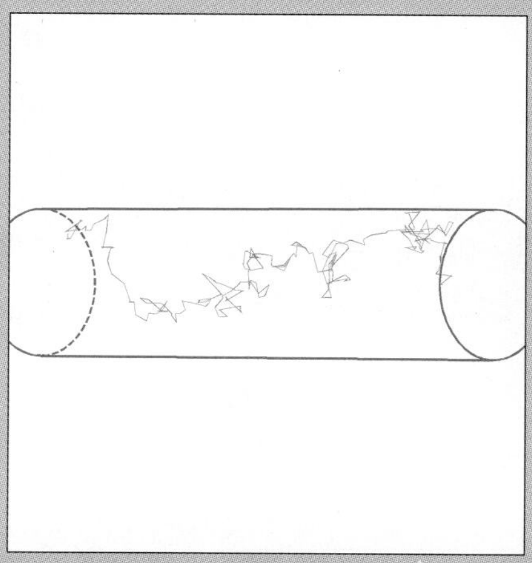

A 3D DRAWING ILLUSTRATING AN ELECTRON MOVING THROUGH A COPPER WIRE. DUE TO NUMEROUS COLLISIONS, AN ELECTRON'S PROGRESS IS SLOW. THE ELECTRONS "DRIFT" ALONG IN THE DIRECTION THEY ARE PUSHED RATHER THAN ACCELERATE TO HIGHER SPEEDS.

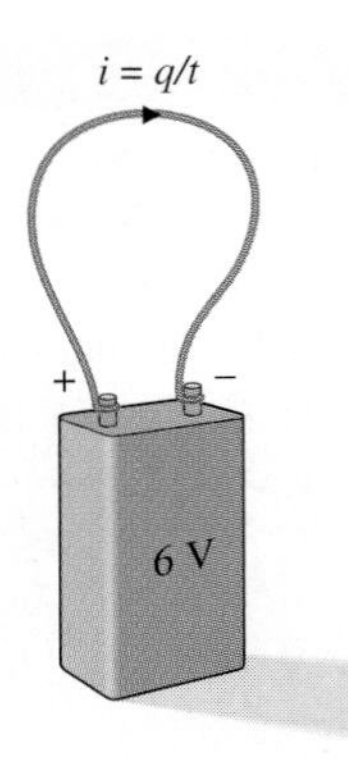

FIGURE 18-4 CHARGE FLOWS THROUGH A COPPER WIRE WHEN IT IS CONNECTED ACROSS AN ELECTRIC POTENTIAL DIFFERENCE. IF A SILVER WIRE OR AN ALUMINUM WIRE OF THE SAME LENGTH AND CROSS SECTION WERE USED INSTEAD OF COPPER, THE CURRENT WOULD BE DIFFERENT DUE TO THE DIFFERENT ATOMIC PROPERTIES OF THOSE METALS AND THEIR DIFFERENT NUMBERS OF CONDUCTION ELECTRONS PER ATOM.

other moving electrons, and such collisions usually change the direction and speed of the colliding electron. On the average, all the electrons move in the direction of the current, but the collisions keep them from gaining much speed. A collision may stop an electron's forward progress entirely or even send it backward against the flow at any time. A typical conduction electron in a copper wire creeps along at a speed of much less than a millimeter a second. We say the wire *resists* the flow of the electrons.

MEASURING ELECTRICAL RESISTANCE: OHM'S LAW

A copper wire connected to the terminals of a battery (Figure 18-4) conducts a certain current, i. Another wire of the same length and diameter but made of silver conducts a larger current from the same battery. A similar aluminum wire draws less current than the copper wire does. Each of these metal wires has a different electrical **resistance** to the flow of charge for two reasons. One, in different conductors a different percentage of electrons are free to move. Two, the distance an electron can travel before a collision changes its direction depends on the atomic properties of the conducting material.

The current in a conducting wire increases if the potential difference between the ends of the wire increases. But the larger the resistance of the wire, the smaller the value of the current for a given voltage across the two ends. That is, current = voltage/resistance, or,

PUTTING IT IN NUMBERS—OHM'S LAW

potential difference = current × resistance

$$V = iR$$

where R is the symbol for the resistance of the conducting path to the current i. The SI unit of resistance is the ohm (Ω). Georg Simon Ohm, 1789–1854, discovered this relationship, often called Ohm's law, while working as a high school math and physics teacher in Germany.

Example: If a potential difference of 6 volts across the ends of a wire draws a current of 2 amperes, what is the resistance of the wire? $R = \frac{V}{i} = \frac{6 \text{ volts}}{2 \text{ amps}} =$ **3 ohms** $= 3\Omega$.

JUST AS RUNNERS TRAVEL AT DIFFERENT SPEEDS ON DIFFERENT SURFACES FOR THE SAME EFFORT, ELECTRONS MOVE THROUGH DIFFERENT CONDUCTING MATERIALS AT DIFFERENT SPEEDS UNDER THE SAME ELECTRIC FIELD.

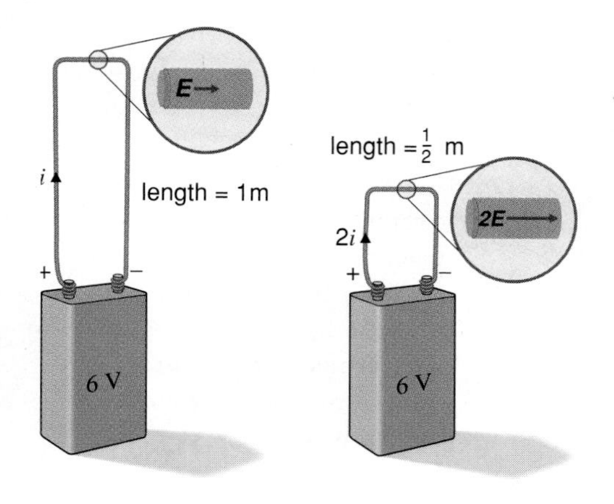

FIGURE 18-5 IDENTICAL BATTERIES WITH DIFFERENT LENGTHS OF THE SAME CONDUCTING WIRE ATTACHED. AN ELECTRON MOVING FROM THE NEGATIVE TERMINAL TO THE POSITIVE TERMINAL OF EITHER BATTERY GETS THE SAME AMOUNT OF ENERGY GIVEN TO IT. SINCE THE FORCE ON THE ELECTRON IS $Q_{ELECTRON} \times E$, AND SINCE WORK = FORCE × DISTANCE, THE LONGER WIRE HAS A SMALLER ELECTRIC FIELD THAN THE SHORTER ONE DOES AND THUS HAS A SMALLER CURRENT.

AS A REMINDER, "VOLTAGE" IS OFTEN USED TO DENOTE ELECTRIC POTENTIAL DIFFERENCE.

A long copper wire connected across the terminals of a battery conducts a certain current. Cut that wire in half, as in Figure 18-5, and one of the shorter lengths connected across the same cell will draw about twice as much current. That is because an electron that is moved from one terminal to the other goes through the same potential difference—hence has the same work (qV) done on it—no matter what distance it must travel. Since $W = F \times d$ and since the work is the same in both cases, a shorter wire means the strength of the electric force qE—and hence the strength of the electric field E—must increase. Since the electric field is greater with the shorter wire, the electrons in that wire have a higher average speed. Therefore, the rate at which charge passes a given point—in other words, the current—is greater in the shorter wire than in the longer one.

A thicker copper wire, one with a larger cross section, offers less resistance than a thinner one of the same length (Figure 18-6). If the same potential difference is applied to each wire, the current is less in the thin wire because there are fewer electrons to move past a point along the wire in a unit of time. As seen from Ohm's law, $V = iR$, if V is the same while i is smaller, the thinner wire has a larger resistance R than the thicker wire does.

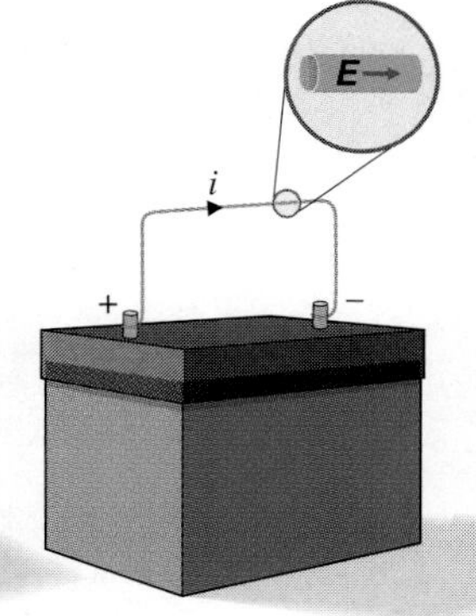

FIGURE 18-6 THICKER CONDUCTING WIRES OF THE SAME LENGTH WITH THE SAME POTENTIAL DIFFERENCE CARRY A LARGER CURRENT THAN SMALLER WIRES. WHILE THE ELECTRIC FIELD INSIDE THEM IS THE SAME, THE THICKER WIRE HAS MORE ELECTRONS THAT CAN PASS A GIVEN POINT IN A SECOND DUE TO ITS GREATER CROSS-SECTIONAL AREA. THIS SITUATION CAN BE COMPARED TO THE RELATIVE FLOWS THROUGH A LARGE-DIAMETER HOSE (SUCH AS FOUND ON FIRE TRUCKS) AND A SMALL GARDEN HOSE, BOTH CONNECTED TO A CITY'S WATER SYSTEM (AT THE SAME WATER PRESSURE).

ELECTRICAL CIRCUITS: DC AND AC

The voltage across the terminals in a car battery can start the engine, light the way, clear the windshield, and bring you music. The voltage across the wires in the walls of a house does an even wider range of tasks. What does the work on the electrons? Energy that comes from the *source* of the voltage. For a battery, it is energy from the chemical reactions. For a home circuit, it is energy from a power plant, which typically comes from burning coal or oil, from water falling at a dam, or from nuclear fission.

To operate an electrical appliance, you need to provide a conducting path for electrons across the potential difference supplied by the battery or wall outlet. Any such conducting path is called an electrical **circuit** (Figure 18-7). A switch lets you turn the appliance on or off. If the switch is *on,* the circuit is complete, or *closed,* and charge flows in the appliance. If the switch is *off,* there is a break in the circuit, and the circuit is *open.* With no conducting path across the potential difference for electrons to move along, no electrons can flow. Notice these terms are backward from what you might expect if you

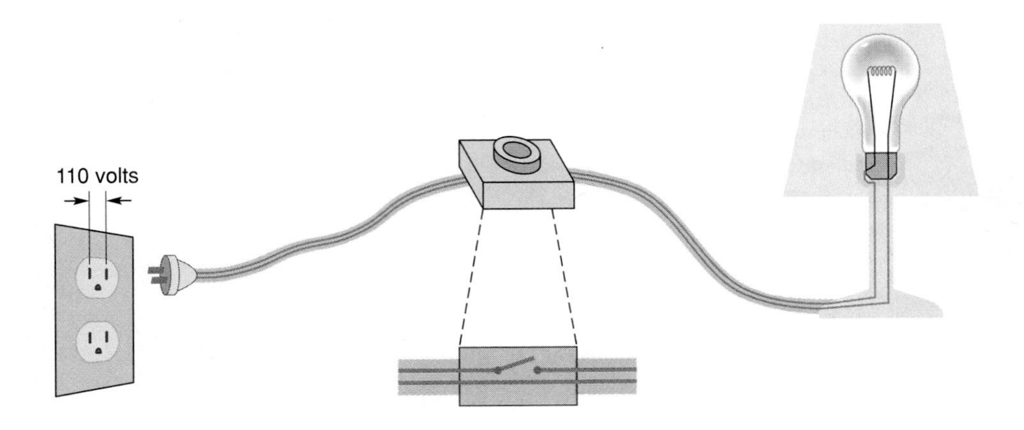

FIGURE 18-7 THE DETAILS OF THE ELECTRIC CIRCUIT OF A LAMP. TWO CONDUCTING WIRES LEAD FROM THE POTENTIAL DIFFERENCE MAINTAINED IN THE HOME CIRCUITS TO OPPOSITE SIDES OF THE LAMP'S FILAMENT. A SWITCH EITHER OPENS OR CLOSES THE CIRCUIT, AS DESIRED.

think of water flowing through pipes and valves to open or close them. If a faucet is open, water flows, but if a circuit is open, charge does not flow.

When current goes in one direction through a circuit, as it does when a battery supplies the voltage, it is called **direct current** (DC). A battery supplies a very steady voltage to a circuit. The current from the electrical outlet in a home, however, is **alternating current** (AC). That's because the voltage that creates the current is not steady; instead, it oscillates, changing the direction of the electric force first one way, then the other. The reason has to do with efficiency, and we'll see why in Chapter 20.

The alternating current used by power companies throughout North America is a 60-cycle-per-second current. Responding to the alternating electric field in a wire, electrons sway back and forth in a very short distance 60 times every second, more quickly than your eye could follow. So when an appliance is connected to a home outlet, the electrons in the appliance just oscillate, swinging back and forth under an average potential difference of 110 to 120 volts.

ELECTRICAL POWER

When a charge of q coulombs travels through a potential difference of V volts, the energy given to the charge is q (coulombs) times V (volts), or qV joules (from Chapter 17). That is, electrical energy $= qV$. To operate an electrical appliance requires not just energy, but also a certain amount of electrical power, P, which is the rate at which energy is delivered, energy/time. So for electrical power,

PUTTING IT IN NUMBERS—CALCULATING ELECTRIC POWER

$$\textbf{power} = P = \frac{qV}{t} = \frac{q}{t}V = \textbf{current} \times \textbf{voltage, or}$$

$$P = iV$$

Power is measured in joules per second, or watts. **Example:** A 60-watt light bulb gets 60 watts of power when it is connected to the potential difference of 110 volts in a home's electrical wiring. What is the current passing through the bulb? $i = \frac{P}{V} = \frac{60 \text{ watts}}{110 \text{ volts}} =$ **0.54 amp.**

ELECTRICAL HEATING

A current of electrons moves along a copper wire that's connected between the terminals of a battery. The electrons continually run into copper atoms in the wire, and each atom that is struck gets a quick push that, on the average, causes it to vibrate faster. Faster vibration of those atoms means *the temperature of the wire goes up when there is a current through it.* If more current is pushed through a conductor, the rate of heating increases. For that reason a short copper wire connected to the terminals of a battery gives off more heat than a long copper wire does. The short wire has less resistance, so it draws more current from the battery, which has a fixed potential difference between its terminals.

A toaster is connected to the 110-volt potential difference in a wall socket by an electrical cord. Inside the cord are two copper wires, kept apart by insulating material. Each is connected to a metal prong in the plug. These copper wires lead to opposite ends of a strip of wire inside the toaster, and this toaster wire has a much higher resistance than the wires in the cord do. The amount of charge flowing through the high-resistance wire is exactly the same as the amount flowing through the wires in the cord, but because the cord's wires have less resistance than the wire inside the toaster, much less heating occurs in them. Essentially all the energy the toaster circuit uses goes into heating the wire in the toaster.

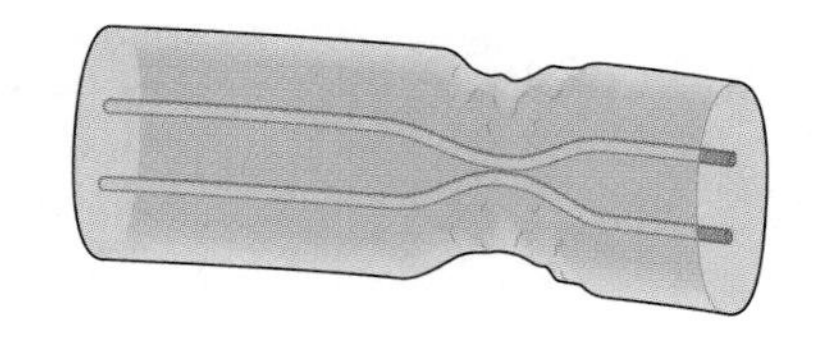

FIGURE 18-8 AN ELECTRICAL CORD THAT HAS BEEN SEVERELY COMPRESSED MAY CAUSE A SHORT CIRCUIT. RATHER THAN SPENDING ELECTRICAL ENERGY IN THE APPLIANCE ATTACHED TO THE CORD, THE CHARGES CAN PASS THROUGH THE POINT WHERE THE TWO WIRES TOUCH, A POINT OF VERY LITTLE RESISTANCE. THIS IS CALLED A SHORT CIRCUIT, AND THE CURRENT THROUGH THE SHORT CIRCUIT WILL BE LARGE ENOUGH TO CAUSE DRASTIC OVERHEATING.

Suppose the insulation in the cord becomes damaged, and the two wires, which are at different electric potentials, touch while the toaster is on, as in Figure 18-8. The new path for the charges, which passes through that point of contact, offers almost no resistance. Most of the moving electrons would go that way. $V = iR$ tells us that if R gets very small while V remains constant (110 volts), the current in the wires that touch must increase. As the current goes up, the rate of heating increases, and that heat can cause a fire. Such a place of low resistance in a circuit is called a **short circuit,** or just a **short.**

A safety feature—either a fuse or a circuit breaker—protects wires from overheating in case of a short circuit or whenever too many appliances are connected to a circuit, increasing the current to unsafe levels and overheating the wires in the wall. In older homes a meltable fuse encased in glass is part of each circuit (Figure 18-9). If an appliance shorts out, that circuit draws a large current. In a matter of seconds, the fuse melts (at a predetermined current noted on the fuse) and breaks the circuit before the short can cause a fire. The dangerous short should be repaired before the melted fuse is replaced.

FIGURE 18-9 A FUSE LIMITS THE AMOUNT OF CURRENT THAT A CIRCUIT CAN CARRY. A METAL STRIP THAT HAS A LOW MELTING POINT WILL MELT IF THE CIRCUIT CARRIES A CURRENT GREATER THAN THAT MARKED ON THE FUSE. ONCE THAT STRIP HAS MELTED, THE CIRCUIT IS BROKEN AND AN OVERLOAD IS PREVENTED.

Newer buildings use circuit breakers, which are bimetallic strips that curl up when they get hot from carrying too much current. They work the same way as the bimetallic thermostats described in Chapter 12. As a strip overheats, it curls and snaps away from a contact to break the circuit. Breakers work a little faster than fuses, so they are safer. And once an overload is corrected and the open bimetallic strip cools, you only need to flip a switch to reset the breaker.

SERIES CIRCUITS

With a few small lengths of copper wire, a couple of flashlight cells, and some flashlight bulbs, you can learn a great deal about how electrical circuits perform. For example, make the two arrangements shown in Figure 18-10(b) and (c) and touch the loose wire to the end of the bulb in each. You will see that the

THE LOWEST RESISTANCE: SUPERCONDUCTIVITY

In 1911 the Dutch physicist Heike Kamerlingh Onnes (1853–1926) cooled small samples of metals to several kelvins and measured their electrical resistance as they cooled. Onnes discovered to his surprise that mercury's electrical resistance *vanished* at 4.2 K. Many other elements and compounds show similar behavior at some low temperature, called their *critical temperature*. Such materials are called *superconductors*. With no electrical resistance, a current passes through a superconductor without losing energy; no heat loss occurs when charges flow. This makes superconductors the most efficient conductors possible.

All superconductors known before 1986 required cooling to extremely low temperatures. Because low temperatures are difficult to achieve (attaining them requires expensive machinery), everyday use of superconductivity was impractical. In 1986, though, physicists Karl Muller and Johannes Bednorz, working for IBM in Germany, created a ceramic compound containing small amounts of lanthanum and barium, which became a superconductor at 35 K. This discovery sparked intense activity among physicists who specialize in the behavior of solids. Since 1987 new compounds have been made with higher and higher critical temperatures. There is speculation that superconductors that work at room temperatures may soon be fabricated, a development that could revolutionize many of the electrical devices we use today.

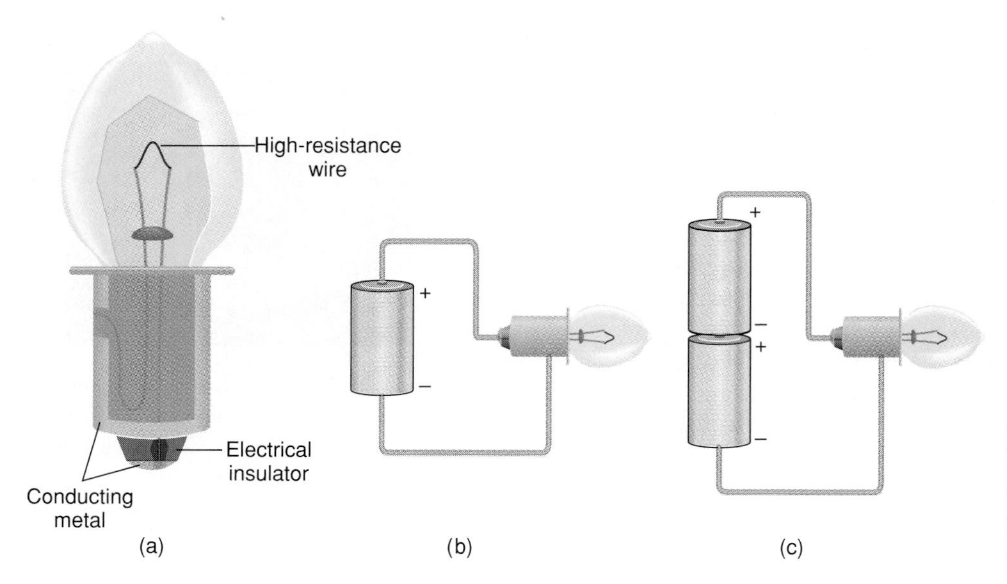

FIGURE 18-10 (A) THE DETAILS OF AN ORDINARY FLASHLIGHT BULB. (B) A BULB LIGHTED BY ONE CELL. (C) A BULB LIGHTED BY TWO CELLS CONNECTED IN SERIES.

FIGURE 18-11 A SINGLE CELL USED TO LIGHT (A) ONE BULB AND (B) THREE BULBS. THE BULBS IN (B) GLOW DIMLY COMPARED WITH THE BULB IN (A). THE BULB IN (A) GETS THREE TIMES AS MUCH CURRENT AS THE (B) BULBS DO SINCE THE (A) CIRCUIT HAS ONLY 1/3 AS MUCH RESISTANCE.

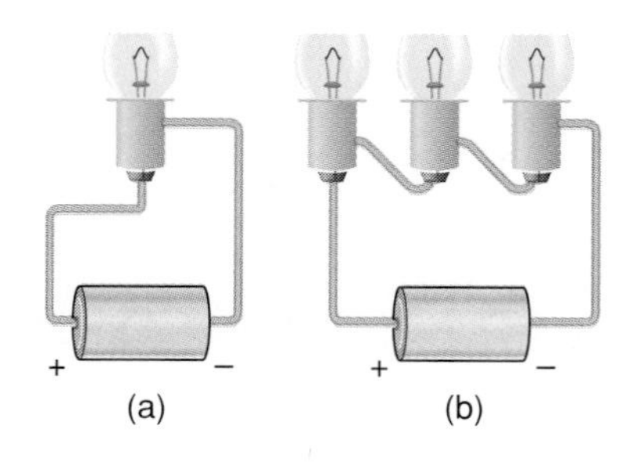

bulb glows more brightly in circuit (c) than in (b). That's because circuit (c) uses the potential difference of two cells—3 volts rather than 1.5 volts. Two cells connected as in circuit (c) are said to be in **series.** In a standard flashlight, the cells are in series, as are the cells in a car battery. When the parts of a circuit are in series, there is only one path for the electrons to take. Since an electron must go through both cells to move through the circuit, it gets a total potential difference of 3V during each time it goes around the circuit.

In both circuits shown in Figure 18-10, essentially all the energy of each electron goes into thermal energy in the bulb's filament, the only place of high resistance in the circuit. The hotter the filament, the more it glows. For an even brighter light, use three or four cells in series. Use a heavy-duty bulb, though, or else you'll melt the filament in a second or two.

Next consider the two arrangements shown in Figure 18-11. The bulb in circuit (a) glows most brightly. Why? To answer this question, let's recount how a bulb glows. The filament offers much more resistance (typically about 10 ohms) to the current than the wires (typically about 0.1 ohm) do. The heating that occurs when the current passes through the high-resistance filament makes a bulb glow. Both circuits get the same voltage but the single

PHYSICS IN 3D/A SIMPLE SERIES CIRCUIT

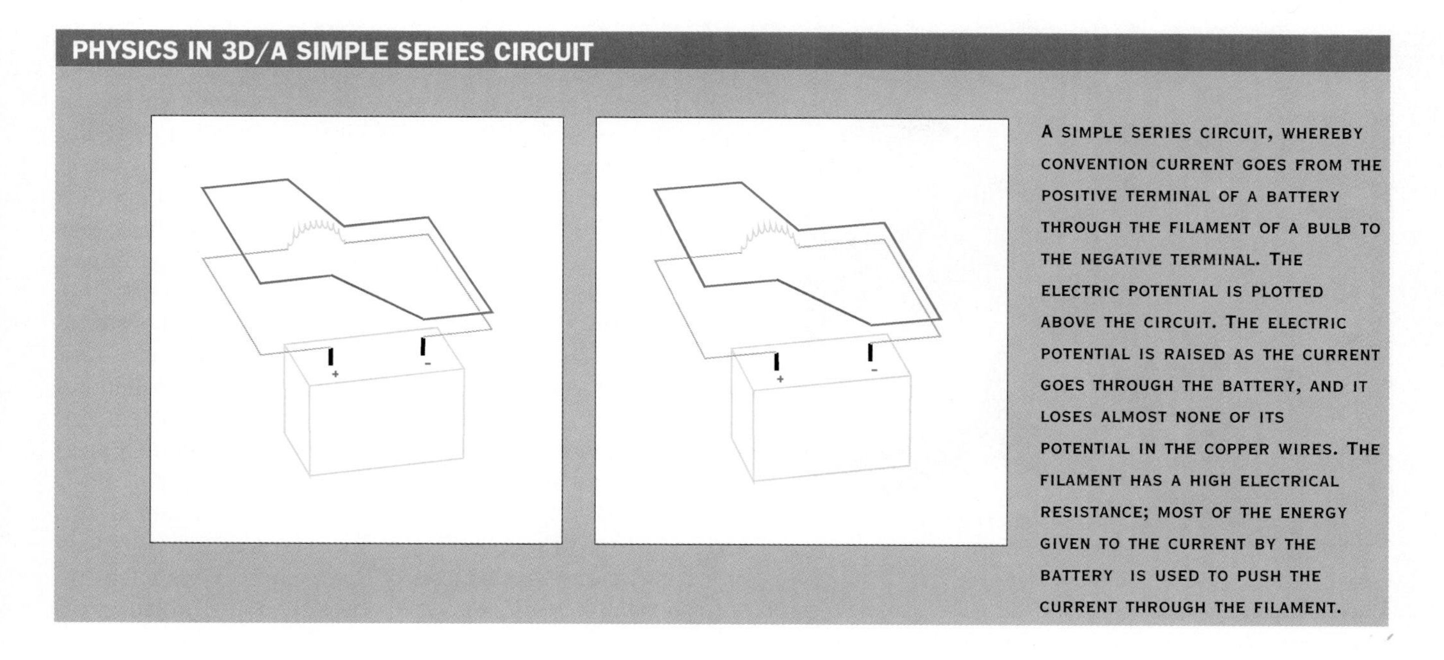

A SIMPLE SERIES CIRCUIT, WHEREBY CONVENTION CURRENT GOES FROM THE POSITIVE TERMINAL OF A BATTERY THROUGH THE FILAMENT OF A BULB TO THE NEGATIVE TERMINAL. THE ELECTRIC POTENTIAL IS PLOTTED ABOVE THE CIRCUIT. THE ELECTRIC POTENTIAL IS RAISED AS THE CURRENT GOES THROUGH THE BATTERY, AND IT LOSES ALMOST NONE OF ITS POTENTIAL IN THE COPPER WIRES. THE FILAMENT HAS A HIGH ELECTRICAL RESISTANCE; MOST OF THE ENERGY GIVEN TO THE CURRENT BY THE BATTERY IS USED TO PUSH THE CURRENT THROUGH THE FILAMENT.

bulb in (a) glows most brightly because there is less resistance in the circuit (hence a larger current, $V = iR$). The bulbs in circuit (b) glow more dimly. The three bulbs offer about three times the total resistance of (a). This means the electrons in (b) each lose about one-third of the cell's potential difference as they move through each bulb, and the current in the 3-bulb circuit is 1/3 of the current in the 1-bulb circuit.

The circuits in Figure 18-11 are series circuits just as those in Figure 18-10 are, with only one path for the electrons to take. Given time, each charge moves through every bulb and cell in the circuit. When the filament of one bulb burns out, the circuit is broken and all the bulbs go dark.

ENERGY IS USED TO PUSH A CURRENT THROUGH A CIRCUIT WITH A RESISTANCE. WHATEVER SUPPLIES THE VOLTAGE TO DO THIS ALSO SUPPLIES THE ENERGY. IF THE VOLTAGE SOURCE IS CONSTANT, AS WITH A FRESH BATTERY, THE CURRENT IS CONSTANT, AND SO IS THE RATE OF USING ENERGY.

PARALLEL CIRCUITS

A **parallel circuit** is a more useful way of connecting electrical appliances to a voltage source (Figure 18-12). Wires lead from a voltage source to the appliances. Each appliance is then connected directly to the (+) and (−) wires. In such a circuit, there is *more than one path* that electrons can take through the circuit.

Parallel circuits have two distinct advantages over series circuits. You can disconnect one appliance without interrupting the current through the others, and every appliance that's connected taps the full voltage of the source. No

IN PARALLEL CIRCUITS, THE VOLTAGE ACROSS EACH APPLIANCE IS THE SAME. BUT IF THE APPLIANCES HAVE DIFFERENT RESISTANCES, THE CURRENTS THROUGH THEM WILL BE DIFFERENT.

FIGURE 18-12 A CAR'S CIRCUIT IS A PARALLEL CIRCUIT, AN ARRANGEMENT THAT ENABLES EACH ELECTRICAL DEVICE TO USE THE FULL VOLTAGE OF THE BATTERY. ALSO, EACH DEVICE CAN OPERATE EVEN WHEN THE OTHERS ARE TURNED OFF, UNLIKE THE CASE FOR A SERIES CIRCUIT.

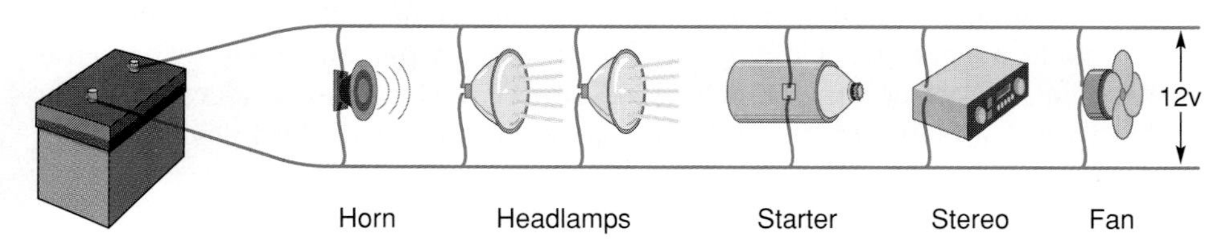

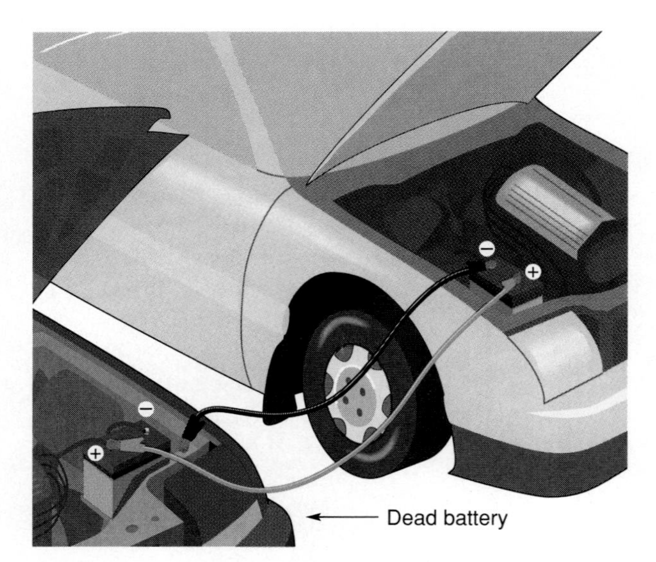

FIGURE 18-13 A PARALLEL CIRCUIT. WHEN JUMP-STARTING A CAR, ALWAYS CONNECT THE POSITIVE TERMINALS OF THE TWO BATTERIES TOGETHER FIRST AND THEN CONNECT THE NEGATIVE TERMINAL OF THE FRESH BATTERY TO THE METAL CHASSIS OF THE CAR WITH THE WEAK BATTERY.

matter which route a particular electron takes, it uses the full electric potential of the source in its trip. This avoids the loss of current, and hence of power, that occurs when additional appliances are placed in a series circuit. Household circuits are parallel circuits.

If you ever jump-start a car, you need to connect another similar battery (fully charged) to the weak one with jumper cables, which are thick copper wires with strong grips to make the contacts, as in Figure 18-13. The fresh battery then lets the starter draw the current it needs. Don't connect the batteries in series! That would increase the voltage difference to a level the car's electrical circuits aren't designed to handle. Connecting the fresh battery in parallel with the weak battery gives the proper voltage to the starter since electrons completing the circuit can go through only one battery. The fresh battery just boosts the capacity for delivering current to the starter. So here's the rule: First connect the positive terminals on the two batteries with one of the cables. Then, since the negative terminal of the weak battery is connected to the engine block, you can connect the cable coming from the negative terminal of the fresh battery to the engine block or the chassis of the car containing the weak battery. That avoids making a spark at the top of the battery, which could be dangerous if there is hydrogen gas emerging from the weak battery.

ELECTRIC SHOCK

CAR BATTERIES ARE RECHARGED BY CONNECTING THEM TO AN EXTERNAL VOLTAGE (FROM THE CAR'S ALTERNATOR) THAT FORCES THE CHEMICAL REACTIONS TO GO BACKWARD. THAT IS, THE CHEMICALS ARE FOR THE MOST PART RETURNED TO THEIR ORIGINAL STATES AND CAN THEN GENERATE CURRENT AS BEFORE.

Figure 18-14 is a diagram of a typical household parallel circuit. The third prong on the electrical plugs of many appliances leads to the grounding device of the home, denoted here by the ⏚ symbol to the left of each socket. This is for the user's protection. Should the insulation around a "live" wire in an appliance fail, the wire might touch a conducting part of the appliance. Someone who then touched the appliance could provide a conducting path to the ground, and charge would flow through that person's body. A ground wire provides a lower-resistance conducting path for the current, thereby protecting the user from a severe shock. Likewise, when the appliance is grounded, any static charge that might build up from friction as the appliance operates goes to the ground, preventing a surprise for the person who touches the device.

An electric shock from a household circuit can be deadly. The damage is done not by the voltage but by the current through the body, which depends on the body's electrical resistance. Dry skin is a good electrical insulator. When the skin is wet, however, its resistance drops to as little as 1/1000 of its value when dry. Inside the body, fluids contain many ions and conduct electricity well; it is there that serious damage can be done. The heart muscles, the muscles used in breathing, and the nerve cells all operate with tiny electrical currents, and disruptions of the normal levels can cause death.

If a few hundredths of an amp passes through a person, muscles tighten involuntarily. At one-tenth of an amp, the heart fibrillates, which means it contracts irregularly. A 6-amp current through a person's chest causes the heart muscles to contract and "freeze." However, if a person's heart is already fibrillating, a high current might be just what is needed. With devices called

ELECTRICAL ENERGY USE IN THE UNITED STATES

On the average, the daily electrical energy consumption of each man, woman, and child in the United States is more than the energy an adult derives from the normal daily intake of food. It is as if there were a silent genie, channeled invisibly through thin copper wires in the walls where you live, at your command 24 hours a day. If the total electrical energy used annually in the United States is divided by the population, that energy amounts to eight times the food energy an adult needs during a year. So each of us benefits in goods and services from the energy equivalent of eight adults laboring full time. People in the United States are used to the comfort and convenience of this energy, but about two-thirds of the world's family dwellings don't have electricity at all.

Physics and Environment

ENERGY USE IN THE UNITED STATES

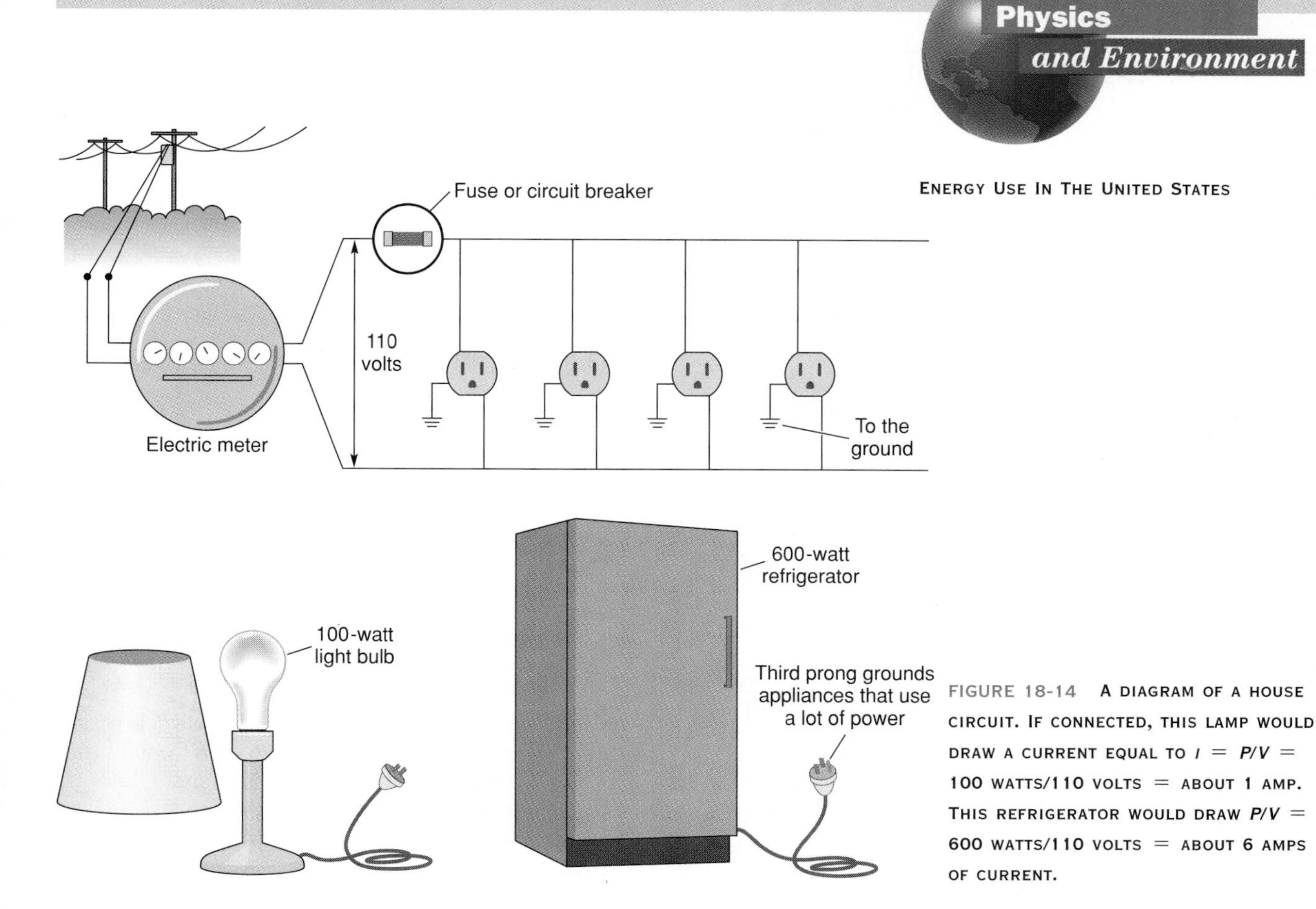

FIGURE 18-14 A DIAGRAM OF A HOUSE CIRCUIT. IF CONNECTED, THIS LAMP WOULD DRAW A CURRENT EQUAL TO $I = P/V =$ 100 WATTS/110 VOLTS = ABOUT 1 AMP. THIS REFRIGERATOR WOULD DRAW $P/V =$ 600 WATTS/110 VOLTS = ABOUT 6 AMPS OF CURRENT.

defibrillators, a large amount of current is quickly passed through the heart by electrodes placed on the person's chest. The heart freezes for an instant and often resumes pumping normally after that.

Controlled shocks are put to use in pacemakers. These devices pulse a tiny current 60 to 80 times a minute to induce the heart to beat at that rate when the natural electrical signals become inadequate for some reason. For example, with one heart condition, the heart pumps at a reduced rate of about 30 times a minute. While the patient won't die because of that level of pumping, often no physical exertion is possible. Pacemakers restore the heartbeat to normal, allowing these patients to lead normal lives.

Summary and Key Terms

electric current–265
amperes–265
chemical electric cell–266
battery–266
electrical resistance–268
ohms–268
Ohm's law–268
direct current–270
alternating current–270
short circuit–271
series circuit–272
parallel circuit–273

Any flow of charged particles, either positive or negative, is an ***electric current***. Current is equal to the *rate* at which charge moves past a location, or $i = q/t$, and it is measured in coulombs/second, or ***amperes***. In a ***chemical electric cell***, reactions take up electrons at one terminal, leaving it positive, while reactions at the other terminal release electrons, leaving it negative. Such a cell is a source of electrical potential difference, or voltage. A ***battery*** is a group of cells joined to produce a voltage that is equal to the sum of the voltages of the individual cells. A conductor has ***electrical resistance***, measured in ***ohms***. ***Ohm's law*** states *potential difference* = *current* × *resistance,* or $V = iR$.

Direct current, like that of a typical cell, flows in only one direction through a circuit. ***Alternating current***, like that drawn from a home outlet, oscillates back and forth rapidly so the electrons never move far. When an electrical circuit is closed, charge flows; when it is open, no charge can flow.

Electric energy = *charge* × *potential difference,* and electric power is the rate at which that energy is delivered. Power, measured in joules/second, or watts, is equal to *current* × *potential difference,* or $P = iV$. As current increases in a conductor, heating increases. A ***short circuit*** is a path of too-low resistance in a circuit that causes it to grow hotter.

A ***series circuit***, like that of a flashlight, provides only one path for a current. A ***parallel circuit***, like those in homes, offers multiple paths for a current. In a parallel circuit, you can disconnect a bulb or any other appliance without breaking the circuit because each appliance is connected directly to the voltage source. In a parallel circuit each appliance taps the full voltage of the source.

EXERCISES

Concept Checks: Answer true or false

1. The quantity of charge that passes a point multiplied by the time is a current.

2. When charge flows through a conducting wire that is connected to a battery, the flow is the result of chemical reactions inside the battery.

3. A conducting wire offers resistance to a flow of electrons because electrons repel each other in the wire.

4. Ohm's law relates the power used in a circuit to the current and the potential difference.

5. The conducting path electrons can take between a supplied potential difference is called an electrical circuit.

6. Direct current and alternating current are both due to steady voltages.

7. Electrical energy is given in terms of the electric potential difference and the charge that goes through that potential difference. It is measured in watts.

8. The temperature of a wire increases when there is a current through it unless the wire is superconducting.

9. A short in a circuit is a point of high resistance.

10. A fuse or a circuit breaker prevents short circuits that could be dangerous in a home.

11. A series circuit has only one conducting path for the electrons; a parallel circuit has multiple conducting paths.

Applying the Concepts

12. A current of 1 ampere can mean (1) there are lots of charged particles per unit volume moving slowly or (2) there are a few charged particles per unit volume moving quickly. True or false?

13. Explain how turning on a flashlight is like turning on the faucet at a sink. (Electricity on tap?)

14. What does an electric current always transport from one place to another? (Careful.)

15. Can you compare drinking a milk shake with two straws rather than one to using a thicker wire rather than a thinner one to draw current from a battery? Discuss.

16. What do you think would happen if someone dropped a

large metal screwdriver across the terminals of a car battery?

17. When you turn on an electrical appliance connected to a wall socket in a room, do any electrons come in or leave from the appliance? Discuss.

18. Two points on an object are at different electric potentials. Does charge necessarily flow between them? Discuss.

19. How does it help to use 220 volts rather than 110 volts to operate an electric clothes dryer?

20. If you made a monstrous flashlight by connecting 73 flashlight cells in series, each of 1.5 volts, what sort of bulb should you use?

21. An electric clock that requires 4 watts to operate uses ______ joules per second.

22. In operating rooms where flammable anesthetics are used, the floors are made of conducting material and the doctors and nurses wear conducting footwear. Why?

23. Most flashlights use several cells in series, so when you buy the packages of cells in a store you are really buying a battery kit. True or false?

24. Explain why it is especially important to have those three prongs (and wires) on appliance cords at any location where electricity could easily pass to the ground through a person.

25. Would you think that a fuse would always be hotter than the conducting wires of the circuit it protects? Discuss.

26. What is an open electric circuit? What is a closed electric circuit?

27. An electric drill operated on a long extension cord will not rotate as fast as one operated on a short cord. Explain.

28. Show that 3600 coulombs pass a point if a current of one amp flows past for one hour.

29. Sometimes people refer to an electric wall receptacle as an "outlet." Why is this not an accurate name?

30. Many chemical reactions proceed much more slowly as the temperature drops, since the atoms or molecules in the reacting substances are then moving more slowly and colliding with less impact. How is this fact connected to the sluggishness of a car's starter on a very cold morning?

31. From your experience in watching cars on the roads at night, are automobile headlamps wired in series or in parallel?

32. A once-frequent cause of electrical fires was a homeowner putting a penny behind a fuse that had burned out to get the electricity restored to the circuits. Discuss.

33. If 6 coulombs of charge passes a point in 1/10 second, what is the current in amps?

34. An example in this chapter showed that a 60-W light bulb operating with a voltage of 110 V draws a current of 0.54 amp. What is the resistance of that light bulb?

35. What is the difference between a cell and a battery?

36. The 60-cycle-per-second alternating current stops instantaneously as it reverses direction. Why don't you notice a flickering of an incandescent lamp as this reversal takes place? (You *can* see this with fluorescent lamps. Swish your hand back and forth under an isolated street lamp!)

37. Why will a small copper wire melt if connected between the wires in a home electrical socket but not if connected across the terminals of a flashlight battery?

38. Why will a 3-volt flashlight bulb quickly burn out if used with a 6-volt flashlight battery?

39. What does a fuse in an electric circuit limit, the potential difference or the current?

40. In the United States, accidental electrocutions occur at the rate of about 1000 per year. Many of these involve electrical appliances placed on the edge of a bathtub, on a shelf over a bathtub, or in a shower stall. Discuss the dangers there.

41. Compare the current used by a 1-watt flashlight bulb that runs on 3 volts and a 40-watt light bulb that runs on 115 volts. Discuss your answer.

42. If a television set uses 300 watts, how much current does it draw?

43. Discuss this statement: "Power companies get all their electrons back, so they should just charge you a rental fee."

44. In 60-Hz alternating current, how many times per second does an electron change its direction? (Careful!)

45. What purpose is served by connecting batteries in series? In parallel?

46. A certain flashlight can use a 10-ohm bulb or a 5-ohm bulb. Which bulb should be used to get the brighter light? Which bulb will discharge the battery first?

47. If a conductor is heated, its resistance usually increases, since the vibrating atoms interfere more with the electrons' motions. In the wire of an extension cord, this heating doesn't matter so much. Think of several cases where a change in the resistance due to an increase in temperature could make a difference.

48. If the copper wires in the electrical cord are electrically neutral even when charge is flowing, why can you get a shock by touching one of the wires if it is bare?

49. A student does a week's wash in an evening. The washing machine works at 300 watts, the electric drier at 700 watts, and the iron at 500 watts. The 100-watt bulb used to see by while ironing was on for 2 hours, the washing machine ran for 1 hour, and the dryer was on for 2 hours. At 10 cents per kilowatt-hour, what was the cost?

50. A 500-watt toaster, a 300-watt coffeemaker, and a 75-watt lamp all operate on the same circuit. When a 400-watt waffle iron is connected to the same circuit, the fuse melts. About what was the amp rating of the fuse?

51. How many amps would flow through a household circuit if the following appliances were being used at the same time: two irons (550 watts each), a hair dryer (1300 watts), and a vacuum cleaner (600 watts)? Would a 30-amp circuit breaker break the circuit under this load?

52. An adult's daily food energy intake is about 3000 Calories, which is equal to about 3.5 kilowatt-hours of energy. Show that if you leave a 100-watt bulb and a 60-watt bulb on for 24 hours, they use more (electrical) energy than an adult uses in the same period.

53. 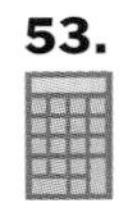 Look up the utility charge per kilowatt-hour in your area (use anyone's electric bill). Then see if electrical energy is cheaper than food energy. In other words, could you feed yourself for a day on the amount 3.5 kWh (3000 Calories) costs?

Demonstrations

1. You can demonstrate the action of a fuse or a short circuit with a flashlight cell. From a pad of steel wool, select the thinnest strand you can find. Then attach two copper wires to the cell, one at each terminal. Use the strand of steel wool to complete the circuit. If you touch the copper wires to the steel strand at points about a centimeter apart, the strand gets warm. If you move the wires closer together along the strand, it gets really hot. A smaller length of the strand offers less resistance, so the current increases. If the points of contact of the copper wires come to within a millimeter or so of each other, the steel will glow white hot like the filament of a light bulb and melt, breaking the circuit.

2. Figure 18-15*a* shows a meter that measures the electrical energy used at a home or apartment. This meter counts the energy consumed in kilowatt-hours. There are five dials (sometimes only four) on an electric meter, (Figure 18.15*b*) one for each digit. The five digits, recorded from left to right, indicate the number of kilowatt-hours used since the meter was last reset to zero. To find the number of kilowatt-hours used during an intermediate period of time, two readings must be taken, and the numbers subtracted. The difference between them is the number of kilowatt-hours used between the dates of the two readings. The interval between readings (often 30 days) and the power company's charge per kilowatt-hour (typically 7 cents to 12 cents per kilowatt-hour in 1995) are indicated on the monthly bill. The cost per kilowatt-hour multiplied by the number of kilowatt-hours used is an estimate of the amount due.

3. A rheostat is a mechanical device that lets someone vary the resistance in a circuit with only the turn of a knob. As the resistance changes, so does the current. Rheostats are commonly used in dimmer switches for home lighting, for the house lights in movie houses and theaters, and in radios and televisions for controlling the volume of the sound. You can make a rheostat with a flashlight cell, a number 2 pencil, and some copper wire. Split the pencil into halves lengthwise to expose the graphite, which is a conductor. Connect a circuit as shown in Figure 18-16. By varying the contact points to use more of the graphite rod in the circuit, you increase the resistance and decrease the current, as you can see by the dimming of the bulb.

(a)

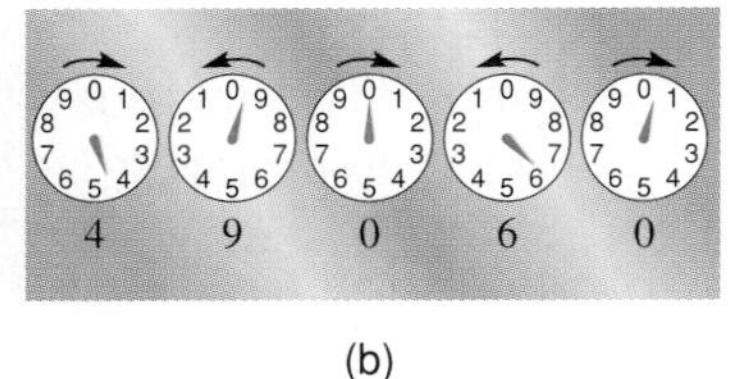

(b)

FIGURE 18-15 (A) A FOUR-FIGURE ELECTRIC METER. (B) THE READING ON THIS FIVE-FIGURE ELECTRIC METER IS 49,060 KWH. TO FIND THE ENERGY CONSUMED BETWEEN TWO DATES, YOU MUST RECORD THE METER READING ON EACH DATE AND SUBTRACT THE FIRST READING FROM THE SECOND.

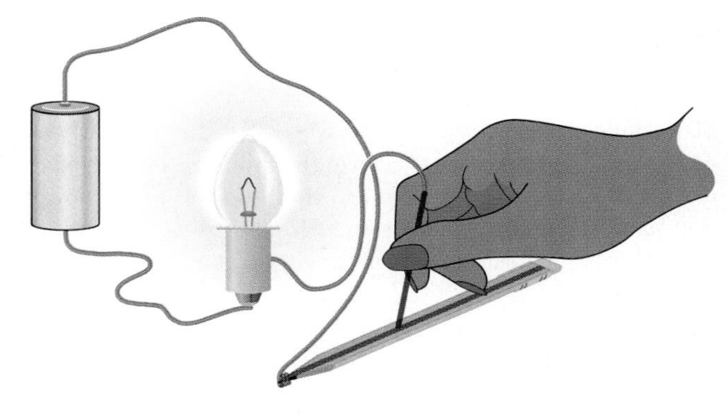

FIGURE 18-16

NINETEEN

Magnetism

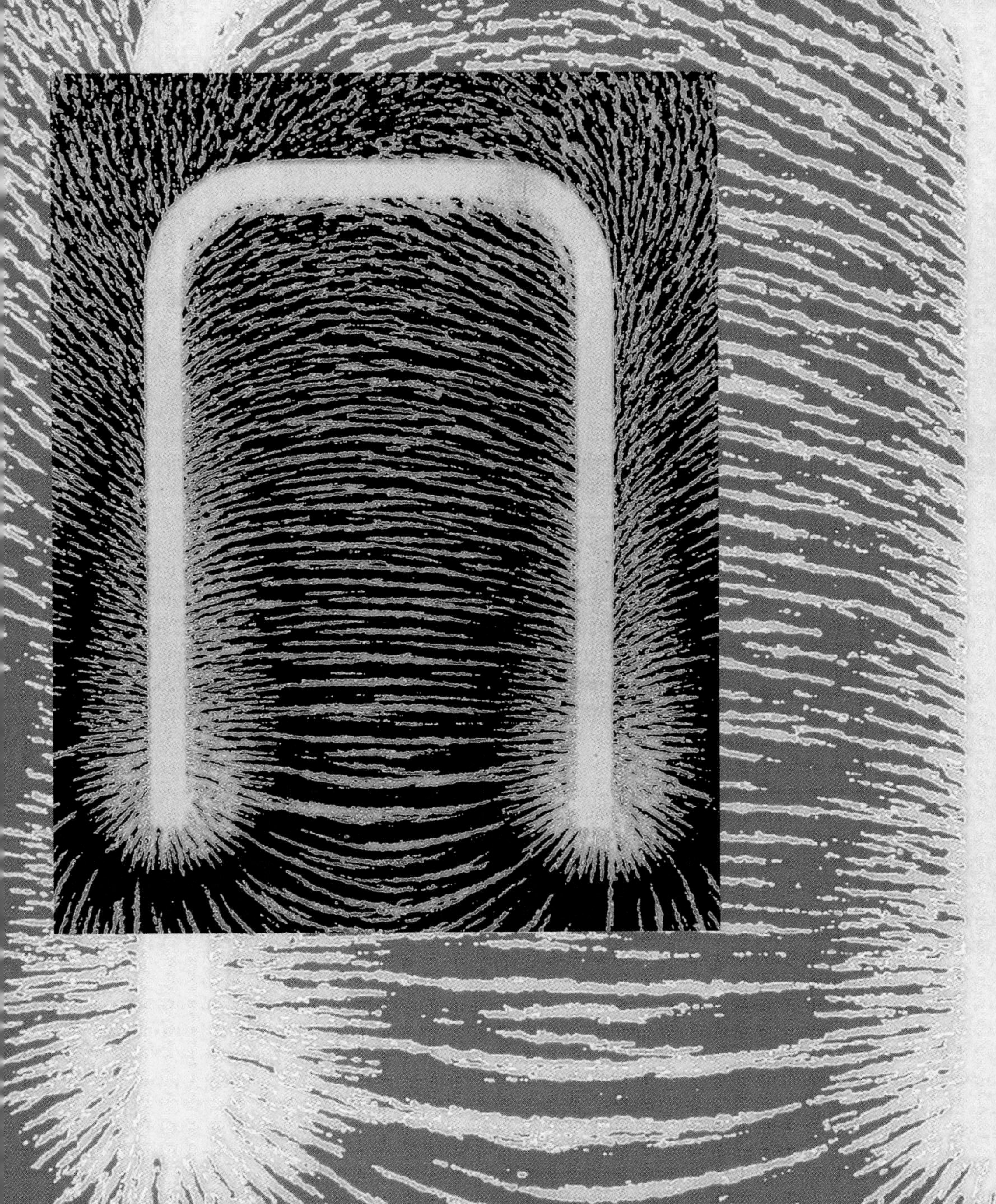

ron filings are shown aligned in the magnetic field of a horseshoe magnet in this computer graphic image.

Rocks of the mineral magnetite (Fe_3O_4) have been mined from early times. These rocks, called lodestones, exert a force that attracts bits of nearby iron; lodestones are natural magnets. Today, manufactured magnets are used in all kinds of ways. In refrigerator doors they provide the last little tug that snaps the door closed; they are used in radio and stereo speakers, in identification strips on credit cards, on video and audio cassette tapes, and in children's toys. Magnets are also used in large particle accelerators, machines that give charged subatomic particles incredible energies so their collisions can be analyzed. Magnetic resonance imaging (MRI) scanners expose the inner bodies of patients to medical doctors today in more detail than was possible before.

MAGNETIC POLES AND THE MAGNETIC FORCE

FIGURE 19-1 THE PLACES OF GREATEST ATTRACTION ON A MAGNET ARE CALLED ITS POLES. THIS BAR MAGNET WAS DIPPED INTO A BAG OF LOOSE IRON FILINGS, WHICH CLING TO THE AREA AROUND ITS TWO ENDS, ITS NORTH AND SOUTH POLES.

Dip a small magnet into a bag of iron filings, and you'll find there are two spots on the magnet where many filings gather (Figure 19-1). These spots are called the **magnetic poles.** The force the magnet exerts on the filings is strongest at these poles.

Magnetic poles come in two kinds, and each magnet has a pole of each type. Float a small bar magnet on a small piece of wood in water, and the magnet (and wood) will slowly turn so that one end of it points approximately toward the north (Figure 19-2). That end is called the magnet's north pole. The other end, called the magnet's south pole, points in a southerly direction. They align that way because the earth itself is a weak magnet, as we'll see later, with its south magnetic pole near earth's geographical north pole. A compass needle is just such a magnet suspended in a holder of nonmagnetic material.

The poles of magnets exert forces on each other, obeying the rule "opposites attract and likes repel." The force is called the **magnetic force,** and it acts very much like the force of gravity and the electric force. That is, the mutual force between two magnetic poles is strong when the poles are close together but gets weak very quickly as the poles are pulled apart.

If you cut a bar magnet along a line halfway between the two poles, as in Figure 19-3, you do not wind up with the north pole isolated on one piece and the south pole isolated on the other; instead you will have two smaller magnets, each with a north and a south pole. For many years physicists have sought a single magnetic pole, or *monopole,* with no success. Magnetic poles, it seems, are always found in N–S pairs. Each N–S pair is called a magnetic dipole, and each has a magnetic field like that of a bar magnet, as discussed in the next section.

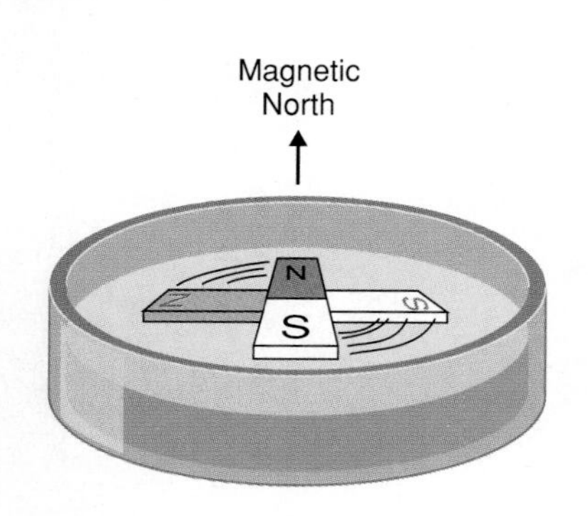

FIGURE 19-2 A BAR MAGNET FLOATING ON A QUIET SURFACE WILL SLOWLY TURN THE RAFT IT FLOATS ON TO ALIGN IN THE NORTH–SOUTH DIRECTION, WITH THE NORTH POLE OF THE MAGNET POINTING APPROXIMATELY TOWARD THE GEOGRAPHIC NORTH POLE.

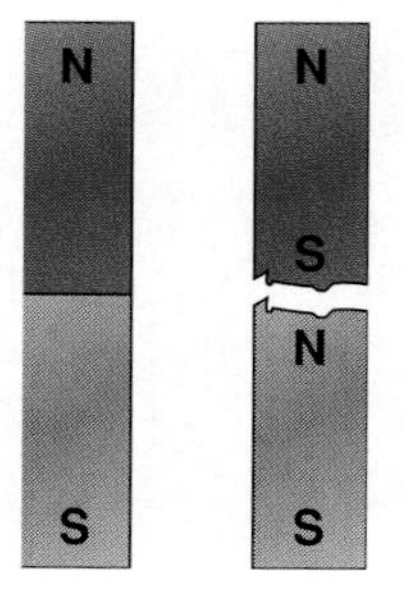

FIGURE 19-3 BREAKING A BAR MAGNET INTO TWO PIECES DOES NOT SEPARATE THE NORTH AND SOUTH POLES OF THE MAGNETS. INSTEAD, EACH PIECE OF THE ORIGINAL MAGNET THEN HAS A NORTH AND A SOUTH POLE.

MAGNETIC FIELDS

Just as for charged particles and their electric fields, we say that magnets have **magnetic fields,** and just like an electric field or a gravitational field, a magnetic field can be visualized by drawing field lines with arrows (Figure 19-4). For magnets, the field lines point away from north poles and toward south poles, and the direction of the field lines at a point in space shows the direction a north pole is pushed if placed at that point. The force on a south pole at the same place would point in the opposite direction from the field lines.

The pattern of a bar magnet's field can be seen by sprinkling some iron filings on a sheet of paper and bringing the bar magnet up beneath the page (Figure 19-5). Tap the paper, and the individual filings rotate to make a clear pattern, one that looks like the electric field of a (+) and a (−) charge, an electric dipole. So the magnetic field around a north and a south magnetic pole resembles the electric field around a positive and negative charge (compare Figure 19-5 with Figure 17-13). The needle of a small compass also lines up with the direction of the filings when the compass is placed on the paper that is over the magnet. The iron filings behave like compass needles because they *become* small compass needles in the presence of a magnetic field. (We'll see why in the section on magnetic induction.) Another way to trace magnetic field lines is to move a compass around in the magnet's field, recording the needle's orientation at each new position.

When two magnetic dipoles or two bar magnets interact, they exert twisting forces, or torques, on each other, as Figure 19-6 shows. The strength of this twisting force gets weaker with distance faster than the electric force or the gravitational force does. (In fact, for distant points the force between dipoles loses strength by the factor $1/d^3$.)

Magnetic fields pass undiminished through most materials, including many metals. As the demonstration with iron filings on paper shows, a magnetic field penetrates paper easily. You could do the same demonstration on a sheet of plastic, a thin piece of wood, or even a piece of aluminum or brass. These materials don't noticeably modify a magnetic field, nor does the field modify them. Iron behaves differently, though, as those filings show, and so do the metallic elements nickel and cobalt. We'll return to these magnetic elements shortly.

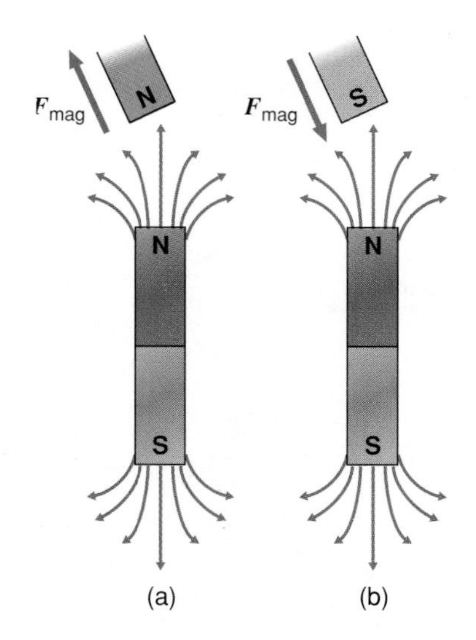

FIGURE 19-4 (A) FIELD LINES POINT AWAY FROM A NORTH POLE AND TOWARD A SOUTH POLE. THE MAGNETIC FORCE EXERTED BY A NORTH MAGNETIC POLE ON THE NORTH POLE OF A SECOND MAGNET IS IN THE DIRECTION OF THE FIELD LINES (LIKES REPEL). (B) THE MAGNETIC FORCE EXERTED BY A NORTH MAGNETIC POLE ON THE SOUTH POLE OF A SECOND MAGNET IS OPPOSITE THE DIRECTION OF THE FIELD LINES (OPPOSITES ATTRACT).

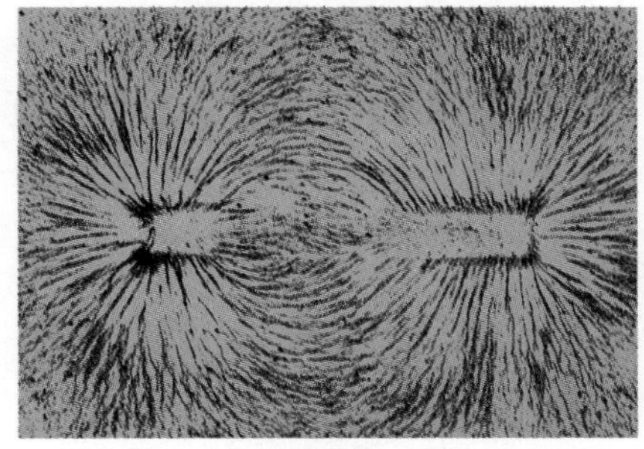

FIGURE 19-5 THE LOOSE IRON FILINGS ON A SHEET OF PAPER RESPOND TO THE MAGNETIC FIELD OF A BAR MAGNET PLACED BENEATH THE PAPER. THE FILINGS TURN, ALIGNING THEIR LONGEST DIMENSIONS WITH THE DIRECTION OF THE MAGNETIC FIELD. COMPARE THIS TO FIGURE 17-13.

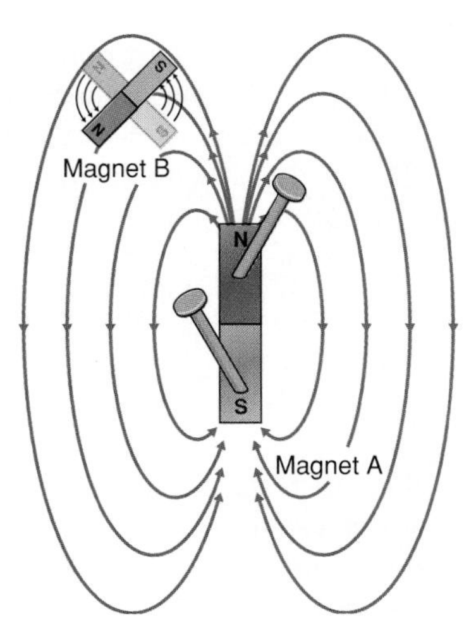

FIGURE 19-6 MAGNET A IS HELD STATIONARY WHILE MAGNET B IS FREE TO ROTATE IN THE MAGNETIC FIELD OF A. MAGNET B TWISTS AND ALIGNS ITSELF WITH THE MAGNETIC FIELD OF MAGNET A.

THE MAGNETIC FIELD CREATED BY AN ELECTRIC CURRENT

In 1681 an English ship sailing to Boston was struck by lightning. After the storm passed, the north pole of the needle on the ship's compass pointed south and the south pole pointed north! Somehow the lightning bolt had reversed the needle's poles.

This incident seemed to show a connection between electricity and magnetism, but early experiments using stationary electric charge and magnets failed to show any forces acting between them. Then Volta's invention of the battery in 1800 provided a way to maintain steady currents, and investigators began to search for a magnetic field related to moving charges. They sent a current through a wire lying on a table, placed a compass on the table beside the wire, and looked for any twisting of the needle. However, no matter where they placed the compass on the table, there was no needle deflection.

In 1820 the connection was discovered accidentally during a physics lecture in Copenhagen. Performing the experiment just mentioned, Professor Hans Christian Oersted was showing his students that the moving charges in the conducting wire did not create a magnetic field. With the wire and compass both on the table, he showed how the current in the wire had no effect on the compass. Part of the wire was aligned north–south, however, and when Oersted picked up the compass and inadvertently held it above that part of the wire, he was amazed to see the needle twist toward the east–west direction (Figure 19-7).

Before the discovery by Oersted, investigators had looked for magnetic field lines pointing straight out from the current-carrying wire the way that electric field lines point out from charges (Figure 19-8). The magnetic field created by an electric current is a field *without poles,* however; this field *circles* the current, as Figure 19-9 shows. In all the experiments prior to Oersted's accidental discovery, the compass needle wasn't deflected because the magnetic field pointed straight up or down, out of or into the tabletop, and the compass needles, being constrained by the horizontal compass housings, could not respond in the up or down direction.

Shortly after hearing of Oersted's findings, a French mathematician-scientist named André Marie Ampère (1775–1836) made a loop in a current-carrying wire and discovered that the magnetic field inside and around that loop looks like the field of a bar magnet (Figure 19-10). That is, an electric current moving in a circular loop creates a magnetic field like those of magnetic dipoles.

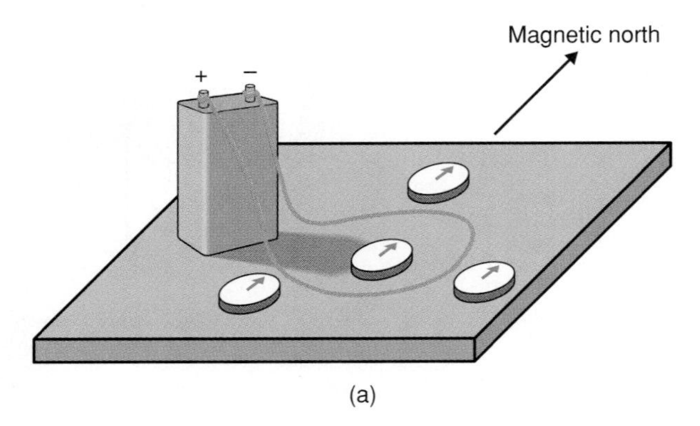

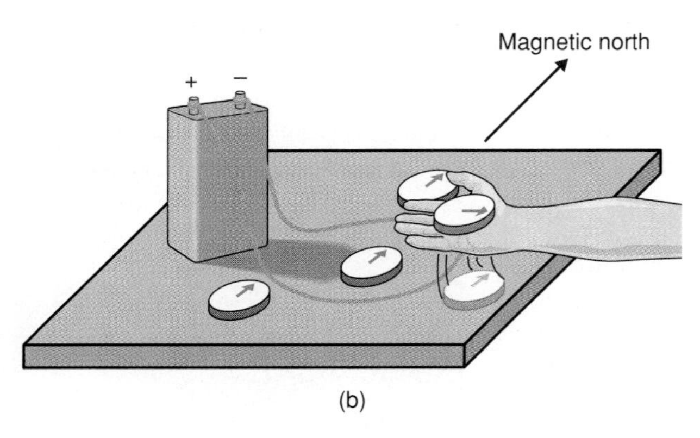

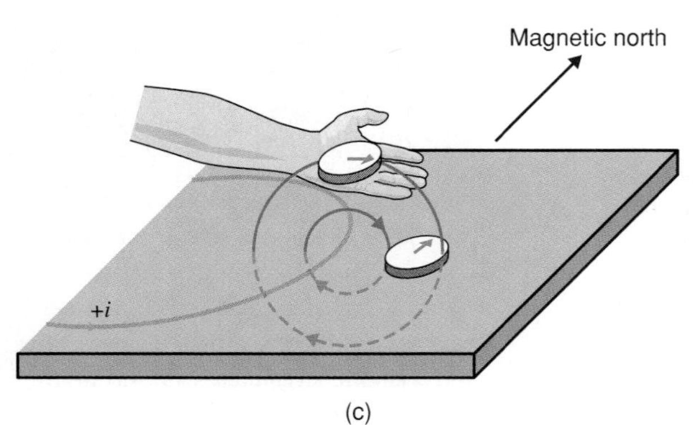

FIGURE 19-7 (A) A CURRENT-CARRYING WIRE LIES FLAT ON A TABLE. COMPASSES SITTING ON THE TABLE NEARBY ARE UNAFFECTED, AND EARLY INVESTIGATORS THOUGHT THAT MEANT ELECTRIC CURRENTS HAD NO MAGNETIC FIELDS. (B) OERSTED ACCIDENTALLY DISCOVERED THAT CURRENTS *DO* GENERATE MAGNETIC FIELDS WHEN HE PICKED UP ONE SUCH COMPASS AND HELD IT *OVER* THE WIRE. (C) THE REASON COMPASSES ON THE TABLE DON'T DETECT THE CURRENT'S MAGNETIC FIELD IS THAT, ON THE TABLE, THE MAGNETIC FIELD IS VERTICAL, AND THE COMPASS NEEDLES COULD NOT MOVE IN THAT DIRECTION.

PHYSICS IN 3D/MAGNETIC FIELD OF A MAGNETIC DIPOLE

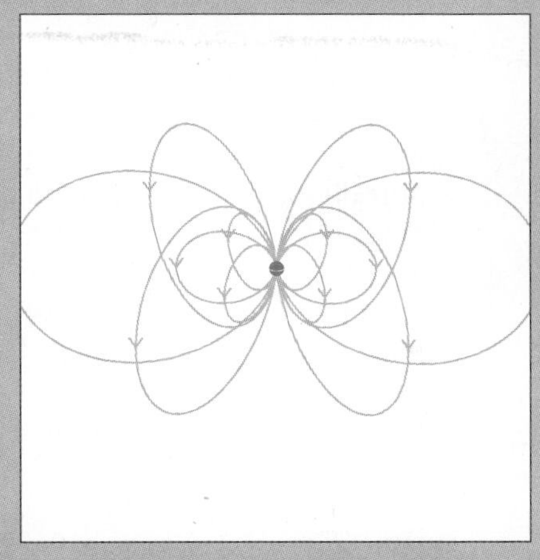

THE MAGNETIC FIELD OF A DIPOLE, WITH THE NORTH POLE AT THE TOP AND THE SOUTH POLE AT THE BOTTOM. AS WITH ELECTRIC FIELDS, THE CLOSER THE FIELD LINES ARE TO EACH OTHER, THE STRONGER THE FORCE IS. THE MAGNETIC FIELD OF EARTH LOOKS MUCH LIKE THIS, EXCEPT THAT THE SOUTH POLE IS IN THE NORTHERN HEMISPHERE.

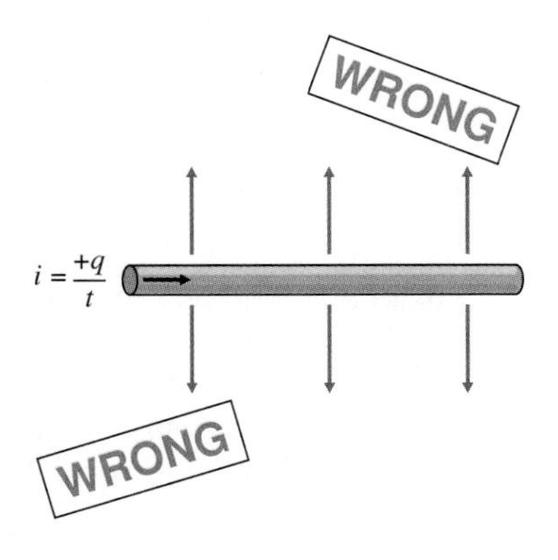

FIGURE 19-8 THIS IS HOW SOME RESEARCHERS EXPECTED MAGNETIC FIELDS TO BE RELATED TO CURRENTS IN WIRES, BUT THAT IDEA PROVED WRONG. FIGURE 19-10 SHOWS THE CORRECT SHAPE OF THE MAGNETIC FIELD LINES.

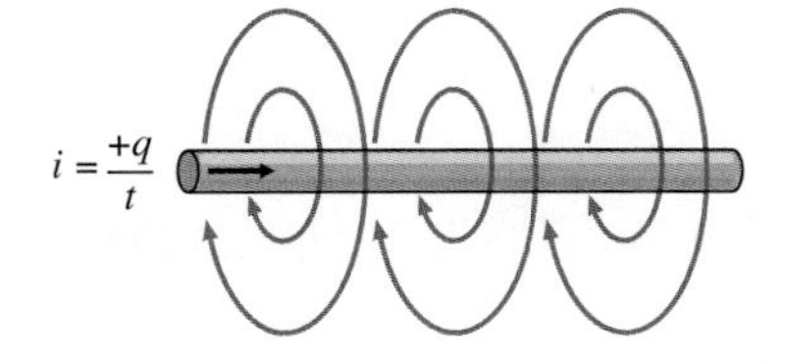

FIGURE 19-9 THE MAGNETIC FIELD CREATED BY A CURRENT CIRCLES THE CURRENT AND HAS NO POLES. THE GREATER THE CURRENT, THE GREATER THE STRENGTH OF THE MAGNETIC FIELD, AND THE FIELD LINES ARE DRAWN CLOSER TOGETHER TO SHOW THIS. IF THE DIRECTION OF THE CURRENT IS REVERSED, THE DIRECTION OF THE FIELD LINES IS REVERSED AS WELL. THE FIELD LINES SHOWN HERE ARE FOR A POSITIVE CURRENT MOVING LEFT TO RIGHT, OR, FOR A CURRENT OF ELECTRONS MOVING FROM RIGHT TO LEFT.

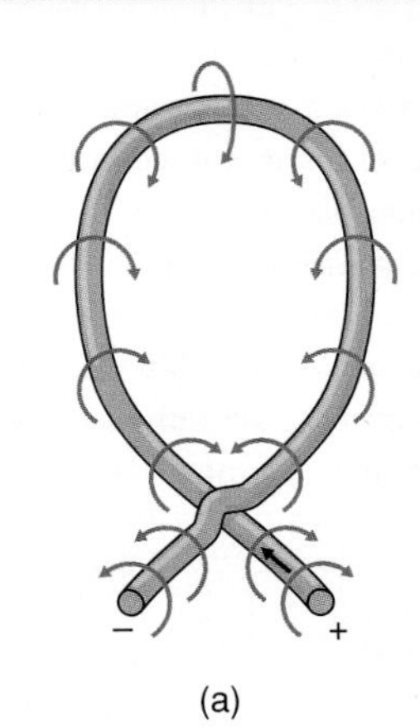

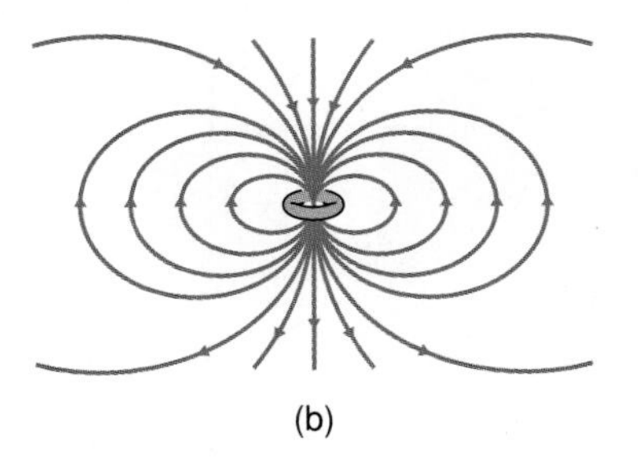

FIGURE 19-10 (A) WHEN AMPÈRE MADE A LOOP IN AN INSULATED CURRENT-CARRYING WIRE, HE SAW THAT THE LOOP'S MAGNETIC FIELD WAS A DIPOLE FIELD LIKE THAT OF A BAR MAGNET. (B) A MORE DETAILED VIEW OF THE MAGNETIC FIELD OF A CURRENT LOOP.

AN ATOMIC VIEW OF MAGNETS: MAGNETIC DOMAINS

Today we know that electrons are the basis for magnetism in iron, nickel, cobalt, and the iron compounds and alloys that make up bar magnets. The magnetic fields caused by electrons come about in two ways. Electrons in orbit around the nucleus constitute a loop of current, and any current loop gives rise to a magnetic field like that of Figure 19-10. Also, electrons (and protons and neutrons, too) have magnetic dipole fields of their own. The **intrinsic magnetic field** is a fundamental property of the electron, just as its mass and charge are. Whether in an atom or drifting freely, the electron has a magnetic field around it. It is this intrinsic magnetic field that is mostly responsible for magnetism in iron, nickel, cobalt, and magnetic compounds. An iron atom has a majority of its electrons' dipole fields aligned in the same direction, giving each iron atom a net magnetic field. That's why you can cut a magnet in half again and again, and there will still be smaller magnets, all the way down to the atomic level.

However, not every piece of iron is magnetized. In ordinary iron, groups of perhaps a million iron atoms align naturally, their magnetic fields adding together. These magnetized groups of atoms are called **magnetic domains.** (Figure 19-11). Iron has a net magnetic field only if many of these domains are aligned with each other, pointing in the same direction, or if the domains naturally oriented in that direction become larger at the expense of others (Figure 19-12). Stroking a piece of iron in one direction with a magnet magnetizes the iron. In fact, just tapping a nail in the presence of a strong magnet magnetizes the nail. However, the nail isn't a permanent magnet. You can demagnetize it—disaligning the domains' directions or causing the size of the aligned domains to decrease—by a sharp blow or two with a hammer or even by just throwing it on the floor.

Some iron alloys hold the domains' size and direction very strongly. If these metals are first taken to very high temperatures and then cooled while in the grip of a strong external magnetic field, the magnetic domains align with that field as a compass needle aligns with the earth's magnetic field. To make permanent magnets, magnetic grains of iron alloys are mixed in epoxy that then hardens while in a strong magnetic field. The result is a permanent magnet that can be machined into any shape.

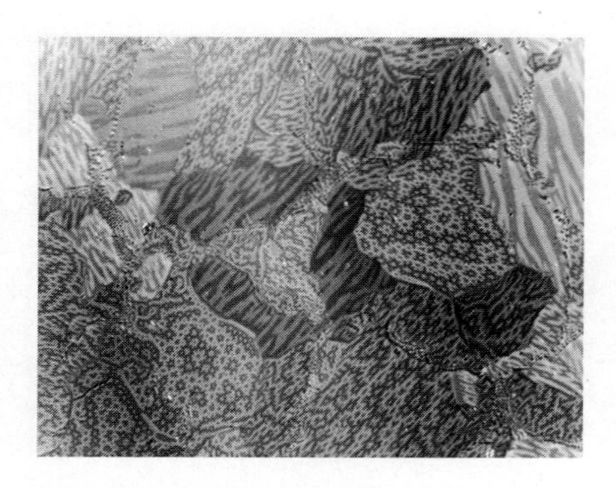

FIGURE 19-11 MAGNETIC DOMAINS IN A COBALT COMPOUND. THE AREAS WITH LINES HAVE ORIENTATIONS PARALLEL TO THE PAGE, AND THOSE WITH STARS HAVE ORIENTATIONS OUT OF THE PAGE.

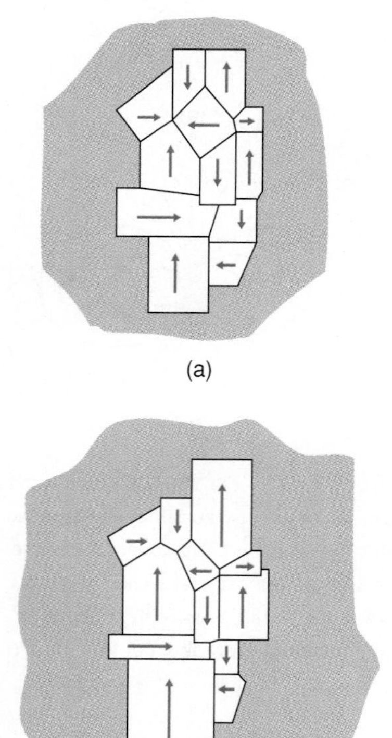

FIGURE 19-12 (A) IN AN UNMAGNETIZED PIECE OF IRON, THE MAGNETIC DOMAINS HAVE RANDOM SIZES AND DIRECTIONS. THE MAGNETIC FIELDS OF THE DOMAINS MORE OR LESS CANCEL. (B) WHEN THE IRON IS IN A MAGNETIC FIELD, THE DOMAINS WHOSE FIELDS ARE IN THE DIRECTION OF THE APPLIED FIELD GROW, INCREASING IN STRENGTH, WHILE THOSE IN OTHER DIRECTIONS SHRINK AND DECREASE IN STRENGTH.

MAGNETIC INDUCTION: ELECTROMAGNETS

Iron or steel becomes temporarily magnetized in a magnetic field as the magnetic domains temporarily turn so that their fields add. This is how long strings of paper clips are able to hang from the pole of a single magnet. Each clip becomes a temporary magnet. We say they have an **induced** magnetic field. Remove the magnet, and their domains disalign immediately because of thermal vibrations. Then the clips separate.

Devices called **electromagnets** use magnetic induction to create magnetic fields of variable strength that can be turned on or off with a switch. An electromagnet is composed of a length of insulated copper wire wrapped around an iron core (Figure 19-13). When there is a current through the wire, each loop generates a magnetic field like that surrounding a small bar magnet,

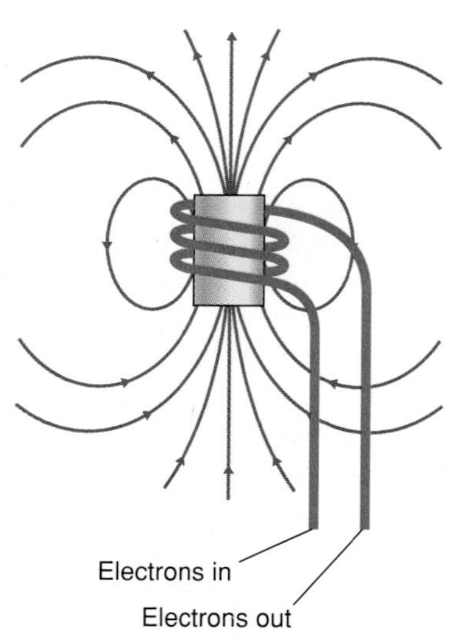

FIGURE 19-13 AN ELECTROMAGNET USES THE COMBINED MAGNETIC FIELD OF MANY LOOPS OF A CURRENT-CARRYING WIRE TO INDUCE A CORE OF IRON TO BECOME MAGNETIC, THEREBY INTENSIFYING THE NET MAGNETIC FIELD (OF THE IRON PLUS THE COILS). THE MORE LOOPS, THE STRONGER THE MAGNETIC FIELD. THE MORE CURRENT CARRIED BY THE LOOPS, THE STRONGER THE MAGNETIC FIELD.

and the fields of all the loops add together. By itself, this field is relatively weak. However, the magnetic domains in the iron core align with the fields of the loops, and the *net* field becomes many times stronger than the field of the current loops alone. The more loops there are, the stronger the field is. Likewise, the greater the current through the wire, the greater the strength of the magnetic field. If the current stops, the domains disalign and the magnetic field disappears.

Earth's weak magnetic field induces a magnetic field in iron. Jostled by normal thermal vibrations, a small percentage of iron's magnetic domains respond to earth's field and align themselves north–south. Place a compass next to your refrigerator, stove, or other iron or steel appliances whose positions don't often change. You will usually see a deflection near these appliances, by fields induced by the magnetic earth.

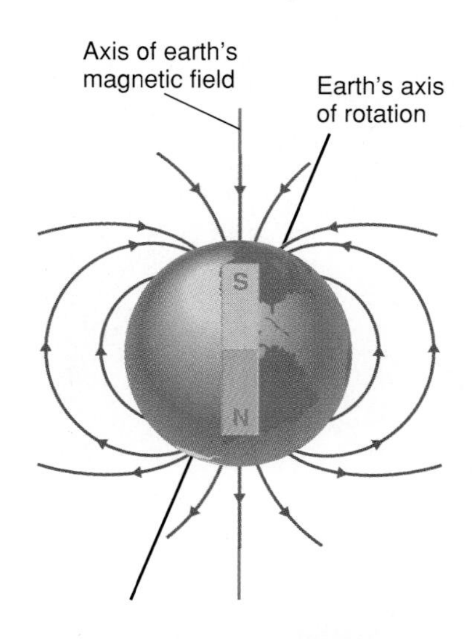

FIGURE 19-14 THE MAGNETIC POLES OF EARTH ARE NOT QUITE ALIGNED WITH THE POLES OF EARTH'S ROTATIONAL AXIS. NOTICE THAT THE MAGNETIC FIELD LINES ARE DIRECTED FROM THE SOUTHERN HEMISPHERE TO THE NORTHERN HEMISPHERE, SINCE EARTH'S NORTHERNMOST MAGNETIC POLE IS ACTUALLY THE SOUTH POLE OF A MAGNET, AS THE IMAGINARY BAR MAGNET INDICATES.

THE MAGNETIC FIELD OF EARTH

Earth's magnetic field is shaped much as if the planet had a great bar magnet at its center. Figure 19-14 shows the orientation of the earth's magnetic field. The magnetic poles of the earth are more than 1000 miles from its geographic poles, which are on the earth's axis of rotation. Notice that earth's *south* magnetic pole is near the *north* geographical pole, which is why the north pole of a compass needle points northward (opposites attract).

It is thought that the earth's magnetic field arises from electric currents in the molten, iron-rich outer core surrounding the solid inner core. The spinning earth carries the conducting material of the outer core around, and convection currents in this rotating fluid are thought to drive the electric currents.

Earth's magnetic field lines usually dip toward or come out from earth's surface at some angle, which is called the **magnetic inclination** or the **magnetic dip** at that location. The dip anywhere in the United States and Canada is very large, as you can see from the field lines in Figure 19-14.

Because the earth's north and south poles do not coincide with its magnetic poles, the "north-pointing" end of a compass needle only points roughly toward the north. The deviation of a compass reading from the direction of true geographical north is called the **magnetic declination** (Figure 19-15). To plot an accurate course with a map, a person must know the magnetic declination in that area. At Daytona Beach, Florida, and Chicago, Illinois, the magnetic declination is close to 0°, so a compass needle at either city points nearly to true geographic north. In Maine the needle points about 20° west of geographic north, while in Oregon it points about 20° east of geographic north.

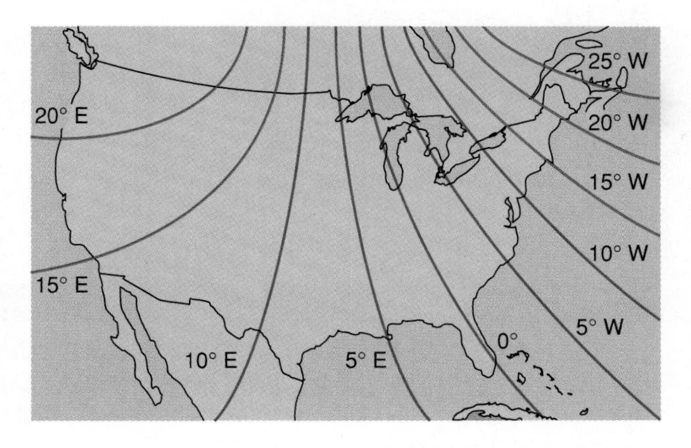

FIGURE 19-15 THESE LINES OF CONSTANT MAGNETIC DECLINATION INDICATE THE APPROXIMATE DEVIATION FROM TRUE NORTH THAT A COMPASS READING HAS AT ANY POINT ALONG THE LINES. THESE LINES HAVE VARIED QUITE A BIT FROM CENTURY TO CENTURY.

CHANGES IN EARTH'S MAGNETIC FIELD

At any location on the earth's surface, the strength of the magnetic field, its declination, and its dip slowly change as time goes by. Variations in magnetic declination were noticed as early as the 1600s. In Paris the magnetic declination was 9° east of true north in 1580, 0° in 1665, and 22° west in 1820. At present the magnetic declination at Paris is approaching 0° again. Such fluctuations are thought to be connected with changes in the electric currents in earth's outer core.

Magnetic minerals are abundant in the rocks of the earth's crust today, and many have retained the effects of the magnetic field present at the time they solidified. Such rocks show that the earth's magnetic poles have reversed, the north magnetic pole becoming the south magnetic pole and vice versa, about 150 times over the past 60 million years. Over the past 5 million years, a compass left at one location would have pointed south more often than north! The magnetic poles also wander as time goes by, but in recent centuries the average position of the wandering north magnetic pole has been very close to the present north geographical pole. A reversing magnetic field isn't exclusive to earth. The sun's magnetic field reverses regularly, with a period of 22 years.

ANIMAL MAGNETISM

MAGNETISM AND LIFE

In 1974 scientists discovered several species of aquatic bacteria that swim almost entirely along earth's magnetic field lines. These bacteria have their own built-in lodestones, perfect little crystals of magnetite put together from iron (taken from the water) and oxygen. A tiny chain of these crystals, apparently of one magnetic domain each, points along the length of the cell. These strong internal compasses respond to the earth's field and orient the bacteria along the field lines through no action of their own. In the laboratory, researchers have reversed the local net magnetic field, and the bacteria's lodestones immediately adjusted the bacteria's direction of travel. Dead specimens also responded to the direction of the field, proof that it is the magnetic force that governs their direction of travel.

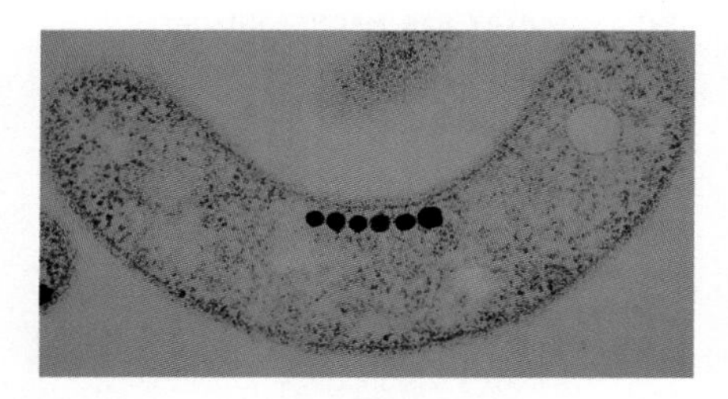

FIGURE 19-16 A CLOSE-UP OF THE CHAIN OF MAGNETIC DOMAINS IN A MAGNETIC BACTERIUM.

In the species of magnetic bacteria native to the Northern Hemisphere, the magnets' north poles point forward, away from the bacteria's tails (Figure 19-16). The best guess for the reason for this orientation concerns survival. If a bacterium is stirred up from the bottom of a pond or saltwater marsh (its natural habitat), traveling north along the magnetic field lines in the Northern Hemisphere takes it back down to its home in the sediment. The magnetic

MAGNETISM AND YOU

Wherever there is an electric current, there is a magnetic field: in a bolt of lightning, in the circuits of a hand calculator, and even in the extremely weak ion currents that travel along the nerve cells in your body. The magnetic fields of the currents along your nerves are very weak, about one-billionth as strong as the earth's magnetic field. Whenever you reach out to touch something, nerves carry an electrical impulse to the muscles you need and cause them to contract. This impulse creates a temporary magnetic field. Two major sources of magnetic fields in the body are the heart (whose powerful contractions are triggered by large synchronous electrical impulses) and the brain.

Magnetism has found important uses in medicine. The lungs of foundry workers sometimes have large magnetic fields due to inhaled iron particles, as do those of arc welders. To detect such particles, the patient is painlessly magnetized in a medium-strength magnetic field, and any magnetic particles in this field align. The external magnet is turned off, and special detectors quickly measure the magnetic field of the aligned particles. The same technique is used to detect iron levels in the liver. With these devices, a medical physicist could tell if you'd just eaten canned beans by the trace amounts of iron in your stomach from the can, as the accompanying figure shows. In Chapter 21 we'll see how magnetic resonance imaging (MRI) gives high-resolution pictures of the tissues inside a patient's body. A strong magnetic field, up to 60,000 times the intensity of earth's magnetic field, is used to align the magnetic fields of the protons of the hydrogen atoms in the patient. The conducting coils used in MRI machines are cooled by liquid helium to make them superconducting, enabling them to carry a steady, high current to produce the high magnetic field needed for the machine's operation.

dip of earth's magnetic field steers the bacterium down as it paddles! Because of its microscopic size and neutral buoyancy, the bacterium would otherwise wander aimlessly; gravity would not help it return to the bottom.

Similar magnets have been found in algae, shrimp, dolphins, and whales. Magnetite crystals have been discovered in the abdomens of honey bees and in the brains of homing pigeons. In 1992 researchers discovered magnetite crystals in human brains, with a density of about 5 million crystals in a thimbleful of gray matter. About a millionth of an inch long, they resemble the crystals found in the magnetic bacteria (Figure 19-17). At the time of this writing, their purpose, or even where they are located in the brain cells, remains unknown.

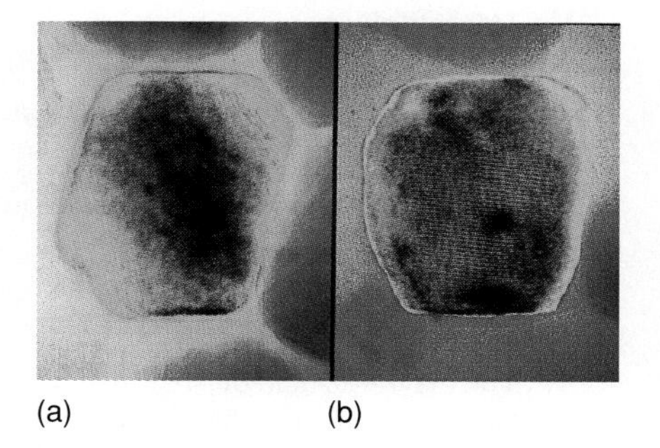

FIGURE 19-17 (A) A CRYSTAL OF MAGNETITE FROM A HUMAN BRAIN, AND (B) A CRYSTAL OF MAGNETITE FROM A MAGNETIC BACTERIUM.

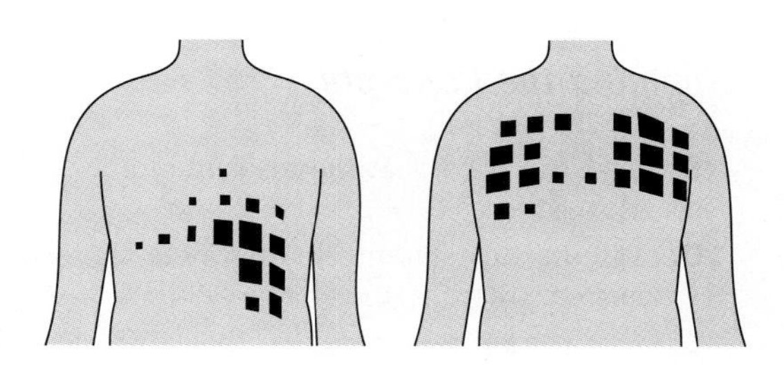

THE MAGNETIC PROFILES OF TWO PEOPLE JUST AFTER AN EXTERNAL MAGNETIC FIELD HAS ALIGNED THE IRON PARTICLES IN THEIR BODIES. SPECIAL MAGNETIC DETECTORS WERE USED TO PICK UP THE PARTICLES' VERY SMALL MAGNETIC FIELDS. THE DARK SQUARES REPRESENT THE INTENSITY OF THE MAGNETIC FIELDS, WHICH RELATES TO THE AMOUNT OF IRON PRESENT. THE SUBJECT ON THE LEFT HAD JUST EATEN CANNED BEANS, WHICH CONTAINED ABOUT 100 MICROGRAMS OF IRON OXIDE ABSORBED FROM THE INTERIOR OF THE CAN. THE LUNGS OF THE SUBJECT ON THE RIGHT, A WELDER, HAD ABOUT 500 MICROGRAMS OF IRON IN THEM.

Summary and Key Terms

Iron filings gather most densely on a magnet where its ***magnetic field*** is the strongest, on its north and south ***magnetic poles***. A magnetic pole attracts an opposite pole and repels a like pole. Like the gravitational force and the electric force, the ***magnetic force*** between two magnetic poles increases rapidly as they come together and decreases rapidly as they move apart. The direction of the magnetic field lines surrounding any magnet shows the direction of the force exerted by that magnet on the north pole of any other magnetic nearby.

magnetic field–281
magnetic poles–280
magnetic force–280

Electric currents can produce strong magnetic fields. The magnetic field around a current in a wire has no poles, it circles the wire. The magnetic field inside and around a loop of a current-carrying wire looks like the field of a dipole.

Electrons are the subatomic basis for the magnetism of iron, cobalt, and nickel. Electrons have ***intrinsic magnetic fields***, dipole fields that are as basic to the particle as its mass and charge. In atoms like iron, most of the electrons' intrinsic fields are aligned in the same direction, giving each atom a net magnetic field. In ordinary iron, large groups of atoms align naturally, with their magnetic fields adding, into ***magnetic domains***. For iron to have a net magnetic field, many of its domains must be aligned in the same direction.

intrinsic magnetic fields–283
magnetic domains–284

A strong magnetic field can temporarily align magnetic domains, and that is called ***magnetic induction***. An ***electromagnet*** has loops of conducting wire wrapped around an iron core. When charge flows through the wire, each loop has a magnetic field, and the fields of all the loops add. This field induces a magnetic field in the iron core, giving a stronger net field.

magnetic induction–284
electromagnet–284

Magnetic dip is the angle at which earth's magnetic field lines dip toward or away from earth's surface. The deviation of a compass reading from the direction of true geographical north is called ***magnetic declination***. At any location on earth's surface, the magnetic field's strength, declination, and dip slowly change. Earth's magnetic poles have reversed many times in the past.

magnetic dip–285
magnetic declination–285

EXERCISES

Concept Checks: Answer true or false

1. Like magnetic poles attract each other, and unlike magnetic poles repel.
2. Magnetic monopoles have not been found, and magnetic poles are found in pairs.
3. The magnetic field of a bar magnet resembles the electric field of an electric dipole.
4. The magnetic field created by a current in a straight wire has no poles; instead, it circles the current.
5. The magnetic fields of some of the electrons in iron atoms add to give iron magnetic properties.
6. Magnetic domains in iron are large groups of atoms whose magnetic fields add. They are the reason every sample of iron always has a large magnetic field.
7. Electromagnets use magnetic induction to create magnetic fields that can be varied in intensity and turned on or off at will.
8. The magnetic field around earth resembles that of a bar magnet with the north pole near earth's geographic north pole.

Applying the Concepts

9. Give two ways that the magnetic force is similar to the electric force.
10. Discuss this statement: Anywhere there is an electric current, there is a magnetic field.
11. If you took a compass to the north geographic pole, which way would its needle point? If you took it to the south pole, which way would the needle point?
12. Which objects would you expect to be magnetized? (a) a sewing needle left in a drawer for a long time, (b) an aluminum baseball bat, (c) the head of a steel hammer, (d) a brass wind chime, (e) an iron nail in a house.
13. A newly developed machine applies a strong magnetic field across a patient's liver and detects how strongly the liver is magnetized. The result tells the doctor how much iron is stored in the liver, a useful thing to know if iron deficiency is suspected. What principle does this machine use?
14. If a paper clip hangs from the north pole of a strong magnet, where is the north pole on the paper clip? The south pole?
15. Exactly what is happening when iron filings line up along the magnetic field lines as in Figure 19-5?
16. To magnetize an iron nail, you can stroke it with one end of a bar magnet. Should you stroke it back and forth or only in one direction? Why?
17. In the cable car system of San Francisco, the miles of steel cable that pass through the powerhouse are inspected continually for wear. A magnetic tester is used. How might such a device work?
18. In 1986 biologists discovered that a yellowfin tuna has as many as 10 million magnetic crystals in its skull bones. These may aid the tuna in navigation. If you were a marine biologist, how would you test these tuna without harming them to see if the magnetite in their skulls does indeed affect their navigation?
19. Suppose someone handed you three similar iron bars and told you one was not magnetic but the other two were. How would you find the one that wasn't (without using the earth's magnetic field)?
20. If you are handed two bar magnets, how would you show that the force between the north pole on one and the south pole on the other decreases as the distance between them increases?
21. Why will either pole of a strong magnet attract an iron nail?
22. Huge magnets at junk yards are useful for moving heavy steel junk around. These are always electromagnets and never permanent magnets. Why?
23. A steel garbage can more than likely becomes slightly magnetized between pick-up days. How could you reverse its magnetic poles?
24. If you performed Demonstration 1 in Chapter 10, you may have noticed something interesting about the floating needle. If oriented in the east–west direction when the tissue paper sank, the needle rotated to the north–south direction. Why are sewing machine needles magnetized?
25. If a large steel bolt is held pointing north and downward at a large angle, about 60° to 70° down from the horizon in the Northern Hemisphere, it becomes temporarily magnetized by earth's magnetic field when struck a few times with a hammer. (Try it! It will pick up several paper clips.) Why should it be angled downward as it is struck?
26. Using Figure 19-14, decide if the north magnetic pole induced in a kitchen stove in Orlando, Florida, will be near the top of the stove or at its base.
27. Shortly after the discovery of magnetic bacteria in the Northern Hemisphere, a similar species was located in the Southern Hemisphere. However, the north poles of those bacteria were near their tails, opposite the orientation for those in the Northern Hemisphere. Explain.
28. Describe how you can use an inexpensive compass to find the magnetic dip in your area.
29. Why can't you feel the $1/d^2$ attraction of gravity as you can with a magnetic force? (See Demonstration 1.)
30. The path of the wandering north magnetic pole of earth for the last few centuries has been traced in studies of

annual deposits of clay from river banks. Explain how such layers might reveal the pole's direction.

31. The strikers of many doorbells are driven by electromagnets. What takes place between an electromagnet and a striker?

32. Sensitive magnetometers (devices that detect and measure magnetic fields) can be carried in airplanes to search for ore deposits beneath the flight path of the plane. Which of these ores are likely to be detected in this way: titanium, copper, iron, magnesium, nickel, gold, silver?

33. At which locations on earth would you not expect to find magnetic bacteria?

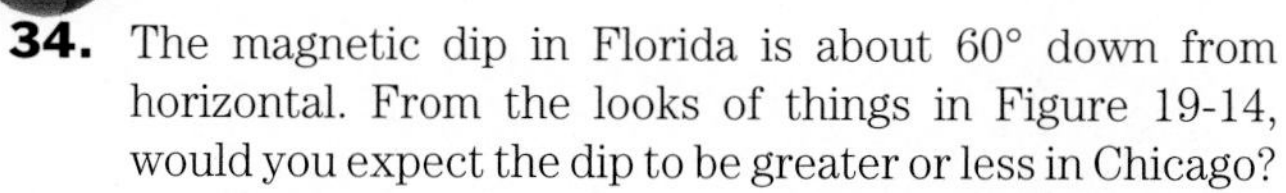

34. The magnetic dip in Florida is about 60° down from horizontal. From the looks of things in Figure 19-14, would you expect the dip to be greater or less in Chicago?

35. Magnetic stirrers are used in chemistry labs and hospital labs. First, a plastic-covered magnet is dropped into a beaker of liquid that needs to be mixed. The beaker is put on a small platform, a switch is turned on, and the covered magnet begins to rotate rapidly. Explain how the device works.

36. Explain why a compass needle tells you that the magnetic pole of the earth in the Northern Hemisphere is a south magnetic pole.

37. Figure 19-18 is a schematic of a radio/television speaker. Tell how a changing voltage across the wire of the electromagnet can help to make a sound wave.

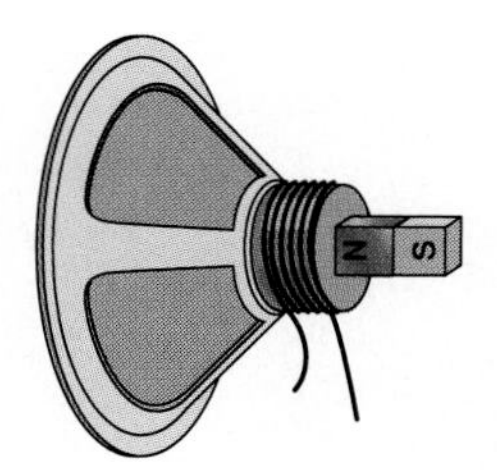

FIGURE 19-18

38. After Ampère discovered that loops of conducting wire made magnetic dipole fields, he suggested that internal currents in iron probably caused its magnetic properties. Another scientist, Augustin Jean Fresnel, pointed out in several letters to Ampère that the "Amperian" currents must be on the atomic or molecular level because of the lack of heating. Discuss.

39. Will the north-seeking end of a compass needle still point north if the compass is taken to the Southern Hemisphere? Describe the magnetic dip of a compass needle in the Southern Hemisphere by observing Figure 19-14.

40. If a bar of iron is bent into the shape of a horseshoe and then magnetized, its north and south poles are close together. What advantage does that offer?

Demonstrations

1. The magnetic force law can let you feel a force that, like gravity and the electric force, varies in strength inversely as the square of the distance between the interacting objects. If you align two identical bar magnets end to end (north pole to north pole) and let them push apart for a short distance, say, less than half the length of the magnets, you feel an inverse-square repulsion with your hands. Then turn one of the magnets around and feel an inverse-square attraction.

2. Put a paper clip into a jar of water and bring a large magnet up to the rim to see if the water or glass interferes with the magnetic field. Put another paper clip on a table and hold your hand an inch or less above it. Then hold the magnet over the top of your hand to see if your hand shields the magnetic field.

3. With a 1.5-volt cell, a length of very thin insulated copper wire, and a large nail, you can make an electromagnet. First, connect the ends of the wire to the cell terminals to see if the wire will pick up a paper clip when there is a current through the wire. (It probably won't.) Don't leave the wire connected for long; the cell will run down quickly. Next, shave about a centimeter of the insulation from the wire somewhere near the middle, reconnect the wire ends to the terminals, and try to pick up a paper clip by touching it to the exposed wire. (You can usually succeed. The magnetic field is stronger closer to the wire.) Now make a coil of the wire by wrapping it around your finger a few times, making sure the bare centimeter of wire lies at the very end of the coil. This time, when you connect the wire to the cell, the bare spot should attract several paper clips at once. Finally, coil the wire tightly around a large nail. When this electromagnet is connected to the cell, the nail will lift even more paper clips.

TWENTY

Electricity and Magnetism

Both high-voltage transmission wires and transformer stations are nearby wherever electrical energy is used today. The physics of electrical power generation and its transmission are subjects of this chapter.

In 1820 Oersted's announcement that electric currents produce magnetic fields was read aloud to the faculty at the Institute of France. In the audience was the person who would make the next steps connecting electricity and magnetism: André Marie Ampère discovered what would become the basis for electric motors. He applied Newton's third law to Oersted's discovery that electric currents exert forces on magnets and asked if magnets exert equal and opposite forces on currents. He quickly showed that the magnetic field from a current-carrying wire could exert a force on a current in another nearby wire. Stringing two straight wires side by side, he passed currents through them at the same time (Figure 20-1). When the currents were in the same direction, the wires drew closer together; when the current in one wire was reversed, the wires repelled each other. This discovery gave people the means to produce physical pushes or pulls with electrical power, but electric motors weren't invented until more than 50 years later (see the last section of this chapter). Today we can explain Ampère's discovery in terms of the forces on the electrons moving in the wires.

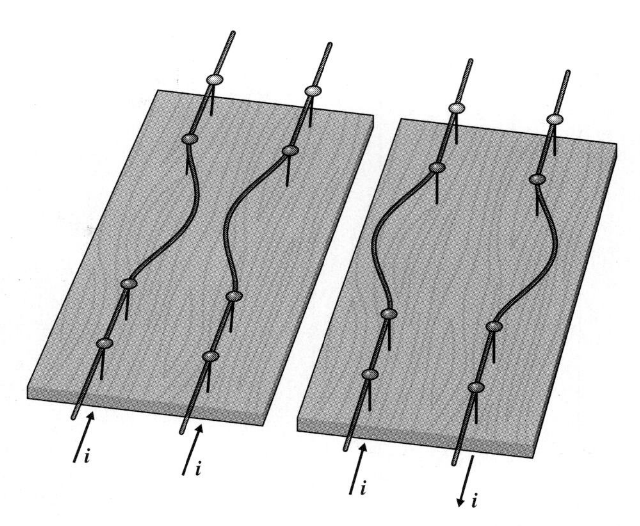

FIGURE 20-1 AMPERE DISCOVERED THAT IF CURRENTS ARE SENT IN THE SAME DIRECTION IN PARALLEL WIRES, THE WIRES ARE DRAWN TOWARD EACH OTHER. IF CURRENTS ARE SENT ALONG THE WIRES IN OPPOSITE DIRECTIONS, THE WIRES REPEL EACH OTHER. THE REASONS HAVE TO DO WITH THE WAY ELECTRONS MOVING IN A MAGNETIC FIELD ARE DEFLECTED, AS WE SHALL SEE.

THE EFFECT OF A STEADY MAGNETIC FIELD ON A CHARGED PARTICLE

Early experimenters found that a steady magnetic field exerts no force on a charged particle at rest in that field. If the particle moves *parallel* to the magnetic field lines, as in Figure 20-2a, there is still no force on the charge. If the particle moves *perpendicular* to the field lines, however, as in Figure 20-2b, it is deflected by the field. The charged particle gets a push that is perpendicular to the field lines and perpendicular to the velocity of the particle. Because the force is perpendicular to the particle's velocity, the magnetic field exerts only a *turning force* on the charged particle. It is a centripetal force (Chapter 4), as Figure 20-3 shows. It doesn't change the particle's speed.

Figure 20-4 shows what happens when a charged particle moves at an angle across the magnetic field lines. The particle's component of velocity parallel to the field is not affected, but the velocity component perpendicular to the field lines turns continuously, so the particle follows a helical path.

This effect of a magnetic field also acts on currents of electrons in conducting wires, and it explains why parallel wires push or pull on each other when there are currents through them. Each current generates a magnetic field that

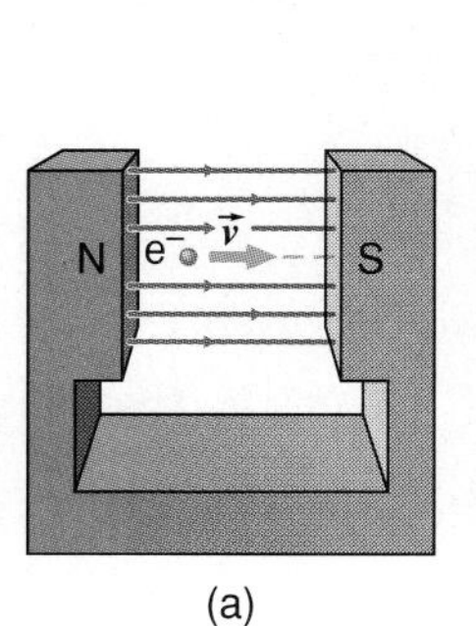

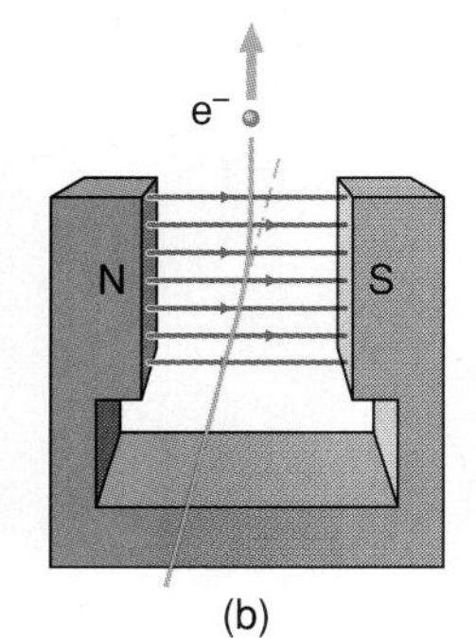

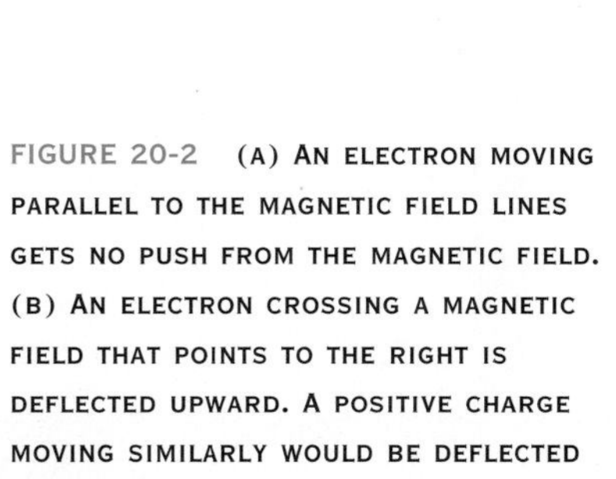

FIGURE 20-2 (A) AN ELECTRON MOVING PARALLEL TO THE MAGNETIC FIELD LINES GETS NO PUSH FROM THE MAGNETIC FIELD. (B) AN ELECTRON CROSSING A MAGNETIC FIELD THAT POINTS TO THE RIGHT IS DEFLECTED UPWARD. A POSITIVE CHARGE MOVING SIMILARLY WOULD BE DEFLECTED DOWNWARD.

FIGURE 20-3 AN ELECTRON WITH A VELOCITY EXACTLY PERPENDICULAR TO THE MAGNETIC FIELD LINES GETS A PUSH PERPENDICULAR TO BOTH THE FIELD LINES AND ITS OWN VELOCITY VECTOR. THAT TURNING FORCE (A CENTRIPETAL FORCE) TURNS THE ELECTRON IN A CIRCULAR PATH.

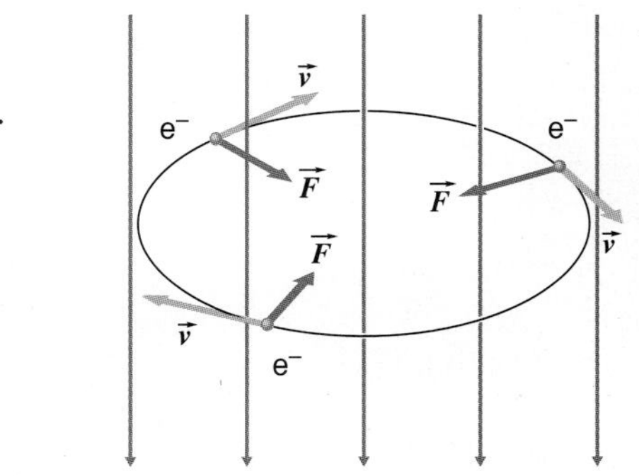

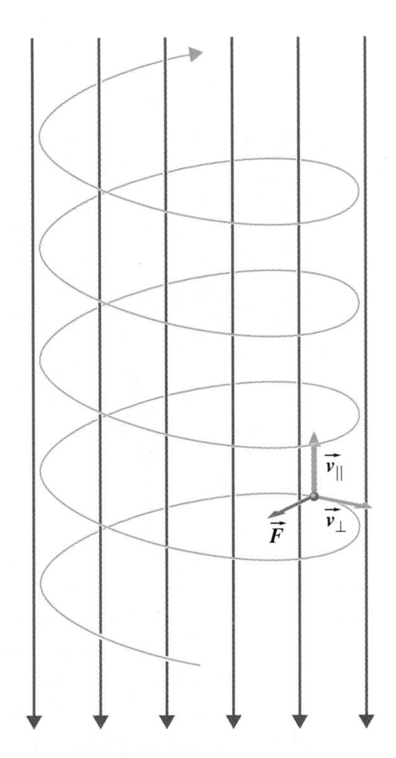

FIGURE 20-4 THIS PARTICLE HAS A VELOCITY WITH ONE COMPONENT PARALLEL TO THE MAGNETIC FIELD LINES AND ONE COMPONENT PERPENDICULAR TO THE FIELD LINES. THE PARALLEL COMPONENT IS NOT AFFECTED BY THE MAGNETIC FIELD; THE PERPENDICULAR COMPONENT TURNS IN A CIRCULAR FASHION, AS BEFORE. THE RESULT IS A HELICAL PATH AS IT RISES.

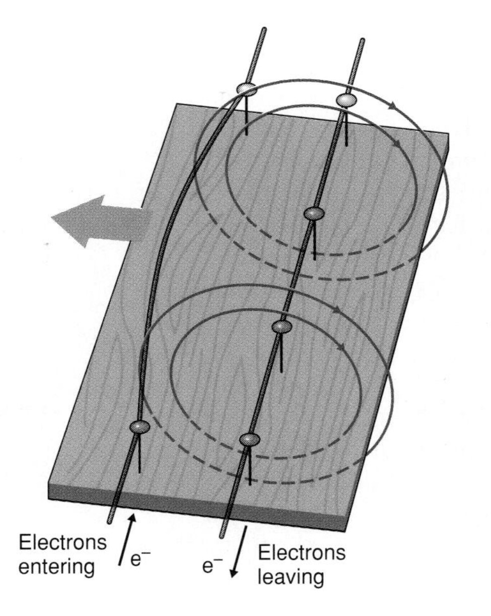

FIGURE 20-5 AMPÈRE'S PARALLEL WIRE EXPERIMENT. ELECTRONS MOVING IN THE RIGID WIRE ON THE RIGHT CREATE THE MAGNETIC FIELD SHOWN. THE MOVING ELECTRONS IN THE OTHER WIRE ARE DEFLECTED TO THE LEFT.

penetrates the other wire. The moving electrons deflect, turning to one side. In their collisions with metal atoms, the electrons give the wire a sideways push as in Figure 20-5. In other words, these featherweights, whose mass accounts for less than one-tenth of 1 percent of the mass of the conducting wire, push the wire around.

In homes, magnetic fields deflect electrons in television picture tubes to guide them to the screen; on a global scale, deflection of charged particles traveling in earth's magnetic field creates "radiation belts" around the earth (see box on pp. 294–95).

PHYSICS IN 3D/ELECTRON MOTION IN A UNIFORM MAGNETIC FIELD

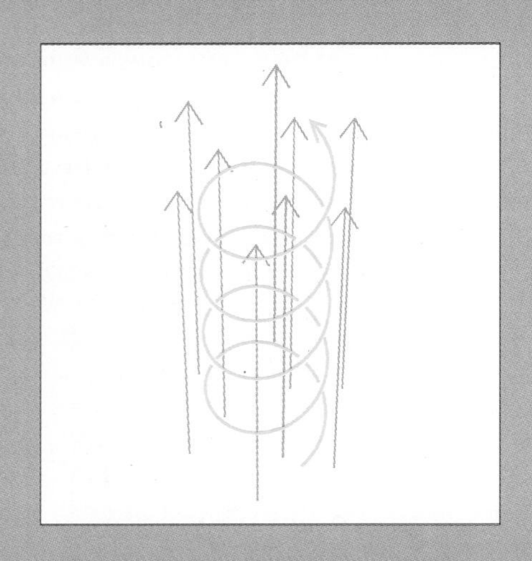

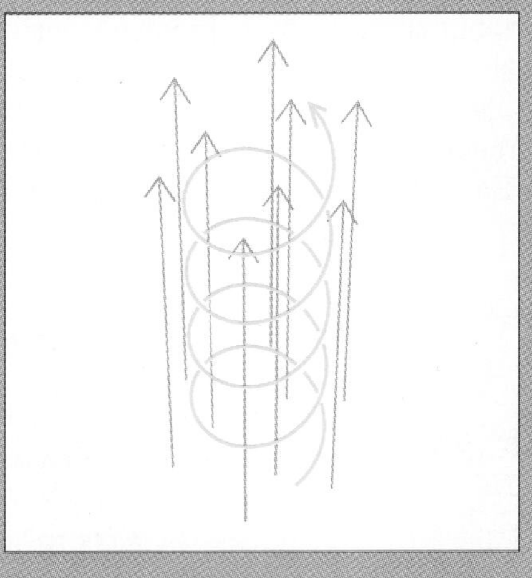

AN ELECTRON SPIRALING UPWARD IN A UNIFORM MAGNETIC FIELD, ALONG A HELICAL PATH. ITS COMPONENT OF VELOCITY PARALLEL TO THE FIELD LINES IS UNAFFECTED, BUT ITS COMPONENT PERPENDICULAR TO THE FIELD LINES IS CONSTANTLY TURNED.

ELECTRICITY FROM MAGNETISM: MICHAEL FARADAY

The next important connection between electricity and magnetism was found in London by Michael Faraday. Faraday had been a bookbinder's apprentice, but after listening to some public lectures by the famous chemist Humphrey Davy (Chapter 18), he decided at age 21 to change trades. He made careful notes of Davy's lectures, bound them in leather, and sent them to Davy. Soon he was working in Davy's lab as a bottle-washer.

Faraday had no formal education and never acquired a working knowledge of mathematics; he has been called a mathematical illiterate. Instead of using mathematics to describe physics, he composed physical pictures to help him understand things. It was Faraday who first used imaginary field lines to describe the magnetic field and the electric field, and he is regarded as one of the finest experimenters the world has ever known. He even carried a small magnet and a coil of wire in his pocket to tinker with at odd moments, hoping to "convert magnetism to electricity."

In August 1831, Faraday made a breakthrough, discovering how to use magnetism to generate electric currents. He coiled two separate insulated wires around the two halves of an iron ring, as in Figure 20-6. The first coil he connected to a battery, and the second he connected to an ammeter (a meter that measures electric current). When the connection with the battery was made, a current was set up in the first coil, and this current created a magnetic field in the iron ring. This field was amplified as the magnetic domains in the iron snapped into alignment with the first coil's field, and induction throughout the ring guided the magnetic field through the second coil.

Faraday noticed that at the instant when he connected the first coil to the battery, the needle of the ammeter attached to the second coil moved. It jumped quickly to one side and just as quickly returned to zero, indicating a brief current in the second coil. When he disconnected the first coil from the battery, the needle moved again, but to the other side, indicating a current in the opposite direction. Whenever the current through the first coil was either steady or zero, the meter registered no current in the second coil.

Faraday reasoned that the first deflection of the needle took place while the current was increasing in the first coil, before it reached its steady rate. As the current increased, so did the magnetic field it was creating, and the *growing magnetic field spread through the second coil.* This was when the needle jumped, showing a flicker of current in the second coil. When the current in the first coil reached a steady value, the magnetic field in the second coil was no longer increasing, and the needle returned to zero. Then as Faraday

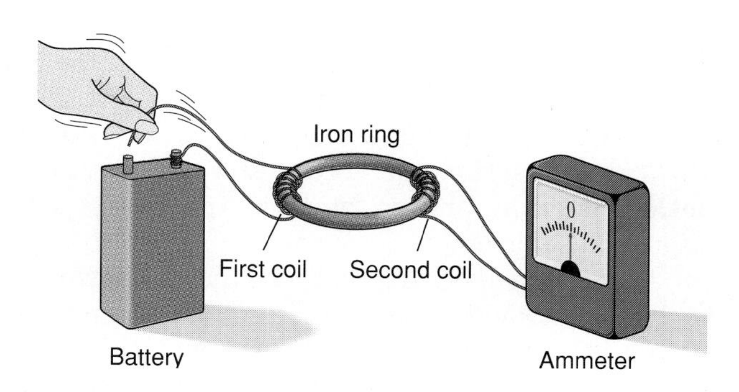

FIGURE 20-6 FARADAY'S EXPERIMENT. ONLY WHEN THE CURRENT IN THE FIRST COIL CHANGES DOES THE MAGNETIC FIELD CHANGE ACROSS THE SECOND COIL AND INDUCE A CURRENT THERE. IF THE CURRENT IN THE FIRST COIL IS STEADY, ITS MAGNETIC FIELD IS STEADY ALSO, AND NO CURRENT IS INDUCED IN THE SECOND COIL. IN SUCH AN ARRANGEMENT WITH A BATTERY, THE CURRENT CHANGES ONLY BRIEFLY WHENEVER THE CIRCUIT IS CLOSED OR OPENED.

THE VAN ALLEN RADIATION BELTS

Earth's magnetic field has a strong effect on incoming charged particles from space. About 90 percent of these particles are electrons and protons that stream outward from the sun, making up what is called the *solar wind.* The other 10 percent come from interstellar space; these are called *cosmic rays.*

As the speeding particles approach, they encounter earth's magnetic field, which grows in strength as the particles approach the earth. The charged particles that cross the field lines are deflected into complicated corkscrew paths, resulting in many particles being trapped largely in two doughnut-shaped regions centered on the earth's magnetic equator, as the top figure shows. These regions are called the Van Allen radiation belts. (James Van Allen, an American physicist, deduced their existence in 1958.) The inner one is centered about 3200 km (2000 mi) above earth's surface; the outer one is some 16,000 km (10,000 mi) overhead.

The concentration of electrons and protons in these belts is thousands of times greater than the concentration in the solar wind in earth's vicinity. Some of the particles in the inner belt have enough energy to penetrate several inches of lead. No space station meant for human occupation will ever be sent to orbit in these regions.

As you can see in the drawing, the portions of the belts near the magnetic equator are at higher altitudes than the portions near the magnetic poles. At the poles, incoming charged particles aren't confined to the belts; instead, they

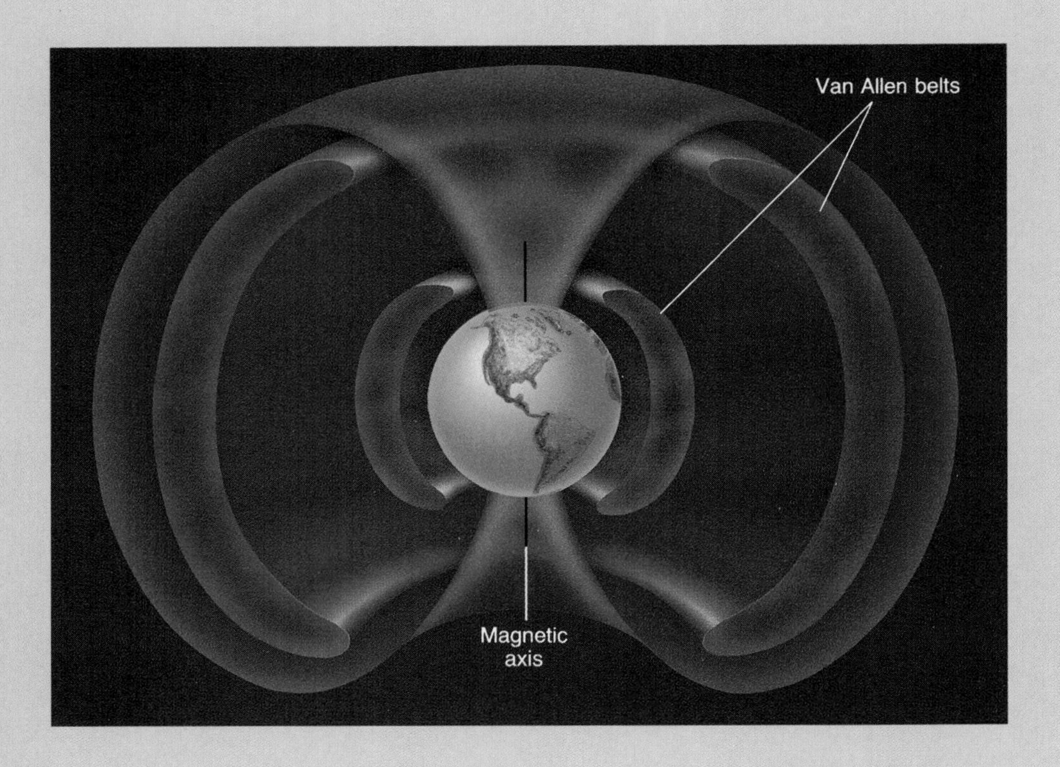

disconnected the battery, the magnetic field of the first coil collapsed across the second coil and the needle registered a flash of current in the other direction. Faraday had discovered that *a changing magnetic field in a conductor induces a current* (Figure 20-7). His discovery is called **electromagnetic induction,** and he later "explained" this action with his picture of magnetic field lines.

As we saw earlier, Ampère found that a charged particle crossing magnetic field lines gets pushed to one side. Faraday found that if magnetic field lines cross a charged particle, the particle gets a similar push. It doesn't matter whether the particle moves across the magnetic field or the magnetic field moves across the particle. Either way, the particle gets a push.

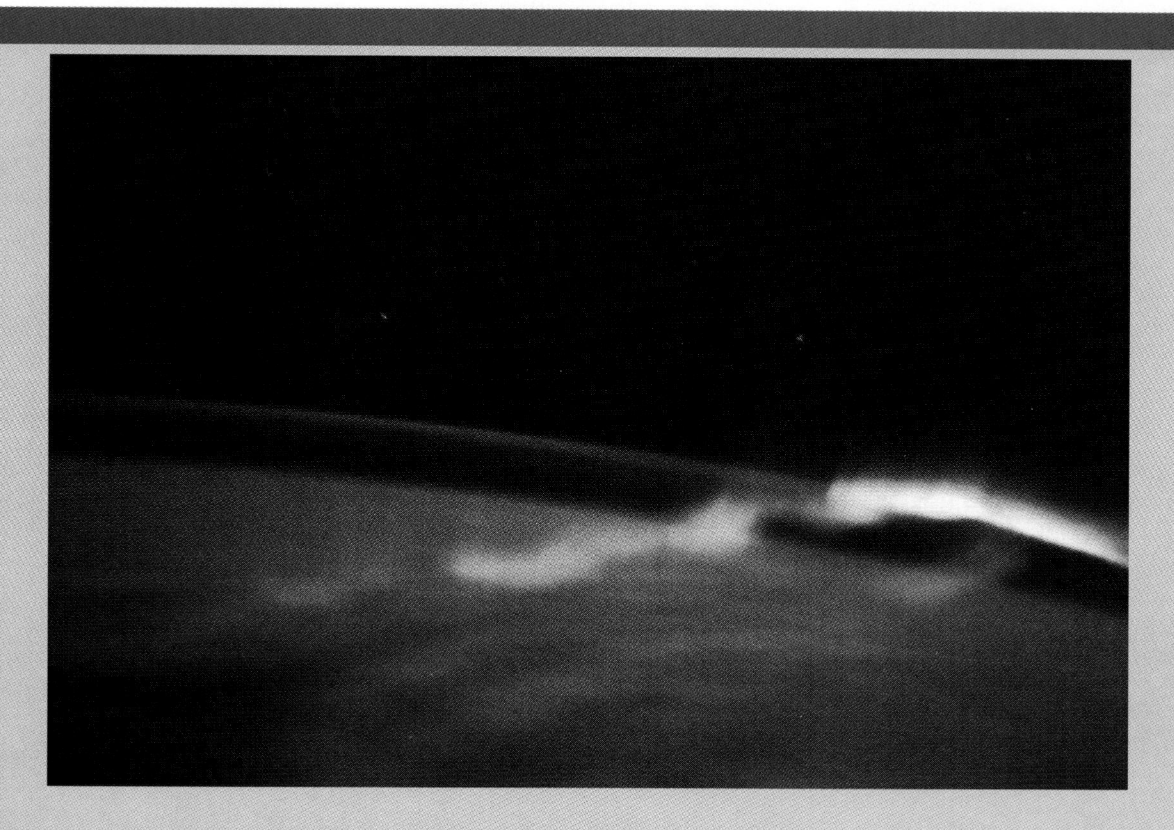

can spiral downward into the atmosphere. As a consequence,the number of charged particles from space that reach the earth's surface is much greater near the poles than in equatorial regions.

The intensity of the solar wind picks up whenever a magnetic "storm" breaks out on the sun's surface, signaled by the appearance of many sunspots or a nearly invisible solar flare. Then earth's sun-side magnetic field compresses as the increased solar wind (which is a current of moving charged particles) adds its own magnetic field to earth's, and the belts dip lower at the poles. If the compression is severe enough, the belts spill charged particles into the upper atmosphere. At 100 km (60 mi) up, those particles (mostly electrons) stimulate atmospheric atoms and molecules to emit light in an action called luminescence (Chapter 24). That light is called an *aurora*. When this happens in the Northern Hemisphere, it is an aurora *borealis* (or northern lights); in the Southern Hemisphere, it is an aurora *australis.*

In the last chapter we saw that earth's magnetic field has reversed itself many times in the past. When it reverses, the field goes to zero, and the change apparently takes place over a period of 4000 to 8000 years. With no magnetic field to deflect them, the number of solar wind and cosmic radiation particles reaching earth's surface increases dramatically. This ionizing radiation is harmful to life, and if the magnetic field keeps changing at the present rate, we may be only 2000 years away from the next reversal.

COSMIC AND SOLAR RADIATION AND EARTH'S MAGNETIC FIELD

Push a pole of a magnet through the center of a coil of wire (Figure 20-8) and a current appears in the coil in one direction; pull the magnet back out and a current appears in the other direction. The faster the magnet is inserted or withdrawn, the greater the current will be. Keep the magnet steady, so that its field doesn't change across the conducting coil, and there is no induced current.

To summarize, *change the net magnetic field in any way across a closed conducting circuit, and a current is produced.* In fact, if you merely wave a coil of copper wire in the air as Faraday once did, the weak magnetic field of the earth induces enough current in that coil to deflect the needle of a sensitive ammeter. Or just take a class ring or a gold wedding band and use

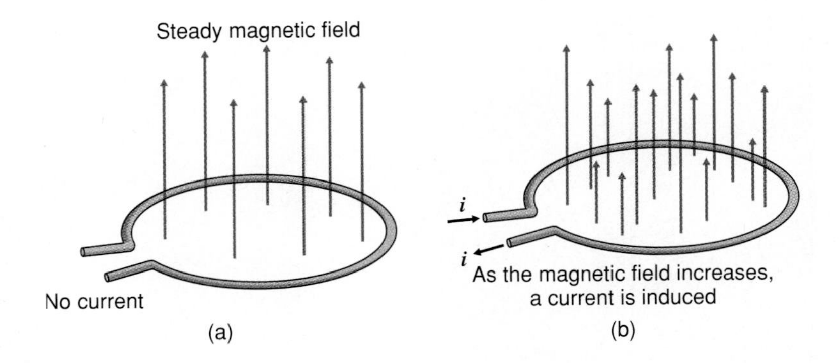

FIGURE 20-7 A CHANGE IN THE MAGNETIC FIELD THROUGH THE LOOP INDUCES AN ELECTRIC CURRENT IN THE LOOP. (A) A STEADY MAGNETIC FIELD INDUCES NO CURRENT. (B) AS THE FIELD STRENGTH GROWS, A CURRENT IS INDUCED IN THE COIL.

one finger to stand it vertically on a tabletop. Flick it on one side with another finger to set it spinning like a top and a current goes around the circle of the ring because of earth's magnetic field.

FIGURE 20-8 (A) A CLOSED CIRCUIT IN A STATIONARY MAGNETIC FIELD. THERE IS NO CURRENT IN THE CIRCUIT, SO THE METER READS ZERO. (B) A CHANGING MAGNETIC FIELD ACROSS THE LOOP INDUCES ELECTRONS TO MOVE AS SHOWN. (C) WHEN THE DIRECTION OF THE MAGNET IS REVERSED, SO IS THE DIRECTION OF THE ELECTRON MOTION.

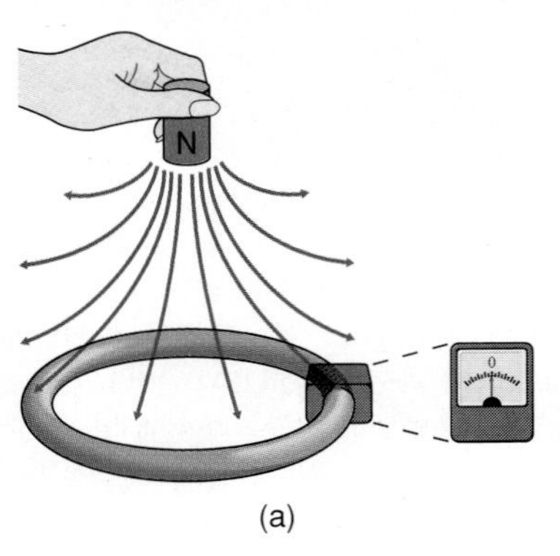

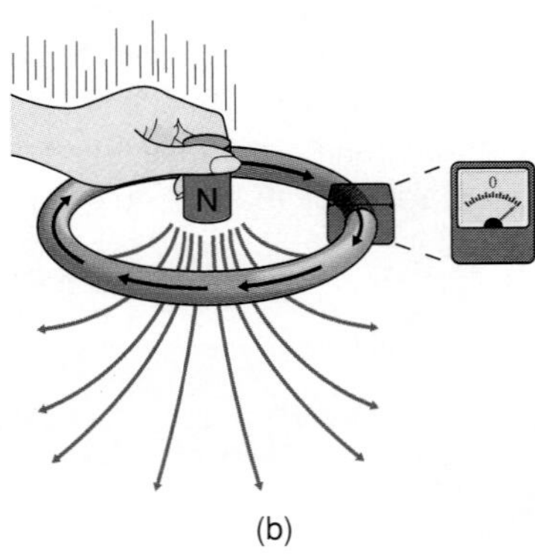

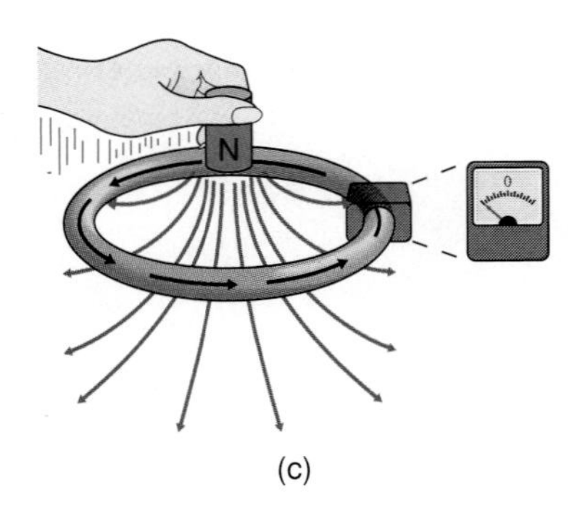

GENERATORS OF ELECTRICITY

Faraday saw how he could use electromagnetic induction to produce a steady electric current, and he built a hand-cranked **electrical generator** (Figure 20-9). He mounted a solid copper disk on a conducting axle and brought a magnet up close to the disk. When he turned the disk, the conduction electrons in the copper passed through the magnetic field lines and received a sideways push, toward the edge of the disk, as in Figure 20-9b. To provide a closed circuit for the moving charges, he touched one end of a copper wire to the axle and the other end to the edge of the spinning disk. The conduction electrons in the disk then had a complete conducting path and a push along that path, so charge moved through the wire. The current was steady so long as the disk turned at a steady rate. As Faraday turned the crank, much of the mechanical energy he put into rotating the copper disk went into the energy carried by the electric current.

Faraday's hand-cranked electric generator soon evolved into larger machines to convert the energy of falling water or high-pressure steam to electricity. Of special interest are the generators that supply the AC electricity you use in home appliances. We can see how they work with an example. If you put a strip of aluminum foil (a good conductor) around the edges of this book as shown in Figure 20-10 and then spin the book in the earth's magnetic field, an alternating current will appear in the foil strip. When the face of the rotating book is perpendicular to the field lines, as in step 1, the conduction electrons in the foil move parallel to the lines, so they feel no force and there is no current. As you rotate the book, however, the movable electrons in the foil cut the field lines as in step 2 and generate electron motion as shown. (Notice that edges A and B have opposite velocities, so the charge in them flows in opposite directions, as it must if there is to be a loop of current.) In step 3 the face of the book becomes perpendicular to the field lines again, and there is again no current. Next, in step 4, the long edges once again cut field lines and charge flows, but in the opposite direction from step 2. The current in the foil is thus an alternating current, and part of the mechanical energy you use to rotate the book goes into the electrical energy carried by the current. Rotating coils of wire (instead of single loops) in magnetic fields is a common way to produce alternating current.

Figure 20-11 shows how mechanical power delivered to a rotating conducting coil (only one loop is shown) sends AC to a light bulb. Either falling water or rushing steam pushes the blades of a turbine to turn an axle. A coil attached

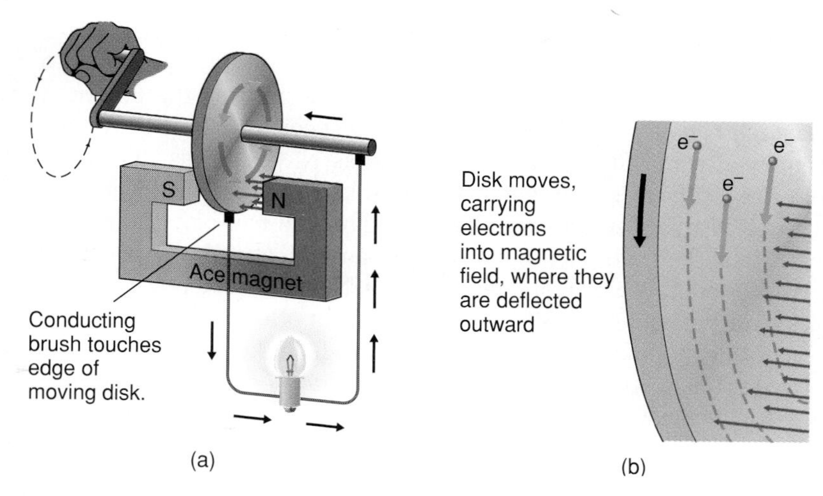

FIGURE 20-9 (A) IF A METALLIC DISK IS TURNED IN A MAGNETIC FIELD, THE ELECTRONS IN THAT DISK GET A SIDEWAYS PUSH AS THEY ARE CARRIED PERPENDICULARLY TO THE MAGNETIC FIELD LINES. IF A COMPLETE CIRCUIT IS FORMED, ELECTRONS WILL FLOW THROUGH IT AS SHOWN. SUCH A DEVICE IS CALLED AN ELECTRIC GENERATOR. (B) A CLOSE-UP OF THE AREA ON THE DISK WHERE THE MAGNETIC FIELD IS LOCATED. CONDUCTION ELECTRONS MOVING THROUGH THE FIELD ARE DEFLECTED TO THE EDGE OF THE DISK.

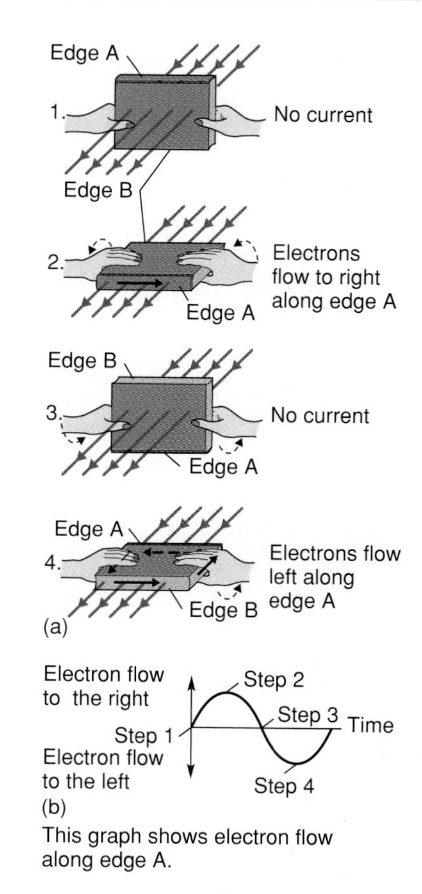

FIGURE 20-10 (A) GENERATING AN ALTERNATING CURRENT, STEPS 1 THROUGH 4. (B) A GRAPH OF THE CURRENT GENERATED BY THE SPINNING BOOK VERSUS TIME.

to this axle is in a region where there is a magnetic field. The movable electrons on one side of the loop get a push one way, while those on the other side get an opposite push, just as with the foil strip around the book. If the ends of the loop are connected to a circuit outside the generator, charge will flow. Some of the mechanical energy of the water or steam is converted to the electrical energy carried by the current.

Commercial electrical generators in North America deliver an alternating voltage that oscillates 60 times a second. If you visit Europe, you will need an adapter for your hair dryer or electric razor, for there electrical outlets are 220 volts and generators oscillate at 50 Hz. Commercial clocks that use 60-Hz voltage to keep time in the United States register only 50 minutes for every hour when in Europe.

DELIVERING ELECTRICAL POWER EFFICIENTLY

TRANSMITTING ELECTRICITY: TRANSFORMERS

Years after the breakthroughs of Faraday and Ampère, other scientists and inventors learned to make practical use of the power produced by electrical generators. The first commercially produced light bulb appeared in 1880, and in no time, New York City had the world's first electric power utility plant to sell electrical energy from its generators to the public. But as conducting wires from that plant were taken to farther and far-

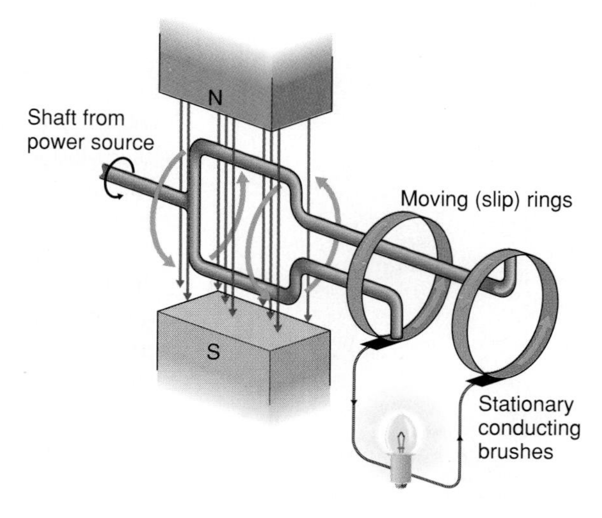

FIGURE 20-11 A SCHEMATIC OF A MODERN GENERATOR (SHOWING ONLY A SINGLE LOOP OF WIRE) THAT GENERATES AN ALTERNATING CURRENT AND VOLTAGE. HALFWAY THROUGH EACH REVOLUTION OF THE LOOP, THE CURRENT REVERSES DIRECTION AS THE SIDES OF THE REVOLVING LOOP CHANGE THEIR DIRECTIONS ACROSS THE MAGNETIC FIELD. STATIONARY BITS OF GRAPHITE, OR BRUSHES, PRESS AGAINST THE ROTATING RINGS TO PROVIDE A COMPLETE CIRCUIT. A STEAM TURBINE USUALLY PROVIDES THE POWER TO TURN THE COILS THROUGH THE MAGNETIC FIELD, ALTHOUGH FALLING WATER OR WIND CAN BE USED.

ther points, the longer lines meant more resistance. As a result, the wires turned much of the electrical energy produced back at the utility plant into heat, meaning much of the energy was wasted in transmitting the power. A more efficient method of transmission was needed, especially to take advantage of more remote natural energy resources, such as Niagara Falls. Its falling water could drive generators at minimal cost, but the great distance from population centers was a serious drawback. Today, you can play a tape, fry an egg, or wash clothes with electricity only because electrical power can be transmitted in conducting wires for long distances. Next we'll see how alternating currents made possible this efficient transmission of electric power.

$P = IV$ THROUGH THE TRANSMISSION LINE, WHICH HAS A RESISTANCE R. BUT $V = IR$, AND USING THIS WE FIND $P = IV = I(IR) = I^2R$. BY KEEPING I VERY SMALL, THE ENERGY LOST PER UNIT OF TIME TO HEATING THE RESISTANCE (I^2R) IS MINIMAL.

There is nothing a power company can do to reduce the resistance of the long copper wires (called *powerlines* or *transmission lines*), but it can reduce the current the lines carry to minimize energy loss. How does a generating plant deliver power with low currents? The answer lies in the formula for electrical power, power = current × voltage. To deliver substantial power at low current requires a high voltage.

Commercial AC electric generators produce electrical power at a modest voltage, perhaps 2200 volts. Then the voltage is increased to 120,000 volts or perhaps up to 500,000 volts before it is applied to the long-distance transmission lines. (Remember, electrons aren't transmitted from the power plant to a home; they only vibrate back and forth. Only energy is transmitted.) In the vicinity of the consumers' homes, the high voltage is lowered to 7200 volts or so. Then even closer to home, usually at a power pole a block or less away, the voltage is lowered to a level that is safer to use, normally 220 volts or 110 volts. Transformers, devices much like the one Faraday used to discover electromagnetic induction, change the voltages up and down.

A **transformer** is two separate coils of insulated conducting wire wrapped around an iron core (Figure 20-12). If one coil (called the primary) carries a current, the magnetic field this current produces inside the iron core is shared almost totally with the other coil (called the secondary). The primary carries an alternating current, which surges back and forth and therefore creates an alternating magnetic field. The moving magnetic field lines cut across the secondary, inducing an alternating force on the conduction electrons there.

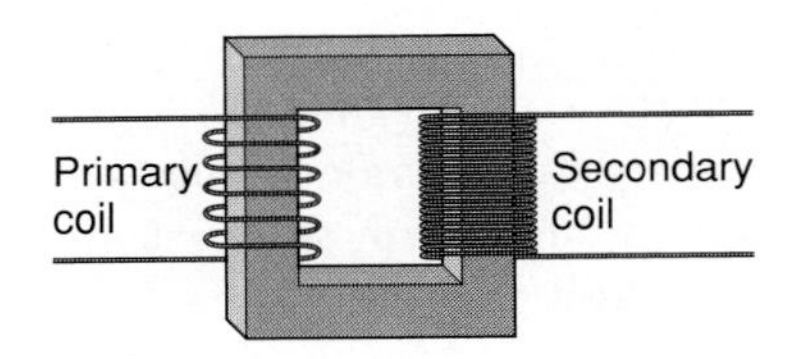

FIGURE 20-12 A TRANSFORMER WITH 5 LOOPS IN THE PRIMARY COIL AND 20 LOOPS IN THE SECONDARY. THE MAGNETIC FIELD INDUCED IN THE IRON CORE IS SHARED EQUALLY BY EACH LOOP OF WIRE AROUND THE IRON, AND THE SAME VOLTAGE WILL PASS THROUGH EACH LOOP.

The net magnetic field that goes through each loop on the primary passes through each loop on the secondary. Faraday and others discovered that this equal magnetic coupling of all the loops *causes each loop to have the same voltage across it.* No matter what the voltage is across the primary or how many loops it contains, the voltage *per loop* in the primary is the same as the voltage *per loop* induced in the secondary. A direct current would not generate a changing magnetic field, so the voltage from a DC generating plant could not be stepped up or down with transformers unless the direct current changed frequently.

Here is what happens in a transformer that has 5 primary and 20 secondary loops, as in Figure 20-12. If there are 10 volts across the primary, giving it 2 volts per loop, the secondary will have 2 volts per loop induced for each of its 20 loops, for a total of 40 volts across it. This transformer therefore steps up the voltage by a factor of 4. In a step-down transformer, the secondary has fewer loops than the primary.

Transformers increase or decrease voltage, but energy is conserved, so the power out cannot be any more than the power in. Transformers waste very little electrical energy. In commercial transformers, the electrical energy coming from the secondary is more than 90 percent (sometimes up to 99 percent) of the energy going through the primary. This means the power into the primary is approximately equal to the power from the secondary.

PUTTING IT IN NUMBERS—CALCULATING TRANSFORMER OUTPUTS

$i_p \times V_p \approx i_s \times V_s$

where $\approx$ means "approximately equal to."

Example: A neighborhood step-down transformer takes the voltage from 7200 V to 120 V. If the current used in a home taking power from that transformer is 40 amps, what is the current in the primary of the transformer? $i_p \approx (i_s \times V_s)/V_p = (40\text{ A} \times 120\text{ V})/7200\text{ V} = \mathbf{0.67\text{ A}}$

The next time you press the button on a doorbell, think of a transformer. The voltage coming from the house wiring is 120 volts, enough to shock you. However, a small transformer reduces the voltage between doorbell and button so that a wet finger on the button is less likely to receive a serious shock. Likewise, a toy electric train runs at reduced voltage to protect the children's fingers (and the dog's nose) should they complete a circuit by touching both rails of the track while the switch is on. Those high voltage electrical lines carried by high towers across the landscape from power plants lose only 4 to 5% of their energy to heating the wires over long distances, thanks to step-up transformers.

MOTORS

Washing machines, refrigerators, vacuum cleaners, blenders, dishwashers, and fans all use an **electric motor,** which is a device that converts electrical energy to mechanical energy (Figure 20-13). This is how they do it. A coil of wire wrapped around an axle sits in a magnetic field. You flip a switch, and the ends of this coil complete a conducting path with the electric circuit in your home. There is then a current in the coil. As electrons in the wire move through the magnetic field, they get a sideways push that also gives the wire a sideways push. That push on the wire causes the coil to rotate and turn the motor's axle.

A motor is essentially an electrical generator that works in reverse. Whereas a generator turns mechanical energy into electrical energy, a motor turns electrical energy into mechanical energy.

There's an interesting bit of history behind motors. Faraday's first generator was made in 1831, but the electric motor did not come into being until 1873, at an exhibition in Vienna. An unknown worker there made a mistake

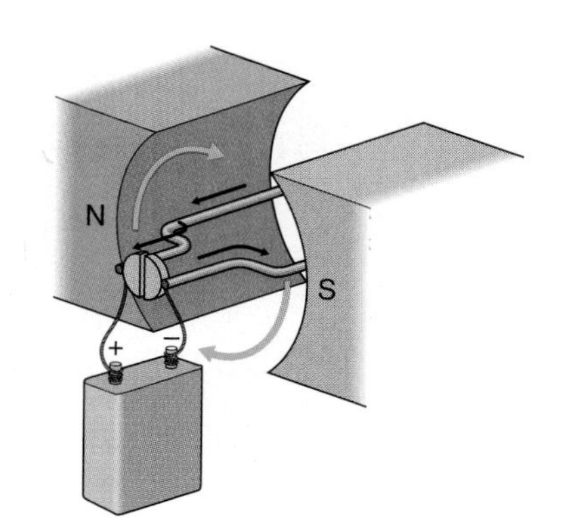

FIGURE 20-13 A SCHEMATIC OF ONE LOOP OF A COIL IN AN ELECTRIC MOTOR OF A HOUSEHOLD APPLIANCE. THE ELECTRONS MOVING THROUGH THE LOOP GET SIDEWAYS DEFLECTIONS FROM THE MAGNETIC FIELD, AND THESE DEFLECTIONS CAUSE THE LOOP TO TURN. THE COILS OF SUCH A MOTOR ARE WOUND AROUND AN AXLE (NOT SHOWN) THAT IS USED TO DRIVE THE BLENDER, VACUUM CLEANER, REFRIGERATOR COMPRESSOR, ETC.

THE REWARDS OF SCIENTIFIC RESEARCH

Faraday probably never imagined the ultimate effect his generator of electric current would have. Indeed, he knew he couldn't foresee its consequences, as we can tell by his replies to inquisitive people who asked the question, "Of what use is it?" The prime minister of England saw Faraday's demonstration and asked what use such an invention might possibly have. Faraday replied, "Why, Sir, someday you will be able to tax it." That was a prophetic answer, as anyone who pays electricity bills knows. The same question is often asked of people who do research in science today. "Why should the government (or industry, or private foundations) support scientific explorations that may never pay off (especially by fiscal year 1996 or 1998)?" "What could society miss by not sponsoring scientific research?" is a better question. Historically the answer is clear: A lot.

in connecting some conducting wire from one electrical generator to another. When the first generator was switched on, the coils of the second began whirling around at a great speed, transforming the electrical power of the first generator back into mechanical power. Needless to say, it was a significant accident!

Summary and Key Terms

A magnetic field exerts no force on a charged particle that is either at rest or moving parallel to the magnetic field lines. If a charged particle moves across the field lines, however, the field pushes the particle *perpendicular* to both the field lines and its velocity, exerting a turning force. Faraday showed that changing magnetic fields produce electric currents in a closed circuit, an action called ***electromagnetic induction***.

electromagnetic induction–294
electrical generators–296

Today we get our electricity from ***electrical generators***, which convert mechanical energy to electrical energy. These generators use falling water or high-pressure steam to push the blades of a turbine, which turns a conducting coil within a magnetic field, generating an alternating current. As powerlines carry this current to remote areas, the resistance of the conducting wires to the current turns electrical energy into heat. Power companies minimize power loss by lowering the current and increasing the voltage in the lines. The ***transformers*** used to step the voltage up and down consist of two separate coils of insulated conducting wire wrapped around an iron core.The primary coil carries an alternating current, which creates an alternating magnetic field. The net magnetic field that moves through each loop on the primary passes through each loop on the secondary, causing *each loop to have the same voltage difference across it.* Little energy is lost to heat, so the power out is about equal to the power in: $i_{primary} \times V_{primary} \approx i_{secondary} \times V_{secondary}$.

transformers–298

electric motors–299

Electric motors convert electrical energy to mechanical energy. In a motor, a coil of wire wrapped around an axle sits in a magnetic field. When there is a current in that coil, the electrons in the wire move through the magnetic field, getting a sideways push that is transferred to the wire, causing the coil and its axle to rotate.

EXERCISES

Concept Checks: Answer true or false

1. A charged particle moving perpendicular to magnetic field lines experiences a force that is perpendicular to the field lines and to the velocity of the particle.
2. Faraday discovered that a changing magnetic field through a conducting circuit induces a current in that circuit.
3. Ampère found that a charged particle crossing a magnetic field gets a sideways push; Faraday saw that magnetic field lines in motion across a charged particle produce a similar push.
4. An electric generator is a device that converts mechanical energy into electrical energy.
5. In commercial power plants in North America, electrical generators deliver an alternating voltage that oscillates 60 times a second.
6. A transformer transforms current into voltage.
7. Power companies use high-voltage lines to transmit electrical power in order to lower the energy lost during transmission. The voltage is then stepped down so the power can be used at lower current levels.
8. An electric motor is a device that converts electrical energy into mechanical energy.

Applying the Concepts

9. An electron at rest in a steady magnetic field experiences no force from the field. True or false? Does that electron exert a force on the magnet that gives rise to the steady magnetic field? Why or why not?

10. The underwater lights you see in the sides of swimming pools use a step-down transformer to operate with a voltage lower than 120 volts. Why should such lights need to operate at a reduced voltage?

11. Describe the magnetic field around a long, straight, current-carrying wire.

12. Did the copper disk that Faraday used to make the first generator create alternating current or direct current?

13. How can a magnetic field be used to move a charged particle that's initially at rest?

14. As your car moves along the road, it cuts through lines of the earth's magnetic field. The metallic parts of the car are conductors, so there is a voltage generated across the car as a result. What does speeding up do to this induced voltage?

15. Is the current through the coil of the generator in Figure 20-9 alternating or direct?

16. At airports today you can see security guards passing a metal loop over the body of passengers who have tripped a signal from a larger metal loop they just passed through. What's going on?

17. Could you hold the coil in a generator stationary and rotate the magnet to obtain a current?

18. Explain how energy can be transmitted in a power line that sits still day and night with nothing visible going on. Why don't the wires move? Why don't electrons flow into one end and out the other?

19. Would a transformer work if a direct current were used rather than alternating? (Careful!)

20. If two coils are placed side by side and there is an alternating current through one of them, will a similar alternating current be induced in the other? Discuss. (Simple phone taps work like this.)

21. In a television tube, electromagnets deflect a fast-moving beam of electrons that sweeps across the fluorescent screen, producing visible light. Do these electromagnets change the speed of the electrons or only their direction?

22. Any coil of wire in the orbiting space shuttle has a small urrent in it, even if no potential difference is applied to its ends. Explain.

23. A transformer has an input of 10 volts and an output of 30 volts. If the input is changed to 15 volts, what will the output be?

24. Why is it more accurate to say electrical power is transmitted in conducting wires rather than transported?

25. A small device mounted next to the rim of some bicycle wheels powers a headlight for use at night. What's that device called? Where does it get its energy?

26. How does coiling a wire (making loops that lie on loops) intensify the magnetic field produced by the wire?

27. Higher voltage lowers the current for long-distance transmission of electrical power. So why don't power companies step up the voltage from 120,000 volts to, say, 100 million volts? (*Hint:* Glance over the last section of Chapter 17.)

28. If home circuits used 1500-V lines rather than 120-V lines, they would operate more efficiently. Why won't power companies do this? Why did automobile companies switch from 6-V to 12-V systems in the 1950s?

29. Fluorescent lamps use a high voltage to create a plasma hat gives off light. You can often hear their transformers humming as the currents inside oscillate at 60 Hz. Why should parts of the transformer vibrate, causing that 60-Hz sound wave?

30. Give three or four ways you could induce a current in a loop of copper wire.

31. Recording cassettes contain a plastic ribbon (the tape) coated with iron oxide or chromium oxide, both of which have magnetic domains. To record, sound waves create a varying electrical current in a microphone. An amplifier magnifies this current, and the amplified current is then used to operate an electromagnet in the recording head. As the tape passes through that head, the fluctuating magnetic field aligns the domains in the tape. The magnetic pattern thus created can be used to make a sound very much like the original sound. Describe how the tape recorder can use the same magnetic head to replay the tape.

32. A ski lift carries a passenger to the top of a slope. The energy to do this comes from the swarm of ultralight electrons that sway back and forth a fraction of a millimeter even as they jitter around with their normal thermal motion within the copper wires of the lift's motors. Explain how such tiny motions can deliver so much energy.

33. Not far from San Francisco, there is a geyser field where ground water passes through heated rock and turns into steam. This steam is then used to power steam turbines, which turn electrical generators. Meanwhile, a 750-horsepower electric motor in San Francisco uses some of that energy to move a cable car. Trace all the steps in the harnessing, transmission, and use of the heat energy from the earth's interior.

34. The input to the transformer in Figure 20-14 is 110 volts. What is the output? If 1 amp is the current in the primary, about what is the current in the secondary?

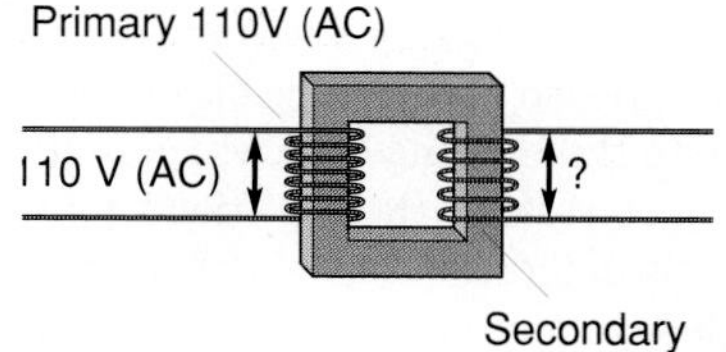

FIGURE 20-14

35. Some traffic signals detect an approaching car and quickly change to let the car pass. If you look, you can often see a thin cut in the asphalt road in the shape of a rectangular loop just in front of the intersection. A conducting wire is beneath that cut. Describe what is happening.

36. In car radios there are power converters that use voltages higher than the 12 volts supplied by the car's battery. A small transformer steps up the voltage—but wait! Battery current is direct, not alternating, so how can a transformer possibly work? Think about this for a moment, and then read the answer at the back of the book.

37. The small lamps that provide light for the dials on stereos, radios, and TVs may require only 4 or 6 volts. Using the 110-volt alternating current from a home circuit, you could light these lamps in two ways. A step-down trans ormer could provide the voltage they need, or a large resistor could be placed in series with the lamps. Most of the 110 volts across the circuit would be used to push the current through the large resistor, leaving only 4 or 6 volts across each bulb. Which method is more energy-efficient?

38. James Clerk Maxwell, whose accomplishments you will read about in the next chapter, was quoted in a popular magazine in 1883 as saying that the reversibility of the electric generator into a motor was "the greatest scientific discovery of the last quarter of a century." Take the view that this was probably a misquote and explain why.

39. Do the copper wires in your home lead to other wires that form an unbroken conducting link all the way to the power company? Explain. Note that we use AC voltages.

40. If the voltage in the secondary of a transformer is 4 times the voltage in the primary, show that the current in the secondary is 1/4 the current in the primary.

41. In the late 1800s, there was a great debate about whether AC or DC should be distributed by power companies. DC lost. What arguments can you think of to support this historical decision?

42. Look at Figures 20-2 and 19-14 and tell in which direction an eastward-moving beam of electrons at the earth's magnetic equator would be deflected.

43. Would a metal ring that is spun like a top in the earth's magnetic field have a current in it if a small slice is cut from the ring so that it doesn't make a complete circle?

44. Explain how you could move a loop of conducting wire through a magnetic field without inducing a voltage in the loop.

45. What happens to the voltage produced by a electric generator if the rate of spin of the coils is increased?

46. When a baseball player swings an aluminum bat, is there a voltage induced between the ends of the bat? Is there a current through the bat? Explain why or why not.

47. When the metal end of the second hand on a wall clock rotates, is there necessarily a voltage induced between the ends of that hand? Discuss.

48. The voltage used to cause a spark across the spark plug of a car is about 15,000 volts. However, the source of electrical power is a 12-volt battery. A transformer in the car called the "coil" converts 12 volts to 15,000 volts. How many loops does the secondary in that transformer have for every loop in the primary?

49. In Figure 20-15, a bar magnet is dropped through the center of a vertical coil of wire that is connected to an ammeter. As the magnet passes through the coil, the meter indicates a current through the coil. At the same time, the magnet accelerates downward with an acceleration less than g. Why? (Air resistance is *not* the major cause. Remember conservation of energy!)

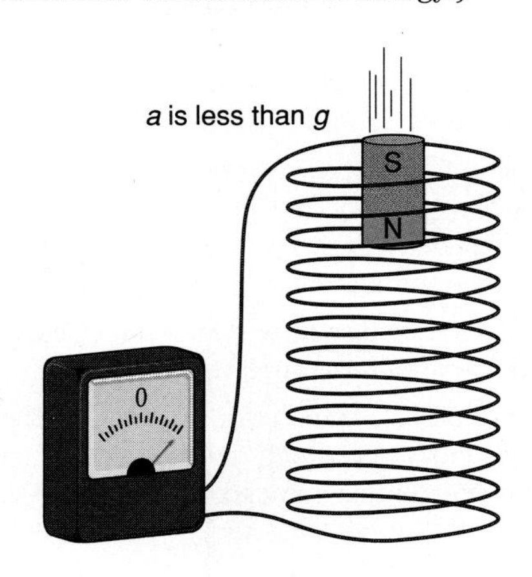

FIGURE 20-15

Demonstration

This demonstration is worth your while. Ask your instructor for the materials. You'll need two coils of thin insulated copper wire (each with numerous loops), two magnets, and two short lengths of insulated copper wire.

Connect the two coils with the wires. Place one coil on a table and suspend the other over the table's edge. Use a book to pin the connecting wires to the table so that the hanging coil doesn't pull on the other one. Bring a magnet up close to the hanging coil (which is at rest), and hold the magnet steady. Have a friend move the second magnet back and forth past the coil on the table, as close as possible without touching. The changing magnetic field this movement causes induces a current in the table coil. Electrons move in the connecting wires and in the hanging coil as well. The current in the hanging coil gets a push from the magnetic field of the magnet you are holding next to the coil, and the coil swings back and forth as a result.

The table coil and the moving magnet represent an electrical generator. The relative motion between the magnetic field and the movable electrons in the coil creates an electric current. The two connecting wires are the transmission lines. The hanging coil and stationary magnet represent an electric motor, which uses the push from a magnetic field on a current to convert electrical energy into mechanical energy.

TWENTY-ONE

Light: Electromagnetic Radiation

A beam of white light enters a prism and disperses into the colors of the spectrum. Newton's experiments with prisms like this one led to discoveries about light's properties.

When you read a book, watch TV, or just enjoy the scenery, your eyes detect light that comes from those things before you. Some of the early Greek philosophers thought light radiated from a person's eye, struck an object, and returned to the eye, a point of view that seems strange to us now. (Pull the curtains in a room and it gets dark, so your eyes certainly aren't flooding the scene with beams of light!) Today we know that the source of most of the light around us is the sun. The energy sunlight brings keeps the earth warm, the oceans liquid, and the atmosphere aloft.

In the early 1700s, there were two ideas about what light was. Newton argued that light was composed of tiny, fast-moving particles; other people thought that light was vibrations, waves of some sort. In 1802 an experiment showed that light has a wave nature, and in 1865 the gifted Scottish physicist James Clerk Maxwell (1831–1879) formulated the first successful theory of the wave nature of light. The predictions of Maxwell's theory were astounding, leading to the understanding of an enormous amount of phenomena concerning both visible and invisible light, all of which is called *electromagnetic radiation*—the subject of this chapter. We shall concentrate on the wave nature of light in this chapter, but in the early 1900s discoveries showed that light also has a particle nature that becomes apparent at the atomic and subatomic levels (Chapter 24).

YOUNG'S EXPERIMENT: THE WAVE NATURE OF LIGHT

Thomas Young, an Englishman, did an experiment in 1802 that proved light has wave properties. He showed that two similar beams of light can meet and *disappear* at certain points, a difficult thing to imagine with particles. Remember the discussion of waves on a string in Chapter 15. If two similar waves traveling in opposite directions on a string met (see Figure 15-9b), the two cancel perfectly when a crest of one wave passes some point just as a trough arrives from the other wave, and the string lies flat for that instant. Those waves *interfere destructively*. The same interference occurs with sound waves and water waves, and Young's experiment proved that this also happens with light, showing that light acts like a wave.

Young's experiment is illustrated in Figures 21-1 and 21-2. He sent a bright beam of light through a small vertical slit in an opaque screen. Some distance from the side of the screen opposite the light source, he placed a second screen that had two parallel vertical slits to let the light through. He reasoned that, if light were indeed a wave, the waves from the slit in screen 1 would diffract (spread out) much as ocean waves do when they pass through a hole in a jetty (see Figure 15-15), and that the wavefronts would become parallel as they approached the two slits in screen 2. In this way crests would emerge simultaneously from each slit in screen 2, followed by troughs, then by crests, and so on. Because there was only one light source (the slit in screen 1), the crests of the two waves had the same amplitude. When the light from each slit in screen 2 emerged, it diffracted just as it did at the slit in screen 1. So light from both slits fell onto the entire area of a third screen, and when it did Young discovered bright bands (lots of light, constructive interference) alternating with dark bands (little or no light, destructive interference), as shown in Figure 21-3.

FIGURE 21-1 LIGHT FROM A SOURCE PASSES THROUGH A SINGLE SLIT IN A SCREEN. AS IT DOES, IT DIFFRACTS, SPREADING OUT INTO THE REGION BEHIND THE SCREEN. NOTICE THAT, THE FARTHER THE LIGHT TRAVELS FROM THE SLIT, THE MORE PARALLEL THE WAVEFRONTS BECOME. SO WHEN THIS LIGHT STRIKES A SECOND SCREEN, THIS ONE HAVING TWO SLITS (SEE FIGURE 21-2), THE LIGHT ARRIVING AT THE TWO SLITS WILL BE "IN STEP" (CRESTS ARRIVING AT THE SAME TIME) AND WILL HAVE THE SAME AMPLITUDE.

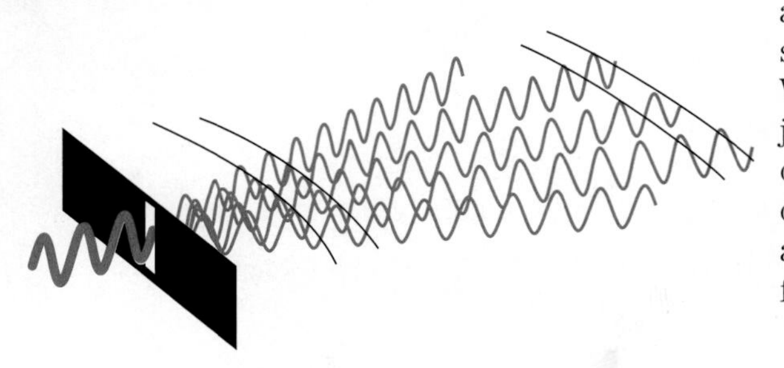

Even though there was just as much light from each slit

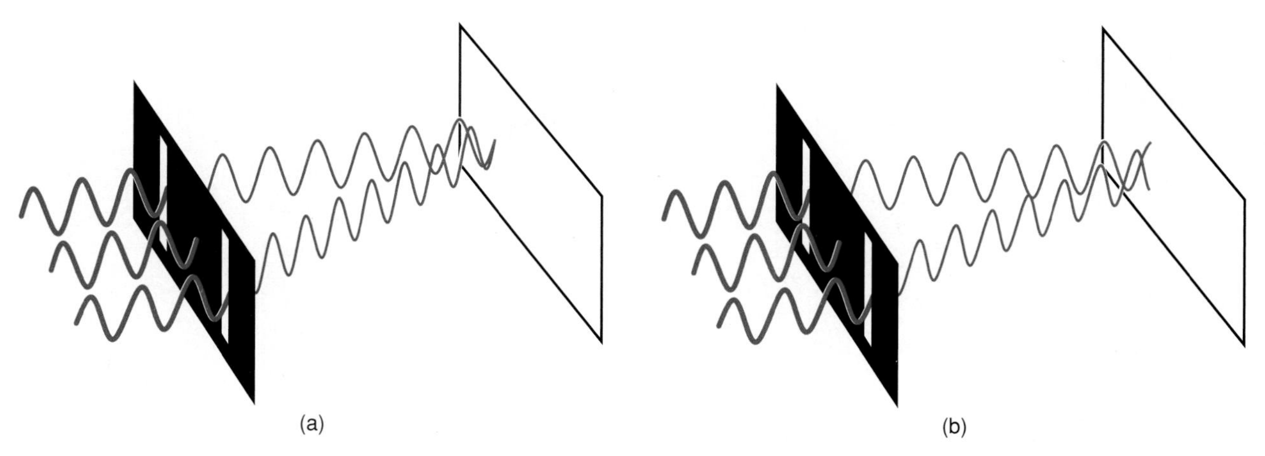

FIGURE 21-2 (A) AT ANY POINT ON THE THIRD SCREEN LIGHT ARRIVES FROM BOTH SLITS OF THE SECOND SCREEN SINCE DIFFRACTION OCCURS AT BOTH SLITS. IF A CREST OF ONE WAVE ARRIVES AT THE SAME TIME AS A CREST OF THE OTHER, THE WAVES ADD CONSTRUCTIVELY AND THERE IS A LOT OF LIGHT AT THE POINT WHERE THEY HIT THE THIRD SCREEN. (B) AT POINTS ON THE SCREEN WHERE ONE WAVE TRAVELS EXACTLY $\frac{1}{2}$ WAVELENGTH, OR $\frac{3}{2}$ WAVELENGTHS, OR $\frac{5}{2}$ AND SO ON, FARTHER THAN THE OTHER, THE LIGHT WAVES *CANCEL* SINCE A CREST OF ONE WAVE ARRIVES AT THE SAME TIME AS A TROUGH OF THE OTHER. ALTHOUGH LIGHT FROM BOTH SLITS REACHES THESE POINTS, IT IS DARK THERE.

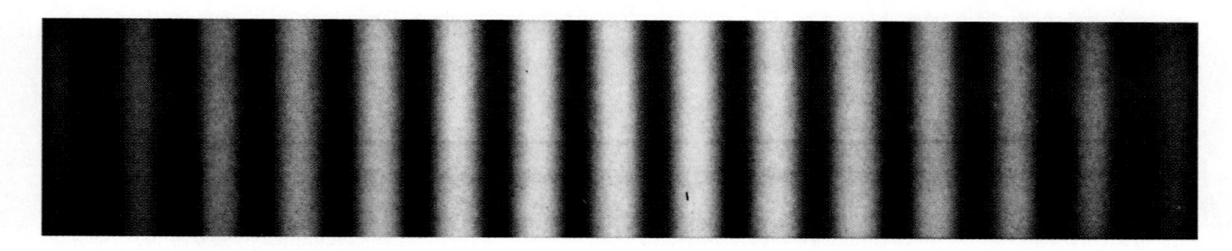

FIGURE 21-3 AN INTERFERENCE PATTERN FROM A DOUBLE-SLIT EXPERIMENT, AS IN FIGURE 21-2.

PHYSICS IN 3D/DIFFRACTION FROM A SINGLE SLIT

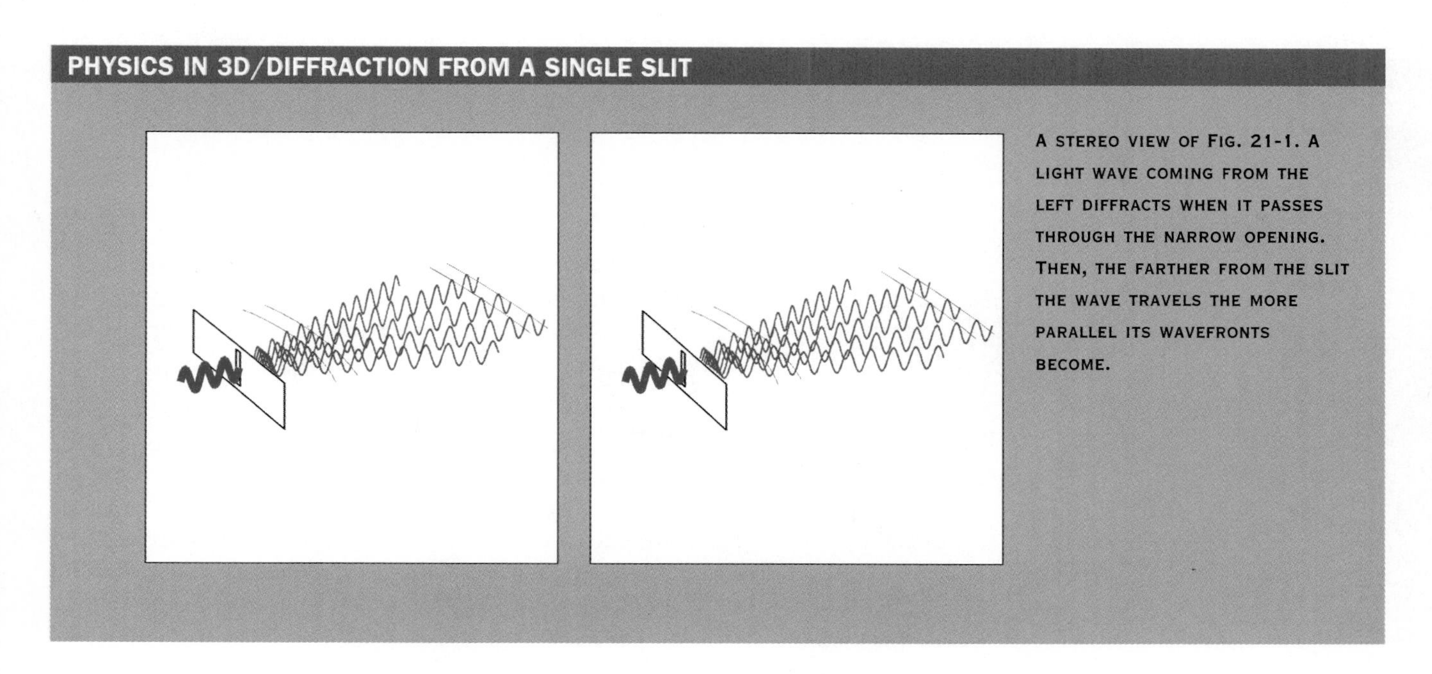

A STEREO VIEW OF FIG. 21-1. A LIGHT WAVE COMING FROM THE LEFT DIFFRACTS WHEN IT PASSES THROUGH THE NARROW OPENING. THEN, THE FARTHER FROM THE SLIT THE WAVE TRAVELS THE MORE PARALLEL ITS WAVEFRONTS BECOME.

PHYSICS IN 3D/A POINT OF CONSTRUCTIVE INTERFERENCE IN YOUNG'S EXPERIMENT

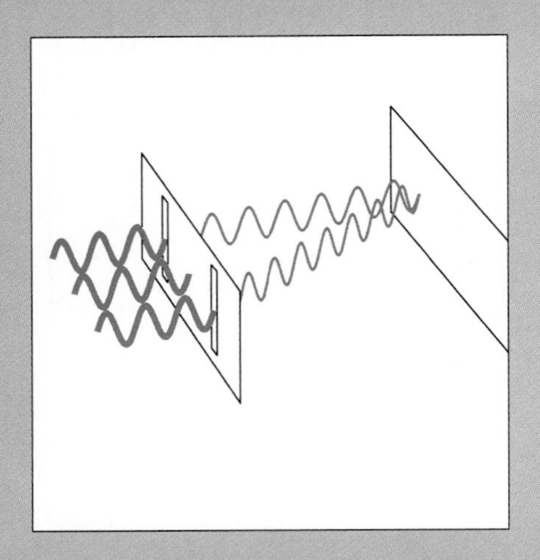

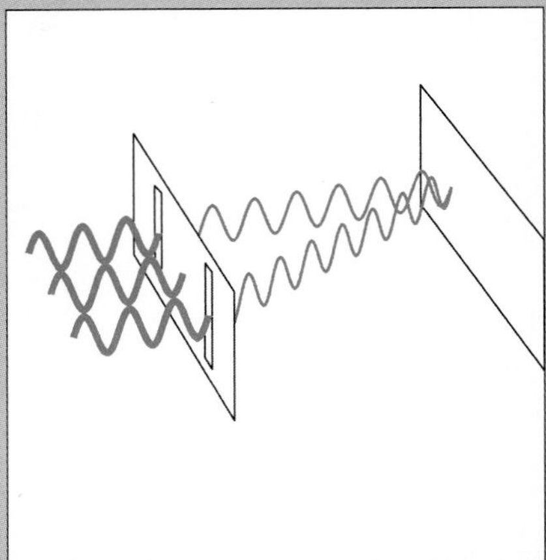

A STEREO VIEW OF FIG. 21-2(A), SHOWING A POINT OF CONSTRUCTIVE INTERFERENCE ON THE SCREEN ON THE RIGHT.

PHYSICS IN 3D/A POINT OF DESTRUCTIVE INTERFERENCE IN YOUNG'S EXPERIMENT

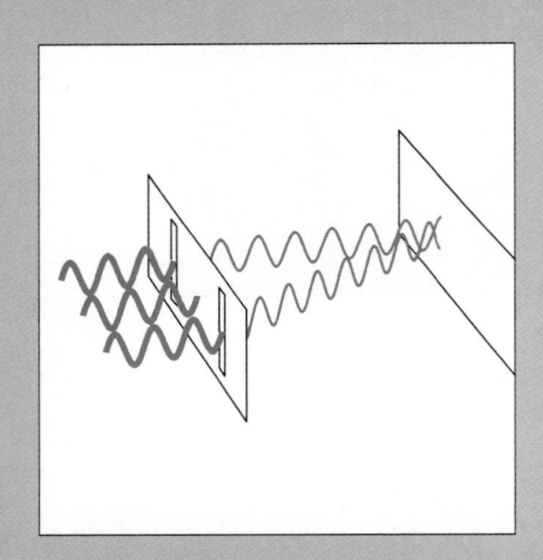

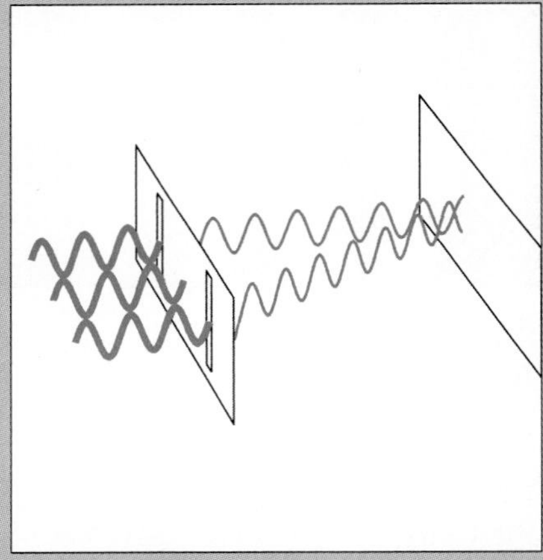

A STEREO VIEW OF FIG. 21-2(B), SHOWING A POINT OF DESTRUCTIVE INTERFERENCE ON THE SCREEN AT THE RIGHT.

falling onto the areas of the dark bands as onto the areas of the bright bands, the light waves that fell in the dark bands canceled almost perfectly. The pattern Young saw is called an **interference pattern,** and this pattern established that light has the properties of waves.

THE SPEED OF LIGHT

In Galileo's time, light's speed was a mystery. He proposed (and apparently performed) an experiment to see how much time light required to travel from one point to another. Holding covered lanterns, he and an assistant stood some distance apart (less than 2 km) in the dark of night. Galileo uncovered his lantern, and the assistant, upon seeing Galileo's lantern, uncovered his own.

The time light took to travel to the assistant and then back to Galileo, divided by twice the distance between the two men, would give the speed of light. Galileo concluded that the experimenters' reaction times far exceeded the time light needed for the round trip, and he was right.

In modern experiments to determine the speed of light, human reaction time has been eliminated. These experiments show that the speed of light in a vacuum, called c, is 299,792.46 km/s, (186,282.03 mi/s). In round numbers,

$$c = 300{,}000 \text{ km/s, or } 186{,}000 \text{ mi/s}$$

In air, light slows down a tiny amount, to about $0.99997c$ at sea level. In about three-hundredths of a second, the blink of an eye, light could go from New York to Los Angeles, be reflected from a mirror and travel back to New York—if it had a clear and straight path.

MAXWELL'S WAVES

When Maxwell investigated electricity and magnetism, he combined the discoveries of Oersted, Ampère, and Faraday with insights of his own. The result was four equations that express the connections between electricity and magnetism. In the last chapter we saw Faraday's discovery that a changing magnetic field creates, or *induces,* a current in a conductor. The reason that current appears is that a changing magnetic field at any point in space induces an electric field at that point, and that electric field can push electrons along in a conductor. Maxwell showed induction was even more general. He found that whenever an electric field changes at some point in space, a magnetic field is induced there. That is, *a changing magnetic field always induces an electric field and a changing electric field always induces a magnetic field.* This is called **electromagnetic induction.** The induced electric and magnetic fields are always perpendicular to each other.

Then Maxwell made an incredible discovery. His equations enabled him to predict that the electric field lines of any accelerated charged particle carry ripples on them, oscillations that look (in the mathematical sense) exactly like the transverse waves that travel down a stretched string. As these ripples move along, the changing electric field produces an accompanying magnetic field, perpendicular to the electric field. So the waves Maxwell predicted came to be known as **electromagnetic waves,** or **electromagnetic radiation.** He also predicted with his equations that these waves travel at the speed of light.

MAXWELL TRIED TO EXPLAIN WHAT MEDIUM HIS ELECTROMAGNETIC WAVES MOVED THROUGH BECAUSE THE ONLY WAVES KNOWN IN HIS DAY NEEDED A MEDIUM—SOLID, LIQUID, OR GAS—TO MOVE THROUGH. HE POSTULATED AN "ETHER" IN SPACE THAT WAVES OF THE ELECTRIC AND MAGNETIC FIELDS COULD MOVE IN. IT TURNED OUT THAT NO ETHER WAS NECESSARY, THAT THE WAVES MOVE THROUGH A VACUUM JUST FINE. (IN CHAPTER 26 WE'LL SEE THAT EINSTEIN ELIMINATED THE POSSIBILITY OF AN ETHER IN 1905.)

Maxwell discovered that light is traveling oscillations of electric and magnetic fields that arise because of the *acceleration* of a charged particle. According to Maxwell's equations, light from any source—the sun, a fire, an electric bulb, a firefly—comes from charged particles that have accelerated for some reason. Electrons normally emit the most radiation because if an electron and a proton are pushed by the same electric force, the electron accelerates 1856 times as fast as the proton due to its smaller mass.

The 3D illustration on page 308 shows a "snapshot" of a light wave—oscillating electric and magnetic fields—as the wave travels along at the speed of light. When such a wave passes a point in space, there is a rapidly oscillating electric and magnetic field at that point, building up in one direction, shrinking to zero, and building up in the other direction, in very rapid succession as the wave passes. As with other waves, the more rapid the oscillations (the higher the frequency), the more energy they carry, and the higher the amplitude, the greater their energy.

PHYSICS IN 3D/AN ELECTROMAGNETIC WAVE

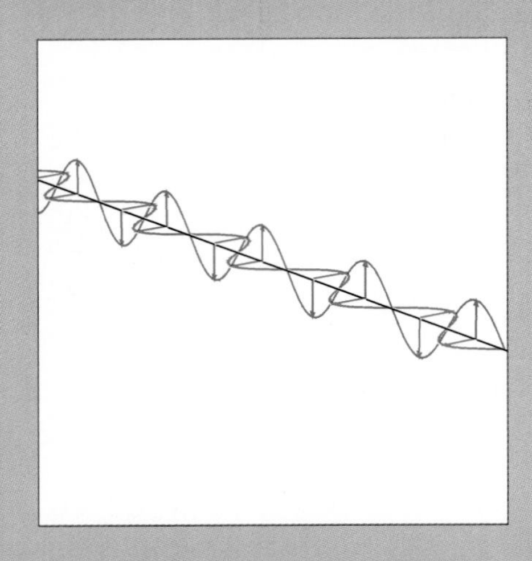

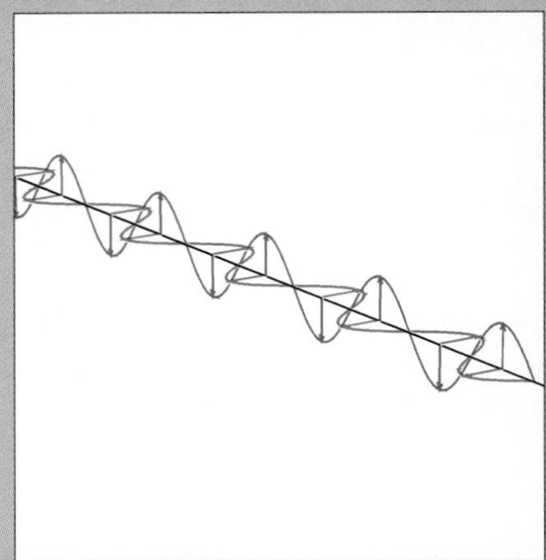

A LIGHT WAVE IS A COMBINATION OF OSCILLATING TRANSVERSE ELECTRIC AND MAGNETIC FIELDS THAT TRAVEL AT THE SPEED OF LIGHT.

THE ELECTROMAGNETIC SPECTRUM

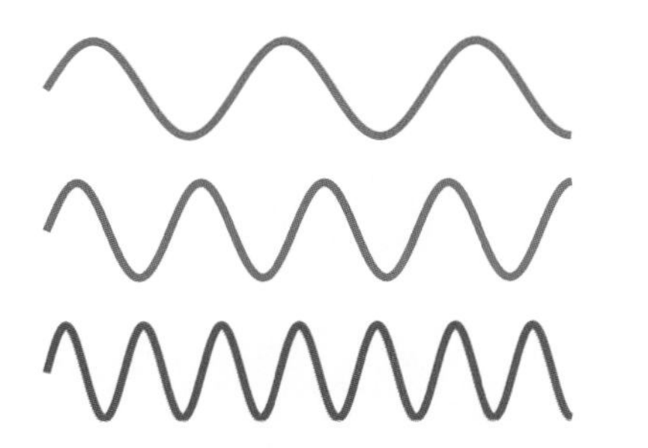

FIGURE 21-4 THE RELATIVE WAVELENGTHS OF RED, GREEN, AND BLUE LIGHT. BLUE HAS THE SHORTEST WAVELENGTH (AND HIGHEST FREQUENCY) OF THESE THREE LIGHT WAVES.

Newton used a prism to separate sunlight into the rainbow of colors our eyes can see. He showed that these colors are fundamental. For example, red light passed through a prism remains red; it separates into no other colors. Newton went on to recombine all the colors to show that they made white light when they came together. (Details of how a prism separates the colors of light are in the next chapter.) What these colors represent remained a mystery before Maxwell's theory, however. Then it became clear that *the visible colors are different frequencies of electromagnetic radiation.* The rainbow of colors in white light is called the **spectrum of visible light.** Figure 21-4 shows the relative wavelengths for blue, green, and red light.

In Chapter 15 we saw that a wave's speed is equal to its frequency times its wavelength. This is also true for electromagnetic waves;

PUTTING IT IN NUMBERS—WAVE SPEED, FREQUENCY, AND WAVELENGTH

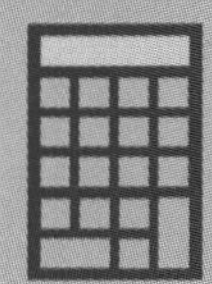

$$c = f \times \lambda$$

Example: An experiment shows yellow light has a wavelength of about 575×10^{-9} m, or 575 nanometers (nm). What is the frequency of that light?

$$f = \frac{c}{\lambda} = \frac{3 \times 10^{8}\ \text{m/s}}{575 \times 10^{-9}\ \text{m}} = \mathbf{5.2 \times 10^{14}\ Hz}$$

The word **light** usually refers to the frequencies of electromagnetic radiation we can see (Figure 21-5). Yet Maxwell's equations placed no bounds on the frequencies of electromagnetic waves. Every frequency is possible, even frequencies that we cannot see. In 1800 William Herschel, an English astronomer, placed a thermometer in the various colors of the spectrum made by a

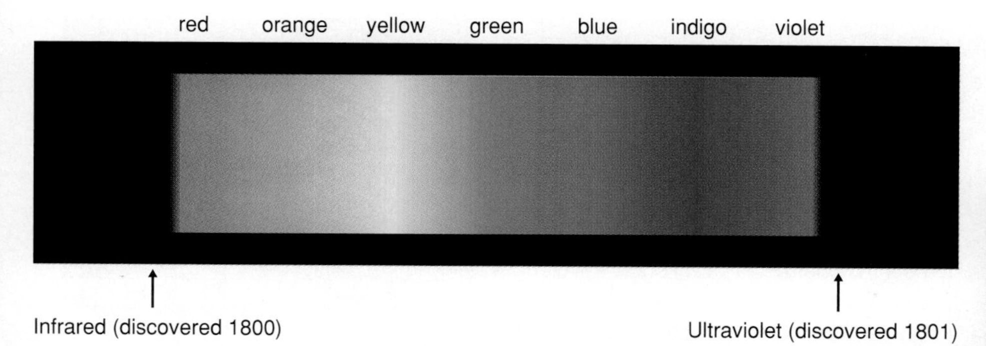

FIGURE 21-5 THE VISIBLE SPECTRUM PLUS THE POSITIONS OF THE FIRST TWO TYPES OF INVISIBLE "LIGHT," OR ELECTROMAGNETIC RADIATION, TO BE DISCOVERED.

A PAINTING OF HERSCHEL CARRYING OUT HIS EXPERIMENT THAT DETECTED IR FOR THE FIRST TIME.

prism. He was startled to see the temperature rise the most when the thermometer was off to the side of the red band of the spectrum, where there was no visible light. The prism was directing invisible rays to this area, rays that were highly effective in heating ordinary matter. These rays are called **infrared** rays, and their wavelengths are longer (and their frequencies lower) than the waves of the visible spectrum. Since they are so effective at heating matter, infrared rays are also called **heat rays.** The next year, J. W. Ritter, of Germany, detected other invisible rays to the side of the violet region of the spectrum. He exposed paper that had been soaked with a silver compound to the spectrum of sunlight that passed through a prism. Wherever light fell, the silver compound turned black, but the greatest blackening occurred just beyond the violet end of the spectrum. These **ultraviolet** rays have shorter wavelengths and higher frequencies (hence higher energy) than visible radiation. In humans, overexposure to these waves causes sunburn, premature aging of the skin, and skin cancer.

FIGURE 21-6 THE ELECTROMAGNETIC SPECTRUM.

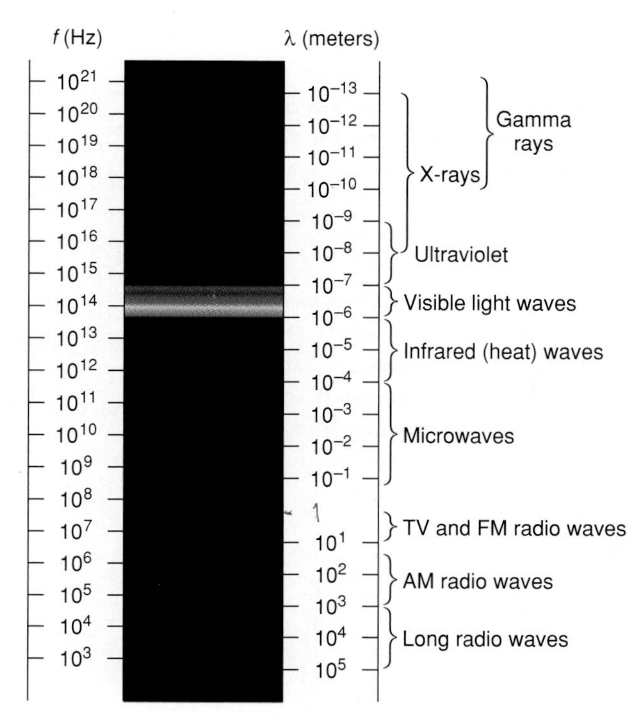

Today we know there can be any frequency of electromagnetic wave, as Maxwell suspected. The electromagnetic waves with the longest wavelengths are called *radio* waves. Next are *microwaves,* which carry TV and telephone signals and cook food. Then, with even shorter wavelengths, come *infrared, visible,* and *ultraviolet rays,* followed by *X-rays* and finally *gamma rays,* which have the shortest wavelengths, and the highest frequencies, of all. All of these waves travel at the speed c in a vacuum. Figure 21-6 shows the entire **electromagnetic spectrum.**

We saw in Chapter 17 that a rubber rod stroked with fur becomes charged. If the charged rod is shaken back and forth, the charge is accelerated, and invisible electromagnetic waves travel out from it at about 3×10^8 m/s. If you flip the rod back and forth 4 times a second, the wave's frequency is 4 Hz and its wavelength is about 75,000 km. To make a wave with a wavelength comparable to earth's diameter, put the charged rod on the end of a table and "twang" it—20 Hz is all the rod needs for that wavelength. If you could oscillate those charges at 10^{14}–10^{15} times a second, you could make electromagnetic waves of about 5000 atomic diameters, and they would be visible to your naked eye!

X-RAYS

Next to almost every dentist chair, there's an evacuated glass tube partially shielded with lead and attached to the end of a flexible arm. Inside the tube are two bits of metal, called electrodes, a centimeter or two apart. Next to one electrode lies a heating coil. With the press of a button, a high-voltage power supply is turned on, causing the heated electrode to become negatively charged and the other to become positively charged. Suddenly there is a potential difference of 60,000 volts between the electrodes. Thermally agitated electrons evaporate from the surface of the heated electrode, and they accelerate through the vacuum toward the positive electrode. The very strong electric force gives them enormous accelerations. After only a centimeter or so of travel, the electrons are moving at over one-third the speed of light!

As they collide with atoms in the positive electrode, 99 percent of the electrons' energy turns into heat. However, a few penetrate close to the nuclei of the metal atoms, where they get a very strong pull. They turn sharply around the nuclei and, as they do, emit radiation of a very short wavelength called X-rays. Directed through a tooth and onto a small sheet of film behind the tooth, the X-rays are absorbed partially by the tooth and partially by the film. In the dental X-ray shown here, the X-rays are completely blocked by metal fillings.

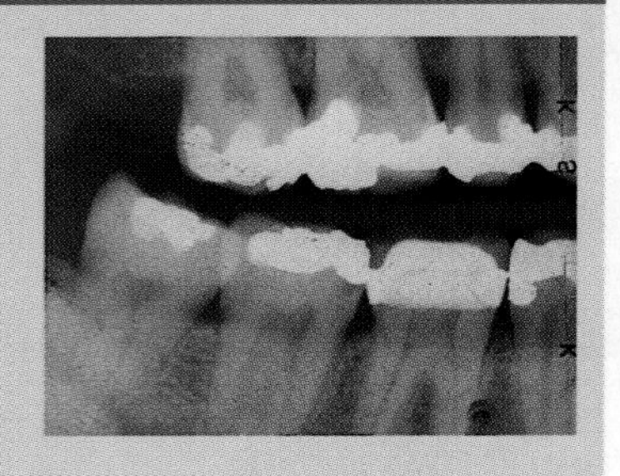

The electrons in the appliances and wiring where you live don't accelerate nearly as much as electrons do in an X-ray machine. Nor would you want them to. The last thing anyone needs is X-rays from a toaster.

MAKING AND DETECTING RADIO WAVES

Radio and television sets operate with electromagnetic waves generated by commercial and public stations. Here's a brief explanation of how a radio wave is broadcast at one location and received at another. A sound wave from a voice or musical instrument enters a microphone at the radio station, as shown in Figure 21-7. The variations in air pressure the wave causes inside the microphone mechanically generate an electrical current called the *signal.* That signal is used to modify, or **modulate,** an intense *carrier* radio wave that is made by the station's transmitter.

FIGURE 21-7 SOUND WAVES MOVE A DIAPHRAGM IN A MICROPHONE, AND THIS MOVEMENT CREATES AN ELECTRICAL SIGNAL THAT MODULATES A CARRIER WAVE BROADCAST FROM THE RADIO STATION.

The carrier wave comes from fast oscillations of charge in a conducting rod called the *antenna.* In essence, one end of the antenna is positive and the other is negative, setting up a dipole electric field around the antenna. When the charges in the station's antenna are reversed hundreds of thousands of times a second, the dipole field follows suit, and oscillating field lines move outward over the local landscape at the speed of light. If the signal produced at the microphone modulates the amplitude of the carrier wave but not its frequency, the station is broadcasting an AM (amplitude-modulated) radio wave. If the carrier wave's amplitude is steady but its frequency is modified by the signal from the microphone, the broadcast wave is FM (frequency-modulated).

As it leaves the transmitting antenna, a radio wave spreads out in all directions (Figure 21-8). Its energy too spreads out along the expanding wavefronts; we say the signal weakens. Movable electrons in a distant radio's metal antenna are pushed back and forth by the electric field of the incoming radio wave. These electrons oscillate, dancing a jig that's a miniature version of the electron motion in the transmitting antenna. Radios that aren't too far away can pick up the distance-weakened signal, make a more intense copy of it (the *amplifier* in the radio does this), and use that copy to reconstruct the original sound from the radio station (the *speaker* takes care of this part). Each metal pot and pan in the kitchen is a receiving

FIGURE 21-8 THE OSCILLATION OF CHARGE ON THE ANTENNA OF THE RADIO/TV STATION'S TOWER CAUSES ELECTROMAGNETIC RADIATION. AS SHOWN HERE, THE WAVEFRONTS MOVE OUTWARD FROM THE TRANSMITTING TOWER IN ALL DIRECTIONS.

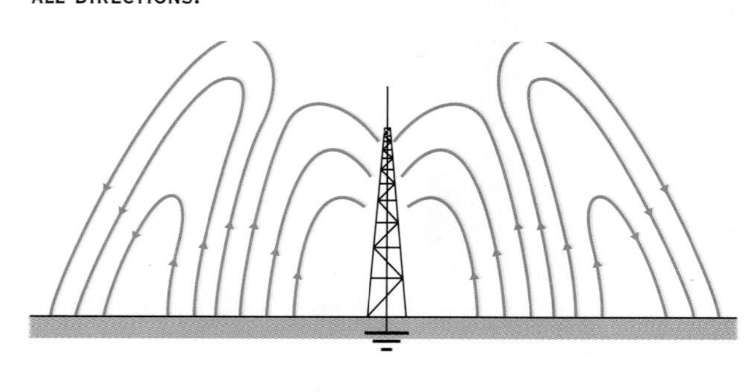

antenna for these waves, too, but they have no amplifiers, circuits, and loudspeakers for processing those electrical signals to reproduce the sound.

Because radio waves come in from every nearby station at the same time, you must adjust, or *tune,* the radio to follow the particular carrier signal you want. When tuned to a certain station, the radio resonates with and amplifies only that one carrier wave from the various signals picked up by the antenna. When an electrical storm is in the area, accelerated charges in the lightning send out radio waves of all frequencies. Some will be in the frequency range of the station a radio is tuned for, and the radio amplifies that signal (called *static*) along with the broadcast signal it is receiving.

Radio waves (as well as many other electromagnetic waves) move electrons only at the surface of metal; electric fields don't penetrate metal, as we saw earlier. Instead, the oscillating electrons at the surface absorb a wave and then, because of their own accelerations, re-emit it in the action called *reflection.* Metals reflect radio waves just as a mirror reflects light. A radio won't play inside a closed metal box. And yet waves pass right through our bodies all day, every day. (That is easy to prove. Put a small portable radio on the ground, lie over it, and it still plays.) Radio waves don't greatly disturb the electrons in any nonconducting matter because the waves carry very little energy and electrons in a nonconductor aren't free to move around. Compared to metals, human bodies are very poor conductors. Nonconducting matter

MRI: MAGNETIC RESONANCE IMAGING

Today, magnetic resonance image scanners are used by medical doctors to provide high-resolution pictures of the tissues inside a patient's body. The scanner creates a strong magnetic field with superconducting electromagnets, and the protons that are the nuclei of the hydrogen atoms in the body all respond by aligning with that field. Unlike a compass needle aligning in earth's magnetic field, the proton and its tiny magnetic field wobble around the direction of the intense applied field, much as a child's spinning top wobbles slowly around a vertical line. (Protons, like electrons, have a "spin" property along with angular momentum, as well as a magnetic field.) Then the wobbling protons are slammed with a burst of radio waves tuned perfectly to push the proton's axis sideways, perpendicular to the large magnetic field. Once this happens and the radio waves have passed, the protons quickly spiral back into their wobbling patterns around the direction of the large magnetic field. As they do, they emit a very faint radio wave, or signal, which is picked up by sensors on the superconducting magnets. Analyzed by a computer, these signals reveal varying densities of hydrogen atoms in the body and their interaction with the surrounding molecules. A hydrogen atom bound in cartilage emits a different frequency from one bound in muscle or one bound in any organ of the body. The image resulting from these signals makes a clear distinction between the different types of tissue, fluid, and bone. (See the figures on page 12 of Chapter 1 for a look at an MRI machine and an MRI image.)

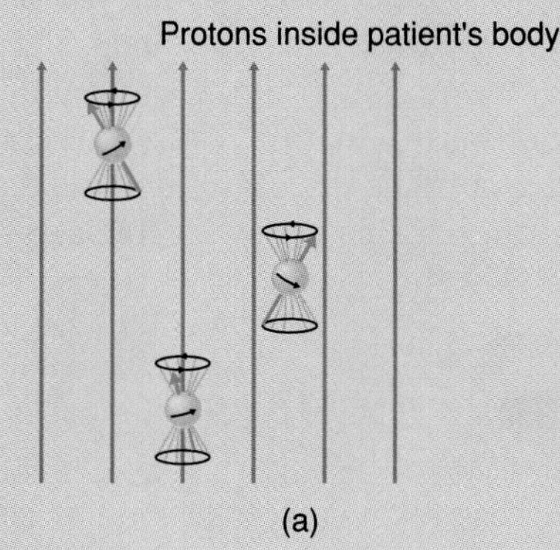

(a)

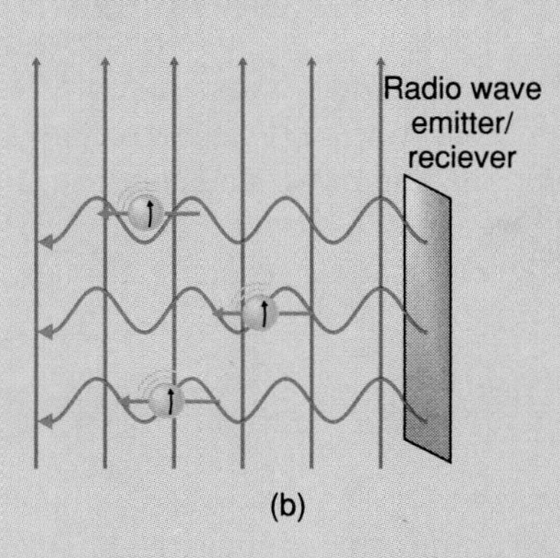

(b)

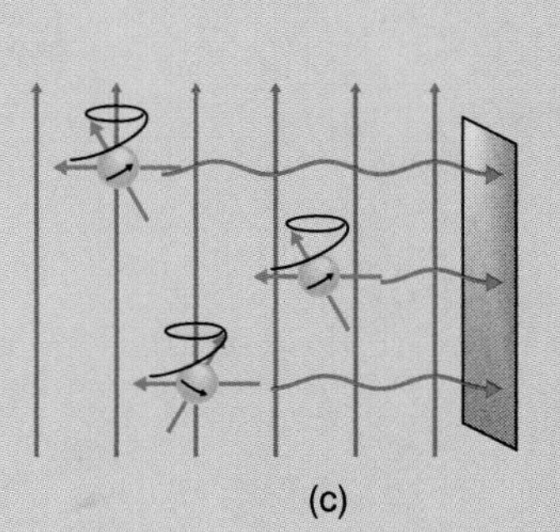

(c)

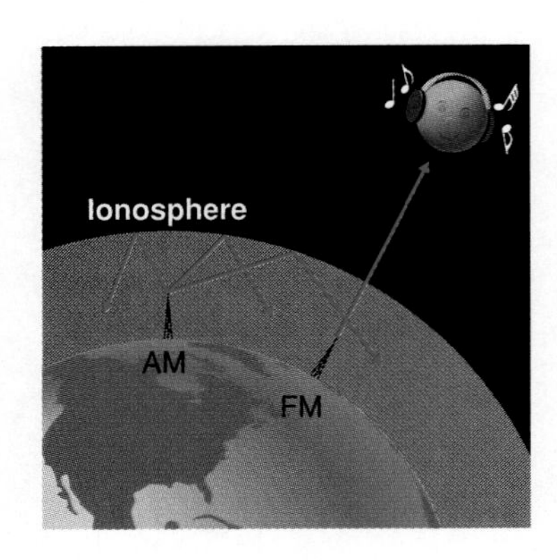

FIGURE 21-9 **FM RADIO WAVES (AND TELEVISION FREQUENCIES) AREN'T REFLECTED EFFECTIVELY BY THE IONOSPHERE, ALTHOUGH AM RADIO WAVES ARE.**

doesn't disturb the incoming waves much, letting them pass through a roof, a wall, or a person without much absorption.

In the thin gas of the earth's ionosphere, above the ozone layer, strong ultraviolet radiation from the sun ionizes some atoms and molecules (Figure 21-9). These ions respond to the oscillating fields of AM radio waves rising from the earth's surface. As the ions accelerate, they send out a perfect copy of the radio waves, just as if the ions were miniature antennae. Part of those signals return to earth. This is why shortwave radio transmissions and some AM radio stations may be received over great distances without a straight line of sight to the transmitters. The ionosphere does *not* reflect FM radio and TV signals, however. Their frequencies are so high that the ionosphere's electrons can't respond fast enough to the changes in the electric fields of those waves to reflect them back to earth.

RADAR

DETECTING EARTH'S RESOURCES WITH RADAR

During World War II, British and American scientists developed radar to detect approaching enemy aircraft. Short bursts of low-frequency microwaves are aimed by an antenna, they reflect from metal surfaces, and the time it takes the reflected signal to return indicates the distance to the reflecting object. Modern air traffic is monitored by radar at airports, and because radar frequencies are also reflected by rainclouds, weather radar reports on TV news are familiar to everyone. Radar waves scatter from droplets in the cloud, and the return "echo's" strength indicates the rainfall's intensity. Now special radar units can detect the Doppler shift in the returning signals, measuring the horizontal speed of the winds in a rainstorm. This Doppler radar quickly detects severe storms and tornadoes and measures the wind speeds in hurricanes. It also indicates when conditions are right for severe wind shear, winds that can come *down,* a very dangerous condition for aircraft that are landing or taking off.

Different radar wavelengths penetrate different types of matter. Radar aboard spacecraft have accurately mapped the elevations of the terrain of Venus, which cannot be seen from above the planet due to its dense cloud cover. In the 1980s the space shuttle carried a 6-cm-wavelength radar unit that bounced off water or solid rock, but penetrated clouds and sand. In 1981 this system exposed three long-dry rivers and their tributaries beneath 5 meters of sand in the desert of Western Egypt. One was as wide as the Nile Valley. Since water often seeps downward from rivers into aquifers, the Egyptian government drilled for and found water some 100 meters down, and now crops are grown on over 20 square kilometers from this water supply. The shuttle also discovered ancient roads below the desert sands. The newest version of shuttle radar operates with 3-cm, 6-cm, and 23-cm radar waves. The 3-cm waves reflect better from low density matter than the 6-cm waves, while the 23-cm waves penetrate solid ground better. By combining data from all three, much sharper images are produced. This is now being used to study the geology and hydrology of many sites world-wide.

GUIDING ELECTROMAGNETIC SIGNALS

We use telephones to talk to people across town or across country as if they were in the same room with us. Telephones use traveling oscillations of electric fields much like radio waves except that in the case of telephones, the waves are guided by copper wires rather than broadcast by antennas. While the electrons that make up currents in copper wires have speeds of less than a millimeter per second, the electric field *changes* in the wire that push those electrons travel at almost *c*. The electric fields flash through the wire whenever there are changes in the voltage across the closed circuit at the telephone, and these voltage changes occur whenever someone talks into the handset.

Imagine trying to carry on a conversation with a friend 2 miles away by shouting through a megaphone. Your words would need about 10 seconds to reach your friend. Allowing another 10 seconds for the reply to reach you would make even a short conversation very long. With telephones, though, only overseas calls have any noticeable lag, and that is because the electronic signal is beamed to a satellite almost 35,000 km up and then back down. (Microwaves perform that part of the connection.) Even at the speed of light, the signal takes about 1/4 of a second to get to the satellite and back.

MICROWAVE OVENS

A conventional oven cooks by conducting heat from the outside of the food to the inside, but a microwave oven cooks by generating heat throughout the food's volume. Microwaves travel through food very well, reacting mostly with the water molecules, which are permanent electric dipoles, and not with bones or with protein molecules. In the presence of a microwave's oscillating electric field, the positive end of the water molecule is pulled along the field line and the negative end is pulled in the opposite direction. These actions cause the molecule to rotate—it has work done on it. As the microwave electric field oscillates, the molecule turns back and forth, gaining rotational energy. The microwaves have just the right frequency to oscillate the molecule so that it resonates, absorbing a great deal of energy. Through collisions with neighboring molecules, its rotational kinetic energy is changed into translational kinetic energy, raising the temperature of the food all throughout its volume at the same time. The heat-producing process goes on wherever there are water molecules. A large portion of the microwaves that enter the food travel all the way through it, but the oven's metal liner reflects them back for another pass and another. Each time, more of the microwave's energy is absorbed. Often in just a matter of minutes, heat from the flipping water molecules cooks the food. Dry paper, styrofoam plates, or other containers recommended for use in microwave ovens contain no water, and microwaves pass through them with no effect. The same is true for ice, because the water molecules are locked into their positions in the ice and cannot turn back and forth.

There are other situations where heating by microwaves is useful. For

LIMITS TO MAXWELL'S EQUATIONS

The importance of Maxwell's discoveries can hardly be overstated. His equations explained everything known about light at the time, all electric and magnetic phenomena, and they correctly predicted many things later verified by experiments. For his tremendous accomplishment in explaining electromagnetic behavior, Maxwell is remembered as one of the giants in the history of physics. By 1900, however, physicists had found several features of the way light is emitted by matter that could not be explained with Maxwell's equations. These puzzling features were not fully explained until the 1920s, when a new theory of matter that included the details of the nature of atoms was developed. Individual atoms and molecules, it turns out, absorb and emit light only in lumps of energy. These individual quantities of light energy are called *quanta* (singular, *quantum*), and the bit of light that carries one such lump of energy is called a *photon*. We'll study these ideas more closely in later chapters. In the next two chapters, however, we'll look at some of the effects around us that can be explained with the wave nature of light and Maxwell's theory.

instance in diathermy, the heating of tissue for medical therapy, low-intensity microwaves penetrate outer layers of fat, which contain little water, and warm the underlying muscles, which contain a higher percentage of water.

ULTRAVIOLET RADIATION AND LIFE ON EARTH

Physics *and Environment*

MONITORING UV IN THE STRATOSPHERE

The sun emits ultraviolet radiation, which in large amounts is deadly to living things, destroying both protein molecules (as in sunburns) and DNA molecules. The amount of this energetic and ionizing radiation that reaches this planet would destroy life here if it reached earth's surface. For at least 500 million years, however, a thin veil of ozone, O_3, in the stratosphere has intercepted the incoming UV. Less than 1 percent of the incoming UV reaches to sea level.

As mentioned in Chapter 13, CFC molecules, manufactured for use in refrigerants and propellants, are now known to reach the stratosphere and form ClO molecules. In complicated reactions, they destroy ozone. Each chlorine atom may remain in the stratosphere for 40 to 100 years and destroy tens of thousands of ozone molecules. Satellites now monitor the loss of ozone, as do balloon-carried and ground-based instruments. Chlorine freed in the Northern Hemisphere is concentrated by a large atmospheric eddy (vortex) over Antarctica every spring (fall in the Northern Hemisphere). The severe depletion of ozone there is called the "ozone hole." (See the figure on page 12 in Chapter 1.) In 1990 scientists measured a 6 to 12 percent decrease in production of phytoplankton in Antarctic waters—the very bottom of the food chain—as a result of the increased UV. In the same year Antarctic explorers and scientists reported skin burns and swelling wherever they left skin exposed to the sunlight, and the intense UV caused nylon outer garments to disintegrate in weeks. When the vortex breaks up in November, patches of ozone-depleted air spread outward, toward the equator. In Australia, short-term 20 percent increases in UV have been measured, and Australian TV stations carry daily UV measurements and warnings to keep out of the sunlight when the levels rise. Scientists are investigating possible ways to help reverse the ozone loss, and by 1992 some 92 countries (responsible for 90 percent of CFCs) had signed the Montreal Protocol, which now calls for a complete ban of CFC production by the year 2000.

Summary and Key Terms

interference pattern–306

Newton thought that light was composed of tiny, fast-moving particles, but Young later proved that light behaves like a wave—under certain conditions intersecting beams of light can create an ***interference pattern***.

electromagnetic induction–307

Maxwell developed a theory to describe the connections between electricity and magnetism, and this theory gave ***electromagnetic induction*** a general definition: *A changing magnetic field always induces an electric field, and a changing electric field always induces a magnetic field.* Maxwell's equations described light waves as ***electromagnetic waves*** (or ***radiation***), traveling oscillations of electric and magnetic fields. *Electromagnetic waves always arise when charged particles accelerate.*

electromagnetic waves–307

light–308
spectrum of visible light–308
electromagnetic spectrum–309

When we use the word ***light***, we usually mean the particular frequencies of electromagnetic radiation that we can see. The ***spectrum of visible light*** is only a small part of the ***electromagnetic spectrum***, which includes all possible wavelengths of electromagnetic waves. The electromagnetic waves with the longest wavelengths are radio waves, and the next down in size are microwaves. Below that are ***infrared rays***, or ***heat rays***, which appear beside the red band of the visible spectrum and have longer wavelengths than visible light. ***Ultraviolet rays***, which appear beside the blue-violet band of the visible spectrum, have shorter wavelengths than visible light. X-rays are shorter still, and gamma rays have the shortest wavelengths of all. In a vacuum, all the frequencies of electromagnetic radiation travel at the same speed *c*.

infrared rays–309

ultraviolet rays–309

A radio transmitter uses a sound wave to generate an electric signal, which then ***modulates*** a carrier radio wave transmitted by the station's antenna. If the signal modulates the carrier wave's amplitude while its frequency remains steady, the station is broadcasting an AM signal. If the signal modulates the wave's frequency while its amplitude remains steady, it is an FM signal.

modulation–310

Microwaves of the frequency that cause resonance with water molecules cook food in microwave ovens. Materials without water in them are unaffected.

EXERCISES

Concept Checks: Answer true or false

1. Young's experiment showing interference of light proved that light has wave properties.

2. Visible light waves, heat rays, X-rays, and radio waves all travel at the same speed in a vacuum.

3. Waves of light, according to Maxwell's equations, are traveling oscillations of electric and magnetic fields created by accelerations of charged particles. These oscillations are known as electromagnetic waves or electromagnetic radiation.

4. Light travels at the same speed in air as in a vacuum.

5. Light that we can see is only a small part of the spectrum of electromagnetic radiation.

6. An AM radio wave carries a signal by having its amplitude modulated. This type of wave is not affected by the ionosphere.

7. TV signals have higher frequency than AM radio signals.

8. Lightning sends out radio waves that a radio receiver can pick up. They arrive at the same time as thunder does from the lightning strike.

Applying the Concepts

9. Which have the shortest wavelength? Which has the highest frequency? (a) radio waves, (b) infrared waves, (c) ultraviolet waves, (d) green light. (*Hint:* See Figure 21-9.)

10. When someone "looks out the window" or "casts a glance," what is going on in the physical sense? (Does something material travel out of the window or in the direction of the glance?)

11. We sometimes say that black and white are colors, but neither is found in the spectrum of sunlight. Are they really colors? Discuss.

12. Why are sounds that consist of a large range of frequencies, such as the sounds made by waves at the seashore or by a waterfall, sometimes called "white noise"?

13. Explain why it is a good guess that the light from a candle's flame comes from electrons rather than protons.

14. Often it is possible to tune in AM radio stations from a distant city, and yet you can't pick up TV broadcasts from that far away. Why?

15. What is the frequency of a 2-cm microwave traveling through a vacuum?

16. If a radio station's signal is weak in your area, you can sometimes improve the reception by placing the radio by an open window. Why should that have an effect?

17. TV and radio broadcasts are often said to be "on the air." What is right and what is wrong with this statement?

18. Arrange the following electromagnetic waves in order of increasing wavelength; ultraviolet rays, X-rays, infrared rays, AM radio waves, microwaves, green light.

19. NASA uses microwaves rather than radio waves to communicate with astronauts in space. What makes this necessary?

20. The sun's visible "surface" is at a temperature of almost 6000 K. Explain how Maxwell's theory predicts that the random collisions of charged particles there produce white light.

21. What is the wavelength of an AM radio station that broadcasts at 625 kHz?

22. If you charge a plastic comb by rubbing and then wave it back and forth in the air, do you create an electromagnetic wave?

23. The ozone layer in the earth's atmosphere blocks not only most of the ultraviolet radiation and X-rays from the sun but also the small amounts of X-ray and gamma radiation from interstellar space. Why is this blocking important for life?

24. What sort of time lag could you expect in conversations with someone in Europe just from the speed of light traveling to a geostationary satellite and back for each side of the conversation? Would you expect to be able to detect that lag?

25. Fair-skinned beachgoers, wary of sunburn, sometimes go into the water up to their necks to "get out of the sun," but a doctor might tell you that being under a little water doesn't help much. What does that tell you about the transmission of ultraviolet light by the water?

26. In terms of the energy an electromagnetic wave carries, explain how the light from a large, luminous star in the night sky pales against the light from your desk lamp.

27. Would you expect microwaves in a microwave oven to penetrate a metal cup in the oven? Why or why not?

28. The U.S. National Park Service has used infrared detectors mounted on helicopters to spot hidden cave entrances in the area around Carlsbad, New Mexico. Discuss how that is possible.

29. FM radio stations broadcast on frequencies between 88 and 108 megahertz. What are the longest and shortest FM radio waves? AM radio stations use carrier waves with frequencies between 550 and 1600 kilohertz. What are the maximum and minimum wavelengths of those waves?

30. Because we are warm from the heat released by chemical processes in our bodies, we emit infrared radiation. That is why infrared detectors can reveal a person's presence at night. How do you think clothing affects a person's visibility where an infrared detector is concerned?

31. Snow, sand, and ice are excellent reflectors of ultraviolet light, the electromagnetic waves that cause tanning and sunburn. Discuss the implications for fair-skinned beachcombers and alpine skiers. (Small doses of ultraviolet light are beneficial to humans, however, producing vitamin D precursors in the skin that are essential for survival.)

32. Atlanta, Georgia, had the second radio station licensed in the United States, WSB (Welcome South, Brother). When it signed on in 1922, it broadcast with 100 watts of power. Today it puts 50,000 watts into its broadcasts. What good does this increase in energy do?

33. A long time ago, an atomic bomb test on an island in the Pacific was telecast to the United States. TV viewers in the United States heard the explosion before observers stationed nearby heard it. Explain.

34. Why is a fair-skinned person in more danger of sunburn high in the mountains than at sea level?

35. The longest electromagnetic wave broadcast by TV stations has a frequency of 54 megahertz (1 megahertz = 1 million Hz). Find its wavelength.

36. Are TV waves longer or shorter than the waves we detect with our eyes? Than microwaves?

37. What type of electromagnetic wave has a wavelength of 1 centimeter?

38. Although the sun emits most of its energy in visible light, it also emits radio waves and microwaves and UV waves. Every autumn, cable TV reception in the Northern Hemisphere is affected by the sun for a few weeks. For half an hour each day, the sun is almost directly behind the geostationary communications satellites that broadcast the TV signals, as shown in Figure 21-10. Why should this position of the sun affect reception at earth stations?

39. Suppose a beam of light could travel in a circle around the equator of the earth. Show that it would pass someone standing at one point on the equator 7 times in a single second. (The circumference of the earth is about 40,000 km.)

40. A hydrogen atom is roughly 1×10^{-10} meter across. How many hydrogen atoms would fit across one wavelength of green light of frequency 5.5×10^{14} Hz?

FIGURE 21-10

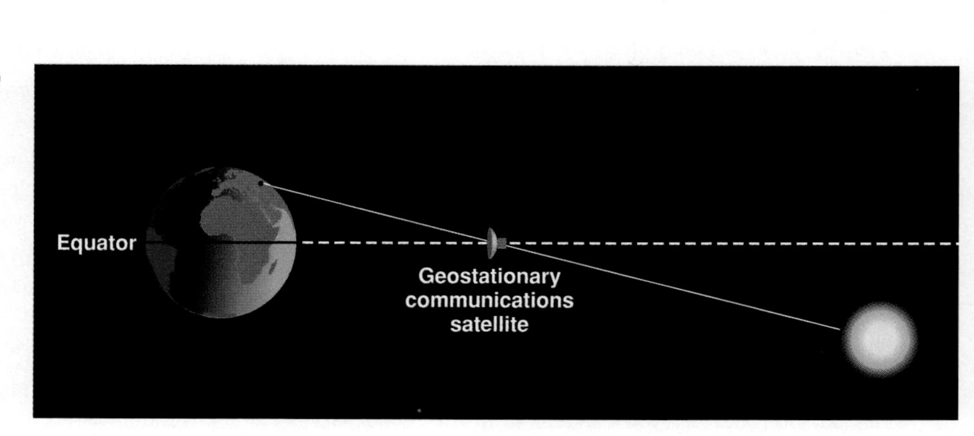

41. Alternating current accelerates electrons back and forth with a frequency of 60 Hz. These accelerated charges therefore give off electromagnetic waves of frequency 60 Hz. Calculate their wavelength. (Since the late 1970s claims have been made that long-term exposure to such low-frequency electromagnetic waves may cause cancer in humans. Early studies did not account for all the variables, and at the time of this writing the verdict is still out. Some electric blanket manufacturers have already designed and produced electric blankets whose 60-Hz waves essentially cancel out, just in case there is a correlation.)

42. If a frozen dinner was frozen solid, could a microwave oven cook it? Discuss.

43. If an electron moves at a constant velocity, is an electromagnetic wave produced? If an electron moves in a helical path, is an electromagnetic wave produced?

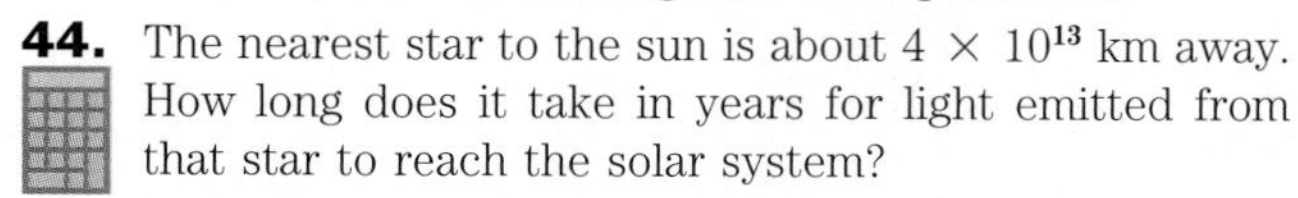

44. The nearest star to the sun is about 4×10^{13} km away. How long does it take in years for light emitted from that star to reach the solar system?

45. Refer to Figure 21-6 for the frequencies of visible light and microwaves. Compare their frequencies by forming a ratio and show that visible light oscillates about a million times faster than microwaves.

46. The moon's average distance from earth is about 384,000 km (240,000 mi). When you look at the moon, then, how old is the image that you see?

Demonstrations

1. Some night when you are in a car, tune the AM radio to the farthest radio station you can hear. (In Florida, for example, radios sometimes pick up stations in Louisiana and Ohio at night.) Then leave the dial untouched and return to the car the next morning. You won't be able to receive the station. What you are indirectly demonstrating is the daytime/nighttime change in ion concentration in the ionosphere. The solar UV radiation that forms the ions in the daytime is absent at night, and the ions begin to recombine into neutral atoms. This atom formation happens fastest where the ions are closer together, at the lower (and hence more dense) edge of the ionosphere. As night falls, it is as if the lower boundary of the ionosphere rises, and the reflected AM radio waves go farther since they are reflected at greater heights (Figure 21-11).

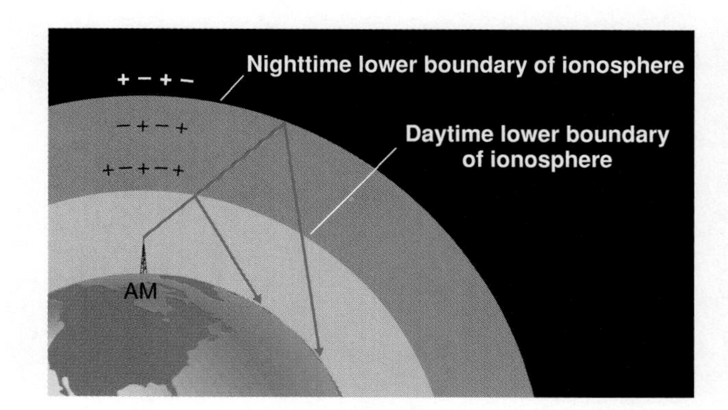

FIGURE 21-11

2. Turn on a portable radio and lie over it—it will still play, showing you that radio waves go right through your body, as mentioned in the text. Then take a sheet of aluminum foil and wrap it around the radio—it will stop playing! The free electrons in the metal oscillate with the incoming wave and cancel its electric field inside the metal. If you ship an audio tape across country to a friend, you might wrap it in aluminum foil to keep any stray electric fields from damaging the tape in transit.

TWENTY-TWO

Properties of Light

The droplets of water on the leaves of a garden plant form clear, sharp images of the flowers behind them. Rounded by surface tension, the droplets serve as tiny wide-angle "fisheye" lenses.

People have long made practical use of many of light's properties. Glass lenses to correct vision came into use in the thirteenth and fourteenth centuries, and today over half of the adults in North America depend on eyeglasses or contact lenses for good vision. The lenses of early microscopes let biologists see objects much smaller than could be seen with the unaided eye, and modern astronomy began with Galileo's use of lenses to build a telescope. This chapter discusses the properties of light used in these and other applications.

LIGHT RAYS AND THE PINHOLE CAMERA

In a darkened theater or concert hall, you can see the beam of a spotlight when some of the light is scattered by smoke and dust in the air. The beam travels in a straight line, and the lighting technician can narrow its width until the whole beam disappears. Early philosophers spoke of infinitely thin beams of light and called them **rays.** Although there is no such thing as an infinitely thin beam of light, the idea of rays is useful in tracing the paths light takes through space.

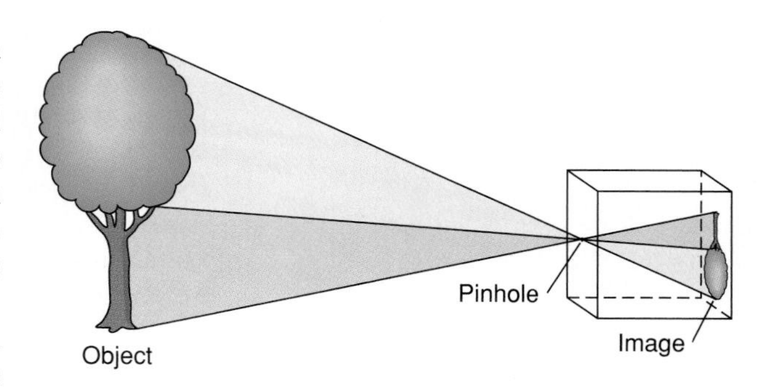

FIGURE 22-1 A DIAGRAM OF THE IMAGE PRODUCED BY A PINHOLE IN A BOX. RAYS OF LIGHT FROM THE OBJECT TRAVEL IN STRAIGHT LINES AND ENTER THE PINHOLE. A RAY LEAVING THE TOP OF THE TREE, GOES THROUGH THE PINHOLE AND STRIKES THE LOWEST POINT AT THE IMAGE POSITION ON THE BACK OF THE BOX. A RAY FROM THE BASE OF THE TREE STRIKES THE HIGHEST POINT AT THE IMAGE POSITION. AS A RESULT, THE IMAGE IS UPSIDE DOWN RELATIVE TO THE OBJECT.

Rays can show us how an image is formed inside a closed box that has a pinhole in one side (Figure 22-1). An **image** is a likeness of an object or scene (latin *imago,* meaning an imitation or copy). Tracing rays of light through the pinhole to the back of the box shows an upside-down image is formed there. This device was mentioned by Aristotle, and much later pinhole images were traced by some artisans to make photograph-like drawings before photography was invented. A *pinhole camera* is just such a light-tight box having a pinhole opening, with photographic film added inside the box opposite the pinhole.

REFLECTION OF LIGHT

From the time we wake up in the morning until we go back to sleep at night, light enters our eyes to form images of the world around us. When you look at anything except a light source, such as a lightbulb or a star in the sky, the light you see is not "straight from the source" but rather is light that has been reflected and scattered by the objects you are looking at. For example, you see this page because light from a lamp or the sun strikes the page and reflects in all directions, and some of that light comes to your eyes (Figure 22-2). The printing on the page reflects much less light (it absorbs more) than the white paper does, so the letters appear darker than the paper they are printed on.

FIGURE 22-2 YOU SEE A PAGE OF A BOOK BECAUSE LIGHT REFLECTS FROM EACH PART OF THE PAGE IN ALL DIRECTIONS, SO THAT SOME OF THE LIGHT RAYS FROM EACH PART OF THE PAGE ENTER YOUR EYE. BECAUSE ALMOST NO LIGHT IS REFLECTED BY THE PRINTED WORDS, YOU "SEE" THEM AS BLACK AREAS.

Most objects reflect light in an infinite number of directions because, as Figure 22-3 shows, the vast majority of surfaces found in the everyday world are far from smooth on the microscopic level. Such surfaces **diffuse** the light, meaning that they scatter it in many directions. In contrast to these materials, a smooth surface of silver reflects a ray of light in only one direction. Therefore rays that come perpendicularly into a mirror or other smooth reflective surface reflect straight back. Light that comes in at an angle to a perpendicular line drawn from the surface reflects at an equal angle, as shown in Figure 22-4. The angle between the incoming ray and the perpendicular is called **angle of incidence,** and the angle between the perpendicular and the reflected ray is

SILVER IS SUCH A *GOOD* REFLECTOR BECAUSE IT HAS MANY MOVABLE (CONDUCTION) ELECTRONS. THESE ELECTRONS MOVE WHEN THE ELECTRIC FIELDS OF THE LIGHT WAVES HIT THE SURFACE, CANCELING OUT THE ELECTRIC FIELD INSIDE THE METAL. THAT MEANS THE LIGHT (LIKE ANY OTHER ELECTRIC FIELD) CANNOT PENETRATE THE SILVER AND THEREFORE IS ALMOST 100 PERCENT REFLECTED AT THE SILVER SURFACE.

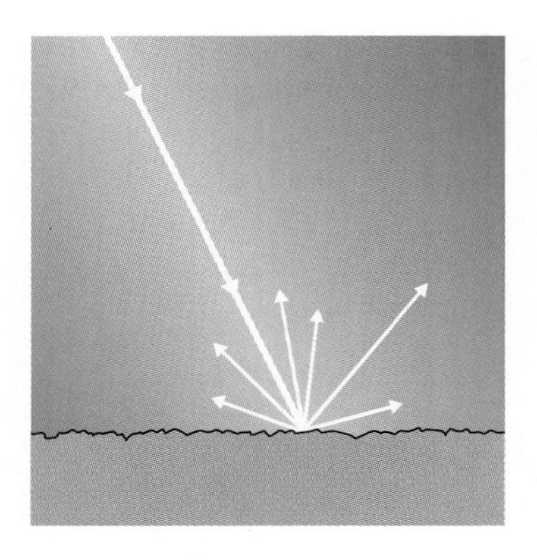

FIGURE 22-3 A MICROSCOPIC VIEW OF A SURFACE SHOWS IT TO BE NOT SMOOTH BUT ROUGH. REFLECTION FROM SUCH A SURFACE IS DIFFUSE. THE UNEVENNESS OF THE SURFACE CAUSES BEAMS OF LIGHT TO REFLECT IN EVERY DIRECTION AWAY FROM THE SURFACE.

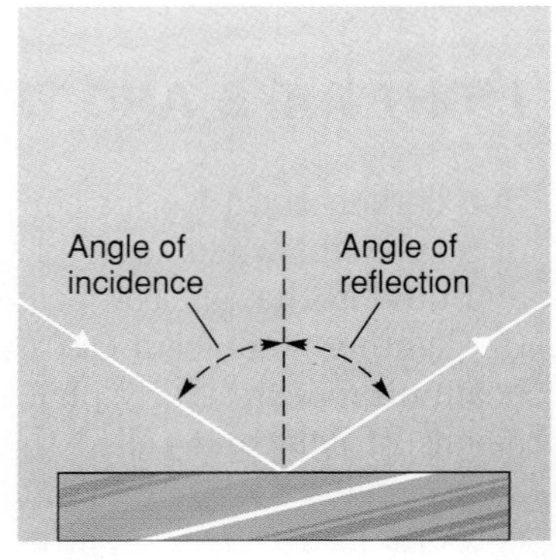

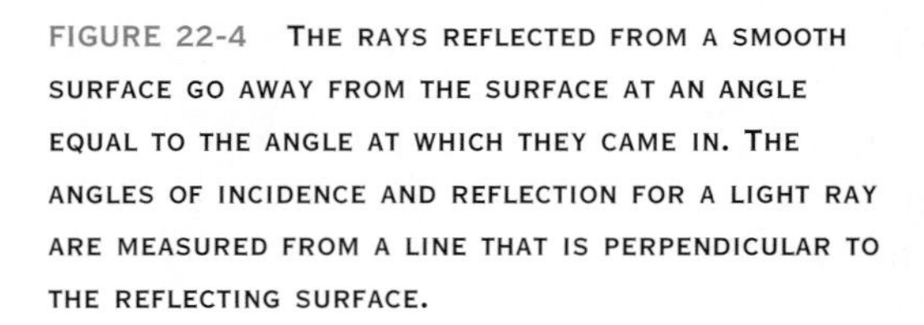

FIGURE 22-4 THE RAYS REFLECTED FROM A SMOOTH SURFACE GO AWAY FROM THE SURFACE AT AN ANGLE EQUAL TO THE ANGLE AT WHICH THEY CAME IN. THE ANGLES OF INCIDENCE AND REFLECTION FOR A LIGHT RAY ARE MEASURED FROM A LINE THAT IS PERPENDICULAR TO THE REFLECTING SURFACE.

called the **angle of reflection.** The **law of reflection** states that, for reflection from a smooth surface,

the angle of reflection equals the angle of incidence.

Rays of light, then, bounce off a very smooth surface just the way a pool ball bounces off the cushion of a pool table. The reflected rays from a *flat* smooth surface such as a mirror make an image that to an extent resembles the object those rays came from.

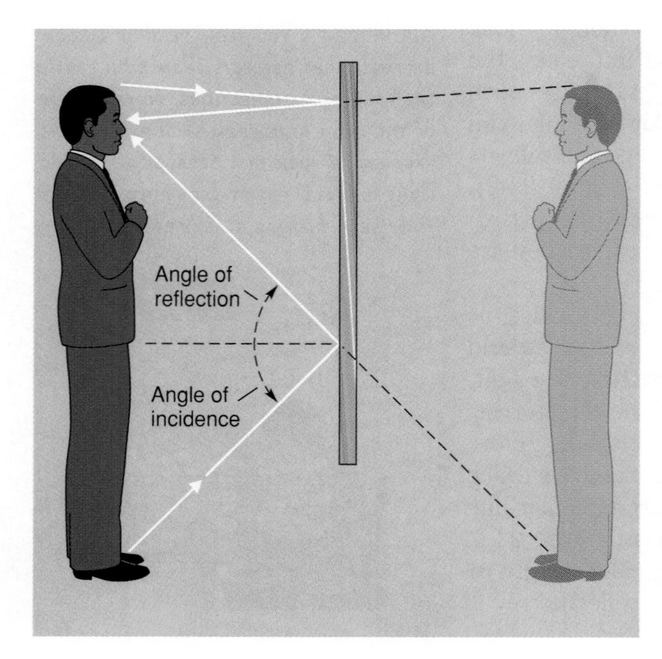

IMAGES FROM REFLECTIONS

The image produced in a pinhole camera is formed by rays of light that actually reach the image location; that type of image is called a **real image.** The images you see on TV screens and at the movies are real images. Your eyes and brain detect the light as coming from the direction the ray had just before striking the eye.

When you look at a mirror, however, all the light rays you see have been reflected at the mirror surface, changing their direction. So when you stand in front of a mirror, you see an image formed by light reflected from the mirror surface that

FIGURE 22-5 SINCE THE ANGLE OF INCIDENCE EQUALS THE ANGLE OF REFLECTION, YOU CAN TRACE LIGHT RAYS THAT STRIKE A MIRROR AND SEE WHERE AN IMAGE IS FORMED. YOUR EYES DETECT THE LIGHT RAYS AFTER THEY HAVE BEEN REFLECTED, SO THE DASHED LINES SHOW WHERE THE RAYS *APPEAR* TO ORIGINATE, FROM YOUR POINT OF VIEW. SINCE THEY DON'T PASS THROUGH THE IMAGE YOU SEE, IT IS CALLED A VIRTUAL IMAGE.

appears to be *behind* the mirror (see Figure 22-5). The dashed lines show the direction you see the light coming from; tracing the light rays like this lets you see where the image is located. The image is a distance behind the mirror that is equal to your distance in front of the mirror, as the dashed lines in Figure 22-5 indicate and as Figure 22-6 shows from an oblique angle. Such an image, formed by light rays that don't actually pass through the location of the image, is called a **virtual image.**

When you look into a mirror, the image you see looks almost like a perfect picture of you, *but it isn't.* Blink your right eye, and the image that faces you blinks its left eye. The image is a left–right reversal of the object it comes from, and the rays traced in Figure 22-7 show why. Images with this property are called *mirror images.* Ambulances and some other vehicles have mirror images painted on their front surfaces so that a driver in front of them will see the signs correctly through the rear-view mirror.

Reflections from a curved mirror give distorted images, as you can see by looking into the bowl of a metal spoon. Mirrors can be specially curved to reflect light in a certain way. The mirror in a flashlight or a spotlight directs the rays of light from a bulb or a carbon arc into a parallel shaft of light (Figure 22-8). When light from a star falls onto the curved mirror of a reflecting telescope, the mirror directs all the light to a single point. Concentrating the light this way makes stars that are too dim to be seen with the unaided eye visible through the telescope.

FIGURE 22-6 SUSAN AND HER IMAGE, SEEN FROM AN OBLIQUE VIEW WITH A CAMERA. THE METER STICK SHOWS THAT HER IMAGE IS EXACTLY AS FAR BEHIND THE MIRROR AS SHE IS IN FRONT OF THE MIRROR. NOTICE THAT HER LEFT HAND IS THE RIGHT HAND OF THE IMAGE.

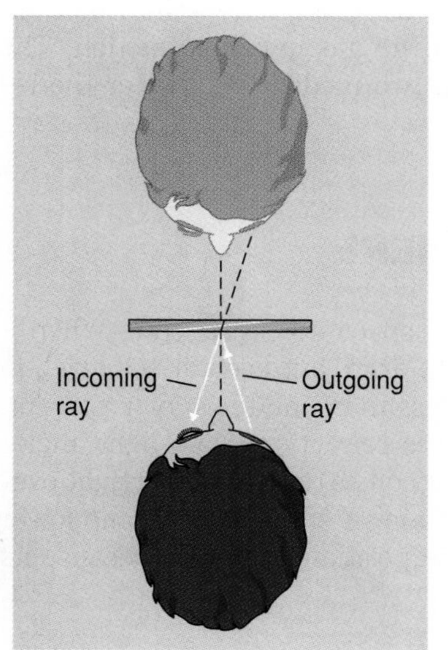

FIGURE 22-7 A PERSON LOOKING AT A MIRROR, AS SEEN FROM ABOVE. A RAY FROM THE CLOSED RIGHT EYE TRAVELS A DISTANCE TO THE MIRROR, REFLECTS, AND TRAVELS AN IDENTICAL DISTANCE TO THE OPEN LEFT EYE. THE OPEN EYE FORMS AN IMAGE OF THE CLOSED EYE FROM MANY SUCH RAYS. THE IMAGE OF THE CLOSED EYE APPEARS TO BE *BEHIND* THE MIRROR AT A DISTANCE EQUAL TO THE DISTANCE TRAVELED BY THE INCOMING RAY.

A MIRROR-IMAGE ADVERTISEMENT MEANT TO BE READ BY DRIVERS IN CARS IN FRONT OF THE TRUCK. VIEWED THROUGH A REAR-VIEW MIRROR, THE MESSAGE IS READABLE.

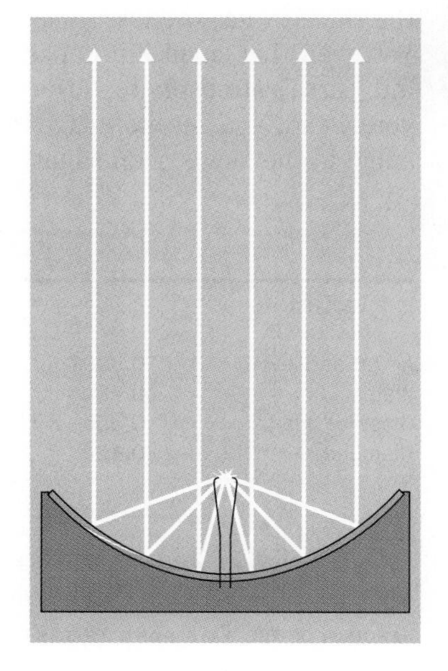

FIGURE 22-8 MIRRORS IN SEARCHLIGHTS AND FLASHLIGHTS ARE CURVED TO CONCENTRATE THE LIGHT FROM A SOURCE INTO PARALLEL RAYS.

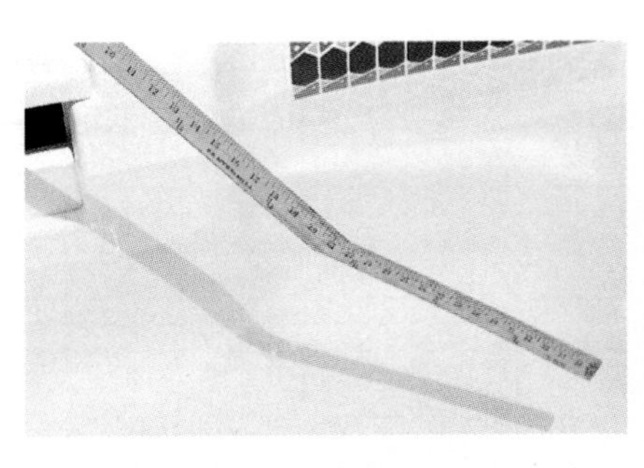

FIGURE 22-9 THIS YARDSTICK LOOKS BENT, BUT IT ISN'T. THE RAYS OF LIGHT COMING FROM THE UNDERWATER PORTION BEND AS THEY LEAVE THE WATER AND ENTER THE AIR. CONSEQUENTLY, THE IMAGE THEY FORM IN YOUR EYES OR A CAMERA IS DISPLACED RELATIVE TO THE TRUE POSITION OF THE STICK.

REFRACTION

On a clear day, your sharp shadow on the ground agrees with the old saying, "light travels in straight lines." However, if you dip one end of a ruler or some other object into water at an angle to the surface, the submerged part looks bent (Figure 22-9). Its image is *displaced* because the light coming from the underwater portion of the object *changes direction* as it leaves the water. This change of direction when light crosses a boundary between two media at an angle is called **refraction.** The refraction is caused by the increased time it takes light to travel through water compared to air (Table 22-1).

Figure 22-10 shows what happens when wavefronts of light traveling through air enter water at an angle to the water's surface. The first part of the wavefront that enters the water slows down while the part of the wavefront still in the air continues to move at the original speed. This difference in speed causes the wavefront to *bend,* or *refract,* at the water's surface. Wavefronts (as discussed in Chapter 15) are always perpendicular to the wave's direction of motion, so as the wave enters the water it travels in a different direction as a result of its refraction. If the light enters the water perpendicular to the water's surface, it doesn't bend because in that case all points of the wavefront slow at the same time.

Light is also refracted if it travels in the other direction, from water into air. If the wave leaves the water at an angle to the water's surface, the emerging part of the wavefront picks up speed and bends away from the perpendicular line to the surface, causing displaced images such as the bent ruler in Figure 22-9 or the shortening of someone's legs as in Figure 22-11.

Whenever light crosses from one medium—air, glass, water—into another, its speed in the second medium determines the amount of refraction. The greater the difference in light's speed in the two media, the greater the refraction.

REFRACTION IN THE ATMOSPHERE

Incoming sunlight or starlight travels through space at the speed *c.* When it enters earth's atmosphere, its speed begins to drop, but even at sea level, in the thickest air, the loss in speed is slight—less than one-tenth of 1 percent. Yet even this small difference in speed causes refraction when light angles into earth's atmosphere. If the light from a star comes down from straight over your head, you see its image in its true position. A star seen at some lower angle in the sky is seen slightly out of its true position, however, because its

Table 22·1 Light's Speed in Various Materials

In a vacuum	c	(about 300,000 km/s or 186,000 mi/h)
Air (at sea level)	$0.9997c$	(about 299,900 km/s)
Water	$0.75c$	(about 225,000 km/s)
Glass or quartz	$0.65c$	(about 195,000 km/s)
Diamond	$0.42c$	(about 126,000 km/s)

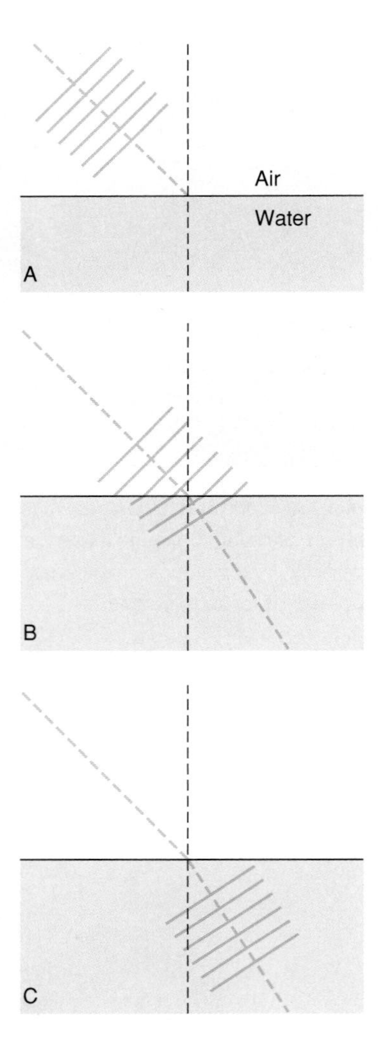

FIGURE 22-10 AS WAVEFRONTS ANGLE INTO THE WATER, THE PART OF EACH WAVEFRONT THAT IS IN THE WATER TRAVELS MORE SLOWLY THAN THE PART STILL IN AIR. THE RESULT IS A TURNING, OR BENDING, OF THE WAVEFRONT. THIS IS CALLED REFRACTION. THE DASHED LINE PERPENDICULAR TO THE WAVEFRONTS IS THE PATH A RAY OF LIGHT WOULD TAKE. (THE *REFLECTED* PORTION OF THE WAVE IS NOT SHOWN.)

(a)

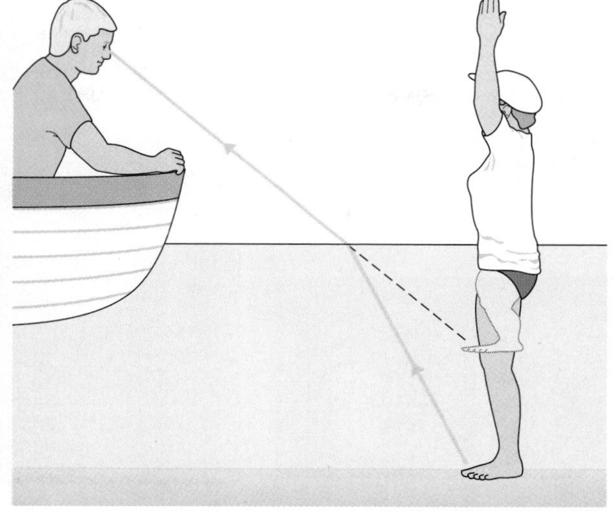

(b)

FIGURE 22-11 (A) SALLY'S LEGS SEEM VERY SHORT HERE COMPARED WITH HER SIZE FROM THE WAIST UP. THE REASON IS REFRACTION, AS SEEN FROM A SIDE VIEW IN (B). LIGHT FROM HER FEET IS REFRACTED AS IT COMES INTO THE AIR FROM THE WATER, AND A PERSON IN THE BOAT SEES AN IMAGE WITH SALLY'S LEGS SHORTENED.

light refracts downward as it comes through the air. You see the image of that star as being a little higher in the sky than the star really is (Figure 22-12). The greater the angle of incidence of the starlight entering the atmosphere, the more bending occurs.

Rays coming through the atmosphere nearly horizontally, as with sunlight each time the sun rises or sets, are bent the most. At sunset, the globe of the sun seems to flatten a bit just before it touches the horizon. It flattens because the rays from the bottom of the sun pass through a bit more of the thickest part of the atmosphere and bend more than do the rays from the top (Figure 22-13). So the bottom of the sun's image is raised a bit more than the top of the sun's image, and the sun appears compressed. In fact, the sun is actually just *below* the horizon at the moment you are seeing its image just *touch* the horizon. See Figure 22-14 for a similar effect with the moon viewed from a spacecraft in orbit around the earth.

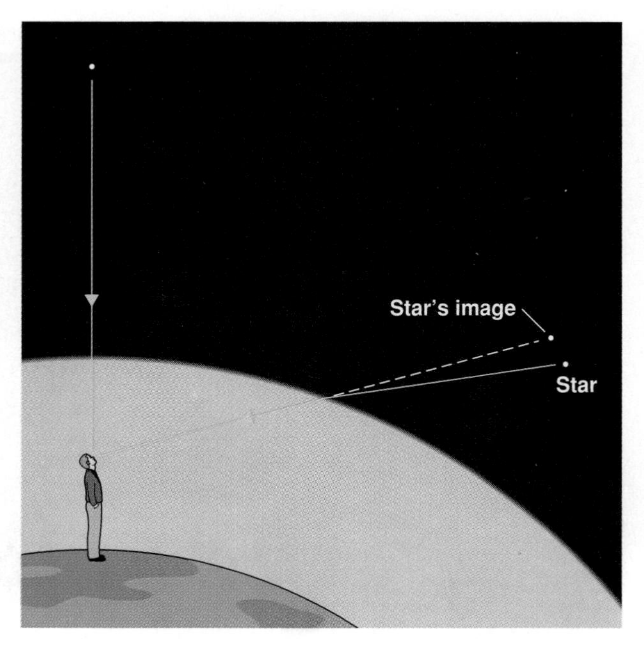

FIGURE 22-12 THE STAR DIRECTLY OVERHEAD IS SEEN IN ITS TRUE POSITION. HOWEVER, THE ATMOSPHERE REFRACTS STARLIGHT THAT COMES IN AT AN ANGLE, AND WE SEE THOSE STARS SLIGHTLY OUT OF POSITION, THAT IS, HIGHER ABOVE THE HORIZON THAN THEY REALLY ARE. (THE DASHED LINE SHOWS THE DIRECTION OF THE DISPLACED IMAGE.)

(a)

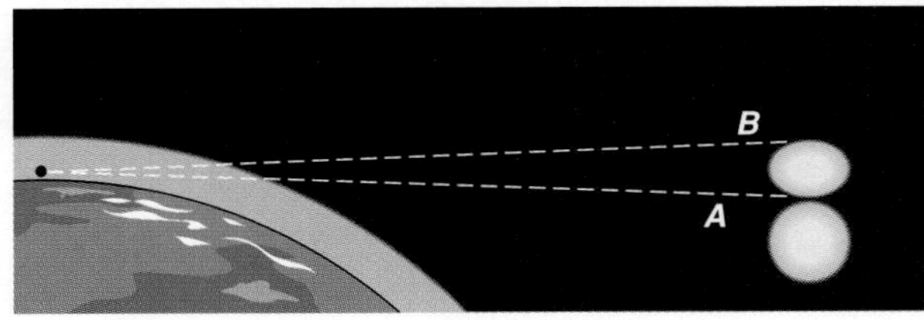

(b)

(c)

FIGURE 22-13 (A) JUST AFTER THE SUN'S TOP EDGE HAS DISAPPEARED BELOW THE HORIZON, YOU CAN STILL SEE THE SUN BECAUSE ITS LIGHT IS REFRACTED BY THE ATMOSPHERE. BECAUSE OF THIS REFRACTION, RAYS THAT WOULD PASS OVER YOUR HEAD IF THE AIR WAS NOT THERE ARE BENT DOWNWARD AND YOU CAN SEE THEM. (B) ON EARTH THE ATMOSPHERE AND CURVATURE OF THE HORIZON ARE SUCH THAT THE SUN'S IMAGE AT SUNRISE OR SUNSET APPEARS ABOUT ONE DIAMETER OVER ITS TRUE POSITION. THE RAYS FROM THE BOTTOM OF THE SUN PASS THROUGH MORE AIR AND ARE REFRACTED MORE, HENCE APPEAR HIGHER, THAN THOSE FROM THE TOP PORTION. (WE'LL SEE IN THE NEXT CHAPTER WHY THE SUN'S IMAGE IS RED AT SUNRISE AND SUNSET.) (C) "COMPRESSED" IMAGE OF THE SUN NEAR SUNSET. THE COMPRESSION IS MORE OBVIOUS IF YOU ROTATE THE BOOK 90°.

(a)

(b)

FIGURE 22-14 TWO VIEWS OF THE FULL MOON AS SEEN FROM SKYLAB IN THE 1970S. (A) NO PART OF EARTH'S ATMOSPHERE IS BETWEEN THE SPACECRAFT AND THE MOON, SO THE VIEW IS UNDISTORTED. (B) WHEN THE MOON IS SEEN THROUGH THE EDGE OF THE ATMOSPHERE, THE BOTTOM PORTION OF THE MOON IS COMPRESSED SEVERELY BY REFRACTION. RAYS FROM THE BOTTOM PORTION OF THE MOON ARE BENT DOWNWARD MORE BECAUSE THEY PASS THROUGH MORE AIR THAN DO THE RAYS FROM THE TOP PORTION, AND THIS EXTRA BENDING MAKES THE BOTTOM APPEAR CLOSER TO THE TOP.

MIRAGES

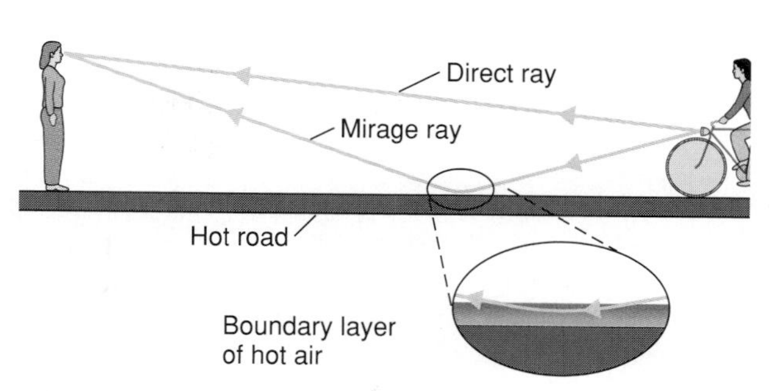

FIGURE 22-15 LIGHT THAT ARRIVES AT THE ROAD SURFACE NEARLY TANGENT TO THE BOUNDARY LAYER OF HEATED AIR NEXT TO THE ROAD CAN BE REFRACTED UPWARD. THIS UPWARD BENDING KEEPS THE LIGHT FROM STRIKING THE ROAD AND BRINGS A MIRAGE—A "MISPLACED" IMAGE—TO YOUR EYES. YOU SEE THE OBJECT TWICE, ONCE FROM THE DIRECT RAYS THAT COME STRAIGHT TO YOUR EYES FROM THE OBJECT AND AGAIN FROM THE REFRACTED RAYS. THIS MIRAGE APPEARS BELOW THE OBJECT. (THE ANGLES ARE NORMALLY MUCH SMALLER THAN THOSE SHOWN HERE.)

Mirages can be seen on any sunlit highway, where they look like water puddles when viewed from a distance but grow thin and disappear as you get closer to them. In a mirage, you see the light from trees, clouds, sky, and so on as if a mirror were lying on the road. Mirages are not reflections of light on the road surface, however; they are *images caused by refraction.* A road heated by sunlight warms the air that touches it, forming a thin boundary layer of hot air next to the pavement. Light that enters this hotter air travels faster than it travels through the cooler air above it because there are fewer molecules per unit of volume in the warmer air to slow the light. Light that angles into this hot layer picks up speed and bends away from the road. The bending is slight, but light rays that almost graze the road can refract and turn upward enough to travel to your eye (Figure 22-15). The mirage comes from light from the sky or from any objects that are located just above the road and behind the mirage. A mirage forms over any smooth dip or bump in the hot road where the light grazes an area nearly parallel to its path. If the surface of a road is wavy, you might see a series of puddles, as in Figure 22-16.

Look for a mirage over a road and walk or drive toward it. The "puddle" will get narrow and finally disappear. Since the light from a mirage is only slightly bent, the rays drop just below your line of sight as you approach the image. If you stoop down while looking at such a mirage, you see more of the refracted rays, and the mirage gets wider. Mirages form over vertical surfaces just as they do over horizontal surfaces. Look directly along the side of a sun-warmed wall. Any object that's just to the side of your line of sight along the wall will appear as a mirage on the wall's surface, as Figure 22-17 shows. Move your head slowly side to side, and you'll see an image of the object appear and disappear.

FIGURE 22-16 MOST OFTEN A MIRAGE IS THE IMAGE OF THE SKY NEAR THE HORIZON. SUCH A MIRAGE USUALLY LOOKS LIKE A PUDDLE OF WATER ON THE HIGHWAY. THIS PHOTOGRAPH SHOWS A SERIES OF MIRAGES FROM THE RUNNING LIGHTS OF A SPORTS CAR. RAYS FROM THE CAR'S LIGHTS TRAVEL NEARLY TANGENT TO THE ROAD AT A SERIES OF DIPS AND EACH TIME REFRACT UPWARD IN THE HOT BOUNDARY LAYER OF AIR THERE.

TOTAL INTERNAL REFLECTION

To see light refract, take a flashlight to a swimming pool at night. Holding the light out to the side at arm's length and just above the surface of the water, aim the beam at the water's surface in front of you. At the surface, some of the light is reflected back into the air and some enters the water and is refracted

FIGURE 22-17 A MIRAGE FROM A WARM WALL. THE LOWER PART OF THE WALL IS A DARKER COLOR THAN THE TOP, SO THE LOWER PART GETS WARMER IN THE SUNLIGHT AND CREATES A WIDE MIRAGE. THE LIGHTER PORTION OF THE WALL REMAINS COOLER AND, AS YOU CAN SEE JUST ABOVE THE MAN'S HEAD, CREATES ALMOST NO MIRAGE AT ALL.

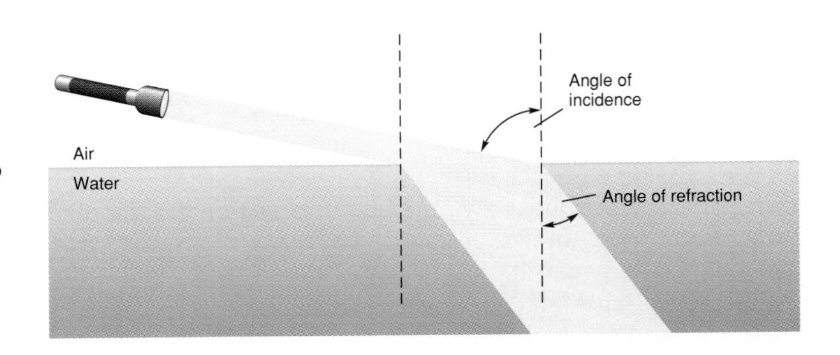

FIGURE 22-18 LIGHT REFRACTS DOWNWARD AS IT ENTERS THE SURFACE OF A SWIMMING POOL AT AN ANGLE. THE GREATER THE ANGLE OF INCIDENCE, THE GREATER THE ANGLE OF REFRACTION. (NOT SHOWN HERE IS THE REFLECTED PORTION OF THE BEAM.)

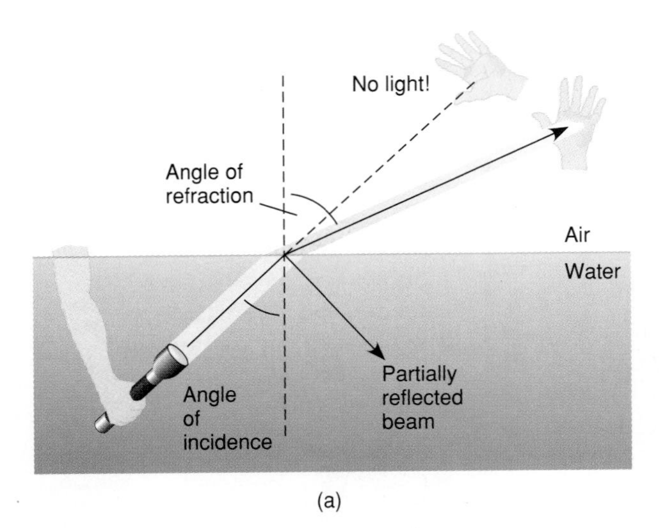

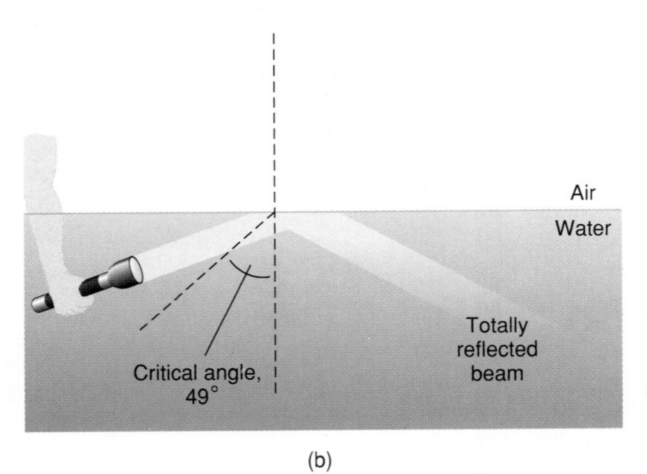

FIGURE 22-19 (A) LIGHT FROM A FLASHLIGHT PICKS UP SPEED WHEN GOING FROM WATER TO AIR, SO ITS ANGLE OF REFRACTION IS GREATER THAN THE ANGLE OF INCIDENCE. YOU CAN SEE THIS BY FINDING THE BEAM WITH YOUR HAND. SOME OF THE LIGHT IS ALSO REFLECTED BACK INTO THE WATER BY THE SURFACE. (B) LIGHT STRIKING THE AIR–WATER INTERFACE AT ANY ANGLE GREATER THAN THE CRITICAL ANGLE IS TOTALLY REFLECTED.

(Figure 22-18). With a flashlight wrapped in a clear, waterproof plastic bag, you can see refraction take place in the opposite direction. Holding the light just under the water, tilt the beam upward at an angle, as in Figure 22-19a. Place your other hand above the water's surface, directly in front of the flashlight. There's no light there! Then slowly lower that hand until the refracted light hits it. If you tilt the flashlight even more, so that the beam strikes the surface at a larger angle of incidence, you'll eventually have to put the hand that is not under water right down to the water surface to find the refracted beam as it travels nearly parallel to the surface. Tilt the light a tiny bit more and your hand goes dark. The refracted light vanishes. At that **critical angle** no light escapes from beneath the surface (Figure 22-19b). The critical angle for a water–air interface is 49°, and light coming to the surface at any angle greater than that is *totally* reflected back into the water. This is called **total internal reflection.** For any angle of incidence greater than the critical angle, the water–air surface acts like a mirror for light coming from below.

A diver sees the effects of total internal reflection when looking up at the surface. Light coming from straight above the diver's head comes straight down, while light coming in at an angle is bent. As you can see from Figure 22-20, even the light from near the horizon angles in for the diver to see. Past the angle of 49° from the vertical, however, the diver sees only reflected light from beneath the surface. Because this happens in all directions around the diver's position, the diver sees a bright circle centered above his or her head. This circle's edge is at the critical angle.

LENSES: BENDING LIGHT TO MAKE IMAGES

Light angling in through a window refracts at both surfaces of the glass (Figure 22-21a), but we still get a fairly accurate view of the outside scenery because those surfaces are flat. In addition, some of the light is reflected back and forth numerous times (Figure 22-21b). Unless the object is extremely bright, however, as in Figure 22-22, these multiple reflections are rarely noticed.

When light comes through a *curved* piece of transparent material rather than a flat piece, refraction bends the rays in different directions. A simple **lens** is a thin, clear piece of glass or plastic with curved surfaces. If the curvature causes parallel light rays to focus to a point, as in Figure 22-22a,

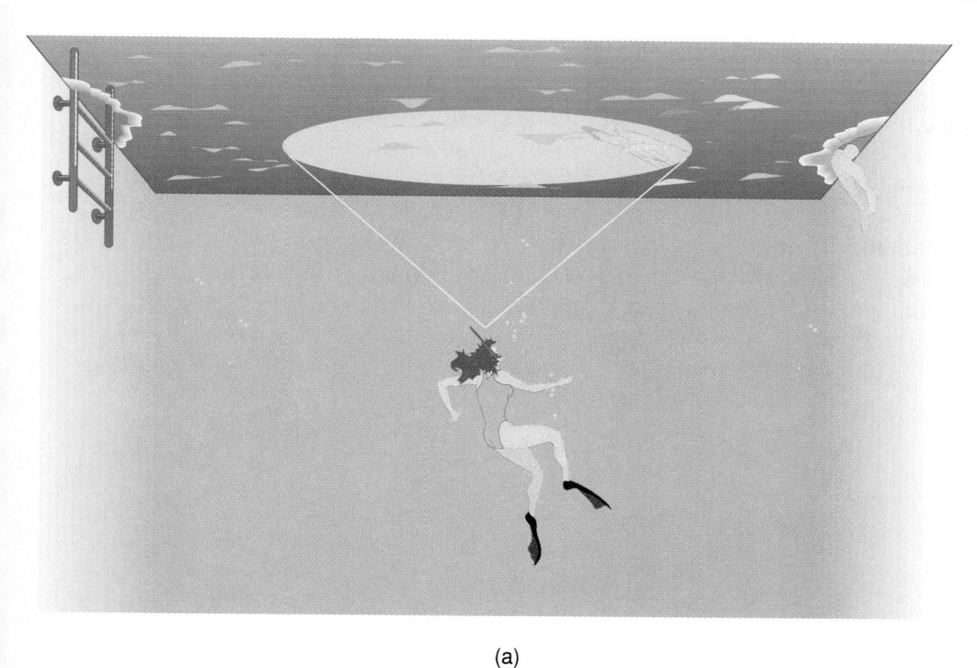

(a)

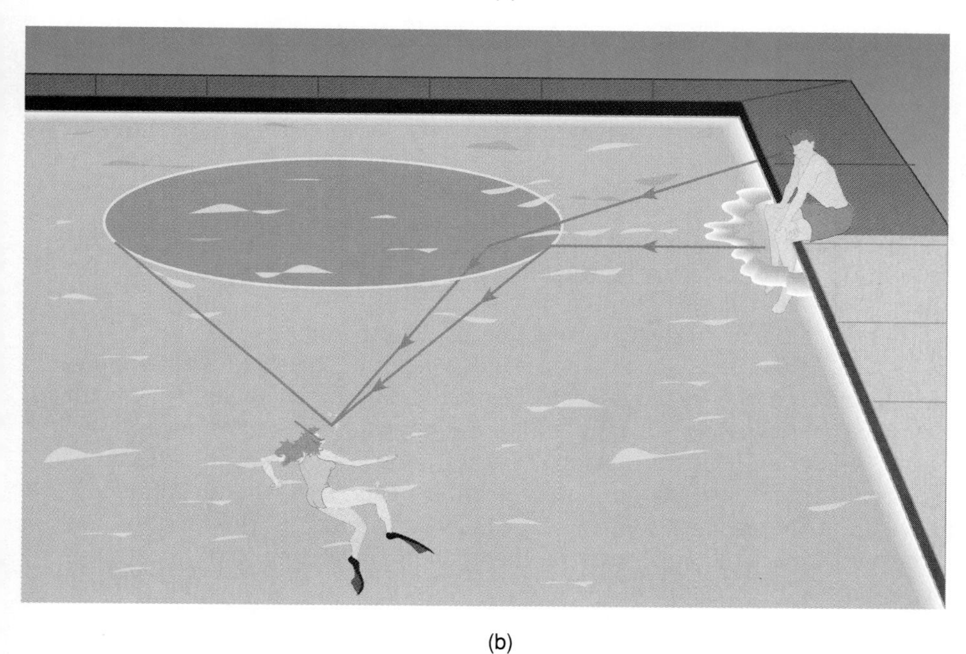

(b)

FIGURE 22-20 (A) WHAT A DIVER SEES BELOW THE SURFACE OF THE WATER. THE IMAGE OF SOMEONE SITTING ON THE EDGE OF THE POOL WITH LEGS IN THE WATER IS SPLIT. THE DIVER CAN SEE THE LEGS BY DIRECT RAYS, BUT THE TORSO APPEARS MUCH HIGHER DUE TO REFRACTION. ALL RAYS COMING INTO THE DIVER'S VIEW FROM ABOVE THE WATER'S SURFACE ENTER THROUGH THE CIRCULAR AREA SHOWN, WHICH IS DEFINED BY THE CRITICAL ANGLE. (B) A VIEW OF THE REFRACTION FROM ABOVE THE WATER.

FIGURE 22-21 (A) AN IMAGE OF AN OBJECT VIEWED THROUGH A PANE OF GLASS IS NOT DISTORTED, BUT IT IS SLIGHTLY DISPLACED WHEN VIEWED AT AN ANGLE. WHEN VIEWED BY AN OBSERVER INSIDE THE ROOM, THE RAY IN THIS DIAGRAM APPEARS TO COME FROM THE DIRECTION INDICATED BY THE DASHED LINE. (B) MULTIPLE REFLECTIONS AND TRANSMISSIONS TAKE PLACE WHEN LIGHT PASSES THROUGH GLASS, BUT THEY USUALLY ARE NOT VERY NOTICEABLE. FIG. 22-22 SHOWS A CASE WHERE THEY ARE, HOWEVER.

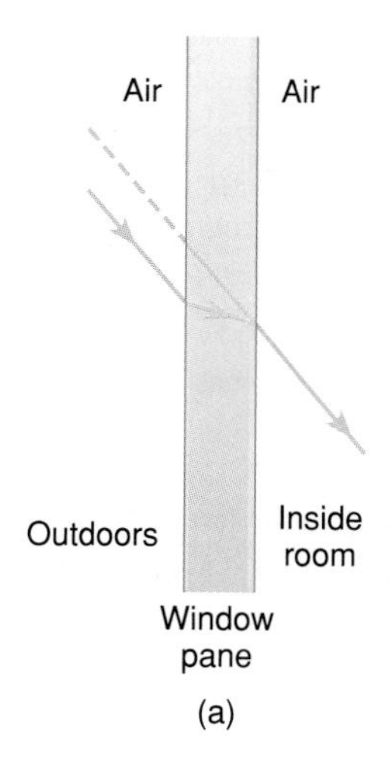

(a)

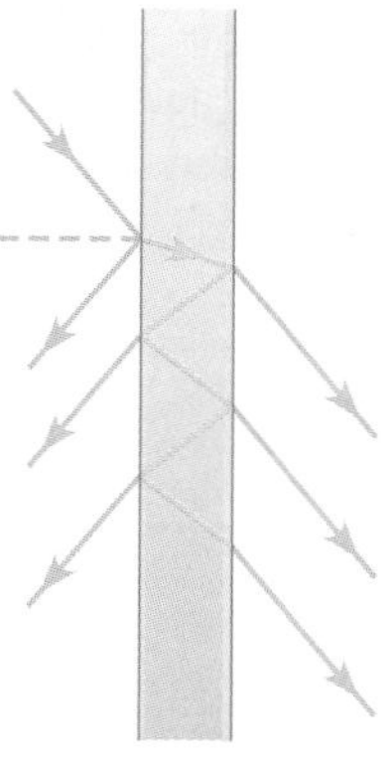

(b)

the lens is called a **converging lens.** If the curvature causes light to be spread out, as in Figure 22-22b, the lens is called a **diverging lens.** A converging lens, also called a **convex** lens, is thicker at its center than at its sides, while a diverging lens, called **concave,** is thinner at its center. The surfaces of lenses, including those of eyeglasses, are normally portions of perfect spheres or cylinders because more complex surfaces are difficult to shape or polish accurately (Figure 22-22c).

Parallel light rays passing through a converging lens meet on the side of the lens opposite the light source at a point called the **focal point** of the lens. The distance from the center of the lens to the focal point is the **focal length** of the lens. Because of the sun's great distance, the rays of sunlight that fall on the earth are very close to parallel. If you have ever used a magnifying glass to focus sunlight to a point and burn a hole in a piece of paper, you've found the focal point of that converging lens. A diverging lens has a focal point, too, as shown in Figure 22-22b. It is the point from which parallel rays passing through the lens seem to originate. Both lenses pictured are symmetric from right to left, and this means they both have two focal points, one on each side of the lens, because parallel light rays coming from either side would be brought to a focus (converging lens) or appear to diverge from a point (diverging lens). For a situation where both types of lenses occur all the time, see Figure 22-27 later in this chapter.

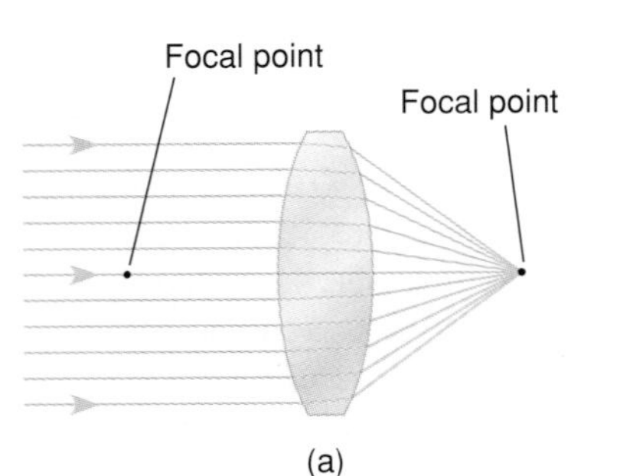

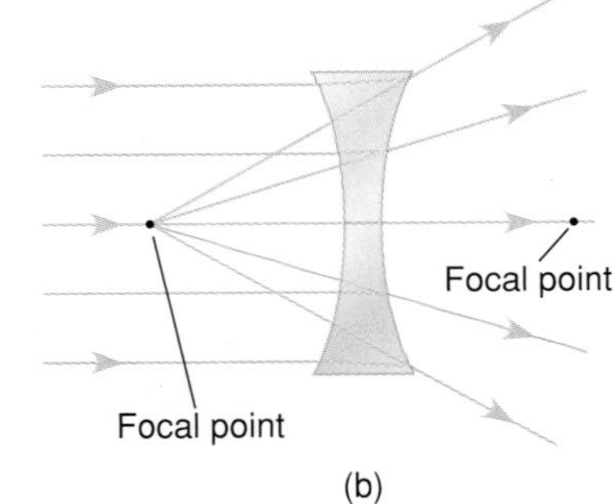

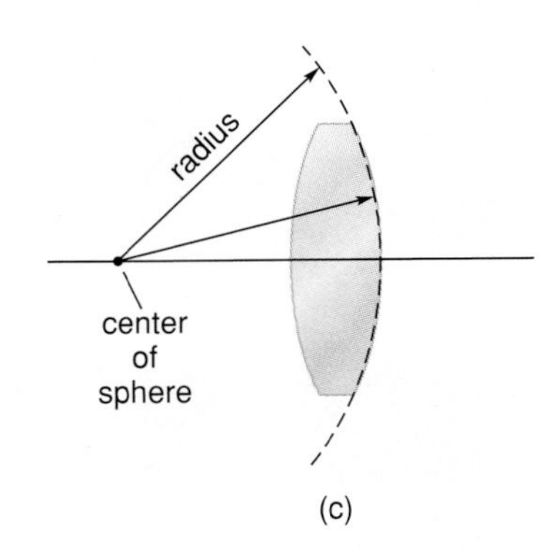

FIGURE 22-22 (A) THE LIGHT RAYS ON THE LEFT ARE TRAVELING PARALLEL TO THE AXIS OF A CONVEX LENS. (THE AXIS OF A LENS IS THE CENTER LINE PERPENDICULAR TO THE FACE OF THE LENS.) THE RAYS REFRACT AT BOTH SURFACES OF THE LENS AND CONVERGE AT A POINT ON THE OPPOSITE SIDE OF THE LENS CALLED THE FOCAL POINT. ANY LENS HAS A FOCAL POINT ON EACH SIDE, SINCE PARALLEL RAYS COMING IN FROM EITHER DIRECTION WILL COME TO A FOCUS. (B) PARALLEL LIGHT RAYS COMING INTO A CONCAVE LENS REFRACT AND DIVERGE SO THAT THEY APPEAR TO COME FROM A POINT LOCATED IN FRONT OF THE LENS. THAT POINT IS THE FOCAL POINT OF THE CONCAVE LENS. (C) MOST LENSES HAVE SURFACES THAT ARE A PART OF A SPHERE (OR A CYLINDER). HERE A SPHERICAL LENS IS SHOWN WITH THE RADIUS OF THE SPHERE IT IS A PART OF. (THE RADIUS OF THE SPHERE IS *NOT* AT THE FOCAL POINT.)

Converging lenses can form real images, like the images slide projectors make on a screen. Real images appear only when the object is more than one focal length from the converging lens. Figure 22-23 shows how this works with a candle that's two focal lengths from a convex lens. The rays shown in this figure are chosen because they are easy to trace. Ray A from the bottom of the candle passes straight through the center of the lens without refraction. Ray B from the candle's top travels parallel to the axis of the lens, so it is refracted through the focal point on the far side. Ray C from the candle's top passes through the center of the lens. (It refracts upon entering but refracts the same amount in the opposite direction when leaving.) Ray D passes through the focal point on the object side and so emerges parallel to the axis of the lens on the image side. Where rays B, C, and D meet is the position of the top of the candle's image. As seen in the figure, the candle has a life-sized real image two focal lengths behind the lens, but it is upside down, or *inverted.* This is why the slides must go in upside down in a slide projector.

Move the object farther away from the lens, and the inverted image moves in toward the lens and becomes smaller (Figure 22-24). As shown in Figure 22-25, a scene on the far horizon forms a small image on a plane behind a converging lens, called its *focal plane.* In a typical 35-mm camera, the inverted image made by the lens falls on the film to expose it. To take a picture of a distant scene, the lens must be one focal length from the film. To focus on something closer, the camera's lens must slide toward the object.

To use a converging lens to magnify, the object must be closer to the lens than the focal point, as in Figure 22-26. Ray C passes through the center of

FIGURE 22-23 A REAL IMAGE IS FORMED BY A CONVEX LENS WHEN THE OBJECT IS MORE THAN A FOCAL LENGTH AWAY. THE LOCATION OF THE IMAGE IS FOUND BY USING RAYS THAT ARE EASY TO TRACE. RAY A GOES THROUGH UNREFRACTED, SO THE BOTTOM OF THE CANDLE'S IMAGE LIES ALONG THE AXIS OF THE LENS ON THE RIGHT. RAY B IS PARALLEL TO THE LENS AXIS ON THE LEFT AND SO GOES THROUGH THE RIGHT-SIDE FOCAL POINT. RAY C PASSES THROUGH THE CENTER OF THE LENS AND IS ESSENTIALLY UNDEFLECTED. RAY D PASSES THROUGH THE FOCAL POINT ON THE LEFT SIDE AND THUS TRAVELS PARALLEL TO THE LENS AXIS ON THE RIGHT SIDE.

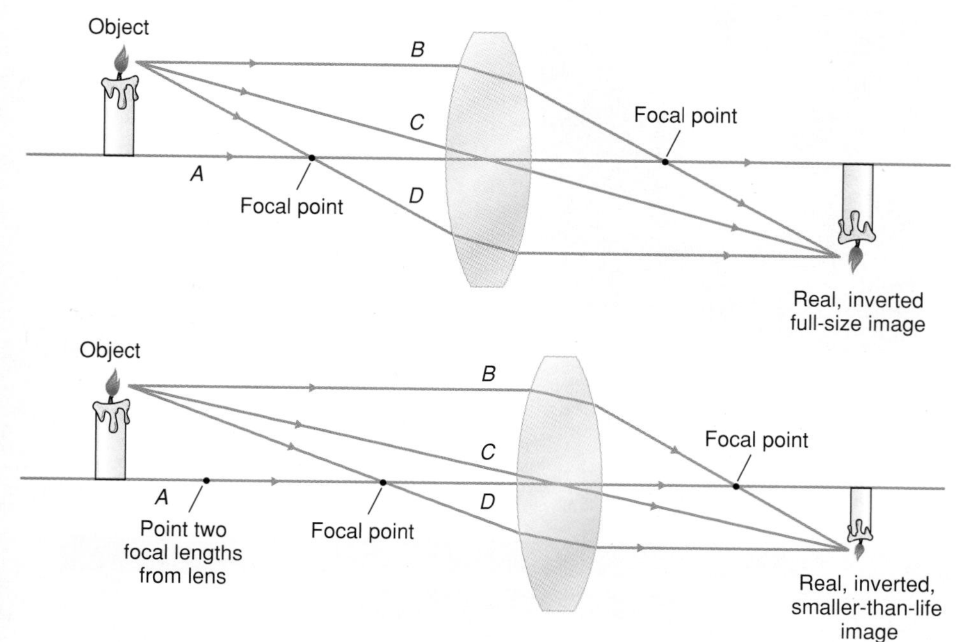

FIGURE 22-24 A SITUATION SIMILAR TO THAT OF FIGURE 22-25, BUT WITH THE OBJECT FARTHER AWAY FROM THE LENS. THE IMAGE REMAINS INVERTED AND BECOMES SMALLER.

the lens with no change in direction. Ray B moves parallel to the axis before entering the lens, so it refracts to go through the focal point on the right side. The dashed lines at the left of the lens show the directions from which the rays appear to have come. Your eyes see the top of the image located where the two rays from the top of the tack seem to originate where the dashed lines converge. The image is upright and also virtual because light rays do not pass through it.

Focal point
Focal plane

FIGURE 22-25 PARALLEL LIGHT RAYS THAT COME INTO A CONVEX LENS AT AN ANGLE COME TO A FOCUS IN A PLANE PERPENDICULAR TO THE LENS'S AXIS AT THE LENS'S FOCAL POINT. THIS IS CALLED THE *FOCAL PLANE* OF THE LENS.

FIGURE 22-26 USING A CONVEX LENS AS A MAGNIFIER. THE LARGE THUMBTACK AT THE LEFT IS THE IMAGE OF THE SMALL TACK SEEN THROUGH THE LENS BY AN OBSERVER. TWO RAYS ARE ENOUGH TO SHOW THE IMAGE POSITION HERE. RAY C PASSES THROUGH THE LENS ESSENTIALLY UNDEFLECTED, AS DID RAY C IN FIGURE 22-25. RAY B IS PARALLEL TO THE AXIS BEFORE ENTERING THE LENS AND THEREFORE PASSES THROUGH THE FOCAL POINT ON THE RIGHT. THE DASHED LINES SHOW WHY THE IMAGE APPEARS WHERE IT DOES. THE IMAGE IS VIRTUAL BECAUSE THE LIGHT RAYS DON'T REALLY PASS THROUGH IT OR COME FROM IT. NOTICE THE TACK IS NEARER TO THE LENS THAN THE LENS' FOCAL POINT.

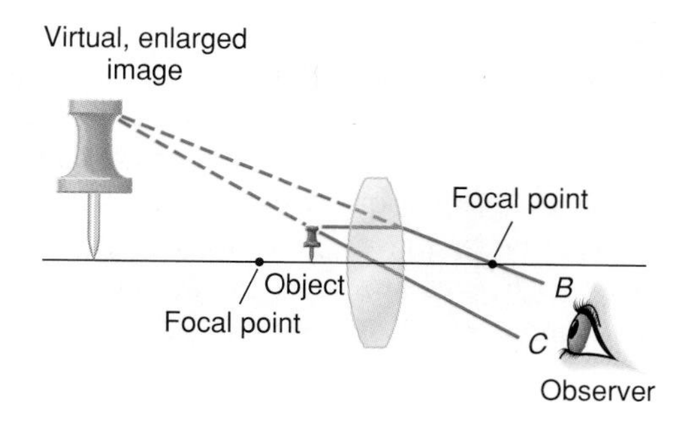

(A) A building with a wine glass on a table in front of it. (B) The image of the building behind the wine glass is inverted, because the glass acts like a convex lens. (Compare to Figure 22-26.)

OPTICAL FIBERS

As we saw in Chapter 21, transmitting equipment puts information onto radio waves and microwaves by changing either the amplitude or the frequency of the wave.

Waves of visible light have a much higher frequency than do radio waves, so more information can be sent per second with light beams than with radio waves or microwaves. And light can be guided by the thin glass or plastic fibers known as *optical fibers.* Total internal reflection keeps the light inside these fibers.

Light comes in at one end of a light pipe and goes straight until it hits the inner wall. If its angle of incidence with this wall is less than the critical angle, some of the light passes through the transparent material that forms the pipe and is lost. However, the light is totally reflected if its angle of incidence on the wall is greater than the critical angle. Then the totally reflected beam travels in a straight line until it hits the inner wall again. A thin optical fiber can be twisted into loops or even knots and still carry the light. Laser light traveling through these fibers can eclipse the information rate of, say, ordinary computer transmissions over telephone lines. Via telephone lines, it would take an hour to transmit the text of the Bible, whereas an optical fiber can carry that much text in 1/16 of a second.

If optical fibers are distorted by pressure, the amount of light they transmit is affected. An optical fiber housed in a protective cable and placed on a road can "weigh" cars and trucks as they run over it—a heavy truck deforms the fiber more than a car, so less light passes through. Optical fibers with sensors embedded in concrete walls and column supports of new buildings detect tiny cracks or even small movements. Optical fiber cables installed on and in dams can monitor the water pressure on the dam, the water flow through the spillway, and the vibrations caused by the turbines and electrical generators.

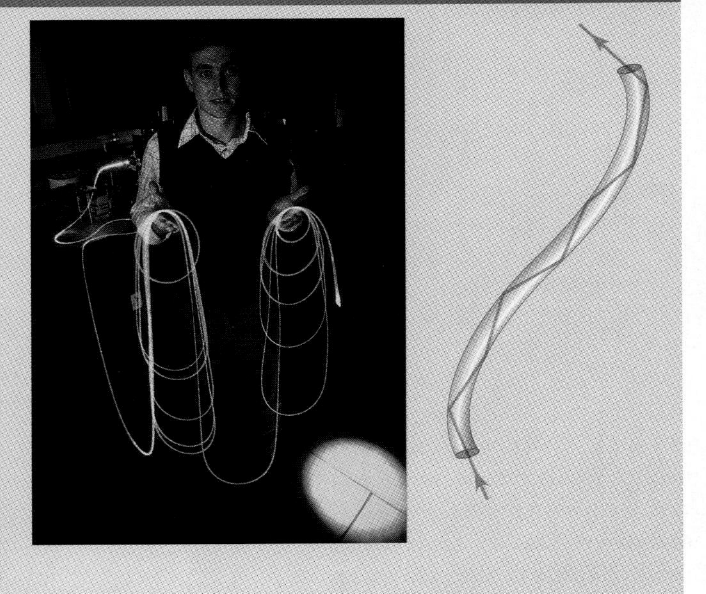

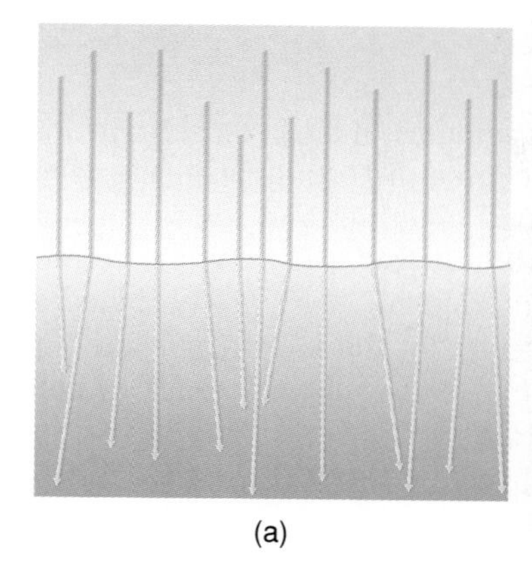

(a)

(b)

(c)

FIGURE 22-27 (A) WHEN A WATER SURFACE IS RIPPLED WITH WAVES, LIGHT TRANSMITTED THROUGH THE SURFACE IS CONVERGENT OR DIVERGENT, DEPENDING ON WHERE IT ENTERS. THE CURVES OF THE SURFACE REFRACT LIGHT MUCH AS CONVEX AND CONCAVE LENSES DO. SINCE THE WAVES MOVE, THE LIGHT BELOW MOVES ALSO. (NOT SHOWN ARE THE REFLECTED WAVES.) (B) A DOLPHIN IS ILLUMINATED WITH LIGHT THAT IS "LENSED" BY THE WAVES IN THE WATER SURFACE OVER ITS HEAD. (C) SMOOTH WAVES, HERE IN A RIVER, ALSO REFLECT LIGHT AS CURVED MIRRORS WOULD, CONVERGING OR DIVERGING THE RAYS.

FIGURE 22-28 THE BRIGHT IMAGE OF THE FILAMENT OF AN UNFROSTED LIGHT BULB REVEALS THE MULTIPLE REFLECTIONS AND TRANSMISSIONS THROUGH A PANE OF GLASS THAT USUALLY ARE SO FAINT WE DON'T NOTICE THEM. (USUALLY ABOUT FOUR PERCENT OF THE LIGHT IS REFLECTED EACH TIME LIGHT ENTERS OR LEAVES A PANE OF GLASS.)

DISTORTION OF IMAGES

Light coming through the edge of a spherical lens focuses a little closer to the lens than light passing through its center, distorting the image. Modern cameras have series of spherical lenses glued together, and such combinations of lens compensate for this **spherical aberration** with multiple refractions. These lenses, usually made with different kinds of glass, are designed with the help of computers. Guided mathematically by the known refractive properties of the glass in the lenses, a computer sends imaginary rays to bend through a set of hypothetical lenses. Adjusting the curves of the lens surfaces with the computer lets the designer minimize aberrations.

PHYSICS IN 3D/SPHERICAL ABERRATION OF A LENS

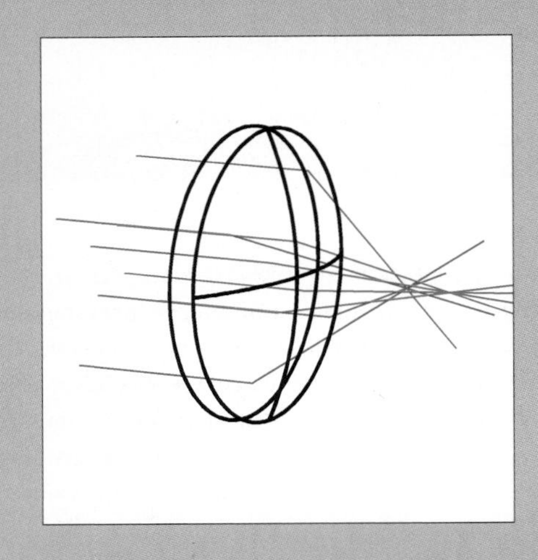

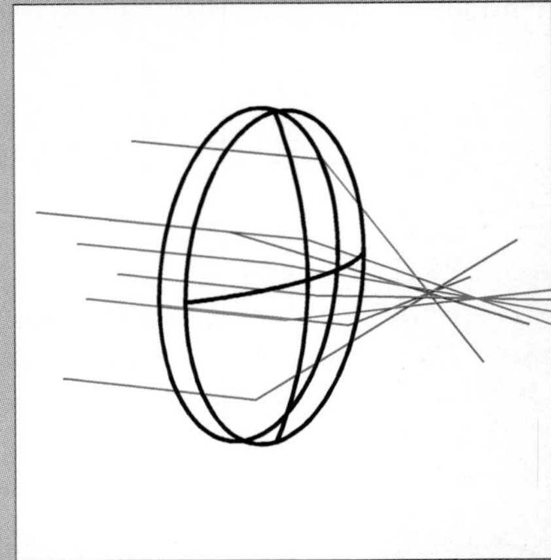

SPHERICAL ABERRATION. RAYS REFRACTED FROM THE EDGE OF A SPHERICAL LENS FOCUS A LITTLE CLOSER TO THE LENS THAN RAYS THROUGH ITS CENTER.

PHYSICS IN 3D/CHROMATIC ABERRATION OF A LENS

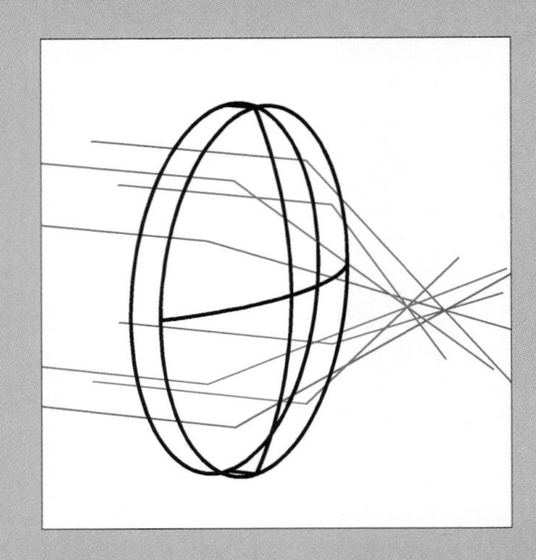

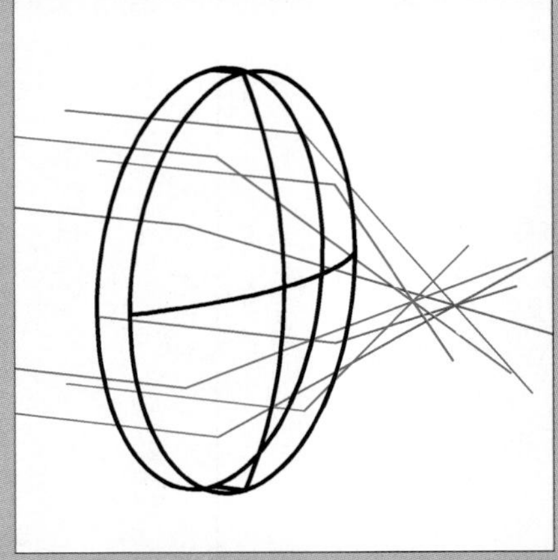

CHROMATIC ABERRATION. SINCE DIFFERENT FREQUENCIES (HENCE COLORS) OF LIGHT TRAVEL THROUGH GLASS AT SLIGHTLY DIFFERENT SPEEDS, THEY HAVE SLIGHTLY DIFFERENT ANGLES OF REFRACTION AND COME TO A FOCUS AT SLIGHTLY DIFFERENT POINTS. HERE RED AND BLUE LIGHT RAYS, ALL INCIDENT AT THE SAME DISTANCE FROM THE LENS CENTER, SHOW CHROMATIC ABERRATION.

THE SIMPLE TELESCOPE

A simple telescope (Galileo built the first one) consists of two convex lenses. The front lens, called the *objective,* brings parallel light rays from a distant object into focus, and a second lens, called the *eyepiece,* magnifies that image. If a distant object takes up angle A when seen by the naked eye, it will take up the larger angle B when seen through the telescope.

As you can see from the figure, the image formed by a simple telescope is inverted. In small commercial telescopes, this image is inverted again by another lens so that you see the image upright. Binoculars use prisms to invert the image (see Exercise 36). Modern telescopes use curved mirrors to form the intermediate image, avoiding the distortions caused by a lens and eliminating the light lost to absorption within the lens.

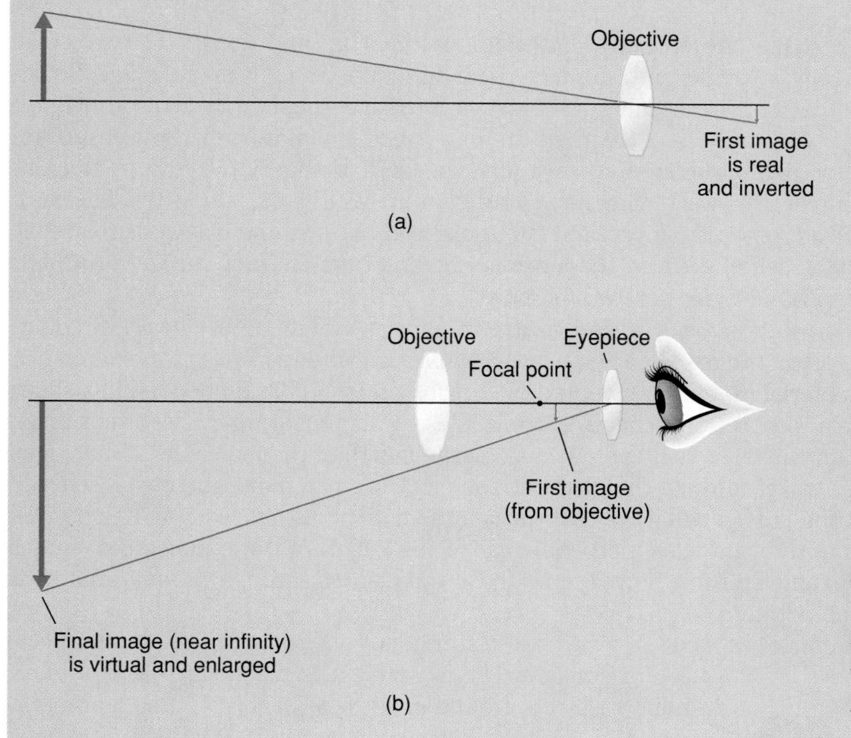

IMAGE FORMATION BY A SIMPLE TELESCOPE. (A) A CONVERGING LENS CALLED THE OBJECTIVE FORMS A SMALL, REAL, INVERTED IMAGE OF A DISTANT OBJECT. (B) A SECOND CONVERGING LENS, CALLED THE EYEPIECE, MAGNIFIES THE IMAGE FORMED BY THE OBJECTIVE TO GIVE A FINAL IMAGE THAT IS VIRTUAL AND LARGER THAN THE OBJECT APPEARS TO THE UNAIDED EYE.

Chromatic aberration is another source of image distortion. Different frequencies of light travel at slightly different speeds in a glass lens, so they are refracted at slightly different angles. Multiple lenses of different kinds of glass that correct for this are called *achromatic* lenses.

INTERFERENCE OF LIGHT WAVES

The shimmering colors in soap bubbles or in oil films on water come from the interference of light waves. When light strikes a thin film, some is reflected from the outside surface and some is transmitted into the film (Figure 22-29). Once inside the film, the transmitted portions of the waves partially reflect from the other surface of the film, return, and escape. Among these escaping waves, some waves meet waves of the same wavelength reflecting from the front surface in just the right manner for them to cancel. Other wavelengths from inside the film exit in such a way as to add constructively. Those added waves are the bright colors you see from the glimmering bubble or film.

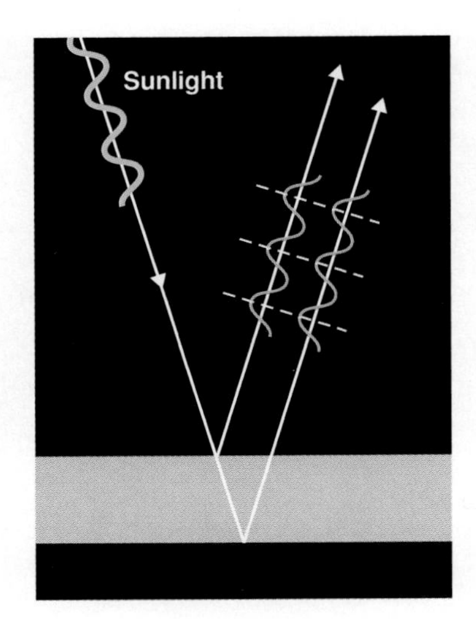

FIGURE 22-29 INTERFERENCE IN A THIN FILM, SAY, OF OIL OR SOAP. IF THE PART OF A WAVE THAT IS REFLECTED FROM THE SECOND SURFACE ADDS CONSTRUCTIVELY WITH LIGHT OF THE SAME WAVELENGTH COMING FROM THE FIRST SURFACE, THAT COLOR IS SEEN BY THE OBSERVER. IF THE TWO PARTS ADD DESTRUCTIVELY (IF THEY ARE OUT OF PHASE), THAT COLOR IS NOT SEEN. THE THICKNESS OF THE FILM DETERMINES WHETHER A GIVEN WAVELENGTH INSIDE EMERGES IN STEP WITH THE SIMILAR WAVE REFLECTED AT THE SURFACE.

POLARIZATION

If you shake one end of a clothesline straight up and down, the wave that scoots away wiggles only in the vertical direction, which is in one plane. Shake the same line horizontally, and the wave you make wiggles only in the horizontal plane. Waves whose oscillations are in a single plane are called **plane-polarized** or just **polarized.** If you jerk the end of the clothesline in random directions, the wave that moves away from your hand along the line isn't polarized. Likewise, most light isn't polarized because the electrons that emit the light, whether from the sun's surface or from the hot surface of a light bulb's filament, accelerate randomly.

Some substances absorb light waves that vibrate in a certain direction while letting those vibrating in a perpendicular direction pass. An example is the material *Polaroid,* invented by Edwin Land in 1938. Embedded in a sheet of Polaroid are long chains of molecules in alignment, and these molecules contain electrons that can move along the length of the molecules. When light waves travel through the material, the electrons can move and absorb energy from the part of the wave that pushes them along the molecules. The oscillations of the light wave perpendicular to the length of these molecules cannot move the electrons, however, so these waves pass through the material with less absorption.

Sunglasses made of polarizing material cut out some of the strong reflections of sunlight. Direct sunlight is not polarized, but when sunlight reflects from a road or a lake (or other non-metal horizontal surface), it becomes partially polarized. The light we call glare is mostly reflected light with waves that oscillate in the horizontal plane. Polarizing lenses can block the horizontally polarized glare, while any vertically oscillating light waves pass through the lenses. Figure 22-30 shows why the vertically oscillating waves don't reflect well. The electrons in the atoms in the surface layers are accelerated almost perpendicularly to the observer's eyes, so almost no light comes to the observer.

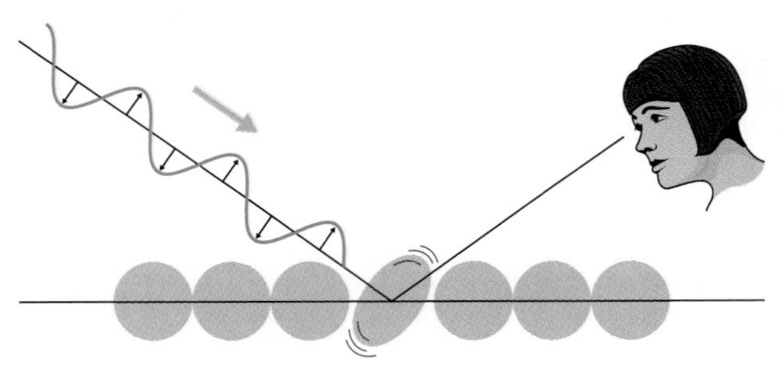

FIGURE 22-30 THE VERTICALLY OSCILLATING WAVE INDUCES LITTLE REFLECTED LIGHT OF THIS POLARIZATION. THINK OF THE ATOMIC ELECTRONS AS BEING AT THE END OF A ROPE IN THIS FIGURE, WITH A FIELD LINE POINTING TOWARD THE PERSON. IF THEY WERE SHAKEN SIDE-TO-SIDE, A TRANSVERSE WAVE WOULD TRAVEL TO THE PERSON'S EYE. BUT SHAKEN TOWARD AND AWAY FROM THE EYE, LITTLE OR NO TRANSVERSE WAVE WOULD APPEAR.

DIFFRACTION OF LIGHT

Water waves that pass through an opening in a barrier will *diffract,* which means that they spread into the region behind the barrier (Figure 15-15). Diffraction of light occurs whenever a portion of a wavefront is blocked or limited in some way. A light wave diffracts to some extent if a portion enters a tiny pinhole in a piece of aluminum foil or if the wavefront passes a sharp edge of any solid body.

In Chapter 21 we saw the double-slit interference patterns that were the first proof of light's wave nature. Due to diffraction, however, interference occurs with just a single slit or even a single edge. Christian Huygens, about 1678, explained such interference by considering each point along a wavefront to be a point source of new waves traveling only in the forward direction. Notice what this implies for a wavefront at a single opening in a screen, as in Figure 22-31. While only three point sources are drawn, you can see what will happen. Each source acts like a single slit, sending out a diffracted wave. At various distances and angles to the right, then, we should expect to find bright and dark patterns where these waves interfere. You can see the interference caused by diffraction in Figure 22-32. Also see the Demonstrations.

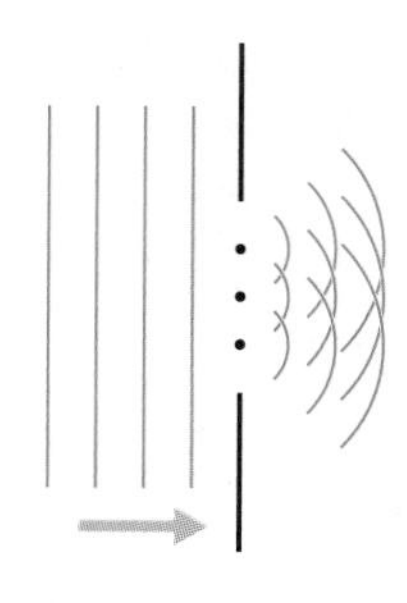

FIGURE 22-31 DIFFRACTION OF LIGHT FROM A SINGLE HOLE. POINTS WITHIN THE HOLE ACT AS SMALL OSCILLATORS THAT DISPERSE THE RADIATION IN ALL FORWARD DIRECTIONS. THE WAVES FROM THESE POINTS WILL INTERFERE WITH EACH OTHER AT POSITIONS TO THE RIGHT.

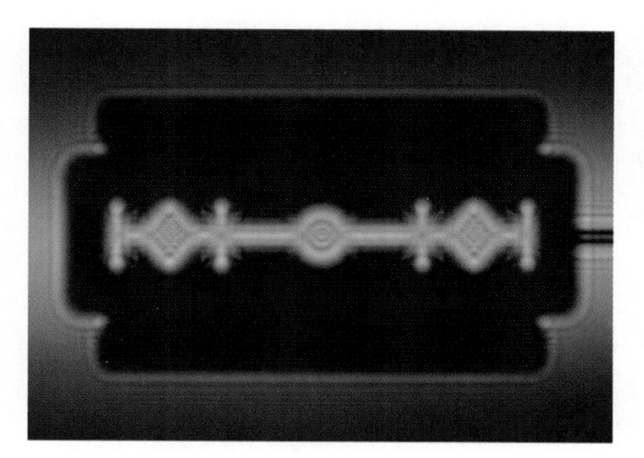

FIGURE 22-32 DIFFRACTION OF LIGHT AROUND THE EDGES OF A RAZOR BLADE.

Summary and Key Terms

Images are likenesses of the objects around us formed by the light they reflect. We can imagine that light consists of infinitely thin beams called ***rays***. Surfaces like this page, which are microscopically rough, ***diffuse*** light by scattering the rays in many directions, but a smooth surface reflects light rays so that the ***angle of incidence*** equals the ***angle of reflection*** (the ***law of reflection***). A mirror's image appears behind the mirror; it is a ***virtual image***, one formed by rays that don't pass through the image location. A ***real image*** is formed by rays that reach the image location.

When light angles into a different medium, its speed changes and it ***refracts***, changing direction. The greater the change in light's speed as it enters a new medium, the more it refracts. If light moves into a medium in which it has a greater speed at an angle greater than the ***critical angle, total internal reflection*** occurs and no light escapes.

A ***converging lens*** is ***convex***, bringing passing rays together, while a ***diverging lens*** is ***concave***, spreading passing rays apart. The ***focal point*** of a converging lens is the point at which parallel rays passing through the lens meet on the other side, while that of a diverging lens is the point from which parallel rays entering the lens seem to come from. The ***focal length*** is the distance from the focal point to the center of the lens. ***Spherical aberration*** can distort an image as light passing through the edge of a spherical lens focuses closer to the lens than light passing through its center. ***Chromatic aberration*** can distort an image as different frequencies of light, traveling at different speeds in the lens, refract at slightly different angles.

image–319
rays–319
diffuse reflection–319

angle of incidence–319
angle of reflection–320
law of reflection–320
virtual image–321
real image–320
refraction–322

critical angle–326
total internal reflection–326
converging (convex) lens–327
diverging (concave) lens–327
focal point–328

focal length–328
spherical aberration–332

chromatic aberration–333

plane-polarized–334

Sunlight that reflects from nonmetallic surfaces, called glare, has waves that oscillate largely in a single plane. We say that glare is ***plane-polarized***. The lenses of Polaroid sunglasses are made of a *polarizing* material oriented to absorb the horizontally oscillating glare while transmitting any light waves that oscillate vertically.

EXERCISES

Concept Checks: Answer true or false

1. When light focuses to form a likeness of an object, that likeness is called an image.
2. A pinhole camera works because light travels in straight lines through the air. The image formed in the camera is real and upright.
3. If light reflected by a surface is scattered in many directions, it is diffused.
4. The angle of incidence and the angle of reflection for a ray of light reflecting from a flat, highly reflective surface are equal.
5. The difference between a virtual image and a real image is that you cannot see a virtual image.
6. Light refracts, or bends, when it changes speed upon entering a different medium perpendicular to its surface.
7. The speeds of light through air, water, and glass are slower than the speed of light through a vacuum.
8. Refraction due to earth's atmosphere can change the position of a star's image in the sky.
9. A mirage is caused by diffraction of light.
10. A converging lens brings parallel rays of light to a focus one focal length away from the center of the lens.
11. A diverging lens brings parallel rays of light to a focus one focal length away from the center of the lens.

Applying the Concepts

12. Sometimes people don't think they look like themselves in photographs. Can you suggest a reason for that? (*Hint:* How do people ordinarily "see" themselves?)
13. Industrial spies have been known to leave innocuous-looking, apparently empty cardboard boxes lying around on a table for a day and then collect them. Play sleuth and guess what these boxes are.
14. The sea animal known as the chambered nautilus has a pinhole opening into a water-filled chamber. Opposite the pinhole, the wall of the chamber is lined with light-sensitive nerves. In other words, the pinhole and chamber act as an eye. Discuss the advantages and disadvantages of this arrangement for this mollusk.
15. Kingfishers most often dive into the water parallel to the rays of the sun. In the morning, they angle in from the east, near noon they dive straight down, and in the afternoon they angle in from the west. How do these flight plans help these birds fish for food?
16. Astronomers who concern themselves with accurate star positions must take refraction into account as they measure angles between stars. Why?
17. Discuss this claim: When you walk toward a mirror, your image seems to come closer to you at twice the speed with which you are walking.
18. Does the diameter of the circle of vision above a diver's head (through which light enters the water from above) grow or shrink as the diver descends? (It helps to make a drawing.)
19. When buying sunglasses, how can you tell if the lenses transmit mostly plane-polarized light?
20. Stars seen through even the largest telescopes still look like points of light. What does this tell you about the rays of a star's light?
21. If rays of light come into glass and diamond surfaces at identical angles, which rays will be bent more by refraction? (Look at Table 22-1 to get the answer.)
22. When you look at a plane flying at an altitude of 35,000 feet and located at a large angle from the point directly over your head, do you see it in its true position? Discuss. (Recall that half of the atmosphere lies below the 18,000-foot level.)
23. "Twinkle, twinkle, little star . . ." Often stars "twinkle" at night, meaning their images seem to wiggle around and even change brightness a tiny bit. Describe how currents of air of different temperatures and densities moving in the upper atmosphere can cause this behavior. (*Hint:* See Figure 22-24.)
24. The archer fish (sometimes called the rifle fish) squirts water with its tongue and pallet. The volley explodes through the water's surface to knock unsuspecting bugs into the water. Its range is about 1 meter. It almost always fires from straight beneath its prey. Why does this technique make sense?
25. If a pinhole camera has a too-small pinhole, diffraction of light will smear the image. What if the pinhole is too large? Draw a diagram to explain what happens in the too-large case.

26. An entire classroom of students can see an instructor's face at the same time. Would you say that fact derives mostly from (a) refraction (b) diffraction (c) total reflection (d) diffuse reflection (e) absorption of light?

27. Describe a plane-polarized wave.

28. The first electric light bulbs were made of clear glass. Today most of them have frosted glass. Why? (The first frosted light bulbs, by the way, were frosted on the outside. Why is interior frosting more desirable?)

29. The fact that we normally see our environment with very little distortion depends on (a) the wave nature of light (b) the speed of light (c) diffusion (d) the fact that light travels in straight lines through the air (e) all of the above.

30. The best mirrors are made by coating very smooth surfaces with a thin layer of aluminum or silver. Why is metal a better reflector of light than most other materials?

31. Make a drawing of a person standing in front of a full-length mirror that is just as tall as the person. Then use a ruler and a protractor to draw rays that reflect from the shoes to the eyes. (Remember the angle of incidence equals the angle of reflection.) Use that drawing to convince yourself that a mirror needs to be only half as tall as the person to form an entire image.

32. Explain why a shaft of sunlight streaming through a break in storm clouds seems to have such sharp edges, whereas light passing through a small slit diffracts.

33. The smoother the surface before it is coated, the better the quality of the mirror. The best flats of glass are smooth to within 20 atomic diameters over appreciable areas. Explain why these flat pieces of glass reflect visible light so uniformly even though they contain variations of as much as 20 atomic widths. (*Hint:* Recall from Chapter 21 what the wavelengths of visible light are in terms of atomic diameters.)

34. Explain why water that is being heated (as in a glass teapot) produces wavering images if you look through the water.

35. The cleaved surface of a single crystal of graphite may be perfectly flat for distances of several thousand atomic diameters. Why can't this material be used to make the most perfect mirrors in the world?

36. The image made by a camera lens on the film is inverted, but the image in a typical camera's viewfinder is not. A prism of glass in the viewfinder is responsible. The prism's end, or cross-section, is a triangle that has angles of 90°, 45°, and 45°. The light enters one edge of the prism perpendicularly and exits at an angle of 90° both to the original beam and to another side of the prism. Draw a ray diagram to show what must happen.

37. Light from high-powered lasers can be focused with a lens and used to weld delicate metal parts. Such lasers can also drill (or melt) smooth holes in ultrahard metals to one-millionth of an inch. Explain why such a powerful laser beam doesn't crack the lens as it travels through the glass.

38. An ordinary rock becomes much darker if you wet it, besides being more reflective. Guess why that is. It is for the same reason that a concrete sidewalk looks darker when it is wet with rain.

39. A near-invisible animal or insect (such as a jellyfish, baby eel, or a South American glass-winged butterfly) casts no shadow. Discuss the advantages and disadvantages of being nearly invisible.

40. Beer and other beverages are often sold in 6-packs with a hard clear plastic sheet of "rings" connecting them for easy handling. Wherever careless people let these 6-pack rings reach the ocean, they are often eaten by sea life such as sea lions and whales, which die as their digestive tracts are blocked by the plastic. They are fooled since the plastic looks almost clear in water, like a jellyfish. (The speed of light in the plastic is not too different from the speed of light in seawater, so little reflection occurs.) Think of several ways this product could be changed to reduce this risk.

41. Penguins swimming in the ocean are black when seen from above and light-colored when seen from below. What purpose could this color difference serve?

42. Why does a high-flying plane cast no shadow although a low-flying plane does cast one? Draw a ray diagram as part of your explanation.

43. Draw a top view of a pinhole camera and convince yourself that the image is reversed left to right (as a mirror's image is) as well as inverted.

44. Flat mirrors reverse the image of an object, exchanging right for left and vice versa, as you can see from Figure 22-7. Why, then, don't mirrors exchange the top of your face for the bottom? Draw a ray diagram to show why a flat mirror doesn't form an inverted image.

Demonstrations

1. Take a large cardboard box outside on a sunny day and punch a pinhole in one side. Use a sharp pencil or an icepick. A hole that's 0.5 millimeter to about 0.75 millimeter across works well. Cut a small peephole along one side of the box, and use a cloth to prevent much light from entering this as you look inside. If light comes in other than through the pinhole, use tape to seal the leaks. Look at the wall opposite the pinhole. On that wall you'll see a faint inverted image of the scene outside the box, as Figure 22-1 shows. The rays of light traced from the scenery through the pinhole show why this image appears as it does.

2. On a still, sunny day, find a shade tree and stand in the shadow cast by its branches. Occasionally a small beam of sunlight may emerge from between the leaves to strike the ground. If the opening it comes through is small, a small circular image appears on the ground even when the opening isn't circular. The tree, the hole, and the shadow are like a natural pinhole camera, and the image you see is an image of the sun. Because the image is round or slightly elliptical, depending on the time of day, it isn't especially striking. You might try looking for an image of the moon the same way—a crescent moon should give a crescent image. Investigate!

3. Dip a wire loop or the hole in a scissors-handle into a soap solution to form a film across the hole. (Use a few drops of dishwashing liquid in a glass of water.) Hold the loop vertically and watch as the film drains downward, getting thin at the top and thick at the bottom. Notice how the interference causes colors that move downward as the liquid drains. Near the top, the film eventually gets so thin that no interference can take place. That thinnest area reflects much less light than the lower part of the film and so looks darker.

4. With a pin or needle, make a small hole in a sheet of aluminum foil. Use this to look at a distant streetlamp at night with your eye close to the foil, and you will see circular interference patterns caused by diffraction.

5. With your forefinger and thumb held parallel and close together, form a small slit. Place this slit two inches from your eye and look at a bright light from across the room or a distant streetlamp at night. Then look carefully for the interference pattern of alternating dark and bright lines. Once you've spotted these "fringes," and you know what to look for, you can see them just by holding your fingers up to an open window in a room.

6. Here's a helpful study tip. Place this book on a table with a study lamp behind it as shown in Figure 22-33a. Move the book around until you see glare coming from the ink of a section title or a black region of a photo. Then put on a pair of Polaroid sunglasses and the glare will vanish. Rotate the Polaroids 90° and the glare reappears. You get glare from the book just like the glare from a highway. The study tip is this: To avoid the glare without using sunglasses, always place the lamp to the side of the book, as shown in Figure 22-33b, rather than behind it.

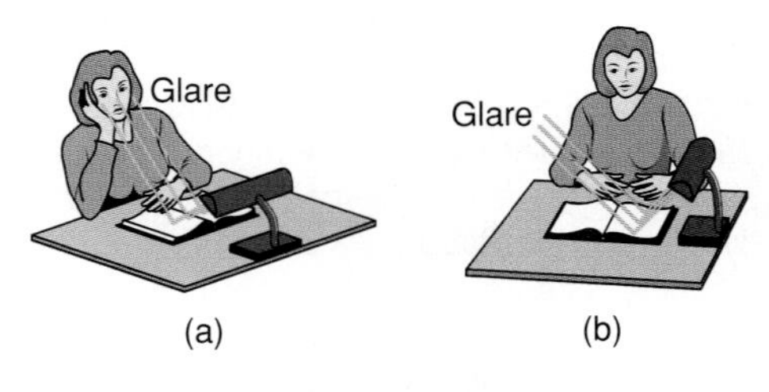

FIGURE 22-33

TWENTY-THREE

Colors and Vision

Alligators, like cats and many other nocturnal creatures, have reflective layers behind their retinas. Light that the retinas don't absorb is reflected back through to give the retina a second chance to absorb it, increasing the efficiency of the eye for night vision. The reflectors are caught here in the powerful flash lamp of the photographer's camera.

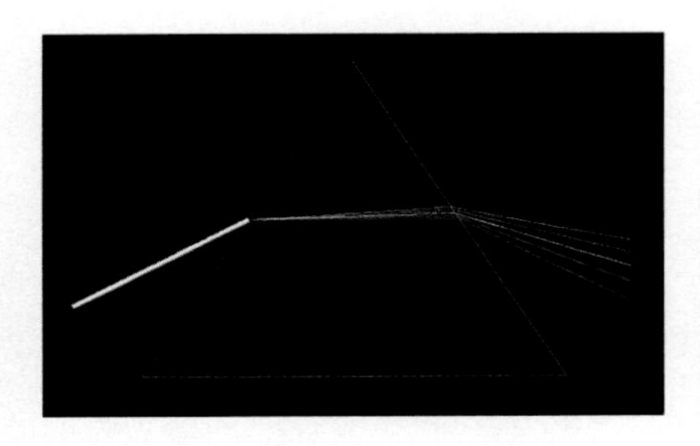

FIGURE 23-1 WHEN VISIBLE FREQUENCIES OF LIGHT ENTER GLASS, THE LOWER FREQUENCIES LOSE LESS SPEED AND ARE REFRACTED LESS. VIOLET, THE HIGHEST VISIBLE FREQUENCY, SLOWS THE MOST AND IS BENT AT A LARGER ANGLE BOTH WHEN IT ENTERS AND WHEN IT LEAVES THE GLASS. THIS COMPUTER-GRAPHIC IMAGE SHOWS THE SEPARATION OF THE COLORS IN WHITE LIGHT AT BOTH SURFACES OF A PRISM.

The beautiful colors in our world come to us when light is reflected from opaque objects such as flowers and butterflies or transmitted through objects such as a stained glass. When white sunlight reflects from trees, fields, and buildings, we are treated to rich greens, yellows, and hundreds of other hues. Sometimes light transmitted through a clear substance is dispersed into the colors of the rainbow, as when sunlight comes through a cut-glass pendant or a glass prism.

THE COLORS OF WHITE LIGHT

We call the separation of the colors of sunlight by a glass prism **dispersion**. The cause of dispersion is that different frequencies of light travel at slightly different speeds in the glass. For that reason, they refract by different amounts when entering or leaving a prism. All transparent matter—diamond, quartz, clear plastics, water—is dispersive to some extent.

Figure 23-1 traces some of the rays of light through a prism. As the light passes into the glass, the violet portion is slowed more than the red portion. Violet light, then, refracts more than red light, and the other colors refract at angles in between red and violet. Notice that refraction takes place where the rays leave as well as where they enter, spreading them even farther apart.

FIGURE 23-2 (A) THE DISPERSION OF A BEAM OF WHITE LIGHT IN A SPHERICAL RAINDROP. WHEN THE DIFFERENT COLORS ENTER, THEY ARE BENT BY DIFFERENT AMOUNTS SINCE THEY TRAVEL AT SLIGHTLY DIFFERENT SPEEDS. REFLECTING FROM THE BACK OF THE DROP (SOME OF THE LIGHT EXITS THERE, NOT SHOWN IN THIS DRAWING), THE COLORS RETURN TO EXIT FROM THE FRONT. WHEN THEY LEAVE, REFRACTION SPREADS THEM EVEN FARTHER. (B) TWO DROPS FROM A RAINBOW-PRODUCING CLOUD OF DROPS. THE PERSON SEES RED LIGHT FROM THE TOP DROP, VIOLET FROM THE LOWER DROP, AND THE OTHER SPECTRAL COLORS FROM DROPS IN BETWEEN THESE TWO. (C) THE RAINBOW APPEARS AS AN ARC. EVERY POINT ALONG THE RED PORTION OF THE ARC IS 42° AWAY FROM THE RAYS OF SUNLIGHT. (D) THE FULL RAINBOW AS SEEN FROM THE GROUND.

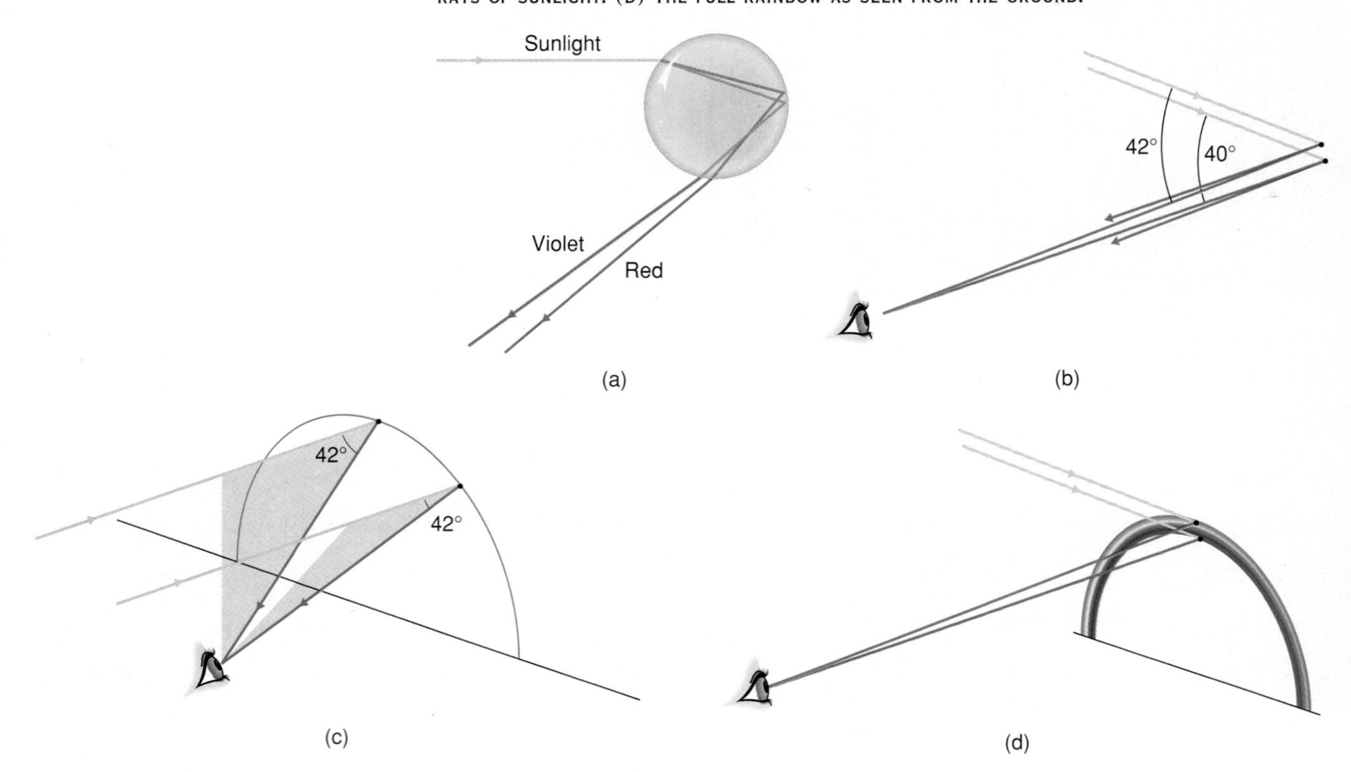

The spread of colors—red, orange, yellow, green, blue, indigo, and violet—is called the **spectrum** of visible light, and they have been remembered by generations of students by the mnemonic ROY G BIV.

RAINBOWS

A rainbow is a dispersion spectrum formed when sunlight falls on a rain cloud or on the mist of a lawn sprinkler or a waterfall. When sunlight strikes a drop of water, some passes into the drop (Figure 23-2a). The rays that enter the drop reflect from the back of the drop and return out through the front surface. The colors of the sunlight are dispersed first as they enter the drop and then even more as they leave it. Red light exits at an angle of about 42° from the direction line of the sun's rays, while violet leaves at an angle of about 40° from that line. The other colors exit at angles between these two values.

You see a color from a drop only if a straight line from your eye to the drop makes an angle between 40° and 42° with the sun's incoming rays (Figure 23-2b and c). The rainbow appears as an arc for that reason, with red on the outside and violet on the inside (Figure 23-2d). If you are on the ground, the arc ends at the ground, where there are no more drops to reflect the light. From an airplane flying past this same cloud of droplets, however, you'd see a full circle of rainbow, as shown in Figure 23-3.

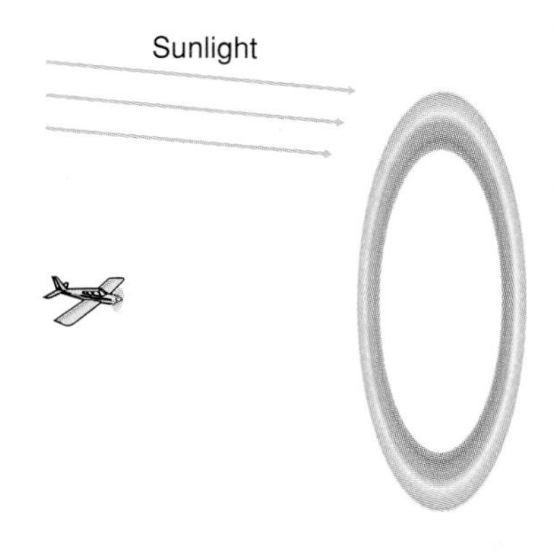

FIGURE 23-3 PASSENGERS IN AN AIRPLANE ARE SOMETIMES TREATED TO THE FULL CIRCLE OF A RAINBOW RATHER THAN JUST AN ARC, SINCE THE GROUND DOES NOT INTERRUPT THE CLOUD OF DROPLETS.

Sometimes you can see a companion rainbow that circles farther out than the bright primary rainbow (Figure 23-4). This one, called the *secondary rainbow*, comes from rays that undergo two reflections at the back of each raindrop and exit at a wider angle, about 50° from the incoming sunlight. The second reflection changes the order of the emerging colors, so the secondary rainbow's colors have their order reversed relative to the colors of the primary rainbow.

FIGURE 23-4 THE PRIMARY AND SECONDARY RAINBOWS. NOTE THE COLORS ARE REVERSED IN THE SECONDARY RAINBOW.

PHYSICS IN 3D/DETAILS OF A RAINBOW

A STEREO VIEW OF A RAINBOW. THE PERSON SEES THE ARC OF SPECTRAL COLORS SINCE ALONG THAT ARC THE ANGLE BETWEEN THE SUN AND THE PERSON'S EYE IS CONSTANT, BETWEEN 40° AND 42°. FOR SIMPLICITY HERE, ONLY THE RED RAYS ARE SHOWN, WHICH COME TO AN OBSERVER'S EYES AT 42° FROM THE INCIDENT SUNLIGHT. (THE LOWER ARC IS WHERE BLUE WOULD APPEAR.)

THE COLOR OF THINGS

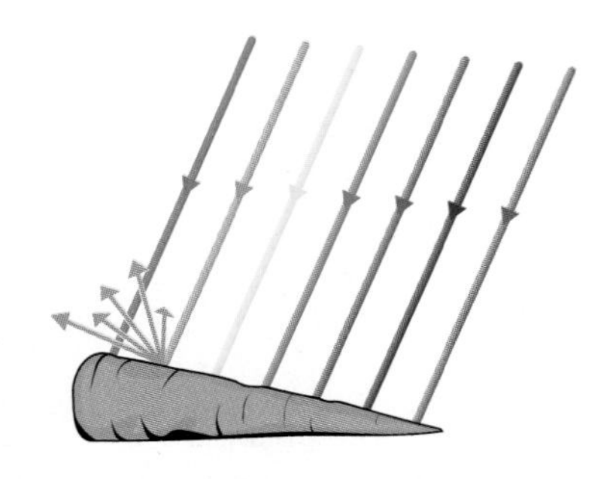

FIGURE 23-5 ALL THE COLORS OF WHITE LIGHT STRIKE THIS CARROT, AND ALL BUT ORANGE WAVELENGTHS ARE ABSORBED. THE ORANGE LIGHT IS REFLECTED, SO THE CARROT APPEARS ORANGE.

Most of the grand variety of colors we see around us is due to **selective absorption** and **selective reflection**. When sunlight strikes a yellow shirt, for instance, all the colors of the visible spectrum shine on the material, but mostly the yellow wavelengths are reflected to our eyes. The other wavelengths are largely absorbed by the material. (Just which wavelengths are absorbed by any material depends on how the electrons are bound in the molecules and atoms of the material.) A carrot is orange in sunlight (or under electric light) only because it absorbs most of the other colors in the spectrum and reflects mostly orange light (Figure 23-5). Held under a source of pure blue light with no other light reaching it, the carrot would look black. It would absorb the blue light, and there would be no orange light to be reflected.

In the same way, the reflected colors seen in a painting are the colors that the painting has not absorbed. Any color not in the visible spectrum is a mixture of different wavelengths from the spectrum. When an artist mixes two or more paint pigments to get a desired hue, the color we see is a mixture of whatever wavelengths of visible light those various pigments reflect. When sunlight falls on this page, white light is reflected to your eyes except where the black ink absorbs the sunlight, causing the letters to appear black. Neither white nor black is considered to be a color, however; white is a mixture of all visible colors and black is the absence of all of them.

Just as opaque matter absorbs some wavelengths and reflects others, transparent matter absorbs some wavelengths and transmits others. This process is called **selective transmission**. If red light enters a piece of green glass, it is absorbed, for example, while green light that enters is transmitted. Some of the red light is reflected, however, as well as the green, causing the reflected green light to be less pure than the transmitted green light (Figure 23-6). This

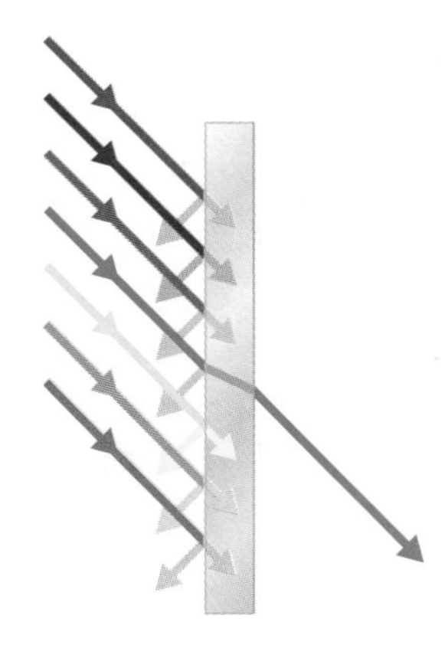

FIGURE 23-6 A PIECE OF STAINED GLASS TRANSMITS ONLY CERTAIN WAVELENGTHS, BUT IT REFLECTS AND ABSORBS MANY MORE. THE REFLECTED LIGHT CONTAINS MORE COLORS THAN THE TRANSMITTED LIGHT AND SO APPEARS DRAB WHEN COMPARED TO THE TRANSMITTED LIGHT. (SEE FIGURE 23-7.)

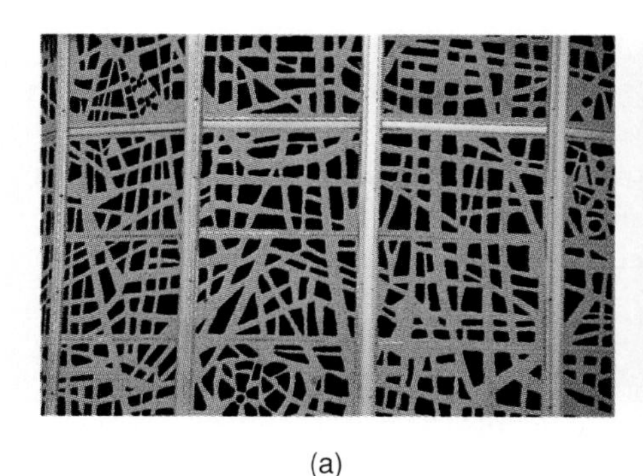

(a)

(b)

FIGURE 23-7 (A) STAINED GLASS VIEWED FROM OUTSIDE. NOTICE HOW DULL THE COLORS ARE. (B) THE SAME GLASS FROM THE INSIDE. THE TRANSMITTED COLORS ARE MUCH MORE VIVID.

is one reason for the stunning effect of sunlight coming through stained glass windows. The transmitted color is purer than the reflected color. Seen from outside during the day, the windows are less vivid than they are when viewed from inside (Figure 23-7).

The sunlight that enters an emerald (a green gemstone) or a ruby (red) often undergoes several internal reflections before emerging. As the light travels back and forth through the gem, it becomes even more green or red as the other wavelengths are absorbed more completely. Likewise, multiple reflections on an opaque surface remove more of the absorbed colors than a single reflection can. The colors coming from deep inside the folds of draperies or from deep between the petals of a rose are purer than those reflected from the outer areas (Figure 23-8) because much of the light coming away from those deeper regions underwent several reflections on the way in.

FIGURE 23-8 NOTICE HOW DEEP THE COLOR OF THIS ROSE BECOMES FARTHER DOWN IN THE PETALS. MULTIPLE REFLECTIONS, WITH THEIR ACCOMPANYING SELECTIVE ABSORPTIONS, HAVE INTENSIFIED THE COLOR.

BLUE SKIES AND RED SUNSETS

When light interacts with objects of dimensions much smaller than the wavelength of the light, the light is said to be scattered rather than reflected. The electrons of such a small object are all shaken up and down at the same time by the electric field of the light wave, and they radiate that frequency of light

LIGHT IN THE OCEANS

As sunlight travels down through clear water, the lowest-frequency waves are absorbed best, while the higher frequency waves penetrate farther. IR from sunlight is absorbed within a hand's breadth of the surface, and only blue light reaches many tens of meters down. As a scuba diver descends, red light disappears from view first, followed by orange, yellow, and green. The blood of a diver who is cut while underwater appears drab green due to the absence of red, orange, and yellow light. Colorful fish all appear uniformly blue to divers who descend 10 meters or so; the colors emerge only when their pictures are taken with an underwater flash. Penguins don't see red at all, their eyes being adapted to hunting at depths where only blue and green survive. Some deepwater fish sense UV, which transmits well through water and also reflects well from their prey's scales.

Phytoplankton, one-celled plants, capture energy from sunlight and combine CO_2 molecules with H_2O molecules to form carbohydrates and oxygen in the process called photosynthesis. These reactions can take place only near the ocean's surface, where the sunlight penetrates with sufficient intensity. As mentioned in previous chapters, the oceans absorb about half of the CO_2 emitted each year from the burning of fossil fuels. The photosynthesis-supporting zone, called the *photic* zone, is fairly shallow due to the limited penetration of sunlight. Nutrients, in addition to sunlight, are critical to production, as are organic trace materials, such as vitamins and hormones produced by bacteria. Pollution washed in through rivers, spilled from ships, or due to acid rain can prohibit plant production, and this is yet another front where scientific studies are proceeding at an increasing rate. Satellites now take images of earth's surface in many wavelengths to monitor this pollution.

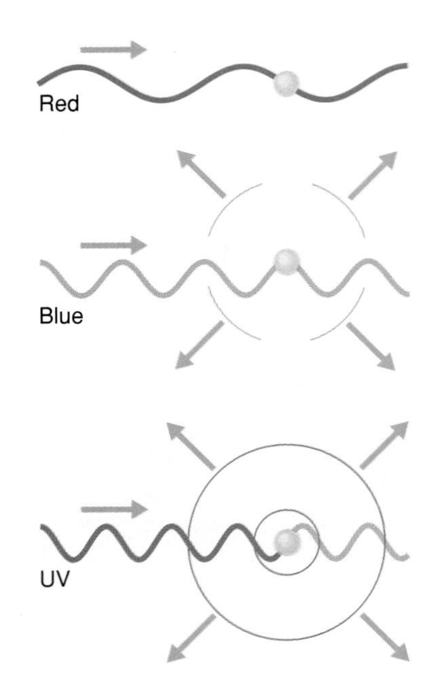

FIGURE 23-9 THE GREATER THE FREQUENCY OF A LIGHT WAVE, THE MORE ENERGETICALLY IT OSCILLATES THE ELECTRONS OF THE ATMOSPHERE'S MOLECULES. BECAUSE BLUE LIGHT HAS A HIGHER FREQUENCY THAN RED LIGHT, FOR EXAMPLE, AIR MOLECULES SCATTER THE BLUE LIGHT FROM SUNLIGHT MUCH MORE EFFECTIVELY THAN THEY SCATTER THE RED LIGHT. ULTRAVIOLET IS SCATTERED EVEN BETTER THAN BLUE LIGHT SINCE IT IS OF A HIGHER FREQUENCY. HOWEVER, THERE IS LESS ULTRAVIOLET LIGHT IN SUNLIGHT THAN THERE IS BLUE LIGHT, AND OUR EYES DON'T SENSE ULTRAVIOLET ANYWAY.

in all directions. The diameter of most molecules is much smaller than the wavelengths of visible light.

The electrons in molecules of air vibrate when an electromagnetic wave comes through. If the wave has a very long wavelength, as radio waves do, the electrons act like corks that rise and fall gently on a lake as a wave passes. As a result, the molecules don't absorb or scatter much of the wave, and we say the air is very *transparent* to radio waves. Shorter wavelengths of electromagnetic radiation, such as visible light, carry more energy, however, and their more rapid oscillations (remember, shorter wavelength means higher frequency) cause a greater disturbance of the electrons in the air molecules. These electrons become miniature antennae—they oscillate with the passing wave and send light out in all directions. From the IR through the UV frequencies, the greater the frequency of the light, the more the electrons respond. Said another way, the shorter the wavelength, the more light is scattered. Figure 23-9 pictures this effect.

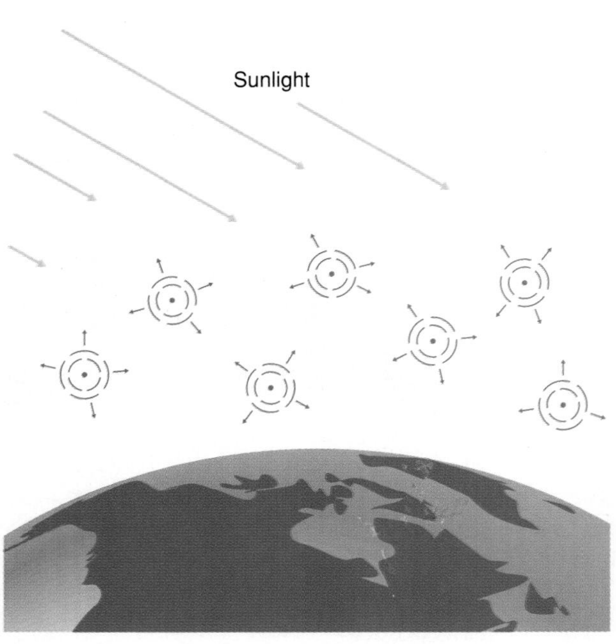

FIGURE 23-10 THE DAYTIME SKY IS BLUE IN EVERY DIRECTION, WHEREVER THERE IS SUNLIT AIR, SINCE AIR'S MOLECULES SCATTER THE BLUE LIGHT MOST.

Of the visible wavelengths, then, air molecules scatter blue light the most, which is why the sky looks blue (Figure 23-10). Though indigo and violet are scattered more than blue, much less of the sunlight's energy is carried by those two colors, and our eyes are in effect overwhelmed by the intense blue light. Ultraviolet light, with its shorter wavelength, is scattered even more than blue light is, but our eyes can't detect UV. A fair person's skin can, however, which is why light-skinned people can get a mild sunburn on a sunny day even if they stay in the shade. Of the UV that gets through the ozone layer, half is in the sun's direct rays and half is from the blue sky.

Scattering explains why the sun is red at sunset or sunrise. When the sun is low on the horizon, its light must pass for many kilometers through the densest part of the atmosphere before it reaches your eyes. During this trip, the blue, yellow, and green wavelengths are effectively removed by scattering. When the sun's rays reach your eyes or reflect from nearby clouds, almost all that is left of the visible wavelengths is the least scattered, which is red.

Scattering also explains why a blue jay's feathers are blue. About 100 years ago, experiments showed that there are no blue dyes or pigments in the feathers—in other words, no molecules that absorb all colors but blue from sunlight. Instead, the feathers are blue because small cells (called alveolar cells) in their barbs scatter light just the way air molecules do. Being much smaller than a wavelength of visible light, these cells scatter blue most effectively. Many of the blues and greens of animals and insects come from the same effect. In fact, when you see someone with blue eyes, the color comes from the *lack* of the pigment that makes most other eyes brown. The blue comes from the scattering of light from particles in the clearer tissue of the iris.

THIS POST ON A PIER IS SCATTERING SMALL-WAVELENGTH WATER WAVES MUCH AS AIR MOLECULES SCATTER LIGHT WAVES, IN ALL DIRECTIONS.

When light hits objects somewhat larger than a wavelength, the light reflects much the same as it does from large surfaces of that substance. A typical droplet of water in a cloud in the sky is many times larger than the wavelengths of visible light and so reflects some of every color that strikes it. That's why a cloud illuminated by sunlight appears white. On the other hand, opaque dust particles larger than visible-light wavelengths selectively absorb colors (Figure 23-11). For that reason, the color of a dust cloud made by a passing car on a dry dirt road is the same color as the road.

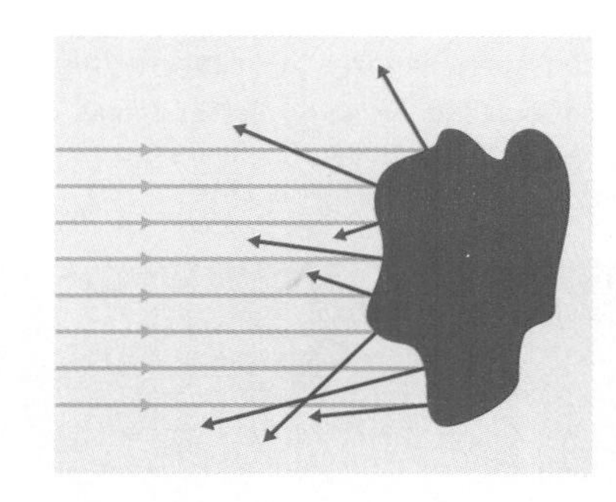

FIGURE 23-11 LARGE PARTICLES OF DUST OR SMOKE IN THE SKY SCATTER LIGHT BY SELECTIVE REFLECTION IN ALL DIRECTIONS. IF ENOUGH OF THOSE PARTICLES ARE IN THE AIR, THE SKY TAKES ON THE COLOR (OR MIXTURE OF COLORS) THEY REFLECT. THE NORTHERN HEMISPHERE HAD MANY EXTRA-COLORFUL SUNSETS FOR SEVERAL YEARS AFTER MOUNT PINATUBO IN THE PHILLIPINES ERUPTED EXPLOSIVELY IN 1991.

IRIDESCENCE

When light is reflected from a regular pattern of tiny objects, interference causes colors to appear. Figure 23-12 shows how light striking a reflective surface with regularly spaced grooves in it, called a *grating*, can interfere constructively, causing intense reflection of particular wavelengths at certain angles. If the viewer changes angles with respect to the grating, some other wavelength interferes constructively—the color seen depends on the angle of observation, just as with a rainbow. Manufactured reflective gratings are also called *diffraction* gratings.

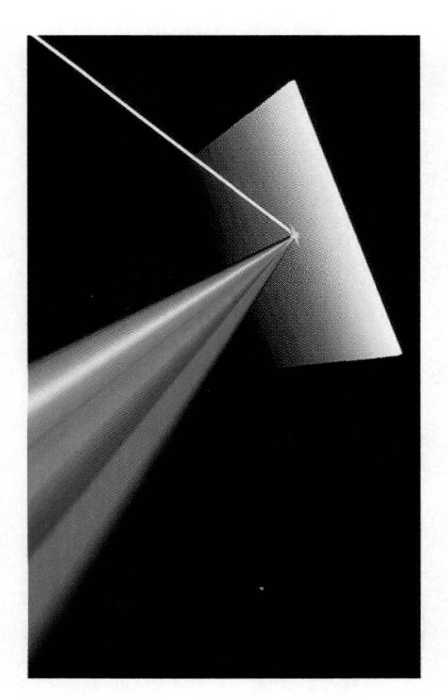

AN ILLUSTRATION OF A REFLECTIVE DIFFRACTION GRATING.

FIGURE 23-12 WHEN WAVES REFLECT FROM A GRATING, AT CERTAIN ANGLES SOME WAVELENGTHS INTERFERE CONSTRUCTIVELY, GIVING A BRIGHT REFLECTION OF THAT COLOR. MANUFACTURED REFLECTING GRATINGS WITH VERY SMALL SPACINGS TO DIFFRACT LIGHT ARE CALLED REFLECTIVE DIFFRACTION GRATINGS.

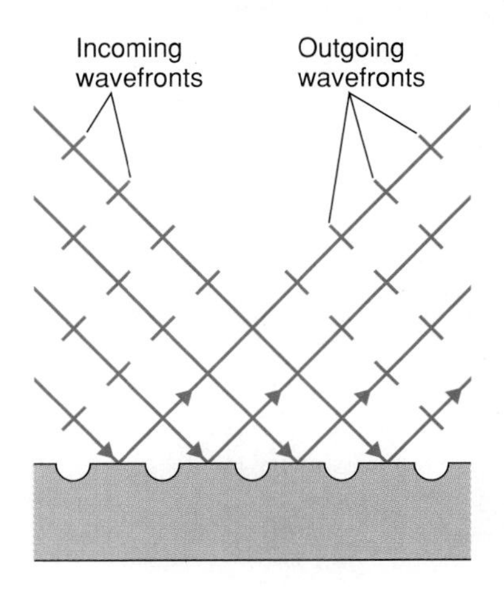

FIGURE 23-13 A CATERPILLAR HUNTER BEETLE'S BACK IS IRIDESCENT DUE TO MICROSCOPIC GROOVES ALIGNED WITH THE DEEP MACROSCOPIC GROOVES SEEN HERE ON ITS BACK.

Such colors from interference are called **iridescence** (iris: Latin for rainbow). The shells and wings of some wasps and beetles have parallel grooves that produce iridescence (Figure 23-13). Iridescent butterflies have scales that act as reflective gratings.

Iridescence also occurs from multiple reflections from layers of material. The iridescent seaweed in Figure 23-14 has thin layers of material at its surface, and the speed of light is about the same in every *other* layer, so light changes speed at each layer—meaning some light reflects at the surface of each layer (Chapter 15). All the reflections of certain wavelengths are in phase as they emerge from the surface of the seaweed (Figure 23-15). Those iridescent colors can be much brighter than the colors from a soap film, which is a single thin film. The multiple reflections ensure that almost all of the light of an iridescent color is reflected (Figure 23-16). The brilliant iridescent blues and greens from these seaweeds and from mother-of-pearl (abalone shells) come from constructive reflections from layers of thin films.

FIGURE 23-14 IRIDAEA, AN IRIDESCENT SEAWEED FOUND IN TIDAL POOLS ALONG THE CENTRAL AND NORTHERN COASTS OF CALIFORNIA. ALTERNATING SURFACE LAYERS ARE ABOUT 200 NM THICK. IF THE LIGHT EMERGING FROM THE VARIOUS REFLECTIONS IS IN PHASE, IRIDESCENCE IS SEEN. FIGURE 23-15 SHOWS THE DETAILS.

THE HUMAN EYE

The mechanisms for seeing colors are our eyes and brains. These "detectors" let us distinguish about a million different colors, with only a small percentage of them being single wavelengths of light from the visible spectrum. Before we discuss color perception, however, we need to discuss the instrument that

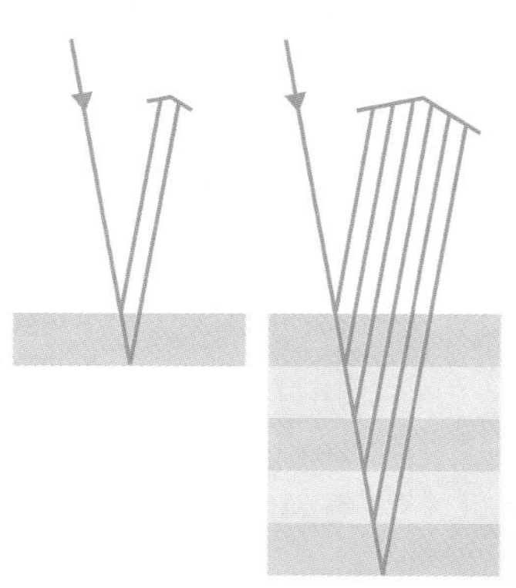
FIGURE 23-16 HOW LIGHT IS INTENSIFIED BY MULTIPLE REFLECTIONS FROM ALTERNATING TRANSPARENT LAYERS OF MATTER. THE SPEED OF LIGHT IN THE TOP LAYER IS $0.43c$, WHILE IN THE SECOND LAYER IT IS $0.82c$, AND THE LAYERS ALTERNATE AS SHOWN. AFTER 9 LAYERS, 99.99% OF THE LIGHT CAN BE REFLECTED.

FIGURE 23-15 AS LIGHT ENTERS THE MULTIPLE LAYERS, SOME IS REFLECTED AT EACH LAYER. WHATEVER COLORS OF THAT LIGHT RE-EMERGE FROM THE TOP SURFACE IN PHASE, ADDING CONSTRUCTIVELY, ARE SEEN WITH GREATER INTENSITY THAN ORDINARY REFLECTIONS.

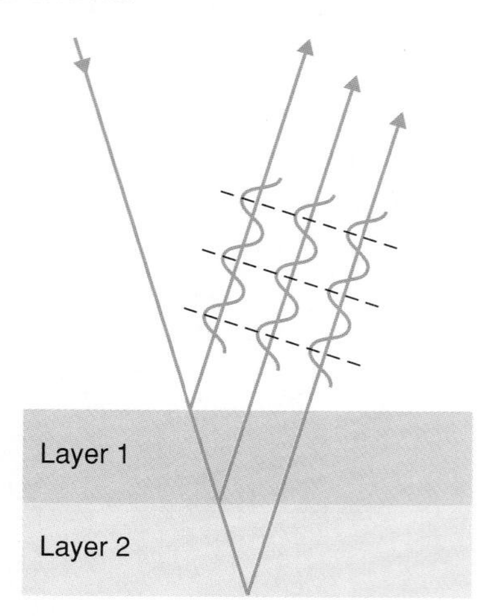

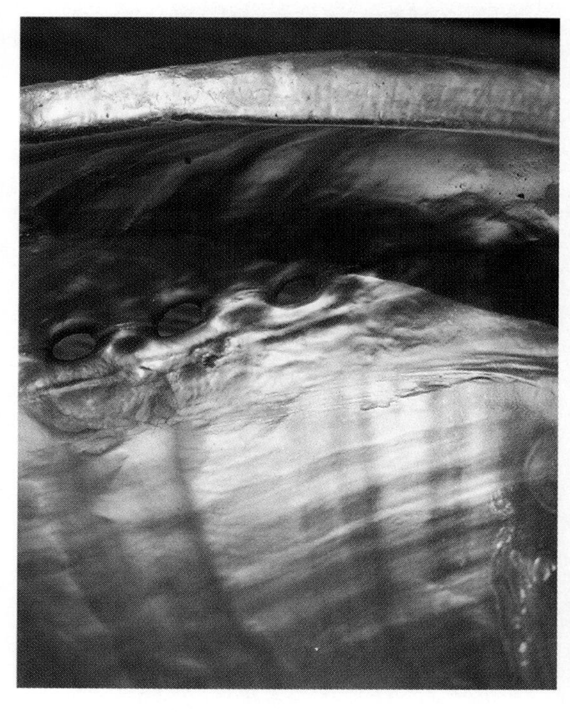
IRIDESCENCE FROM AN ABALONE SHELL.

makes it possible: the human eye (Figure 23-17). The curved front surface of the eye is called the cornea, and it converges the rays of light. Just under the cornea, a convex lens converges the light a little more. When light passes into the eye, a tiny, inverted, real image is formed on the curved *retina*, a hair-thin layer of tissue. The retina's light-sensitive cells continuously sample the light of the image and send electrical signals to the brain.

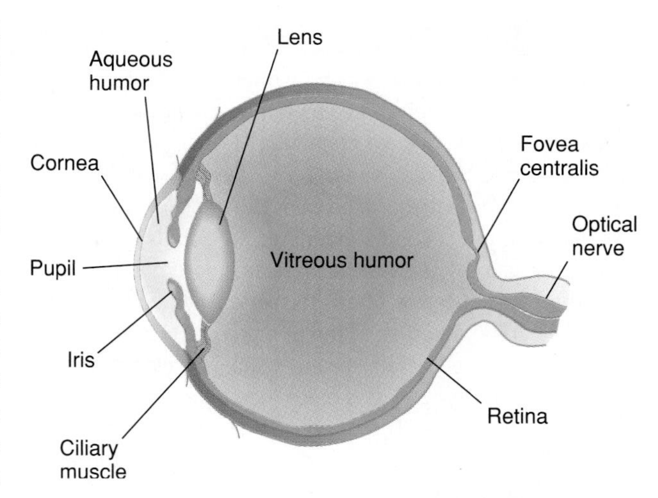

FIGURE 23-17 THE HUMAN EYE.

To focus at different distances, the ciliary muscle around the elastic lens squeezes down to change the shape of the lens. When this muscle is relaxed, the lens is thinnest and the eye is focused on things far away. When you read this page, your ciliary muscles contract, giving your lens more curvature to keep the image focused on your retina. The lenses tend to harden with age; usually sometimes after age 40 the ciliary muscle can no longer curve the lens enough to bring the print in a book or newspaper into focus. When this happens, a person needs to wear corrective lenses.

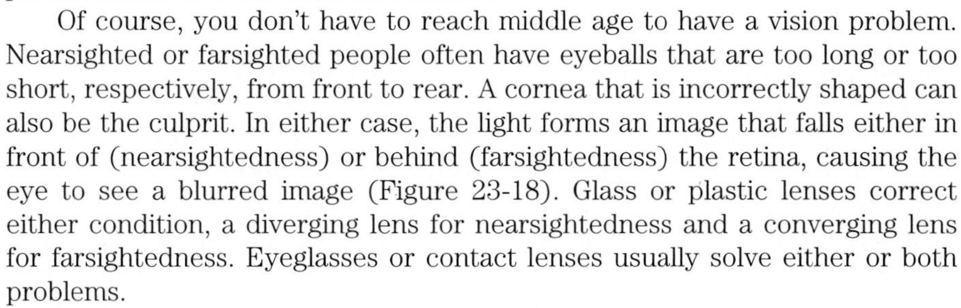

Of course, you don't have to reach middle age to have a vision problem. Nearsighted or farsighted people often have eyeballs that are too long or too short, respectively, from front to rear. A cornea that is incorrectly shaped can also be the culprit. In either case, the light forms an image that falls either in front of (nearsightedness) or behind (farsightedness) the retina, causing the eye to see a blurred image (Figure 23-18). Glass or plastic lenses correct either condition, a diverging lens for nearsightedness and a converging lens for farsightedness. Eyeglasses or contact lenses usually solve either or both problems.

The cells in the retina that sense light are called *rods* and *cones*. Cones are concentrated at the spot called the *fovea centralis*, and they increase the eye's resolving ability. For example, if you see some friends a block away, their images on your retina are a mere two-hundredths of a centimeter across, but the concentration of cones at the retina's center lets you see (or resolve) enough detail to tell them apart clearly. The cones are also responsible for giving us color vision. The rods, which outnumber the cones about 20 to 1,

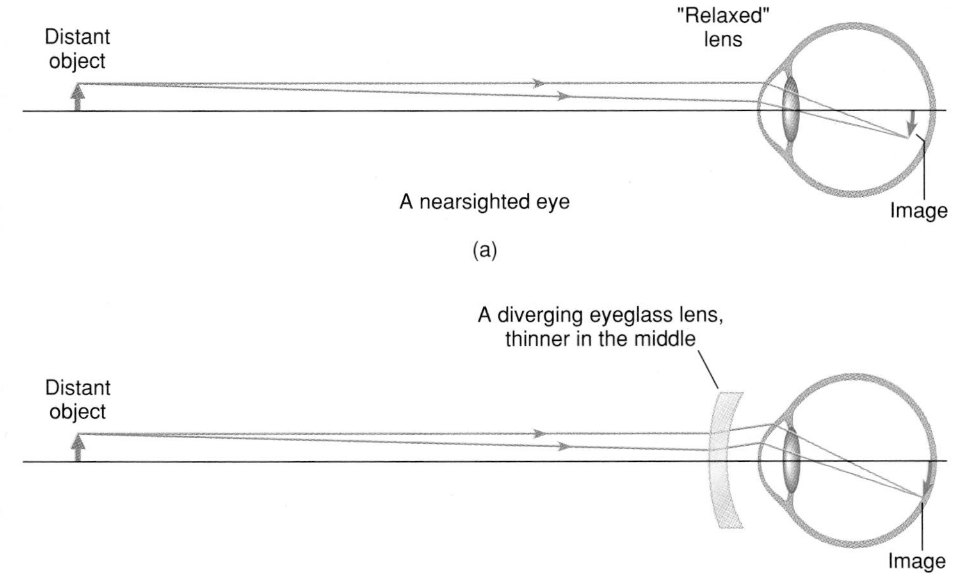

FIGURE 23-18 (A) IF AN EYE IS NEARSIGHTED, THE IMAGE OF A DISTANT SCENE FORMS IN FRONT OF THE RETINA. THE PERSON SEES THE SCENE BLURRED, OR OUT OF FOCUS. (B) A DIVERGING LENS IN FRONT OF THE EYE CAN MOVE THE IMAGE BACK TO THE RETINA, AND THE PERSON SEES THE DISTANT SCENE CLEARLY. (FOR FARSIGHTED EYES, THE IMAGE FORMS BEHIND THE RETINA AND A CONVERGING LENS BRINGS IT ONTO THE RETINA.)

SEEING CLOSE-UP

You see them on the beach—small, short-beaked birds that dart in close behind a wave as it recedes and turn to race just ahead of the next wave coming in. They stay busy, their bills dabbing into the wet sand. Try to see what they are eating. Get down and put your face near the sand, and you still won't see it. Sandpipers and sanderlings feed on tiny larvae, brought to the beach by the waves, that are too small for us to see. In the eyes of these birds, the lens is curved to focus light on the retina, much as in the human eye. However, the bird's eyes have exceptionally strong ciliary muscles that can apply pressure to curve the lens more than ours and bring very close objects into focus. The sandpiper curves its lens enough to get microscopic vision, focusing on objects at the end of its beak. This ability gives these tiny birds a real advantage in hunting; shore birds with longer beaks cannot get close enough to the sand to bring the tiny larvae into focus.

SANDPIPERS SCURRYING TO THE NEXT MEAL. THEIR STRONG CILIARY MUSCLES LET THEM FOCUS ON MICROSCOPIC LARVAE AT THE END OF THEIR BEAKS.

account for night vision—but they are not color-sensitive. A rod can detect light with as little as 1/100,000 of the light energy a cone needs for activation. The pupil opens wide in the dark, allowing almost all of the areas of the lens and retina to collect the light. The rods are mostly concentrated away from the retina's central axis. Try looking slightly to the side of a faint star rather than directly at it. Then more light falls on the rods, and you see the star better.

The rods and cones send electrical impulses that travel inside the optic nerve to the brain. While the image on the retina may change continuously, there is a lag in the brain's reception and interpretation, which is why 30-frames-per-second television broadcasts look so real to us. A part of this lag is because when the rods and cones flash an electrical response to the light they receive, they must then "recharge" briefly before they flash again.

SEEING COLORS

Pinks and purples, gold and silver, browns, iridescent greens and fluorescent reds, there are too many colors for each to have a unique name. That is not to say people won't try to name them! For example, there is a color called "crevette." It is described as a strong yellowish pink that is redder and slightly darker than average salmon, redder and darker than salmon pink, and deeper than melon. Such colors, not found in the rainbow, are all due to the way our eyes and brains perceive mixtures of colors. What enters your eyes when you look at a scene is a jumble of lightwaves of different intensities and wavelengths. What you "see" depends on how your visual system (including the visual center of your brain) processes the information delivered to the retina.

FIGURE 23-19 APPROXIMATE RESPONSE CURVES FOR RED-SENSITIVE CONES, GREEN-SENSITIVE CONES, AND BLUE-SENSITIVE CONES. NOTICE THAT EACH CONE SENSES MANY WAVELENGTHS OF LIGHT, HENCE MANY COLORS. BECAUSE THESE THREE RESPONSE CURVES OVERLAP, WE CAN DISTINGUISH MANY MORE COLORS.

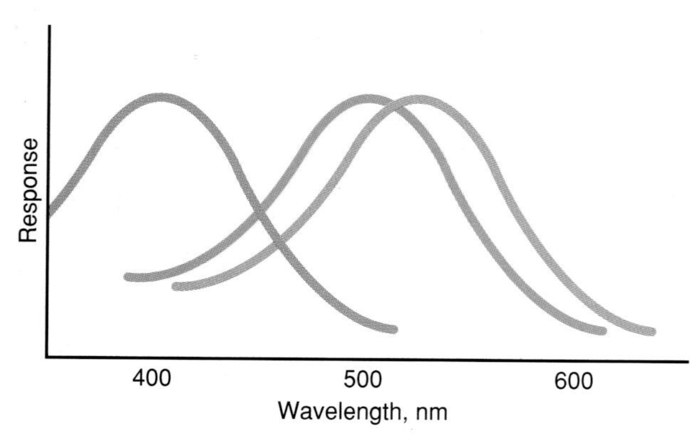

Experiments show there are three types of the color-sensitive cones, each responding best to a different region of the visible spectrum. One detects red light best, another green, and the third type blue (Figure 23-19). Each also responds to the other colors somewhat. The sensation of color arises from the simultaneous responses of the three types of cones. For example, if you shine circles of red, green, and blue lights from three slide projectors onto a reflective screen so that the beams overlap, as in Figure 23-20, you won't see any of those colors. The mixture of those wavelengths stimulates all three types of cones at once, and your visual system "sees" white, not a pure color at all. Any light perceived in this way is called an **additive mixture** of colors. Additive mixtures of red, green, and blue light in various intensities give us most of the colors we can distinguish, and they are called the **additive primary colors**.

Notice from Figure 23-20 that similar amounts of blue and green give the color cyan, blue and red give magenta, and green and red give yellow. On a color TV screen, tiny bars of three types of phosphors give off red, green, or blue light when struck by electrons. When you sit a normal distance from the screen, your eyes cannot resolve the individual dots, so each part of the screen you see carries a mixture of the primary additive colors. You see that mixture as the color of the image (Figure 23-21). Some large-screen televisions, however, project three separate images, one of each primary color, onto a reflective screen. When green light and red light reflect from the same area, for instance, you see that part of the image as yellow.

Any two colors that add together to give white light are called **complementary colors**. Since we perceive a mixture of red, green, and blue light as white, and since red and green together appear as yellow, if a blue light and a yellow light reflect from the same screen, they appear as white light. Thus yellow and blue are complementary, as are cyan (blue + green) and red, and magenta (red + blue) and green.

Colors can be produced by means other than additive mixtures. Certain colors subtracted from white light produce other colors. For example, when a shirt appears yellow in sunlight, it is because all the other colors of the spectrum are absorbed by the cloth and only the yellow is reflected. This subtraction can be done by transmitting light through filters. A red filter used with a theater spotlight or a camera absorbs (filters out) the green and the blue from white light. In other words, the green and the blue are subtracted by a red filter, which is just a piece of red glass. If all the pigments on an artist's palette are mixed, the result is not black but dark brown or dark purple. Mixing of paints, then, is different from mixing of colored lights. If you mix two colors of paint, the new color you get is from a subtractive process.

When an additive primary color is subtracted from white light, a **subtractive primary** color remains. Thus white minus red is blue plus green, or cyan; white minus green is red plus blue, or magenta; white minus blue is green plus red, which is yellow. Yellow, magenta, and cyan are the three subtractive primary colors. Printers used these three primary subtractive colors to make the colored photographs in this book. (Use a magnifying glass to see this!) When white light shines through a cyan filter, the filter subtracts the red light. If that light is then passed through a magenta filter, the green light is absorbed, and only blue light is left. Pass this through a yellow filter, and no light comes through. The sensation caused by the absence of all light is what we call black. See Figure 23-22 for the ranges of wavelengths in the primary additive and subtractive colors.

Not everyone can see colors equally. About 8 percent of men and about 0.5 percent of women are color blind—which means they confuse certain colors. That condition arises when one of the three types of cones is missing. It is very rare for a person to have no color separation at all, meaning two of the three types of cones are missing.

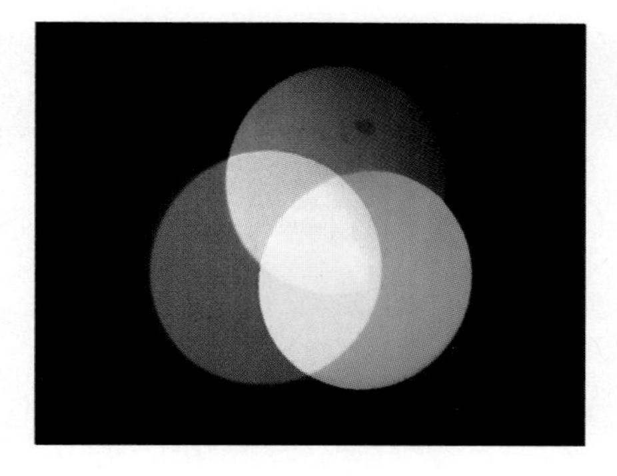

FIGURE 23-20 WHEN THE PRIMARY ADDITIVE COLORS ALL REFLECT FROM THE SAME AREA, YOU SEE WHITE LIGHT. YOU SEE GREEN AND BLUE ALONE AS CYAN (C), RED AND BLUE AS MAGENTA (M), AND RED AND GREEN AS YELLOW (Y).

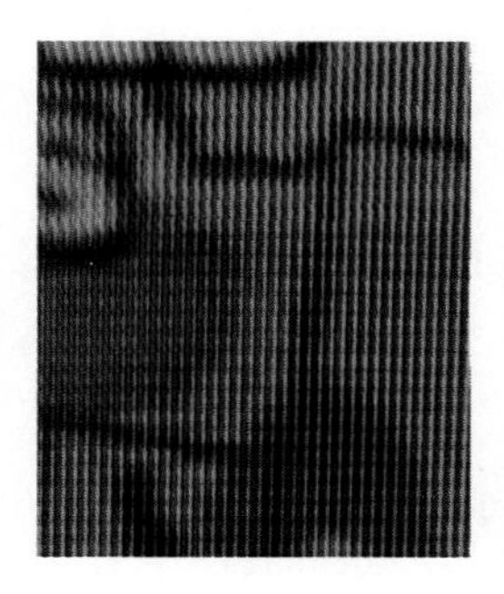

FIGURE 23-21 CLOSE-UP PHOTO OF COLOR TV DURING A WEATHER REPORT. ALL THE COLORS YOU SEE COME FROM TINY BARS OF RED, BLUE, AND GREEN PHOSPHOROUS SO CLOSE TOGETHER THAT YOUR EYES CANNOT RESOLVE THEM AT NORMAL VIEWING DISTANCES.

FIGURE 23-22 THE APPROXIMATE RANGES OF WAVELENGTHS WE SENSE AS THE ADDITIVE PRIMARY COLORS AND THE SUBTRACTIVE PRIMARY COLORS.

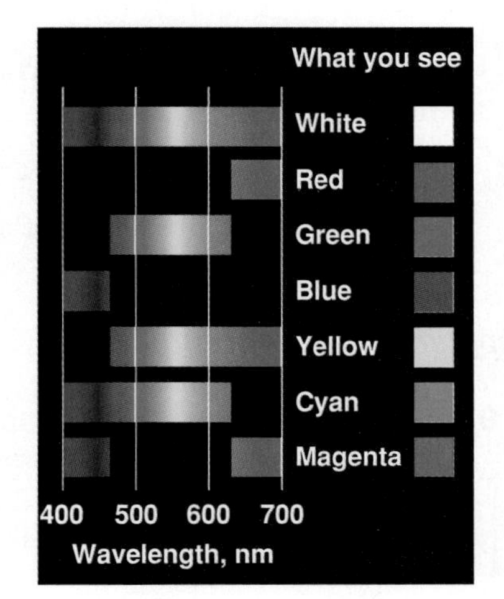

Summary and Key Terms

spectrum–341
dispersion–340
selective reflection–342
selective absorption–342
selective transmission–342
iridescence–346
additive primary colors–349
additive mixtures–349
complementary colors–349
subtractive primary color–349

Sunlight (white light) angled in to a prism separates into a ***spectrum*** due to ***dispersion***: The various wavelengths of waves in white light refract by differing amounts due to their different speeds within the glass. The colors of the spectrum, in the order ROY G BIV, are identical to those of a rainbow. Any color not in the spectrum is a mixture of different wavelengths from the spectrum.

Most of the colors we see are due to ***selective reflection***, in which only certain wavelengths are reflected while all others are absorbed (***selected absorption***). When white sunlight strikes a green shirt, the electrons in the green dye reflect only the green wavelengths, absorbing the others. Colors in transparent matter are due to ***selective transmission***, in which only certain wavelengths are transmitted while all others are absorbed. Of sunlight's visible frequencies, air's molecules scatter blue light the most, so our sky looks blue. The sky looks red at sunset because the sun's light passes through so much of the densest part of the atmosphere, scattering out most of the blue, yellow, and green light, leaving mostly red light. ***Iridescence*** is the appearance of color due to the constructive interference of particular wavelengths in sunlight as they reflect from alternating layers of thin films or reflect from gratings.

Our eyes consist of a curved cornea and a convex lens that act together to converge light onto the thin retina, where it forms a tiny inverted image. The color-sensitive cones are concentrated around the retina's center and give us color vision. The more abundant rods account for night vision. There are three types of cones, one that best detects each of the ***additive primary colors***, red, green, and blue. Most of the colors we can distinguish are ***additive mixtures***, colors produced by the combination of the wavelengths of light that come to our eyes simultaneously. Any two colors that when added together give white light are ***complementary colors***. When an additive primary color is subtracted from white light, a ***subtractive primary color*** remains.

EXERCISES

Concept Checks: Answer true or false

1. The spectrum of white light is every frequency of electromagnetic radiation we can see with our eyes.

2. Refraction, dispersion, and reflection act to separate sunlight into its spectrum with a prism.

3. You see the colors from an apple and an orange because of selective reflection of light.

4. Our eyes and brains perceive the mixture of light frequencies coming from an object as its color. If a color is not found in the rainbow, it is not a pure spectral color.

5. Indigo light is scattered more by the atmosphere than is orange light. Blue light is scattered more than green light.

6. Constructive and destructive interference of light explain both iridescence and scattering.

7. Two colors that when added together give white light are called primary colors.

8. When an additive primary color is subtracted from white light, the result is a subtractive primary color.

Applying the Concepts

9. If the various colors of visible light traveled at the same speed while passing through glass, would there be any dispersion? If they traveled at the same speed in glass as they do in a vacuum, would there be any refraction?

10. When white light enters a face of a cut diamond, brilliant splashes of color may emerge at some other point. Explain why.

11. What is the order of the colors in the secondary rainbow?

12. About 150 colors of paint are used on American cars and trucks each year. Those with metallic or pearly sheens have flecks of mica, graphite, aluminum, titanium, or special pigments spread throughout the coat of paint. To reflect the colors of the flecks, the base paint (in which the flecks are embedded) must be (a) highly reflective, (b) highly transmittive, (c) highly absorptive.

13. What happens to the energy of the light absorbed by an opaque object?

14. Why do distant snow-covered mountains have a yellowish tinge on a bright, clear day?

15. A lobster seen on an underwater reef looks almost black, but when brought to the surface, it looks red. What does this tell us about how water absorbs light?

16. Explain why white as a color is very impure.

17. Explain how we can see colors that aren't in the spectrum of visible light.

18. At times, the light from a full moon casts sharp shadows and is bright enough to read by. Moonlight is reflected sunlight and so contains the spectrum of visible frequencies. Nevertheless, anything most of us see by moonlight alone appears in black, shades of gray, and white. Explain why.

19. Rainbows can be seen in sprays of salt water, but light travels at a slightly different speed in salt water than in fresh water. Exactly what difference would you expect to see in such a rainbow?

20. The skin of frogs and anole lizards and the feathers of parrots contain a yellow pigment, but they also have cells that scatter blue light. The blue and yellow mixed additively give these animals their green color. In addition, the anole lizard has specialized cells that can open and close over the yellow pigments. Tell how the working of these special cells allows the lizard to change its color somewhat.

21. An apple and a rose appear black if held under a streetlamp at night with no other light around. What must be happening?

22. What color would the sky be if the earth had no atmosphere? What color do the astronauts see from orbit when they look away from earth?

23. Why does the snow, which looks white in sunlight, look blue when it is in a shadow, as in Figure 23-23?

24. A laser eraser aims a pulse of intense light at the unwanted letter, and "poof." The black ink absorbs the brief pulse of light so rapidly the ink is vaporized, but the paper below is not scorched. Give two reasons why the paper isn't harmed.

FIGURE 23-23

25. Explain why sunlight is reflected so well from clouds while radio waves pass almost unimpeded through them.

26. Explain why, in daylight, you see things that are straight ahead so much more clearly than things that are slightly to the side of where your eyes are pointing.

27. Tired of reading? Look up and focus your eyes on something across the room for a minute. That usually helps, but why should it?

28. Notice the Big Dipper in the sky of Figure 23-24. Its stars are trailing—meaning this is a long exposure! The photographer took this picture by only the light of the full moon at his back. Now—why is the sky so blue?

FIGURE 23-24

29. Look at Figure 23-2 and decide where the sun must be in the sky to create a rainbow of the greatest possible height.

30. Explain how ROY G BIV can let you decide which colors travel faster in glass than others.

31. What color would a red car appear to be at night if it were under a lamp that emitted only blue light?

32. The park rangers for New Cave, near Carlsbad, New Mexico, take visitors to a location far into the dark cave and have them all extinguish their flashlights. Ten minutes later, the visitors are still blind. Why don't their eyes adjust to the dark—or do they?

33. Have you ever seen a part of a rainbow, that is, only part of the normal arc? Explain why it's possible.

34. The haze in the atmosphere over industrial centers and large cities comes from particles much larger than molecules and closer to the size of the wavelengths of visible light. How would you expect those particles to scatter light?

35. Would you expect to be able to see a rainbow by moonlight rather than sunlight? Explain.

36. At the *fovea centralis* in your eye there are about 250,000 cells, mostly cones, per square millimeter. Explain why this high concentration of cells is needed to let your brain resolve a clear image of a faraway scene. (*Hint*: The eye's lens is a converging lens—remember the discussion of converging lenses in Chapter 22.)

37. Why does the sky sometimes appear so much bluer after a rainstorm?

38. After an extensive forest fire in British Columbia, the full moon looked blue to people in the area, and after the volcanic explosion of Krakatoa, many people reported a green moon. Discuss.

39. Why does the color of cloth sometimes depend on whether it is viewed in fluorescent light, incandescent light, or sunlight?

40. Before Viking 1 landed on Mars on July 20, 1975, many scientists suspected that Mars had a pale blue sky, just as earth does. However, Viking found the Martian sky to be pinkish to orange because of microscopic dust particles from Mars's red surface kept up in the air by winds. What process would you say colors Mars's sky? (a) molecular scattering, (b) dispersion, (c) total reflection, (d) selective reflection, (e) refraction.

41. At fairs, circuses, and promotionals for shopping centers, searchlights are sometimes used, throwing their beams into the night sky so they will be seen far and wide. Would this work on the moon? Discuss.

FIGURE 23-25

42. If you ever drive through mountains, notice how the most distant range appears the bluest (Figure 23-25). Explain why.

43. When very fine black pigment particles are dilutely mixed into white paint, the paint becomes tinted blue. Explain.

44. In autumn, a tree whose leaves turn red will early-on have a mixture of green leaves and red leaves. From a distance, however, that tree will look brownish yellow. Discuss.

45. When a red spotlight strikes the same area as a blue spotlight, what color do you see?

46. A pigeon's photoreceptors are only cones, while a cat's are nearly all rods. Discuss.

47 What colors are complementary to (a) red, (b) green, (c) magenta?

48. Among acrobats, yellow is considered an unlucky color. It is interesting to speculate why. What would happen if the acrobat's swinging hoops or bars were yellow and a blue spotlight was used on their act? Or, what if a performer wore blue clothes and she or he were suddenly illuminated by only a yellow light?

49. White paper looks red if red light shines on it, but black, green, and blue paper do not. Explain.

50. There are two varieties of chlorophyll, called chlorophyll *a* and chlorophyll *b*, and they use the energy from light of two different wavelengths, 680 and 700 nanometers, to drive life processes in plants. What *color* of light do they use? If one plant was put under red light, an identical plant was put under green light, and a third identical plant under blue light, which would survive?

51. Snow is white because its large grains scatter all visible wavelengths of light equally. If you poke a hole in a snowbank, however, you can see that, just like water, ice (even in the form of snow crystals) absorbs red light much better than blue. The deeper you look into the hole, the bluer it looks. Explain.

52. White ice in the Arctic Ocean is white because it is full of air bubbles, while the blue ice seen in glaciers has been remelted and compressed and is almost free of bubbles. Explain how the difference in how far incident light can go before it is reflected is responsible for the two very different "colors" of white and blue ice.

Demonstrations

1. Here's a way to make your own blue skies and reddish sunsets. When a flashlight beam penetrates a pitcher of clear water, there's little change in the color of the beam. Add a few drops of milk to the water, however, and you'll see the beam of light turn reddish orange (Figure 23-26). The milk's molecules scatter the blue light (and some green and yellow, too) in all directions before it can reach your eyes, just as the air's molecules do for the rays of sunlight at sunset. Now look through the side of the pitcher, perpendicular to the beam. Presto! There's the blue light, scattered to the sides (and in all directions), just as the air scatters blue light from sunlight to give us blue skies.

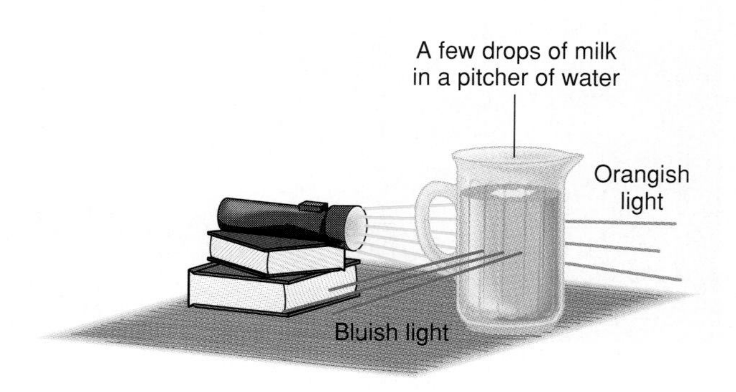

FIGURE 23-26

2. Where the optic nerve is connected to the retina, there is a blind spot in your vision. You can find it with the help of Figure 23-27. Close your left eye and look straight at the black square. As you stare at the square, move the book toward you. You'll see the cat disappear as its image falls on your blind spot. Move the book even closer as you continue to stare at the black square, and the cat reappears but the mouse disappears (but not its tail, which is focused outside the blind spot).

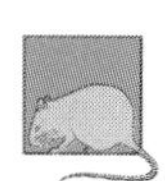

FIGURE 23-27

3. Starlight can show that earth's atmosphere not only refracts light but disperses it. With a pair of binoculars on a clear night, look at a bright star near the horizon. The spectrum of starlight you see shows that air slows the various frequencies of light by different amounts, causing them to bend at slightly different angles as they come through the atmosphere. Look at the stars overhead with the binoculars and you won't see a spectrum.

4. Your brain retains an impression of the image that falls on the eye for about 0.1 second after that image is removed. That *persistence of vision* is why the images in movies (24 frames per second, or 0.04 s per frame) and on television (30 frames per second, or 0.03 s per frame) seem to move smoothly. To test your persistence of vision, you need only to blink your eyes. Ordinarily you don't notice a gap in the image when you blink because your eyelid is shut for only a tenth of a second or less. A lazy blink of a half second could be risky in a game of catch or driving in very fast, close traffic.

5. You can make your own rainbow—and even a moonbow, when the moon is full. Experiment with a garden hose that makes a fine spray. (If you want to read about it, look up Helen Perry's article in *The Physics Teacher*, **13**, p. 175, 1975, in a library.) Do this near sunset (or sunrise) and notice the brightness of the spray when the sun reflects from it back to your eyes as well as when the sunlight comes through it; it is easy to see the greater forward scattering of light.

6. Motion pictures, we've seen in Demonstration 4, take advantage of the persistence of human vision so that 24 still pictures flashed before our eyes each second can simulate motion. The same thing happens with colors. Your brain can't instantaneously turn off an image or its color. When you look at a color and turn away quickly, there is a very brief afterimage. As a matter of fact, you can trick yourself into seeing white even through you are gazing at colors. Cut a circle from posterboard about 10 centimeters in diameter and color it so there are three pie-shaped segments, one red, one green, one blue. Punch two holes on either side of the center, pass a long cord through them, and tie the cord so about 30 centimeters of cord is on each side of the disk. Holding opposite ends of the cord, twirl the disk in one direction so that the cord winds around itself; then pull sharply against the ends of the wound-up cord, and the colored disk will twirl as the cord unwinds. As it does, your brain is stimulated by all three colors simultaneously, and the twirling disk appears to be white.

NEW BEGINNINGS: PLANCK, EINSTEIN, AND BOHR

Early investigators of electricity learned how to make a spark jump between two pieces of metal, or *electrodes.* Today anyone who has jumped-started a car has seen such sparks move through the air, caused by the passage of an electric current. But in the early days the nature of the current was unknown. When scientists put two electrodes in a sealed glass tube and evacuated the tube with a vacuum pump, their ammeters showed that a current traveled through the wires leading to the electrodes as the air pressure dropped. Studies with these *discharge* tubes led to the discovery of the nature of the electrical current, which flowed better through a vacuum than through air.

In 1839 Michael Faraday wrote of a phosphorescent glow in a discharge tube, and, quick to follow, other investigators soon discovered more. When the vacuum was very good, the end of the glass tube near the positive electrode glowed with a soft green light. Apparently something was striking the tube to produce that light. Moreover, solid obstacles placed between the two electrodes cast shadows upon the glowing end of the tube. Such sharp shadows meant that the invisible cause of the green glow traveled much as rays of light—that is, in straight lines. Named *cathode rays,* because they travel from the cathode (negative terminal) to the anode (positive terminal) and never backward, these radiations were soon found to have a property not shared by light: When a magnet was brought up to the discharge tube, the shadows moved, proving that the cathode rays were some sort of charged particles in motion.

By 1895 physicists had shown that the cathode-ray particles could penetrate thin metal foils, meaning they were probably smaller than atoms. In the years 1897 to 1899, the English physicist J. J. Thomson managed to determine an important property of the charged particles. He found that the ratio of the amount of charge the particle carried to its mass to be about 1000 times greater than the same ratio for ionized hydrogen (hydrogen is the lightest of the atoms). Presuming that the amount of charge on the cathode-ray particle is not much different from that of ionized hydrogen, this indicated that the cathode-ray particles had very little mass. For this evidence that cathode-ray particles (now known as *electrons*) were the first of the particles to be discovered that were smaller than atoms, Thomson was later given the Nobel Prize. (Despite the progress Thomson made with his investigations, it has been said that he always left the second decimal place to someone else to discover, as he did in this case: The more accurate ratio of charge to mass for the electron, e/m, is about 1837 times the same ratio for ionized hydrogen, or protons as they are now called. But the assumption about the amount of charge was correct. Protons and electrons have equal but opposite charges.)

In the wake of this development, discoveries by Max Planck (see "Max Planck's Model for Blackbody Radiation"), Albert Einstein ("Another Puzzle: The Photoelectric Effect"), and Neils Bohr ("The Bohr Model of the Hydrogen Atoms"), were the beginning of the understanding of the atom's structure. Their findings are in Chapter 25, but here we'll take a brief look at the scientists themselves.

Max Planck was quiet, very religious, and musical. His views were typical of many physicists at the end of the 1800s, especially in Germany; he thought all of natural behavior could be traced in principle to Newtonian mechanics. Many of his colleagues were of the opinion that all the great discoveries of physics had already been made. When his revolutionary discovery came along, he could not quite believe it himself. He was over 40 years old in 1900 when he discovered the radiation law that predicts how matter radiates because of its temperature. It earned him the Nobel Prize in physics in 1918. His discovery was the first major step in a new understanding of matter's behavior, so different that the physics done from the time of his work or is often called "modern" physics, while pre-1900 discoveries are called "classical" physics.

A scientific contemporary of Planck's, Albert Einstein was only 26 in 1905 when he explained the details of the photoelectric effect, the knocking of electrons from the surface of a metal by light. He won the 1921 Nobel Prize in physics for that work. According to his sister, at his birth his mother was frightened because of his large head. His grandmother's first words when she saw him were, "much too thick, much too thick." Between 2 and 3 years of age, the not-as-yet-talking Albert decided to speak—in entire sentences, quietly rehearsing each before saying it out loud. When he was 4 or 5 his father showed him a magnetic compass, and he was so struck by it he trembled. He later said of the event, "I experienced a miracle." When he was 12 someone gave him a geometry book, and he soon called it his "holy geometry book." For the next 4 years he studied math on his own, and in time he became an excellent student.

Einstein graduated from college with a diploma that allowed him to teach in high school. After a short time as a substitute teacher he settled in Bern, Switzerland, and tutored students until a friend's father got him a job at the Swiss patent office. A student during that period wrote of him: "about five feet ten, broad-shouldered, slightly stooped, a pale brown skin, a sensous mouth, black moustache, nose slightly aquiline, radiant brown eyes, a pleasant voice." He seeemed to be able to be deep in thought without appearing to be aloof from those around him. It was while he was at the patent office, where he could easily do his regular duties and still have time to think about physics, that he published three papers (in 1905) in different areas of physics, any one of which was worthy of a Nobel prize.* In the same year, he completed a Ph.D. thesis. His first, submitted in 1901, had been rejected.

Einstein remained at the patent office until 1908, when he left to become a junior, then an associate, then a full professor at various Swiss, German, and Austrian universities. During this period he worked on the general theory of relativity, which deals with motion in accelerated frames of reference (see Chapter 26). Early in 1919 Einstein divorced his first wife, and he agreed that she would have the proceeds of a Nobel prize should he win one. He promptly married a cousin, Elsa, herself divorced. He was awarded the 1921 Nobel Prize in physics and his former wife received all the money, as promised.

Einstein continued to work on fundamental problems of physics for the rest of his life. He attempted to unify the forces of nature into one grand theory, the unified field theory. In 1932, several months before the Nazis came to power in Germany, he was appointed professor at the Institute for Advanced Study at Princeton, New Jersey. There his only duty other than research was to attend faculty meetings (he disliked teaching courses). Despite his position as a pacifist, first stated during World War I in Germany, he sent a letter to President Roosevelt in August, 1939, to draw attention to possible military uses of nuclear energy. This helped influence Roosevelt to set into motion the effort that resulted in the development of nuclear weapons. Upon Einstein's death in 1955, his ashes were scattered by friends in an undisclosed location.

Niels Bohr was born in Copenhagen in 1885, and his parents were progressive and took part in every phase of their children's education. Though overshadowed as a student by his brother Harald, who became an eminent mathematician, Bohr earned early fame as a thoughtful, thorough scientific investigator. Upon finishing his Ph.D. thesis in 1911, Bohr went to England to study with J.J. Thomson. Because his thesis had been on the actions of electrons in metals, Bohr wanted to work with the discoverer of the electron. But Thomson had other interests by then, so a disappointed Bohr went to Manchester where Ernest Rutherford had established an active laboratory and only the year before had discovered the nucleus. There Borh attacked the problem of how the electron behaved about the nucleus of the atom. After 6 months he returned to Copenhagen, where he worked out a model for hydrogen that explained how the atom absorbed and emitted light with the motions of its electron. This model of the hydrogen atom brought Bohr the Nobel Prize in 1922. Several years later the Danish government built a laboratory for Bohr on land donated by some of his friends. The Institute for Theoretical Physics became a focal point for a revolution in physics over the next few decades.

The revolution resulted in the *quantum theory* of atoms and molecules. It explained the structure and the behavior of those tiniest units of matter. First introduced by Planck, Einstein, and Bohr, the Quantum Theory would come into full bloom in 1925 and the years afterward. This theory overturned the classical notions of the behavior of matter (see Chapter 26), and in doing so it divided the opinions of the three who made the earliest contributions to its development. Bohr became the champion of the quantum theory, and Einstein and Planck, among others, became its antagonists. Again and again, Bohr and Einstein argued the theoretical points, with Bohr and the quantum theory the ultimate winners. Bohr came to use these arguments with Einstein as private daily mental exercises, providing a constant identity with the new theory. Even after Einstein died, Bohr was heard to argue against Einstein as though he were still alive.

When the Nazis occupied Denmark during World War II, Bohr refused to leave Copenhagen. He stayed to exert whatever influence he could for exiled scientists in Denmark. He was also concerned about retaliation against his familiy, friends, and colleagues at the Institute should he leave. At some point, however, Bohr and his brother Harald were warned that they were on a list to be deported to German concentration camps (their mother was Jewish). Bohr arranged for his family and himself to escape to neighboring Sweden. Once there, he was invited by Winston Churchill to come to England to help the war effort, and he accepted. Later he came to the United States to act as a consultant for the international team of scientists involved in the development of the nuclear bomb.

*Chapter 26 is devoted to his special theory of relativity, which explains motion as viewed from frames of reference with different speeds, the subject of one of those 1905 papers. Yet another, concerning Brownian motion, is claimed by some to have settled once and for all the question of whether there really are atoms and molecules.

TWENTY-FOUR

Light and Atoms

Lasers are devices that could be made only after physicists understood a great deal about atoms and light and their interactions. Today they are finding widespread uses in fields from medical operations to communications and earthquake monitoring.

In the late 1800s, physicists were trying to understand how light was emitted and absorbed by matter. Maxwell's equations showed that radiation comes from accelerated charges, but the nature of atoms, other than that they contain two types of electric charge, was unknown. What these physicists eventually discovered is a realm so different from everyday experience that the events that occur there go against all common sense. Anyone can relate to the pushes and pulls of Newton's laws, but no one can ever experience what goes on inside an atom.

Developed largely in this century, the physics of the atom is part of what is usually called *modern* physics, while pre-1900 physics is often called *classical* physics. This chapter and the four that follow describe the world of modern physics. This part of nature provides a great test for human investigative skills and imagination and has been the scene of great triumphs of understanding. Modern physics has led to revolutionary progress in chemistry, biology, medicine, and astronomy as well as in technology, manufacturing, communications, and energy production.

SPECTRA: THE FINGERPRINTS OF ATOMS

In 1800 William Wollaston, a London physician, became partially blind and decided to retire. His hobby was science, and he soon devised an instrument to get a closer look at the visible spectrum of sunlight. Like Newton a hundred years earlier, Wollaston passed sunlight through a prism, but then he sent the dispersed light through a small telescope to enlarge the spectrum for a closer look. (An instrument that spreads wavelengths of light for observation through a telescope is called a **spectroscope**, shown in Figure 24-1. If the instrument uses a camera to photograph the spectrum, it is called a **spectrograph**.) This magnification revealed thin dark spaces in the spectrum where some wavelengths seemed to be absent. Wollaston thought these dark regions might be some sort of boundaries between the colors. In 1814 an apprentice optician, Joseph von Fraunhofer, made a better spectroscope and found that Wollaston's fuzzy regions were actually very sharp, distinct dark lines (Figure 24-2). That is, specific wavelengths of light were missing. He measured those wavelengths for almost 600 of the darkest lines of the sun's spectrum, but why these wavelengths were missing, he couldn't say.

In the 1840s, the German physicist Gustav Kirchhoff passed white light from a bright lamp through a window and into a container of gas, sending the light that emerged from a window on the opposite side into a spectroscope (Figure 24-3). For each element he used in the container—hydrogen gas, oxygen gas, and others—different dark lines appeared in the spectrum of the light that had passed through the gas; each element left its own "fingerprint" on the spectrum of white light. The gaseous atoms in the container *absorbed* certain wavelengths of light, letting all the others pass. Such spectra containing dark lines, or missing wavelengths, are now called **absorption** spectra.

Kirchhoff next placed a second spectroscope at another window on the side of the container of gas, out of the path of the light beam from the source.

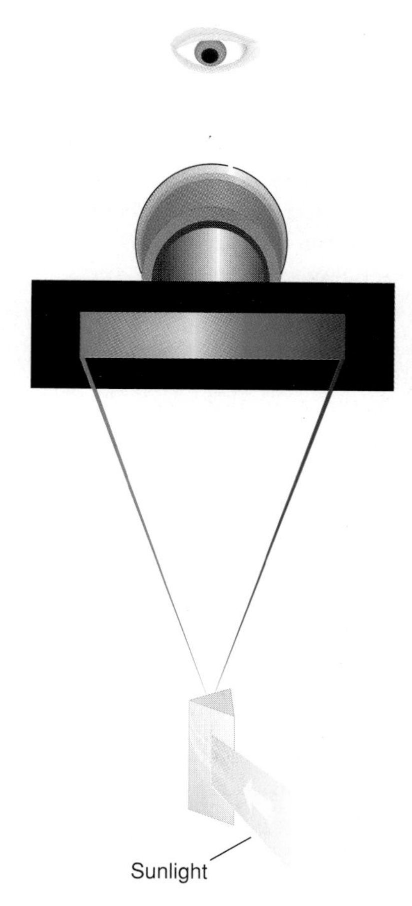

FIGURE 24-1 A SPECTROSCOPE IS AN ARRANGEMENT TO ENLARGE SMALL PORTIONS OF THE SPECTRUM OF LIGHT (HERE, SUNLIGHT).

FIGURE 24-2 A SMALL PORTION OF THE SPECTRUM OF A BEAM OF SUNLIGHT THAT HAS PASSED THROUGH A SLIT. THE DARK LINES SHOW THAT CERTAIN PRECISE WAVELENGTHS (OR FREQUENCIES) OF LIGHT ARE MISSING IN THE SUNLIGHT.

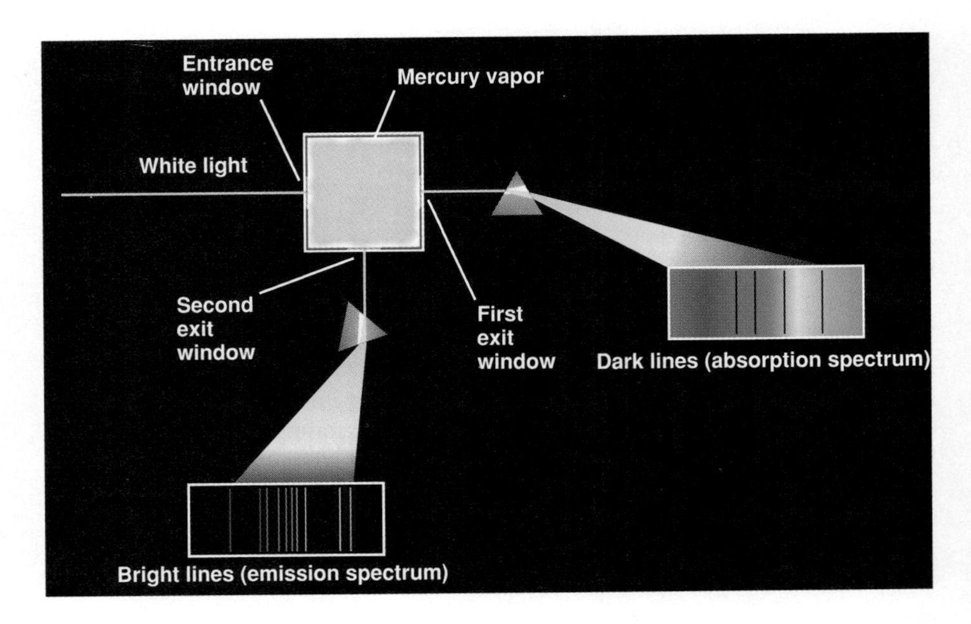

FIGURE 24-3 OBSERVING THE ABSORPTION AND EMISSION SPECTRA OF A GAS. MERCURY VAPOR ABSORBS SOME WAVELENGTHS FROM THE WHITE LIGHT, CAUSING DARK LINES IN THE SPECTRUM OF THE BEAM WHEN IT EXITS FROM THE CONTAINER. A SPECTROSCOPE AT THE SIDE OF THE GAS GETS NONE OF THE RAINBOW SPECTRUM OF THE DIRECT BEAM, HOWEVER. INSTEAD, THE SPECTRUM HAS BRIGHT LINES ON A DARK BACKGROUND. SOME OF THESE ARE IDENTICAL TO THE WAVELENGTHS ABSORBED FROM THE DIRECT BEAM.

When he looked into this second spectroscope, he saw sharp lines of color in an otherwise dark field of view, and the wavelengths missing in the absorption spectra matched up perfectly with some of the wavelengths of these bright lines. (We'll see why all the lines don't match in a later section.) The gas atoms absorb light of certain wavelengths, and this absorbed light adds energy to the atoms. (We say the atoms are *excited*.) The atoms then lose that energy almost instantly by emitting light of identical wavelengths (plus some others) randomly in all directions. The bright-line spectra are called **emission** spectra.

Why the atoms in a gas emit and absorb light this way was still a puzzle in the late 1800s. Nothing known in physics at that time could predict such behavior.

RADIATION AND SOLIDS

Another puzzle at the end of the 1800s was how solids interact with light. Solids emit and absorb radiation over wide ranges of wavelengths, rather than at just specific wavelengths as gases do. In general, darker-colored bodies absorb more of the white light that falls on them than lighter ones do. A black-colored body absorbs light best of all. An ideal **blackbody** absorbs *all* the electromagnetic radiation that strikes it, reflecting none. Yet energy does leave the blackbody—a blackbody in thermal equilibrium emits as much radiant energy as it absorbs each second. The *blackbody emission spectrum* for two temperatures is shown in Figure 24-4. The hotter the blackbody is, the more radiation it emits at all wavelengths (see Demonstration 1), and the wavelength at which it emits the most energy is shorter with increasing temperature. At "room" temperatures, its radiation is mostly at infrared wavelengths. When a blackbody gets as hot, say, as an electric stove element on high, it emits not only infrared but visible light as well. Notice that a blackbody doesn't emit just single wavelengths in its emission spectrum as a gas does. It emits a continuous spectrum, which means it emits some radiation at every wavelength.

FIGURE 24-4 EMISSION SPECTRA FOR BLACKBODIES AT 5800 K AND AT 2800 K. A BLACKBODY EMITS SOME LIGHT AT EVERY WAVELENGTH, AND THE CHANGE IN INTENSITY FROM ONE WAVELENGTH TO THE NEXT FOLLOWS SMOOTH CURVES LIKE THE ONES ON THIS GRAPH. A BLACKBODY AT 5800 K (THE TEMPERATURE OF THE SUN'S SURFACE) EMITS MOST OF ITS RADIANT ENERGY IN THE WAVELENGTHS OF VISIBLE LIGHT. AT 2800 K, THE TEMPERATURE OF A LIGHT BULB'S FILAMENT, MOST OF THE ENERGY IS EMITTED AT WAVELENGTHS LONGER THAN THOSE OF VISIBLE LIGHT—THAT IS, AS INFRARED, OR HEAT, RAYS. (A TYPICAL INCANDESCENT LAMP EMITS LESS THAN 10 PERCENT OF ITS RADIANT ENERGY AS VISIBLE LIGHT.)

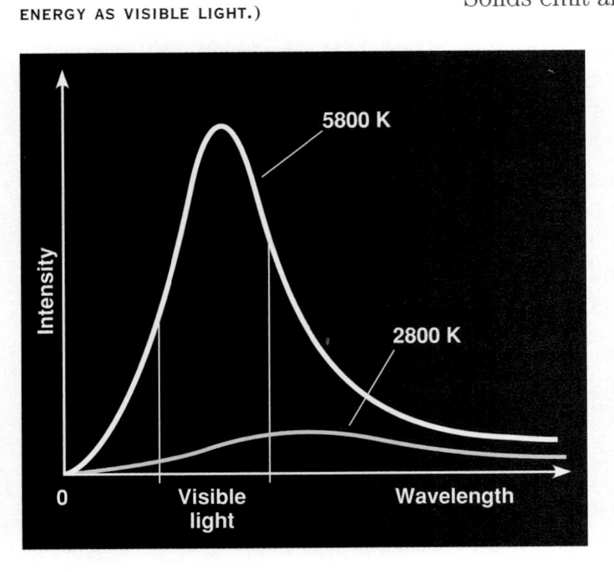

At least a small fraction of any light that strikes any solid is reflected, so not even the darkest solid is a perfect blackbody. However, if a closed box has a tiny hole punched in it, any light that strikes the hole will pass into the interior and be absorbed by the material inside the box (sometimes after several bounces). The hole acts like an ideal blackbody. All matter at any temperature above absolute zero emits radiation, and the inside surfaces of the box do, too. Detectors on the outside can obtain the spectrum of the emitted radiation as some of it escapes from the hole. Physicists in the late 1800s found that, *no matter what material is lining the box*, the spectrum of the radiation emerging from the hole is the same. So blackbodies show how atoms of *any* kind in the solid state emit and absorb radiation when reflection and transmission are not present.

Physicists tried to predict the shape of blackbody emission curves. An atom's thermal vibrations, according to classical ideas, should be random, so physicists thought that the charged particles that emitted electromagnetic radiation would vibrate randomly as well. But the answers never came out right; the laws of classical physics weren't predicting the facts found by the experiments, just as they weren't explaining the spectral lines of gases.

EVERY TIME YOU LOOK INTO A FRIEND'S EYES, YOU ARE LOOKING AT BLACKBODIES! LIGHT ENTERS THE EYE THROUGH THE PUPIL, AND ONLY 0.01 PERCENT OF THAT LIGHT REFLECTS BACK THROUGH THE PUPIL FROM THE INTERIOR OF THE EYE. THAT MAKES A PUPIL A GOOD APPROXIMATION TO A BLACKBODY. THE SPECTRUM OF ELECTROMAGNETIC WAVES EMERGING FROM SOMEONE'S PUPIL IS THAT OF A BLACKBODY AT 37°C, OR 310 K, WITH THE GREATEST INTENSITY IN THE INFRARED. (YOU CAN SEE REFLECTIONS WHEN YOU LOOK AT THE PUPIL, BUT THESE COME FROM THE SURFACES OF THE CORNEA AND THE LENS OF THE EYE.)

PLANCK'S MODEL FOR BLACKBODIES

In 1899 Max Planck investigated the radiation emitted by a hot blackbody. Realizing that he had to accept the experimental curves shown in Figure 24-4 as his answers, he looked for a mathematical formula that would fit those curves at every point. He found a set of curves that fit the blackbody emission spectra for any blackbody temperature. A constant number appearing in the formula, which we today symbolize by the letter h, is now known as *Planck's constant,* and the formula is called *Planck's radiation law*.

Planck still had to explain *how* molecules or atoms give up this emitted light. Electrons were discovered just about the time he was doing this work, and as yet no one knew about the atomic nucleus. He knew radiation occurred whenever charged particles accelerate. The spectrum of the light emitted by randomly accelerated charges did not match the blackbody spectrum, but Planck discovered a model that would. The atoms themselves do not change places in solids, so he figured the charges inside them wouldn't either. He assumed that the charged particles oscillate in place, swinging back and forth, their acceleration causing them to radiate. Their frequency of oscillation, f, would be the same as the frequency of radiation they emitted. What Planck found was this: Whichever frequency f an oscillator had, it gave up or absorbed only definite amounts of energy that were equal to a whole number times Planck's constant times the frequency of the oscillator. In other words, the radiation is emitted in lumps, or bundles, and each bundle has an energy hf. Planck called that amount of energy a **quantum** (plural = *quanta*).

Despite his model's perfect prediction of blackbody radiation, Planck was disturbed. To emit only quanta of energy meant that the blackbody's charges could not swing back and forth in a continuous motion like a mass oscillating on a spring. They could have only certain amplitudes of swing, and they could not move with other amplitudes. Imagine a child's swing having to follow such a rule when you gave it a push. Rather than swinging smoothly, from the vertical up to an angle of 15 degrees or so, it would jump abruptly from, say 0° to 3°, then from 3° to 6°, and so on, never to be found at any angles in between. Nothing in our everyday world moves that way, and yet Planck's oscillating charged particles did! Planck's model made no physical sense to him or anyone else at that time, but because it explained the blackbody radiation curves perfectly, he published it in 1900. For the next five years, no more was heard of the quantum, though Planck spent a lot of time trying to find another way to explain his radiation law. He couldn't bring himself to believe that his strange model really described what was going on in a radiating solid.

ANOTHER PUZZLE: THE PHOTOELECTRIC EFFECT

About the same time Planck was working with blackbody radiation, other physicists were experimenting with another effect between light and solids: the **photoelectric effect**. Ultraviolet light hitting a plate of zinc causes electrons to pop from the surface, much like water molecules evaporating from a sunlit lake. Like those water molecules, the electrons need a certain amount of energy just to pull away from the metal's surface. For water, that energy is called the latent heat of vaporization. For electrons in metals, it is called the *work function.*

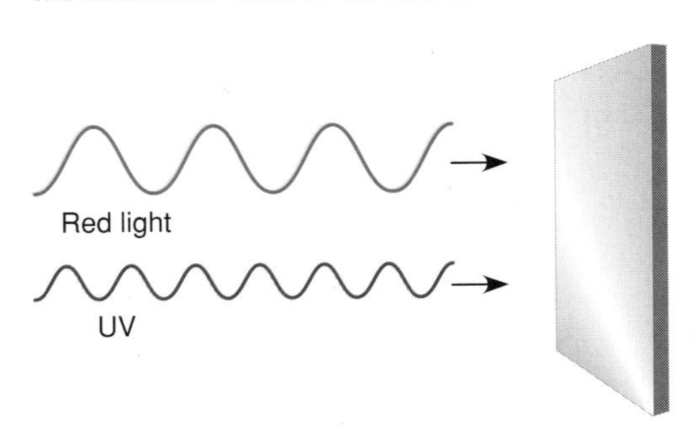

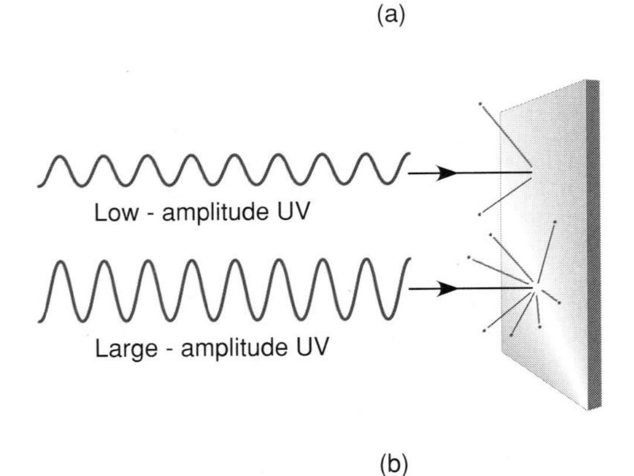

FIGURE 24-5 (A) AN INTENSE (HIGH-AMPLITUDE) RED LIGHT WAVE CARRIES THE SAME ENERGY PER SECOND AS THE WEAK (LOW-AMPLITUDE) UV WAVE, BUT ONLY THE UV WAVE CAN FREE ELECTRONS FROM THE ZINC PLATE. (B) UV WAVES OF SMALL AND LARGE AMPLITUDES ARE INCIDENT ON A ZINC PLATE. THE LARGER-AMPLITUDE WAVE FREES MORE ELECTRONS PER SECOND, BUT THE MAXIMUM KINETIC ENERGY OBSERVED FOR THESE ELECTRONS IS THE SAME AS THAT OF THE ELECTRONS FREED BY THE WEAKER WAVE.

Classical physics could not account for the freeing of these electrons. The wave theory of light says that, when an electromagnetic wave strikes the surface of a metal, the movable electrons begin to oscillate along with the wave, taking energy from it with each push. Whenever an electron at the surface gains more energy than the work function, it should escape. According to classical physics, that much energy could be gained from any frequency of light just so long as the electrons are exposed long enough to gather it. Experiment did not agree with the theory, however! No photoelectric effect was observed when the zinc was illuminated with red light, even very intense (bright) red light. The result was the same for green and blue light. With even extremely weak UV radiation, however, electrons left the zinc surface the instant the light struck the surface. There was no detectable delay, no time for an electron to oscillate and gradually gain energy (Figure 24-5). Moreover, when the experimenters sent in a single frequency of ultraviolet light and varied the wave's intensity (the energy per second arriving at the metal surface), corresponding to different intensities of visible light, the number of electrons escaping per second was doubled if the light intensity doubled, but their speed remained unchanged. The predictions of wave theory were wrong.

EINSTEIN ANSWERS THIS PUZZLE

In 1905 Albert Einstein studied the photoelectric effect and decided that the known facts of physics couldn't explain the actions of the electrons. Einstein saw that the electrons react to the ultraviolet light waves exactly as if the waves *were a hail of pellets of radiant energy*. That is, the electrons absorb energy from the electromagnetic wave *only in precise lumps* of energy, or *quanta* of energy. Einstein named these pellets of light **photons**. Each photon carries a quantum of energy in the amount $E = hf$, where f is the frequency of the light. So the bundles of energy that Einstein found in light waves corresponded exactly to the difference in energy that Planck's blackbody oscillators could take on. Together, these findings of Planck's and Einstein's revolutionized the understanding of light.

PUTTING IT IN NUMBERS—ENERGY OF A PHOTON

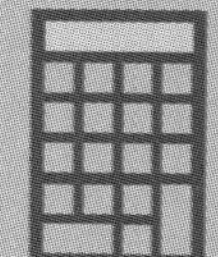

energy of a photon of light = Planck's constant times the light's frequency *f*

$$E = hf$$

Example: Blue light has a frequency of about 6.4×10^{14} Hz. The value of h is 6.626×10^{-34} joule-seconds, so a single photon of blue light has an energy of hf, or about 4.2×10^{-19} joules. In other words, a single quantum of blue light carries a *tiny* amount of energy.

Photons give us a completely new picture of light. When we say a steady beam of red light of frequency f falls on a metal's surface, we mean a number of photons of energy $E = hf$ are incident on that surface each second. If we say the beam is made brighter (by its source), or more intense, we mean only that more photons of energy $E = hf$ arrive per second. The fact that red light does not eject electrons from zinc's surface means the photons in that light carry too little energy. To be ejected from zinc, an electron must absorb a photon whose energy $E = hf$ is greater than the work function of zinc. An increase in intensity of the red light won't help; a higher frequency of light is needed (Figure 24-6).

The interference experiments of Young in 1802 (Chapter 21) proved that light acts like a wave on a scale that we can see; the interference patterns in Young's experiment are visible to the eye. On the scale of atoms, however, light is a rain of photons each carrying energy hf. The oscillators that Planck had described, with their curious jumps in energy levels, are nothing more than electrons absorbing and emitting photons. The electrons in matter cannot absorb or give up energy in any amounts other than whole quanta of energy. These findings were only the first revelations about the very strange world of electrons and atomic structure.

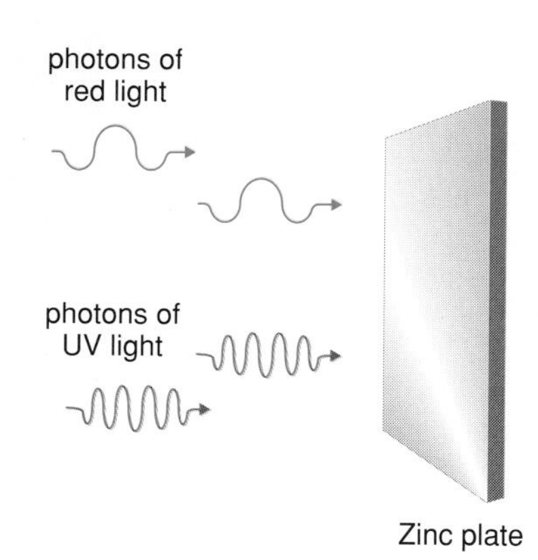

FIGURE 24-6 THE MODERN PHYSICS VIEW OF FIGURE 24-5. BEAMS OF LIGHT OF A SINGLE FREQUENCY ARE MADE UP OF PHOTONS OF THAT FREQUENCY, SMALL BUNDLES OF ENERGY. ELECTRONS INTERACT ONLY WITH WHOLE PHOTONS, AND ONLY THE UV PHOTONS HERE HAVE ENOUGH ENERGY TO EXCEED THE WORK FUNCTION FOR ZINC, RELEASING ELECTRONS.

THE BOHR MODEL OF THE HYDROGEN ATOM

The discoveries by Planck and Einstein concerning light and solids offered no help toward explaining the absorption and emission spectra of gases. The origin of those patterns of precise wavelengths was still a mystery. Then in 1911 Ernest Rutherford discovered the nucleus, and suddenly all the volume in the atom around the tiny positive nucleus was seen to be the domain of electrons (Chapter 8).The next year a young Danish physicist, Niels Bohr, took on the problem of how electrons behave in atoms to try to explain how an atom emits and absorbs light.

Bohr knew that the electric force kept the negative electron and the positive nucleus together. He decided the electron must orbit the nucleus, something like a planet orbits the sun. Otherwise he thought the two particles would just fall together under their mutual attraction. The hydrogen nucleus (a single proton) is almost 2000 times as massive as an electron, so he reasoned

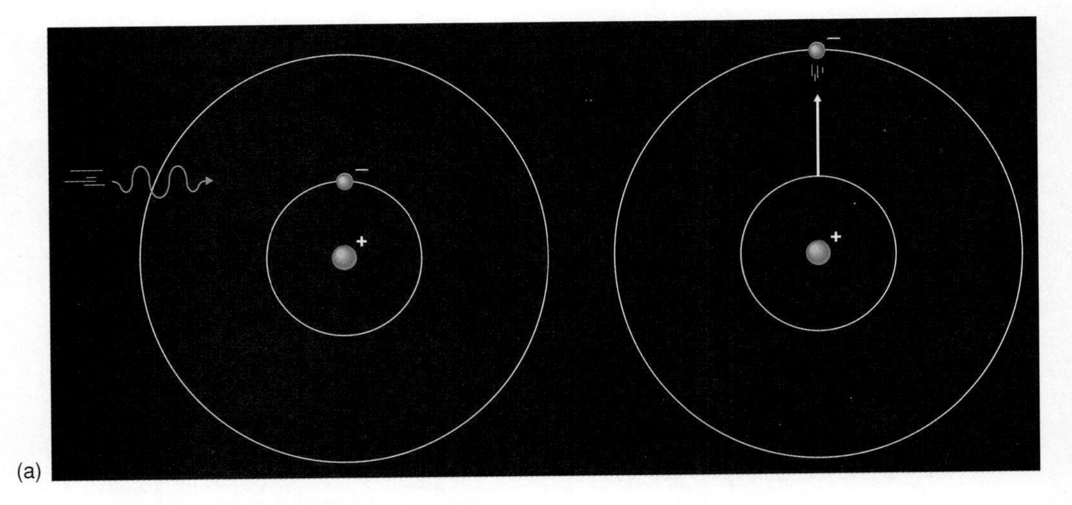

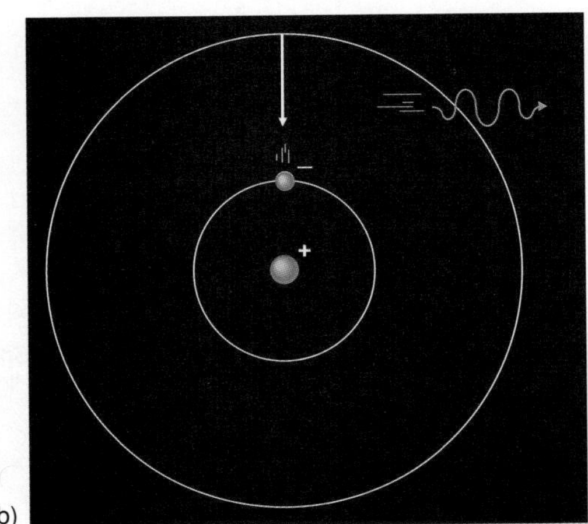

FIGURE 24-7 (A) BOHR REASONED THAT, WHEN THE ENERGY OF A PHOTON OF LIGHT WAS ABSORBED BY A HYDROGEN ATOM, THE ELECTRON WOULD MOVE FARTHER FROM THE NUCLEUS, GAINING POTENTIAL ENERGY, BUT STILL ORBIT THE NUCLEUS IN AN ALLOWED ORBIT. (B) WHEN THE ATOM EMITS THAT ENERGY, THE ELECTRON WOULD RETURN TO ORBIT CLOSER TO THE NUCLEUS, LOSING POTENTIAL ENERGY, WHILE THE PHOTON CARRIED OFF THAT SAME AMOUNT OF ENERGY.

it moves relatively little as the lightweight electron sweeps around it. That accounted for the large volume of the atom that electrons occupy.

Next, Bohr tackled the problem of how the hydrogen atom absorbs or emits light. When a photon of light strikes a hydrogen atom and is absorbed, the atom takes on the photon's energy. The electron, being the lighter particle in the atom, is affected more by this absorbed energy than is the proton. In Bohr's model, the absorbed energy causes the electron to pull away from the proton to orbit at a greater distance, as shown in Figure 24-7, just as an artificial satellite moves to a higher orbit around the earth if its energy is increased by booster rockets. Likewise, if that excited hydrogen atom emits light, the electron gives up the energy that is taken away by the photon. Because it gives up this energy, the electron falls toward the proton to orbit at a closer distance. The atom absorbs or emits *only the frequencies of light found in its absorption and emission spectra.* To account for that fact, Bohr decided the electron must move *only on a certain set of orbits* (Figure 24-8). In a move from one of these orbits to another, the electron's energy changes by $E = hf$, the difference in the energy of the two orbits, where f is precisely one of those frequencies found in the hydrogen spectra. Bohr deduced where the orbits would be. When the electron is in its lowest-energy orbit, closest to the nucleus, it is said to be in its **ground state**. The possible orbits farther from the nucleus are called **excited states**.

To relate to the electron energies in Bohr's model, think of walking up or down a flight of stairs (Figure 24-9). It takes energy to go up a step, and you

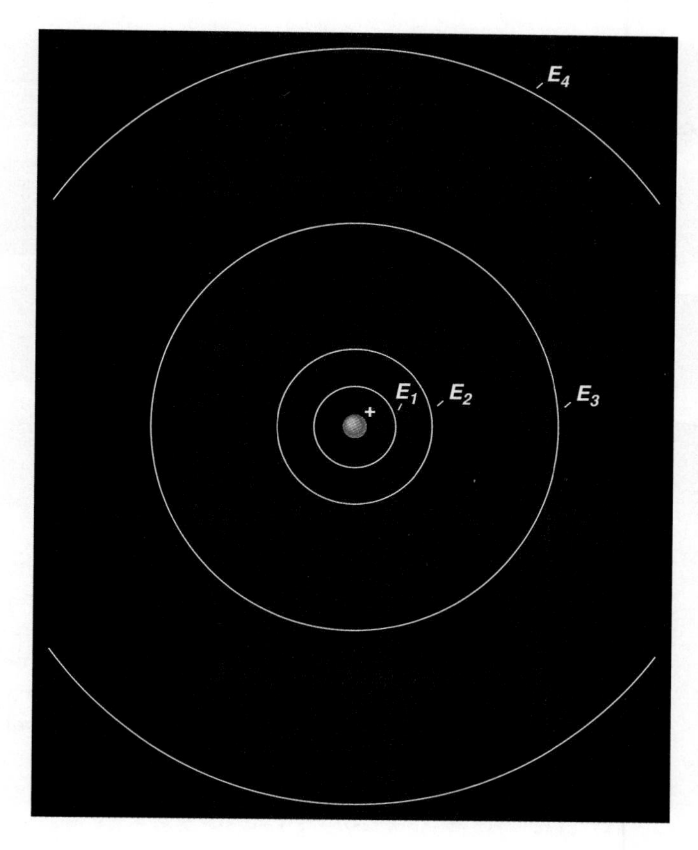

FIGURE 24-8 BOHR PREDICTED CIRCULAR ORBITS FOR THE ELECTRON IN A HYDROGEN ATOM SUCH THAT IF THE ELECTRON JUMPED FROM ONE ORBIT TO ANOTHER, IT WOULD ABSORB OR EMIT AN AMOUNT OF ENERGY EXACTLY EQUAL TO THE ENERGY OF THE WAVELENGTHS SEEN IN HYDROGEN'S ABSORPTION AND EMISSION SPECTRA. THE FOUR BOHR ORBITS LOWEST IN ENERGY ARE SHOWN HERE (NOT QUITE TO SCALE).

lose potential energy if you step down. A step down (toward the nucleus) means a loss of energy, and a step up (away from the nucleus) means the electron gains energy. The "steps" on an atom's energy "staircase" are not all the same size, however. A different amount of energy is lost or gained between any two steps. Just as you can't stand between two steps, neither can the electron land between its energy steps, or **energy levels**. The electron, like you, can take steps two or three at a time (or even more), skipping one or more energy levels, but while it is part of the atom, it must always stop at one of its energy levels. In classical physics, particles can lose or gain any amount of energy, but the electron in a hydrogen atom cannot. The sharp spectral lines of all the elements show that their electrons also have definite energy levels.

Despite Bohr's success with the hydrogen atom, no one could explain why the electron could not orbit anywhere between the allowed orbits. Also, Maxwell had shown 50 years before that any accelerated charge radiates energy, and an electron moving along a circular path accelerates toward the center of the circle. Yet Bohr's electron did not radiate while in its orbit; it radiated only when it made a **transition**, that is, when it jumped from one orbit to another. Clearly something was wrong. Then Bohr found his model did not predict the observed spectral lines of helium or any other atom con-

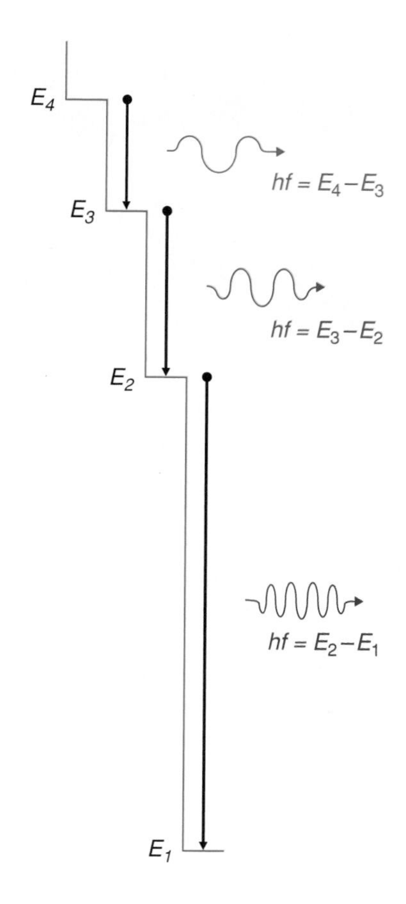

FIGURE 24-9 SOME OF THE "STEPS" OF ALLOWED ENERGIES FOR A MODEL ATOM. (IN THE BOHR THEORY, THESE STEPS CORRESPOND TO THE ENERGIES OF THE ELECTRON IN THE FOUR ORBITS CLOSEST TO THE NUCLEUS.) IF THE ELECTRON LOSES ENERGY ONE STEP AT A TIME, AS SHOWN HERE, IT WILL EMIT THREE PHOTONS ON ITS WAY TO THE GROUND-STATE ENERGY, E_1. THE GREATER THE ENERGY LOSS OF THE ELECTRON, THE GREATER THE ENERGY (HENCE FREQUENCY) OF THE EMITTED PHOTON.

COMPARISON OF THE EMISSION SPECTRUM FOR A BLACKBODY AT 6000 K (TOP) WITH SEVERAL EMISSION SPECTRA OF ELEMENTS.

taining more than one electron. A dozen years later, the reasons for these failures would be known, and we shall see the explanation in the next chapter. Bohr's model *was* wrong. It turns out that the hydrogen energy levels were correct in his model, and all electrons in atoms do have precise energy levels—but his ideas of electron orbits had to be abandoned. Next we'll look at some atomic phenomena using the idea of electron energy levels.

WHY WE GET WHITE LIGHT FROM THE SUN

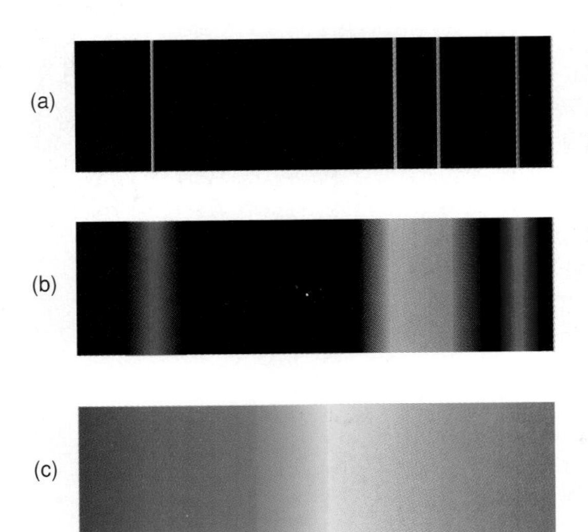

FIGURE 24-10 (A) A RELATIVELY COOL, LOW-DENSITY GAS EMITS SHARP SPECTRAL LINES. (B) IF THE DENSITY AND/OR TEMPERATURE OF THE GAS IS INCREASED, THE SHARP LINES BECOME BROADENED DUE TO COLLISIONS AND THE DOPPLER SHIFT. (C) AT EXTREMELY HIGH DENSITIES AND/OR TEMPERATURES, THE BROADENING IS SO GREAT THAT THE GAS EMITS ALL FREQUENCIES OF LIGHT; THE SPECTRUM IS THEN CALLED A *CONTINUOUS* SPECTRUM.

A hot, low-density gas in the chamber of a spectroscope sheds energy by giving off photons, and these emissions appear in the spectroscope as a few sharp, bright lines of color. If the density of gas in the chamber is increased, the spectrum begins to change. The once-sharp bright spectral lines begin to broaden, spreading out to become small *bands* of color. The gas pressure and density can be increased until those bands of color spread to overlap completely (Figure 24-10). The light from any gas under these conditions gives a *continuous* spectrum when seen through a spectroscope. In simpler terms, a hot, dense gas gives off white light, and a major reason involves collisions.

In a dilute gas, the atoms rarely collide while in the process of emitting photons. In a dense, hot gas, though, the chances are good that an excited atom will collide with another atom *while the excited atom is emitting a photon*. During the collision, the other atom exerts pressure on the excited atom, and this change in pressure alters the energy of the electron levels in the excited atom. The energy of any emerging photon, then, is different from that emitted by a similar atom that is not under pressure. Since the energy of the photon is different, its frequency is also, since $E = hf$. The different frequencies emitted by colliding atoms are seen through the spectroscope: A once-sharp spectral line becomes broadened. The hotter the gas, the more violent the collisions can be, and the denser the gas, the more collisions per second. Both temperature and density influence how wide the spectral lines are.

The Doppler effect (Chapter 16) also comes into play at high temperature. If an atom moving away from your position emits a photon, the frequency of that photon relative to you is lower—much like the lower pitch you hear from a receding train whistle. If the atom is approaching you as its photon emerges, the light you see has a higher frequency. In a hot gas, some atoms have great speeds of approach and recession, and the photons that appear have a spread

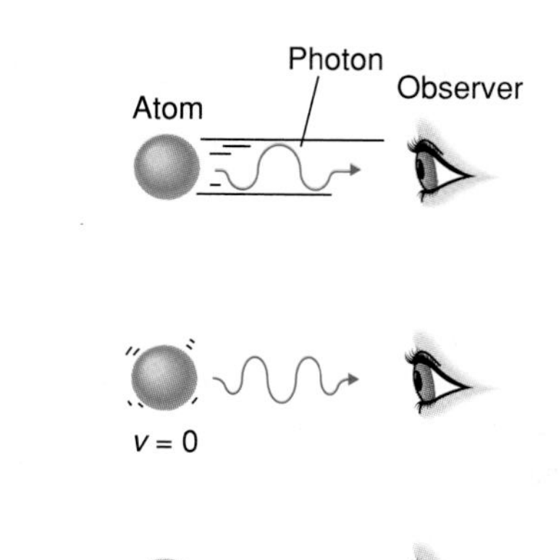

FIGURE 24-11 THE EFFECTS ON THE FREQUENCY OF A PHOTON DUE TO THE EMITTING ATOM'S MOTION. THE FREQUENCY OF A PHOTON IS REDUCED IF AN ATOM IS SPEEDING AWAY FROM THE OBSERVER (HERE TO THE RIGHT) DURING AN EMISSION. AN ATOM THAT IS RELATIVELY STILL DURING AN EMISSION EMITS A PHOTON WHOSE FREQUENCY BELONGS TO THE ATOM'S NORMAL SPECTRUM. IF AN ATOM APPROACHES THE OBSERVER WHILE THE EMISSION TAKES PLACE, THE PHOTON'S FREQUENCY IS HIGHER THAN NORMAL. THE SHIFT IN FREQUENCIES DUE TO THE RELATIVE MOTION OF THE ATOMS IS CALLED THE DOPPLER EFFECT.

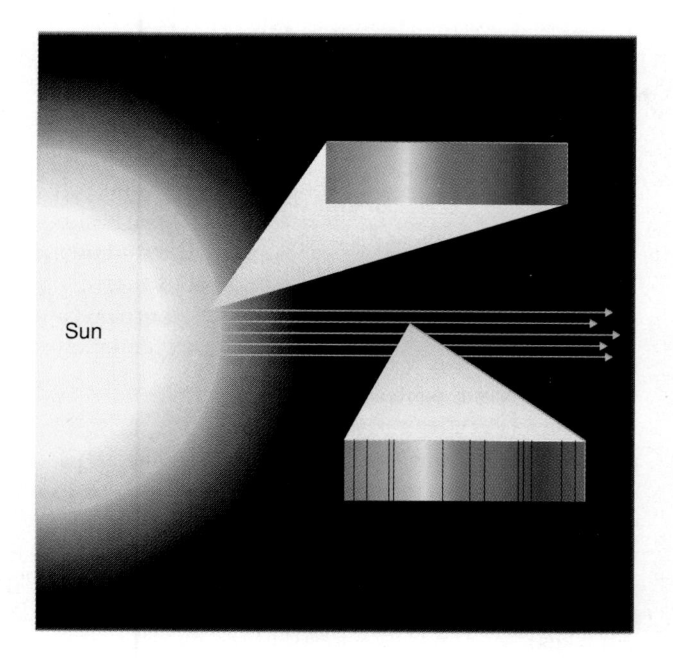

FIGURE 24-12 AS THIS LIGHT PASSES THROUGH THE SUN'S ATMOSPHERE, CERTAIN FREQUENCIES ARE ABSORBED BY ATOMS AND MOLECULES THERE. SO THE SUNLIGHT WE RECEIVE ON EARTH CARRIES—IN THE FORM OF ABSORPTION LINES—THE "FINGERPRINTS" OF THE GASES FOUND IN THE SUN'S ATMOSPHERE. (NOT SHOWN IN THE ABSORPTION SPECTRUM ARE "BANDS" OF CLOSELY-SPACED LINES DUE TO MOLECULES.)

of frequencies. This effect is called *Doppler broadening* (Figure 24-11) of the spectral lines.

Sunlight is white for all these reasons. The gases (mostly hydrogen and helium) at the sun's surface are hot, excited by radiation that comes from the sun's hot interior. The incessant collisions at high speeds and the Doppler effect spread the spectral frequencies of the atoms until they overlap completely and become a continuous spectrum containing every frequency of light. Sunlight contains all the frequencies of the rainbow as it leaves the sun's surface, but it then travels through the sun's rarefied atmosphere. There, atoms (and ions) of many elements absorb their spectral wavelengths from the passing light, while letting the others through. This is why, when he turned his spectroscope toward the sun, Fraunhofer saw thousands of dark absorption lines within the rainbow spectrum of sunlight (Figure 24-12).

FLUORESCENCE AND PHOSPHORESCENCE

Suppose an atom has been excited. Perhaps another atom has struck it, causing an outer electron to go to an energy level higher than its ground-state energy level (Figure 24-13a). Or perhaps the atom absorbed a photon from an incident

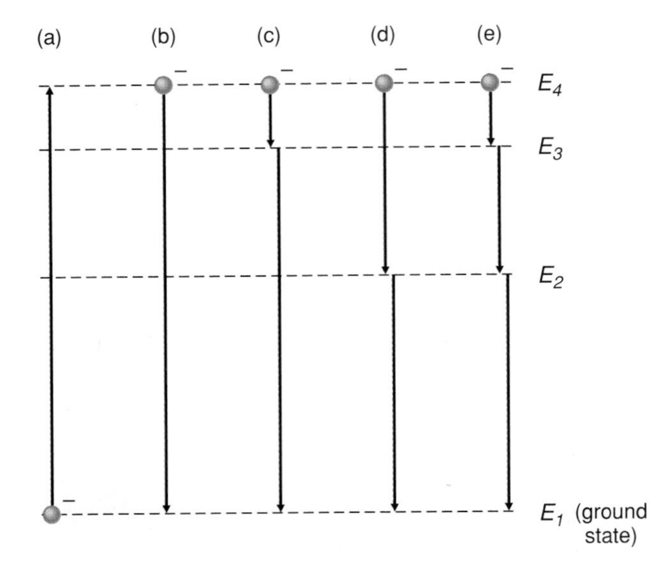

FIGURE 24-13 (A) AN ELECTRON IN A MODEL ATOM IS EXCITED TO ITS FOURTH ENERGY LEVEL BY A COLLISION WITH ANOTHER ATOM OR BY ABSORPTION OF A PHOTON. THAT EXCITED ELECTRON CAN RETURN TO THE GROUND STATE (B) IN ONE JUMP, EMITTING ONE PHOTON, OR (C) OR (D) IN TWO JUMPS, EMITTING TWO PHOTONS WHOSE TOTAL ENERGY EQUALS THAT OF THE PHOTON EMITTED IN THE SINGLE JUMP IN (B). (E) THE ELECTRON CAN ALSO RETURN TO THE GROUND STATE IN THREE JUMPS, EMITTING THREE PHOTONS WITH A TOTAL ENERGY EQUAL TO THE SINGLE PHOTON OF PROCESS (B).

light beam to cause the transition. However it happened, the excited atom will eventually lose the extra energy. It does this *spontaneously*. The electron may return to its lowest energy state with one transition, emitting a photon that carries away all its extra energy (Figure 24-13b). Or, by emitting a photon of lesser energy, it may jump to an intermediate energy level and later, by emitting a second photon, make another jump and return to its ground state (Figures 24-13c and 24-13d). Once an atom has been excited, it usually emits one or more photons within about 10^{-8} second. This way of giving off light energy is called **fluorescence**. Most atoms fluoresce, and the short stay at excited levels influences their absorption spectra (for details see Figure 24-14). Sometimes, however, the electron finds itself in an energy level, or state, where the transition to a lower level takes more time. It may need anywhere from 1/1000 second to days or even weeks to make the light-producing jump to the lower-energy state. This slower process is called **phosphorescence**, and the state where the electron is hung up temporarily is a **metastable state**.

The soft glows of "neon" signs come from a gas of neon or other atoms. These atoms are struck by oscillating electrons in the tube and become excited. Excited neon atoms have no transitions that give photons of blue or green light, but they emit red light intensely by fluorescence, and other atoms fluoresce at other wavelengths. Auroras are natural displays of fluorescence. Nitrogen and oxygen molecules and oxygen atoms at high altitudes near the north and south

FIGURE 24-14 THE EMISSION AND ABSORPTION PROCESSES OF A MODEL ATOM, SHOWING THE EMISSION AND ABSORPTION SPECTRA THAT RESULT. THERE ARE FEWER ABSORPTION LINES THAN EMISSION LINES. FOR EXAMPLE, THERE IS NO ABSORPTION LINE FROM LEVEL $E_2 \rightarrow E_3$, WHILE THERE IS AN EMISSION LINE FROM $E_3 \rightarrow E_2$. THAT IS BECAUSE IN A COOL, LOW-DENSITY GAS, MOST ELECTRONS ARE IN THE GROUND STATE, AND WHEN ONE IS EXCITED TO LEVEL E_2, IT TYPICALLY REMAINS THERE ONLY 10^{-8} SECOND—NOT MUCH TIME FOR ANOTHER PHOTON TO HAPPEN BY AND KICK IT UP TO LEVEL E_3. CONSEQUENTLY, THERE IS NO DARK LINE IN THE ABSORPTION SPECTRUM CORRESPONDING TO THAT ENERGY JUMP.

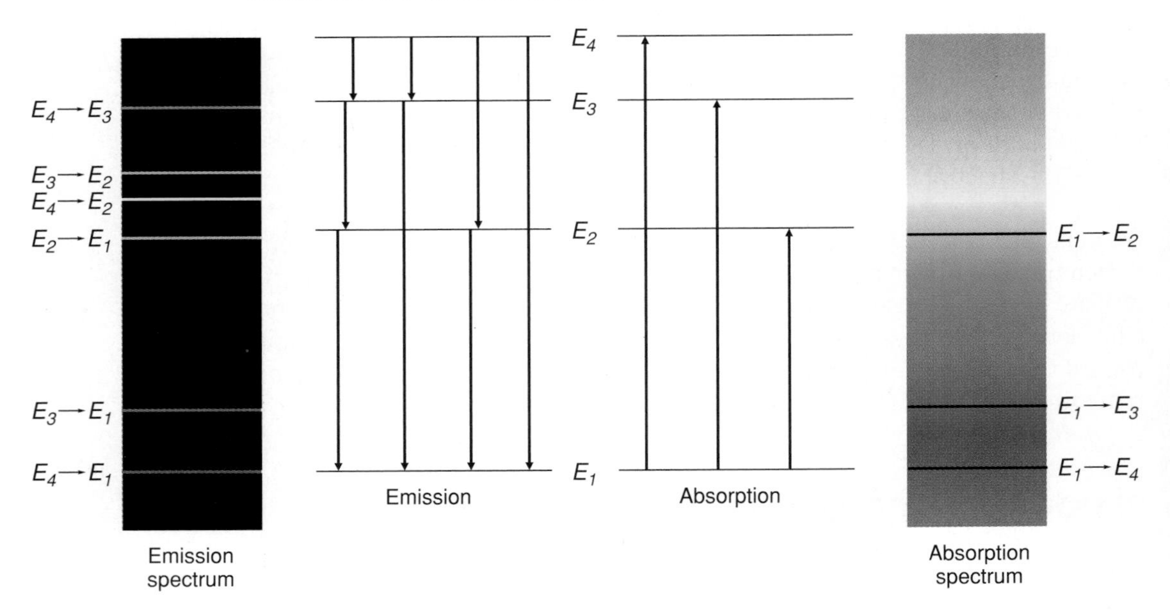

magnetic poles of the earth collide with ions from the sun, become excited, and fluoresce in visible colors.

Examples of phosphorescence are found in living creatures. The firefly uses a chemical reaction to emit light so it can be seen. However, certain squid emit visible light to become invisible! To fool predators below, these squid carefully regulate their brightness to match the intensity of the sunlight at their depth—they use light for camouflage.

In a fluorescent light, oscillating electrons excite mercury atoms, which then give off intense ultraviolet light, photons that we cannot see (Figure 24-15). The inside surface of the lamp is covered with a *phosphor* (a phosphorescent material), however, and this material first absorbs the ultraviolet photons and then emits photons of visible light. The excited electrons in this phosphor take several jumps to come back down to their original energy level, splitting the energy of those ultraviolet photons into several parts, and these less energetic photons have frequencies in the range of visible light.

AURORA AUSTRALIS (SOUTHERN LIGHTS) SEEN IN THE SOUTHERN HEMISPHERE.

LASERS

Excited atoms emit photons spontaneously in due time, but in 1917 Einstein determined that excited atoms can also be *stimulated* to emit photons. If an excited atom is struck by a photon of exactly the same frequency as one that the atom can eventually emit, the chances are very good that it will emit that photon immediately. In the light from the sun or a light bulb, the photons are of almost all frequencies and go in all directions. They are emitted independently from one another and arrive at any surface like raindrops splashing on a sidewalk. In stimulated radiation, however, an atom gives up a photon that travels in the same direction as the stimulating photon. The two photons also travel *in phase* so that the crests and troughs of the electromagnetic waves these photons belong to coincide, in perfect constructive interference (Figure 24-16). This sort of radiation is called **coherent**, whereas the photons from spontaneous emissions are **incoherent**, meaning the photons occur independently and at random.

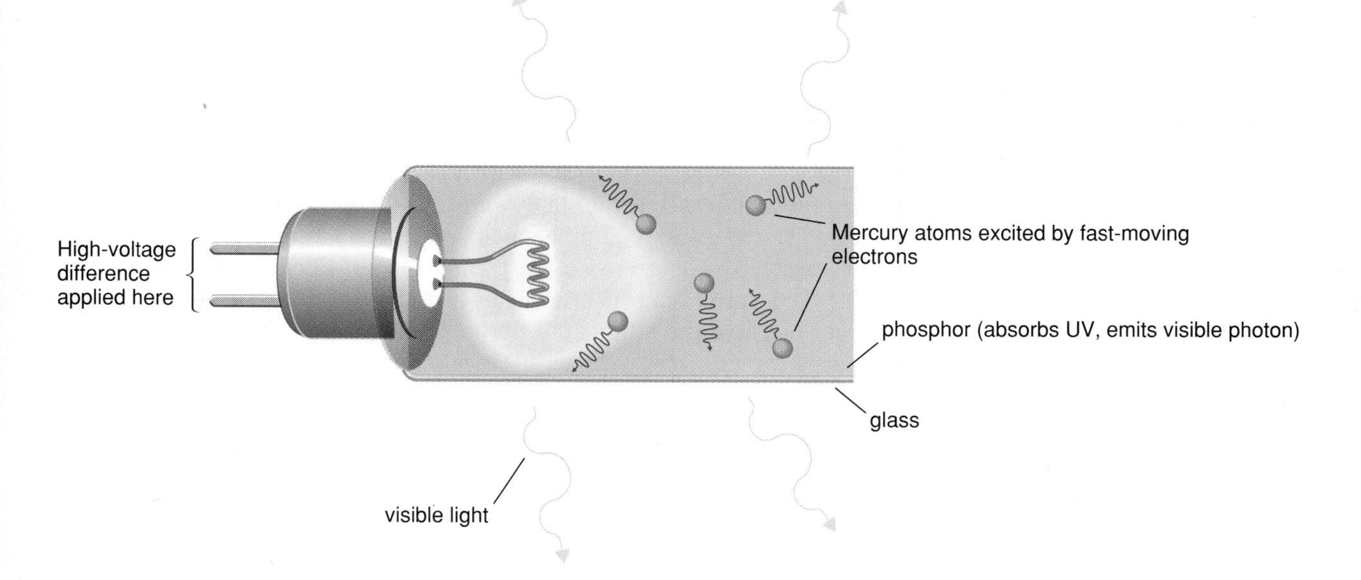

FIGURE 24-15 IN A FLUORESCENT LAMP, MERCURY ATOMS INSIDE ARE EXCITED BY ELECTRONS THAT STRIKE THEM. THE ATOMS EMIT UV PHOTONS, WHICH ARE ABSORBED BY PHOSPHORS ON THE INSIDE OF THE GLASS. THESE PHOSPHORS EMIT THAT ENERGY IN SEVERAL TRANSITIONS RATHER THAN ONE, AND THE PHOTONS THAT EMERGE FROM THE LAMP ARE IN THE VISIBLE REGION OF THE SPECTRUM RATHER THAN THE ULTRAVIOLET REGION.

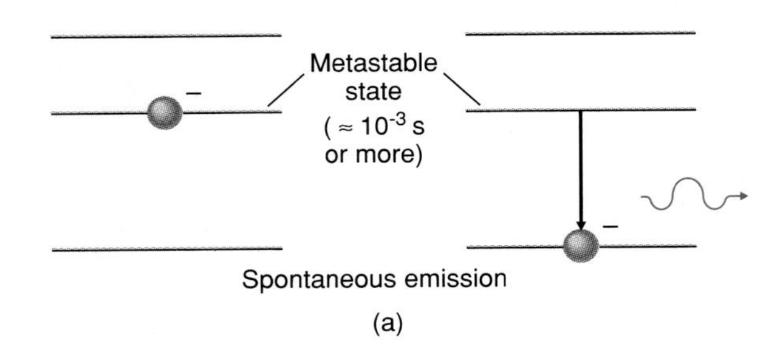

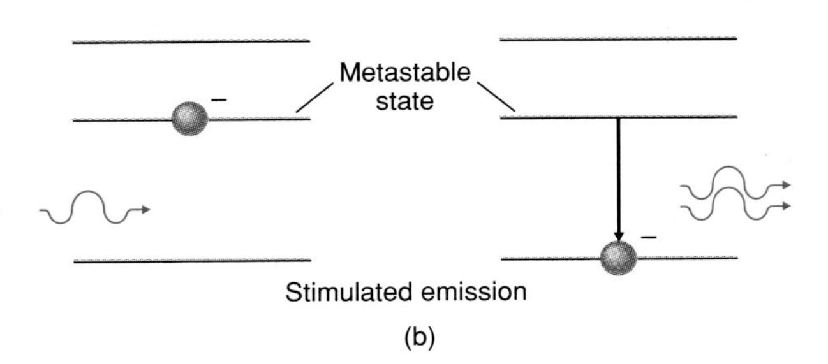

FIGURE 24-16 (A) SPONTANEOUS EMISSION BY AN EXCITED ELECTRON FROM A METASTABLE STATE. (B) STIMULATED EMISSION OF AN ELECTRON IN A METASTABLE STATE. THE PHOTONS IN STIMULATED EMISSION ARE IDENTICAL IN FREQUENCY AND DIRECTION AND TRAVEL IN PHASE.

Whenever two coherent photons happen upon another excited atom of the same type, the chances are even better than in the first episode that this atom will be stimulated to add its photon to the first two. The more stimulations that occur, the better that group of photons is at stimulating other photons. With each stimulation, the pulse of light gains energy—the light is *amplified.* **Lasers** are devices that amplify light of a single frequency in this way. The name is an acronym formed by the first letters of the words that describe their purpose: **l**ight **a**mplification by **s**timulated **e**mission of **r**adiation.

In a helium-neon laser, the kind you have probably seen in grocery stores or department stores at checkout counters, a low pressure mixture of helium and neon is in a tube with small mirrors at its ends. A high voltage is applied to electrodes at each end; the current flowing through the tube continually excites the helium atoms, which collide with the neon atoms, putting them into a metastable state. The neon atoms return to the ground state, emitting photons of red light. If a photon happens upon another excited neon atom, a stimulated emission occurs. Many of these photons pass through the side of the tube, but some reflect back and forth between the mirrors inside. One mirror reflects essentially all of the light that reaches it. The other end reflects about 99 percent, so 1 percent passes through. The mirrors cause most of the photons to reflect repeatedly through the gas tube, stimulating more emissions with each pass (Figure 24-17). The light that emerges through the partially reflecting end is the laser beam, highly amplified coherent light. Today there are many types of lasers, including small solid-state lasers that use electrical energy to excite the electrons to metastable states. Because the photons in the laser's beam are coherent, the beam's net electric field is astonishingly strong. And because the stimulated photons all travel in the same direction, a laser beam spreads very little as it travels. These unique properties make laser light invaluable for many applications in science, medicine, and communications.

Various lasers produce light throughout the range of visible, infrared, and ultraviolet frequencies. The cross section of a typical laser beam might expand

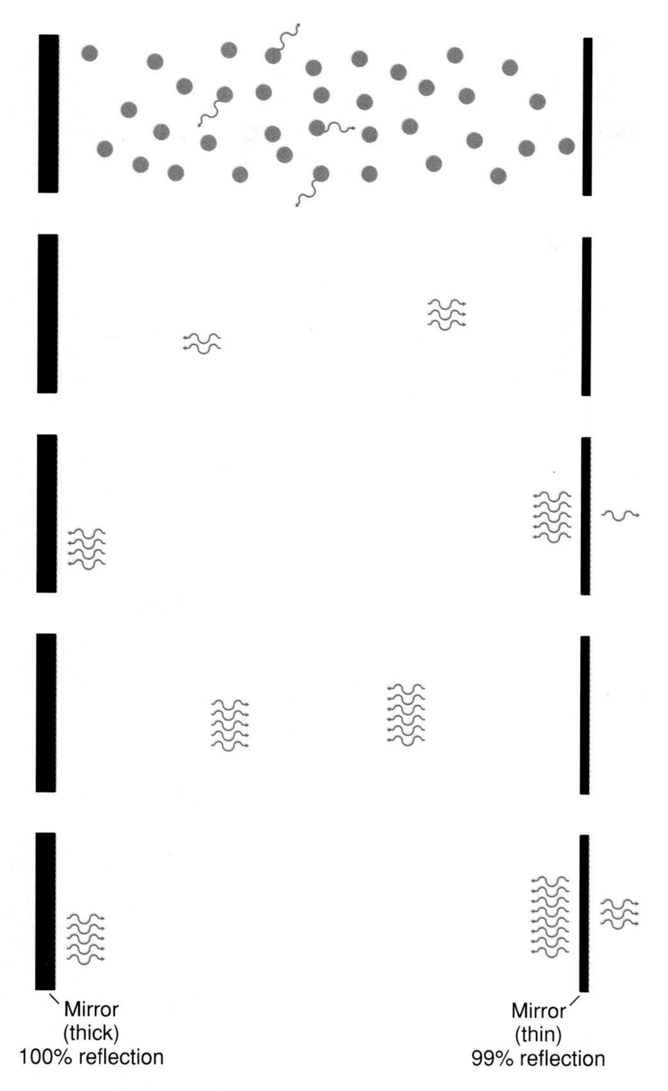

FIGURE 24-17 In a helium–neon laser, a current excites the helium atoms. These collide with the neon atoms, putting them in a metastable state. As the neon atoms return to their ground state they emit photons in random directions, as seen in the top figure. The lower figures omit the atoms for clarity. Those photons that reflect from the mirrors can make multiple passes through the laser, stimulating many other photons to travel with them. Repeated passes build up the coherent radiation, a small amount of which passes out with each reflection from the thin-mirror side and constitutes the laser beam.

less than one centimeter for every two kilometers it travels, so the compact laser beam serves today as an instant straight line for use in surveying, the construction of skyscrapers and pipelines, and also for the precise alignment and calibration of machinery (Figure 24-18).

Today laser beams aimed at the moon reflect from special reflectors left by Apollo astronauts in the 1970s. The time of flight is measured with an atomic clock accurate to one part in a hundred million. Multiplying this time by the speed of light reveals the round-trip distance to the moon with unprecedented accuracy, with an uncertainty of under 10 centimeters (Figure 24-19). Lasers are used in computerized cash registers at the checkout counters in many grocery stores (Figure 24-20). The beam illuminates (and the computerized detector reads) the identification code placed by the manufacturer on almost all grocery products. Laser printers are found in many offices, and lasers are used to play compact-discs (Figure 24-21). Some nationally circulated newspapers are printed simultaneously in a dozen or more cities across the country with laser assistance. Guided by a satellite transmission, a laser exposes

FIGURE 24-18 LASERS ARE USED TO MEASURE DISTANCES ACCURATELY AND FOR STRAIGHT-LINE ALIGNMENTS IN SURVEYING AND CONSTRUCTION.

(a)

(b)

FIGURE 24-19 (A) RETROREFLECTORS LEFT BY AN APOLLO CREW ON THE MOON. (B) MCDONALD OBSERVATORY IN FORT DAVIS, TEXAS. WHEN THE POWERFUL LASER BEAM SENDS A BURST OF LIGHT TO THE MOON, SOME OF THE PHOTONS ARE RETURNED ALONG THEIR ORIGINAL PATH BY THE REFLECTORS.

FIGURE 24-20 LASER SCANNERS READ THE BAR CODES ON STORE ITEMS.

FIGURE 24-21 A COMPACT DISC (CD) HAS A SPIRAL TRACK OF LENGTH OF ABOUT 3.5 MILES WITH UP TO 3 BILLION PITS OF VARIOUS LENGTHS THAT A LASER BEAM REFLECTS FROM. WHEN THE BEAM HITS A PIT, DESTRUCTIVE INTERFERENCE WITH THE PART OF THE BEAM NOT HITTING THE PIT CAUSES A FLUCTUATION IN BRIGHTNESS OF THE REFLECTED BEAM. THE AMOUNT OF LIGHT REFLECTED, BRIGHT OR DIM, IS CONVERTED TO ELECTRICAL SIGNALS AND THEN TO SOUND BY AN AMPLIFIER. THE SPIRALS OF THE TRACK ARE SO CLOSE TOGETHER THAT THE CD ACTS AS A REFLECTION DIFFRACTION GRATING IN THE SUNLIGHT, AS SEEN HERE.

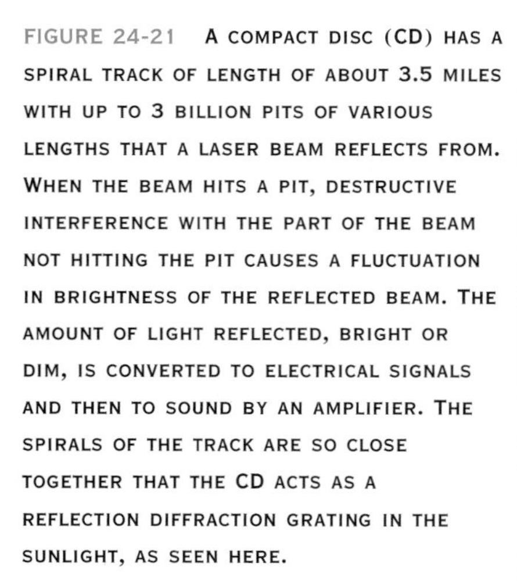

the photographic film that is used to make the printing plates. The light used in fiberoptics communication is laser light.

Lasers have been used in medicine since the 1960s (Figure 24-22). A surgeon working on the human eye can send a perfectly controlled laser beam through the cornea and lens and bring the beam into focus at the edge of a rip in the thin retina. At the point of focus, the concentrated laser light increases the temperature, coagulating the proteins there and welding the retina into place.

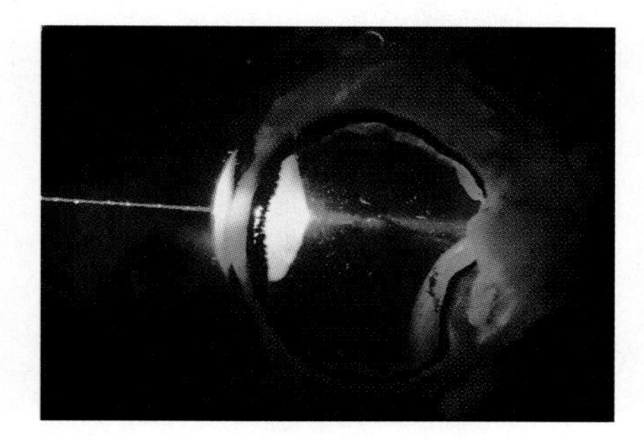

FIGURE 24-22 A LASER BEAM DIRECTED INTO AN EYE WHICH IS COVERED BY A CONTACT LENS. THE LASER BEAM'S TARGET WAS A MELANOMA WHICH WAS TREATED SUCCESSFULLY.

Summary and Key Terms

When a ***spectroscope*** or ***spectrograph*** magnifies the spectrum of white light that passes through a gas, it reveals dark lines that show the frequencies the gas absorbs, its ***absorption spectrum***. The excited gas also has an ***emission spectrum***, a dark field with bright lines showing which frequencies the gas emits. Classical physics could not explain why gases absorb and emit only certain frequencies. Nor could it explain the emission spectra of ***blackbodies***, solid objects that absorb all the light that strikes them. Blackbodies emit more radiation at shorter wavelengths as their temperature increases. Planck found that blackbodies emit or absorb radiation only in definite amounts of energy, or ***quanta***. The energy of one quantum is $\boldsymbol{E = hf}$, where h is Planck's constant.

spectroscope–357
spectrograph–357
absorption spectrum–357
emission spectrum–358
blackbody–358
quanta–359

In the ***photoelectric effect***, a metal absorbs light and emits electrons at its surface. To explain this effect, Einstein deduced that light travels in bundles, now called ***photons***, each carrying a quantum of energy. Electrons absorb or emit energy only in whole quanta.

photoelectric effect–360
photons–360

Since hydrogen absorbs or emits only certain frequencies of light, Bohr speculated that its electron moves only on a certain set of orbits, which had certain ***energy levels***. In a move from one orbit to another, a ***transition***, the electron absorbs or emits a photon, gaining or losing a quantum of energy that corresponds to one of the frequencies in the hydrogen spectrum. The energy level closest to the nucleus is the ***ground state***, and the other possible states are ***excited states***. Bohr's theory was wrong concerning electron orbits, but the idea of energy levels proved correct.

energy levels–363
transition–363
ground state–362
excited states–362

Excited electrons normally move to a lower state by spontaneously emitting photons, a process called ***fluorescence***. If excited electrons first move to an intermediate ***metastable state***, where transition to the ground state takes longer, they give off light by ***phosphorescence***. An excited atom can also be stimulated to emit a photon if it is struck by a photon of the same frequency as one it can emit, as in a ***laser***. The two photons move off together, in phase. That is ***coherent radiation***. Radiation from random, spontaneous emissions is ***incoherent***.

fluorescence–366
metastable state–366
phosphorescence–366
laser–368
coherent radiation–367
incoherent radiation–367

EXERCISES

Concept Checks: Answer true or false

1. A spectroscope produces a magnified spectrum of a light beam for visual observation.
2. Dark spectral lines appear in a continuous spectrum when certain wavelengths of light are absorbed as white light passes through a gas.
3. In a thin, cool gas, atoms absorb light of certain wavelengths and emit only light of identical wavelengths.
4. The light emitted by a low-density gas of excited atoms appears in a spectroscope as bright lines of color on a dark field. This is called an emission spectrum.
5. Unlike a gas, a solid generally absorbs and emits light over a wide range of wavelengths.
6. A blackbody absorbs all the light that strikes it; it reflects none and emits none.
7. An atom gives off light when one of its electrons makes a transition to a lower energy level or to a higher level.
8. In Bohr's theory, a transition from one orbit to another means the electron has lost or gained a quantum of energy.
9. The photoelectric effect is used by lasers.
10. A photon carries the amount of energy equal to Planck's constant, h, times the frequency f. A light beam of a single frequency is just a large number of photons of that frequency.
11. An atom can emit photons by two different processes, spontaneous emission and stimulated emission.
12. A laser is a device that amplifies light by spontaneous emission. Its radiation is coherent.

Applying the Concepts

13. Helium is very scarce on earth and was discovered first on the sun and only later on earth. How could that happen?
14. A star that appears red emits most of its visible light in the red region of the spectrum, and a blue star emits more visible light in the blue region. Which star is hotter?
15. If you just barely open the door to a dark room, the crack—if it is just a fraction of a centimeter across—is an excellent approximation to a blackbody. When you open the door a fraction more, though, you begin to see reflected light from inside the room. What are several other common examples of blackbodies?
16. Since the hydrogen atom contains only one electron, why does its spectrum have so many lines?
17. Why does the bulb in a flashlight appear red when the flashlight cells get weak?
18. How do some materials and paints glow under a black light, an electric bulb that emits ultraviolet light?
19. A gamma-ray photon has a frequency of 10^{22} Hz. What is its energy in joules?
20. Indicate whether the following statements are true or false. (a) The distribution of radiation in the spectrum of a blackbody depends only on the temperature of the blackbody. (b) The peak of the spectrum of a blackbody is at shorter wavelengths for higher temperatures. (c) To obtain a greater intensity of red light from a blackbody at 5800 K, you could lower its temperature to 2800 K. (*Hint:* See Figure 24-4.)
21. Ultraviolet radiation, X-rays, or gamma rays that pass through a gas can ionize some of the atoms or molecules and make the gas conduct electricity. Would you expect the greatest ionization from (a) a low intensity of high-frequency light or (b) a high intensity of low-frequency light?
22. How can a technician in a crime lab determine the atomic and molecular composition of a flake of automobile paint found at the scene of a hit-and-run accident? Or a toxic compound found at a waste site?
23. In a photography darkroom, you can find red lights of very low power that let the photographer see to work yet don't expose the light-sensitive chemicals in the photographic paper. Comment on whether any other color would do just as well.
24. Astronomers today use the absorption lines in the spectrum of the light from a star or planet to tell about that body's atmospheric composition. Explain how.
25. If an astronomer gave you photographs of the spectra of a dozen types of stars, some of the lines would be precisely the same in each spectrum, provided the exposures were identical. Guess the origin of those lines.
26. If you turn off a TV in a dark room, why does the screen continue to glow for a few seconds? (*Hint*: Recall from Chapter 23 that a color TV has three types of phosphors in its screen.)
27. What might the spectrum of hydrogen look like if the electron could move between orbits of any radius in Bohr's model?
28. About how many photons per second are emitted by a 100-watt bulb if the average frequency of the photons is 5×10^{14}Hz? (Typically 7 percent of the energy each second goes away as visible light. Heat takes away most of the energy.)
29. How would sunlight be different if collisions between atoms and the Doppler effect had no influence on the photons the atoms emit?
30. Suppose you are looking through a spectroscope at the emission spectrum of hydrogen gas. Describe what hap-

pens to the lines if you turn on a heater that greatly increases the temperature of the gas.

31. Think of a solid as a condensed gas and give one reason why you might expect a solid's emission spectrum to be continuous, containing almost all frequencies of light (although different frequencies are emitted with different intensities).

32. An astronomer can use a spectroscope to observe the sun's absorption spectrum. What would happen if he or she were watching this spectrum when a total solar eclipse occurred, where the moon moves between the sun and the observer, leaving only the sun's atmosphere (called the corona) visible? The continuous part of the solar spectrum would suddenly disappear, but that's not all. The dark lines would become bright lines. Explain.

33. High-pressure sodium-vapor lights emit an amber-pink glow. Low-pressure sodium-vapor lights emit a deep yellow light and are brighter and less expensive to operate. Both are used as street lamps that scatter light into the atmosphere in every direction. Astronomers want the cities close to observatories to use the type of light that interferes less with their spectral analysis of starlight. One of these lights emits a broad continuum of light, while the other emits its light mostly at two discrete frequencies, allowing the astronomers to work in the remainder of the spectrum with their spectrographs. Which lights are the ones astronomers prefer?

34. Lasers can be used to detect flaws inside solid objects. A laser beam is focused on the solid's surface and literally blows off a small amount of the material, causing only minor cosmetic damage. As the vaporized material blows off the surface, it exerts a strong, sudden pressure on the surface. It creates a sound wave that begins at that point and travels through the solid. Tell how that large-amplitude, short-wavelength sound wave might be used to detect flaws.

35. How could the photoelectric effect be used in a device to open doors when someone approaches?

36. Broadening of the characteristic spectral lines of an element's atoms takes place because of (a) collisions, (b) high temperatures, (c) the Doppler effect, (d) all of these.

37. Broadcasts via microwave transmissions let people watch the astronauts walking and riding on the moon and sent pictures to earth from Mars, Jupiter, and Saturn. Knowing this, what can you infer about the absorption spectra of the oxygen and nitrogen molecules in earth's atmosphere?

38. Could you say that light from an incandescent bulb resembles a rain shower falling on a sidewalk, while light from a laser is more like all the drops falling at once?

39. Think of a reason fluorescent lamps are more efficient than incandescent lamps in turning energy into light. That is, fluorescent lamps give equal amounts of light with much less energy lost as heat.

40. Ultraviolet and infrared laser beams cannot be seen with the eye. Why does that make them more dangerous to use in a laboratory than visible beams?

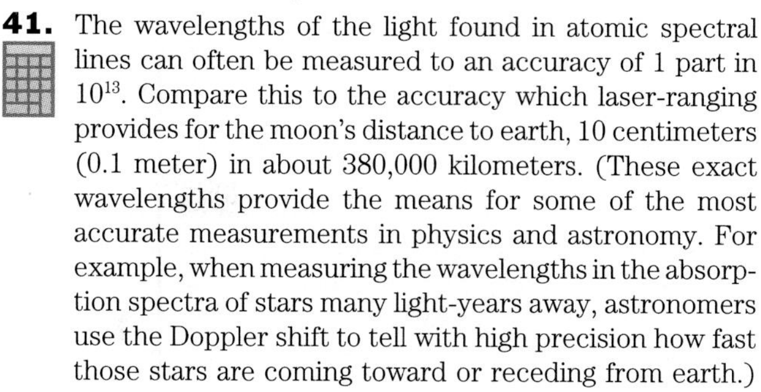

41. The wavelengths of the light found in atomic spectral lines can often be measured to an accuracy of 1 part in 10^{13}. Compare this to the accuracy which laser-ranging provides for the moon's distance to earth, 10 centimeters (0.1 meter) in about 380,000 kilometers. (These exact wavelengths provide the means for some of the most accurate measurements in physics and astronomy. For example, when measuring the wavelengths in the absorption spectra of stars many light-years away, astronomers use the Doppler shift to tell with high precision how fast those stars are coming toward or receding from earth.)

42. Could materials that do not have any metastable states produce coherent beams of light by laser action?

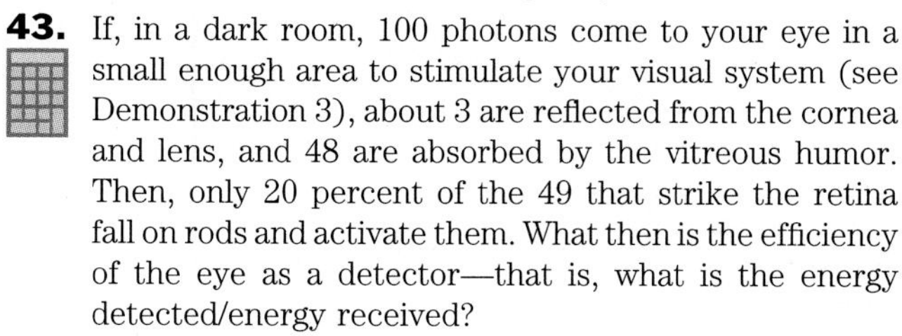

43. If, in a dark room, 100 photons come to your eye in a small enough area to stimulate your visual system (see Demonstration 3), about 3 are reflected from the cornea and lens, and 48 are absorbed by the vitreous humor. Then, only 20 percent of the 49 that strike the retina fall on rods and activate them. What then is the efficiency of the eye as a detector—that is, what is the energy detected/energy received?

44. An argon ion laser can emit powerful pulses of blue-green light. Could a flash tube that emitted only red light be used to power that laser? Explain.

45. If you know the work function of a metal and the frequency of the incident radiation, could you predict the maximum kinetic energy you would observe for the escaping electrons? (*Hint:* Where else could the energy of the absorbed photon go? Some electrons will lose some energy just traveling through the metal.)

46. "Grow-lamps" for house plants give off most of their energy in the red portion of the spectrum, as seen in Figure 24-23. Why should they? (*Hint:* See Exercise 23-50.)

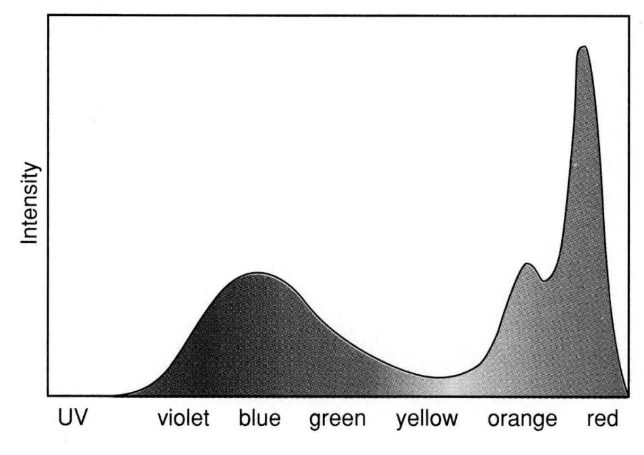

FIGURE 24-23

47. Considering the revolution in understanding that the unexplained experimental data from gas spectra and blackbody radiation brought about after 1900, could you make an argument for supporting physics research in the year 2000 with taxpayer monies? Do you think Congress or scientists should make the decisions on which projects to support, and where?

48. If you have two lamps on your desk and both are turned on, would you expect to see interference patterns on your desktop? Why or why not?

49. A laser and a photon detector mounted in an airplane have been used to chart the depths of shallow coastal waters. A brief burst of light from the laser strikes the water's surface, where some of the beam is reflected and returns to the detector. However, a portion of the beam enters the water and travels to the bottom, where once again a small amount of light is reflected. Even in 10 meters of turbid water, a few photons from the light that made it to the bottom makes it back to the detector on the plane. The time lag between the light reflected from the surface and the light reflected from the bottom can be used to find the depth. What would the time lag be for a water depth of 10 meters? (The speed of light in water is $0.75c$.)

50. For a certain metal to begin to emit electrons from its surface, the incident radiation must have a frequency of at least 5×10^{16} Hz. What is the work function of this metal in joules?

51. If an electron makes two jumps on its way to its ground state and loses in the first jump exactly half of the energy it loses in the second jump, how do the frequencies of the two emitted photons compare?

52. When light goes through a centimeter of glass, it travels at $0.65c$, or about 2×10^{10} cm/s. Since fluorescence typically occurs in 10^{-8} s, how likely do you think it is that the slowing of light in glass is due to absorption and re-emission of the light from each individual atom the light crosses in its path? (There would be about 10^{8} atoms across one centimeter of glass.)

Demonstrations

1. Some night, turn off the lights in a kitchen that has an electric stove. Then turn one of the stove element controls to the setting for simmer or warm. In daylight you wouldn't be able to see the element glow at this setting, but in the dark you can because of your night vision. Hold a hand close to the side of the element; what you feel is "light" that you cannot see, infrared radiation. Slowly turn the control to the low, medium, and high settings, feeling the infrared radiation at each step and noting the colors of the element. As the temperature increases, the color of the most intensely emitted visible wavelength goes from dark red to lighter red to red-orange; in other words, to shorter wavelengths, just as the peaks of the blackbody radiation curves do in Figure 24-4. The different infrared intensities you feel also correspond to the curves of blackbody radiation. As the element becomes hotter, the infrared intensity rises, just as the blackbody curves indicate. The element isn't a perfect blackbody, but its emission spectrum follows the same general pattern as that of a blackbody.

2. Even at the threshold of seeing, your eyes are sensitive detectors of light. A single rod will detect a single photon, but your visual system responds and you see a flash of light only when 2–10 photons are absorbed by rods very close together within 0.1 second. Single-photon events are counted as background "noise" and don't register. Otherwise, we would see the triggering of rods (cones, remember, take much more energy) due to random chemical reactions, changes in pressure caused by eye motion, breathing, and even heartbeats. To demonstrate such "noise" in an exaggerated form, close your eyes in a very dark room and *gently* tap your eyelid with a finger. You'll see bright flashes of light. Many people can also see such flashes if, with eyes closed, they swing their eyes back and forth as if following a ping pong match.

3. You can make an excellent blackbody with a coffee can that has a removable plastic lid. Spray paint the top of the lid with dull black paint. Then punch a small hole in its center and replace it on the can. Even though the inside of the can is very shiny (reflective), the hole will appear very black when compared with the dull black lid.

TWENTY-FIVE

The Wave Nature of Particles

A quantum "corral" of iron atoms placed in a circle on a copper crystal by a scanning tunneling microscope. The waves inside are standing waves of electrons ordinarily free to move at the copper's surface. This chapter discusses the wave nature of atomic particles and atoms.

As we saw in the last chapter, something was wrong with Bohr's model for the hydrogen atom. The classical laws of physics say the electron should accelerate in its circular orbit, and therefore radiate light and lose energy, and spiral inward to crash into the proton—but it doesn't. And although his model explained the emission and absorption spectra of hydrogen, neither Bohr nor anyone else could say why the electron should move only on a certain set of orbits. It took a new theory to solve this mystery, one that shattered established notions of what particles are like and how they behave. The radical ideas needed came from a number of physicists, and, like Planck with his theory of blackbody radiation, some of them could not believe in the end what they had collectively discovered. The first step was taken by a member of the French nobility, Prince Louis Victor de Broglie (1892–1987).

DE BROGLIE'S IDEA

While a physics student at the University of Paris, de Broglie was fascinated with Einstein's revelations about light. The fact that light acts like a wave in some experiments and like a collection of particles in others led de Broglie to pose this question: Can particles of matter have a dual nature, too? In other words, can particles ever act like waves? No experiment had showed such behavior, and no wave nature was observed in the behavior of macroscopic objects, such as baseballs and elevators. Despite this lack of evidence, he persisted in searching for answers to his questions. He compared the energy of a light wave with the energy of a particle moving at a constant speed, both as given by Einstein's theory of relativity (Chapter 26), and he deduced a connection between the momentum mv of a particle and a wavelength λ for that particle:

PUTTING IT IN NUMBERS—PARTICLE WAVELENGTHS

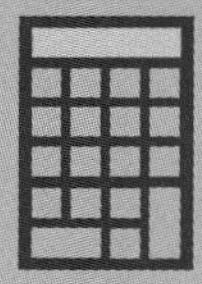

momentum of particle = Planck's constant/wavelength

$$mv = \frac{h}{\lambda}, \text{ or } \lambda = \frac{h}{mv}$$

Example: The diameter of a typical atom is 10^{-8} cm. What is the speed of an electron that is represented by a de Broglie wave having a length equal to this typical atomic diameter?

$v = \frac{h}{m\lambda}$, where $\lambda = 10^{-8}$ cm $= 10^{-10}$ m, $h = 6.6 \times 10^{-34}$ J·s, and $m_e = 9.1 \times 10^{-31}$ kg.

So $v =$ **7.3×10^6 m/s,** which is about 2 percent of the speed of light.

De Broglie's speculation could have been a blind alley, except for one thing. He noticed that the electron in Bohr's hydrogen atom model moved uniformly on each of its orbits—that is, with a fixed amount of momentum for each orbit. Using his formula, de Broglie calculated what the wavelength for the electron would be for each orbit, and there he had success.

In each case a *whole number* of wavelengths (of length $\lambda = h/mv$) fit exactly on Bohr's orbits (Figure 25-1). If electrons behave as if they have

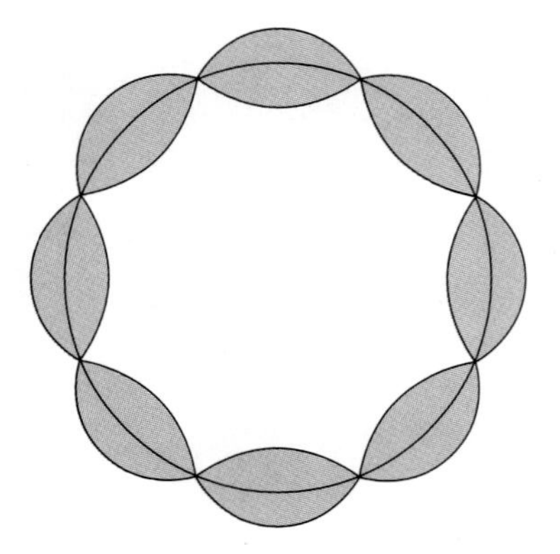

FIGURE 25-1 AN EXAMPLE OF A WAVE THAT HAS A WHOLE NUMBER OF WAVELENGTHS ON A BOHR ORBIT. CONSECUTIVE PASSES OF THE WAVE AROUND THE ORBIT JUST RETRACE THIS PATTERN.

wavelengths of these sizes, this would explain why the electron could stay only on certain orbits around the nucleus. Since the wave calculated from de Broglie's equation fit perfectly on the electron's orbit, the wave might represent a *standing* wave whose pattern repeats perfectly, its wavelength and amplitude never changing. Otherwise, if the wavelength for an orbit did not fit perfectly, the wave would interfere destructively as it went around and disappear after a number of turns. Representing the electron by a standing wave could help explain why that particle didn't lose its energy (as classical physics predicts) and spiral into the nucleus. In standing waves, energy is "trapped" by reflections and can therefore remain constant.

Encouraged, de Broglie presented his ideas in his Ph.D. thesis. With no experimental evidence to support his ideas, though, his professors were skeptical. One sent de Broglie's thesis to Einstein for an outside opinion, and Einstein replied that the ideas certainly looked crazy, but he thought they were very important and "sound." The prince received his Ph.D. degree in 1924 and in 1929 was awarded the Nobel Prize in physics for the work in his thesis, which, by then, was confirmed by experiments.

Before de Broglie's thesis was published in 1924, Clinton Davisson and Lester Germer, working at Bell Laboratories in the United States, aimed beams of electrons at crystal faces and monitored the electrons as they rebounded at various angles. A large number rebounded straight back along the incoming beam's path, but the others scattered off to the sides. The scattered electrons bounced off mostly at one certain angle and rarely at any other angle. Davisson and Germer reported the strange pattern in a physics journal. After de Broglie's work appeared, several European physicists made a remarkable observation: Davisson and Germer's patterns of electron scattering resembled an interference pattern, *as if the electron beam had the properties of a wave*. These physicists convinced Davisson to do the experiment again, carefully measuring both the momentum of the electron and the angles in the interference pattern. Soon he and Germer showed that the wavelength de Broglie's formula predicts for electrons agreed with the patterns they observed. This experiment was the first to confirm the wave nature of matter. Davisson and Germer soon confirmed de Broglie's wavelengths with 19 different crystals. Many more such experiments followed. In 1930 other workers scattered whole hydrogen and helium atoms from crystal faces and found the same effect. Both subatomic particles and whole atoms have a wave nature in their actions.

AN EQUATION FOR MATTER WAVES

De Broglie's formula relating the momentum of a particle to a wavelength for the particle tells us little about the wave nature of the particle. For example, the value the formula gives for the wavelength says nothing about the amplitude of the wave. Nor does it tell us what the amplitude means or even what the wave really is. The picture in 1924 was incomplete.

Early in 1925 Erwin Schrödinger, an Austrian physicist, began to search for an equation that would describe the *amplitude* of the wave of an electron in a hydrogen atom. He assigned the wave's amplitude a symbol, ψ (Greek letter psi, pronounced "sigh"). Within a year he found an equation that predicted three-dimensional standing waves having precisely the energies of the electron in Bohr's orbits—the energies that give the spectra of hydrogen when the electron makes transitions between the energy levels. (His equation involves advanced math, so we won't work with it.) In Schrödinger's picture, the wave of the electron changes from one standing wave pattern to another in order to emit or absorb a photon. His wave equation also shows what happens

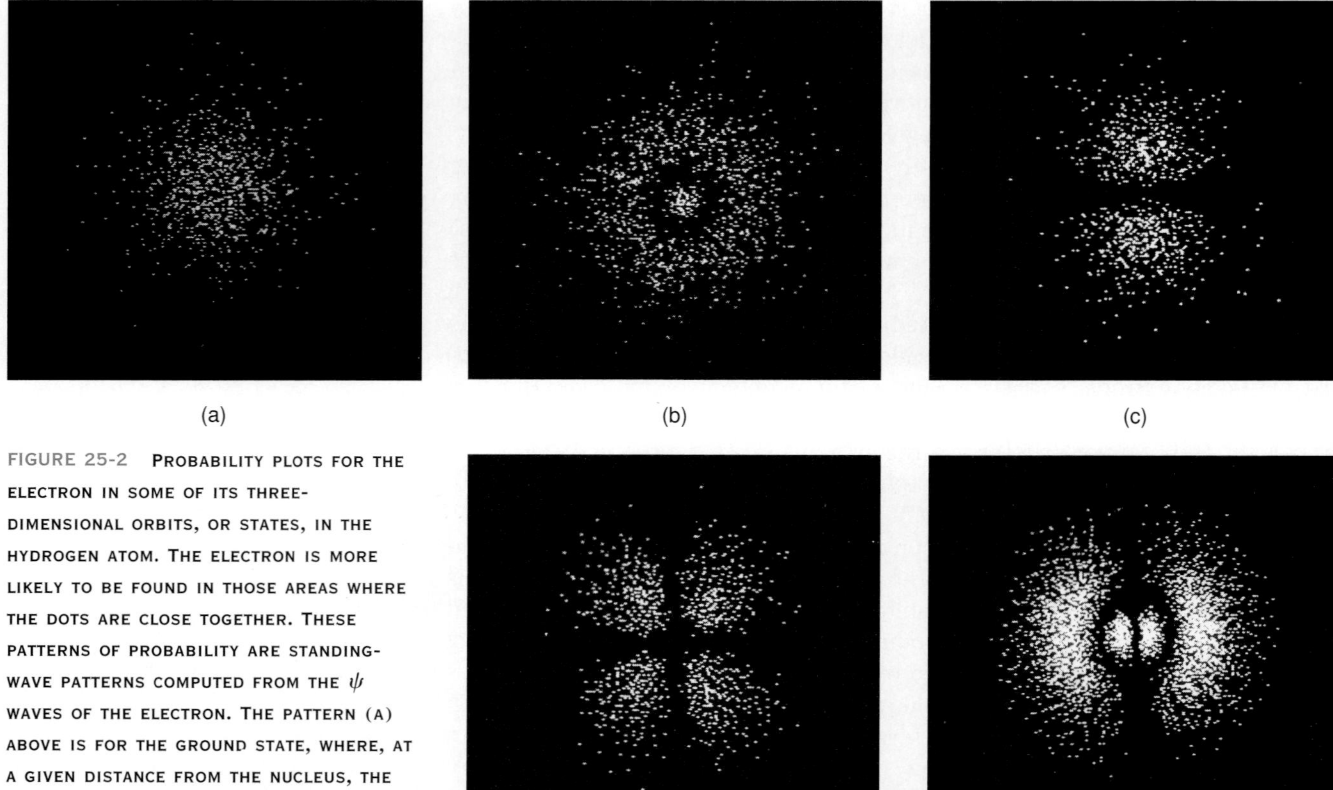

FIGURE 25-2 PROBABILITY PLOTS FOR THE ELECTRON IN SOME OF ITS THREE-DIMENSIONAL ORBITS, OR STATES, IN THE HYDROGEN ATOM. THE ELECTRON IS MORE LIKELY TO BE FOUND IN THOSE AREAS WHERE THE DOTS ARE CLOSE TOGETHER. THESE PATTERNS OF PROBABILITY ARE STANDING-WAVE PATTERNS COMPUTED FROM THE ψ WAVES OF THE ELECTRON. THE PATTERN (A) ABOVE IS FOR THE GROUND STATE, WHERE, AT A GIVEN DISTANCE FROM THE NUCLEUS, THE ELECTRON HAS EQUAL PROBABILITIES FOR BEING FOUND AT ANY ANGLE. THE OTHER PATTERNS, MOVING FROM LEFT TO RIGHT, ARE FOR INCREASINGLY EXCITED STATES FARTHER FROM THE NUCLEUS. FOR SIMPLICITY, ALL OF THESE PROBABILITY PLOTS ARE SHOWN WITH THE SAME MAXIMUM DIAMETER.

to the wave for an electron (or any other particle) moving under the influence of a force.

Schrödinger thought these waves meant the electron was smeared out, as if a child pulled on the edges of a wad of bubble gum and stretched it thin. In his view, the amplitude of the wave would be big where most of the particle was and small at the particle's edges where the particle was thin. He thought the square of ψ, which corresponds to the intensity of a classical wave, might indicate the density (mass/volume) of the smeared-out electron. Figure 25-2 shows a few of the wave "clouds" of psi-squared predicted by Schrödinger for the hydrogen atom. Where the dots are thickest, the value of psi-squared is largest. However, Schrödinger's idea about the smeared-out electrons was not correct, as we'll see in the next section.

Schrödinger called his equation the **wave equation**. The behavior of the waves of matter is known as **wave mechanics** to distinguish it from classical mechanics, which uses Newton's laws to describe the behavior of larger particles.

WHAT MATTER WAVES MEAN

Max Born, another German physicist, saw that Schrödinger's idea that an electron might be spread out along its wave was wrong because electrons are always found whole, at some point, and not smeared out over an area or volume. Experiments never find a fraction of an electron. Born suggested that the density of the wave clouds, psi-squared, represents only the *probability* for finding the electron at various places. The electron has a good chance of being found in a region only if the square of the wave amplitude is large there. Where that quantity is small, an experiment has little chance of finding the electron. In other words, the wave clouds in Figure 25-2 show the *probability density* for the electron.

Strange as this idea sounds, it agrees with the experimental facts. Born's idea about Schrödinger's wave clouds has been verified by many experiments since the 1920s. The electron is a point-like particle; it is not a wave, but the probability of the electron appearing at any point in space is calculated by a wave equation. The wave ψ has no mass or energy of its own—a rain of photons cannot scatter from this psi wave to give an image, as they do from a cloud of smoke over a fire or from a billow of dust above a dirt road. However, the *square* of the wave at any point gives the electron's probability for being at that point, and that probability is detected in experiments. The picture of the atomic "corral" at the beginning of this chapter was made from detecting psi-squared for the electrons in those atoms, by means we'll discuss later.

PHYSICS IN 3D/ELECTRON PROBABILITY DISTRIBUTION

A 45° SLICE OF THE PROBABILITY DISTRIBUTION FOR THE ELECTRON IN THE 1ST EXCITED STATE OF THE HYDROGEN ATOM. (THINK OF SLICING AN ORANGE FROM THE TOP INTO EIGHT EQUAL PIECES.) THE NUCLEUS IS AT THE LEFT EDGE OF THESE DOTS, AT THE CENTER OF THE EDGE. THIS 3-D DISTRIBUTION IS FOR THE 2-D DISTRIBUTION IN FIG. 25-2(B). NOTICE THE EMPTY "SHELL" NEAR THE NUCLEUS WHERE THE STANDING WAVE HAS A NODE.

PHYSICS IN 3D/ELECTRON PROBABILITY DISTRIBUTION

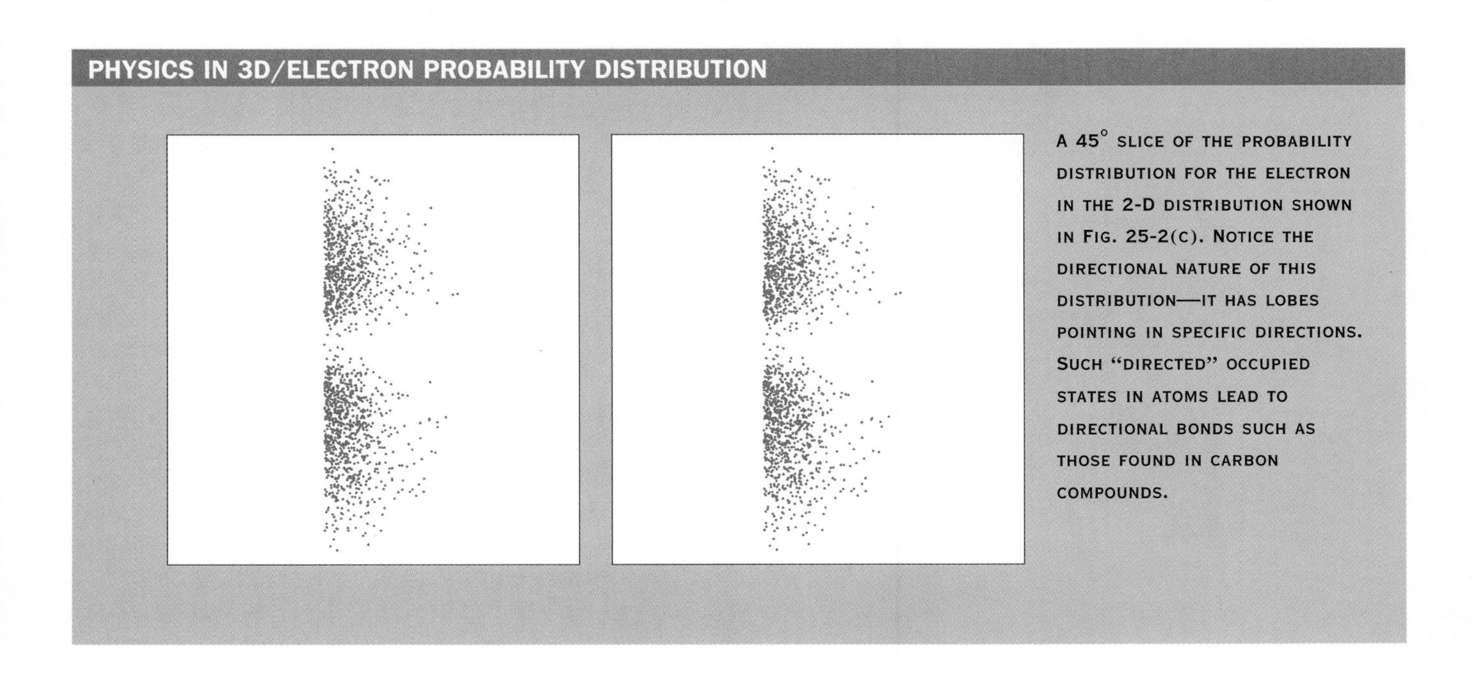

A 45° SLICE OF THE PROBABILITY DISTRIBUTION FOR THE ELECTRON IN THE 2-D DISTRIBUTION SHOWN IN FIG. 25-2(C). NOTICE THE DIRECTIONAL NATURE OF THIS DISTRIBUTION—IT HAS LOBES POINTING IN SPECIFIC DIRECTIONS. SUCH "DIRECTED" OCCUPIED STATES IN ATOMS LEAD TO DIRECTIONAL BONDS SUCH AS THOSE FOUND IN CARBON COMPOUNDS.

Here is the interpretation of the hydrogen atom according to wave mechanics. Bohr's orbits are gone. A path for the electron is not to be found in wave mechanics. The only thing wave mechanics tells us is the probabilities for the electron's appearance. If at some instant the electron in a hydrogen atom *absorbs* a quantum of energy, its pattern of probability density changes to another pattern of probability density representing the higher energy and increasing the electron's probability of being found farther from the nucleus. If the atom *emits* a photon, the electron abruptly loses energy and is then represented by a probability density that is concentrated closer to the proton. The standing wave clouds of probability that replaced Bohr's orbits are still called **orbitals** despite the fact that they have nothing to do with orbits.

THE STRUGGLE TO UNDERSTAND WAVE MECHANICS

Bohr's institute at Copenhagen, Denmark, soon emerged as a center for interpreting the new wave mechanics. He and his colleagues came to the belief, called the **postulate of wave mechanics**, that the wave ψ for a particle, as calculated by Schrödinger's equation, contains *all* the information we can obtain about the particle's position, momentum, and energy.

At this point, Einstein, Schrödinger, Planck, and de Broglie all objected. They could not believe it is impossible to predict precisely where to find an electron in an atom, that probabilities for finding the positions of such small particles are all anyone can ever know. They thought there must be another interpretation of the theory of wave mechanics, or perhaps more to be learned. Schrödinger's equation replaced Bohr's jumping electron with a progression of standing waves, but the "densities" of his waves were only patterns of probability. The electron was a pointlike particle that must still "jump" when its probability pattern changed, emitting or absorbing a photon in the process. Schrödinger said to Bohr, "Had I known that we were not going to get rid of the damned quantum jumping, I never would have involved myself in this business." As for the probabilities that the wave formula predicted, Einstein protested, "God does not play dice . . ." Bohr's answer was, "How do you know?"

Bohr's interpretation of the wave cloud accurately predicts the results of experiments at the subatomic and atomic levels. Let's look now at a hypothetical electron-scattering experiment to illustrate how the theory of wave mechanics makes predictions about the behavior of particles.

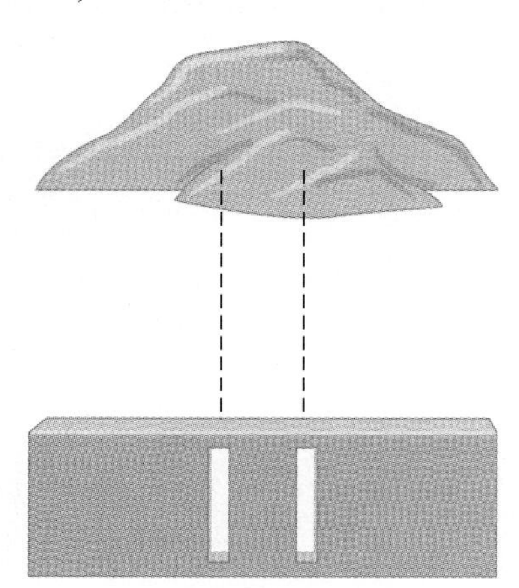

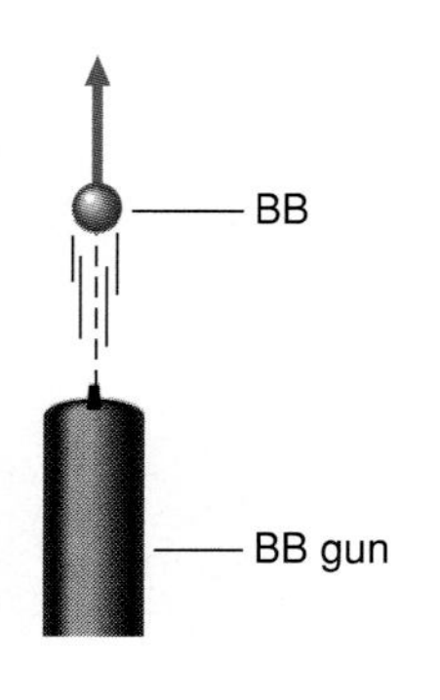

FIGURE 25-3 A BB GUN FIRES BBS AT A PLATE HAVING TWO SLITS IN IT. BBS PASSING THROUGH THE PLATE BURY THEMSELVES IN A PILE OF SAND, WHERE THEY CAN BE FOUND LATER AND THEIR NUMBERS PLOTTED ON GRAPHS, AS IN FIGURE 25-4.

A "THOUGHT" EXPERIMENT WITH ELECTRONS

This "thought" experiment lets you see how strange the behavior of particles on the atomic level is when compared with the behavior of objects such as books and baseballs. Proposed by the physicist Richard Feynman (1918–1988), it uses electron ψ waves in an analogy with Young's experiment showing the wave nature of light (Chapter 21). Recall from Figures 21-1 through 21-3 that a coherent light wave, diffracting as it goes through two slits, causes an interference pattern on a screen some distance away. Feynman's example sends both macroscopic objects and electrons through the slits.

First, imagine a steel plate having two slits in it. In front of this plate is a BB gun, and behind it is a large pile of sand (Figure 25-3). Random shooting of many BBs at the plate with the left slit closed results in the BBs distributed in the sand as shown by the graph in Figure 25-4a. If the shooting occurred with the right slit closed instead, you'd find the distribution shown in Figure 25-4b. If both slits were open for the entire shooting, you'd find the sum of these two distributions of BBs in the sand, as in Figure 25-5.

Next, imagine a submicroscopic setup like this with an electron gun firing one electron at a time toward a double slit. Replace the pile of sand with a screen that emits a flash when and where an electron hits it. With the left slit closed, you'd find a distribution like that in Figure 25-4a for the many electrons arriving at the screen, and with the right slit closed, you'd find a distribution like that in Figure 25-4b. The wave equation says that the psi wave for an electron will diffract at each slit, spreading out behind the slit. If both slits are open while many electrons are "fired," however, the distribution of electrons arriving at the screen would not look like that in Figure 25-5. Instead, it would look like the distribution shown in Figure 25-6: an interference pattern just like Young found for light waves.

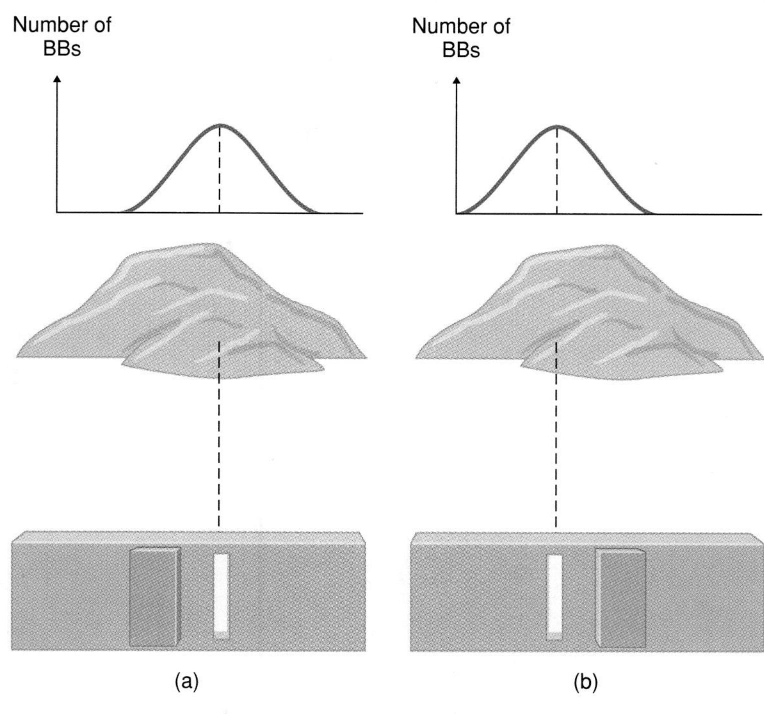

FIGURE 25-4 (A) A PLOT OF WHERE THE BBS ARE FOUND AFTER MANY PASSED THROUGH THE RIGHT SLIT WHILE THE LEFT SLIT WAS CLOSED. (B) A PLOT OF THE BBS' NUMBERS AND POSITIONS AFTER MANY PASSED THROUGH THE LEFT SLIT WHILE THE RIGHT SLIT WAS CLOSED.

Here is how wave mechanics explains this result. There is no "trajectory" for any electron leaving the gun, just as there is no orbit for an electron in an atom. The ψ wave representing each electron can be found with Schrödinger's equation, and behind each slit the ψ wave diffracts as the light waves in Young's experiment did. When the diffracted psi waves from each slit arrive at some point on the screen, they interfere. Where the value of psi-squared on the screen is high, many electrons will arrive there if hundreds or thousands are "fired" at the double slits one by one. Where psi-squared is zero or near zero, none or few electrons show up there. The point is this: The pattern of Figure 25-6 gives the probability of a *single* electron arriving at the screen; one cannot predict where within that pattern the electron will strike ahead of time, however. This probability distribution tells us everything we can know about the electron's path (or energy or momentum) in advance of its arrival. Once it arrives and strikes the screen, we "measure" its position (we see the flash) and perhaps measure its energy (from the intensity of the flash). However, we can know only probabilities for these quantities before the measurement is made. If a similar experiment is performed with single photons, a similar interference pattern appears; both light and particles exhibit wave behavior.

THE UNCERTAINTY PRINCIPLE

The postulate of wave mechanics, which in effect says that probability governs the world of subatomic particles, left most physicists hoping for a way out of that interpretation. They thought they should be able to track an electron

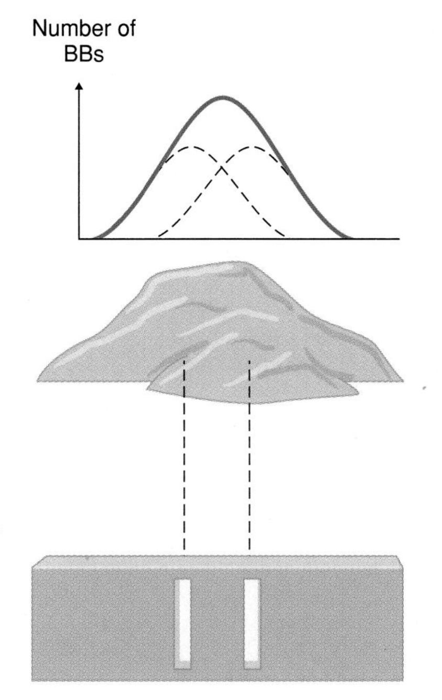

FIGURE 25-5 IF BOTH SLITS ARE OPEN AND A LARGE NUMBER OF BBS COME THROUGH, THE DISTRIBUTION IS THE SAME AS THE SUM OF THE TWO DISTRIBUTIONS OF FIGURE 25-4.

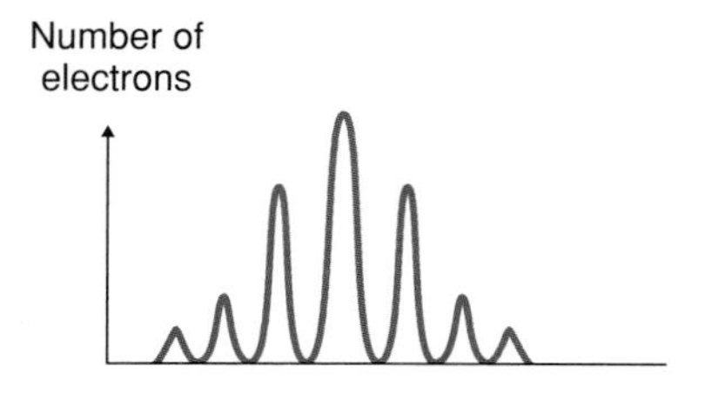

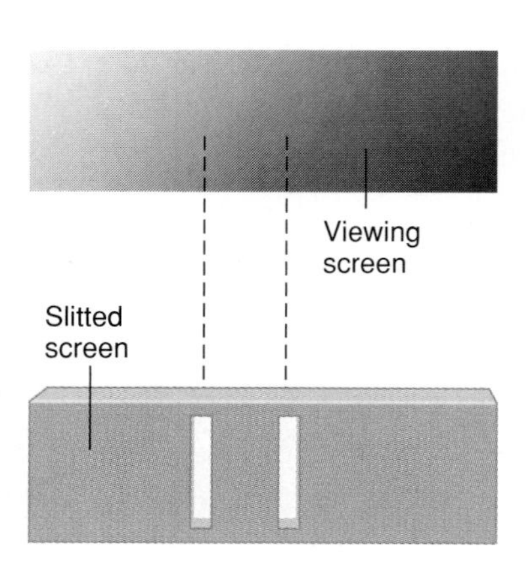

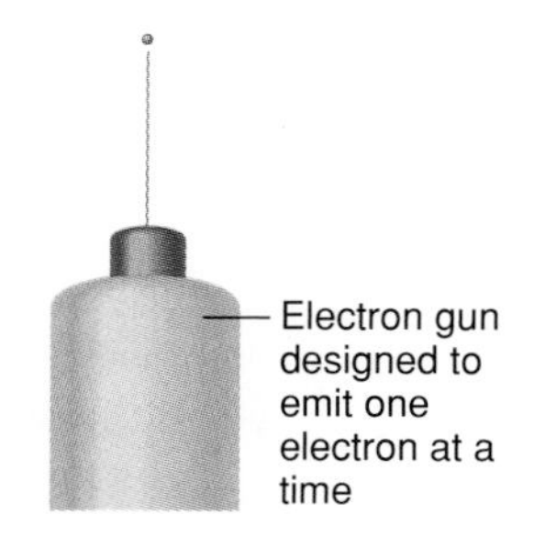

FIGURE 25-6 IF AN ELECTRON GUN FIRED ONE ELECTRON AT A TIME TOWARD THIS SCREEN WITH SLITS, THE DISTRIBUTION OF THE ELECTRONS ARRIVING AT THE VIEWING SCREEN WOULD APPEAR IN AN INTERFERENCE PATTERN MUCH LIKE THE ONE YOUNG FOUND WHEN HE SENT LIGHT THROUGH A DOUBLE SLIT. THAT PATTERN IS MUCH THE SAME AS THE PROBABILITY DENSITY AT THE SCREEN FOR A SINGLE ELECTRON'S ψ WAVE, SQUARED. ELECTRONS WILL NOT GO TO THE AREAS WHERE THE DISTRIBUTION IS ZERO—NO FLASHES WILL APPEAR THERE. AN ELECTRON'S WAVE INTERFERES DESTRUCTIVELY WITH ITSELF AT THOSE POSITIONS. WHILE THE ELECTRON MUST PASS THROUGH ONE SLIT OR THE OTHER, ITS WAVE AS DETERMINED BY THE SCHRÖDINGER EQUATION TAKES INTO ACCOUNT THE POSSIBILITY THAT THE ELECTRON COULD PASS THROUGH EITHER SLIT ON ITS WAY TO THE VIEWING SCREEN.

along a single trajectory, since electrons are always detected at a single point, never "spread out." Werner Heisenberg, a German physicist, closed the door to those hopes. He found through both experimental and theoretical results that there is a *fundamental uncertainty* in a particle's behavior.

Consider an electron, perhaps one that made it through one of those slits in our previous thought experiment. We know photons interact with electrons (Chapter 24), so suppose we flood the area with light to try to track the electron as it goes through the slit. If a single photon "bounces" from the electron, and if we can then detect that photon, we should know the direction of the electron's position. However, that is not the case. The electron and photon both obey the laws of conservation of momentum and energy, so if the photon is deflected by its "collision" with the electron, *so is the electron*! The act of finding the electron disturbed it, so we don't know its position very well at all. If it had been heading for the slit, the very act of finding it deflects it so that it might not make it through. Heisenberg found there is always an uncertainty in measured quantities such as position, momentum, and energy whenever a measurement is made on an atomic system, and that those uncertainties are paired. For example, the position and momentum of a particle in one dimension have uncertainties given by:

uncertainty in position times uncertainty in momentum is greater than or equal to Planck's constant

$$\Delta x \cdot \Delta mv \geq h$$

The more certain the position is (smaller Δx), the greater the uncertainty in the particle's momentum (Δmv), and vice-versa. This **uncertainty principle** comes from the nature of the waves that must be used to predict the motion of matter's smallest particles. The more accurately you know a particle's position, the less you can know about its momentum (hence its speed). Likewise, if you know its momentum almost exactly, you can have almost no idea where the particle is.

AN EXAMPLE OF THE UNCERTAINTY PRINCIPLE

Consider an experiment that tries to aim the electrons in a beam at a single atom on the face of a crystal (Figure 25-7). If two plates, each with one small pinhole in it, block the path of the beam, the electrons that get past both plates have traveled along a straight line toward a single atom on the crystal (if the pinholes are small enough). Even then, however, many electrons that make it through both pinholes miss the target atom and hit another, as we can see by the following analysis.

The momentum that moves an electron along in the line of the beam can be known very accurately because the electron's position along the line is never known. (There aren't any detectors along the way to pin down its location.) However, each electron must pass through the second pinhole to reach the crystal. At the instant it does, we know its horizontal position to within Δx, the width of the pinhole. The uncertainty principle tells us what has to happen to the electron's ψ wave.

The small value of Δx causes the momentum *in that direction* to have a large uncertainty. That's because the electron's ψ wave spreads out in the x-direction, or diffracts, as it travels toward the crystal face, as Figure 25-8 shows. When the ψ wave arrives at the crystal, it encounters all the atoms in the vicinity of the target atom. The electron has a probability of hitting any one of them! The attempt to aim the electrons is foiled by the uncertainty principle.

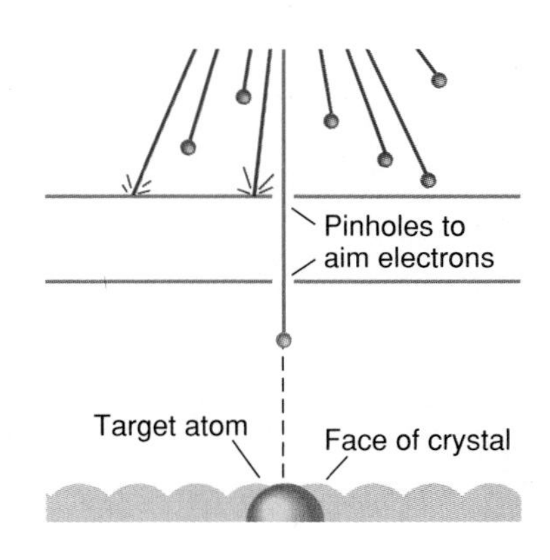

FIGURE 25-7 PINHOLES IN A PAIR OF PLATES ALLOW ELECTRONS TO PASS ONLY IF THEY ARE TRAVELING IN THE APPROPRIATE DIRECTION. THIS SETUP IN EFFECT "AIMS" THE ELECTRONS AT A POINT ON THE SURFACE OF ATOMS SHOWN HERE.

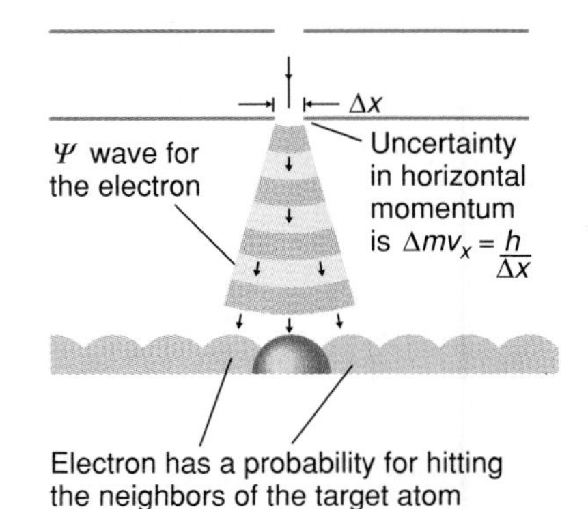

FIGURE 25-8 THE UNCERTAINTY PRINCIPLE INTERFERES WITH AN ATTEMPT TO AIM AN ELECTRON AT AN INDIVIDUAL ATOM. KNOWING THE POSITION OF THE ELECTRON WELL IN THE HORIZONTAL DIRECTION MEANS ITS MOMENTUM (HENCE SPEED) IN THAT DIRECTION HAS A LARGE UNCERTAINTY. THAT IS, THE ψ WAVE DIFFRACTS AT THE OPENING, AND THE ELECTRON MAY TRAVEL AT A CONSIDERABLE ANGLE TO ITS ORIGINAL STRAIGHT-LINE PATH. THE LIGHT AND DARK SHADED AREAS REPRESENT THE CRESTS AND TROUGHS OF THE ELECTRON'S ψ WAVE.

The sizes of the uncertainties in momentum or position are connected by Planck's constant, h. This is such a small number that, for large collections of atoms such as planets or people or even bacteria, the uncertainties cannot be measured. For individual electrons and protons, though, and sometimes even for whole atoms, uncertainty plays a large part in their behavior. One cannot know the exact position and momentum of such a particle simultaneously.

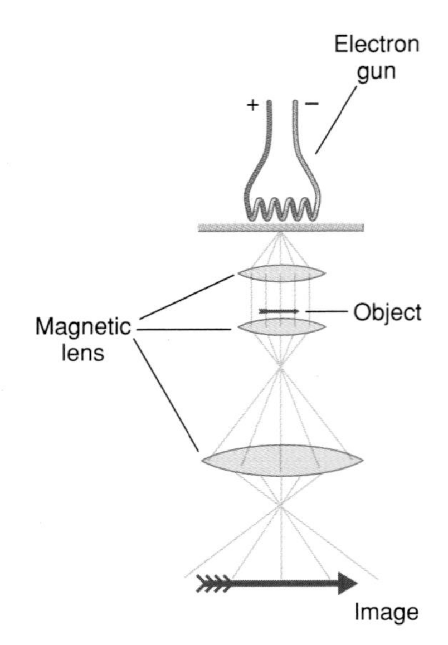

FIGURE 25-9 A SCHEMATIC OF A TRANSMISSION ELECTRON MICROSCOPE.

USING AND SENSING THE WAVE PROPERTIES OF ELECTRONS: ELECTRON MICROSCOPES

If a target is bombarded with many electrons all having the same de Broglie wavelength, the distribution of electrons that results from scattering can be used to form an image of the target much as the distribution of photons that strike the film in a camera is used to form an image. This is how a *transmission electron microscope* works. The electron microscope must furnish a vacuum for the electrons to move through or else the lightweight electrons would scatter from air molecules, and magnets rather than glass lenses must focus the electrons, as Figure 25-9 shows. The short wavelengths of the electrons, typically 0.005 nm (1 nanometer $= 10^{-9}$ m) or so, make it possible to form a sharp image at high magnification. Light waves, X-rays, and electron waves cannot reveal any details of objects that are about the size of a wavelength or less, because diffraction occurs in addition to reflection, smearing the image. Because of the short wavelengths of the electrons, a transmission electron microscope can magnify up to 50,000 times, whereas optical microscopes have a useful magnification of around 500× (Figure 25-10).

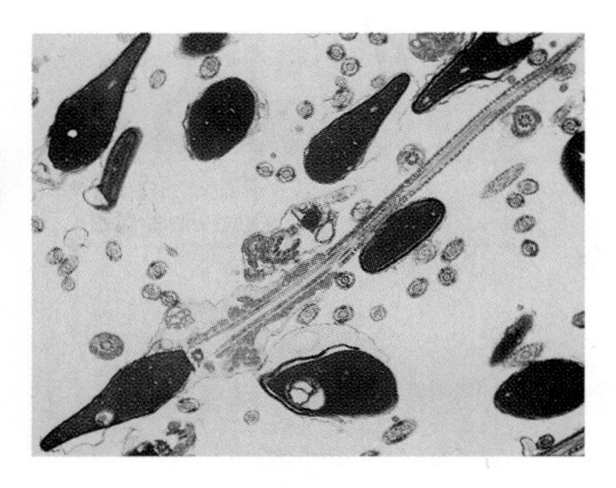

FIGURE 25-10 TRANSMISSION ELECTRON MICROGRAPH OF HUMAN SPERM IN VARIOUS SECTIONS. THE CENTRAL SPERM HERE SHOWS ITS HEAD, NECK AND TAIL REGIONS.

The *scanning electron microscope* scans the surface of the target with an electron beam. The beam knocks electrons from the surface of the target, and those "secondary" electrons are detected and used to form an image that is almost three-dimensional (Figure 25-11). Both types of electron microscopes are giving researchers in the life sciences extremely realistic pictures of the microscopic world.

A third type of microscope—this one invented in the 1980s and called the *scanning tunneling microscope*—uses an extremely sensitive electrical device to move a sharp tip (one or a few atoms across) just above the surface of a specimen. A tiny voltage induces electrons to leave the specimen and come to the tip. The tip is moved up and down as it makes parallel passes over the specimen in a way to keep the current, called a *tunneling* current, constant as it moves (Figure 25-12). The tip in essence "surfs" over the electron probability waves at the surface, the constant current meaning it stays a constant distance from a certain value of the electron probability density there. The opening illustration of this chapter, then, is a plot of the surface probability densities of the electrons on the copper surface. The tip can be stopped, the voltage changed, and a surface atom can be tugged along by the tip to another surface location. The IBM researchers who made that "quantum coral" of iron atoms in the opening illustration at the same time made a barrier for conduction electrons that can move over the surface. You can see standing waves of their probabilities trapped inside the circle of atoms. The same sort of pattern can be seen in a full coffee cup; place it on a counter and hit the counter gently with the heel of your hand. The standing waves you see are caused by reflections from the circular boundary of the cup.

FIGURE 25-11 SCANNING ELECTRON MICROGRAPH OF SPERM AND EGG. MAGNIFICATION IS ABOUT 3000X.

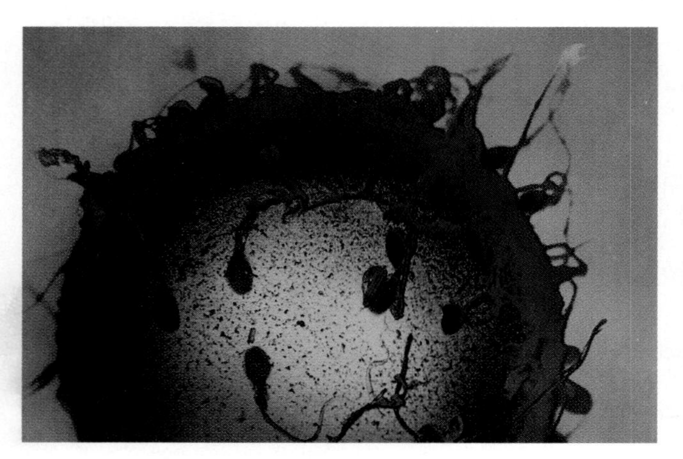

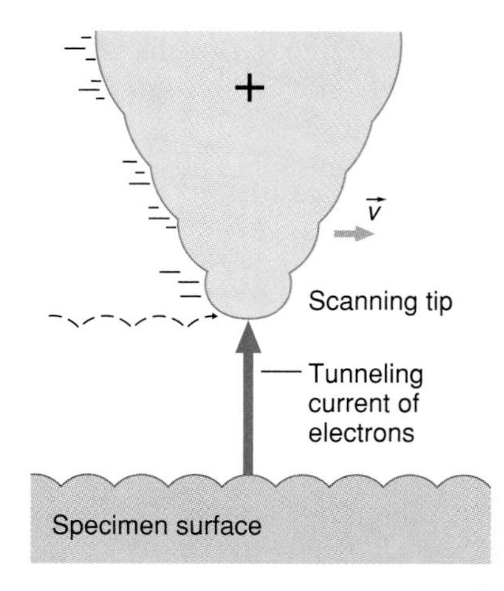

FIGURE 25-12 THE TIP OF A SCANNING TUNNELING MICROSCOPE. AS THE TIP PASSES OVER, OR SCANS, THE SURFACE IN PARALLEL LINES (MUCH AS A TV PICTURE IS PRODUCED ONE LINE AT A TIME), THE TIP MOVES UP AND DOWN, KEEPING THE TUNNELING CURRENT CONSTANT. IN THIS WAY A THREE-DIMENSIONAL IMAGE OF THE "SURFACE" IS FORMED.

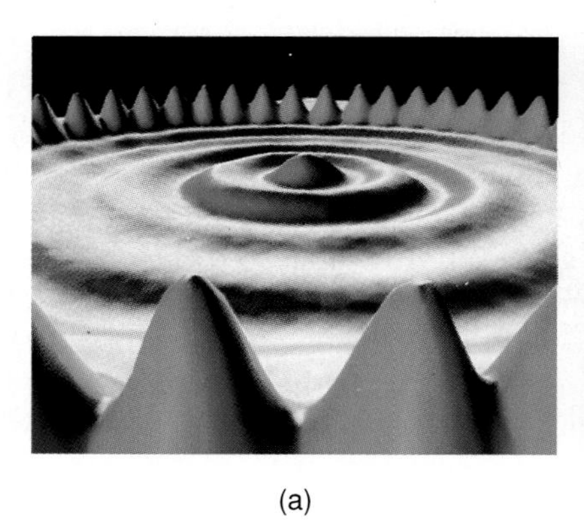
(a)

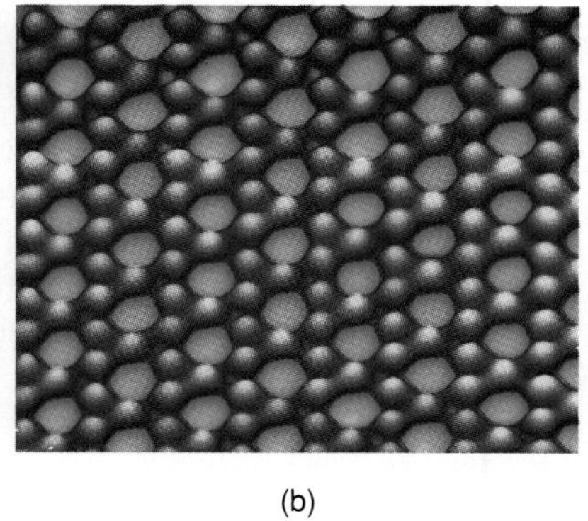
(b)

(A) AN OBLIQUE VIEW OF THE QUANTUM CORRAL ON THE CHAPTER-OPENING PHOTOGRAPH, SHOWING THE SURFACE WAVES OF ELECTRON DENSITY "TRAPPED" BY THE CIRCLE OF IRON ATOMS ON THE COPPER SURFACE. (B) IODINE ATOMS (GREEN) REGULARLY SPACED ON THE CRYSTAL OF PLATINUM (PINK).

QUANTUM MECHANICS

The **state** of an electron in an atom describes its physical properties, and all of them are *quantized*. That is, they can be represented by numbers called **quantum numbers**, which have a set of allowed values just as the old Bohr orbits did. In fact, one part of the state of an electron in an atom is its ψ wave's "shell"; electrons in the same shell have about the same average distance from the nucleus and about the same energy. Another part of each electron's state is the angular momentum the electron can have in its "orbital" around the nucleus, and also the orientation of that angular momentum. (For a two-dimensional example of orientation, think of clockwise or counterclockwise.) Another part involves the electron's magnetic properties—the electron has a tiny magnetic field. The quantized values for the shell, the angular momentum, and the angular momentum orientations are all predicted by the Schrödinger equation. The magnetic orientation, called *spin*, is not, but information taken from spectral lines revealed that it has only two values, which are arbitrarily called *up* and *down*.

IN 1927, LESS THAN TWO YEARS AFTER SCHRÖDINGER FOUND HIS EQUATION, P.A.M. DIRAC, AN ENGLISH PHYSICIST, COMBINED QUANTUM MECHANICS WITH RELATIVITY (CHAPTER 26) AND THE RESULTING *DIRAC EQUATION* PREDICTED THE SPIN OF THE ELECTRON.

A major failure of the Bohr theory was that it could not predict spectra of elements other than hydrogen—that is, atoms containing more than one electron. The Austrian physicist Wolfgang Pauli studied the spectral lines of multi-electron atoms, and in 1925 he found another principle that let wave mechanics predict the electron states, and hence the spectral lines, of those atoms. Called the **Pauli exclusion principle**, it says:

No two electrons in an atom can be in the same state at the same time.

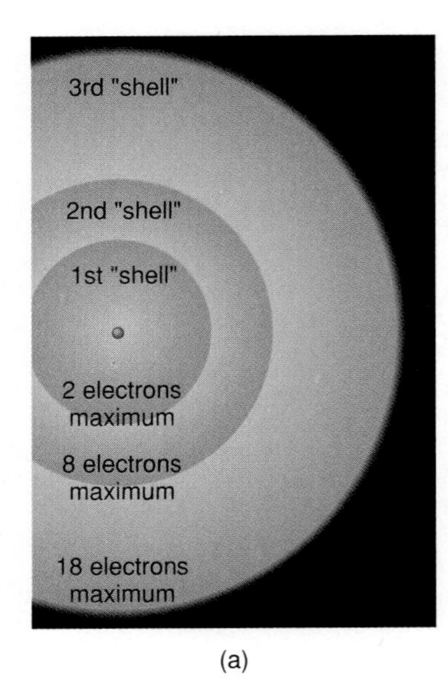

(a)

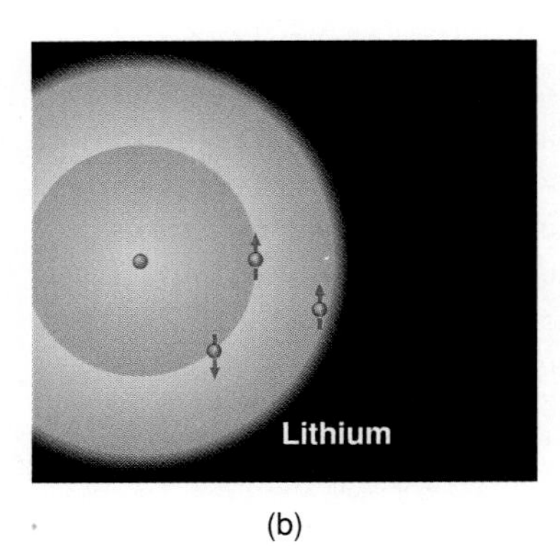

(b)

FIGURE 25-13 (A) THE MAXIMUM NUMBER OF ELECTRONS IN EACH SHELL, AS DETERMINED BY THE EXCLUSION PRINCIPLE. NO TWO ELECTRONS (IN SUCH A CLOSE SYSTEM WHERE THEY INTERACT) CAN HAVE ALL THEIR QUANTUM NUMBERS THE SAME. (B) THE GROUND-STATE "CONFIGURATION" OF THE ELECTRONS IN A LITHIUM ATOM. THE ELECTRONS ARE SYMBOLIZED WITH ARROWS THROUGH THEM TO SHOW THEIR "SPINS" AS UP OR DOWN.

Whereas two identical photons can travel precisely together, electrons cannot. When one electron occupies a state, all other electrons are excluded from that state.

Because the physical properties of an electron in an atom are quantized, the complete theory of electrons in atoms is called **quantum mechanics**. We won't elaborate on the quantum states here except to note that the possible states in each shell are limited. For example, the shell closest to the nucleus has two possible electron states (the electrons must have opposite spins to ensure that the states are not the same). The second shell has eight possible states because the electrons can have different values of angular momentum, and the third shell has eighteen (Figure 25-13a). Helium, with two electrons, has a full inner shell when in its ground state. Lithium, with three electrons, has a full inner shell and one electron in its second shell (Figure 25-13b). Neon, with ten electrons, has the first two shells filled when in its ground state. The chemical properties of the elements are determined by the electrons in the outermost shell. If the outermost shell is totally occupied, the element is chemically stable.

The exclusion principle keeps all of the electrons in an atom from occupying, for example, the lowest energy shell in the atom, the one closest to the nucleus. Once the electrons are assigned to their proper quantum states in an atom, the Schrödinger equation gives (after complicated calculations) the observed energy levels and the atomic spectrum when electrons change states by either emitting or absorbing photons. Today chemists and physicists use quantum theory to understand the chemical bonds that hold atoms together in molecules, liquids, and solids. Many effects have been predicted that would have gone undiscovered (or at least unexplained) without this remarkable theory.

BOHR BRINGS TWO THEORIES TOGETHER

Bohr realized that the predictions of any new theory of nature at the subatomic level had to apply to macroscopic particles also. Otherwise, nature would be operating with two sets of laws, one for small particles and another for large particles built from the smaller ones. If this were the case, at some point in between the physics would change abruptly.

After wave mechanics with its powerful, accurate predictive powers came into being, Bohr was able to show that, for everyday objects, where the mass and energy of an object are large, Schrödinger's equation predicts the same thing as Newton's laws do. Newtonian mechanics charts with great precision the motions and positions of moons and planets, pendulums, and falling apples. However, Newton's mechanics is an upper limit of a theory that says we cannot know an object's speed and position with complete certainty. Bohr's demand that wave mechanics and classical mechanics be logically consistent with each other was only a part of a general principle he promoted that is called the **correspondence principle**: *Any two theories or hypotheses whose areas of prediction overlap should agree at those places; otherwise at least one of them is wrong.*

A LOOK BACK

Newton's laws of motion and his law of gravity were the very foundations of physics. Successful predictions followed whenever they were applied in the 1700s and 1800s, and the universe seemed to run like a clock at every level.

For example, once the positions of the planets and their velocities were known at some instant, Newton's laws let physicists trace planetary histories both forward and backward in time. (The sole exception is a tiny precession of Mercury's orbit, which was explained by Einstein's general theory of relativity in 1916.) Even historical eclipses could be predicted to check with old records. Physicists and philosophers alike naturally assumed atoms and molecules would follow the same rules.

The advent of wave mechanics, which contradicts this classical, Newtonian view, brought with it a reminder: Human beings produce the theories of physics. They use imagination to build abstractions that they pit against the facts of nature observed by themselves and other human beings. Scientific theories are upheld only if they agree with the known facts and predict others. The abstract models of wave mechanics and the quantum theory of light passed tests that Newton's laws of mechanics did not. Historians of science have marveled that this revolution in understanding came to the most fundamental of the sciences in so short a time.

Summary and Key Terms

Matter, like light, has a dual nature, behaving both like particles and like waves. De Broglie found that a particle's wavelength, λ, is equal to Planck's constant, h, divided by the momentum of the particle: $\lambda = h/mv$. Schrödinger's ***wave equation*** predicts the amplitude, ψ, of the waves for electrons in an atom or in other situations. An atomic electron's ψ wave changes from one standing-wave pattern to another to emit or absorb a photon. An ***orbital*** is a three-dimensional standing-wave pattern whose amplitude squared at a point gives the electron's probability for appearing at that point.

wave equation–378

orbital–380

Wave mechanics describes the wave behavior of subatomic particles, and its ***postulate*** assumes that a wave describes as completely as possible a particle's position, momentum, and energy. Probabilities for finding an electron's position (and momentum and energy) are all we can know. If we know an electron is localized between two points along direction x, then the distance between those points, Δx, is the uncertainty of the electron's position. A particle's wave nature prevents us from knowing its precise position and momentum simultaneously, a fact known as the ***uncertainty principle***. The product of uncertainty of position and uncertainty of momentum is greater than or equal to Planck's constant: $\Delta x \cdot \Delta mv \geq h$.

wave mechanics–378

postulate of wave mechanics–380

uncertainty principle–383

Wave mechanics can predict the electron ***states***, and hence the spectral lines, of atoms other than hydrogen as a result of the ***Pauli exclusion principle***: No two electrons in an atom can be in the same state at the same time. Electron states describe quantized physical properties, including energy, angular momentum, and spin, which are represented by ***quantum numbers***. The complete theory of electrons in atoms is called ***quantum mechanics***. When the electrons are placed in their proper quantum states according to the exclusion principle, the wave equation gives their observed energy levels and corresponding atomic spectra when they emit and absorb photons.

state–385

Pauli exclusion principle–385

quantum numbers–385

quantum mechanics–386

The ***correspondence principle*** says that any two theories whose areas of prediction overlap should agree at those places, or at least one of them is wrong. Bohr proved that, for large objects, wave mechanics and Newton's laws give the same results. Newton's laws do not make accurate predictions at the atomic level, however, while wave mechanics agrees with the known facts and predicts others.

correspondence principle–386

EXERCISES

Concept Checks: Answer true or false

1. De Broglie's equation for the wavelength associated with a particle of momentum mv tells us that the larger the momentum, the larger the wavelength.
2. Schrödinger's equation predicts the amplitude of matter waves, ψ, even when the particles are moving under the influence of a force.
3. Schrödinger proposed that the density of the matter wave for a particle represents the *density* of the particle. Born proposed that the density of matter waves represents the *probability* of finding the particle. They were both correct.
4. The Davisson-Germer experiment was the first to reveal the actions of matter waves by showing constructive and destructive interference in the reflection of electrons from atoms in a crystal.
5. As a consequence of matter's wave nature, a particle's momentum and position cannot be known simultaneously with absolute certainty. The uncertainty in these quantities is connected by h, a tiny constant, meaning cars and trees and pets are not significantly affected by these uncertainties.
6. A fundamental difference between wave mechanics and Newton's laws led Einstein to say, "God does not play dice. . . ," but the evidence today indicates the modern theory is correct.
7. The postulate of wave mechanics elaborated by Bohr states that the ψ wave tells everything about the state of an electron in an atom—its precise energy, position, and momentum, for example.
8. The exclusion principle plays a role in the theory of atomic structure; it excludes electrons from occupying the same shell in an atom.
9. For objects such as people and planets and even bacteria, Schrödinger's equation predicts the same motions as Newton's laws do.

Applying the Concepts

10. Which of the following are true? (a) Bohr's model for the hydrogen atom predicted the hydrogen spectrum but failed to explain why the orbiting electron did not radiate energy. (b) Bohr couldn't explain why the electron should be found only in certain orbits. (c) De Broglie set out to find a better theory for the hydrogen atom. (d) De Broglie waves explained why the electron was found only in special orbits, or orbitals. (e) Electrons on orbitals in an atom don't radiate energy because the probabilities for finding them are given by standing-wave patterns that don't change unless the electrons make transitions between orbitals.
11. Describe an experiment from the previous chapter that shows light behaves like a stream of particles when it interacts with electrons. Describe an experiment from this chapter that shows electrons have wave properties.
12. Any molecule has a wavelength associated with it according to de Broglie's equation. To increase the wavelength of the molecule, must one (a) slow it down or (b) speed it up?
13. Does de Broglie's formula give any clue as to the amplitude of the wave associated with a particle?
14. Is an electron a particle or a wave? Explain.
15. Is it more correct to say that an electron is a particle found in atoms or that an electron is a wave cloud around an atom? Explain.
16. Sound waves move through the medium of matter. What medium do ψ waves move through?
17. Give several facts that indicate a ψ wave doesn't experience friction and die out the way a wave on a clothesline does.
18. If a proton and an electron have identical momenta, how do their wavelengths compare? If they have identical speeds, how do their wavelengths compare?
19. If one neutron travels twice as fast as another, which has the longer wavelength?
20. A helium atom ($m = 4$ u) travels at the same speed as a hydrogen atom ($m = 1$ u). Which has the shorter wavelength?
21. An electron is confined to a shoebox. Can we know its speed with absolute certainty?
22. Explain why it is easier to predict future positions for a more massive particle than for a less massive particle.
23. Discuss what happens when atoms in a solid cool to near absolute zero. With less thermal motion, their positions in the solid become less uncertain. (Atoms in a solid can never give up all their motion.)
24. The physicist John Archibald Wheeler has said (as a warning against acceptance of any theory), "There is no law except the law that there is no law." Discuss this statement as it pertains to how we have come to accept the theories of wave mechanics in the face of so complete a view as Newton's mechanics was thought to be.
25. It is a result of the uncertainty principle that electrons

in the orbital closest to the nucleus cannot collapse onto the nucleus under its attractive electrical force. Explain why it would require extra energy to compress an electron in the ground state of the hydrogen atom into a smaller volume containing the nucleus. (So the uncertainty principle helps make the atom huge compared to the size of its constituents, the electrons and the nucleus.)

26. The photoelectric effect is used to turn streetlamps on at night. Would you say such lamps use a quantum effect? Do the "photoelectric cells" detect light or the absence of light?

27. Some smoke alarms use photoelectric cells (see Exercise 26), as do those elevator doors that stay open while passengers are boarding. Discuss. Do these devices use the quantum properties of light or the classical properties?

28. How would you go about telling the difference between an electron and a photon that have the same wavelength?

29. The air in the room around you doesn't give off visible light. What does this tell you about the energy of the present states of its electrons?

30. Light diffraction patterns can be seen with just a small slit between your fingers (see Demonstration 5 in Chapter 22). Why are much smaller "holes" (such as the spaces between atoms on the surface of a crystal) necessary to show similar diffraction of matter waves?

31. Compare the Schrödinger and the Bohr theories of the hydrogen atom.

32. Some heavy metal atoms emit X-ray photons in their spectra. Would you expect these photons to come from transitions between states close to the nucleus or from transitions between states far away from the nucleus?

33. If a proton and an electron had the same speed, which would have the shorter wavelength? Shorter wavelengths mean a better image can be formed for small objects; can you think of any reasons you might not want to use the particles with the shorter wavelength in a microscope?

34. At what speed would an electron have to move in order to have the same wavelength as a photon of red light (about 650 nanometers)?

35. Find the de Broglie wavelength for a 100-kg running back moving at 8 m/s. Do you think this wavelength would be useful to illustrate diffraction? Discuss.

36. Explain why you shouldn't expect the uncertainty in the position of a car in a parking lot due to its wave nature to affect the owner's ability to find it.

37. Find the wavelength of an oxygen molecule in the air that is traveling at 450 m/s. Compare this with the molecule's diameter, about 1.2×10^{-10} m.

38. What would happen if you walked through a doorway in a universe where h was 100.0? (*Hint*: calculate your de Broglie wavelength, and calculate your uncertainty in a momentum. You would diffract!)

TWENTY-SIX

Special Relativity

Albert Einstein discovered that measurements of length and time are different when made by observers who are moving with respect to each other. He also showed how momentum and energy differed from the classical formulas, and that mass itself is a form of energy.

During a class your ears and eyes get messages carried by the sound and light coming from the lecturer at the front. You aren't at all distracted by the fact that light and sound travel at very different speeds, as you might be if you were sitting high in a stadium at halftime, where the distance to the band is so much greater. There you can see a drummer hit a drum and then hear the sound later. Light moves nearly a million times faster than sound, making light appear to travel instantaneously. Just as the sound from an airplane you see high overhead seems to come from a different direction, however, the image you see of the plane does, too—even though its displacement is very, very small.

Aside from its great speed, light's motion is *fundamentally* different from other types of wave motion. In 1905, Albert Einstein published a theory based on the properties of light—the special theory of relativity—that was as revolutionary as the wave theory of matter would be. Just as de Broglie's matter waves are outside our everyday experience, so are Einstein's remarkable discoveries. His special theory of relativity, the subject of this chapter, deals with objects moving relative to each other. (Einstein's other theory of relativity, called the general theory, deals with how gravity affects time and space, as we saw in Chapter 7.)

WHAT DOES LIGHT MOVE THROUGH?

When Maxwell showed in the 1860s that light waves are oscillations of electric and magnetic fields, he was puzzled about the medium through which those waves traveled. A wave moves along a stretched Slinky because the spring is elastic; its tension tends to restore the spring from any displacement. Maxwell assumed his electromagnetic waves also moved through an elastic medium, and he called this medium the *luminiferous* (light-bearing) *ether*. In the late 1800s, physicists could only guess about the properties of this medium because no one had detected it. Since light arrives at earth from distant stars, the ether had to be almost everywhere. And the ether must be thin, they reasoned, for otherwise its drag on the earth and other planets would cause them to slow and fall into the sun. Because light's speed is so great, Maxwell and his contemporaries thought the ether must be very tense, or stiff; if you pluck a guitar string, the wave you make travels fastest when the tension is greatest. The universe, it seemed to them, was filled with an extremely thin, astonishingly stiff, invisible stuff! To clearly see their dilemma, we can look at some properties of waves in another medium.

If you've ever spent time sitting by a pool of quiet water in the woods, you've probably seen small waterbugs moving over the surface. Each kick of a bug's legs sends a wave outward in a growing circle, as Figure 26-1 shows. The water wave travels outward in every direction with the same speed—let's call it s. In a slow-moving stream rather than a still pool, a bug can kick just enough to stay in one place in the stream (Figure 26-2), and you, sitting on the stream bank, can see the effect of the current on the circular wave created by each kick. From your point of view (and that of the stationary waterbug), the edge of the wave going downstream has a speed $(s + v_{\text{current}})$ while the edge going upstream has a speed $(s - v_{\text{current}})$. The current—that is, *the motion of the medium that the wave is traveling through*—makes quite a difference to the speed of the water waves as you see them from the bank.

FIGURE 26-1 A WATERBUG KICKS ITS LEGS AND A WAVE TRAVELS OUT IN EVERY DIRECTION OVER THE STILL POND WITH THE SPEED *s*.

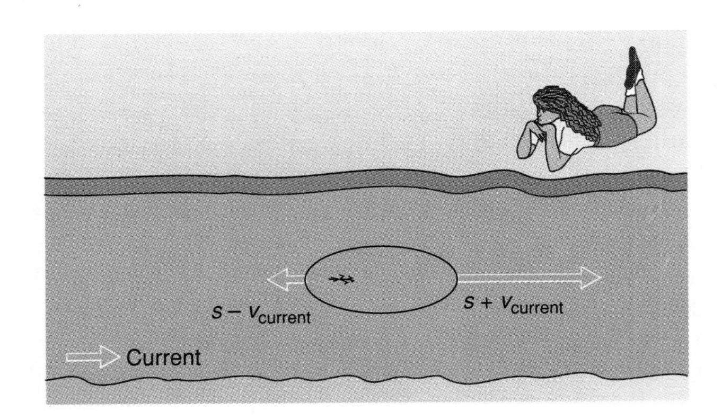

FIGURE 26-2 WAVES IN WATER, SOUND WAVES IN AIR, AND WAVES IN SOLIDS ALL MOVE EITHER ON OR IN A MEDIUM. IF THE MEDIUM HAS A MOTION OF ITS OWN, THE WAVES ARE CARRIED WITH IT. HERE, AS THE WATERBUG KICKS TO REMAIN STILL RELATIVE TO THE BANK, THE SPEED OF THE CURRENT INCREASES THE SPEED OF THE WAVE IN THE DIRECTION OF THE CURRENT AND DECREASES IT IN THE DIRECTION OPPOSITE THE CURRENT.

Physicists thought the same would be true for light waves moving through the ether; the differences in light's speed when it travels in opposite directions should show how fast the ether moves and in what direction (Figure 26-3).

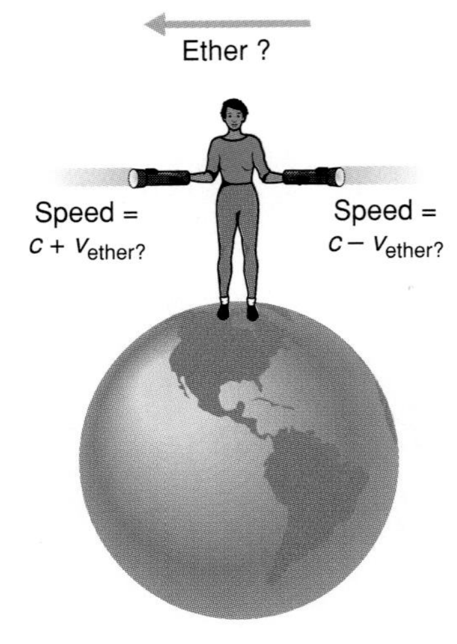

FIGURE 26-3 THE IDEA BEHIND AN EXPERIMENT TO DETECT A MEDIUM THAT LIGHT TRAVELS THROUGH. JUST AS THE SPEED OF THE WATER AFFECTS THE RELATIVE SPEED OF THE WAVE IN FIGURE 26-2, IF THE MEDIUM FOR THE LIGHT WAVES (THE SO-CALLED ETHER) WAS IN MOTION, BEAMS SHINING IN OPPOSITE DIRECTIONS SHOULD HAVE DIFFERENT SPEEDS. SUCH EXPERIMENTS HAVE NEVER FOUND A DIFFERENCE IN THE SPEED OF LIGHT TRAVELING IN DIFFERENT DIRECTIONS IN A VACUUM.

THE SEARCH FOR THE ETHER

In 1887, two Americans, Albert Michelson and E.W. Morley, built a device to measure the difference in the speeds of two light beams traveling in different directions. Any motion of the ether would affect the speed of the two beams differently. The results of their experiments were a shock; they found *no* difference in the speed of light beams moving in different directions. Light's speed was *not* changed by the motion of an ether. It seemed improbable that Michelson and Morley could have caught the ether "at rest," but they repeated the experiment six months later, when the earth was going in the opposite direction along its orbit around the sun than it was six months earlier. They even looked at light coming from a star as the orbiting earth speeded first toward the star and then away from it. Again they found no difference in the speed of light in any direction. This behavior of light is very different from anything in our experience, as the following thought experiment shows.

WHAT THE MICHELSON-MORLEY RESULTS MEAN

Imagine you are in interplanetary space and you shine a flashlight in some direction (Figure 26-4a). Measure the speed of the beam as it leaves the flashlight, and you find it is traveling at speed c. Next, leaving the flashlight at rest, you could jump into a spaceship and accelerate alongside the beam until you move at 30 percent of this speed, $0.3c$, with respect to the flashlight (Figure 26-4b). Then while coasting at $0.3c$, you could use your measuring apparatus to measure the speed of the beam again. Common sense says you should find the light traveling relative to you at $0.7c$, and all of the physics known to Galileo and Newton agrees with you. However, the results of Michelson and Morley tell us that

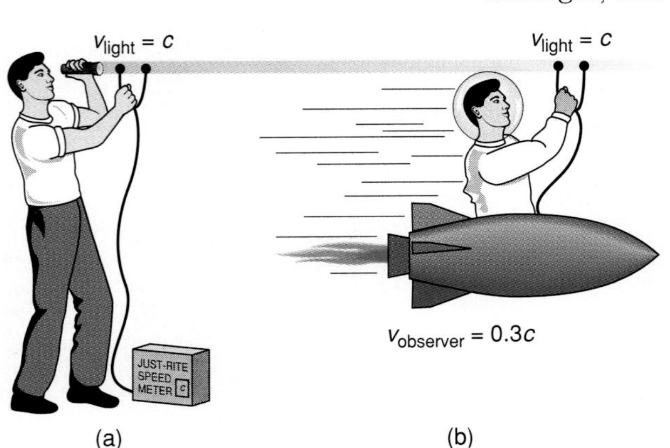

FIGURE 26-4 THE OBSERVER IN THIS THOUGHT EXPERIMENT FINDS THE LIGHT BEAM TRAVELS AT THE SAME SPEED c NO MATTER WHETHER HE IS (A) AT REST RELATIVE TO THE FLASHLIGHT, OR (B) MOVING ALONGSIDE THE LIGHT BEAM AT A SPEED EQUAL TO $0.3c$.

the light speed you measure when you are traveling at $0.3c$ is NOT $0.7c$. The beam you are traveling alongside *still* moves with the speed c. In other words, you cannot reduce the relative speed of the beam by speeding along with it. Nor can you increase the relative speed of light by turning around and speeding toward the flashlight. *The relative speeds of the source and observer do not influence the speed of light.*

Something seems terribly wrong with this statement the first time you hear it. A car doing 55 miles per hour that passes you while you are doing 50 miles per hour surely does not appear to be moving at 55 miles per hour as you watch it through your windshield—it moves ahead at only 5 miles per hour relative to *you*. No matter how fast you chase a beam of light in space, though, it will still outrun you with a relative speed c. You have never experienced anything like that, but that is what the experiments show light does.

FRAMES OF REFERENCE

Long ago, Galileo reasoned that people traveling in a ship with a perfectly constant velocity over smooth seas could not tell by on-board experiments if they were in motion. If they rolled a heavy ball, for example, it would go in a straight line with a constant speed so long as there was no net force on it (Newton's first law); it would behave exactly as if it were on land. From the passengers' point of view, the ship could be at rest with the ocean moving past it.

Suppose the ship passes an island while the on-board experiment is taking place. The islanders could watch the ball and they would agree that it follows Newton's first law. The only difference in what the islanders and the sailors see would be the addition of the relative speed between the island and the ship. The sailors and islanders see (or measure) what is happening from different perspectives, that is, from different frames of reference. However, Newton's first law of motion holds (in the horizontal direction) for both groups of observers because the two frames of reference are both *nonaccelerated*. Whenever Newton's first law holds—whenever an observer doesn't accelerate (speed up, slow down, or change direction)—we say the observer is in an **inertial frame of reference**. It is called this because in it Newton's law of inertia, his first law of motion, holds true.

Galileo's observation can be put into simple words: *All inertial frames of reference are equivalent; that is, Newton's laws predict the motion of things equally well in any inertial reference frame.* This idea agrees with our experience. Suppose a train traveling at 55 miles per hour on smooth and level railroad tracks passes a car that is stopped at a crossing. At the moment the train passes, the passengers in the car and the train can look out and see each other behaving normally. Passengers on the train could be walking in the aisles, or even pouring a cup of tea, because the laws of physics are exactly the same aboard that steadily moving train as they are for the passengers in the stopped car.

THE TWO POSTULATES OF RELATIVITY

The following statement is the **first postulate of relativity**, known to Galileo and Newton and restated by Einstein in 1905:

The laws of physics are identical in all inertial frames of reference.

Einstein's **second postulate of relativity** did away with the need for an ether:

The speed of light in a vacuum has the same value, *c*, in all inertial frames of reference.

Unlike other waves, light (and all other electromagnetic waves, too) travel through space without a supporting medium of any sort. Apparently there is no medium at all in the vacuum of space. Einstein realized that the unchanging speed of light meant something was wrong with some of our "common sense" ideas. Then he showed how light's unchanging speed modified the old ideas of time and space, since speed is a measured distance divided by a measured time. Thought experiments can show us the results of his **special theory of relativity**, which relates distances and time intervals (and more) between different inertial frames of reference.

STRANGE RESULTS

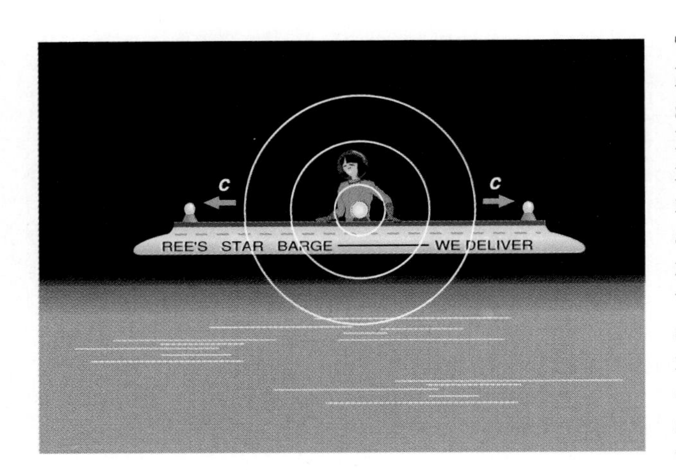

FIGURE 26-5 FROM THE CENTER OF HER SHIP, REE SEES THE LANDSCAPE BELOW HER MOVING AT A CONSTANT SPEED. A BURST OF LIGHT TRAVELS OUTWARD FROM THE FLASHBULB AT THE SHIP'S CENTER AND REACHES THE LIGHT DETECTOR-CLOCKS ON THE ENDS AT EXACTLY THE SAME TIME.

Two people take part in this thought experiment, call them Ree and Ray. Ree is positioned at the center of a long spaceship moving low over level ground on an airless planet, as Figure 26-5 shows, and Ray is standing on the surface, watching the ship pass. At each end of the ship, at equal distances from Ree, are light detectors connected to very accurate atomic clocks that keep precisely the same time (we'll say more about this shortly). Her ship moves over the ground with a uniform speed that is an appreciable fraction of the speed of light, say 50 percent or more. If Ree fires a flashbulb, it sends a burst of light out in every direction. The detector-clocks at each end of the ship record the time when this light strikes them. Later, she checks their records, and she finds that each detector received the light at the same instant. The light, traveling at speed c, covered the equal distances in equal times.

Suppose there is a row of identical detector-clocks on the ground beneath the flight path, with their detectors turned upward. Imagine a flagpole in the middle of the row. As Ree flies over, she could cause another bulb to flash directly above the tip of the flagpole (Figure 26-6). Once again the light spreads in all directions at the speed c and illuminates the two detectors on the ship. The ground detectors closest to those on the ends of the ship could record the exact time the light arrived at the ship detectors. Ray could then walk around and check their records. He'd discover that, according to the ground detectors, the light did *not* arrive at the ship's two detectors at the same instant. According to the ground detectors' clocks, the rear detector on the ship was illuminated before the forward detector was.

How did this happen? When the flash goes off at the point above the flagpole, even though it is given off by a bulb moving along at a speed v_{ship},

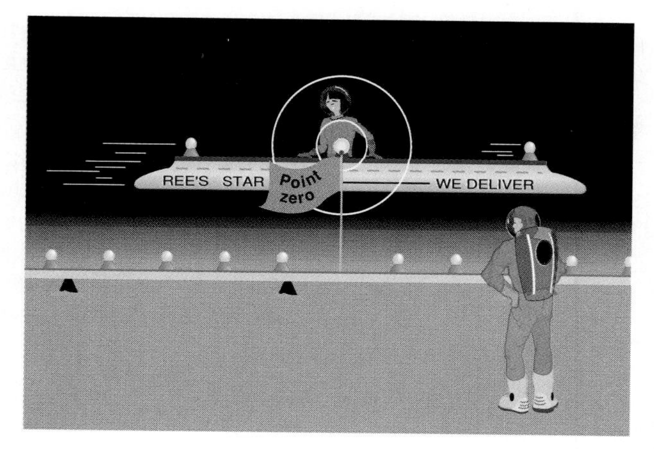

FIGURE 26-6 AN OBSERVER ON THE GROUND MEASURES THE BURST OF LIGHT CREATED ON THE SHIP AS IT PASSES THE FLAGPOLE. THAT LIGHT TRAVELS OUTWARD AT SPEED *C* IN EVERY DIRECTION AS MEASURED BY THE DETECTORS ON THE GROUND, ACCORDING TO THE SECOND POSTULATE OF RELATIVITY.

the burst of light passes the ground detectors at the speed c in each direction. As seen from the ground, the ship's rear detector rushes toward this uniformly expanding shell of light and the ship's forward detector moves away from it (Figure 26-7). The light hits the ship's rear detector first because the light goes a shorter distance than it does to catch the forward one. Hence, the ground detectors see the ship's rear detector illuminated first. Yet the two on-board detectors register the light simultaneously as seen on their clocks. This lack of agreement is due to the second postulate of relativity. *Light travels at the same speed over the ground as it does over the ship.*

At this point you may say, "Some of the clocks must be wrong," or "The clocks weren't set to the same time to begin with." So let's consider just what it means to tell time at different places and to set clocks to read the same.

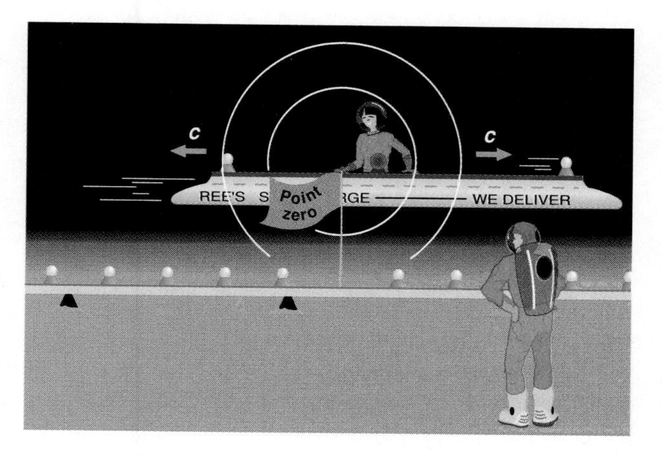

FIGURE 26-7 AFTER A FRACTION OF A SECOND, THE SHIP HAS MOVED TO THE RIGHT AND THE BURST OF LIGHT HAS SPREAD OUTWARD IN ALL DIRECTIONS. AS DETECTED FROM THE GROUND, HOWEVER, THE CENTER OF THE BURST IS STILL OVER THE FLAGPOLE. NOTE THAT THE SPREADING WAVEFRONTS HAVE REACHED DETECTORS THAT ARE AT EQUAL DISTANCES FROM THE POLE. THE GROUND DETECTORS ALSO RECORD THE TRAILING END OF THE SHIP AS RECEIVING THE WAVEFRONTS BEFORE THE LEADING END OF THE SHIP DOES. COMPARE THIS TO FIGURE 26-5. OBSERVERS IN DIFFERENT FRAMES OF REFERENCE SEE THE LIGHT REACH THE END POINTS OF THE SHIP AT DIFFERENT TIMES *BECAUSE LIGHT'S SPEED IS INDEPENDENT OF THE SPEED OF THE LIGHT SOURCE OR THE SPEED OF AN OBSERVER.* (*NOTE:* THE DRAWINGS REPRESENTING LIGHT WAVES AND CLOCKS, FIGURES 26-7, 8, 9, AND 10, ARE EXAGGERATED TO SHOW THE EFFECTS OF RELATIVITY.)

WHAT IT MEANS TO KEEP THE SAME TIME

Suppose you and a friend put your watches side by side. If the watches read the same time at the same instant and stay precisely together as they tick off the seconds, we say they are keeping the same time. Then suppose your friend goes 3 km across town, to a point where you could see her or him with a telescope. How could you be sure that the two watches still agree? When you stood together and compared watches, you both were in the same inertial reference frame. Because your friend moved away at some speed, he or she *changed* inertial frames, however. When the relative motion stopped, your friend rejoined your inertial frame, but from what we have just seen about moving clocks, we might wonder if the two watches still agree.

Here is Einstein's prescription for making sure two separated clocks keep the same time. (He took great care in defining this notion when he published his theory in 1905.) Suppose you use a telescope to look at your friend's watch. You'd find it is ticking off seconds identically to your own. If that watch is really keeping the same time as yours, however, it will show a time different from what your watch says at any given instant. This is because the light you see from your friend's watch must travel 3 km to the telescope, and this

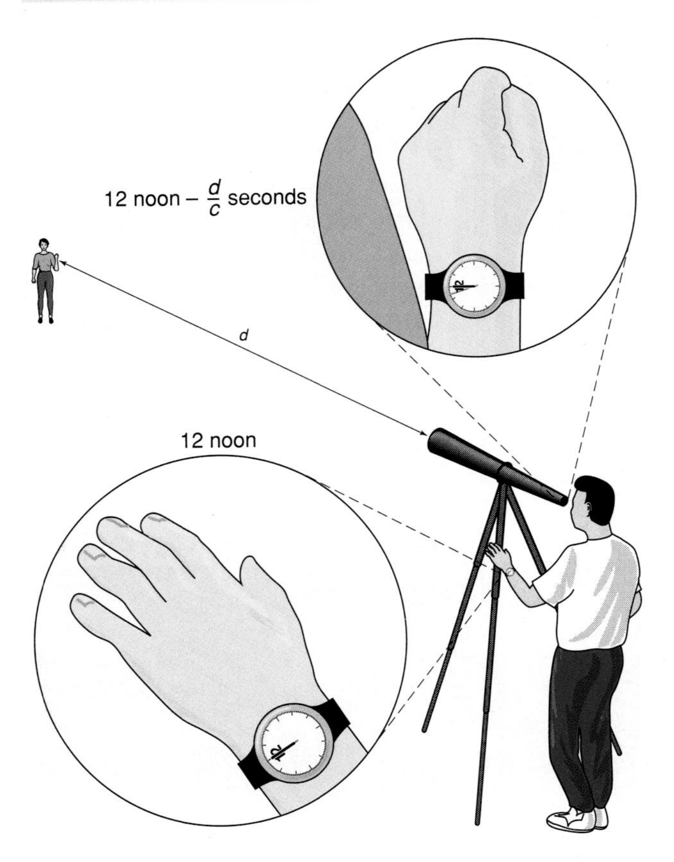

FIGURE 26-8 IF CLOCKS AT REST BUT SOME DISTANCE APART ARE TO KEEP THE SAME TIME, THE DISTANCE BETWEEN THE CLOCKS MUST BE KNOWN. THE CLOCK AT THE TELESCOPE MUST BE EXACTLY D/C SECONDS AHEAD OF THE CLOCK A DISTANCE D AWAY, SEEN THROUGH THE TELESCOPE, SINCE IT TAKES THE LIGHT A TIME D/C TO TRAVEL FROM THE FARAWAY CLOCK TO THE TELESCOPE.

traveling takes time. To make certain the watches really agree, you must know the distance and correct for the time the light needs to go that far. The distance covered is equal to the speed multiplied by the time of flight, or $d = ct$. So if light travels a distance d, it takes an amount of time $t = d/c$ to do so. This means that, in order for the watches to be keeping exactly the same time, a watch a distance d away from you should be set so that it appears through a telescope to read d/c seconds *behind* your watch (Figure 26-8). We say the watches are then synchronized or "in sync."

THE DILEMMA

Let's return to the experiment with the flash of light from the spaceship. Ray could use Einstein's method to set all the clocks on the ground to read the same time, and Ree could do the same for the clocks on her spaceship. Even then, the results of the experiment would be the same. Remember what happens as the two spaceship detectors move over the ground. The ground detectors see the ship's rear clock receive the light first, and then a little later the rays catch the ship's forward clock. We know, however, that when Ree checks those two detectors on her ship, they have recorded the flash at exactly the same time! That means that if the ground detectors also record the *times* on the faces of the ship detectors as they are lit, the two ship detectors must be seen to have exactly the same time on their faces.

There is only one possible explanation. *Detectors on the ground "see" the two clocks passing overhead as being out of sync, with the rear spaceship clock* ***set ahead*** *of the forward clock*. That is the only way these separate observations can agree.

The detectors aboard the spaceship can record the same sort of thing. At the instant the light strikes these detectors, they could record the time displayed on the face of the closest ground clocks. Ree could then say, "No wonder you have trouble, *your* clocks were out of sync. The clock on the ground that was closest to my forward clock was set ahead of the ground clock that was nearest my rear clock. That is why you think our forward clock was struck at a later time." *Each observer finds that the clocks in the other frame of reference are out of sync, with the rear clock (or clocks) always ahead of the forward clock (or clocks) in time*. Clocks that are synchronized in one inertial frame of reference are out of sync when seen from any other frame of reference (Figure 26-9).

It is the constant speed of light that fouls up the notion that time should be the same for everyone, even for people in different reference frames. Both Ree and Ray measure the same speed c for the same beam of light even while they are moving past each other. Einstein saw what this meant. Either that beam of light must be in two places at once or else we have to accept new ideas about time and distance. He showed in 1905 that *time and distance change when seen from different inertial frames of reference*. The predictions he made based only on the two postulates of relativity have since been proved accurate by uncounted experiments.

TIME IS RELATIVE

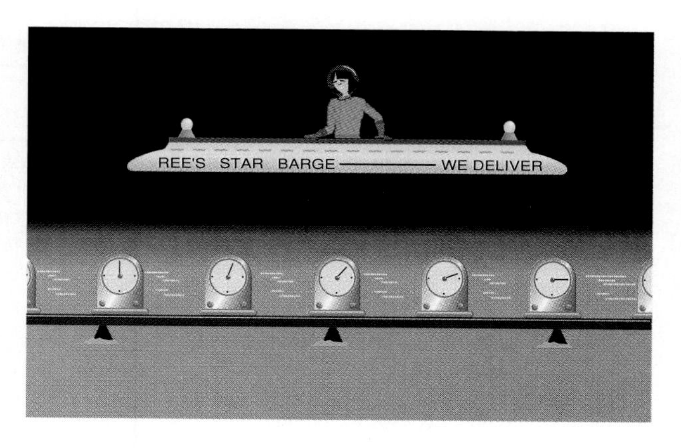

FIGURE 26-9 A SNAPSHOT OF AN INSTANT IN TIME, SHOWING WHAT MEASUREMENTS FROM THE FLYING SPACESHIP WOULD REVEAL. THE CLOCKS MOVING TO THE LEFT BENEATH THE SHIP APPEAR OUT OF SYNCHRONIZATION, WITH THE TRAILING CLOCKS (THOSE TO THE RIGHT) BEING SLIGHTLY AHEAD OF THE LEADING CLOCKS. AN OBSERVER ON THE GROUND WOULD FIND THAT THE CLOCKS WERE IN SYNC. THIS ILLUSTRATION SHOWS WHAT MEASUREMENTS COULD REVEAL, BUT NOT WHAT REE COULD ACTUALLY *SEE* AT THAT INSTANT, SINCE LIGHT FROM THOSE GROUND CLOCKS WOULD TAKE DIFFERENT AMOUNTS OF TIME TO REACH HER EYES!

One of the flying clocks could be monitored by the detectors on the ground for a period of, say, 1 second on the ship. The ground detectors could record by their own clocks precisely when the flying clock's second started and when that second ended. (Of course, the ground detector aligned with the flying clock at the end of the second would be *very* far from the one that saw its beginning!) All the detectors on the ground keep the same time, though, and once again Ray could look at the ground clocks' records. He would find that the moving clock *ticks more slowly than his own*; that is, more than one second passes on the ground, in the inertial frame of reference where the observer is at rest, for every second that passes on the moving clock. This slowing of clocks that move past an observer's reference frame is called **time dilation**.

Why does time dilation happen? Remember that, from the moving clock's frame of reference, each ground clock it passes is set *ahead* of the previous ground clock. Naturally, then, a second of time in the frame of the moving clock would be shorter than a second on those ground clocks, which were out of sync. Said another way, at the end of the spaceship clock's second, the ground clock closest to it will show that more than one second has passed on the ground. Likewise, if a line of clocks on a very long spaceship were used to observe a single ground clock for one second, it would be seen to run more slowly than the spaceship clocks! That ground clock would be passing the line of clocks on the ship, and each trailing clock would appear to the ground clock to be set ahead of the one in front of it. Measurements of a clock moving with respect to any observer's frame of reference always find the clock running more slowly than the observer's own clocks.

Einstein worked out the formula for time dilation. He found

PUTTING IT IN NUMBERS—TIME DILATION

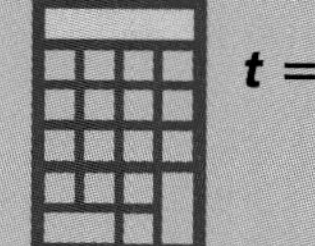

$$t = \frac{t_0}{\sqrt{1 - \frac{v^2}{c^2}}}$$

where t is the time needed in the observer's frame of reference for a time interval t_0 to pass in a frame of reference that's moving past the observer with a speed v. In other words, if t_0 is, say, 1 hour that passes on a clock in a spaceship moving past the earth at speed v, then t is the number of hours that elapse on earth during that time. Table 26-1 lists the factor $1/\sqrt{1 - v^2/c^2}$ for some values of v.

Example: If a clock moves past you at a speed of $0.5c$, then using a calculator or using the value from Table 26-1 for $v = 0.5c$, we find $t = t_0/0.870 = 1.15t_0$. So 1.15 seconds would elapse for you for each second that elapsed on the moving clock. Likewise, for every hour passing on the moving clock, 1.15 hours would pass for you.

Notice in Table 26-1 that the faster the moving clock travels relative to an observer, the greater the slowing, or dilation (think of *stretching*), of its time as seen by the observer. The results of the theory of special relativity become large only when relative speeds approach the speed of light.

Table 26·1 The Relativity Factor for Time Dilation

v	$\frac{1}{\sqrt{1 - v^2/c^2}}$
Speed of space shuttle in orbit (29,000 km/h or 18,000 mi/h)	1.00000000036
Speed of earth along its orbit (29 km/s or 18 mi/s)	1.0000000047
0.1c	1.005
0.5c	1.15
0.8c	1.67
0.9c	2.29
0.95c	3.20
0.99c	7.09
0.999c	22.37
0.9999c	70.7
0.99999c	223.6

LENGTH IS RELATIVE

Our two experimenters can next compare lengths in their respective frames of reference. Suppose Ree stretches a chain between herself and the rear detector on her ship (Figure 26-10). The detectors on the ground could mark where each end of that chain is at a predetermined time, such as 10:00 A.M. Then Ree could stop her ship and bring the chain back and lay it down beside those marks on the ground. The marks would be closer together than the ends of the chain, meaning that, from Ray's point of view, the *chain is shorter when it is moving than when it is at rest!*

We can see why from Ree's point of view. She and the chain are at rest relative to each other and the ground detectors are rushing backward past them. As with any line of synchronized clocks that are moving past an observer, the trailing clocks are ahead (in time) of the leading clocks. So she sees the forward end of her chain marked by a (trailing) ground detector when the clock on that ground detector reaches 10 A.M. *before* the (leading) ground

THE TWIN EXPERIMENT

Suppose twins Marvin and Trudy were 20 years old when Trudy left in a spacecruiser. After cruising at 0.99c for a year (by its clock), the ship turned around and came back to earth at 0.99c. When Trudy stepped out after landing, she was 22 years old. **How old was Marvin?** Referring to Table 26-1, $t_{\text{Marvin}} = 2 \text{ years} \times 7.09 = 14.18$ years. So Marvin was 20 years + 14.18 years = **34.18 years old!**

From Trudy's point of view, Marvin did the moving, so Marvin's clock ran more slowly than Trudy's while the relative speed was 0.99c. So why does Marvin age more while Trudy is gone? It seems paradoxical. However, the cruiser and Trudy accelerated at takeoff, decelerated to turn around, accelerated to come back, and decelerated to stop and land, so her frame of reference changed during those (lengthy) times. Meanwhile, Marvin and the earth kept essentially the same frame of reference for the whole time. The experiences of the twins aren't reversible because each experienced something different. Trudy felt the accelerations and changed frames of reference, but Marvin didn't.

An analysis of a line of clocks placed along Trudy's path would show that Marvin *lived extra time* during Trudy's accelerations and decelerations, and the difference in elapsed time isn't entirely defined until they get back together again and compare clocks. We won't do that analysis, but abundant experiments in atomic physics have proved that the predictions of special relativity such as these are accurate.

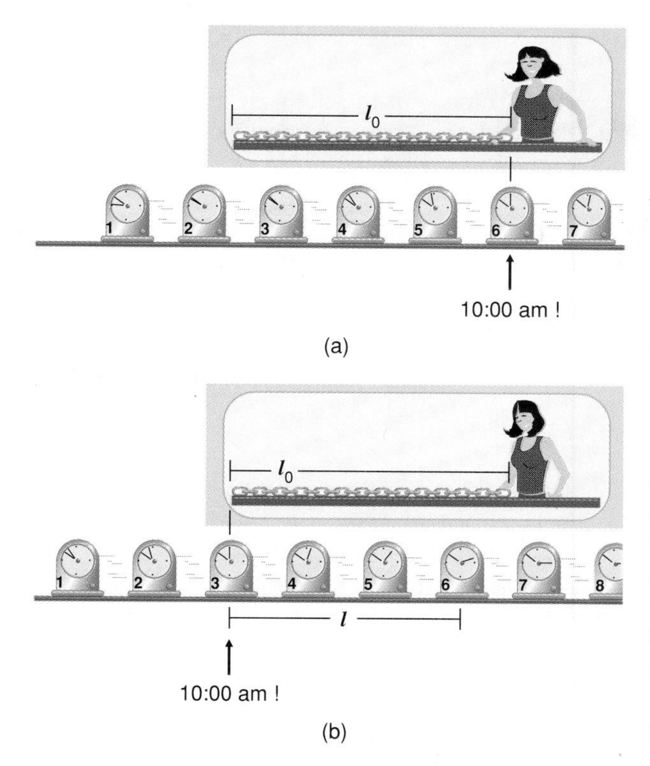

FIGURE 26-10 IN THIS ILLUSTRATION, THE RATE AT WHICH THE HANDS OF THE CLOCKS MOVE HAS BEEN GREATLY EXAGGERATED TO HELP SHOW HOW LENGTH CONTRACTION OCCURS. THE DETECTOR CLOCKS ON THE GROUND CAN PINPOINT THE ENDS OF THE CHAIN AS IT FLIES PAST BY SEEING WHERE THE ENDS ARE AT PRECISELY 10 A.M. THEN RAY CAN CHECK THEIR RECORDS AND MEASURE THE LENGTH OF THE CHAIN—IT IS THE DISTANCE BETWEEN THE TWO CLOCKS THAT DETECTED ENDS OF THAT CHAIN AT 10:00 A.M. SHARP. THE DISTANCE WILL BE LESS THAN THE LENGTH OF THE CHAIN MEASURED WHEN IT IS AT REST RELATIVE TO THE MEASURING DEVICE, AND REE CAN TELL WHY. BECAUSE THE TRAILING CLOCKS LEAD THE FORWARD CLOCKS *AS SEEN FROM HER FRAME OF REFERENCE*, (A) A TRAILING CLOCK WILL MARK THE FORWARD END OF THE CHAIN BEFORE ANY CLOCK NEAR THE REAR OF THE CHAIN REACHES 10:00 A.M. (B) THE CHAIN THEN MOVES A SHORT DISTANCE BEFORE IT CAN ENCOUNTER A LEADING CLOCK THAT REGISTERS 10:00 A.M.

detectors near the rear of the chain reach 10 A.M. and mark the position of that end. Meanwhile, the rear of the chain has moved forward a little bit. So she would agree that marks on the ground should be closer together than the length of the chain at rest. Objects shrink, or contract, along their direction of motion as seen by an observer in another frame of reference. And if Ree measured an identical chain that was resting on the ground as she flew over it, she would find it to be shorter than the chain at rest in her spaceship.

Einstein found the **length contraction** of a moving object along its direction of motion is

PUTTING IT IN NUMBERS—LENGTH CONTRACTION

$$\ell = \ell_0 \times \sqrt{1 - \frac{v^2}{c^2}}$$

where ℓ is the observed length if the object is passing an observer at a speed v and ℓ_0 is the length of the object if it was at rest relative to the observer.

Example: If a 1000-meter long spaceship (as seen by the crew) passed over a colony on the moon at a speed of 0.95c, what ship length would careful measurements by the colonists reveal? $\ell = 1000 \text{ m} \times \sqrt{1 - \frac{v^2}{c^2}}$. Using your calculator or the value for this square root listed in Table 26-2 (p. 403), we get $\ell = 1000 \text{ m} \times 0.312$, which means ℓ is about **312 m**.

FORCE AND ENERGY IN RELATIVITY

More than three centuries ago, Newton defined inertial mass with his second law. For motion in one dimension, $F = ma$. Expressing this in terms of momentum, mv, we have

$$F = m\frac{v_f - v_i}{t} = \frac{\Delta mv}{t}.$$

Almost a century ago, Einstein's special relativity modified Newton's second law, which became

$$F = \frac{1}{\sqrt{1 - \frac{v^2}{c^2}}}\frac{\Delta mv}{t}$$

This tells us that the faster an object moves, the harder it is to speed it up! In special relativity, then, the relationships between momentum and speed are different at different speeds.

The formula tells us something else. An object cannot reach the speed of light—the value of the force approaches infinity as the object's speed approaches c. *No material object can accelerate up to the speed of light.* Also notice that when v is small relative to c, the relativity factor is essentially 1 and the equation above gives us Newton's second law as we met it in Chapter 3. Relativity did not overthrow the old physics so much as modify it.

If the work = change in kinetic energy formula is calculated with the relativistic force, this equation is found:

$$KE = \frac{mc^2}{\sqrt{1 - \frac{v^2}{c^2}}} - mc^2$$

Notice what this says. If $v = 0$, the kinetic energy is zero, as we would expect. However, the classical term $\frac{1}{2}mv^2$ is replaced by the difference between a term that depends on v and one that doesn't. The term mc^2 is energy the object has while it is at rest, its rest energy. This means an object has energy in the form of its mass!

This is no doubt the most widely known equation in physics. If an amount of mass m disappears in some reaction, the energy that appears in other forms is equal to m times the speed of light squared.

PUTTING IT IN NUMBERS—THE ENERGY–MASS RELATIONSHIP

$$\boldsymbol{E = mc^2}$$

Example: Calculate the energy represented by the mass of a single gram of water. $E = (0.001 \text{ kg}) (3 \times 10^8 \text{ m/s})^2 = \mathbf{9 \times 10^{13}}$ **J.** If this is set equal to *mgh*, you can show this is the amount of energy needed to lift well over a million mid-sized cars 1000 m (about 3300 ft) into the air! Since c^2 is such a large number, even a tiny mass represents an enormous amount of energy.

The term $mc^2/\sqrt{1 - v^2/c^2}$ in the equation above for KE is the total energy of an object of mass m. As v approaches c, a material object's energy approaches infinity. We cannot arrange to give an object an infinite amount of energy, so no mass ever moves at c with respect to us. Light, of course, does move at c, but if you stop a photon to "weigh" it, it disappears. It has energy but no "rest" mass m.

In the nuclear processes at work in the sun, in other stars, and in nuclear power plants, the release of large amounts of energy is accompanied by a relatively small decrease in mass. (In Chapter 28 we'll have a closer look at how this is done.) More than that, however, Einstein's special theory of relativity applies to every process in nature, not just nuclear processes and not just those that happen at extremely high speeds. For example, in any chemical reaction that releases heat, if the initial mass of the material is determined with the greatest precision and then compared with the mass of the final products, it is found that a tiny fraction of mass has changed into energy, energy that can be detected as heat.

Summary and Key Terms

Early researchers speculated that light waves must travel through an extremely thin, tense elastic ether that exists everywhere. They assumed that the ether's motion past an observer would change light's relative speed, showing the ether's speed and direction. However, Michelson and Morley found that *the relative speeds of the source and observer do not influence light's speed.* Einstein's ***special theory of relativity*** shows how this constant speed of light modifies the old ideas about space and time.

special theory of relativity–394

Einstein's ***first postulate of relativity*** is *the laws of physics are identical in all **inertial** (non-accelerating) **frames of reference**.* His ***second postulate of relativity*** is *the speed of light through space is the same in any inertial frame of reference.* Light moves with no supporting medium. Einstein showed this means time and distance change when seen from different inertial frames of reference.

first postulate of relativity–393
inertial frames of reference–393
second postulate of relativity–394

If a spaceship moves over the ground at a large fraction of light's speed, clocks that are synchronized in the reference frame of the ship are out of synchronization when seen from the reference frame of the ground, and vice versa. Also, the rear clocks always lead the forward clocks in time. If detectors on the ground monitor a clock on the moving ship, they will find that more than one second passes on a ground clock for every second that passes on the ship's clock. The faster the clock travels, the more slowly it runs as seen from the ground. This is called ***time dilation***.

time dilation–397

If ground detectors mark the length of the ship as it speeds by, and then the ship returns, stops, and parks beside the marks, the marks will be closer together than the ends of the ship. *So the ship is shorter when it is moving past the observer than when it is at rest.* This is called ***length contraction***: Objects shrink along their direction of motion as seen by an observer in another reference frame.

length contraction–399

Einstein also found an ***energy–mass relationship***: If an amount of mass m disappears in a reaction, an amount of energy equal to m times the speed of light squared, or $E = mc^2$, is released. Since c^2 is such a large number, even a tiny mass represents a huge amount of energy. The energy emitted by the sun and in nuclear power plants is accompanied by a decrease in the mass of the interacting matter.

$E = mc^2$–400

EXERCISES

Concept Checks: Answer true or false

1. The speed of water waves through water depends on whether or not the water is moving.

2. Michelson and Morley found the ether's speed, and it was zero.

3. The speed of light in a vacuum does not depend on the motion of anything; this speed is a constant no matter what inertial frame of reference it is viewed from.

4. Relativity shows that mass, length, and time are altered for any object that has speed relative to an observer.

5. The length of an object moving relative to the measuring device is shorter than its length measured when the object is at rest relative to the measuring device.

6. Whenever a second of time passes on the clock of a moving object, a time less than one second passes for an observer at rest.

7. The relationship between energy, mass, and the speed of light for an object is $E = mc^2$.

8. According to special relativity, no material object can move as fast as the speed of light as seen by any observer.

9. Light waves are not material objects and have no mass if stopped, so their moving at c does not contradict relativity.

10. The consequences of special relativity seem far removed from everyday experiences because everyone and everything we know shares almost the same reference frame—the surface of the earth. Relative speeds here are usually a few kilometers per hour, and such low speeds don't reveal the effects of relativity.

11. All starlight passes the earth at speed c regardless of whether the earth is moving toward or away from the star.

Applying the Concepts

12. Which postulate of relativity is the more difficult for you to believe? Which have you had personal experience with?

13. In Newton's time, after the revelation that earth circles the sun at a great speed, people wondered if there was a point of *absolute rest* in the universe. Newton wrote "... it may be that there is no body really at rest, to which the places and motions of others may be referred." Discuss this statement in light of the postulates of relativity.

14. An NFL quarterback is running backward from his pursuers at the rate of 3 meters per second, and he throws the ball over their heads with a speed of 10 meters per second relative to him. How fast will the ball travel downfield as seen by a lineman at rest on the ground? If the quarterback instead shined a light beam in the same direction, how fast would it pass that lineman?

15. When astronauts went to the moon, the finite speed of light introduced itself to the people watching on TV. The microwaves used for communications took about 2.5 seconds for a round trip, causing a noticeable delay in the conversations between Mission Control in Houston and the moonwalkers. From this information, calculate the approximate distance to the moon.

16. Does Newtonian mechanics place a limit on how fast things can travel? Does special relativity?

17. Can any everyday object move at the speed of light past some other everyday object? Give a reason for your answer.

18. From the results of special relativity, would you say that space, time, and distance are absolute quantities?

19. As a train approaches an intersection with a road, the light traveling down the track from its headlight travels at (a) $c + v_{\text{train}}$, (b) $c - v_{\text{train}}$, (c) c.

20. Suppose you could look through a telescope and see the watch of a friend who is 5 kilometers away. If that watch and your own are keeping time simultaneously as defined by Einstein, by how many seconds should your watch be set behind your friend's?

21. Suppose you could somehow "beat" special relativity, travel across your state with a speed $2c$, then quickly stop and turn around. Describe what you could see, according to *classical* physics.

22. If the laws of physics weren't the same in all inertial frames of reference, discuss how a car trip might be very different from one you might ordinarily take.

23. The effects of relativity are not apparent in our world because the relative speeds we encounter are so much less than the speed of light. Discuss how things would be different if you could find a place where the speed of light in a vacuum was 10 meters/second, about the speed of a competitive sprinter.

24. In a motorboat on a lake, you can catch up to a wavefront from behind and pass over it. Can you do this with a light wave? Explain.

25. As monitored by someone on the ground, is the time between heartbeats of people on a moving train (a) less than, (b) greater than, (c) the same as the time between heartbeats of the same people when the train is stopped at the station?

26. Discuss this statement: Mass is the most inert form of energy.

27. Two ruffians sit 5 meters apart by the side of some

railroad tracks. After synchronizing their watches, they simultaneously fire peas through peashooters, aiming straight at the side of a moving boxcar. Later, when the car is at rest, they measure the distance between the marks made by the peas. If they could make extremely accurate measurements, what would they discover?

28. When Einstein was young, he mused over how the world might appear to him if he were riding on a light beam. In light of his later discoveries, is such an observation possible?

29. Evaluate this statement: The distance from New York to San Francisco depends on how fast you travel.

30. A spacecraft moving between two stars a fixed distance apart travels at 0.999*c*. Which statements are true? (a) Observers located at rest near the stars see the travelers age more slowly than they themselves do. (b) The travelers feel they are aging more slowly than the observers. (c) The travelers see the distance between the stars as shorter than the observers do. (d) The observers see the distance between the stars as shorter than the travelers do. (e) The observers see the travelers moving in slow motion relative to themselves.

31. While the speed of light doesn't change from one inertial reference frame to another, its frequency does. Discuss how changing frequency might be an aid to astronomers who wish to know the relative motion between stars and the earth.

32. If you live to be 100 years old, how long would that seem to someone on another planet circling a star that moves at 0.99*c* with respect to earth? Use Table 26-1.

33. From Table 26-2, find how fast a car would have to move (as measured by someone standing on the road) to be only 60 percent of its present length.

Table 26·2 The Relativity Factor for Length Contraction

v	$\sqrt{1 - v^2/c^2}$
Speed of space shuttle in orbit (29,000 km/h or 18,000 mi/h)	0.99999999964
Speed of earth along its orbit (29 km/s or 18 mi/s)	0.9999999953
0.1*c*	0.995
0.5*c*	0.866
0.8*c*	0.600
0.9*c*	0.435
0.95*c*	0.312
0.99*c*	0.141
0.999*c*	0.0447
0.9999*c*	0.0141
0.99999*c*	0.00447

34. You are preparing to take a 0.5*c* trip with an extraterrestrial you met last night. A friend tells you to buy a smaller belt since you'll be skinnier along your direction of motion. Is your friend correct? Explain.

35. A bug that finds itself on a ceiling fan crawls along one of the revolving blades until it reaches the tip. How is time different for you and that bug? If both you and the bug could measure lengths with great accuracy, how would you appear to the bug and how would the bug appear to you?

36. Find the energy in joules stored in the mass of 1 kg of potatoes.

37. When a uranium nucleus splits in a process called fission, about 0.001 of its mass *m* becomes energy that is carried away by the fragments. About how many joules of energy appears if one kilogram of uranium undergoes fission?

38. Speculate what you might see if the speed of light traveled only at 1 mm/day while all the laws of physics remained unaltered for you.

39. Is the space we move in over earth's surface an inertial frame of reference? Discuss.

40. Incoming cosmic rays travel at a large percentage of the speed of light and collide with nuclei of air molecules. High-speed muons, unstable particles that quickly (in 2×10^{-6} second) decay into other particles, emerge from the interaction, among other things. These muons travel downward at a speed greater than 0.99*c*. According to classical physics, they would decay after traveling about 600 m, but they don't. Instead, most travel about 9000 m before decaying. Discuss, considering what happens to lengths in relativity.

41. Sirius A is a star 8.8 light-years from the sun. That is, it takes light traveling at *c* about 8.8 years to make a one-way trip between Sirius and the earth. Find how far in light-years that distance would seem to someone in a spaceship traveling at 0.9999*c*. Use Table 26-2.

42. Discuss this statement: If the speed of a moving object is small relative to the speed of light, the time dilation is extremely tiny. If this were not so, an ordinary car trip across the continent would cause watches to disagree noticeably.

43. The closest star to the sun is a little more than 4 light-years (see exercise 41) away. If the energy per second emitted by that star in visible radiation suddenly changed today, when could we detect it? What would we detect between now and then? Astronomers who use telescopes to gather light from distant stars are looking back into the past.

MORE DISCOVERIES AT THE TURN OF THE CENTURY

In the physics laboratories of the 1890s, some physicists had bad results with photographic plates. When developed, the plates were sometimes "fogged," just as if light had reached them from some source other than through the shutter of the camera. Wilhelm Roentgen, a German physicist, wondered if the electrical discharge tubes in his lab could be exposing the film. (Discharge tubes were in common use in the physics labs of that time; see "New Beginnings," following Chapter 23.) In 1895 he directed the cathode rays in his discharge tubes at different minerals and found that the rays caused many to fluoresce. Then he darkened the room to make the faint glow of the cathode rays easier to see. When he activated the tube, something else in the room caught his eye. A piece of paper coated with fluorescent material and lying some distance away was glowing brightly in the dark. He carried the paper into the next room, and the glow continued. Roentgen had discovered a kind of radiation that did more than just fog photographic plates; *it could pass through a wall!*

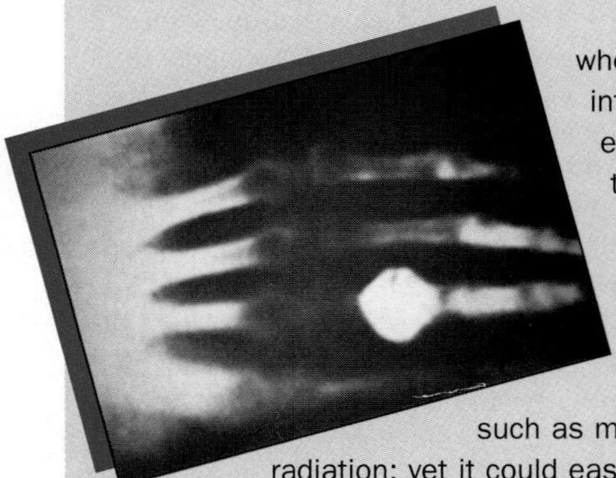

The cathode rays, when they slammed into the matter at the end of a discharge tube, produced the X-rays that traveled through the room. Roentgen soon learned that heavier atoms, such as metals, stopped the radiation; yet it could easily pass through lighter atoms, such as those of glass or water. He directed these unidentified ("X") rays through his hand onto a photographic plate, and when he developed the plate he saw images of the bones of his hand (bones are about 20 percent calcium, which is a metal).

On January 20, 1896, the discovery of X-rays was announced at a meeting of the French Academy of Sciences. Henri Becquerel, a French physicist, was in the audience. He knew about the fluorescence of certain minerals when they were exposed to sunlight, and he wondered if Roentgen's X-rays might be found in that fluorescent glow. For a month he placed all kinds of naturally fluorescent minerals in the sunshine to cause them to glow. Beneath them he put photographic plates wrapped in black paper to keep out the sunlight. When he developed the plates he found no fogging—until he happened to use a mineral that contained uranium. Placing coins under the uranium samples, Becquerel repeated the experiments and saw outlines of the coins on the developed plates. The metal coins had stopped the radiation being given off by the uranium crystal. Becquerel, believing the sunlight had caused the radiation and that the radiation was the same as Roentgen's X-rays, excitedly announced his results to the French Academy in February 1896. Both assumptions were wrong, but a real discovery was soon to come.

For several days after his report, the Paris skies were cloudy, and Becquerel could not do his experiments without sunlight. He tossed the minerals and paper-covered plates into a dark drawer to wait for a sunny day. Four days later he returned to his laboratory and decided to check some of his plates before resuming the experiments. When he developed them, he was amazed to find not fogged spots but rather black areas where the crystals had been touching the plates in the drawer. The uranium mineral was sending out rays without the aid of sunlight. Becquerel had accidently discovered *natural* radiation, or *natural radioactivity*.

The natural radioactivity of uranium proved to be too weak to make pictures of bones, as Roentgen's X-rays could do, but Becquerel pressed on with his research. When he found that a uranium ore called pitchblende affected photographic plates far more than it should have for the percentage of uranium it contained, he suspected that another, unknown radioactive element was present. He passed this idea on to a young Polish graduate student, Marie Curie. By early 1898 she and her husband, Pierre, managed to separate the element polonium (named in honor of her native country) from the pitchblende. Later in 1898 they also separated radium.

The discovery of polonium and radium, elements that are far more radioactive than uranium, attracted the attention of Ernest Rutherford, then a student at Cambridge University. Rutherford found two types of charged radiation in the rays given off by polonium. The positively charged radiation ionized the air very strongly; the negatively charged rays ionized weakly by comparison. He named them alpha and beta rays, respectively. Gamma rays, with no charge, were discovered the next year, so three types of natural radioactivity had been identified. Today we know both cathode rays and beta rays are electrons. Rutherford soon discovered the details of the rate of decay of radioactive atoms and proposed that radioactivity caused an internal change in an atom. Rutherford and a chemist, Bertram Boltwood, identified three chains, or families, of radioactive elements.

In 1908 Rutherford and Hans Geiger discovered that an alpha particle would make a faint flash when it struck a screen coated with zinc sulfide. Every flash showed precisely where an alpha particle landed, and Rutherford realized he could use such a screen to investigate how the alpha particles interacted with atoms. He sent a beam of alpha particles straight into a thin gold foil, and he placed a zinc sulfide screen on the opposite side to intercept them. By observing how the alpha particles scattered as they passed through the gold atoms, Rutherford hoped to find clues to the structure of the atom. Geiger handled the counting because Rutherford was too impatient to sit still and record the tiny flashes. Geiger found that almost all the alpha particles cut through the gold atom in the gold atoms in the foil like a hot knife through butter, deflecting only a degree or less. Wider deflections of, say, 5° were rare. This wasn't unexpected. Only the year before, Rutherford had shown that alpha particles were totally ionized helium atoms, that is, helium atoms minus electrons. These alpha particles outweighed an electron by about 8000 to 1, so an energetic alpha particle wouldn't be slowed much when it hit an electron or two. Rutherford didn't drop the experiment, however; in the spirit of "trying any damn fool thing," he asked a 19-year-old student, Ernst Marsden, to spend some time looking for alpha particles deflected by as much as 45°. The apparatus was modified, and Marsden began his observations.

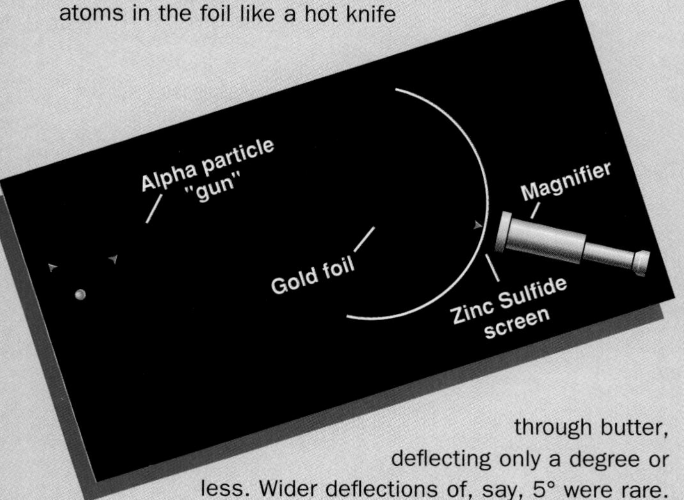

By their accounts, the job was tedious. The equipment was in the cold, wet basement of a building because the faint flashes were best seen in the dark. Marsden had to sit in the dark for half an hour before his eyes were fully adjusted to the dark. Then he could peer through a telescope to see flashes on the screen. Swinging the screen and telescope 45° from the direction of the incoming alpha particles, Marsden saw a few pinpoints of light in the course of these sessions. Then he swung the screen and telescope around past 90° and found that a few alpha particles were coming back *from the front side of the gold foil.* When Rutherford heard about this, he was astonished. He later said, *"It was quite the most incredible event that has ever happened to me in my life. It was almost as if you had fired a 15-inch shell at a piece of tissue paper and it came back to hit you."*

Geiger and Marsden counted over a million flashes in order to record enough of the wide-angle events to form a statistically correct pattern of the scattering, and Rutherford tried to find a model of the atom that would predict what they had observed. Knowing that the alpha particles would pass through several hundred gold atoms on their way through the foil, Rutherford knew the scattering was a statistical process. Though he was a brilliant experimentalist, he was not an accomplished mathematician, so he went to a statistics class. The students in that class in Manchester, England, were startled to find a Nobel Prize winner sitting with them and taking notes. Early in 1911, Rutherford proclaimed to Geiger, "Now I know what the atom looks like!"

Only an extremely tiny, highly repulsive, massive object could explain the small number of deflections at wide angles. (Only 1 in 20,000 alpha particles was deflected by as much as 90°.) Indeed, the gold nucleus could send any alpha particles that came directly at it straight back, but the nucleus was so small compared to the size of the gold atom that the chance of an alpha particle coming straight at it was *extremely* small. Rutherford estimated the size of the newly discovered nucleus, and the understanding of atomic structure had begun.

TWENTY-SEVEN

The Nucleus

Two invisible gamma rays come from the bottom of this photo, have close encounters with protons, and create electrons (colored red) and positrons (green) which curve in opposite directions in the magnetic field of this detector. At the bottom, an atomic electron (colored white here) takes some of the gamma ray's energy and flies off to the top right.

In 1911, Rutherford discovered the speck at the center of the atom that holds more than 99.9 percent of its mass, the nucleus. If an atom were the size of the earth, the nucleus would be a sphere with a diameter of less than 150 yards. The nucleus is incredibly dense; if a cubic inch of nuclear matter could be assembled on earth, it would weigh about a billion tons! Nuclei could not be packed this way on earth, of course, because of the great repulsion of their positive charges. But if the atoms in your body could be stripped of their electrons and the nuclei packed into a box, that box would need to be only 0.0004 centimeter across. That means your entire bodily mass, except for 10 to 20 grams of electrons, would fit into the tiny cube shown in Figure 27-1. You could not see that cube with the naked eye.

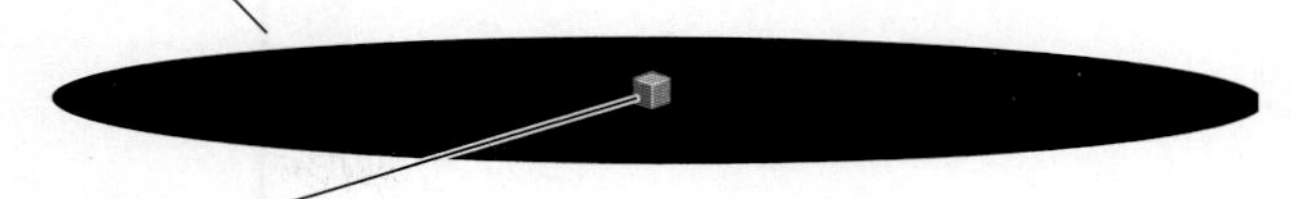

FIGURE 27-1 AN ATOM IS IMMENSE WHEN COMPARED WITH THE TINY NUCLEUS AT ITS CENTER. IF THE PROTONS IN THE NUCLEI DID NOT REPEL EACH OTHER, YOUR BODY'S NUCLEI, AND HENCE 99.9 PERCENT OF YOUR BODY'S MASS, COULD FIT INTO THE CUBE SHOWN HERE. THINK OF WHAT THIS MEANS. WHEN YOU LOOK AT SOMEONE, YOU SEE THAT PERSON ONLY BECAUSE OF LIGHT REFLECTED FROM HER OR HIS ATOMS. AND THE ATOMIC ELECTRONS BOTH REFLECT THAT LIGHT AND ACCOUNT FOR THE VOLUME OF ATOMS AND MOLECULES AND HENCE THE VOLUME OF THAT PERSON. HOWEVER, IT IS THE TINY NUCLEI THAT GIVE THAT PERSON INERTIA, THAT ACCOUNT FOR THAT PERSON'S MASS.

A REVIEW OF THE NUCLEUS

The nucleus of the simplest atom, hydrogen, is a single proton, but the nucleus of any other element is a tight collection of protons and neutrons. The number of protons in a nucleus is the *atomic number* of that atom, and the number of protons and neutrons together is its *mass number*. For example, helium has 2 protons and 2 neutrons in its nucleus. Its mass number is therefore 4, it has a mass of about 4 u, or $4 \times 1.66 \times 10^{-27}$ kg, and its symbol is ${}^{4}_{2}\text{He}$. The number of protons in any nucleus determines how many electrons are in the neutral atom, and the electrons in the outermost shell govern the chemical and most of the physical properties of an atom. In a copper atom, ${}^{63}_{29}\text{Cu}$, the nucleus contains 29 protons and is surrounded by 29 electrons. Take away a proton, and it becomes a nickel nucleus, Ni, with atomic number 28. Add one proton, and the copper nucleus becomes a zinc nucleus, Zn, with atomic number 30. (Remember, the periodic table on the back cover shows both the atomic number and the average atomic mass for all the elements.)

THIS SECTION IS A BRIEF REVIEW OF SOME OF THE MATERIAL OF CHAPTER 8. FOR MORE DETAILS, YOU CAN REVIEW THE SECTIONS OF CHAPTER 8 THAT DEAL WITH THE NUCLEUS.

Unlike the number of protons, the number of neutrons contained in the nucleus can vary for a given element. For example, the copper nuclei in a copper tea kettle or in copper wiring have various numbers of neutrons. On the average, 7 of every 10 copper nuclei have 34 neutrons, so that their mass number is $29 + 34 = 63$. However, the other three copper nuclei have two additional neutrons and a mass number of 65. These copper nuclei with different numbers of neutrons are called *isotopes* of copper.

SOMETIMES AN ISOTOPE OF AN ELEMENT IS REFERRED TO BY GIVING ITS NAME FOLLOWED BY ITS MASS NUMBER, SUCH AS CARBON-12 OR URANIUM-238.

It would be impossible for a nucleus to contain more than one proton—the electric force would cause the protons to leave the nucleus explosively—if there were not another, stronger force that acts to keep the protons there. This force within the nucleus, called the **nuclear force** or the **strong interaction**, is an attraction between all nucleons (a *nucleon* is a neutron *or* a proton), but it is a force with very short range. Experiments show that it acts only when nucleons are next to each other, or, in classical terms, close enough to "touch."

SINCE THE STRONG INTERACTION HAS THE SAME MAGNITUDE BETWEEN TWO NEUTRONS AS IT DOES BETWEEN TWO PROTONS, AND ALSO BETWEEN A NEUTRON AND A PROTON, THESE PARTICLES ARE BOTH CALLED *NUCLEONS*.

UNSTABLE NUCLEI

Most of the atoms found on earth are stable (unchanging) in their natural state. Of the elements found on earth, there are about 190 stable isotopes, all of which have atomic numbers between 1 (hydrogen) and 83 (bismuth). The atoms of these stable nuclei give the matter around us its permanence. How-

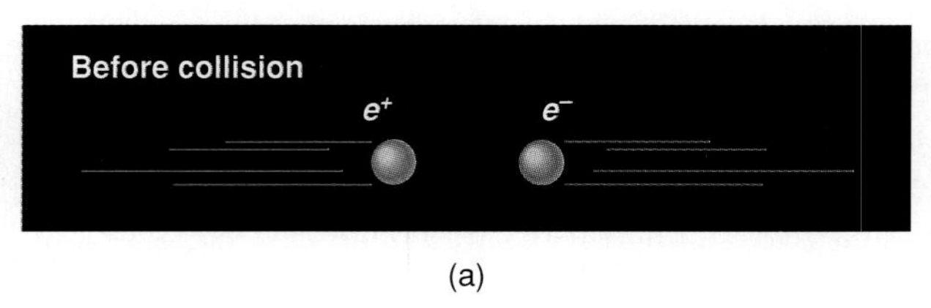

(a)

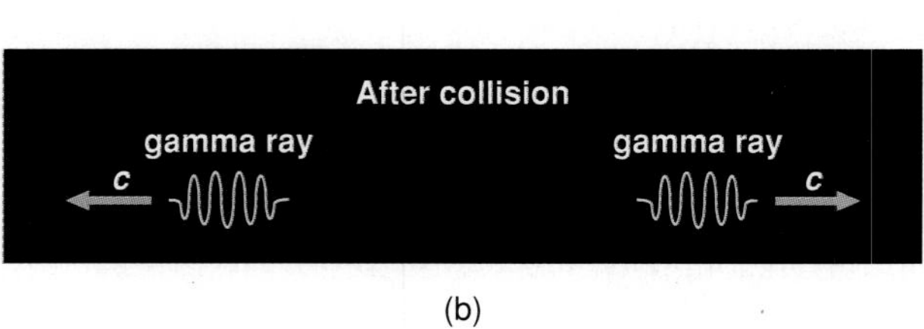

(b)

FIGURE 27-2 (A) AN ELECTRON AND A POSITRON BEFORE A HEAD-ON COLLISION. (B) ONE OF THE POSSIBLE RESULTS, DEPENDING ON THEIR SPINS, IS ANNIHILATION PRODUCING TWO GAMMA-RAY PHOTONS.

ever, a tiny fraction of the atoms naturally occurring on earth are unstable; that is, they change, or *decay*, into different nuclei by emitting an energetic particle of some kind. These unstable nuclei are called **radioactive**.

As you saw in *More Discoveries at the Turn of the Century*, the first three types of decay products to be identified from radioactive nuclei were alpha particles (helium nuclei), beta particles (electrons), and gamma rays (high-energy photons). Later some of the beta particles were found to be **positrons**, or *positive electrons* (symbol e^+). The positron is called the antiparticle of the electron (symbol e^-) because if a positron meets an electron, the two are annihilated, and as they disappear gamma rays carry off the energy of their mass, according to $E = mc^2$ (Figure 27-2). The existence of antimatter was postulated by the British physicist P. A. M. Dirac in 1928 and discovered experimentally in 1932. Physicists are still uncertain as to why the universe seems to be made mostly of normal matter because the equations of physics don't seem to prefer matter over antimatter.

ANTIPROTONS, ANTINEUTRONS, AND EVEN ANTIATOMS HAVE BEEN CREATED IN EXPERIMENTS, AND POSITRONS ARE FOUND WHEN COSMIC RAYS STRIKE NUCLEI OF MOLECULES IN OUR ATMOSPHERE.

THE SHELL MODEL OF NUCLEAR STRUCTURE: GAMMA EMISSION

In Chapter 25 we saw that the electron states correspond to shells, or layers, around the nucleus. Electrons in the same shell have about the same energy, full shells tend not to react chemically, and an atom absorbs or emits photons by transitions of its electrons from one shell to another. Nucleons, like electrons, follow the rules of quantum mechanics. In 1949, using electron shells as an analogy, the Polish-American physicist Maria Goeppert Mayer predicted shells of nucleons in the nucleus. Each shell accommodates a certain number of nucleons, and nuclei with full outer shells are particularly stable.

The shell model helps to explain gamma ray emission from a nucleus. If a particle such as a fast proton from a particle accelerator or a cosmic-ray particle strikes a nucleus, one or more nucleons may go into an excited state. As with electrons in excited states, nucleons returning to their ground states emit a photon that takes away energy. Because the nuclear force is so strong, nucleon transitions are of very high energy. For example, a typical gamma-ray photon from an excited nucleus may carry a million times more energy than a visible photon does. And, like atoms, the photons emitted by a nucleus have distinct frequencies, so the spectrum of gamma rays is like a set of fingerprints, making nuclei that emit gamma rays easy to identify.

BETA DECAY

The stable nuclei in nature follow a distinct pattern. For low atomic mass atoms, stable nuclei have almost equal numbers of protons and neutrons, but high-mass nuclei need more neutrons than protons in order to be stable. (We'll say more about this shortly.) A graph of number of neutrons versus number of protons reveals a *line of stability* for nuclei, shown by the dashed line in Figure 27-3. Often an unstable nucleus emits an electron or positron in the action called **beta decay** to become a new nucleus that is on or closer to the line of stability.

Beta decay occurs because of the *weak interaction*, another force that acts only at the extremely short ranges found in the nucleus. The weak interaction is much weaker than the electromagnetic force but stronger than gravity. In the 1960s, physicists showed that the weak interaction and the electromagnetic force are two parts of one interaction, called the **electroweak force**. (Before this unification, there were thought to be four fundamental forces in nature: gravity, electromagnetic, strong, and weak. Now there are thought to be only three: gravity, electroweak, and strong. Physicists who work on "grand unification theories," GUTs, are trying to reduce these three to one truly fundamental force of nature.)

A carbon-14 nucleus undergoes beta decay because its neutron-to-proton ratio is too high. It is to the left of the line of stability in Figure 27-3. Such a nucleus is said to be *neutron-rich*. In essence, a neutron inside that nucleus decays into a proton, an electron, and an antineutrino (which we discuss later), as shown in Figure 27-4. The antineutrino and the electron escape from the nucleus immediately because neither is affected by the attractive nuclear force. The proton remains, however, bound by the surrounding nucleons. The new nucleus, called the **daughter**, has almost the same mass as its predecessor, called the **parent**, because the electron is light and the neutrino is apparently massless (the jury is still out on this). The atomic number of the daughter is one higher than the atomic number of the parent because of the extra proton. In equation form,

$$^{14}_{6}\mathrm{C} \rightarrow {}^{14}_{7}\mathrm{N} + \mathrm{e}^{-} + \text{antineutrino}$$

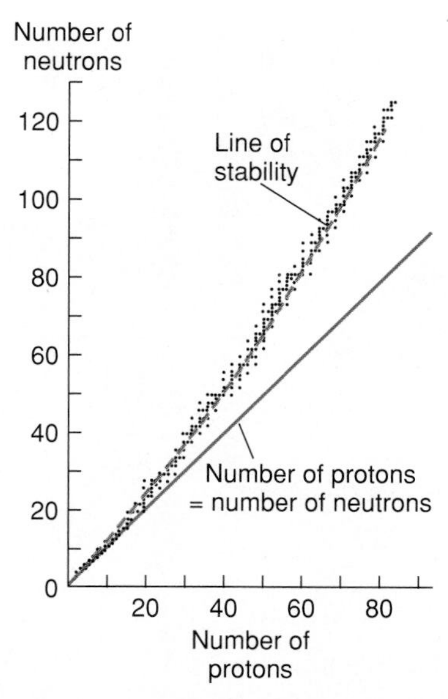

FIGURE 27-3 THE NUMBER OF NEUTRONS VERSUS THE NUMBER OF PROTONS FOR THE STABLE ISOTOPES. THE SOLID LINE IS WHERE A NUCLEUS CONTAINING EQUAL NUMBERS OF PROTONS AND NEUTRONS WOULD LIE. ALL HEAVY AND MIDDLEWEIGHT NUCLEI THAT ARE STABLE CONTAIN MORE NEUTRONS THAN PROTONS, SO ALL THESE STABLE ISOTOPES LIE TO THE LEFT OF THE SOLID LINE, ALONG A "LINE OF STABILITY" INDICATED BY THE DASHED LINE. UNSTABLE NUCLEI TO THE LEFT OF THAT LINE TEND TO UNDERGO BETA DECAY; THE RESULTING LOSS OF A NEUTRON AND GAIN OF A PROTON MOVES THE DAUGHTER NUCLEUS CLOSER TO THE LINE OF STABILITY. ELEMENTS TO THE RIGHT TEND TO EMIT EITHER ALPHA PARTICLES (IN THE CASE OF HEAVY ELEMENTS) OR POSITRONS, SO AGAIN THE DAUGHTER NUCLEUS IS CLOSER TO THE LINE OF STABILITY THAN ITS PARENT NUCLEUS.

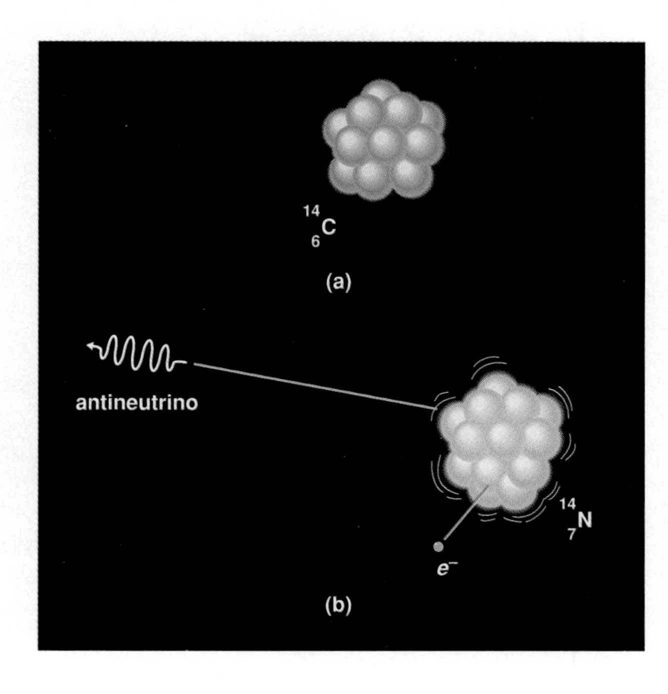

FIGURE 27-4 (A) CARBON-14 IS RADIOACTIVE AND UNDERGOES BETA DECAY. (B) THE END PRODUCTS ARE A NITROGEN NUCLEUS, AN ELECTRON, AND AN ANTINEUTRINO.

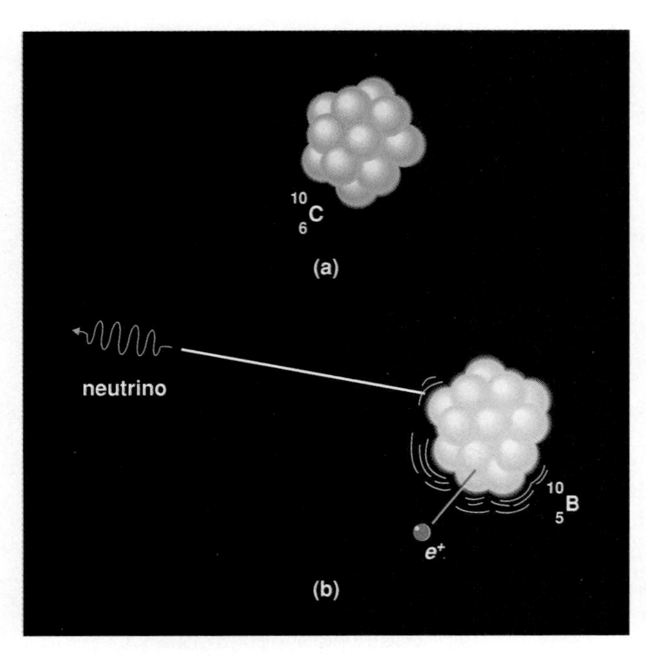

FIGURE 27-5 (A) CARBON-10 IS RADIOACTIVE AND UNDERGOES BETA DECAY. (B) THE END PRODUCTS ARE A BORON NUCLEUS, A POSITRON, AND A NEUTRINO.

The nitrogen nucleus has a lower neutron-to-proton ratio than its parent ($\frac{7}{7}$ versus $\frac{8}{6}$). The nitrogen nucleus is on the line of stability.

As another example, carbon-10 is a nucleus whose neutron-to-proton ratio is too low; such a nucleus is said to be *neutron-poor*. It is to the right of the line of stability in Figure 27-3. Carbon-10 decays by emitting a positron and a neutrino (Figure 27-5). In symbols,

$$^{10}_{6}\text{C} \rightarrow {}^{10}_{5}\text{B} + \text{e}^{+} + \text{neutrino}$$

where B is the symbol for boron (atomic number 5). In this decay, a proton in the carbon-10 nucleus and extra energy supplied by work done by the nuclear force become a neutron,

THE LITTLE NEUTRAL ONE

When a nucleus undergoes radioactive decay, a set amount of nuclear energy is lost in the process. The laws of conservation of energy and momentum predict that, if a nucleus decays by emitting only one particle, that particle and the daughter nucleus will share the energy and the momentum exactly the same way each time. However, experiments done in the early days of nuclear physics showed that the energies of the electrons emerging from beta decay are not equal. Instead, the electrons have a *range* of energies as they emerge. When these energy differences were discovered, it seemed as if conservation of energy might not hold true for beta decay.

Wolfgang Pauli decided that was unthinkable. In 1930 he hypothesized that, during beta decay, an additional particle must be emitted along with the electron. If the decay energy is shared among three particles, the electrons could indeed have a range of energies. Called the **neutrino** ("little neutral one"), it apparently has no mass (like photons, it travels at *c*) and no charge and only very rarely reacts with matter. Neutrinos are therefore very hard to detect and were not found by experiments until 1956, even though nuclear reactions in the sun and in nuclear reactors on earth emit them in staggering numbers. These particles come close to being unstoppable; they can easily pass through the earth without deflection—in fact, if 10^9 neutrinos were aimed at the earth's center, all but one would emerge unchanged on the other side.

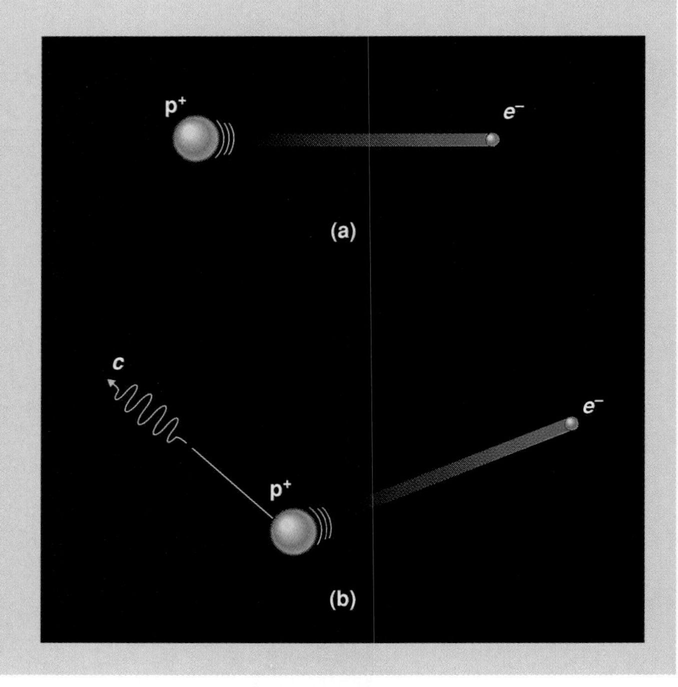

(A) IF A NEUTRON DECAYED ONLY INTO A PROTON AND AN ELECTRON, CONSERVATION OF MOMENTUM AND ENERGY WOULD ENSURE THAT ALL THE ELECTRONS WERE EMITTED WITH THE SAME AMOUNT OF ENERGY—BUT THEY AREN'T! (B) PAULI SUGGESTED THAT THIS UNEQUAL-ENERGY RESULT WAS EVIDENCE FOR ANOTHER PARTICLE, ONE THAT IS MASSLESS, CHARGELESS, AND VERY HARD TO DETECT. THE NEUTRINO WAS EVENTUALLY DETECTED IN 1956; IF IT ACCOMPANIED BETA DECAY, THE ELECTRON COULD EMERGE WITH A LOT OF DIFFERENT ENERGIES, AS WAS OBSERVED.

USING ANTIMATTER IN MEDICAL PHYSICS: POSITRON EMISSION TOMOGRAPHY (PET)

The half-life of $^{11}_{6}C$ is about 20 minutes, and when it decays it emits a positron. If this occurs inside a patient, the positron travels only a millimeter or so before it collides with an electron and the two annihilate. Two energetic gamma ray photons appear, and they leave the scene traveling in opposite directions (see Figure 27-2). A circular array of gamma ray detectors surround the area of the patient being imaged, and when two such photons are detected in opposite directions, the slight difference in time when the photons arrived is used by a computer to calculate where the annihilation occurred. The source is located more precisely than in any other kind of nuclear medicine procedure, to within one-half of a centimeter. Oxygen-15, with a half-life of about 2 minutes, and nitrogen-13, with a half-life of about 10 minutes, are also used in PET. (Tomography refers to the computer process where multiple events are used to form a three-dimensional image of the area where the positrons annihilate.)

Extremely small amounts of the positron-emitting nuclei are needed, so the dose of radiation received by the patient is extremely small. The radioactive nuclei are produced by a particle accelerator in the hospital and immediately added to the appropriate tracer molecules, which in turn go to the specific area being studied. With PET, researchers can monitor blood flow, blood volume, and consumption of oxygen and other chemicals. For an example of its use, when chess players are introduced to a chessboard with a game in progress, PET activity shows first the various areas of the brain that are used to distinguish colors and spacial arrangements, then where in the brain the player remembers the rules of the possible motions of the piece in question, and finally where the player actually plans (thinks about) the next move. For the first time medical researchers can actually map the areas of the brain where these various processes are taking place.

RADIOACTIVITY IN MEDICINE

a positron, and a neutrino. A neutron is more massive than a proton; this means that energy must be added to the proton if a neutron is to appear. The neutron stays in the nucleus, while the positron and neutrino escape. The details of these decay processes are understood today in terms of particles (called *quarks*) that make up protons and neutrons, a topic that is discussed in a later section.

ALPHA EMISSION

Alpha particle emission most often occurs for heavy unstable nuclei, those having a mass number of about 140 or greater. For an example, let's look at radium, which decays into radon and an alpha particle:

$$^{226}_{88}\text{Ra} \rightarrow {}^{222}_{86}\text{Rn} + {}^{4}_{2}\text{He} \text{ (the helium nucleus is an alpha particle)}$$

The radon nucleus and the alpha particle repel each other since they are both positive, which sends the alpha particle away with a lot of kinetic energy. An alpha particle significantly ionizes the matter it plows through before it is stopped.

The details of why nuclei emit alpha particles are complex, but one result of alpha emission is to form a daughter nucleus having a higher neutron/proton ratio than the parent nucleus. This means the protons in the daughter nucleus are "padded out" a little more by the neutrons than they were in the parent. Keeping the protons, which repel each other, a fraction farther apart makes the nucleus more stable. (This increase in stability will be discussed in more detail in the next chapter.)

THE LIFETIME OF RADIOACTIVE NUCLEI

When the nucleus of a radium atom decays, the alpha particle that escapes tears through the neighboring atoms, freeing some electrons and exciting others. As those excited electrons jump back down into lower-energy states, they emit photons of visible light, ultraviolet light, or even X-rays. As a result, a specimen of radium glows steadily in the dark. When Marie Curie first isolated radium from other elements, its steady rate of energy emission presented a puzzle. It appeared that the source of the radiation was constant and perpetual—a state of affairs that would violate the principle of conservation of energy. The solution to this mystery came only after other radioactive elements were studied. Rutherford and other physicists isolated several radioactive elements that lost their radioactivity in days or even hours, and he found that nuclear decays follow a statistical process: *A large collection of radioactive nuclei always loses the same fraction of its parent nuclei in the same amount of time.*

As an illustration, suppose it takes a year for a specimen containing 10,000 nuclei of some radioactive isotope to lose 10 percent of its nuclei. Years later, when there are only 1000 of the radioactive nuclei left, the specimen will still be losing about 10 percent of that number during the next year. The more nuclei there are, the more closely the sample follows this statistical rule.

A convenient benchmark in the life of a sample of radioactive material is the time it takes for half of the nuclei to decay. This time is called the **half-life** of the material. If the half-life of an unstable isotope is 2 days, for example, a sample of a million such nuclei will lose about 500,000 through decays in the first 2-day period. Over the next 2 days, it would lose about 250,000 (which is half of the remaining 500,000 nuclei), and so on. Figure 27-6 shows a plot of how samples of three unstable nuclei having different half-lives decay as time goes by.

FIGURE 27-6 A PLOT TO COMPARE HOW THREE RADIOACTIVE SUBSTANCES WITH DIFFERENT HALF-LIVES DIMINISH WITH TIME. (RED CURVE = 1-MINUTE HALF-LIFE, GREEN CURVE = 2-MINUTE HALF-LIFE, BLUE CURVE = 50-MINUTE HALF-LIFE.) THOUGH THE CURVE FOR EACH SPECIMEN DIFFERS, THE QUANTITY OF NUCLEI IN ANY GIVEN SPECIMEN TAPERS OFF BY THE SAME FRACTION OVER EQUAL INTERVALS OF TIME. FOR EXAMPLE, THE SUBSTANCE WITH A 1-MINUTE HALF-LIFE LOSES HALF OF ITS INITIAL NUMBER OF PARENT NUCLEI IN THE FIRST MINUTE (SEE THE BLACK DASHED LINES THAT ARE PARALLEL TO THE AXES). DURING THE SECOND MINUTE, IT LOSES HALF OF THE NUMBER THAT REMAINED AFTER THAT FIRST MINUTE. LIKEWISE, AFTER 3 MINUTES, ONLY HALF OF THE RADIOACTIVE NUCLEI THAT WERE PRESENT AT THE END OF 2 MINUTES REMAIN, AND SO ON. THE LONGER THE HALF-LIFE, THE MORE SLOWLY THE SUBSTANCE LOSES ITS NUCLEI AND THE MORE LEVEL THE CURVE THAT REPRESENTS ITS DECAY.

Rutherford then solved the mystery of why radium seems to give off radiation at a constant rate. He found the half-life of radium (^{226}Ra) to be 1620 years; radium has such a slow rate of decay that the *change* in the energy emitted per second was difficult to detect. Radium glows in the dark not because the number of decays per second is extremely large but because each individual decay emits so much energy. The decay of a single radium nucleus releases as much energy as a million or more visible photons carry.

A ROCK CONTAINING A URANIUM-BEARING MATERIAL, TOGETHER WITH A PHOTOGRAPH MADE IN THE DARK. THE RADIOACTIVE AREAS OF THE ROCK EXPOSED THE FILM.

The statistical nature of nuclear decays reflects the statistical nature of quantum transitions from one energy level to another. That is, the instant at which any given radioactive nucleus will decay is *unpredictable.* For the same reason, it is impossible to know exactly when an excited electron in an atom will return spontaneously to its ground state, emitting a photon.

SECULAR EQUILIBRIUM OF ISOTOPES

Imagine what would happen if a machine could turn out nuclei of a radioactive isotope at a constant rate, say 100 per second. After 10 seconds, there would be 1000 nuclei minus those that decayed during that time. Of course, the number that decay in 10 seconds depends on the half-life, but it also depends on how many nuclei are in the sample because a given fraction will always decay in a given time. As time passed and the machine kept producing nuclei, the total number of these unstable nuclei would increase. Eventually there would be enough nuclei so that 100 would decay every second. At this point, the rate at which nuclei were added would be equal to the rate at which they would disappear, and *the net amount of radioactive substance would be constant.* Sometimes in nature a radioactive isotope is created at a constant rate, and such an isotope is therefore present in a constant amount. This condition is called **secular equilibrium** (Figure 27-7).

Because of radium's long half-life, the number of decays per second from a specimen of radium is essentially constant over a human lifetime. That means radon, with a half-life of a mere 3.8 days, is created at a constant rate and is in secular equilibrium with its parent radium. The total amount of radon present on earth because of radium decay is constant.

SECULAR = LONG-TERM

THE RADIOACTIVE SERIES

All isotopes of the elements having atomic numbers greater than 83 are radioactive, emitting (almost exclusively) alpha and beta particles. These atomic species are disappearing. For example, radium nuclei that were present when the earth was formed from materials orbiting the young sun 4.5 billion years ago are long gone; they have all decayed. Radium's half-life of 1620 years guarantees that. Yet there is a constant amount of radium in uranium ore today, in secular equilibrium with its long-lived parent uranium. Uranium's half-life is very close to the age of the earth, so about half of the initial uranium atoms are still here.

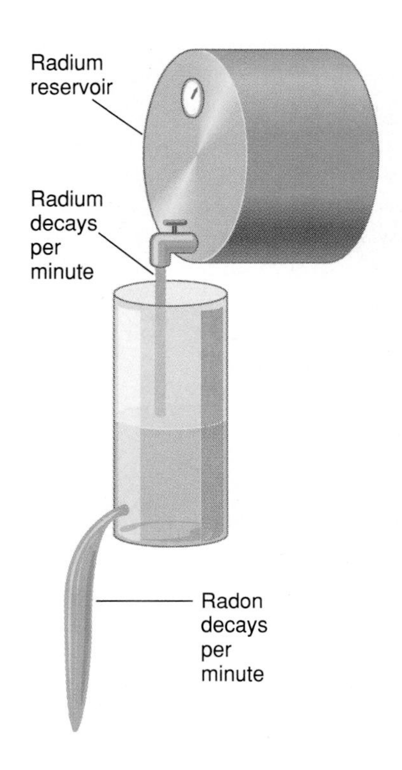

FIGURE 27-7 TO UNDERSTAND HOW THE SECULAR EQUILIBRIUM OF A DAUGHTER NUCLEUS COMES ABOUT, THINK OF WATER BEING POURED SLOWLY INTO A CONTAINER THAT HAS A HOLE NEAR THE BOTTOM. THE AMOUNT OF WATER IN THE CONTAINER INCREASES UNTIL THE PRESSURE AT THE BOTTOM FORCES JUST AS MUCH WATER OUT PER SECOND AS COMES IN AT THE TOP. THE AMOUNT OF WATER IN THE CONTAINER THEN REMAINS CONSTANT. IN THIS ILLUSTRATION, A SPECIMEN OF PURE RADIUM HAS A NEARLY CONSTANT NUMBER OF DECAYS PER SECOND, CREATING RADON AT A CONSTANT RATE, ANALOGOUS TO THE STREAM OF WATER. THE SIZE OF THE HOLE IN THE CONTAINER CORRESPONDS TO THE HALF-LIFE OF THE RADON. RADON BEGINS TO ACCUMULATE, AND THE AMOUNT OF RADON PRESENT GROWS UNTIL JUST AS MANY RADON NUCLEI DECAY EACH SECOND AS ARE CREATED BY NEW RADIUM DECAYS. THEN THE NET AMOUNT OF RADON PRESENT IS CONSTANT IN TIME, OR IN SECULAR EQUILIBRIUM.

CARBON DATING OF ORGANIC MATERIALS

Cosmic rays (mostly hydrogen nuclei) crash into our atmosphere from space 24 hours a day, raining down at a more or less constant rate. Most of these rays eventually hit the nuclei of atmospheric molecules and blast them apart. Among the debris are neutrons, which can strike the nuclei of nitrogen atoms—plentiful in our atmosphere as molecular nitrogen, N_2. When this happens, the neutron can knock a proton out of the nitrogen nucleus ($^{14}_{7}N$) as the neutron is absorbed, thereby creating a carbon-14 nucleus ($^{14}_{6}C$). The new-born carbon atoms combine with oxygen molecules to form molecules of carbon dioxide, CO_2. Carbon-14 is radioactive, with a half-life of 5730 years. Created at a reasonably steady rate in the process just described, it is in approximate secular equilibrium, so the amount of it in the air remains relatively constant. In the carbon dioxide in the air we breathe, there is roughly one carbon-14 atom for every trillion (10^{12}) carbon-12 atoms.

Plants use carbon dioxide from the atmosphere (and emit oxygen) in their life processes, so, while it is alive, any plant has the same ratio of carbon-14 to carbon-12 atoms as does the atmosphere. Because all animals (including humans) depend on these plants through the food chain, they, too, have carbon-14 and carbon-12 in this same ratio. When a plant or animal dies, the percentage of carbon-14 in the dead body gradually diminishes—since no more carbon-14 is being absorbed. The carbon-14 activity from a sample of old organic matter compared with the carbon-14 activity in the atmosphere at present indicates when the exchange of carbon dioxide stopped—in other words, when the organism died. Thus, the age of the wood from a sunken Spanish galleon, of a scrap of bone from an ancient campfire, of seeds from an Egyptian tomb, or of material from a Dead Sea scroll can be determined with fair precision—up to an age of about 50,000 years, after which time the carbon-14 activity becomes too small to measure reliably with today's instruments.

RADIOACTIVE DATING IN ARCHAEOLOGY

The heavy radioactive isotopes found in minerals today occur in patterns, or series. The only way these isotopes lose significant mass is by alpha decay, when they lose four nucleons at once. Very often the large daughter nucleus that remains is also unstable and emits an alpha particle or a beta particle. If the second daughter is unstable as well, the process continues until what remains of the original nucleus is finally stable. For example, a nucleus of uranium-238 decays through the sequence of nuclear-mass change:

$$238 \rightarrow 234 \rightarrow 230 \rightarrow 226 \rightarrow 222 \rightarrow 218 \rightarrow 214 \rightarrow 210 \rightarrow 206$$

In this way, uranium-238 nuclei evolve into a stable isotope of lead, $^{206}_{82}Pb$, ending the radioactive series. See Table 27-1 for the details.

Table 27·1 Consecutive Nuclear Isotopes in the Uranium-238 Series of Radioactive Elements

Element	Isotope	Half-Life	Types of Radiation
Uranium	$^{238}_{92}U$	4.55×10^9 years	Alpha
Thorium	$^{234}_{90}Th$	24.1 days	Beta, gamma
Protactinium	$^{234}_{91}Pa$	6.6 hours	Beta, gamma
Uranium	$^{234}_{92}U$	2.48×10^5 years	Alpha
Thorium	$^{230}_{90}Th$	7.6×10^4 years	Alpha, gamma
Radium	$^{226}_{88}Ra$	1622 years	Alpha, gamma
Radon	$^{222}_{86}Rn$	3.8 days	Alpha
Polonium	$^{218}_{84}Po$	3.05 minutes	Alpha
Lead	$^{214}_{82}Pb$	26.8 minutes	Beta, gamma
Bismuth	$^{214}_{83}Bi$	19.7 minutes	Alpha, beta, gamma
Polonium	$^{214}_{84}Po$	1.6×10^{-4} second	Alpha
Lead	$^{210}_{82}Pb$	19 years	Beta, gamma
Bismuth	$^{210}_{83}Bi$	5.1 days	Beta
Polonium	$^{210}_{84}Po$	138 days	Alpha, gamma
Lead	$^{206}_{82}Pb$	Stable	—

DATING ROCKS

The half-life of uranium-238 is about 4.5 billion years. Uranium that is frozen in the solid structure of rocks decays, but the decay products are trapped at the same location. So the fraction of uranium in a rock compared with the fraction of its decay products gives an accurate age of the rock—the length of time since it formed as a solid.

Using this technique, geologists have learned that those rocks containing the oldest fossil shells of sea animals are about 500 million years old, and the rock surrounding the bones of our earliest ancestors dates back about 4 million years. First prize for the most ancient mountain formations goes to the Appalachians of the eastern United States. The rocks of Grandfather Mountain in North Carolina, for instance, are 1 billion years old. However, the oldest rocks on earth—some 4 billion years old—lie in Greenland. Moon rocks collected by the Apollo astronauts date back 4.6 billion years, only slightly predating meteorite material that is 4.55 billion years old.

GRANDFATHER MOUNTAIN IN NORTH CAROLINA, JUST OFF THE BLUE RIDGE PARKWAY. ROCKS ON THE FACES SEEN HERE ARE 1 BILLION YEARS OLD. THE APPALACHIANS, OF WHICH THE BLUE RIDGE IS ONE PART, WERE ONCE HIGHER THAN THE HIMALAYAS ARE TODAY; EROSION HAS GRADUALLY WORN THEM DOWN.

Physics *and Environment*

RADIOACTIVE DATING IN GEOLOGY

Notice from Table 27-1 that each decay in the uranium-238 series skips three atomic masses. That is, between 238 and 234 there are three possible isotopes, those having the masses 237, 236, and 235. If the atomic mass of each of *these* nuclei were reduced by four through an alpha emission, then four again, and so on, each of the three nuclei could be part of other series that don't coincide with the uranium-238 series. Two of these chains are found in the elements of earth. Uranium-235 decays through the sequence 235 $\rightarrow$ 231 $\rightarrow$ 227 $\rightarrow$ 223 . . . and becomes stable at the isotope of lead ^{207}Pb. A natural isotope of uranium with a mass of 236 is nonexistent today, but thorium, ^{232}Th, is present and passes along the sequence 232 $\rightarrow$ 228 $\rightarrow$ 224 $\rightarrow$ 220 and so on to end with yet another stable isotope of lead, ^{208}Pb.

The radioactive chain that would include neptunium-237 is not found on earth except for its end product, the stable nucleus bismuth-209. The reason is that the longest-lived isotope in that chain, neptunium-237, has a half-life of only 2 million years, a very short time compared to the earth's age. Whatever isotopes the earth may once have had from that chain are gone, having passed through too many half-lives to survive.

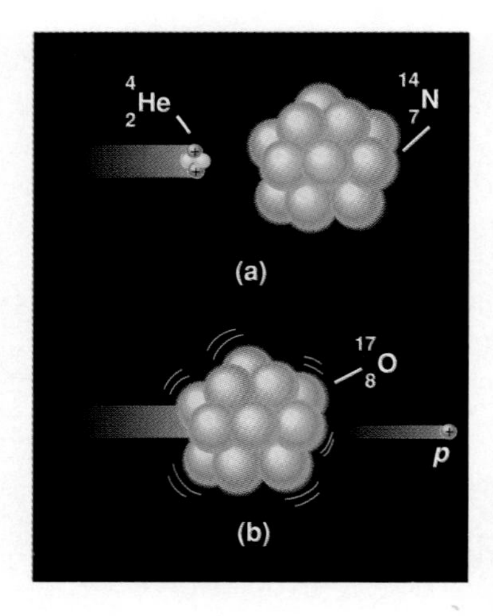

FIGURE 27-8 THE FIRST TYPE OF NUCLEAR REACTION PRODUCED BY EXPERIMENT. (A) AN ALPHA PARTICLE, WHICH IS A HELIUM NUCLEUS ($^{4}_{2}$HE), COLLIDES WITH A NITROGEN NUCLEUS. (B) THE NITROGEN NUCLEUS ABSORBS THE ALPHA PARTICLE AND EMITS A PROTON, TRANSMUTING THE ORIGINAL NUCLEUS INTO AN ISOTOPE OF OXYGEN. IN SYMBOLS, $^{4}_{2}\text{He} + ^{14}_{7}\text{N} \rightarrow ^{17}_{8}\text{O} + ^{1}_{1}\text{H}$.

NUCLEAR REACTIONS

Both in the spring and at the "fall of the leaf," Newton's secretary wrote, the great genius spent about 6 weeks in uninterrupted, intense investigation into the chemical changes of matter. He, like other alchemists of his day, tried to *transmute* chemical elements, that is, to turn one into another. Unlike Newton's work in physics, however, these sessions never paid off. What he had no way of knowing is that the changes he and the other alchemists sought come not from chemical reactions but from *nuclear* reactions.

Rutherford ushered in a new field of physics when he first produced a nuclear reaction in 1918 (Figure 27-8). He sent alpha particles through nitrogen gas, and a few of the alpha particles struck the nitrogen nuclei and were absorbed. As soon as this absorption took place, the new nuclei ejected protons. Rutherford saw that given the proper subatomic particles for projectiles, he could probe nuclei for more knowledge and perhaps transmute the isotopes of the elements throughout the periodic table. He had only the radiations from natural radioactive isotopes to use for this purpose, but he began to think of a machine designed to accelerate protons to such speeds that they could reach a nucleus despite the strong electric repulsion it exerts.

In 1932 two of Rutherford's students, John Cockroft and Ernest Walton, managed to build a device to accelerate protons. When aimed at lithium nuclei, the protons from their machine caused the lithium to turn into a beryllium isotope not found in nature. These experiments began the modern investigations into the properties of the nucleus. The first crude accelerators led to larger, more powerful machines whose subatomic projectiles can penetrate or even shatter a nucleus. Today gigantic particle accelerators are the biggest machines on earth. More than 100 stable isotopes have been created with accelerators, bringing the total number of stable isotopes known to about 300. However, an even greater number of unstable nuclei have been created with particle accelerators. More than 1300 artificial radioactive isotopes have been studied, and thousands of others are known to exist.

HIGH-ENERGY NUCLEAR REACTIONS AND ELEMENTARY PARTICLES

At one time, atoms were thought to be the most elementary particles, but about 100 years ago the electron was discovered (1899), followed by Rutherford's discovery of the proton in 1919. When he later predicted the existence of the neutron, it seemed that there were four elementary particles: photon, electron, proton, and neutron. But the same year the neutron was discovered (1932), the positron was also discovered. Shortly after that came particles called *pions* and *muons*. Experimenters found that, when a nucleus is smacked with very high-energy particles, many particles emerge. Several hundred such particles, all unstable, have been found to date. These particles decay much more rapidly than most unstable nuclei. Many of them have half-lives of about 10^{-8} second. Light travels only about three meters in that short time. Others have half-lives more than a trillion times shorter than that! The study of these particles has led to a remarkable theory.

In this theory, there are two families of elementary particles, **quarks** and **leptons**. All the particles that interact by means of the strong interaction seem to be made of collections of quarks. Protons and neutrons, for example, are each made of three quarks. Leptons include the electron and the neutrinos; they seem to be point-like, with no internal structure—they are not made of quarks. It is thought that there are six quarks and six leptons (plus their antiparticles), and together they "explain" the other particles that have been found to date. They even helped predict some of those particles before they were found, which is an important test for a theory's validity.

An isolated quark has never been detected. Quarks are held together to form all the nuclear particles—the protons, neutrons, pions, and so forth. At present, it appears that the force keeping the quarks together must *increase* if the quarks begin to separate. It could be that we may never be able to add enough energy to a nucleon to pull the quarks apart! If so, we may *never* see an isolated quark. However, at this time there is experimental evidence for all of the 6 quarks, gained from details of collisions in particle accelerators. The frontier of understanding of the elementary particles of nature is being pushed back by research teams at the high-energy particle accelerators. In the coming years, you will no doubt read of their discoveries in the news.

RADIATION AND YOU

RADIATION IN THE ENVIRONMENT

Your body is built from the food you eat and drink and the air you breathe, and because you come from the earth, you share its radioactivity. Everywhere in the rocks and soil and water and air of our planet are traces of radioactive elements. (Lighter radioactive elements, such as hydrogen-3, carbon-14, and potassium-40, are created when incoming cosmic rays shred the nuclei of air molecules.) This natural radioactivity is called **background radiation**. It is as much a part of your environment as sunshine and rain, and if our bodies couldn't tolerate it, we wouldn't be here.

Yet radiation damages molecules. When alpha and beta particles and gamma rays penetrate matter, they ionize the molecules along their paths, an action that changes chemical properties, playing havoc with molecules. If a cosmic ray causes a transmutation in a nucleus, the atom suddenly becomes an atom of a different element, again affecting the parent molecule. Therefore we should be concerned about any additional radiation we receive. Both alpha and beta particles have very short paths in air before they are stopped, and much shorter paths in liquids or gases, so they are harmful only if the radioactive atoms that emit them are within your body—that is, in the food you eat or the air you breathe. Gamma rays have much greater penetrating ability.

For most of us, about 85 percent of the radiation we get each year comes from background radiation. Most of the other 15 percent comes from diagnostic X-rays, color television, and extra cosmic radiation received during jet airplane travel, when you are above much of the atmospheric shield. Two percent or so comes from fallout from atmospheric tests of nuclear weapons decades ago. (Since 1963 such tests have been banned by most of the countries that have

Table 27·2 Approximate Percentages of an Average Person's Annual Radiation Dose from Natural Causes

Source	Percent
Food	15
Cosmic rays (sea level)	10
Ground and buildings	10
Radon	65

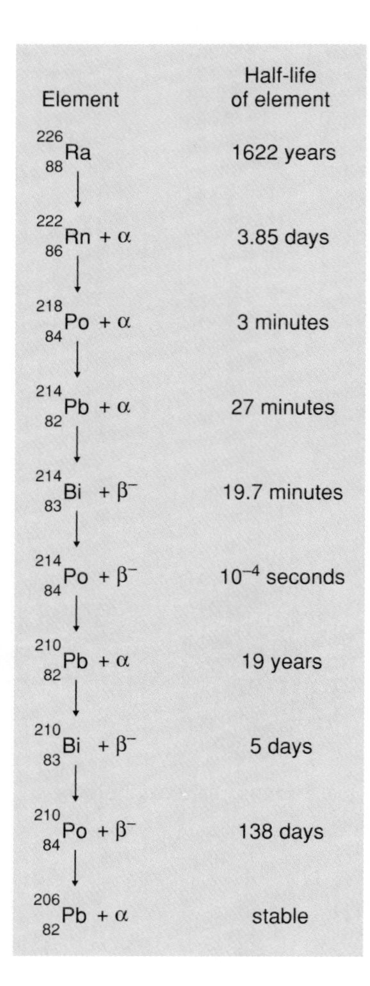

FIGURE 27-9 **RADIUM AND ITS DAUGHTERS ARE ALL RADIOACTIVE. AFTER RADON DECAYS, THERE ARE SEVEN MORE RADIOACTIVE ELEMENTS BEFORE THE REMAINING NUCLEUS BECOMES STABLE AT LEAD-206.**

nuclear weapons.) Nuclear power plants contribute a tiny fraction of the 15 percent of nonnatural radiation we receive.

Here is a typical breakdown of the 85 percent of exposure that comes from natural radiation. About 15 percent comes from food, and at sea level about 10 percent comes from cosmic rays. About 10 percent comes from the ground and buildings because construction materials come from the earth and therefore contain traces of radioactive elements. When you are inside a building, these materials surround you, and you are exposed to this radiation from all directions. For this reason, in a tent you get less than half of the radiation you get while inside a wooden building. Buildings made of concrete block or brick are a bit more radioactive than wooden buildings, and buildings of stone are the most radioactive of all. However, if you are in a mine or in a basement or even in a house that has been closed up for awhile, the air you breathe will have accumulated radon, the daughter of radium decays. Radon accounts for about 65 percent of the natural radioactivity we receive on the average. Radon is an inert gas, and therefore when it is created in the ground or in the brick or cement blocks of a home, it escapes as a gas into the air. Radon has a short half-life, and there is only a small trace of radon in the air you breathe. However, when a radon atom decays inside your body, the energetic alpha particle ionizes a huge number of molecules in its path. Even worse, radon's daughter, granddaughter, and so on, are also radioactive (Figure 27-9). For this reason, it is best to exchange the air in a home or basement with outside air as regularly as possible.

At sea level the atmosphere is a good protective blanket against cosmic rays, but not if you live high above sea level. A person living in Denver, Colorado, is above much of the atmosphere and gets more than twice as much cosmic radiation as someone who lives near sea level. Traveling by jet at 11,000 m (36,000 ft) puts you above 80 percent of the atmospheric shield and into the thick of the cosmic radiation. A round-trip flight between New York City and Los Angeles exposes you to the equivalent of half the radiation dose of a chest X-ray. Airline personnel are limited to a certain number of hours of air time a year because of this extra radiation.

Summary and Key Terms

nuclear force–407
strong interaction–407

Protons and neutrons stay together in the nucleus because of the ***nuclear force***, or ***strong interaction***, an extremely strong attraction between nucleons that are close enough to "touch." Nucleons, like electrons, follow the rules of quantum mechanics, forming shells that accommodate a certain number of nucleons. A nucleus with a nucleon in an excited state emits a gamma ray when the nucleon returns to the ground state. Some nuclei are unstable, or ***radioactive***. Radioactive ***parent nuclei*** decay into ***daughter nuclei***, releasing a significant amount of energy.

radioactive–408
parent nuclei–409
daughter nuclei–409

All nuclei having an atomic number above 83 are unstable. These nuclei can spontaneously emit alpha particles (helium nuclei). The stability of a nucleus depends on the ratio of neutrons to protons; a plot of stable nuclei reveals a *line of stability*. A light nuclear isotope that is neutron-rich or neutron-poor can undergo ***beta decay***. A ***beta particle*** is either an electron or a ***positron***, the *antiparticle* of the electron. Electron emission is accompanied by an antineutrino, and positron emission by a neutrino. ***Neutrinos*** are massless, chargeless particles that travel at the speed of light. Beta decay is due to the weak interaction, which is unified with the electromagnetic force in the ***electroweak force***.

beta particle–409
beta decay–409
positron–408
neutrinos–410
electroweak force–409

The time of decay of any given radioactive nucleus is unpredictable, and nuclear decays follow a statistical process: *A collection of radioactive nuclei always loses the same fraction of its parent nuclei in the same amount of time.* The time it takes for half the nuclei in a radioactive sample to decay is its ***half-life***. If nuclei of a radioactive isotope are produced at a constant rate, eventually the daughter will reach ***secular equilibrium***, the point at which nuclei are produced and disintegrate at the same rate.

half-life–412
secular equilibrium–413

Background radiation comes from radioactive minerals and from incoming cosmic rays. Whenever a nucleus is struck with very high-energy particles, many unstable particles emerge. Subatomic particles are made up of ***quarks*** and ***leptons***, which are the most elementary of particles.

background radiation–417
quarks–416
leptons–416

EXERCISES

Concept Checks: Answer true or false

1. The nucleus contains the atom's protons and neutrons, also known as the nucleons. For the lighter elements that are stable, protons and neutrons are found in about equal numbers in the nuclei.
2. Individual atoms of an element always contain the same number of nucleons. Isotopes of an element have different atomic numbers.
3. A short-range attractive force holds the nucleus together; it is called the electroweak interaction.
4. Because the protons in the nucleus have like charge, they also experience a force other than the nuclear force.
5. The more massive radioactive elements found naturally on earth emit radiation in the form of alpha particles.
6. In a specimen of radioactive nuclei, a precise number of nuclei always decay each second.
7. Secular equilibrium is the process where two or more nuclei decay to give the same daughter nuclei at a constant rate.
8. Natural background radiation from the earth and the atmosphere accounts for more than 3/4 of the radiation received annually by an average person.
9. The nucleus always loses a neutron in beta decay, whether the beta particle is an electron or a positron.
10. Both positrons and neutrinos were predicted before they were discovered.

Applying the Concepts

11. How long would you likely have to wait to watch any sample of radioactive atoms completely decay?
12. Which type of natural radioactivity leaves the number of protons and the number of neutrons in the nucleus unchanged?
13. Which of the radiations—alpha, beta, or gamma—is emitted from a radioactive nucleus much like a photon is emitted from an excited atom?
14. When you look into a mirror, does anything of what you see come from the nuclei of atoms? When you push off to take a step, does that action have anything to do with the properties of the nuclei in your molecules?
15. Which particle can more easily approach a nucleus at close range, a proton or a neutron? Why?
16. Why do most alpha particles pass through small thicknesses of any kind of matter without a great deflection in their path? How would you expect electrons of the same energy to behave?
17. Would you say that the decay of a radioactive nucleus is subject to the laws of classical (Newtonian) physics or modern physics? Why?
18. Why is the nucleus the seat of the most energy in an atom? What has to be done in order to convert that energy to another form?
19. Early experimenters put samples of radioactive isotopes under high pressure to see if the increased pressure would change the rate of decay. It didn't. Discuss.
20. How much of a 1-gram sample of pure radioactive matter would be left after four half-lives?

21. If a proton and a neutron have exactly the same speed and are aimed at the same target, which will travel farther into the target? Why?

22. Which facts about atoms prevent you from causing nuclear reactions on your kitchen stove?

23. Tritium, ${}^3_1H$, is radioactive. It decays by emitting an electron. What is the daughter nucleus?

24. The nucleus is made of protons and neutrons, so how can electrons be emitted from the nucleus?

25. The principal cause of damage to living cells by nuclear radiation is from the ionization of the matter that the rays pass through. Why would this ionization be more harmful than the nuclear reactions these radiations might cause?

26. A rare type of radioactive decay involves a proton in the nucleus capturing one of the atom's electrons to become a neutron. When the isotope ${}^{190}_{79}Au$ undergoes such an electron capture, what does it become?

27. Gamma rays can be used to sterilize food, which then keeps fresh at room temperature for times exceeding its safe storage time in refrigerators. Discuss how an intense gamma-ray source might be put to use in an underdeveloped country with little or no electricity available and no refrigerated trucks.

28. Ashes from a campfire deep in a cave show a carbon-14 activity of only one-eighth the activity of fresh wood. How long ago was that campfire made?

29. The only stable isotope of sodium is ${}^{23}Na$. What sort of decay would you expect the neutron-poor ${}^{22}Na$ to undergo? What sort of decay should the neutron-rich nucleus ${}^{24}Na$ undergo?

30. Fill in the decay products. See the periodic table on the inside cover at the back of this book to identify any missing elements.

$${}^{254}_{100}Fm \rightarrow {}^{250}_{98}Cf + \_\_\_\_\_\_$$
$${}^{212}_{83}Bi \rightarrow \_\_\_\_\_\_ + {}^4_2He$$
$${}^{39}_{17}Cl \rightarrow \_\_\_\_\_\_ + e^- + \text{antineutrino}$$
$${}^{13}_{7}N \rightarrow \_\_\_\_\_\_ + e^+ + \text{neutrino}$$

31. Why can we still find short-lived isotopes on earth today?

32. Isotopes can be produced in any substance if it is exposed to radiation from a source such as a particle accelerator (or a nuclear reactor). Often the isotopes have short half-lives, making them valuable as tracers. Explain how such tracers could be used to locate leaks in an underground plumbing system, to monitor the wear of piston rings in a car, or to follow medical drugs through someone's body after ingestion.

33. Uranium-238 has a half-life of 4.5×10^9 years. It so happens that the earth appears to be about 4.5×10^9 years old. How much uranium-238 exists in the earth today compared with the amount present when the earth was formed?

34. Sodium-24 is radioactive, with a half-life of 15 hours. Because of its short half-life, it is useful as a tracer for some purposes in the human body. Once in the bloodstream, its absence (detected by lack of decays) in some area reveals circulation blockages. Because of this short half-life, however, ${}^{24}Na$ is not readily kept in stock. Suggest a way to make ${}^{24}Na$ from ${}^{23}Na$ at a hospital.

35. There is a nuclear emission and absorption spectrum from a nucleus just as the electron cloud of an atom has an emission and absorption spectrum. What similarity in structure with atoms do these nuclear spectra come from?

36. One of the best techniques for identifying atomic species in a sample of matter is called neutron-activation analysis. The sample is irradiated with medium-energy neutrons, which merely bump into the sample's nuclei and excite them. When a nucleus spontaneously loses its extra energy, it emits gamma rays of frequencies characteristic only to its spectrum. These nuclear "fingerprints" allow samples of pollutants in the air to be traced to their source, identify flecks of paint from hit-and-run accidents, and so on. Think of a way any product could be secretly protected against counterfeiting by a manufacturer using this technique.

37. Some nuclei will absorb a gamma ray and become so excited that some of the neutrons are "evaporated" from them. Is such a gamma ray found in the emission or absorption spectrum of that nucleus? Discuss.

38. Even though someone living in Denver gets about twice as much cosmic radiation as someone living in New York, the total background radiation in Denver is only about 20 percent higher than the average at sea level. Explain the difference.

39. Nuclear radiation spreads out, just as light does, to become less intense with a greater distance from the source. How would tripling your distance from a gamma-ray source affect the number of gamma-ray photons your body would intercept per second? Discuss.

40. Show that if a nucleus has a diameter 10^{-5} times the diameter of an atom, the nucleus occupies about $1/10^{15}$ of the atom's volume.

41. The isotope ${}^{238}_{92}U$ decays by emitting an alpha particle, ${}^4_2He$. From the ratio of neutrons to protons for ${}^{238}_{92}U$ and a similar ratio for ${}^{234}_{90}Th$, show that the thorium nucleus has a larger ratio of neutrons to protons than the uranium nucleus does.

42. The heat in the earth's interior today comes from the decay of radioactive nuclei there; the decaying atoms are very few in number compared with the number of stable atoms, of course, so why would earth's interior be so hot just from radioactivity?

43. In 1994 the EPA estimated that indoor radon causes between 7000 and 30,000 of the 130,000 deaths in the United Sates each year from lung cancer, putting it second only to smoking as the cause of that disease. In areas where the mineral content of the ground is higher in uranium, the radon level is higher. The EPA suggests putting pipes in the foundations of the homes there to circulate the air, mixing it with outdoor air. The data the EPA uses extrapolates from the very high lung cancer rates of uranium miners, who were exposed to radon levels hundreds or thousands of times higher than found in outdoor air. Not all scientists agree with the EPA estimate. What would be the difficulty in doing an experiment to check the estimates? Discuss.

TWENTY-EIGHT

Nuclear Energy and Our Energy Future

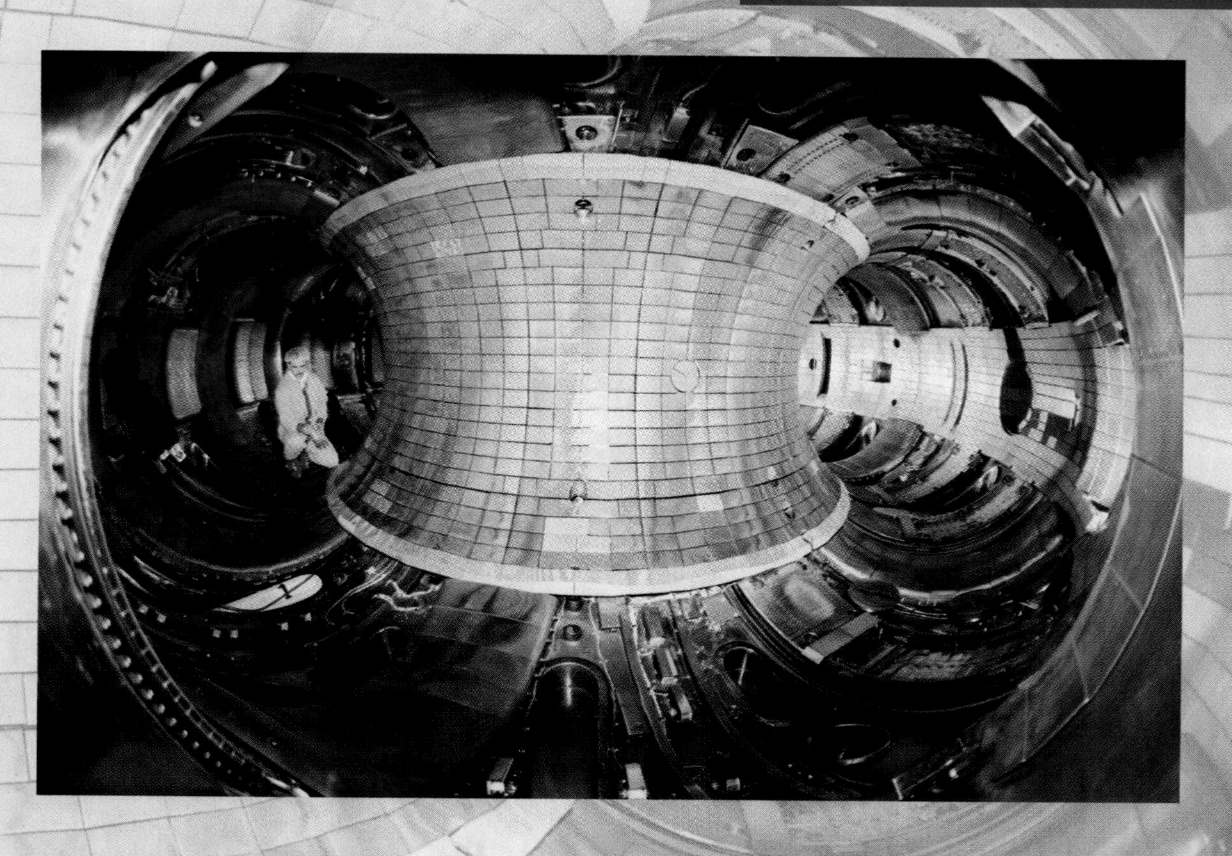

The Tokamak Fusion Test Reactor at Princeton. The doughnut-shaped interior is evacuated and then almost filled with a hot plasma of deuterium and tritium, kept from the walls by magnetic fields.

THE ELECTRON-VOLT IS THE UNIT OF ENERGY MOST OFTEN USED FOR ATOM-SIZED PROCESSES. ONE ELECTRON-VOLT IS THE ENERGY GIVEN TO AN ELECTRON AS IT MOVES THROUGH A POTENTIAL DIFFERENCE OF 1 VOLT AND IS EQUAL TO 1.6×10^{-19} JOULES.

Our distant ancestors who first learned to start a fire brought about a revolution. In time, early people learned to use fire to smelt ores, bringing about first the Bronze Age and then the Iron Age. Thousands of years later their descendants used the heat from fires to drive steam engines, and today with coal and oil for fuel, we use fires to generate electricity. In the late 1930s and early 1940s, we learned to start another kind of fire. Its heat comes from changes in the nuclei of atoms rather than from rearrangements of their electrons in a chemical reaction, and it releases far more energy per kilogram of fuel. In this chapter we discuss the details of nuclear energy—its generation, its prospects, and its problems. Then we look briefly at other sources of energy and their problems.

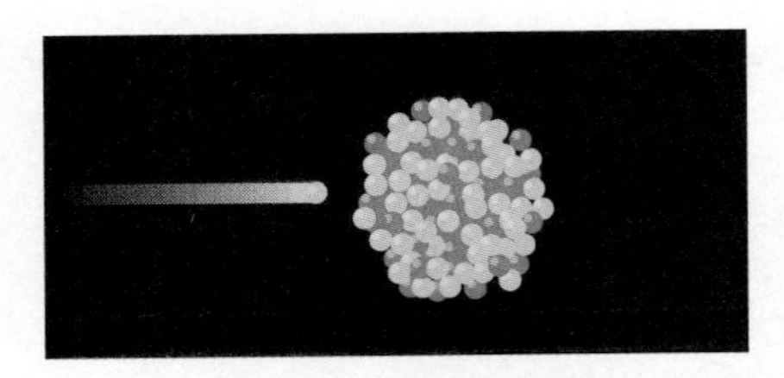

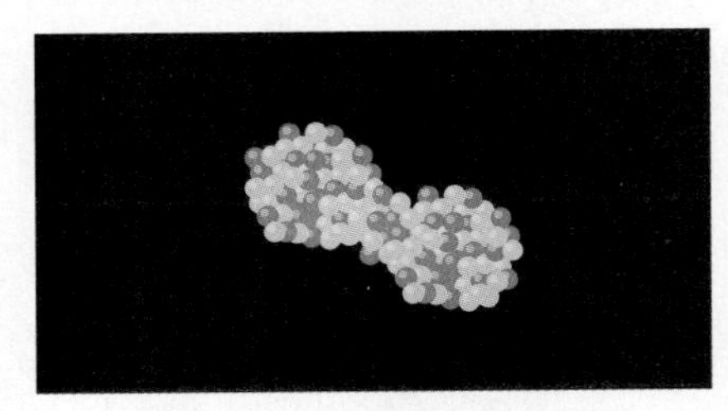

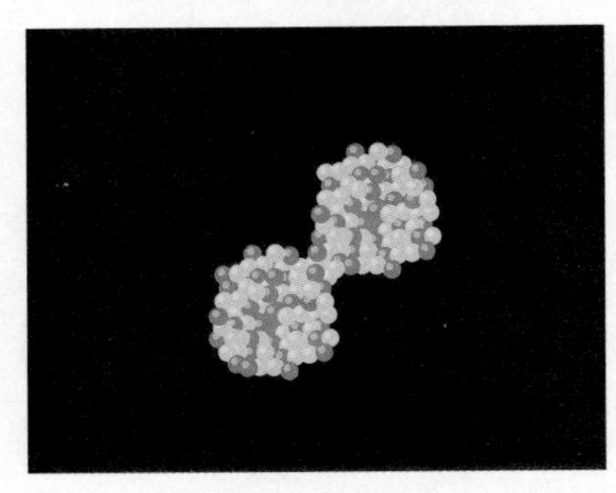

THE DISCOVERY OF NUCLEAR FISSION

In the autumn of 1938, physicist Lise Meitner was forced to flee from the Nazis in Germany, where she had worked in Berlin with a chemist, Otto Hahn, for over 30 years. These two scientists were famous for their studies of radioactivity and nuclear transmutations, and at the time of Meitner's flight to Sweden, they were studying the particles formed when the heaviest known natural element, uranium, was bombarded with neutrons. Unlike alpha particles and protons, neutrons are not repelled by the positive charge of the nucleus. Even slow-moving neutrons can enter a nucleus to *create a different nucleus*, an isotope that may or may not be stable.

That winter, her nephew Otto Frisch, a young physicist at Bohr's Institute in Copenhagen, came to visit with her. On the first morning of his visit, Frisch found his aunt going over a letter from Berlin. Hahn thought that uranium, upon absorbing a neutron in his experiments, was emitting (at least part of the time) the element barium. This result was totally unexpected. *The uranium nucleus had split almost in half!* The only nuclear transmutations known until then involved emissions of protons, electrons, alpha particles, or neutrons—all small particles. Why should a tiny nudge from a single neutron cause the uranium nucleus, with more than 200 nucleons, to split into two large nuclei?

Meitner and Frisch calculated what had happened. The two large nuclei from the split uranium nucleus would push apart because of their large positive electric charges, giving them *enormous* kinetic energy. A comparison with an ordinary chemical reaction shows us how energetic the splitting of a uranium-235 nucleus is. When wood burns in a fireplace, carbon atoms chemically unite with oxygen molecules to make molecules of carbon dioxide. The rearrangement of the electrons that makes the bonds in a single carbon dioxide molecule releases 4.1 electron-volts of energy (see the note in the column). In contrast, the splitting, or **fission**, of a single uranium nucleus releases some *200 million* electron-volts of energy. This was no fire in a fireplace; it was an enormous release of energy from a very small amount of matter.

Frisch went back to Copenhagen and gave Bohr the news just before Bohr was to leave for a visit to the United States. Bohr realized at once how this reaction could take place because of an earlier model of the nucleus that he had put forth, as we'll see in the next section. (Bohr exclaimed, "Oh, what idiots we have all been!") Then Bohr sailed to America, and Frisch did the

FIGURE 28-1 AN ARTIST'S CONCEPTION OF THE FISSION OF A URANIUM-235 NUCLEUS. THE SLOW NEUTRON IS ABSORBED AND DEFORMS THE NUCLEUS, WHICH OSCILLATES ENOUGH FOR TWO LOBES TO APPEAR AND SEPARATE. THREE NEUTRONS APPEAR IN THE BOTTOM FRAME, ALMOST LIKE SPLINTERS, WHEN THIS URANIUM-235 NUCLEUS SPLITS.

experiments that proved uranium nuclei do indeed absorb neutrons and split. That is how the news of nuclear fission came to America, and within a week American physicists had duplicated the experiments.

THE DETAILS OF NUCLEAR FISSION

In a drop of water, the short-range cohesive attraction among the molecules pulls the surface molecules inward, creating a surface tension that pulls the drop into a sphere (see Figures 10-16a and b). Bohr reasoned that, since nucleons are also held together by a short-range attraction, the nuclear force, the nucleus should act like a liquid drop. If a neutron entered this nuclear drop, the drop's surface might elongate and oscillate back and forth (Figure 28-1). If conditions were right, the quivering nucleus might even stretch thin at its middle, and the two positively charged lobes could repel each other enough to overcome the surface tension (caused by the nuclear force) and tear the nucleus apart.

In the United States, Bohr and the American physicist John Wheeler used this liquid-drop model to work out the details of nuclear fission. They found that uranium-238, the most common isotope of uranium, can absorb a neutron and become uranium-239 without splitting, but that the rarer isotope uranium-235 has up to an 85 percent chance of splitting if it absorbs a slow neutron. A fast neutron can also cause fission in ^{235}U, but it is not as likely to start the necessary oscillations; tap a bowl of Jello and it wiggles, but shoot a bullet through it and it hardly moves.

n ^{235}U

(a)

n $^{92}_{36}$Kr $^{141}_{56}$Ba *n* *n*

(b)

FIGURE 28-2 DETAILS OF ONE MODE OF FISSION OF URANIUM-235. BARIUM AND KRYPTON ARE COMMON PRODUCTS OF THIS FISSION.

CHAIN REACTIONS

When a uranium nucleus splits, neutrons are certain to be emitted. Suppose a ^{235}U nucleus splits, and its large fragments are a barium nucleus and a krypton nucleus, as shown in Figure 28-2. The uranium nucleus has more neutrons than the total number of neutrons two medium-sized nuclei ordinarily have. (Recall from Chapter 27 that heavier elements have a higher neutron-to-proton ratio than do lighter elements.) So neutrons appear in the energetic fission of the uranium nucleus. Under the right conditions, these fission neutrons can go on to cause nearby ^{235}U nuclei to fission. Those nuclei can release more neutrons that can then cause more fissions (Figure 28-3a). Any process like this that once started keeps itself going is called a **chain reaction**.

^{235}U nuclei typically emit either 2 or 3 fast neutrons per fission (the average is 2.5). If on the average only one neutron released during each fission eventually causes another fission, a **self-sustaining** chain reaction takes place. If fewer than one neutron per fission on the average causes another fission, the chain reaction dies out. Think, though, about what happens if *more* than one neutron per fission goes on to cause another fission. The number of uranium

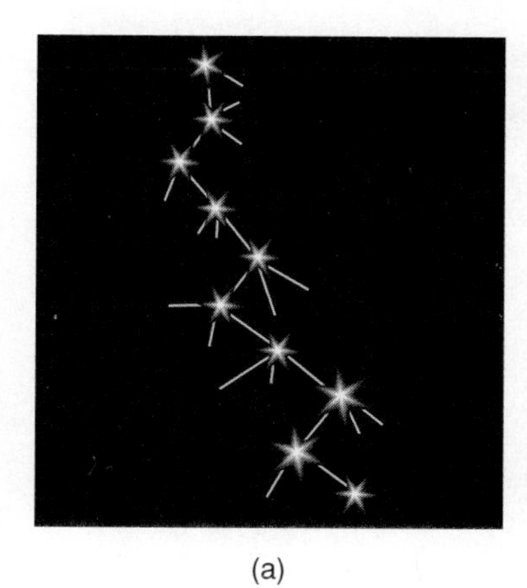

(a)

(b)

FIGURE 28-3 (A) A SELF-SUSTAINING CHAIN REACTION. TO SIMPLIFY THE DRAWING, THE FISSIONS ARE ALL GATHERED IN THE DOWNWARD DIRECTION HERE. THE GOLD LINES INDICATE FISSION NEUTRONS. ON THE AVERAGE, ONLY ONE FISSION IS INDUCED FOR EACH FISSION THAT OCCURS. EACH FISSION EVENT PRODUCES EITHER 2 OR 3 NEUTRONS, FOR AN AVERAGE OF 2.5. (B) A RUNAWAY CHAIN REACTION. MORE THAN ONE FISSION IS INDUCED PER FISSION, AND THE NUMBER OF FISSIONS GROWS EXTREMELY QUICKLY.

nuclei splitting each second *grows*, leading to a **runaway** chain reaction, as in Figure 28-3b. Because so much energy is released so fast, an explosion takes place. Just such a runaway reaction is the basis for nuclear weapons.

Uranium ore can be refined, just as iron ore can, to obtain very pure uranium. A sample of pure uranium metal contains about 99.3 percent ^{238}U nuclei and 0.7 percent ^{235}U nuclei, and the ^{238}U nuclei are very good at absorbing fast neutrons, the kind released during fission. To achieve a runaway chain reaction, then, a sample of uranium must meet two criteria. First, the sample must contain enough ^{235}U nuclei per unit of volume to allow the fission neutrons to cause more fissions rather than be absorbed by the ^{238}U nuclei. This means the uranium sample should be nearly pure ^{235}U nuclei. Finding a way to concentrate ^{235}U (to about 97 percent) was a tough technical challenge. This problem was worked out in the Manhattan Project, the code name for the World War II project to make an atomic bomb.

The second criterion has to do with the quantity of concentrated ^{235}U needed to initiate a runaway chain reaction. If the specimen is too small, many fission neutrons will escape from its surface and fewer than one neutron per fission will cause another fission—hence no chain reaction. Yet the instant a certain **critical mass** of nearly pure ^{235}U is assembled in the right shape and the first fission occurs, the reaction runs away almost instantly, causing an explosion.

ENERGY FROM NUCLEAR FISSION

CONTROLLING THE CHAIN REACTION: NUCLEAR REACTORS

We've seen that in pure uranium metal, non-fissioning ^{238}U is abundant. In uranium metal, a fission of a ^{235}U nucleus cannot lead to a runaway chain reaction—there are too many ^{238}U nuclei around, and they will absorb the fast neutrons released in the fission. Besides, a *slow* neutron needs to impact the ^{235}U nucleus in order for it to have the highest probability to fission. At the same time, ^{238}U absorbs fast neutrons well but *not* slow neutrons. So if something slows the fission neutrons, they are better able to sustain a chain reaction. As the fast neutrons rip through nearby atoms, they do not interact with the electrons since neutrons have no charge and electrons do not feel the nuclear force. If these neutrons are to slow down, they must collide with other nuclei. Fission neutrons slow best when they strike *small* nuclei. A marble shot at a bowling ball bounces off almost as fast as it came in, losing almost no energy. That same marble shot at another marble loses much of its energy, giving this energy to the target marble. To slow a fast neutron typically takes about 1/1000 second.

A nuclear reactor is a very simple machine (Figure 28-4). Uranium metal is shaped into rods, and then enough of these "fuel" rods are brought together to initiate a *controlled* chain reaction. The object is to get an average of one neutron per fission to cause another fission. In a typical reactor, there are movable control rods interspersed among the fuel rods. The control rods contain an element, such as cadmium or boron, whose nuclei *absorb* the neutrons that strike them. So the control rods control the rate of fission reactions. As long as the control rods are in place, the uranium fuel rods, with their large surface areas, lose too many neutrons to the control rods for a chain reaction to take place.

Besides the control rods, there must be a *neutron moderator* to slow the fast neutrons. Ideally, this moderating material should not absorb neutrons,

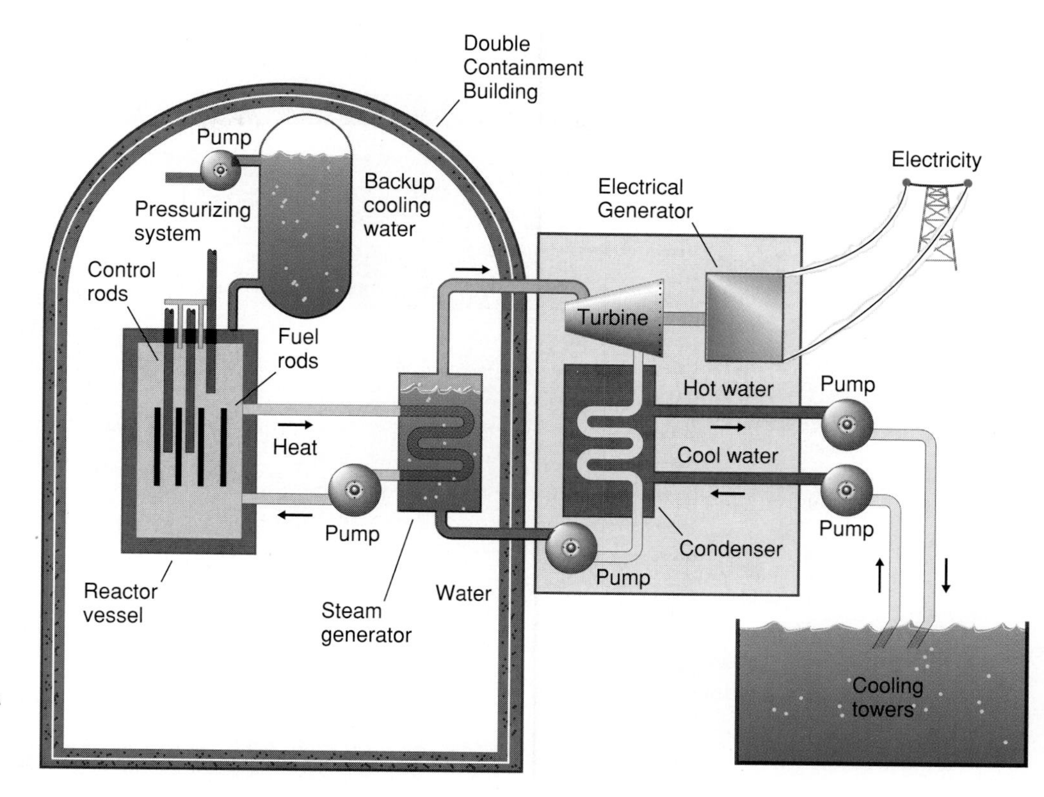

FIGURE 28-4 A SCHEMATIC OF A COMMERCIAL NUCLEAR POWER PLANT.

and its nuclei should be relatively light to absorb a lot of energy when neutrons strike them. When the control rods are slowly withdrawn and the fuel rods are exposed to each other's "moderated" neutrons, a self-sustaining chain reaction proceeds, giving a steady release of energy. When a ^{235}U nucleus fissions, the fragments quickly lose their energy as they rip through the rest of the uranium in the fuel rod, running into many electrons and occasionally another nucleus. This increases the temperature of the fuel rods, and a coolant circulates around the rods, collects the heat, and generates steam. This steam drives turbines that run electrical generators, just as in coal-fueled and oil-fueled power plants.

In the United States, water serves as both the neutron moderator and the coolant for commercial reactors. Pressurized water can be heated to high temperatures without boiling, and its two hydrogen nuclei per molecule slow the fast neutrons efficiently. However, hydrogen nuclei often *absorb* neutrons to become deuterons (^{2}H, or D). To make up for those lost neutrons, the uranium fuel rods used in U.S. reactors must be enriched, containing up to about 3 percent ^{235}U rather than the naturally occurring ratio of 0.7 percent. This small level of enrichment will not support an explosive runaway chain reaction, so a nuclear power plant can never explode like an atomic weapon.

The fuel rods, the control rods, and the moderator/coolant make up the reactor core. A typical pressurized water reactor in the United States uses about 200 tons of enriched uranium per year. In 1994 there were over 200 nuclear power reactors at about 100 plants in the United States, operated by about 40 utility companies. In more than 30 other countries there were more than 300 other nuclear reactors, and in some of these countries nuclear power generates more electricity than any other source. Figure 28-5 shows the percentage of electrical energy generated by nuclear power in countries around the world.

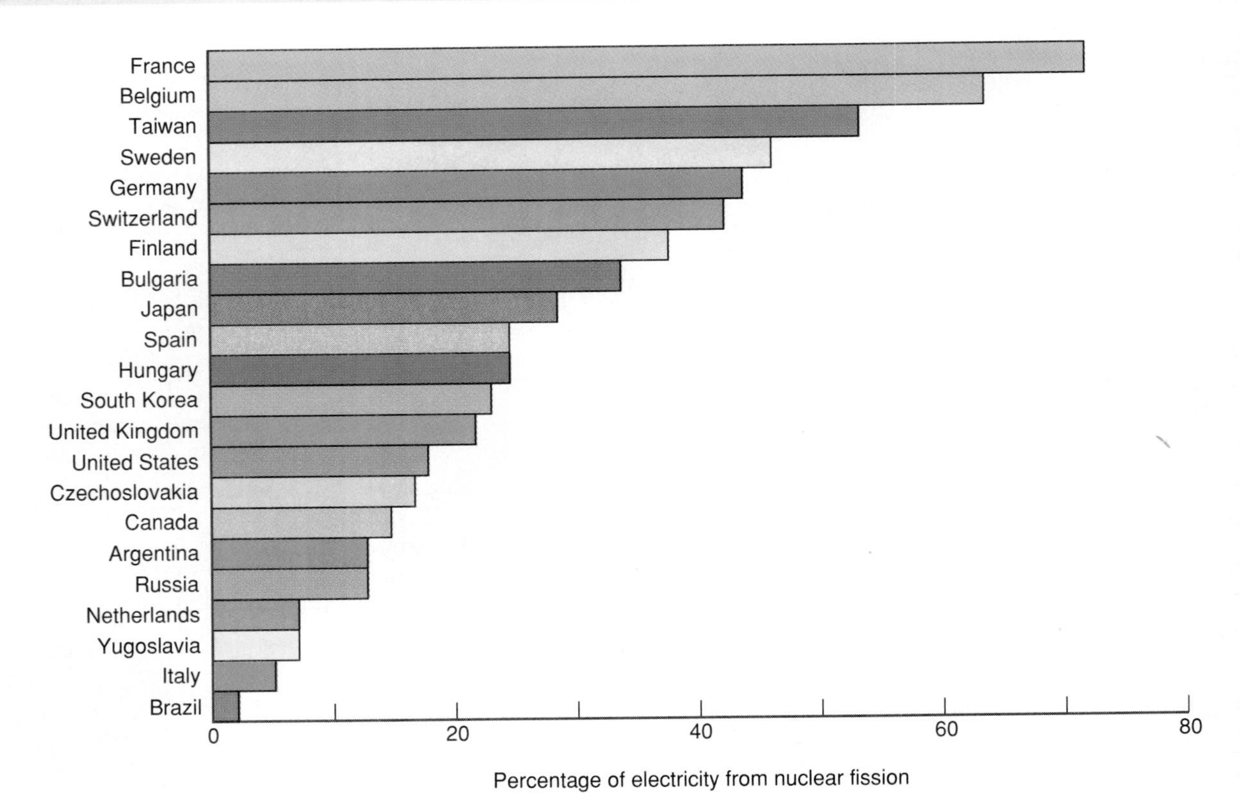

FIGURE 28-5 **PERCENTAGE OF ELECTRICITY GENERATED BY NUCLEAR POWER FOR ALL THE COUNTRIES THAT HAD NUCLEAR POWER IN 1990.**

PROBLEMS WITH NUCLEAR REACTORS

Fuel rods depleted of ^{235}U are treated as waste, and the storage of this waste is a serious problem since many of the highly radioactive isotopes it contains have long half-lives and their release into the environment would be extremely hazardous. As of 1995, reactor wastes are still stored on site at the nuclear power plants. Most countries now plan to store the wastes long-term by geological burial. The United States is studying a deep underground site at Yucca Mountain, Nevada, and may begin storing radioactive wastes there in 2010.

REACTOR SAFETY

Another concern is reactor safety. The uranium used in reactors cannot create a nuclear explosion. However, if the intense heat from reactors in operation is not removed steadily, the temperature can increase enough that the fuel rods can melt and high-pressure steam can burst internal parts. In 1957 uranium metal caught fire in the Windscale reactor in Great Britain, and radioactive gases escaped into the atmosphere. In 1979 a series of malfunctions caused exposure of the fuel rods in a reactor at Three Mile Island in Pennsylvania. Because its surrounding coolant was missing, the core overheated, and superheated steam escaped into the atmosphere, carrying a small amount of radioactive isotopes. In April 1986, a major accident occurred at a reactor at Chernobyl, Russia. During a test, several safety procedures were ignored or overturned by operators and managers, and the reactor generated too much energy too quickly. The fuel rods overheated, broke up, and came into contact with the coolant. The radioactive steam that was generated cracked the single containment vault. (All reactors in the United States have double containment buildings to protect against such accidents.) Graphite, which is pure carbon, was the moderator in the Chernobyl reactor. It caught fire, and a tremendous quantity of radioactive gases and solid material escaped, over 1000 times that released at Windscale. Much of it went high into the atmosphere because of

NUCLEAR RESPONSIBILITY

The reactor accident in Chernobyl is attributed to both poor engineering and an unbelievable disregard for safety. Nevertheless, that disaster took human error to occur. But blatant irresponsibility with nuclear materials in the former USSR came to light in March of 1993, when a committee of Russian scientists released a report to Russian Federation President Boris Yeltsin concerning nuclear waste dumping.

Beginning in 1965 the Soviets dumped 7 nuclear reactors full of fuel into shallow water in the Kara Sea. The reactor cores had been damaged, preventing the removal of the spent nuclear fuel rods. Eleven empty reactors, also highly radioactive, were dumped there, and the Soviet navy dumped two empty reactors in the Sea of Japan, off the east coast of Russia. As late as 1994, low-level radioactive liquid was still being dumped into the Kara and Barents seas. In addition, radioactive material (over twice as much as released at Chernobyl) from a reactor explosion at Mayak has polluted a lake there and is seeping through the ground toward a river nearby. Contaminated fish are only a small part of the problem with such pollution; radioactive contamination of the ocean could affect the entire food chain there.

At the time of this writing no plan has been put forth to clean up these sources of contamination (the extent of the problem is still being studied), and the Russian Navy still lacks any place to store its spent nuclear fuel from either active nuclear-powered ships or dozens of decommissioned nuclear submarines. Irresponsible handling of nuclear materials has the potential to affect everyone; it is a global concern which needs to be addressed by all the developed countries.

Physics *and Environment*

NUCLEAR POLLUTION

the extremely hot fire. Winds quickly spread most of the Chernobyl isotopes over Eastern Europe, Sweden, and Norway. Within a week, a small amount of airborne radiation was detected in the western United States, and a few days later in the eastern United States, underscoring the global nature of concern over such accidents.

Public concern over accidents brought new-reactor construction in the United States to a halt in 1975, eleven years before Chernobyl. However, in some important ways the use of safe reactors is more desirable than the use of fossil fuels. Nuclear reactors create no carbon dioxide or sulfur emissions as coal-fueled and oil-fueled electrical power plants do. These two pollutants, contributors to the greenhouse effect and acid rain, are very real concerns today. The methods of producing electricity are compared in the last section of this chapter.

BREEDER REACTORS

The known reserves of uranium ore in the world are relatively small. In 1980 about 300 uranium mines in the United States produced 20 million tons of ore, but each ton contains less than an ounce of ^{235}U and the richest deposits are being quickly depleted. However, the United States, with its scaled-down nuclear power program (since 1975), has enough uranium stockpiled to fuel its present reactors for their anticipated lifetimes. Other countries are expanding nuclear plants, however, and in the coming decades a shortage of fuel is expected.

One solution to the ^{235}U shortage is to build reactors that produce more fuel as they generate power. For example, plutonium can be made from the plentiful (but nonfissionable) ^{238}U by exposing this isotope to fast neutrons. Plutonium can absorb a neutron and undergo fission, and it releases 2.7 fast neutrons per fission on the average, so it can sustain chain reactions. Plutonium is radioactive as well, but it has a half-life of 24,360 years, so the plutonium in a reactor may be collected and used for fuel. Such reactors produce more fuel than they use, so they are called *breeder* reactors. Currently only France and Japan are developing commercial breeder reactors. If there is to be a long-range future for nuclear fission as a major energy source, these reactors could be it. Breeder reactors could produce all of the world's present electrical energy needs for several centuries.

Breeder reactors have drawbacks, however. They run at very high temperatures and use a coolant that becomes radioactive, so the safety features are more critical than in ordinary reactors. Also, the plutonium they breed can be used to build nuclear weapons, so security is a concern.

BINDING ENERGY AND FUSION

We have seen that large nuclei can break apart to form smaller ones in fission and that in particle accelerators nucleons can be added to nuclei, turning them into larger nuclei. With the addition or removal of nucleons, it is possible to custom-build isotopes (for medical purposes, for instance) and even create elements that, though unstable, are heavier than any elements found in nature. Uranium, with 92 protons, is the most massive atom found on the earth, but elements having as many as 109 protons have been produced artificially.

Nuclear reactions behave, in terms of energy, much like chemical reactions. In chemical reactions the bonding electrons rearrange themselves, *giving up energy* if the end products are more tightly bound, or stable, than the reactants. In the other direction, chemical reactions that create products that are more unstable than the reactants *absorb* energy. Nuclear reactions do the same.

The stability of a nucleus depends on how tightly bound its nucleons are. The **binding energy** per nucleon measures this property. The binding energy tells how much energy is needed to separate a nucleon from its nucleus. Binding energy represents the work done by the nuclear force during the rearrangement of nucleons. Because the binding energy per nucleon is different from one nucleus to another, either breaking a nucleus apart or merging two nuclei together always releases or absorbs energy.

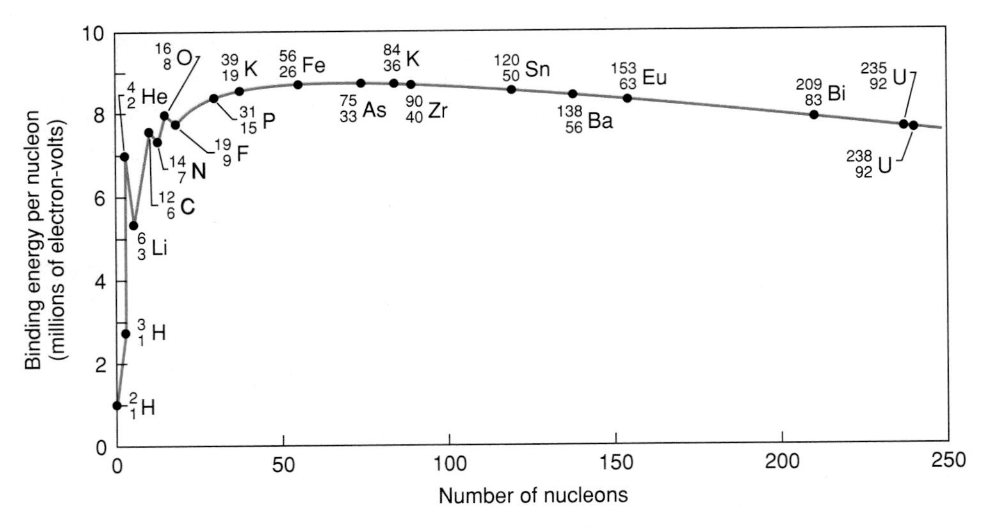

FIGURE 28-6 A GRAPH COMPARING THE BINDING ENERGY PER NUCLEON FOR NUCLEI OF VARIOUS MASSES. THE IRON (FE) NUCLEUS (AT THE HIGHEST POINT ON THE CURVE) IS THE MOST STABLE BECAUSE ITS NUCLEONS ARE ON THE AVERAGE MORE TIGHTLY BOUND THAN THE NUCLEONS IN OTHER NUCLEI. ENERGY IS RELEASED ONLY BY THOSE NUCLEAR REACTIONS WHOSE PRODUCT NUCLEI ARE MORE TIGHTLY BOUND THAN THE INITIAL NUCLEI.

Figure 28-6 shows binding energies for the nuclei. A look at this graph lets us see which reactions give up energy. In a ^{235}U fission, the resulting nuclei (often ^{91}Kr and ^{142}Ba) are more tightly bound than the ^{235}U, so the reaction releases energy. From the binding energy curve, we see that if a large nucleus splits to form several smaller nuclei, energy is released, as long as the nuclei that emerge are not smaller than iron, Fe. However, a look at the left side of the graph shows that elements lighter than iron are less tightly bound than iron as well. In fact, if two light nuclei fuse to form an element that is lighter than iron, in almost every case the nucleus of the fused element is more tightly bound than those it came from.

Such a reaction, where nuclei combine to form larger nuclei, is called **fusion**. *In general, if either a fission or a fusion reaction leaves the nucleons involved more tightly bound, the reaction is exothermic—it gives up energy.*

The fusion of the very light nuclei deuterium, ^{2}H, and tritium, ^{3}H, to form helium, which is illustrated in Figure 28-7, releases about three times as much energy per gram as the fission of ^{235}U does. Of the hydrogen nuclei in water molecules, about 1 in 6500 is a deuterium nucleus. The supply of deuterium in the oceans, though expensive to isolate, would meet the energy requirements projected for the world for millions of years. Tritium, however, has to be made via nuclear reactions and is therefore very expensive.

Aside from expense, there is a great difficulty in producing and controlling fusion, and the problem comes from the electric force. In order for deuterium and tritium to fuse, they must come together at a very high speed. If they do not have sufficient speed (which means very high temperatures), the repulsion between the positive nuclei turn them back before they can merge. Fusion is the reaction used in a "hydrogen" bomb. In these weapons a fission bomb goes off first, raising the temperature of the enclosed deuterium and tritium, and putting them under great pressure. Some of those speeding nuclei overcome their electrical repulsion to merge during their random collisions, releasing an incredible amount of energy very quickly. To be useful in generating electricity for everyday use, however, the energy from fusion must be released in amounts smaller than in a bomb and over longer periods of time.

Because of the high temperatures required for fusion (about one hundred million kelvins), no ordinary material can act as a container for fusion. Since the 1950s, physicists have worked on a magnetic bottling technique to compress a gas of ionized deuterium and tritium while a large magnetic field keeps this plasma (a gas of charged particles) away from the walls of the container. These experiments are coming closer to the critical combination of plasma density and temperature, but it could still be centuries rather than decades before fusion power is harnessed.

Another idea is to use the tremendous power in laser light to fuse tiny amounts of deuterium and tritium. Tiny pellets of these isotopes, encased in a solid material, are centered at the focus of simultaneous laser pulses. That is, they are blasted from all sides at once. The material on the outside of the pellet vaporizes almost instantly and expands in all directions. This expanding gas compresses the deuterium and tritium at the center, dramatically increasing its temperature and making fusion possible (Figure 28-8). The prototypes of these laser fusion devices have initiated fusion but only with very small quantities of deuterium and tritium.

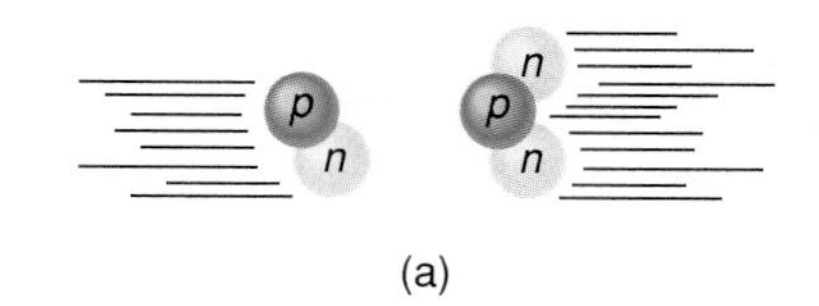

(a)

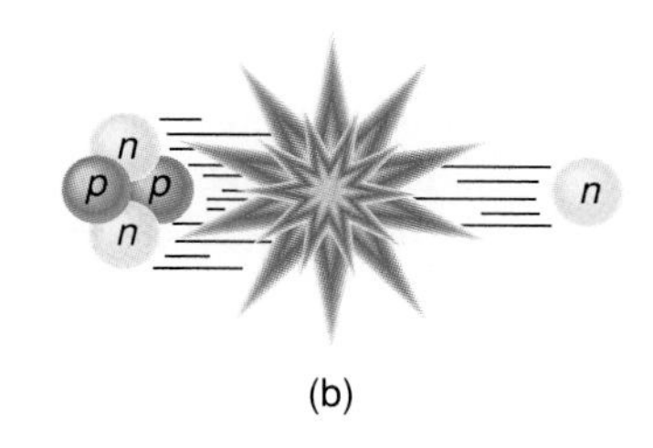

(b)

FIGURE 28-7 THE FUSION OF A DEUTERIUM NUCLEUS AND A TRITIUM NUCLEUS TO FORM HELIUM RELEASES 17.6 MILLION ELECTRON VOLTS OF ENERGY.

ENERGY FROM FUSION?

ENERGY FROM MASS

In nuclear reactions, the reactant nuclei have a certain mass before reaction and the products have a certain mass afterward. If the mass is less after the reaction, energy is released. That energy appears as kinetic energy of the products and as electromagnetic radiation. When a ^{235}U nucleus fissions, for example, almost 0.1 percent of its mass disappears, converted to energy released in the explosion. The same thing is true in ordinary chemical reactions, but the mass lost is ordinarily too small to measure, since the energy given off is on the order of a million times less.

FIGURE 28-8 (A) THESE TINY PELLETS, ALMOST HIDDEN ON THIS PENNY, CONTAIN DEUTERIUM AND TRITIUM. WHEN INTENSE LASER BEAMS ARE SIMULTANEOUSLY DIRECTED AT SUCH PELLETS FROM ALL SIDES, THE SURFACE MATERIAL VAPORIZES AND EXPANDS ALMOST INSTANTLY, CAUSING A SHOCK WAVE TO MOVE TOWARD THEIR CENTER. THE ISOTOPES ARE COMPRESSED TREMENDOUSLY, AND BOTH THEIR DENSITIES AND TEMPERATURES RISE. UNDER SUCH EXTREME CONDITIONS, THE NUCLEI OF THE ISOTOPES CAN COME CLOSE ENOUGH TO FUSE AND FORM HELIUM, RELEASING ENERGY IN THE PROCESS. (B) THE NOVA AT LAWRENCE LIVERMORE LABORATORY—A MACHINE INVESTIGATING LASER-ASSISTED FUSION OF DEUTERIUM-TRITIUM INSIDE SMALL PELLETS. TEN LASER BEAMS CONVERGE ON THE TARGET, WHICH IS SHOWN IN (C).

(a)

(b)

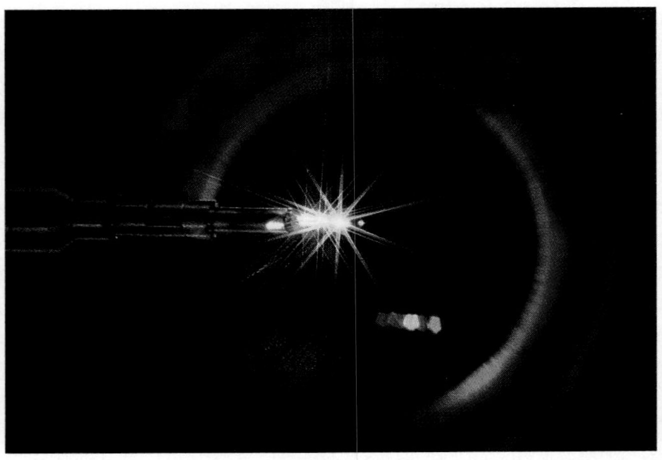

(c)

ENERGY'S COST TO THE ENVIRONMENT

TODAY'S PROBLEMS WITH ENERGY GENERATION

Wood was the fuel used by our ancestors until well after steam engines arrived on the scene in the early 1700s. Coal, discovered in England in the twelfth century and perhaps used by people in other places before then, also became a major fuel. The first producing oil well was sunk in Pennsylvania soil in 1859, and the energy from both coal and oil was used to industrialize civilization.

The human population has grown so much since the Middle Ages that the energy needed to support the earth's peoples today could never again come from wood; trees don't grow fast enough. Besides this fact, wood fires share a problem with the burning of coal and oil: They all produce carbon dioxide, CO_2, which pours into our atmosphere. Up to half of this carbon dioxide may dissolve into the ocean, but the rest remains in the air.

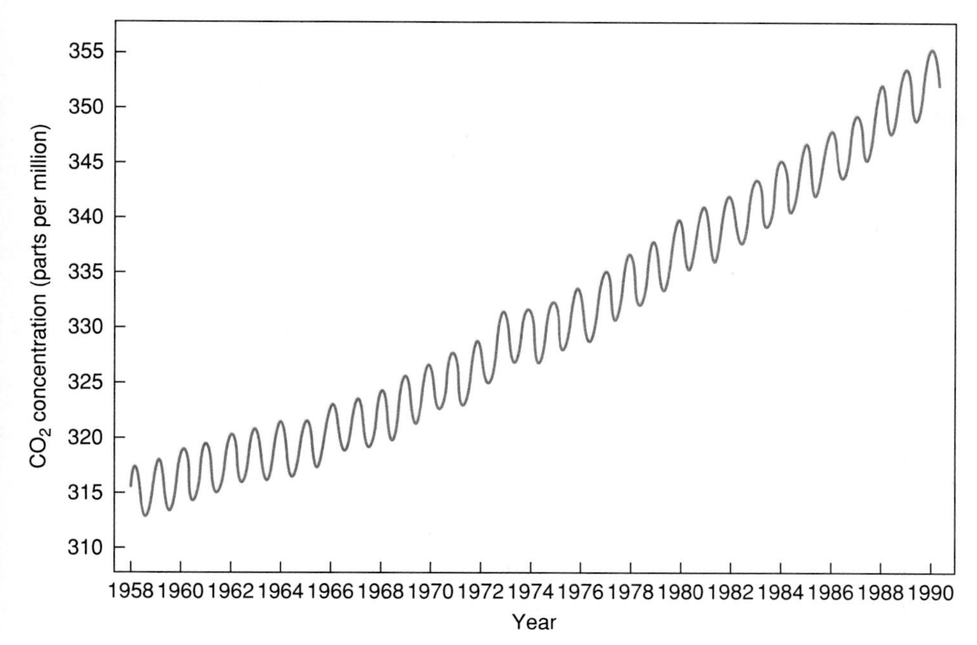

FIGURE 28-9 A GRAPH OF THE CARBON DIOXIDE CONTENT IN THE ATMOSPHERE IN THE NORTHERN HEMISPHERE. THE CONCENTRATION DECREASES IN SUMMER, WHEN PLANT PHOTOSYNTHESIS PROCEEDS AT A HIGH RATE AND CARBON DIOXIDE IS ABSORBED FROM THE AIR. IN WINTER, HOWEVER, MUCH PLANT LIFE DECAYS AND RELEASES CARBON DIOXIDE AND OTHER GASES INTO THE AIR. (DATA TAKEN IN THE SOUTHERN HEMISPHERE REVEALS SIMILAR UPS AND DOWNS IN CONTENT EXCEPT THEY ARE REVERSED SINCE THE SEASONS ARE OPPOSITE.) NOTICE THE CONTINUAL AND INCREASING ANNUAL RISE IN THE CURVE. THAT RISE IS ATTRIBUTED TO THE RELEASE OF CARBON DIOXIDE INTO THE AIR BY HUMAN INDUSTRIAL AND TRANSPORTATION ACTIVITIES. AFTER MT. PINATUBO ERUPTED IN THE PHILIPPINES IN 1991, THE CO_2 LEVEL "LEVELED OFF." SCIENTISTS ARE INVESTIGATING, HOPING TO UNCOVER THE REASONS; CLEARLY WE HAVE A LOT TO LEARN ABOUT THE ATMOSPHERE.

Because CO_2 molecules in the atmosphere absorb heat so well, the earth's atmosphere might warm from this "greenhouse" effect. Environmental scientists monitor the annual rise in the level of carbon dioxide (Figure 28-9), and if the recent trend continues—which seems inevitable as we continue to use fossil fuels—by the year 2025 the carbon dioxide level in our atmosphere will be double the 1980 level. Although no one is certain what this doubling will mean, some computer models suggest that such an increase will cause the earth to warm an average of 1.5 to 3.0 C°, with polar regions having increases of 8 to 12 C°. This warming will lead to a partial melting of the polar ice sheets, and this melting will raise the ocean level, flood coastal cities, and change climate patterns worldwide. In the future, we can expect advances in atmospheric science and computational techniques to make such models more accurate. Meanwhile, natural processes on earth can influence both the CO_2 content of the atmosphere and its temperature. Mt. Pinatubo blew its top in June of 1991, spewing 15 to 20 million tons of sulfur dioxide gas as high as 19 km, where the gas combined with water to form sulfuric acid. These acid droplets absorbed IR from earth's surface, warming the stratosphere sharply, while reflecting sunlight back into space. Ozone concentrations there dropped as well, and the rise in atmospheric CO_2 leveled off. Clearly we still have a lot to learn about all these interacting factors.

WATER VAPOR IN THE ATMOSPHERE ALSO ABSORBS INFRARED ENERGY EFFICIENTLY, AND DEPENDING ON THE LOCATION ITS CONTRIBUTION DWARFS THE ABSORPTION BY CO_2. BUT WORLDWIDE, AND OVER TIME, THE CO_2 LEVEL IS A CONCERN.

Coal, especially, pollutes the air with other damaging impurities besides carbon dioxide. The list includes carbon monoxide (CO), sulfur dioxide (SO_2), various oxides of nitrogen, mercury, hydrocarbons, and even radioactivity from minerals. The sulfur dioxide released through smokestacks at fossil-fueled electrical power plants also combines with water vapor to make sulfuric acid. When it rains, the water that falls is acidic.

Acidic rainwater runs off into lakes and slowly accumulates until the lakewater is unable to support life. Acid rains, largely from coal-burning plants in the American midwest, have killed many lakes in the northeast and Canada. The problem is everywhere. Ten thousand of Sweden's 100,000 lakes are completely dead as a result of sulfur dioxide being carried by the prevailing winds from coal-burning power plants in Germany, France, and Great Britain.

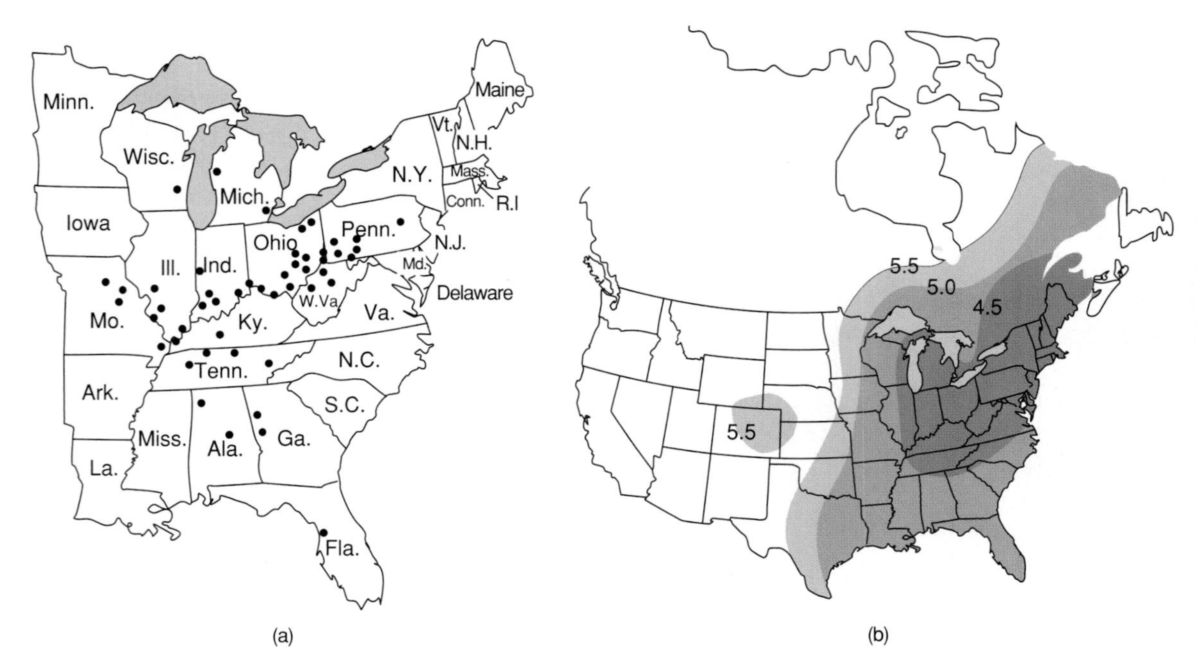

(A) THE LOCATIONS OF THE 50 LARGEST UNCONTROLLED CO_2 EMITTERS, ALL COAL-FIRED PLANTS, IN 1979. EVEN MORE COAL-FIRED POWER PLANTS HAVE BEEN BUILT SINCE THEN. (B) RECENT MEASUREMENTS OF THE PH OF PRECIPITATION. (A PH OF 7 IS NEUTRAL, AND SMALLER NUMBERS ARE INCREASINGLY ACIDIC.)

The surface of both the Lincoln Memorial in Washington, D.C., and the Acropolis in Greece are being dissolved by acid rain. And the damage doesn't stop with acidity: Acid rainwater reacts with minerals in the ground to release heavy metals, some of them poisonous, into the water table. Burning oil to produce energy pollutes less than coal, and natural gas pollutes even less, but these fuels are already in short supply compared with coal. Only 2 percent of the known coal reserves on earth have been used. The threat coal poses to our air and water demands we look elsewhere for power in the near future.

FIGURE 28-10 IN THE MOJAVE DESERT OF CALIFORNIA, 6500 COMPUTER CONTROLLED MIRRORS TRACK THE SUN, HEATING FLUID IN TUBES—WHICH IN TURN BOILS WATER, GENERATING STEAM TO DRIVE TURBINES CONNECTED TO ELECTRICAL GENERATORS.

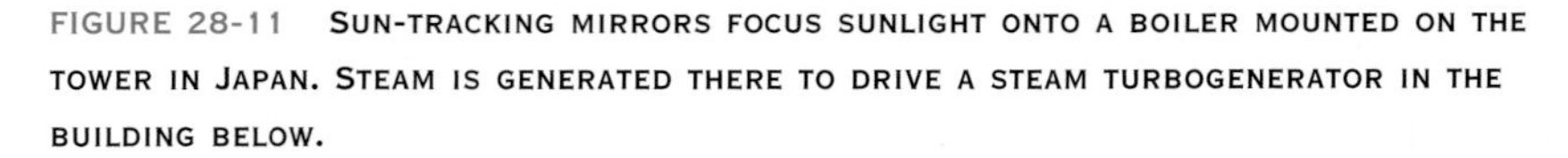

FIGURE 28-11 SUN-TRACKING MIRRORS FOCUS SUNLIGHT ONTO A BOILER MOUNTED ON THE TOWER IN JAPAN. STEAM IS GENERATED THERE TO DRIVE A STEAM TURBOGENERATOR IN THE BUILDING BELOW.

FIGURE 28-12 HOOVER DAM AND LAKE POWELL ON THE COLORADO RIVER IN ARIZONA.

Solar energy would be a perfect solution, with no pollution, no threat of contamination, and no waste products (Figures 28-10 and 28-11). However, up to now there is no economically viable way to convert sunlight to electricity in the amounts needed for industrial activity. Hydroelectric power, where water from dams is used to turn turbines, provides about 12 percent of the energy generated by utility companies in the United States (Figure 28-12). Geothermal power, where steam is generated from hot spots near earth's surface, is in use in a few places, such as Iceland, Italy, and California, but it too contributes little to supplying the total energy needs of the world at present (Figure 28-13). Using the wind to generate electricity has been realized in only a few places, but recent improvements in wind-powered generators have increased the interest in this source (Figure 28-14).

FIGURE 28-13 ELECTRICITY IS GENERATED BY GEOTHERMAL STEAM IN ICELAND.

Nuclear reactors emit no carbon dioxide and no more radiation than emerges from the smokestacks of coal-fueled power plants. However, safe storage of nuclear wastes is a problem, and even worse is the possibility of an accident that would cause considerable radioactive contamination. No one can argue that nuclear energy is perfectly safe, but its cleanliness compared with fossil fuels is a fact worth considering. Should nuclear fusion ever be harnessed for power, it could provide for the world's energy needs for the foreseeable future. Fusion plants should be easier than fission plants to operate safely and should create less radioactive waste. They would not contaminate the atmosphere with carbon dioxide and sulfur dioxide. Unfortunately, the technological problems with small-scale fusion have proved to be tough ones, and progress is slow. The choices of how we get the energy we use are not simple ones, and environmental concerns are certain to be even more important as time goes by.

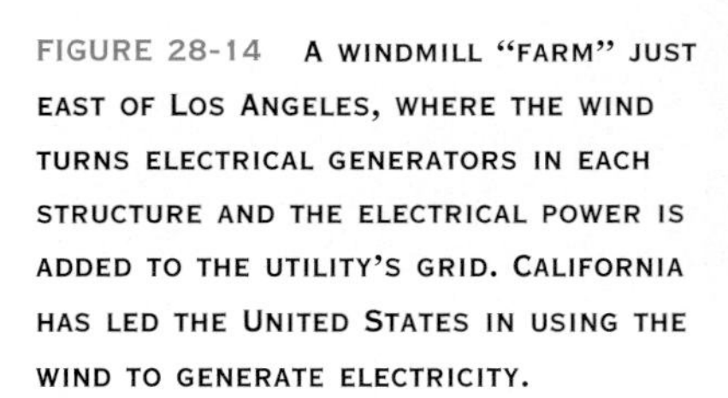

FIGURE 28-14 A WINDMILL "FARM" JUST EAST OF LOS ANGELES, WHERE THE WIND TURNS ELECTRICAL GENERATORS IN EACH STRUCTURE AND THE ELECTRICAL POWER IS ADDED TO THE UTILITY'S GRID. CALIFORNIA HAS LED THE UNITED STATES IN USING THE WIND TO GENERATE ELECTRICITY.

Summary and Key Terms

nuclear fission–422

When a ^{235}U nucleus absorbs a slow neutron, ***nuclear fission*** can occur: The incoming neutron causes the nucleus to split almost in half. The resulting nuclei repel because of their large positive electric charges, giving them enormous kinetic energy.

chain reaction–423

self-sustaining chain reaction–423

When a ^{235}U nucleus fissions, neutrons emerge, and if those neutrons slow by colliding with small nuclei, they can efficiently cause other nearby ^{235}U nuclei to fission, releasing more neutrons, and so on, in a ***chain reaction***. Uranium-235 nuclei usually emit either 2 or 3 neutrons per fission. If one neutron per fission causes another fission, a ***self-sustaining*** chain reaction occurs. If more than one neutron per fission causes another fission, the number of nuclei that fission each second grows, causing a ***runaway chain reaction***. To achieve an explosive runaway chain reaction, a sample of uranium must be nearly pure ^{235}U and must have a certain ***critical mass***.

runaway chain reaction–424

critical mass–424

A nuclear reactor uses nuclear fission to generate electrical power. In the reactor core, uranium fuel rods are alternated with movable control rods, which absorb fission neutrons. A moderator slows the fission neutrons. When the control rods are withdrawn, the uranium rods are exposed to each other's slowed neutrons, and a chain reaction proceeds. As the nuclei fission, the temperature of the fuel rods increases. A coolant circulates around the rods to collect the heat, which generates steam to turn turbines that run electrical generators.

Nuclear reactions in which the end products are more stable than the reactants give up energy, and reactions in which the end products are less stable absorb energy. The stability of a nucleus depends on the ***binding energy*** per nucleon, which is the energy needed to take a nucleon away from its nucleus. If a large nucleus splits to form several smaller nuclei that are heavier than iron, energy is released. If two light nuclei undergo ***nuclear fusion***, uniting to form another element that is lighter than iron, the fused element will likely be more tightly bound than the reactants and the reaction will release energy. Fusion reactions have yet to be tamed for commercial use because of the high temperatures required.

binding energy–428

nuclear fusion–429

Wood, coal, and oil fires all produce pollutants that contribute to acid rain and to the greenhouse effect. Nuclear reactors produce power more cleanly, but there is concern over the safe storage of nuclear wastes and the danger of accidents.

EXERCISES

Concept Checks: Answer true or false

1. In fission, the nucleus splits apart and the fragments are given enormous kinetic energy. This process is not like natural radioactivity because in fission a neutron must be added to cause the nucleus to fission.

2. For a chain reaction to occur in the fission of ^{235}U, each neutron released must go on to cause another fission.

3. A nuclear reactor produces a controlled chain reaction that results in a steady release of energy.

4. The control rods of a nuclear reactor control the speed of the neutrons given off during fission.

5. Breeder reactors create fuel while releasing energy that can be used to generate power.

6. Fusion differs from fission in that the initial nuclei in fission are smaller.

7. The radiation released presents the greatest problem in controlling fusion.

8. If a critical mass of ^{235}U were assembled in a long, thin wire, it would undergo a runaway nuclear reaction.

Applying the Concepts

9. When you start a campfire, are you starting a chain reaction? When you start a car, are you starting a chain reaction? Explain.

10. Draw an analogy between the instability of a large liquid drop that tends to break apart easily and the instability of a large nucleus.

11. Explain why water is effective at slowing neutrons down while lead isn't.

12. In terms of forces, why is it that nuclear processes can give so much more energy than chemical processes?

13. Discuss this statement: A slow neutron that happens to touch a nucleus can fall into that nucleus.

14. Which force limits the maximum size of stable nuclei?

15. Why does burning wood not alter the percentage of carbon-14 in the carbon dioxide of the atmosphere?

16. Explain how the electrical force that resists fusion assists in fission.

17. What facts about the fission of ^{235}U makes a chain reaction possible?

18. Some nuclear reactors use uranium metal, while others use uranium dioxide. Uranium melts at a much lower temperature than uranium dioxide and can also burn in oxygen if the fuel cans are ruptured. This happened at the Windscale reactor in Great Britain in 1957. What advantage does uranium metal have over uranium dioxide as a fuel in a reactor?

19. Fusion is the energy source of the sun, with the hottest temperature and the greatest fusion rate at its center. What confines the fusion reaction, and why doesn't the earth receive gamma radiation from the sun?

20. Explain why a nuclear reactor can't have the type of runaway chain reaction that occurs in a nuclear weapon.

21. The strength of the material in fuel rods is a primary concern with nuclear reactors. Today most reactors boil water to create steam at 350°C, a very low temperature compared with the 600°C used at coal-fueled and oil-fueled power plants. The lower temperature is necessary to prevent overheating of the fuel rods since overheated rods could rupture their containers. What effect does this mandatory lower temperature have on the efficiency of nuclear power plants?

22. When small volumes of ^{235}U are brought together in a bomb to make the critical mass, has the surface area of the uranium per unit of mass increased or decreased? Does this change affect what takes place next?

23. Explain how $E=mc^2$ can be at work in both fission and fusion.

24. Why isn't the water that cools a nuclear reactor's core used to drive the turbine and electrical generator?

25. The fallout of radioactivity from Chernobyl was not evenly distributed over western Europe. "Hotter" spots were found where there were rainshowers as the windborn contaminants drifted over. Why is it important to find all of those hot spots?

26. Strontium-90 is an abundant radioactive fission product that chemically acts like calcium, unfortunately. Discuss the implications of this similarity in view of the fact that calcium ingested in milk goes to our bones.

27. Two fission products are radioactive xenon and krypton, both gases much like neon chemically. Neon has a filled outer shell and doesn't react chemically; it is called an inert gas. Explain why xenon and krypton are relatively safe, at least compared to strontium (see Exercise 26).

28. When a critical mass is brought together in a nuclear weapon, what should its shape be to ensure the most effective chain reaction?

29. Discuss the advantages and disadvantages of using small nuclear reactors to provide power in unpeopled satellites.

30. Deuterium, ^2H, has a stable nucleus. Is its mass greater than or less than the mass of a proton plus the mass of a neutron?

31. Why must an atomic (fission) bomb be used to set off a hydrogen (fusion) bomb?

32. The efficiency with which heat can be transformed into energy is at best equal to $(T_{high} - T_{low})/T_{high}$, where the temperature is in kelvins, as discussed in Chapter 13. On this basis, which type of nuclear power plant can be more efficient, an ordinary fission reactor or a breeder reactor?

33. Carbon particles from coal-fired electric plants are found in the haze above the Arctic Circle. These particles absorb sunlight that might otherwise be reflected by the ice below. How might this absorption affect the amount of ice in that region?

34. Pollution particles that make it into the cold air over the Arctic tend to stay there. Environmental scientists call the Arctic atmosphere a cold trap. Discuss what they must mean.

35. When atmospheric testing of nuclear weapons began in 1945, these devices (via neutron emission) created large amounts of carbon-14 in the atmosphere. By the time of the test ban treaty in the 1960s, the carbon-14 activity of our atmosphere had doubled. What effect will this have on archaeological dating in the distant future?

36. Why wouldn't you expect a chain reaction to proceed in uranium ore residing in surface deposits? Slow neutrons are present at the earth's surface all the time because of cosmic rays. (Such a chain reaction may have once occurred! See *Scientific American* for July 1976.)

37. The first nuclear reactors used bricks of solid graphite, a form of carbon, as the moderator because carbon-12 is relatively light, doesn't absorb neutrons readily, and is very inexpensive and plentiful. The nuclear reactor at Chernobyl also used graphite as a moderator. Discuss the pros and cons of this use of graphite.

38. The fission of 4.5 grams of pure ^{235}U in a reactor today gives the amount of energy used by an average person in the United States in an entire year. About how many kilograms of this isotope would provide the energy needed by the 1/4 billion people in the United States?

39. Canadian reactors use natural uranium metal rather than enriched uranium—a fuel choice made possible by their using pure heavy water (D_2O) as a coolant. The deuterium nuclei don't absorb neutrons, and their small mass slows the fast fission neutrons. Though deuterium atoms are found in water, it is difficult and expensive to separate them from ordinary H_2O. Is there an economic tradeoff?

40. When humans began using coal and oil as their major source of fuel, they began to decrease the percentage of carbon-14 in the air. Coal and oil are so old compared to the half-life of carbon-14 (5730 years) that those fuels contain no carbon-14 nuclei to speak of. The carbon dioxide from the burning of these fossil fuels adds only carbon-12 atoms to the air and so dilutes the carbon-14 in the atmosphere. Since about 1850 the level of radioactive carbon dioxide has decreased by 2 percent for that reason. How would this change affect carbon-14 dating if other carbon-14 pollution didn't interfere?

41. Chemically, ^{238}U and ^{235}U behave in the same way, so nonchemical means had to be used to separate these isotopes in the Manhattan Project. Can you think of ways that separation might be achieved? (Be sure to see the answer.)

42. Is a "population explosion" similar to a runaway chain reaction? Compare a ^{235}U chain reaction with the types of chain reactions possible with the populations in countries where every couple has (a) fewer than two children, (b) two children, (c) more than two children.

43. Show that when ^{2}H (m = 2.0141 u) and ^{3}H (m = 3.0160 u) fuse to form ^{4}He (m = 4.0016 u) and one neutron (m = 1.0087 u), approximately 18 million electron-volts of energy is released.

44. The total energy used in the world in one year could be supplied by about 3500 tons of pure uranium-235. For comparison, the United States alone uses more than 600,000,000 tons of coal annually. Would nuclear reactors have an edge economically in terms of transportation costs?

45. According to some estimates, a large coal-fired plant sends more radioactivity into the atmosphere than a nuclear plant that generates the same amount of electrical energy. Do you think this fact is generally known to the public? Could this be an important fact when decisions on new nuclear plants are in the works?

46. Each ton of coal that is burned releases over 80 pounds of sulfur dioxide into the air. Besides its effect on trees and lakes, the acid that results is harmful to humans. An estimated 8000 people die each year because of its presence, and another 2000 deaths are attributed to the tars in coal emissions. Do you think this is generally known to the public? Could this be an important fact when decisions on new nuclear plants are in the works?

47. One ton of TNT releases about 4×10^9 J. The fission of 1 kilogram of ^{235}U releases about 83×10^{12} J. How many tons of TNT is represented by the energy available in 1 kg (which weighs about 2.2 pounds) of ^{235}U?

48. The mass of a helium nucleus (2 protons and 2 neutrons) is 4.0016 u. The mass of a proton is 1.0073 u, and the mass of a neutron is 1.0087 u. Find the binding energy per particle for the helium nucleus. (*Hint:* Use $E = mc^2$.)

49. Would you rather live next to a coal-burning plant or a nuclear power plant? Why?

50. The two gamma ray photons emitted when an electron and a positron collide during a PET procedure (Chapter 27) carry away little more than the energy of the rest mass of the particles, the mass of 2 electrons. What energy (in MeV) does each photon have?

Appendix One

SIGNIFICANT FIGURES

The number of significant figures in a stated number is the number of digits whose values are known with certainty. When you calculate a number, you may think of your answer as being exact. But measurements of physical quantities of the continuous type such as time and length (see Chapter 2) are never exact! For example, if you measure the length of a pencil and find it to be closer to 17.2 centimeters than to 17.1 or 17.3, but you are not sure what the next digit would be (i.e., such as 17.23), you should state your measurement as 17.2 centimeters, which has only three significant figures. You shouldn't list any digits that are probably in error, and when you are given numbers to use in a calculation, you should assume they are only as accurate as stated. Suppose someone told you that a distance along a road was "about" 100 meters. With nothing else to go on, you would have to assume that the two zeros were *not* significant figures. The actual number might be 110 meters or even 96 meters, or even farther off for all you know! You need to assume there is only one significant figure there, the "1," meaning the distance should certainly be between 0 and 200 meters. On the other hand, if you were told an athlete had run the 100-meter dash at a track meet, you could assume the distance was measured very accurately and use all three digits as significant. If you further knew that the 100 meters was accurate to within a tenth of a meter, you could write that number as 100.0 meters. Because of the zero written after the decimal point, the number of significant digits in this number is four. In any calculation with physical quantities, the significant digits are important.

Here is an example to illustrate the type of calculation you might do in your course. Suppose you need to find the number 130.68×2.2—perhaps the units are meters or seconds, but we won't worry about that here. Your calculator will give the answer 287.496, but this answer is only accurate to 2 significant figures because the number 2.2 is only known to 2 significant figures. So only the first two digits of the number 287.496 are significant, and this answer should be rounded to 290; the third digit, the zero, is not significant here, just as with the 100 meter example above. *When two or more numbers are multiplied or divided, the result is accurate only to the same number of digits as the number in the calculation with the smallest number of significant figures.*

CHANGING UNITS

When you need to change the units of a quantity, a table relating various units is given on the inside front cover. For an example, suppose a car is moving at a constant speed of $12.3\ \frac{\text{yards}}{\text{second}}$. Let's express this speed in the units of meters per second. The first line of "Conversion Units" on the inside front cover tells you that 1 meter = 1.094 yards, and we can use this change the units. Divide both sides of this by 1.094 yards, and you find

$$\frac{1 \text{ meter}}{1.094 \text{ yards}} = \frac{1.094 \text{ yards}}{1.094 \text{ yards}} = 1,$$

since anything divided by itself is 1. Now we can multiply our car's speed (or any other speed) by this fraction without changing its value, since anything multiplied by 1 is unchanged. But this *does* change the units, as we'll see now:

$$12.3 \frac{\text{yards}}{\text{second}} = 12.3 \frac{\cancel{\text{yards}}}{\text{second}} \times \frac{1 \text{ meter}}{1.094 \cancel{\text{yards}}} = \frac{12.3 \text{ meters}}{1.094 \text{ second}} = 11.2 \frac{\text{meters}}{\text{second}}$$

We cancel the $\frac{\text{yards}}{\text{yards}}$ since, as before, anything divided by itself is 1. If you had needed to change a speed from meters per second to yards per second, you could have formed the factor the other way, that is, $\frac{1.094 \text{ yards}}{1 \text{ meter}} = 1$.

POWERS OF 10

When numbers are uncommonly large or small, a shorthand method makes them easier to write and handle. Consider the number one thousand. Because 1000 is the same as $10 \times 10 \times 10$, or 3 factors of 10, we can write 1000 more compactly as 10^3. (Read 10^3 as "ten to the three.") The number 3 is called the *exponent* of 10. It tells how many factors, or *powers*, of 10 are multiplied together to equal the large number. (A single factor of 10 is sometimes called an *order of magnitude*.)

Now consider the number one hundred thousand (100,000). This number is literally 100×1000. Or, factoring 100 and 1000 into powers of ten, we have $(10 \times 10) \times (10 \times 10 \times 10) = 10^5$. But returning to the form 100×1000, we can write $100 = 10^2$ and $1000 = 10^3$. Thus $10^2 \times 10^3 = 10^5$. When powers of 10 are multiplied together, their exponents add.

When powers of 10 are divided, the exponents subtract rather than add as they do in multiplication. For example,

$$1000/100 = 10^3/10^2 = 10 \times 10 \times 10/10 \times 10 = 10^1.$$

Each power of 10 in the denominator eliminates a power of 10 in the numerator; thus the exponent of 10 in a denominator subtracts from the exponent of 10 in the numerator. So we can write

$$10^3/10^2 = 10^{3-2} = 10^1.$$

Notice that the effect of dividing by 10^2 is the same as multiplying by 10^{-2}. So to take a power of 10 to the numerator from the denominator (or vice versa), just change the sign of its exponent. So $10^4 = 1/10^{-4}$ and $10^{-5} = 1/10^5$.

As an example of how to express a number in terms of powers of 10, let's rewrite the approximate number of seconds in a year. In 365.25 days (remember leap years!) there are 31,557,600 seconds. If we factor, say, 3 powers of 10 from this number, we get $31{,}557.6 \times 10^3$. Notice that each factor of 10 removed from a number shifts the decimal point one place to the left. Continuing to factor powers of 10, we find

$$3.15576 \times 10^7 \text{ s} = 31{,}557{,}600 \text{ s} = \text{the number of seconds in 365.25 days.}$$

To help you recognize powers-of-10 equivalents, inspect Table A–1. Prefixes for some of the factors of 10 are given in Table A–2. For example, the average radius of the earth is 6,370,000 meters. This distance can be expressed as 6370 km or 6.37 Mm.

Table A·1

$10^1 = 10$	$10^{-1} = 1/10$ or 0.1
$10^2 = 100$	$10^{-2} = 1/100$ or 0.01
$10^3 = 1{,}000$	$10^{-3} = 1/1{,}000$ or 0.001
$10^4 = 10{,}000$	$10^{-4} = 1/10{,}000$ or 0.0001
$10^5 = 100{,}000$	$10^{-5} = 1/100{,}000$ or 0.00001
$10^6 = 1{,}000{,}000$	$10^{-6} = 1/1{,}000{,}000$ or 0.000001

Also note $10^3/10^3 = 10^{3-3} = 10^0 = 1$

Table A·2

Power of 10	Prefix	Abbreviation
10^{12}	tera	T
10^9	giga	G
10^6	mega	M
10^3	kilo	k
10^{-2}	centi	c
10^{-3}	milli	m
10^{-6}	micro	μ
10^{-9}	nano	n
10^{-12}	pico	p

Appendix Two

DERIVATION OF $d = \frac{1}{2} at^2$ (Chapter 2)

We can relate the distance something covers to its acceleration, if the acceleration is uniform—though the derivation is more difficult than any exercise in this text! An object with an average speed v_{average} covers distance according to $d = v_{\text{average}} \times t$. During a uniform acceleration, v_{average} is the average of the final and initial speeds, or $v_{\text{average}} = \frac{1}{2}(v_{\text{final}} + v_{\text{initial}})$. We will find the distance here for objects that begin at rest, so $v_{\text{initial}} = 0$.

$$v_{\text{average}} = \tfrac{1}{2} v_{\text{final}}$$

and using $at = v_{\text{final}} - v_{\text{initial}}$, we see when $v_{\text{initial}} = 0$, v_{final} is just equal to at, so

$$d = v_{\text{average}} \times t = \tfrac{1}{2} v_{\text{final}} \times t = \tfrac{1}{2} at \times t = \tfrac{1}{2} at^2 \quad \text{or,}$$

$$\boldsymbol{d = \tfrac{1}{2} at^2} \quad \textbf{for objects that start from rest}$$

DERIVATION OF NET WORK EQUALS CHANGE IN KINETIC ENERGY (Chapter 6)

Net work, where the net force is in the direction of the displacement, is $F_{\text{net}} \times d = ma \times d$. Now $d = v_{\text{average}}t$, and the average speed is $v_{\text{average}} = \frac{1}{2}(v_{\text{final}} + v_{\text{initial}})$, while the acceleration a is $(v_{\text{final}} - v_{\text{initial}})/t$. Substituting for d and a we find the work =

$$W = \tfrac{1}{2} m (v_{\text{final}}{}^2 - v_{\text{initial}}{}^2) = \tfrac{1}{2} mv_f^2 - \tfrac{1}{2} mv_i^2 = KE_{\text{final}} - KE_{\text{initial}}$$

Answers to Selected Exercises

Chapter 1

1. F **2.** T **3.** T **4.** T **5.** T **6.** (a) F (b) F

8. Pythagoras, and many natural philosophers and scientists after him, recognized the fact that to understand and predict natural phenomena you have to make measurements, which are expressed in numbers. That is, the understanding must be quantitative rather than qualitative alone. An everyday example is when you follow a recipe to have a better chance of a cooked item turning out the way you want it to turn out.

11. *Observation*: you sense (see or hear) that the stereo doesn't come on. *Hypothesis*: It isn't getting electricity. *Experiment*: Trying another outlet or replacing the electrical plug of the stereo with that of another appliance.

15. (a) No. This statement is testable, you could walk over and check it out. It might even qualify as an educated guess, if you'd seen this happen before. But it is not a statement that you would expect to use for predictions, since on any given day it could easily be interfered with by the driver of a standard car. While it might be a true statement today, it probably has no special predictive powers for tomorrow! A scientific hypothesis is an educated guess to *predict* natural phenomena. (b) This does not qualify as a scientific hypothesis, since if the stuff in the room is undetectable, no experiment can be done to confirm or deny its presence. (c) Suppose you have been watching the weather where you are and that temperature is within the range of temperatures you've been observing. Then this statement qualifies as an educated guess. And since it is testable; you would only have to take a thermometer to school with you on Tuesday. However, it is not a statement that could lead to a theory of *prediction* for the weather since even if it was found to be true once, you wouldn't expect such a prediction to be accurate in every case. A *scientific* hypothesis should offer some hope of *explanation* of a predictable natural phenomena. So this hypothesis would not be a scientific hypothesis. (d) Yes, this is a scientific hypothesis since, although these units are assumed too small to see they can be identified with experiments that people can do. The hypothesis that they are indestructible (not true) can also be tested through a wide variety of experiments today. Moreover, if we apply some imagination we can see (as we will in this text) many other phenomena these atoms could explain if they are real, things as simple as how a gas exerts pressure on its surroundings.

27. Whereas the Babylonians estimated there were 360 days in a year, the Chinese had made more accurate observations and determined there were 365.25 days before the stars repeated their patterns exactly in the nighttime sky and the sun repeated its north-south cycle in the sky. These "cycles" held special significance for these early observers and they used them to divide the circle into parts for measuring.

32. If 6 people peeled 10,000 eggs in 8 hours, then on the average they must have peeled 10,000 eggs/8 hours = 1250 eggs/hour (which is 1250 eggs **per** hour). Since 6 people were peeling, in one hour each person must have peeled (on the average) 1250 eggs/6 people = 208 eggs/person. Per minute, then, a person on the team peeled 208 eggs/60 minutes, or 3.47 eggs/minute. If we round this off to two significant figures, which could mean that the number of eggs peeled was not exact, but that they actually peeled closer to 10,000 eggs than 11,000 or 9,000, then on the average a person on the team peeled only 3 eggs per minute.

34. If 2 million go into higher education, and 80% take a course in natural sciences, then $2{,}000{,}000 \times 0.8 =$ 1,600,000 first-year students are in these courses. If half of 395,000 natural scientists were teaching, and 1/5 of them were teaching a single first-year course, then about $400{,}000 \times 1/2 \times 1/5 = 40{,}000$ such courses were being taught. Divide the students in the courses by the courses, and the result is 1,600,000/40,000 = 40 students/course. This is probably within a factor of ten of the correct answer, i.e., the average class size is probably between 4 and 400! If we assume that all these numbers we used had at least one significant figure, then the answer should be between 30 and 50 students/class. How does that compare to your class?

37. From Exercise 36, 110×10^9 gallons of gasoline were sold. Multiplying this by 20 miles/gallon (which is probably a high estimate), we find 2200×10^9 vehicle miles traveled, and dividing by 100×10^6 we find 22,000, and the statistics say there are 2.5 deaths for each one of these 100 million miles traveled. Thus if the assumption of 20 miles per gallon was close, and the statistics were correct, $2.5 \times 22{,}000$ people should have died. That is 55,000 people. The total number of deaths that year was actually less than 45,000. Discuss where this calculation was probably most uncertain. Can you identify any assumptions made that would throw the estimate off? (i.e., what about gasoline sold for lawn mowers, tractors, and what about those vehicles using diesel fuel rather than gasoline?) Incidentally, when the economy of the US booms, more deaths occur on the highways since more people can afford to travel more.

Chapter 2

1. T **2.** T **3.** T **4.** T **5.** T **6.** T **7.** F **8.** T **9.** F **10.** F **11.** T **12.** T **13.** F

14. (a) exact, (b) exact, (c) continuous, (d) continuous, (e) continuous, (f) exact, (g) ontinuous.

19. The bear's average speed is 500 yd/30 s = 17 yd/s if we keep only 2 significant figures. To compare this to the 10 m/s average speed of an athlete who can do the 100-m dash in 10 seconds, we can convert this just as outlined in Appendix 1, or we can estimate the answer using the fact that a meter is about 9% longer than a yard. That means the 10 m/s speed becomes (10 + .09 × 10) yards/second, or 10.9 yd/s. Keeping only two significant figures, this is 11 yd/s, so the sprinter would see the bear gaining ground at about 6 yards each second!

21. Refer to appendix 1; (a) 100 yd × (1m/1.094 yd) = 91.4 m. (b) 100 m × (1.094 yd/1 m) = 109.4 yd.

22. Yes, the keys' speed is zero instantaneously at the top of their path. That is the point where they stop rising and begin falling, all at the same instant. But they never stop accelerating, their speed changes constantly with the rate g. So you can say that the keys' speed just "passes through" the value of zero as they are turning around.

25. (a) a, c (b) d (c) b, e (d) e

33. 155 mph! Since the ball and the bat are traveling in opposite directions, the speeds must be added to find the relative speed at which they approach each other.

34. (a) a, b, c (b) b, c, d, e (c) a.

42. (b) in two dimensions. He essentially moves in a single plane as he falls, unless a wind would take him to the side of the plane's path (presuming the plane was keeping a straight course).

46. The second day at 9 A.M. you'll start at the top of the axis representing distance. (See the dashed line.) As you walk back, you'll get closer and closer to where you began, so the line that describes your distance at any time will drop down to zero distance at the 5 P.M. mark. Somewhere it will cross (at least once) your "line" from the previous day.

Chapter 3

1. F **2.** T **3.** T **4.** T **5.** T **6.** T **7.** T **8.** T **9.** T **10.** T **11.** F **12.** T

14. You judge inertia by sensing the force you apply compared to the acceleration the object gets in return. (The ratio of force to acceleration, F/a, is the measure of its inertia, m.) Thus, even when you don't know the numerical value for force and acceleration, you do know that if you push hard and the book accelerates little, its inertia is large. If you give it a shove and it accelerates a lot, it has little inertia.

18. Inertia. The greater the mass in motion, the more force required to change its velocity.

20. The bully was wrong! In every push action equals reaction. Perhaps the victorious ruffian had the larger mass, in which case he'd respond to the same force with less acceleration than his opponent. Or perhaps he had a firmer foothold and mustered a greater frictional force to help balance the shove from the loser. Or maybe his push was directed somewhat upward. That would decrease his opponent's contact force from the ground and, thus, the friction on his opponent's feet. At the same time, the reaction force pushes the winner somewhat downward, increases his own contact force, and provides more friction to oppose that push.

21. Let's assume the car is on a level road. To push anything at rest and get it moving, you have to overcome both the object's inertia (its resistance to a change in motion) and any friction or other forces on it that might act to retard motion. But once the car (or other object) is moving at the speed you want, you no longer have to overcome its inertia, only whatever kinetic friction there is.

26. If you are suddenly pushed on your shoulders from behind, your head does not actually snap backward. Your head just tends to stay at rest while your body accelerates forward. For your head to accelerate, there has to be a force, and in this case it must come from your neck. Your neck isn't absolutely rigid, so your head is left somewhat behind as the rest of you accelerates. If your car is struck from the rear, a headrest will give your head a push along with the rest of your back, thus eliminating the whiplash action.

28. Its mass $m = F/a$. Then 1 N/5 m/s^2 = 0.2 kg. (Remember that a newton is a kg m/s^2.)

33. The forces the cars exert are equal (and in opposite directions). In other words, $m_{(\text{car } 1)}a_{(\text{car } 1)} = m_{(\text{car } 2)}a_{(\text{car } 2)}$. If car 1 is much more massive than car 2, then the acceleration of car 2 will be much greater than the acceleration of car 1; otherwise, the products of m and a could not be equal. The passengers in the more massive car will accelerate less than those in the less massive car. So if each car has a 65-kg passenger, the one in the less massive car will receive the greatest force during the collision.

35. No matter where it is, earth or moon, the only quantity a bathroom scale measures is a force. The scale balances your weight with a force from a spring. On the moon the scale would balance your weight there and give an accurate reading of the pull of the moon's gravity on your body. The kilogram reading of a bathroom scale is not measured but derived. The spring measures mg_{earth}, your weight on earth. That kilogram reading on the scale is just the measured value of mg_{earth} divided by g_{earth}. So on the moon that reading of mass would be in error. Why? mg_{moon} divided by g_{earth} won't be equal to your mass in kilograms. The scale would show 1/6 of the astronaut's earth weight, which is true, but it would also say the astronaut's mass was 1/6 of his mass on earth, which is not true. Few astronauts would have gone to the moon if they had thought they would lose 5/6 of their mass in the process!

44. The forces on them are equal but in the opposite direction, so they will accelerate in opposite directions. Since $m_{girl}a_{girl} = m_{boy}a_{boy}$, his acceleration is (40 kg × 1 m/s²)/65 kg = 0.62 m/s².

45. The acceleration of the rocket increases dramatically! Since the force stays constant, and that is equal to ma, then ma must stay constant. When the mass goes down because the fuel is used, the acceleration goes up, keeping ma constant.

48. (a) She is applying 40 pounds of force upward, if the 100-lb rock presses with only 60 pounds against the scale. The net downward force must be (100 lb − her upward force) = 60 lb, so that means her upward force is 100 lb − 60 lb = 40 lb. (b) If she is exerting an upward force of 40 lb, then the scale she is standing on will read her weight plus 40 lb since her body gets the reaction force from the force she applies to the rock. (c) The scale that is underneath all this will read a steady weight equal to her weight, the rock's weight, plus the weight of the two scales. When she begins to lift, the forces act in pairs and don't change the net force downward of all the interacting components.

Chapter 4

1. T **2.** F **3.** F **4.** F **5.** T **6.** T **7.** F **8.** T **9.** F **10.** T **11.** T **12.** T **13.** T **14.** T

20. Because the dictionary moves at a constant speed, there is no net force on it. Since you are pushing with a force of 3 pounds, the friction that opposes the dictionary's motion must also be 3 pounds because the *net* force is zero.

22. We wish to find the acceleration a_{slope} down the slope, which we can find if we know the net force in that direction, since $a_{slope} = F_{slope}/m$ There is a 100 N force *down* the slope and a kinetic friction force of 40 N that points *up* the slope, retarding your friend's downward motion. So the net force down is 100 N − 40 N = 60 N. Since the person's mass is 50 kg, we find a_{slope} = 60 N/50 kg = 1.2 m/s². (If the forces or the mass had only one significant figure accuracy, the answer would be 1 m/s².)

29. When the approach is slower, it is easier to land safely, whether bird or plane. A tailwind (which comes from behind) tends to bring the bird or plane in faster, so a headwind that slows it is more desirable. There's another reason, especially for planes with fixed wings. When the wind's direction is the same as the plane's direction, the relative speed between air and wings is less, giving less lift and drag. Hence, the pilot has slightly less control over the descent rate. However, a headwind increases the air's speed against the wing and increases lift and drag. This is also an important consideration when an airplane takes off.

31. First, a component of the car's weight pulls it down the slope. The backwards pointing friction force from the tires must be greater than the car's component of weight pointing downward if the car is to slow down (negative acceleration). Second, since the car is on a slope, its own contact force with the ground is less than if it is on a horizontal surface. The friction force from the tires depends on the value of the contact force, so the diminished contact force means the friction force is decreased as well, making it more difficult for the friction force to be large enough to overcome the car's component of weight that points down the slope.

36. On a level road, the force the trailer hitch must exert is less than when the car is traveling uphill, when a component of the trailer's weight is opposing the forward motion.

39. The passengers going over a crest feel lighter. The car accelerates downward abruptly and suddenly supports the passengers less. The decrease in the contact force from the car's seats and floorboard means the passengers accelerate downward, along with the car. They feel lighter, just as if they were in an elevator that starts down. When the car turns upward on a slope, the passengers inside accelerate upward with the car. So the contact force from the car on the passengers increases, becoming larger than their weight. Because of this, they feel heavier as the car pushes them upward.

46. The farther mass is from the axis of rotation, the larger the rotational inertia and the greater the torque needed to rotate the upper body. The most difficult is thus when the arms are stretched out, farthest from the axis of rotation.

51. When a car accelerates, gaining speed, the friction forces from the road on the tires point forward. Since the center of mass of the car is higher than the road, these forces exert a torque on the car which tends to rotate its front end upward and its back end downward, just opposite from when the car stops.

56. (a) 600,000 lb; the lift must balance the weight if it is to travel horizontally. (b) 0.6 lb/in² (c) $P_{wings} = 0.04\ P_{atmosphere}$

Chapter 5

1. T **2.** T **3.** T **4.** T **5.** F **6.** T **7.** F **8.** F **9.** T **10.** T

19. The airbag expands quickly, and then as the driver impacts against it, it "gives" as it changes shape and as the air inside compresses, bringing the driver to a halt more slowly than the car stops, decreasing the average force on the driver. The area of the bag is also important, since the larger the area that applies the force to stop the driver, the smaller the average pressure on the driver is, since pressure = force/area.

22. No! The earth gains exactly as much momentum upward as the keys gain downward, and the total momentum of the system of earth plus keys is still the same.

23. The bullet is pushed by the expanding gases for a longer time when it travels through the rifle barrel than when it goes through a pistol's barrel, so its impulse is greater and its change of momentum is also.

27. The soccer player has momentum mv of 50 kg × 7 m/s = 350 kg m/s, while the bullet has a momentum of 0.1 kg × 850 m/s = 85 kg m/s, almost one fourth that of the soccer player. If a 60 kg villain at rest absorbed the momentum of that bullet, the villain's final speed would be (85 kg m/s)/60 kg = 1.4 m/s.

29. The racket is a great deal more massive than the tennis ball. Equal but opposite changes in (mass × speed) mean the racket's speed changes very little compared to the tennis ball's speed.

31. Because the relative speed between the ball and the bat is much greater for the pitched ball, the relative speed after their collision can be much higher, meaning the ball can go much farther.

34. Perfectly elastic means with no deformation at all. Therefore, cars wouldn't bend and crunch during a collision, and repair bills would be less. But the relative speed afterward would equal the relative speed before, and the change in speed (or velocity) during a perfectly elastic collision is greater than for an inelastic collision. The passengers, then, would have greater changes in momentum, which means the forces on them would be larger.

36. The engine has the greatest momentum, mv, 3M × 3 m/s = 9 M m/s. To find the total momentum before the collision, notice that the boxcar on the right has momentum that subtracts from the engine's momentum since it is in the opposite direction. The total momentum is then 9 M m/s − 1 M × 2 m/s = 7 M m/s. Checking to see if that equals the final momentum, we see a total mass of 5 M moving to the right at 7/5 m/s for a total momentum of 7 M m/s, which checks. The relative speed between the engine and the boxcar on the left before the collision is just the speed of the engine, 3 m/s. But the relative speed between the engine and the boxcar on the right is 3 m/s + 2 m/s = 5 m/s. (Since they are traveling in opposite directions their speeds add to give the relative speed.) After the collision they all move off together at 7/5 m/s over the track, meaning their relative speed is zero, so the collision was not just inelastic, but perfectly inelastic.

45. You'll move faster if you catch the ball and throw it back. Here's how to tell. The ball's total momentum change is equal but opposite to yours. It changes by a certain amount if you just catch it and stop it, but if you throw it back, it changes even more. You'll move faster in option (b) and slowest in option (a). Option (c) will give you a speed between (a) and (b). If it drops straight to the ground, the ball's momentum changes more than if you catch it and it continues to go forward at your speed, yet less than if you threw it back.

48. The first reason is to increase the time of impact and so lower the average force that stops the fall. The second reason comes from Chapter 4. The larger the area of the body that takes the impact, the smaller the pressure ($P = F/A$), and thus the less chance of an injury.

54. As the iron (more mass per cubic meter than the surrounding material) sank closer to the axis of rotation, the earth's moment of inertia about that axis would decrease, so conservation of angular momentum tells us the earth's rotation rate would increase.

58. Use impulse = change in momentum, with weight = mg. The answer is 134 seconds.

Chapter 6

1. T **2.** F **3.** F **4.** T **5.** T **6.** T **7.** T **8.** T **9.** F **10.** T **11.** F **12.** T **13.** T **14.** T **15.** T

21. Your first impression of the answer to this question might be no, since the force of static friction is exerted on your feet, and your feet don't move through a distance while that force is acting. Thus the work that starts or stops you seems to come from the muscles of your body as you bend the legs. But remember that your muscles could go through the same sort of motions if you were standing in (or walking through) a puddle of grease and you'd get no work done on you at all. It *is* the external force of friction that is responsible for the work. It is a force that acts on you while you move through a distance, *even though the parts of you the force acts on* (your feet) *are not moving*.

27. The boat doesn't move, so he's *not* doing work on the boat. The water moves as he paddles, however, so he does work on the water.

32. The extra kinetic energy they gain will let the car go farther up the hill before gravity's work can possibly bring it to a halt. The car's engine will have less work to do *while it is on the hill* to get to the top. But the work it does to gain the kinetic energy to begin with will be the same as the work saved while it is on the hill. This technique really saves no work at all.

36. Since the *KE* of the car is four times as great at 60 mph than it is at 30 mph, four times as much work must be done to stop it from the 60 mph speed than the 30 mph speed. If the brakes exert about the same average force both times, then the distance traveled should be four times farther stopping from 60 mph than from 30 mph. Now 162 ft is *about* 4 times 44 feet, so this seems to be very consistent with net work = change in kinetic energy.

38. Potential energy = *mgh*, and ball B is *about* 5 times higher than ball A, so its potential energy is *about* five times as great.

40. If the bike frame bends as the rider pumps hard on the pedals, the rider does work in bending the frame through a certain distance. That work can't go to propel the bike along, so it's wasted effort. (The frame bends back, of course, but most of the energy that goes into stretching the metal turns into heat.) A perfectly rigid frame, one that doesn't bend, wastes none of the biker's energy in that manner.

42. The veins are somewhat elastic. When the heart pumps, they stretch, storing some of the work done to stretch them as potential energy. As the blood pressure drops, they squeeze down a bit as they collapse, helping to maintain a (reduced) pressure and hence flow of blood. This effect is greatest in the aorta, the huge artery leaving the heart. There large quantities of elastic tissue called *elastin* expand greatly with each heartbeat, store work as potential energy, and contract or shorten as soon as the beat is over. This helps to maintain the blood flow even between heartbeats.

47. *The body is less efficient at converting energy into muscular force when the rate of using energy is high.* Training helps. (Untrained muscles have been shown to be only 1/7 as capable of using fatty acids as trained muscles, for example.) There are also the extra energy expenditures a runner has that don't go toward pushing the runner faster. An average male, for instance, inhales about 6 liters (a liter is a volume of about 1 quart) of air per minute while at rest. In a cross-country run the same man may average 150 liters per minute. To breathe that fast, he must force air in and out of his lungs, using perhaps 10 times as much energy per minute as he would for normal breathing. There's some air drag, and the extra extension of tendons and muscles in running uses up some energy. The heart works much harder to get the oxygen to the muscles and also to pump more blood to dilated vessels under the skin for cooling.

55. To find the lost kinetic energy, just find the total *KE* before the collision and the total *KE* after the collision and subtract. To find the energy in joules, first convert 90 km/h and 30 km/h into m/s. Doing this for 30 km/h, we get 30,000 m/h times (1 h/3600 s) = 8.33 m/s, so 90 km/h = 25 m/s. Putting in the numbers, we find the initial $KE = 9.4 \times 10^5$ J and the final $KE = 1.0 \times 10^5$ J, so 8.4×10^5 J was lost, or $(8.4/9.4) \times 100\% = 89\%$ of the initial *KE* was lost.

Chapter 7

1. T **2.** T **3.** T **4.** F **5.** T **6.** T **7.** F **8.** F **9.** T

11. Less, since you are farther from the center of earth and thus farther from its center of gravity.

14. Yes. Suppose a spacecraft is directly between the earth and moon. Their pulls of gravity oppose one another; they point in opposite directions. If the amount of each pull is the same at any point between the surface of earth and the moon, the two forces will cancel. For the earth and moon there is such a point; it lies about 24,000 miles from the moon, or about 216,000 miles from earth.

21. The shuttle is freely falling and travels horizontally over the plane only instantaneously. Lift holds the plane and its contents in a horizontal path, so the passengers get a contact force from the floor for support.

22. Its orbit is nearly the same as the original orbit of the shuttle at the time it was "dropped". As time goes on the tiny amounts of drag it gets from gaseous molecules (and even a small amount of force it gets because sunlight strikes it) will lower its energy and it will burn up upon entering the atmosphere.

27. At apogee the satellite was farthest from earth's center. As it moved outward, the force of gravity from the earth did negative work to slow it. At perigee, the satellite had moved in closer to the earth's center. Gravity did positive work to speed it in this portion of its orbit.

36. 30 km/s, or about 18 mi/s.

42. *Larger*, because the moon's pull would have an even greater percentage change from one side of earth to the other.

50. At the surface the force is equal to your weight, mg. That is equal to $F_g = G(m_{you} \times M_e/R_e/R_e)$. So for (a) 2 radii from earth's center, the denominator of the equation above would be $/(2R_e)^2$, meaning the force is 1/4 as great. (b) 1/9 as great. (c) 200 miles = 322 km, whereas R_e is 6380 km. So $F_{shuttle}/F_{ground} = (R_e/R_{shuttle})^2 = 0.91$, which is 91% of the force on the ground, so in free fall there you would accelerate at 0.91 *g*.

Chapter 8

1. T **2.** T **3.** F **4.** T **5.** F **6.** T **7.** T **8.** F **9.** F **10.** F **11.** T **12.** T

14. The *least* reactive, which either were found in the free state or were easiest to separate from compounds in the simplest chemical reactions. The ten were carbon, copper, gold, iron, lead, mercury, silver, sulfur, tin, and zinc.

23. (a) 342 u (b) Using Cl = 35.5, 154 u (c) Using Cl = 35.5, 74.5 u (d) 100 u

26. ink (a), milk (a), ice (b), gold (c), apple pie (a), copper (c), bread (a), air (a), silver (c), sugar (b), soft drinks (a).

33. About 0.4 lb. That is, since sodium's atomic mass is 23 and chlorine's is 35.5, the molecular mass is 58.5, so 23/58.5 = 0.39, which means about 40 percent of the mass of sodium chloride is sodium.

40. Endothermic, because when nitroglycerine breaks apart the reaction is very exothermic.

50. (a) yes (b) no (c) yes

52. Twenty percent of your weight mg would go up by 13/12, so your weight would increase by 0.2 $mg \times$ 13/12 minus 0.2 mg. If you weigh 120 pounds, for example, your weight would go up by 2 pounds!

Chapter 9

1. T **2.** T **3.** T **4.** T **5.** T **6.** T **7.** T **8.** T

11. These huge stones depend on their weight and accurate fitting to keep them in place. Their shapes interlock somewhat, so there's no easy direction of yield. (Just as the Grand Coulee Dam is called a gravity dam because its sheer weight on its base holds back the water, these Inca walls could be called gravity walls.)

20. When a large force is supported by a small area of contact, the pressure is large. The molecules of each surface that do touch press together with enough force to cause metallic bonding between the atoms in contact.

23. The whole beans have much less surface area per unit of volume than the ground beans will. They won't get stale (lose aroma and oils through their surface area to the air) as quickly as ground beans will.

27. Increasing the surface area of the food helps, because the chemical activity of digestion starts at the surface of the food you swallow.

32. Smaller particles have more surface area per unit of volume than larger ones. So the smallest particles of sand dust have a great surface area compared to their mass. For that reason, they are the easiest for the air to lift and blow around. The skies over southern and central Florida are sometimes whitened for several days at a time with the glare of sunshine on such particles that had been carried by the wind across the Atlantic Ocean from the Sahara Desert.

36. The water comes out of a nozzle that helps to separate it into drops almost immediately. Those drops travel out at nearly the same angle and with essentially the same speed. But the smaller drops slow faster because of air drag, so they travel a shorter distance before hitting the ground. That's why the ground will get wet between the sprinkler and the outer circle where the largest drops fall. You can see this behavior under the stream of a garden hose as well. Why do the small drops slow faster? The smaller droplets certainly have less surface area than the large drops and have less total air drag. But they have more surface area per unit of mass, so *the air drag on a unit of area decelerates a smaller amount of mass in the smaller droplet.*

38. The cardboard box of wrapped eggs will have a huge surface area but very little mass, so its terminal speed will be very slow. The paper wrappings first yield to and then stop the eggs with a force over a large area on the side of the egg, minimizing the pressure.

43. True. The lollipop dissolves only at its surface. When there is more surface area, the rate of dissolving the candy will be greater in proportion.

45. True. The smaller the size of a given shape, the greater its surface area per unit of volume.

55. Just find the pressure a column with a cross section of 1 cm^2 (or any other cross section) exerts on its base and convert to the other units. The column is 1200 cm high, so 1 cm^2 at the base must support the weight mg of the mass of (1.3 g/cm^3 $\times$ 1200 cm^3), or mg = 1.56 kg $\times$ 9.8 m/s^2 = 15.29 N. So the pressure is 15.29 N/cm^2. Since 1 cm^2 = 10^{-4} m^2, P = 152,900 N/m^2. Since 1 atm = 101,000 N/m^2, P = 1.51 atm.

Chapter 10

1. F **2.** T **3.** T **4.** T **5.** T **6.** T **7.** T **8.** T **9.** T **10.** F **11.** T

15. No; the ice floats at a level that displaces its weight in the water. When it melts, it still has the same mass, hence weight, and it still displaces its weight in the water. Its density is the same, so the space or volume it fits into is exactly equal to the volume displaced by the ice cube. The water level will neither rise nor fall.

20. In the middle of the flow, farthest from the solid edges, which slow the flow to zero with friction.

22. The lungs of someone who is snorkeling are about 1 ft below the water's surface. As the swimmer inhales, the diaphragm must work against the pressure of the water that is transmitted through the body to the lungs. Though the pressure at a depth of 1 ft isn't much (1/34 atm = 0.44 lb/in^2), the snorkeler will tire from it after snorkeling for a long time.

25. Surface tension! Air-dried towels are stiff because the wet fibers of the towel on the line group together and stick in mats as a result of the surface tension of the water. As the water evaporates, they tangle, stiffening the fabric. In a clothes dryer the tossing motion and currents of warm air tend to keep the fibers moving and drying independently.

28. The soap reduces the water's surface tension; but so does the extra temperature. At higher temperatures the water molecules move faster and are less able to hold onto one another.

32. Surface tension between the wet fibers of the thread hold them together in a point.

36. The buoyant force helps support them, keeping minimum stress on their bones and muscles.

41. Since the buoyant force must balance its weight, it must displace 1000 tons of seawater.

42. The clay particles stick together with a strong cohesive force.

Chapter 11

1. T **2.** T **3.** T **4.** T **5.** T **6.** F **7.** F **8.** F **9.** T **10.** T

23. As the local air pressure goes up and down during the week, and the temperature changes, the density of the outside air changes. That means the box displaces different weights of air during the week. The buoyant force on the box changes, and it weighs different amounts on the scale. (Since the box is sealed, its volume may well change with different atmospheric pressures. Convince yourself that action would partially correct for the buoyancy lost or gained.)

24. Bernoulli's principle causes the partial collapse of the trachea. Then because the air pushed from your lungs goes through a smaller opening, it picks up speed, exerting a greater direct push on anything in its way. Both effects help to dislodge the particle.

30. With entrances at two different heights, the higher entrance will ordinarily have a bit more air movement over it, so Bernoulli's effect reduces the pressure there and air pushes through from the lower opening to the higher one, exchanging the air in the burrow.

33. The 25 lb of pressure is the amount *over* atmospheric pressure. So there'd be fewer molecules in your tire in Denver to give the same pressure as in Miami at sea level, presuming the same temperature at both locations.

37. At the solid boundary of the fan blades, the air speed is zero even though the blades are pushing air along rapidly.

38. A little more than 11 lb.

41. The birds would weigh less and be easier to support in the air. But the atmosphere would weigh less too. It would bulge outward, leaving the sea-level air much thinner. The birds would have to push much harder to get the same amount of force. The effects, therefore, would at least partially cancel.

45. You must find the weight of the displaced air! That is equal to the buoyant force. The air's mass is 7320 kg, which *weighs* about 72,000 N or 16,000 pounds. (To find the net force upward on the bare balloon, you would subtract the weight of the helium and the balloon.)

47. In space the exhaust gases don't have to push outward against the pressure of surrounding air—over a ton for every square foot of exhaust area at sea level.

48. The deep blue sky isn't very deep or very "thick." The space shuttle orbits in an almost air-free environment, so the atmosphere that presses down all around us is fairly thin in terms of height. The fact that we can see horizontally through it so well means that the molecules are both far apart even near the ground and largely transparent to light.

53. In Boston, most likely. If the hail is traveling at its terminal speed with no extra speed from wind, it will fall somewhat slower in Boston because earth's atmosphere is thickest closer to the ground. At Leadville, the highest incorporated town in the United States, the terminal speed for hail in the thinner air would be higher.

54. Unlike the liquid, a bubble of air can compress as the pressure on the brake fluid increases. The work done on the fluid goes partly to compress the bubble somewhat rather than to push on the brake shoes or brake pads. Therefore, the liquid beyond the bubble in the line won't move as far. (The bubble can't compress all the way and become incompressible liquid air; that would require about 1000 atmospheres of pressure.)

58. 3000 in^2, or about 21 ft^2.

Chapter 12

1. F **2.** T **3.** T **4.** T **5.** F **6.** F **7.** T **8.** T **9.** F **10.** T **11.** T **12.** F

15. It's true! The engine give the car *KE*, and that energy totally changes form each time the car comes to a halt. The brakes heat up as friction there brings the car to a rest, and as the car moves through the air, pushing it aside, some of the energy given to the car is used to do work on the air.

20. The copper veins in the rock both expanded and contracted more than the surrounding rock. The heat made it expand, which broke up the rock, and the water cooled it. It could then be easily removed from the broken rock.

23. 102.2°F

26. As the vapors inside cool, the pressure of those vapors against the lid drop. The air pressure on the outside is undiminished, however, and there will be a large net force that keeps the pot's lid on.

29. Small regions on the surface of the droplet in contact with the pan burst into steam, which has a much greater volume than the liquid water it comes from. Those miniature explosions propel the drop of water one way and then another.

33. Look at Table 12-1. Silver contains much less heat energy than water does at a given temperature.

45. The thermometers reading would cease to change at that temperature, even as more heat is added or taken away.

48. Cooking oil has a much higher boiling point than water does, and the boiling temperature is the highest temperature a liquid can go to unless it is held under pressure. So oil can be used to cook food at higher temperatures than water can.

52. The ice prevents the warmer water underneath from coming into direct contact with the colder air above, and for the water below to lose its latent heat of fusion means that heat must escape through the solid ice (see pg. 178).

54. Muscles in use give off much more heat than the normal resting metabolism does. They decouple their wings and vibrate the muscles very fast.

Chapter 13

1. F **2.** F **3.** T **4.** F **5.** T **6.** T **7.** F **8.** T **9.** T **10.** T **11.** T **12.** T

15. When the tip of your tongue sticks to the scoop, it's because the cold metal conducts heat away from your tongue so fast that it freezes the moisture on your tongue at the point of contact. A child's tongue can be seriously damaged if it's touched to an aluminum light pole or other metal surface in subfreezing temperatures. A wooden spoon conducts heat poorly and won't cause that problem.

18. With no gravity no natural convection currents cause the gases of combustion to rise, bringing in fresh oxygen from the side. Placed under a ventilator that directs forced air, however, the candle could burn in the artificial convection current. Or if the candle was waved around while it burned, fresh oxygen from the air could get to the site of oxidation.

24. The interior of the fireplace gets hot and radiates a great deal of heat energy along with the fire. This increases the surface area radiating heat into the room significantly.

27. The water heater is insulated; without the heater on, the water in the tank stays cool even when the daytime air around it becomes scorching hot. Meanwhile, the water moving through the uninsulated cold water pipes picks up heat from the ground and air and becomes hotter than the water stored in the insulated water heater.

28. When the person stands normally, the insides of the thighs radiate heat to each other and so reabsorb part of that heat radiation. Likewise, the insides of the underarms radiate to the body, and vice versa.

33. Air is a *very* poor conductor of heat compared to water, principally because its molecules are so far apart.

35. Heat must be absorbed through the surface, and a large turkey has *less* surface area per pound of meat. So a large turkey takes *more* time per pound to cook all the way through.

40. When water boils, its volume increases dramatically. When a bubble of steam bursts forth underwater, its molecules occupy roughly a thousand times more volume than they did in the liquid state.

44. It takes energy to move that snow, and you could spend more energy moving snow than the extra insulation could save you in the period of an evening if you pile too much on!

55. If the winedrinker holds the wineglass by the stem, the hand won't warm the wine. (Another possible reason, related indirectly by someone in France, is that when wineglasses were first developed with long stems it was before the days of indoor plumbing and frequent bathing, and winedrinkers, at least those who could afford wineglasses, used perfumes and powders to make mask body odors. The long stem let them sample the bouquet of the wine by keeping the wine farther from their hands and arms!)

Chapter 14

1. T **2.** T **3.** T **4.** T **5.** F **6.** T **7.** T **8.** F **9.** T

12. Infrared radiation that is given off by the ground and objects on it rises and escapes through a clear sky at night. Because the air in an overcast, cloudy sky has a high content of water vapor, however, the radiation warms the air as it travels through it. Water vapor (as does carbon dioxide) absorbs infrared energy and transfers much of the absorbed energy through collisions to other molecules in the air.

15. The long hours of daylight (up to 24 hours) in the area allow plankton to grow at much faster rates than in the middle latitudes.

17. When your warm breath enters the cold air, it cools quickly. This raises its relative humidity to the point of supersaturation, and rapid condensation follows. But the cloud quickly disappears as the small droplets evaporate into the cold air around them which has a lower relative humidity.

20. Friction! The many falling water drops push downward on the air in front of them and sweep it along.

24. When warm moist air strikes a cold can of soda pop, that air loses heat rapidly, and its humidity rises. Because the metal can conducts heat well and because the drink (mostly water) inside requires a lot of heat to warm it, the water vapor cools to the point where it condenses on the surface.

29. The moving air speeds evaporation on the windward side of the finger, making it lose heat faster than the side away from the wind.

32. The upward breeze, called a *valley breeze,* rises as the sun heats the slopes, warming the air next to the slopes. Less dense, this air is buoyed up by the heavier air around it. It lasts from mid-morning to late afternoon. The downward breeze, called a *mountain breeze,* begins as the temperature of the slopes drops in the evening due to heat radiation. The air next to the slopes cools, becomes heavier than the air around it, and flows downward.

34. The vapor in the warm, saturated air condenses on the cool surface of the lens of the glasses. If the lips are pursed, the air expands and cools. This cooling lowers the relative humidity of the air before it strikes the cool lens, and the lens won't fog. (Blow on your hand with your mouth in each shape; feel the difference in temperature of the air.)

38. Large hailstones have less air resistance per unit of mass than small raindrops, for instance, and the hail thus has a greater terminal speed and can descend very quickly. In addition, there will be a larger downdraft because of the hailstones great mass and speed (see Exercise 14-20).

43. Cities contain large areas of streets and rooftops, which absorb solar energy much better than the canopies of leaves and grass in the countryside, so cities are warmer for that reason. Also, vehicles, abundant in cities, emit a good bit of heat from their engines and their exhausts. The cooler air from the countryside is heavier than the warmer air and pushes in along the ground, buoying up the warmer air to cause that circulation shown.

Chapter 15

1. T **2.** F **3.** F **4.** T **5.** T **6.** T **7.** F **8.** T **9.** T **10.** F

12. The wavelength gets shorter when you increase the rate of oscillation, the frequency, but that shorter wavelength doesn't catch and overrun the longer wavelength wave in front of it. The reason? All the frequencies of waves you put on a spring travel at the same speed.

14. Raindrops fall randomly on a lake's surface. Therefore, the waves they make spread from different points, losing amplitude as they do, and the interference of the various waves is just as often destructive as constructive.

18. Such alligators would be close in size to the wavelengths of the wind-generated waves. As a result, they would be considerably disturbed by the passing waves on the water's surface. (See the section "Interacting with Waves," page 227.)

23. The shorter wavelengths, those that nearly match the size of the reeds or weeds along the shore, are both scattered well and absorbed by those weeds. Longer wavelengths pass with less interference (see pg. 227, "Interacting with Waves.")

24. The compressional waves generated in an ice pack by the storm winds pushing it move much faster through the solid ice than the winds themselves can travel.

28. The hairs on the mole can detect vibrations from compressional waves moving through the ground. In a similar way, the blind fish senses a disturbance in its vicinity by a wave traveling through the water.

30. The surface waves that disturb seismographs are caused by the winds that blow over the earth's surface (and the accompanying fluctuating air pressure). These waves resemble ocean waves in that the surface particles perform almost circular motions. However, the amplitudes are quite small and are noticeable only with sensitive seismographs.

32. The sloshing wave you can make in a glass of tea has a wavelength that is twice the diameter of the glass. In a nearly full glass of tea, the bottom will be a wavelength or more below the surface. One wavelength below the surface, the tea is disturbed very little. But if you drank iced tea from a soup bowl or even a pie plate, the bottom of the tea would be much closer to the surface, and that wave would stir the sugar.

37. Marching in step, the troops exerted a large net force with a frequency that was close to a resonant frequency of the bridge. As they marched without interruption, the standing wave (normal mode) grew until the bridge collapsed. If the troops had broken formation while on the bridge, their footfalls would not have applied a large force with a definite frequency, and the resonant mode would not have been excited.

40. Explosions generate large-amplitude compressional waves, or P waves. These waves are used to detect underground nuclear explosion tests.

43. 2 meters—the distance from the undisturbed medium to the top of the wave is 1/2 the distance from the bottom of the trough to the crest.

Chapter 16

1. F **2.** T **3.** T **4.** T **5.** F **6.** T **7.** F **8.** F **9.** F **10.** T **11.** F

12. The conclusion was that the giraffes emit infrasound, very long wavelength sound that would turn (diffract) over the hilltop, and also around the brick wall. Researchers are presently investigating such sounds from large animals with special receivers.

19. The smaller the mass, the less the inertia. Smaller bones can resonate with much higher frequencies because they can move back and forth faster.

21. The high rate of flow through a small opening ensures turbulent flow, and eddies vibrate the valves and pipes and make noise.

25. Softer feathers make less noise when pushed through the air than hard feathers do; harder feathers create more and larger eddies in the air. Quieter flight helps the owl surprise its prey.

27. Much of the energy of the sound waves is reflected at the door's solid surface, since sound travels much faster through the wood than it does through air. The door's great mass, hence great inertia, also means it can't move back and forth very fast as the compressions and expansions arrive.

29. In Chapter 15 we saw that water waves are dispersed more by objects whose physical size is close to a wavelength. Sound waves behave the same way. The tiny pine needles scatter and reflect the high-frequency (short-wave-length) portion of a sound wave best, while the low-frequency portion passes by with less backward scattering.

33. Previously, the sound of the approaching boat reflected from the solid expanse of the rock. Experienced captains used these sounds to judge distance and position when visual conditions were not good.

37. The water that has been pushed aside by the propellers quickly collapses into the cavity that is formed. The sound comes from the impact of the inward-rushing water as the cavity fills.

39. Because of the Doppler effect, a bystander on the ground in front of the motorcycle would hear a higher frequency sound from the horn than the horn actually makes. The sound wave would travel on to reflect from the end of the canyon and return to the bystander, who will hear that higher frequency once again. However, the rider doesn't hear what the bystander does in either case. The rider initially hears the sound of the horn's actual frequency, since the relative speed between them is zero. Later, the rider hears the reflected sound from the wall of the box canyon. Because the rider is moving into that reflected wave, the rider hears a frequency of sound that is even higher than the frequency the bystander hears.

40. Sound waves are scattered best by objects about one wavelength in size, so wavelength = v/f = (340 m/s)/120,000 Hz = 0.0028 m = 0.28 cm. Insects about 1/4 cm in size scatter (or reflect) such sound well.

42. Colder air is more dense and thus more viscous; the eddies that cause sound occur at lower speeds for cold air moving past a sharp corner than they would for warm air.

44. It probably hears that frequency best. Sound at that frequency evidently excites a standing wave with large amplitude in the bat's ear cavity. This increases the response of the ear at that frequency and enables the bat to detect smaller intensities of sound at that frequency.

50. A difference of 10 dB, as seen from Table 16-2, amounts to a factor of 10 in the intensity of the sound. With respect to the *loudness* of the sound as perceived by the human ear, however, a 10 dB increase means the sound is only *twice* as loud. This should be sufficient.

Chapter 17

1. T **2.** T **3.** F **4.** T **5.** T **6.** F **7.** T **8.** T **9.** T **10.** T **11.** T **12.** T

16. Charge separates most easily when *unlike* substances, with different affinities for electrons, are brought into contact.

21. Friction with the air charges the airplane while it is in flight.

22. The paper gets some charge from the comb when it touches, and like charges repel.

29. As shown in Table 17-1, the charge is most likely negative. The chains on older gasoline trucks grounded the truck body, letting the charge that could cause a dangerous spark move off to the ground.

32. Normal matter always attracts other normal matter with the gravitational force. Only if another type of matter occurred that repelled normal matter could gravitational forces be balanced by others. Antimatter, discussed later in this text, may be repelled by normal matter. But experiments to create it and observe its attraction or repulsion for the earth have been inconclusive.

37. The electrons in the neutral atom shift toward the charge if it is positive or away from it if it is negative. This leads to an attraction by way of induction.

40. A beam of electrons in the tube excites the material on the screen and causes it to glow. The screen is negatively charged while in operation. The paper becomes charged by induction and sticks to the screen.

45. The electrons' strong electric attraction to the positive ions they would leave behind prevents the electrons from rushing to the surface because of the electrons' mutual repulsion.

48. Inside the car, a person is surrounded by metal which keeps the charge on its surface and the electric field inside equal to zero. But if the person touches the ground, the metal car is grounded through the person's body and charge will move.

49. St. Elmo's fire is caused by a large difference in electric potential across that region of air where it is seen. It is a small electrical discharge, a small version of lightning. Generally this occurs not because of a charged overhead cloud (which could cause a lightning strike), but because of friction, such as between the boat and the water. If the boat was charged, the charge would be concentrated on the points of the masts, causing a large potential difference between those points and the surrounding air.

Chapter 18

1. F **2.** T **3.** F **4.** F **5.** T **6.** F **7.** F **8.** T **9.** F **10.** F **11.** T

15. Yes; twice the cross-sectional area in the straw or the wire allows twice as much milkshake per second or twice as much charge per second to flow, as the case may be. Whereas the milkshake rises in the straw because of a difference in pressure, the charges flow in the wire because of a different electric potential, or a voltage difference. In either case, twice the cross-sectional area means that the flow is twice as great.

16. A short circuit would occur as a large current flowed through the metal of the screwdriver, which offers little resistance. This would cause rapid heating of the screwdriver. The battery has no circuit breaker, so the current would continue to flow until the circuit was broken or the battery was dead.

21. 4 J/s = 4 W

25. Yes, it has a greater resistance while the same current flows through it.

31. Parallel, since if one headlamp "burns out" the other still operates; if they were in a series circuit, when one broke the other would also go out.

37. The wire would conduct a larger current if connected to the wall socket. $i = V/R$, and V is 110 V for the home circuit but much less for the flashlight battery. The large current would produce a greater rate of heating ($P = i^2R$), and the wire would melt.

44. A 60 Hz current means the electrons perform exactly 60 oscillations in a second. They change directions twice during each oscillation, so they change directions 120 times a second. (Think of an electron at rest before the current is on; then it moves, say, right, stops, zooms back past where it started, stops, and then returns to where it started.)

46. The 5-ohm bulb on both counts. Less resistance means the bulb draws more current, heats more (becoming brighter), and drains the battery faster.

52. $E = P \times t = 160 \text{ W} \times 24 \text{ hr} = 3840 \text{ Wh} = 3.84$ kWh, which is greater than 3.5 kWh.

Chapter 19

1. F **2.** T **3.** T **4.** T **5.** T **6.** F **7.** T **8.** F

11. At the north geographic pole the needle would have a great inclination, or dip, pointing almost straight down. At the south geographic pole it would point almost straight up.

14. The south pole of the clip would touch the north pole of the magnet, and the north pole of the clip would be at the other end, away from the north pole of the magnet.

19. Find the two bars that behave like magnets, such that either end of one bar attracts one end of the other bar but repels the opposite end. The third bar is the nonmagnetic bar. Either of its ends would be attracted to any end of the magnetic bars by induction.

25. Angling the bar downward aligns it more nearly with the earth's magnetic lines of force. Its magnetic domains will, therefore, align all along the length of the bar and make it a stronger magnet.

26. It will be near the base. Remember that the earth's north magnetic pole is actually the south pole of a magnet.

29. We cannot notice or feel a change in the gravitational force because we cannot vary our distance d to the center of the earth by a large percentage. We can, however, vary the distance d between two small magnets.

33. Near the magnetic equator, roughly the equatorial regions of the earth, where the magnetic field has little or no dip.

35. A second magnet is rotated beneath the platform, attached to the axis of a motor. The other magnet, the stirrer, follows the magnet beneath it, its north pole attracted to the moving magnet's south pole, etc.

Chapter 20

1. T **2.** T **3.** T **4.** T **5.** T **6.** F **7.** F **8.** T

12. It made DC. The electrons in the copper disk crossed the magnetic field in only one direction; hence they were pushed in only one direction.

18. Nothing visible goes on because the copper atoms in the wire don't travel; about one electron per atom does the moving. Yet these electrons represent a lot of energy because of the strength of the electric force. Because those moving electrons oscillate back and forth rather than travel in only one direction, the net flow of electrons through an end of a wire is zero.

19. Yes, it could. A transformer only requires the magnetic field to change across the secondary coil. For instance, if AC flows through the primary coil, the magnetic field across the secondary coil oscillates in direction and intensity with that current. On the other hand, if a steady DC flows, the magnetic field across the secondary coil is steady. To use a transformer with direct current, the DC can be regularly interrupted (with an oscillating off-on switch, for example). The DC then changes, and so does its magnetic field. A *changing* DC in the primary coil does generate a voltage across the secondary coil, and current will flow if there is a closed circuit.

21. The magnetic fields of the electromagnets exert only a turning force on the electrons, doing no work on them to change their speed. The electrons are deflected more if the intensity of the magnetic field is greater. By varying the strength of the magnetic field, the beam may be guided to the desired points on the screen.

24. "Transported" implies something material moves from one place to another, and that's not so in this case; the electrons only oscillate back and forth. However, energy is transmitted to points along a conducting wire from the source of the voltage difference across the wire.

28. The 1500-V lines are more dangerous; such high voltages cause insulator breakdowns, even through short air gaps, more readily. The increased voltages for cars were smaller, increasing efficiency without increasing the risk very much.

31. When the tape to be played passes through the magnetic head, the magnetic field of the tape induces in the head a current that can be amplified electronically to reproduce the original sound.

36. Read Exercise 19 and its answer. A switch interrupts the DC voltage, turning it off, then on, then off, and so on.

39. No, the path is not without interruption. Transformers are used at many points along the way, and their primary and secondary coils of wire are not connected to each other. They merely share the same magnetic field.

42. Downward, toward earth. This isn't terribly easy to visualize; compare Figs. 19-4 and Fig. 20-2(b).

49. The currents induced by the magnet in the coils create another magnetic field, one which *opposes* the fall of the magnet. You can figure this out by applying conservation of energy. If the falling magnet induced a current in the wire whose magnetic field *added* to the magnetic field of the falling magnet, the magnet would be attracted by that field and fall even faster than g. If it fell faster, it would induce a greater current in the coil, and be attracted even more, so it would be creating energy—impossible according to the conservation of energy.

Chapter 21

1. T **2.** T **3.** T **4.** F **5.** T **6.** F **7.** T **8.** F

13. The reason lies in the fact that an electron and a proton respond differently to equal forces. Because of its smaller mass, an electron has an acceleration nearly

2000 times as great as a proton that is experiencing the same force. The electric field of the electron, then, becomes much more distorted, and the resulting traveling kinks in the field lines represent the light that we see from the candle's flame in Maxwell's theory.

14. TV signals, which are in the microwave region of the electromagnetic spectrum, pass through the ionosphere rather than reflect from it. Therefore, TV reception depends more on straight-line paths from the transmitter to the receiver than AM radio waves do.

19. Microwaves, like visible light, interact much less with the ionosphere, whereas ordinary radio waves reflect well from the ionosphere. Radio-wave reflections would cause interference in the space-to-earth (and earth-to-space) communications.

28. Cold air emits less infrared radiation than warm air. Cold air (higher pressure) seeps from the cave's entrance into the warmer air (lower pressure) outside, especially when a low-pressure front moves in. A photograph taken with a special camera that uses film sensitive to the infrared wavelengths will reveal a cave's entrance as a dark area because the cool air emits less infrared radiation than the surrounding air does.

30. The inner surface of a person's clothes may reach the surface temperature of the skin. The outer surface, however, will be closer to the air's temperature and thus will emit less infrared radiation per square centimeter than the skin beneath it does.

33. Microwaves (TV signals) and radio waves travel over 186,000 miles (300,000 km) in the time it takes sound to travel about 1000 yards (900 m). There was ample time for microwaves and radio waves carrying the amplified signals from microphones to reach the United States before the sound waves traveling through air reached the local observers.

42. Frozen solid—no, microwaves don't rotate the water molecules in ice since they are held in place in the crystal. Of course, as the surface of the frozen dinner comes into contact with the air, the surface thaws and the liquid water there does absorb heat from the microwaves and the dinner will defrost. But you can have apple pie a la mode warmed in the microwave oven if you put the pie in at room temperature and the ice cream in on top of the pie if it is frozen absolutely solid.

Chapter 22

1. T **2.** F **3.** T **4.** T **5.** F **6.** F **7.** T **8.** T **9.** F **10.** T **11.** F

13. They were pinhole cameras. Tiny pinholes would emit a tiny bit of light to slowly expose the film inside.
15. By coming from the direction of the sun, the kingfisher has two advantages. First, its prey will have poor visibility in the kingfisher's direction, since the direct sunlight is so intense and blinding. Second, this technique keeps the kingfisher from facing the glare of the sun on the water (except at high noon).

23. Light slows a bit upon entering cooler air (higher density) and speeds up when entering warmer air (lower density), so it can refract, changing directions. In the atmosphere currents of moving air, with varying densities, make the light traveling through it move back and forth by small amounts in this way.

28. The frosted bulb diffuses the light. Diffuse light is much more pleasant than light directly from the filament for two reasons. First, if you happen to look at the frosted bulb, the light is less intense than if you looked directly at the filament. Second, the shadows cast by a frosted bulb are not as sharp as those from a clear bulb, making the light seem less harsh, or softer. That is, there is less stark contrast between lit and unlit areas. The smooth outer surface is also easier to clean than a frosted surface would be.

33. Visible light has wavelengths that are thousands of times longer than a molecule of silicon dioxide (SiO_2, glass). The greatest scattering of light occurs when the irregularities are on the order of the size of a wavelength, as we saw in Chapter 15 for waves in general. Hence, the irregularities of about 20 atomic diameters hardly scatter the light striking it at all, and the mirror forms a very clear image.

35. Graphite is very soft, and maintaining the flatness of a single plane of atoms for areas large enough to make a macroscopic mirror would be difficult. A line of graphite a single centimeter long contains about 10^8 atoms, as we saw in Chapter 8. Moreover, the graphite would have to be evenly coated with a metal in order to reflect light as well as possible.

37. The rays of light refract while they travel through the lens and do not come to a focus until they are well outside the lens. Only then is the light energy delivered per cubic centimeter per second great enough to weld and drill the metal parts.

38. The rock is shinier because it reflects more like a mirror. But it actually reflects *less* light when wet, and so appears darker. Here's why. Most of the incident light enters the water, which is quite transparent. The transmitted light is slowed, so when it strikes the rock it experiences less of a change in speed than it would if it traveled through air. Hence less reflects.

Chapter 23

1. T **2.** F **3.** T **4.** T **5.** T **6.** F **7.** F **8.** T

10. The white light spreads or disperses into the colors of the spectrum when it refracts upon entering the diamond. Each time the colors of light reflect internally, they travel through the diamond again, *separating farther as they go*. Finally, when a portion of the light hits a face at less than the critical angle and exits, you see a more narrow range of frequencies, hence a more pure color.

11. They are reversed from those of the primary rainbow due to the extra internal reflection. Thus red is on the inside of the arc, and violet is on the outside.

15. Water absorbs the longer wavelengths (red, for example) better than shorter wavelengths (such as blue).

21. The light from the lamp must not include red light, which is the predominant color these objects reflect.

24. First, the paper is white and reflects much more light than the black spot did. Second, the laser pulse is of extremely short duration and the energy left behind as heat on the paper, though at high temperature, spreads out rapidly into the large mass of the surrounding paper and the temperature drops very quickly.

28. Because moonlight is just reflected sunlight, so it contains all the frequencies of sunlight. Its light scatters in the air just like sunlight does, and the blue frequencies scatter the best. Notice also that the forest floor is green, and the tree bark appears in natural colors as well. While the photographer, Dale, couldn't see all these colors as he stood there with the camera, the film collected the light over 5 minutes or so and was gradually exposed to the colors you see here, much as it would have been in a very short time in full daylight.

29. The sun must be near sunrise or sunset, as low to the horizon as possible.

32. Even with their night vision working full time, there is no light at all for the visitors to see, so the cave remains pitch-black.

39. Materials often have selective absorption of light, where light of one frequency may be absorbed, but that of a nearby frequency is reflected. For example, scorpions are dark in direct sunlight, but fluoresce in ultraviolet, or "black," light, which enables biologists to detect them at night. Sunlight and incandescent light and fluorescent light are all composed of different mixtures of the spectral colors, and a cloth may appear to be different colors under each of them because, for example, a frequency of light that is reflected in sunlight might not appear with as great an intensity in incandescent light or fluorescent light, altering its appearance.

42. The more air there is between you and a distant mountain, the more blue light that is scattered as the sunlight falls on that air. That blue light comes to your eyes as well as light reflected from the mountains.

43. Scattering; just as the molecules of air scatter blue light best, so do those fine black particles.

44. When green light and red light come from a single direction to your eye, you see yellow.

50. Both wavelengths lie in the red portion of the visible spectrum, which lies between about 380 (violet) and 750 (red) nanometers.

Chapter 24

1. T **2.** T **3.** F **4.** T **5.** T **6.** F **7.** F **8.** T **9.** F **10.** T **11.** T **12.** F

21. (a). Photons of higher frequency light carry more energy than those of lower frequency light. As they pass through a gas, its electrons interact with the photons individually, not collectively. As a result, high-frequency light ionizes best, even at lower intensities (photons/second).

22. By performing a spectrographic analysis of the paint, a technician can tell its molecular (and atomic) composition. Molecules, as well as atoms, have definite frequencies (and even bands of frequencies) of light that they absorb and emit well. Identifying those frequencies with a spectrograph identifies the particular compounds used to make the paint and their relative proportions. The compounds and their proportions differ from manufacturer to manufacturer even for the same color of paint.

25. The identical spectral lines come from the gas molecules in the earth's atmosphere. The light of every star we see passes through these gases on the way to us; therefore, these absorption lines appear in every spectrum of starlight. (They are called *telluric* absorption lines.)

27. It would be a continuous spectrum where all frequencies are present.

29. Under these circumstances, sunlight would not contain all the frequencies of light as it now does (with the exception of the absorption lines). We would see only the spectral emission frequencies of the gases at the sun's surface, mostly those of hydrogen and helium.

33. The low-pressure sodium vapor light generally emits light at two frequencies. Using spectrographs, astonomers can still observe starlight in the other portions of the spectrum. However, the high-pressure lamp emits a spread of frequencies due to collision broadening of the spectral lines. This causes more widespread interference with the spectrum.

34. A wave scatters best from imperfections that have dimensions of about one wavelength. Therefore, short-wave sound waves can reveal very small imperfections as they scatter from them.

39. The atoms in the filament of an incandescent bulb emit photons when they absorb enough energy from high-speed collisions with electrons. For every such high-energy collision, the electrons have undergone more lower-speed collisions with atoms. The collisions heat the filament tremendously, so that it loses a great deal of energy as heat. Fluorescent light occurs when the fluorescent atoms absorb photons of higher-than-visible energy, or ultraviolet photons, from the highly excited mercury vapor. They fluoresce by giving up their absorbed energy during several transitions, rather than by a single transition, emitting this absorbed energy as lower frequency photons. No random collision process is involved with the fluorescent atoms as it is with the atoms in the filament of an incandescent bulb, and much less energy is lost as heat.

44. No; the photons of red light don't have enough energy to excite those atoms to the point where they make the higher energy transitions that are necessary to emit green light.

Chapter 25

1. F **2.** T **3.** F **4.** T **5.** T **6.** T **7.** F **8.** F **9.** T

10. (a), (b), (d), (e).

14. An electron is a pointlike *particle* that has a *wavelike* nature in its behavior.

16. Psi is not a physical wave, and it has no medium through which it moves. It is a wave whose amplitude helps to predict probabilities for the actions of small particles.

17. The probability waves for atoms whose electrons are in ground states have constant frequencies and amplitudes, not decaying in time as waves on a clothesline would. If this were not true, all atoms would shrink in time. Also, and more fundamental, if the psi waves for particles did shrink and die out, the probability for finding those particles in the universe would diminish as well. There is no evidence to suggest that particles are disappearing from the universe without a trace.

23. As the thermal motion of an atom decreases, its position becomes more certain. But it can't become completely still, that is, with no x, no uncertainty in its position, because then mv would become infinite. By the uncertainty principle, therefore, each atom, even at absolute zero, has motion. This is called *zero point* motion, and the atom cannot give that motion up as heat energy. $mv^2 = kT$, as discussed in Chapter 12, is not correct as the Kelvin temperature approaches zero.

25. If the electron distribution around the nucleus is compressed, the electrons' positions are known to within a smaller If the electron distribution around the nucleus is compressed, the electrons' positions are known to within a smaller x. That means their uncertainty in momentum, mv, must increase. If the value of v increases, the kinetic energy increases. This means the electrons are more likely to pull farther away from the nucleus, increasing rather than decreasing the volume. In this way the uncertainty principle ensures that the electrons in atoms cannot collapse into the nucleus, even though the nucleus attracts them.

32. The radiating electron(s) must terminate on a state very close to the nucleus in order to lose so much energy.

Chapter 26

1. T **2.** F **3.** T **4.** F **5.** T **6.** F **7.** T **8.** T **9.** T **10.** T **11.** T

19. (c)

25. (b)

27. The distance between marks on the side of the boxcar would be less than 5 meters, which was the distance between the boys' positions.

30. (a), (c), (e).

34. In your rest frame, you would observe both yourself and the belt to look normal. In your friend's rest frame, he or she would see you as skinnier along the direction of motion. But he or she would also see your belt, which moves past at the same speed as you do, shortened by the same amount along the direction of motion. There would be no need for a new belt; your pants or skirt would not fall down.

39. No, earth's surface is not an inertial frame of reference. First of all, points everywhere but the north and south poles all revolve around earth's axis of rotation, so anything on the surface experiences a centripetal acceleration toward that axis. Also the earth performs an orbit about the sun, which gives everything on earth an acceleration toward the sun. (Then the sun itself moves around the center of its galaxy, the Milky Way, and so on. . .) While these accelerations are quite small compared to g, they nevertheless mean earth's surface is not an inertial frame of reference.

Chapter 27

1. T **2.** F **3.** F **4.** T **5.** T **6.** F **7.** F **8.** T **9.** F **10.** T

11. There is no way to know. The behavior of large numbers of radioactive atoms is predictable statistically, but not the behavior of a single atom (such as the last one to decay), or even a small number of those atoms.

14. The nuclei of your body have almost nothing to do with the image you see of yourself in the mirror. That image is formed by reflected light, and that light reflected entirely from the electrons in the molecules and atoms. The nuclear indirectly affect the reflection only in that they order the electron distribution in their atoms—that is, the number of protons establishes how many electrons there are and therefore where the electrons will be situated in the shell structure of the electron probability cloud of the atom (Chapter 25). In turn, that structure affects how the light is reflected.

17. Radioactive decay belongs to the arena of modern physics. With the laws of classical physics, motions and events were in principle predicted precisely; the time of decay of a radioactive nucleus is unpredictable.

19. The pressure that would compress an atom acts on its electrons, located far from the nucleus. Even if the electrons could be so compressed that they had a significantly larger probability of being found within the volume of the nucleus, they would not greatly affect the nuclear processes, since electrons are not subject to the nuclear force.

20. 1/16 gram

23. One of tritium's two neutrons becomes a proton when the electron is emitted. The new nucleus contains 2 protons and 1 neutron. A glance at the periodic table confirms that the daughter nucleus is an isotope of helium.

26. The gold nucleus becomes a platinum nucleus, with 78 protons and 112 neutrons.

29. Neutron-deficient ^{22}Na undergoes positron decay, which raises its ratio of neutrons to protons and brings it closer to the line of stability for nuclei. Neutron-rich ^{24}Na undergoes beta decay, which lowers its ratio of neutrons to protons and brings it closer to stability.

30. (a) $^{4}_{2}$He, (b) $^{208}_{81}$Tl, (c) . $^{39}_{18}$Ar, (d) $^{13}_{6}$C.

36. A manufacturer could include traces of a substance that would leave a clear fingerprint in the gamma-ray spectra of a product. The product could be effectively safeguarded from counterfeiting, since a check with neutron activation analysis could prove whether a specimen was genuine.

37. No; the energy would be too high. Such gamma rays alter the nucleus as ionizing radiation alters the electron structure of the atom, by removing one or more nucleons from the nucleus. Those gamma rays are too powerful to be found in the photons released by transitions between bound states in the nucleus.

Chapter 28

1. T **2.** F **3.** T **4.** F **5.** T **6.** F **7.** F **8.** F

15. Live wood, or seasoned wood that has been cut for only a year or so, has the same percentage of ^{14}C to ^{12}C as the atmosphere, since its carbon came from the air during photosynthesis. Therefore, when it is burned, it releases ^{14}C and ^{12}C in the same ratio as found in the air, and the carbon isotope composition of the air isn't changed.

19. Gravity keeps the sun together; the pressure of the outer layers of the sun "keep the lid on" its fusion reactor at the center. Those same outer layers absorb the gamma rays, preventing earth from being flooded with high-energy, ionizing photons.

22. The larger the shape, the more volume per unit of surface area and the less surface area per unit of volume. The larger volume of uranium, then, has less surface area per unit of volume or mass, and it is less likely for a given fission neutron to be lost to the outside, increasing the possibility that it will cause another fission.

28. Spherical. A sphere has less surface area per unit of volume, so fewer fission neutrons can escape before causing another fission.

30. Its mass is less, some mass being "tied up" as binding energy in the stable nucleus.

34. Cold air is heavy and does not rise as warm air does when it is surrounded by cooler air. Consequently there is little mixing of air in the atmosphere above the Arctic and Antarctic regions. Significant convection currents don't occur there, so pollution is trapped.

41. One way is to put the uranium into a compound that is gaseous at a reasonable temperature and then place the gas in a rotating device, a centrifuge, where the slightly more massive U-238 molecules will be pushed to the outside more than the lighter U-235 molecules, which concentrates the U-235 toward the axis of rotation. Another way is to force the gaseous molecules through porous materials. The U-235 being less massive, they have slightly greater speeds at the same temperature and are able to diffuse though the material faster, concentrating at the "front" of the emerging gas.

43. $E = mc^2$. The mass that disappears is (3.0160 + 2.0141) u − (4.0016 + 1.0087) u = 0.0198 u = 0.0198 $\times$ 1.7 $\times$ 10^{-27} kg = 3.4 $\times$ 10^{-29} kg. Converting this to energy gives $E = mc^2$ = 3.4 $\times$ 10^{-29} kg $\times$ 9 $\times$ 10^{16} m^2/s^2 = 3.1 $\times$ 10^{-12} J. Dividing this by 1.6 $\times$ 10^{-19} J/eV, we find 18.9 $\times$ 10^{6} eV.

Glossary

absolute pressure (161): The amount of pressure that a gas exerts relative to a vacuum.

absolute zero (180): The zero point on the Kelvin scale of temperature; the temperature at which the molecules of a substance have no kinetic energy that can be taken from them.

absorption spectrum (357): A continuous spectrum missing certain definite frequencies of light.

acceleration (21): The rate of change of the speed for a moving body that moves along a straight line. (A negative rate of speed, meaning the object is slowing down, is sometimes called *deceleration*.) More generally, it is the rate of change of an object's velocity, meaning that if a moving object merely turns it is accelerating.

additive mixture (of light) (349): The light produced by combinations of frequencies of light that come to your eyes simultaneously.

additive primary colors (349): red, blue, and green, whose combinations in various intensities produce most of the colors we can distinguish.

adhesion (150): The attraction between molecules of different kinds.

air resistance: (see drag)

alloy (129): A mixture of two or more metals that have been combined in order to form a new material with special properties.

alternating current, AC (270): An electric current whose direction oscillates.

AM radio wave (310): A carrier wave whose amplitude is modified during broadcasting.

amorphous solid (126): Solids like plastics or glass, in which the arrangement of the molecules or atoms is irregular.

ampere (265): The SI unit of electric current; 1 coulomb of charge per second.

amplitude (for a wave) (219): The maximum distance a particle in the medium is displaced from its normal position as a wave passes.

angular momentum (71): The product of the moment of inertia and the angular speed of a rotating body.

Archimedes' principle (145): The buoyant force on an object is equal to the weight of the water it displaces, whether it floats or is submerged.

atmospheric pressure (158): The total pressure exerted by the mixture of gases in the atmosphere at a location.

atom (110): The smallest particle of a chemical element that exhibits the chemical and physical properties of that element. That is, the small units of matter that give matter its chemical and physical characteristics on the macroscopic scale.

atomic mass unit, u (115): 1.67×10^{-24} grams, about the mass of one hydrogen atom.

average speed (18): The distance an object moves in a specific amount of time divided by that time.

axis of rotation (57): A straight line about which a body rotates.

background radiation (417): The natural radiation from radioactive elements and cosmic rays in our environment.

barometer (159): A device used to measure the air's pressure.

battery (266): A combination of two or more cells joined to produce an electrical potential difference equal to the sum of the potential differences of the individual cells.

Bernoulli effect (162): The lessening of fluid pressure at the side of a stream of fluid.

beta decay (409): The process in which either an electron or a positron is emitted from a radioactive nucleus.

beta particle (409): An electron or a positron emitted in natural radioactivity.

binding energy (428): The energy needed to take a nucleon away from its nucleus.

blackbody (358): An object or surface that absorbs all the light that strikes it and emits radiant energy in a way that depends only on its temperature.

bonding force (42): An attractive force between atoms or molecules, strongest in solids, less in liquids, and extremely small in gases.

Boyle's law (161): If the temperature of a confined gas does not change, the product of the volume and the absolute pressure of the gas is constant. $P_iV_i = P_fV_f$

breeder reactor (427): A nuclear reactor which produces nuclear fuel as a byproduct of its fission reactions.

Brownian motion (141): The random movement of microscopic particles in a fluid, caused by collisions from every direction with the fluid's molecules.

buoyant force (144): The upward force exerted by a fluid on an object immersed in it or floating in it.

calorie (174): The amount of heat that raises the temperature of one gram of water by 1 degree Celsius.

capillary action (151): The rising of a liquid in a small vertical space.

cell: (see chemical electric cell)

center of gravity (96): A point at which the net gravitational force acts on an object.

center of mass (59): The balance point for an object. The point about which a free body will rotate.

centripetal force (55): The turning component of a net force on a moving object, perpendicular to the object's actual path. $F_c = mv^2/r$

chain reaction (423): When one event triggers other similar events, which in turn trigger others, and so on.

chemical bond (116): The attractive interactions of the outermost electrons that combine like or unlike atoms to form molecules or compounds.

chemical electric cell (cell) (266): A device that uses chemical reactions to produce a difference in electric potential that can give rise to a current.

chemical element (111): A substance that cannot be broken down into simpler substances by means of chemical reactions.

chemical reaction (116): The combining of either elements or compounds, or the breaking apart of molecules or compounds.

chromatic aberration (333): Light of different colors entering a lens refract by different amounts and so come to focus at different points.

coherent radiation (367): Radiation whose waves (or photons) are in perfect step; the crests and troughs coincide, leading to maximum constructive interference.

cohesion (150): The attraction of like molecules for each other.

complementary colors (349): Two colors that when reflected together give white light.

component vector (26): A vector that is part of a chain of vectors adding to give a net vector.

compressions (218): Regions of higher density, as when a longitudinal wave passes through a gas.

compound (112): A combination of different chemical elements joined by chemical bonds.

condensation nucleus (209): A small particle in the air on which water vapor can condense.

conduction (186): The transfer of heat by molecular vibrations, molecule to molecule; most effective in solids. Even more effective in metals, where free electrons help carry the energy.

conduction electrons (250): Electrons in a conductor that are free to move from atom to atom.

conservation of angular momentum (72): In the absence of external torques, the angular momentum of a system (rotational and/or orbital) is constant.

conservation of energy (88): In every interaction of any kind, the total energy afterward is always equal to the total energy before the interaction, even when energy changes forms during the interaction.

conservation of momentum (70): The sum of the vector momenta of a system of interacting objects remains constant if no external interactions occur.

constructive interference (222): When two or more waves meet and their amplitudes add at some point to give a larger amplitude.

contact force (41): The force of repulsion that occurs when molecules or atoms of matter are pressed together. The contact force is always perpendicular to the surfaces that come together.

continuous quantity (20): A quantity such as time or length that is (apparently) divisible without limit.

correspondence principle (386): The statement that any two theories or hypotheses whose areas of prediction overlap should agree at those places; otherwise at least one of them is wrong.

coulomb: (249): The SI unit of electric charge.

Coulomb's law (249): The law giving the electric force between two charges.

covalent bonds (119): A bond between nonmetal atoms where the outermost electrons are shared between the atoms.

convection (187): The transfer of heat by a current of moving fluid such as air or water.

converging lens (327): A convex lens, one that converges light to a focal point behind the lens.

critical angle (326): The angle of incidence beyond which total internal reflection occurs for a light wave incident on a medium where its speed will increase.

critical mass (424): The minimum mass of a fissionable material such as uranium-235 needed to sustain a runaway chain reaction long enough for an explosion to occur.

crystalline solid (126): A solid whose molecules or atoms are arranged in a pattern that repeats throughout the solid.

daughter nucleus (409): A nucleus that is the product of a radioactive disintegration.

density (132): The ratio of the mass to the volume of an object. Density = M/V. Its units are typically grams/cubic centimeter.

destructive interference (222): When two or more waves meet and their amplitudes oppose at some point to give a smaller amplitude.

dew (210): Condensation of water vapor from the air onto surfaces cooler than the air itself. The air next to the surface must be chilled to or below the dew point, which depends on the relative humidity of the air as well as its temperature.

dew point (208): The temperature for a parcel of air at which that air becomes saturated and the relative humidity is 100 percent.

diffraction (225, 335): The natural spreading of a wave that passes through an opening in a barrier or that passes adjacent to barriers.

diffuse reflection (319, 236): When a microscopically rough surface diffuses light, scattering it in all directions.

direct current, DC (270): An electric current that moves in only one direction.

dispersion (340): The separation of various frequencies of waves in a medium due to their different speeds there.

dissolve (151): When gas molecules enter a liquid's surface, or when molecules or ions of a solid that is immersed in a liquid dissociate from the solid state and enter the liquid, we say they have dissolved in the liquid.

diverging lens (327): A concave lens, one that refracts light to form a virtual image in front of the lens.

Doppler shift (238): A change in wave frequency due to motion of the source or of the listener.

drag (44): The resistive force experienced when a solid object moves through a liquid or a gas.

efficiency (87, 196): The ratio of the work a system does to the energy it uses.

elasticity (42): The property of solid matter that enables it to recover its shape after being deformed. It comes from the actions of contact forces and bonding forces between the object's molecules.

electric cell: (see chemical electric cell)

electric current (265): A flow of charged particles.

electric dipole (256): A positive and an equal negative charge separated by a short distance.

electric field lines (256): Imaginary lines used to visualize the electric force field in the region around a charged particle.

electric force (113): The strong, long-range force associated with the property of matter called charge.

electric force field (255): The electric influence around a charged particle.

electric generator (296): A device that converts mechanical energy into electric energy.

electric motor (299): A device that converts electrical energy into mechanical energy.

electric potential (257): The potential energy *per unit of charge* of a particle due to an electric field.

electric field. Its metric units are joules.

electrical circuit (269): A connected, conducting path for electrons between points of different electric potential.

electrical conductor (250): A material in which charges can easily move.

electrical induction (252): The process of separating charge on an object by bringing it near another charged body.

electrical insulator (250): A material that won't ordinarily carry an electric current; a nonconductor.

electrical potential energy (257): The potential energy a charged particle has due to the presence of an electric field

electrical resistance (268): The opposition to the flow of charge through matter.

electromagnet (284): A device that uses an electric current to create a strong magnetic field.

electromagnetic induction (294, 307): The creation of an electric field by a changing magnetic field, and the creation of a magnetic field by a changing electric field.

electromagnetic spectrum (309): All possible wavelengths of electromagnetic waves.

electromagnetic waves (or radiation) (307): Traveling oscillations in the electric field of a charged particle.

electron (112): The negatively charged subatomic particle that orbits the nucleus of an atom.

electroscope (251): A device used to detect the presence of an electric charge.

electroweak force (409): The weak nuclear interaction unified with the electromagnetic force.

emission spectrum (358): A spectrum of bright lines produced by the emission of light by excited atoms or molecules.

empirical (11): Based on experience with numerical data from measurements.

endothermic chemical reaction (121): A chemical reaction that absorbs energy.

energy levels (of electrons) (363): The energy levels or steps that an electron can occupy in an atom.

energy-mass relation (400): If an amount of mass disappears in some reaction, it yields energy in an amount equal to that mass times the square of the speed of light.

entropy (193): A quantity that measures the state of disorder or randomness of a system. In any process where heat is transferred, entropy increases.

escape velocity (98): The speed that an object must have in order to escape permanently from a planet or a moon. The minimum speed it must have to leave because of gravity.

evaporate (151, 178): When molecules leave the surface of a liquid to go into the gaseous state.

excited states (362): Energy states with more energy

than the ground state, where the particle is still part of the bound system, such as an electron in an atom.

exothermic chemical reaction (121): A chemical reaction that releases energy.

expansions (218): Regions of lower density, as when a longitudinal wave passes through a gas.

experiment (10): A carefully planned arrangement to make quantitative measurements of an action in nature.

first law of thermodynamics (192): The total energy of a system in all of its forms, even though it may change forms during interactions, is a constant. When work is done it takes the form *energy in* = *work done* + *energy out*.

first postulate of relativity (393): All laws of physics are identical in all inertial frames of reference.

fluorescence (366): The light-producing process by which an excited atom loses its extra energy as it quickly moves to a lower-energy state.

FM radio wave (310): A carrier wave whose frequency is modified during broadcasting.

focal length (328): The distance from the focal point to the center of the lens.

focal point (328): The point at which parallel light rays entering one side of a lens will meet (converging lens) or seem to come from (diverging lens).

fog (210): Droplets of water in the air large enough to obstruct the passage of light, occurring when the air cools enough so that the relative humidity exceeds 100%.

force (35): A push or pull on an object.

free fall (40): When gravity is the only force acting on an object, it is in free fall. In free fall objects have no apparent weight, since nothing supports them. This condition is often called "weightlessness."

frost (211): Ice forming from water vapor in the air, directly from the air onto a cold surface.

g (22): The symbol for the value of the acceleration of gravity at earth's surface, which is about 32 feet per second per second, or 9.8 meters per second per second.

gamma emission (408): The process in which a nucleon of a radioactive nucleus moves to a lower-energy state and emits a gamma ray.

gamma ray (408): A high-energy photon, released when a proton or a neutron moves to a lower-energy state in a nucleus.

geocentric (4): *Earth-centered*, as in the Ptolemaic model of the solar system; the Earth was assumed to be at rest, with the sun, moon, and planets revolving around it.

gravitational field lines (97): Imaginary lines drawn to an object which show the direction and value of its gravitational attraction for nearby matter.

gravitational force (95): The mutual attraction of all particles for each other.

gravitational potential energy (85): The measure of the potential of gravity to do work on an object. Near the Earth's surface, gravity's potential energy is equal to mgd, where d is the distance that the object may fall.

ground state (362): The lowest energy state a particle can have, such as an electron in an atom.

grounding (254): Connecting a conductor directly to the earth, which is able to absorb any charge passed to it.

half-life (412): The time needed for one-half of the nuclei of a given specimen of radioactive substance to disintegrate.

heat (87, 186): Energy that flows from one point to another because of a difference in temperature. Heat can move by conduction, convection, or radiation.

heat of sublimation (177): The heat energy needed to change a subliming substance from the solid state directly into the gaseous state, or the energy released as that gaseous substance becomes solid.

heliocentric (4): *Sun-centered*, as in the Copernican model of the solar system; all the planets, including Earth, revolve around the stationary sun.

hypothesis (10): A possible explanation for observations; an educated guess, an unproved theory.

image (319): A likeness of an object or a scene formed by light.

impulse (66): The product of a force on an object and the time it acts. The impulse of the *net* force is a vector that is equal to the object's change in vector momentum.

incoherent radiation (367): Radiation whose crests and troughs come at random positions or times, leading to random interference of light.

induced charge (252): See electrical induction.

inertia (36): The inability of matter to accelerate without a net force acting on it. The measure of inertia is how difficult it is to change something's velocity.

inelastic collision (68): A collision where some relative speed is lost between the objects but not all; the most common type of collision.

inertial frame of reference (393): A fixed system of coordinates from which motion can be viewed and measured *and* where Newton's first law is obeyed.

infrared radiation (188): The invisible radiation that is effective at transferring heat into and out of matter.

infrared rays (309): Invisible electromagnetic waves, with lower frequencies (longer wavelengths) than red light.

infrasonic wave (234): A compressional wave with a frequency lower than 20 Hz, under the limit of human hearing .

instantaneous speed (18): The rate of travel that an object has at a particular instant in time (or at a particular point in space).

intensity (243): The sound-wave energy incident on a given unit of area each second.

interference pattern (306): A pattern caused when two waves in step strike the same area, showing destructive and constructive interference.

internal energy (87): The energy stored within matter in the forms of kinetic and potential energies of its molecules or atoms.

intrinsic magnetic fields (283): A magnetic field that is basis to the particle, such as an electron or proton; it is a property as basic as charge or mass.

ionic bonds (118): A bond between a metal atom and a nonmetal atom, where the metal atom loses its outer electron(s) almost entirely to the nonmetal atom.

iridescence (346): color produced by the constructive interference of particular wavelengths in sunlight.

isotopes (116): The various species of a chemical element that have different numbers of neutrons in their nuclei.

kinetic energy (82): The energy of motion of a body. A measure of a moving body's ability to do work, related to its mass and speed by 1/2 mv^2.

kinetic friction (42): The frictional force between touching surfaces that have relative motion.

latent heat of fusion (176): Heat energy transferred upon freezing or absorbed upon melting that does not change the temperature of the matter.

latent heat of vaporization (177): Heat energy transferred upon condensation or absorbed upon evaporation that does not change the temperature of the matter.

length contraction (399): The length of an object contracts along the direction of motion when the object moves relative to an observer.

leptons (416): Believed to be, together with quarks, the most elementary particles today, the particles that make up all the other known particles.

lever arm (of a force) (58): The perpendicular distance between the axis of rotation (or possible rotation) and the line of a force that acts on the body.

life sciences (11): Those sciences dealing principally with living matter and its interactions with the physical word.

lift (53): the upward component of force on an airplane wing as it pushes through the air.

light (308): The wavelengths of electromagnetic radiation that are visible to the human eye.

longitudinal wave (218): A wave that moves the particles of the medium back and forth along the direction of the wave's motion.

loudness (234, 243): The brain's perception of the intensity of a sound wave.

magnetic declination (285): The deviation of a compass needle from geographic north.

magnetic dip (285): The angle at which the magnetic field lines of earth dip into the earth's surface.

magnetic domains (284): The groups of atoms that align to produce a net magnetic field in a magnetic material.

magnetic field (281): The region of magnetic influence around a magnetic pole or a moving charged particle.

magnetic force (280): The repulsive or attractive force between the poles of magnets.

magnetic inclination: (see magnetic dip)

magnetic induction (284): The creation of a net magnetic field in a material whose magnetic domains align when an external field is present.

mass (34): The measure of the amount of matter in an object; the ratio of force to acceleration when a net force acts on a body. $m = F/a$

mass number (115): The total number of protons and neutrons in the nucleus of an atom.

metals (118): Elements that tend to lose electrons in chemical reactions.

metastable state (366): An excited state in an atom where an electron won't promptly drop to a lower energy level.

modulation (of electromagnetic waves) (310): A modification of either the amplitude or the frequency of an electromagnetic wave, usually in order to carry a signal.

molecule (112): A unit of combined atoms that has its own chemical identity; the smallest unit of a substance of combined atoms that exhibits the chemical and physical properties of that substance.

moment of inertia (58): The measure of a body's rotational inertia. The ratio of the net torque to the rate of change of angular speed of an object.

momentum (66): The vector quantity mv, the mass of an object times its velocity.

natural frequencies (223): Those frequencies of oscillation for an object that excite standing waves in the object.

natural philosophy (4): The examination and explanation of natural phenomena in ancient times. This term was replaced by *science* in the mid-1800s.

net force (35): The resultant force when more than one force acts on an object at the same time; the total force that causes an acceleration.

net vector (26): The single vector that by itself describes the successive or simultaneous addition of two or more (component) vectors.

net velocity (26): The velocity of an object that is found by the vector addition of any independent velocities the object may have.

net work (82): The sum of the works done on an object by all the forces acting. It is equal to the change in the kinetic energy of the object.

neutrino (410): A massless, chargeless particle that is emitted during beta decay and travels at the speed of light.

neutron (112): The neutral subatomic particle contained in the nucleus of an atom.

nonmetals (118): Elements which tend to take on electrons in chemical reactions.

nuclear fission (422): The splitting of a nucleus into two other nuclei.

nuclear force (114): The strong, short-ranged attractive force that keeps the nucleons (protons and neutrons) bound together in a nucleus.

nuclear fusion (429): The building of a larger nucleus from smaller ones.

nuclear reactor (424): A device using nuclear fission to generate heat that may be used to generate electrical power.

nucleus (112, 407): The tiny center of an atom, composed of protons and neutrons (thus having a positive charge) and containing almost all of the atom's mass.

Ockham's razor (8): The principal that a physical theory should be as simple as possible to explain the facts and observations and lead to other accurate predictions. In other words, a correct theory should have no unnecessary assumptions.

ohm (268): The SI unit of electrical resistance; 1 volt per amp.

orbital (380): A three-dimensional standing-wave pattern the amplitude of which defines the physical state and behavior of an electron in an atom.

order of magnitude (appendix 1): A factor of ten.

oscillation (219): One complete cycle of a repetitive motion.

parallel circuit (273): An electrical circuit that provides more than one current path.

parent nucleus (409): A nucleus that decays to form a different nucleus which is called the daughter nucleus.

partial pressure (160): The pressure exerted by a specific gas in a mixture of gases.

Pascal's principle (142): At any point in a fluid, pressure is transmitted equally in all directions.

perfectly elastic collision (68): A collision where the relative speed of the colliding objects is the same before and after the collision.

perfectly inelastic collision (67): A collision where the relative speed between the colliding objects is zero after the collision.

periodic table (116): A table with the chemical elements arranged so those elements in columns have similar chemical properties.

phospher (366): A molecule or compound that absorbs energy from light and re-emits it later at certain frequencies characteristic of that particular substance.

phosphorescence (366): A process identical to fluorescence except that the excited atoms stay in the excited state much longer.

photoelectric effect (360): the emission of electrons from a metal's surface when that surface absorbs light.

photon (360): A quantum of light.

physical law (10): A theory of particularly broad application that correctly explains and predicts many facts, including facts not known before the law's conception.

physical quantity (10): A quantity that can be measured.

physical sciences (11): Those sciences dealing with the physical nature of the universe.

physics (12): The natural science that deals with matter, energy (including radiation), time, and space at the most fundamental levels.

pitch (238): The frequency of a sound wave.

plane-polarized light (334): Light waves that oscillate in a single plane.

polar molecule (253): A molecule that, though neutral overall, has positive and negative areas.

polarization (253): The shifting of charge in an object when the object is exposed to an electric field. Its atoms or molecules become negatively charged on one side and positively charged on the other side.

positron (408): A particle that has the same mass as an electron but carries a positive charge. The antiparticle of the electron.

postulate of wave mechanics (380): The assumption that the psi wave describes the state of any particle, including its position, momentum, and energy.

potential energy (85): Energy of position, the energy a body has because a force acting on the body has the potential to perform work on it.

power (81): The rate of doing work or using energy. Power is the ratio of work done (or energy used) to the amount of time required to do that work (or to use that energy). $P = W/t$ or $P = E/t$

power of ten (appendix 1): The exponent on the (base) ten when a number is expressed in scientific notation.

precession (73): A regular, circular wobbling of the spin axis of an object.

pressure (average) (54): The force on an object divided by the area to which the force is applied. $P = F/A$

prevailing winds (204): Global air currents whose directions are influenced by earth's rotation.

proton (112): The positive subatomic particle contained in the nucleus of an atom.

qualitative (10): Expressible with words or attributes.

quantitative (10): Expressible with numbers or a formula.

quantum (359): A unit of radiant energy.

quantum number (385): A number that designates a possible state for a quantized property of a particle, such as the shell number for an electron in an atom.

quarks (416): Believed to be, together with leptons, the most elementary particles today, the particles that make up all the other known particles.

radioactive nuclei (408): Unstable nuclei that emit either gamma rays or particles in the act of losing energy in order to become more stable.

rate (18): A rate is a quantity divided by a time. A rate tells how fast something happens or how much something changes in a certain amount of time.

ray (319): An idealized, infinitely thin beam of light.

real image (320): The image made by light rays that converge at an image's location.

reflection (221): When part of a traveling wave rebounds upon hitting a medium where it must change speeds abruptly.

reflection, law of (320): The angle of reflection, measured from a normal (perpendicular) line to the reflector's surface, is equal to the angle of incidence measured from that same line.

refraction (226): The change in direction of a wave as it passes into a medium where its speed changes.

refraction, law of (320): The change in light's direction caused by a change in its speed as it enters a different medium at an angle other than 90 degrees.

relative humidity (207): The ratio of the vapor pressure of water in a volume of air at a given temperature to its saturated vapor pressure at that temperature.

relative speed (23): The speed of an object with respect to something else.

resonance (223, 238): The condition where a standing wave grows in amplitude as energy is added by a disturbance. When a natural frequency of an object or a medium is excited.

revolution (57): The turning motion of a body around an axis outside its own body.

rotation (57): The turning motion of a body around an axis within the body.

runaway chain reaction (424): When more than one neutron released per fission on the average goes on to cause another fission; this means the fission rate multiples quickly.

saturated vapor pressure (207): The highest vapor pressure possible at a given temperature before condensation starts.

saturation (151): The point at which the maximum quantity of a gas is dissolved in a liquid for a given temperature and pressure.

science (2): The body of knowledge dealing with natural phenomena.

scientific process, or scientific method (10): A procedure typically followed in scientific investigations, where observations are followed by hypotheses and experiments.

second law of thermodynamics (192): Heat always flows from regions of higher temperature to regions of lower temperature. As it does the energy it transfers becomes more disorganized (or more randomized) and is less useful for doing work.

second postulate of relativity (394): The speed of light through space is independent of any motion of the source or the observer.

secular equilibrium (413): Whenever the amount of a radioactive isotope remains constant because nuclei are being added to the isotope at the same rate at which nuclei are disintegrating.

selective absorption (342): When only certain frequencies of white light are absorbed and the rest are reflected or transmitted.

selective reflection (342): When only certain frequencies of white light are reflected and the rest are absorbed.

selective transmission (342): When only certain frequencies of white light are transmitted and the rest are absorbed.

self-sustaining chain reaction (423): When the neutrons from one nuclear fission cause an average of one more nuclear fission. Fissions then occur at a steady rate.

semiconductor (252): A material that is an electrical insulator in pure form but takes on the properties of a conductor when specific impurities are introduced in small amounts.

series circuit (272): An electrical circuit that provides only one current path.

shock wave (239): A conical wavefront formed behind an object traveling faster than wavefronts can travel in the medium.

short circuit (271): A path of too-low resistance in an electric circuit, leading to a great increase in the amount of current in the (shorted) circuit.

significant figures (appendix 1): The number of significant figures in a stated number is the number of digits whose values are known with certainty.

simple machine (79): An arrangement that trades force for distance to help perform work easier, that is, with less force.

solar system (5): The sun and its nine planets, only five of which can be seen easily with the naked eye. (Also a host of minor planets, called asteroids, and comets, all of which orbit the sun.)

sonic boom (239): The loud noise heard when a passing object travels faster than the speed of sound.

sound wave (234): A longitudinal wave through the air (or any gas, liquid or solid) which has a frequency between 20 Hz and 20 kHz.

special theory of relativity (394): Einstein's theory that modifies the ideas of time and space due to the constancy of the speed of light in any inertial reference frame.

specific heat of a substance (174): The quantity of heat that will raise the temperature of 1 gram of the substance by 1 degree Celsius.

spectroscope (357): An instrument used to magnify the spectrum of light for visual observation.

spectrum (of white light) (341): The continuous range of colors that compose white light: **r**ed, **o**range, **y**ellow, **g**reen, **b**lue, **i**ndigo, and **v**iolet, remembered easily as ROY G. BIV.

specular reflection (236): A reflection that occurs with little distortion of the wave's form.

spherical aberration (332): A light ray entering a spherical lens near the edge of the lens axis comes to focus at a different point than parallel light rays incident closer to the center of the lens.

standing waves (223): Waves formed when a wave's reflection from a boundary interferes with the incoming wave to form a pattern that doesn't move.

state (385): The complete description of a particle's properties in quantum mechanics, all of its quantized properties and its wave function.

static charge (250): Charge not free to move around, such as charges on an insulator.

static friction (42): The frictional force that resists motion between two touching surfaces when they are at rest.

streamlines (157): Lines indicating the speed of a fluid by how close together they are. Where they are closer together, the speed of the fluid is greater.

strong interaction (407): The short-ranged but extremely strong attractive force between the nucleons in a nucleus.

subtractive primary color (349): The color that remains when a primary color is removed from white light.

superconductor (272): An element or compound whose electrical resistance vanishes at or below its critical temperature, allowing current to pass without any loss of energy.

supersaturated air (209): Air whose relative humidity is temporarily above 100 percent; eventually this forces condensation of water vapor.

surface area (131): The area of a surface, which can be either flat or curved. Surface area is measured in terms of a length squared, such as square meters, m^2.

surface tension (149): The tension on the surface of a liquid due to the mutual attractions of its molecules, causing the surface to behave as if it had an invisible "skin" stretched over it.

temperature (173): A measure of the average molecular kinetic energy in a substance.

terminal speed (53): A constant speed that comes about when the pull of gravity on an object is balanced by the air drag it encounters.

theory (10): A scientific explanation or model. It must explain observed facts and predict features beyond those initially leading to the theory.

theory of general relativity (105): Einstein's theory describing the attraction of masses for other mass and even light as a warping of space and time near a mass. This theory did away with the concept of action at a distance.

thermal energy (186): Another term for the energy within a substance that can move as heat. See "heat."

thermal equilibrium (189): The state where an object's temperature is constant; when an object loses just as much energy each second as it gains from its surroundings.

thermodynamics (192): The part of physics dealing with heat and work and matter.

tidal force (102): The stretching and pulling force acting on an extended body that is in the gravity field of a massive object.

time dilation (397): Time passes more slowly for an object in motion relative to an observer than it does for the observer.

torque (of a force) (58): The product of an applied force and its lever arm. A net torque on a body causes a change in the body's rate of rotation.

total internal reflection (326): The total reflection that occurs when light strikes a medium (where it would pick up speed) at greater than the critical angle.

transformer (298): A device used to increase (step up) or decrease (step down) voltage for transmission and/or safety purposes.

transmission (221): When part of a traveling wave enters a medium where it changes speeds.

transverse wave (218): A wave that moves the particles of the medium back and forth perpendicular to the direction of the wave's motion.

turbulence (148): Chaotic flow, as opposed to smooth or laminar flow. When a fluid flow is turbulent, energy is used as in friction between solids—heat is created.

ultrasonic wave (234): A compressional wave with a frequency higher than 20,000 Hz, beyond the limit of human hearing .

ultraviolet rays (309): Invisible electromagnetic waves, with higher frequencies (shorter wavelengths) than violet light.

uncertainty principle (383): The fact that the precise position and the momentum of a particle cannot both be known at the same time.

vacuum (160): A space containing no atoms or molecules.

van der Waals bond (120): A weak attraction that occurs between any atoms due to the polarizing motion of the outer electrons of atoms. Positive and negative areas are created which lead to a net attraction for other neutral atoms nearby.

vapor pressure of water (206): That portion of air pressure due to the water molecules in the air.

vectors (25): Arrows used to represent physical quantities that have directions as well as a size, or magnitude.

velocity (25): A vector that indicates the speed of an object together with its direction.

virtual image (321): An image formed by light rays that don't converge at the location of the image.

visible spectrum: (see *light*)

volt (257): The unit of electrical potential, 1 joule per coulomb.

volume (131): The space enclosed by the surfaces of a substance, or the space enclosed by a designated region in space. Volume is measured in terms of a length cubed, such as cm^3.

watt (81): The SI unit of power, including electrical power; equal to 1 joule per second.

wave (218): A moving disturbance that carries energy through matter or space. A wave in a material medium travels without carrying the particles of the medium with it.

wave equation (378): The equation that predicts the amplitude of the probability waves for a particle in wave mechanics.

wave frequency: The oscillations per second of a particle as the wave passes. Its symbol is f.

wave mechanics (378): The theory describing the wave behavior of subatomic particles.

wavefront (224): A surface where a passing wave affects all the particles of the medium the same way at the same time.

wavelength (219): The distance between two successive compressions or expansions (longitudinal wave) or two successive crests or troughs (transverse wave). Its symbol is λ.

weight (39): The force of gravity on an object. On earth's surface, weight $= mg$. The direction of this force is toward the center of the earth

work (78): A measure of how productive an applied force is. Work is the force component in the direction of motion multiplied by the distance the object move through. $W_{net} = F_{parallel} \times d$. When a force or a force component pushes or pulls an object through a distance (or resists its motion through a distance), it does work on the object.

work function (360): The minimum energy an electron needs to escape from a metal's surface.

Photo Credits

Cover photo, **Fig. 3-4**, **p. 57**, **p. 85** bottom, **Chapter 8** opening photo, **Fig. 8-2a**, **Fig. 9-11**, **Chapter 10** opening photo, **Fig. 11-2a**, **Fig. 11-12**, **p. 187** top, **Fig. 13-7**, **Fig. 13-9b**, **Fig. 13-19**, **Fig. 14-1a**, **Fig. 14-1b**, **Fig. 14-14**, **Fig. 14-16**, **Fig. 15-19**, **p. 238** b, **Fig. 22-13c**, **Fig. 22-16**, **Fig. 22-17**, **Fig. 23-4**, **Fig. 23-23**, **Fig. 23-24**, all by Dale Nichols.

Chapter 1 opening photo, Larry Landolfi; **p. 3**, Peter Buckley; **Fig. 1-4**, Lee Coombs; **p. 5**, top, Ray Renken; **Fig. 1-5b**, NASA/JPL; **Figs. 1-6a**, **b**, Lee Coombs; **Fig. 1-7**, Boston Museum of Science; **Fig. 1-9**, The Bettman Archives; **Fig. 1-10**, Lee Coombs; **p. 10**, Art Resource; **p. 11**, Photo Researchers, Inc.; **p. 12** margins, top, Photo Researchers, Inc., middle, NASA; **p. 12** bottom (a) and (b), Photo Researchers, Inc.

Chapter 2 opening photo, Photo Researchers, Inc.; **p. 20**, Photo Researchers, Inc., **Fig. 2-4**, James Sugar/Black Star, **Fig. 2-5**, Allsport; **Fig. 2-8**, *PSSC Physics*, D.C. Heath and Co.; **p. 33**, Ray Renken.

Chapter 3 opening photo, Photo Researchers, Inc.; **Fig. 3-3**, NASA; **Fig. 3-9**, The Bettman Archive; **p. 40** bottom, Bob Barbour, **p. 41** Hulton Dentsch Collection Limited; **p. 43** Franca Principe Science Museum, Florence.

Chapter 4 opening photo, Photo Researchers, Inc.; **Fig. 4-6**, courtesy of GMNAO Research and Development Center; **p. 53** bottom, Photo Researchers, Inc.; **p. 54** top, Photo Researchers, Inc.; **Fig. 4-7b**, NASA; **p. 55**, Photo Researchers, Inc.

Chapter 5 opening photo, Photo Researchers, Inc.; **Fig. 5-2**, Photo Researchers, Inc.; **p. 68**, NASA; **Fig. 5-5**, The Stock Market; **Fig. 5-8**, Peter Arnold, Inc.

Chapter 6 opening photo, Animals Animals/Earth Scenes; **Fig. 6-10**, courtesy of Gina Metivier and Tarl Spitzer; **Fig. 6-11**, Photo Researchers, Inc.; **p. 85** top, Logan Campbell; **p. 86**, Animals Animals/Earth Scenes; **p. 89**, box, Ron McQueeney; **Fig. 6-16**, Fundamental Photographs.

Chapter 7 opening photo, Lee Coombs; **p. 98** top and middle, NASA; **Fig. 7-9**, NASA; **Fig. 7-10**, NASA.

Chapter 8 **Fig. 8-1**, NASA; **Fig. 8-3**, The Granger Collection; **Fig. 8-18**, Photo Researchers, Inc.; **p. 120** top, John Deisenhofer, University of Texas Southwestern Medical Center at Dallas; **p. 120** middle, Michael G. Rossmann, Purdue University; **p. 121**, Fundamental Photographs.

Chapter 9 opening photo, Tim Bandy/Paul Morgan; **Fig. 9-2a**, GE Research and Development; **Fig. 9-3**, Photo Researchers, Inc.; **Fig. 9-5**, Discover Publications, **p. 130** middle and right, Photo Researchers, Inc.; **p. 133**, Jonathan Meyers; **Fig. 9-12**, Allen Mathews.

Chapter 10 **Fig. 10-12**, Minden Pictures, Inc.; **Fig. 10-17b**, NASA; **p. 150** middle, Courtesy of W.L. Gore and Associates, Inc., Elkton, MD.; **Fig. 10-20**, Photo Researchers, Inc.

Chapter 11 opening photo, NASA; **Fig. 11-1**, Animal Animals/Earth Scenes; **Fig. 11-6a**, The Bettmann Archive; **p. 161**, Sally James; **Fig. 11-10**, John Thorpe, Pacific Northwest Laboratory, Battelle Memorial Institute for the U.S. Department of Energy under contract No. DE-AC06-76LO-1830, previously published in Weatherwise, 39,#6, 1986; **p. 165**, NASA; **p. 166**, NASA; **Fig. 11-11**, Photo Researchers, Inc.

Chapter 12 opening photo, courtesy of Bruce Rowland; **p. 176**, courtesy of Tiny Bird; **Fig. 12-11**, Photo Researchers, Inc.

Chapter 13 opening photo, Ron McQueeney; **Fig. 13-3**, Anthony Bannister Photo Library; **Fig. 13-6**, Photo Researchers, Inc.; **Fig. 13-8**, NASA, **Fig. 13-14**, Grant Heilman Photography.

Chapter 14 opening photo, NASA; **Fig. 14-2**, Photo Researchers, Inc.; **Fig. 14-11**, NASA; **Fig. 14-12**, NASA, **p. 210** Minden Pictures, Inc.; **p. 211** bottom, NASA, **p. 212** top, NASA; **Fig. 14-18**, NASA; **p. 213** middle, NASA; **p. 213** bottom, NASA; **Fig. 14-20**, Paul Doherty.

Chapter 15 opening photo, Allsport; **Figs. 15-11a**,**b**, Paul Doherty; **Fig. 15-13**, International Expeditions; **Fig. 15-15a**,**b**, Paul Doherty; **Fig. 15-15c**,**d**, Fundamental Photographs; **p. 229** top and bottom, NGDC/NOAA.

Chapter 16 opening photo, Four by Five, Inc.; **Fig. 16-4**, Superstock; **Fig. 16-9**, MIT, **Fig. 16-11**, Photo Researchers, Inc.; **p. 241**, Minden Pictures, Inc.

Chapter 17 opening photo, Photo Researchers, Inc.; **p. 252** bottom, Xerox Corporation Photographic Communications; **p. 254**, Sid Clements, Appalachian State University; **p. 259**, NOAA.

Chapter 19 opening photo, Photo Researchers, Inc.; **Fig. 19-1**, Photo Researchers, Inc.; **Fig. 19-11**, General Electric Corporate Research and Development Center; **Fig. 19-16**, Visuals Unlimited; **Fig. 19-17**, Discover Magazine; **p. 295**, NASA.

Chapter 21 opening photo, Photo Researchers, Inc.; **Fig. 21-3**, Springer-Verlag Ohg, Berlin; **p. 309**, NASA/JPL; **p. 310**, courtesy of Frank Diefenderfer.

Chapter 22 opening photo, Charles Krebs; **p. 321** middle, Red Finch; **Figs. 22-14a**,**b**, NASA; **p. 330** top **a** and **b**, Sally James; **p. 330** bottom, Photo Researchers, Inc.; **Fig. 22-24b**, The Stock Market; **Fig. 22-32**, Fundamental Photographs.

Chapter 23 opening photo, Photo Researchers, Inc.; **Fig. 23-1**, Thomas Kimble; **Figs. 23-7a**,**b**, Sally James; **p. 345** bottom, Photo Researchers, Inc.

Chapter 24 opening photo, courtesy of CREOL, Center for Research and Education in Optics and Lasers, University of Central Florida; **p. 367**, NGDC, NOAA; **Fig. 24-18**, Sally James; **Fig. 24-19a**, NASA; **Fig. 24-19b**, University of Texas at Austin; **Fig. 24-20**, Tim Bandy; **Fig. 24-22**, Photo Researchers, Inc.

Chapter 25 opening photo, IBM; **Fig. 25-10**, Photo Researchers, Inc.; **Fig. 25-11**, Photo Researchers, Inc.; **p. 385** middle left, IBM Corporation, Research Division, Almaden Research Center; **p. 385** middle right, courtesy of Digital Instruments, Inc.

Chapter 26 opening photo, The Bettmann Archive.

Chapter 27 opening photo, Laurence Berkeley Laboratory; **p. 412**, Fundamental Photographs.

Chapter 28 opening photo, Princeton University; **Fig. 28-8a**, Los Alamos National Laboratory; **Fig. 28-8b**,**c**, Photo Researchers, Inc.; **Fig. 28-10**, Photo Researchers, Inc.; **Fig. 28-11**, Photo Researchers, Inc.; **Fig. 28-12**, Grant Heilman Photography; **Fig. 28-13**, Island Tourist Board.

Index

D

E

M

XY

PERIODIC TABLE OF THE ELEMENTS

Main groups — Transition metals — Main groups

1A[a] 1	2A 2	3B 3	4B 4	5B 5	6B 6	7B 7	8B 8	8B 9	8B 10	1B 11	2B 12	3A 13	4A 14	5A 15	6A 16	7A 17	8A 18
1 **H** 1.00794																	2 **He** 4.00260
3 **Li** 6.941	4 **Be** 9.01218											5 **B** 10.81	6 **C** 12.011	7 **N** 14.0067	8 **O** 15.9994	9 **F** 18.998403	10 **Ne** 20.1797
11 **Na** 22.98977	12 **Mg** 24.305											13 **Al** 26.98154	14 **Si** 28.0855	15 **P** 30.97376	16 **S** 32.066	17 **Cl** 35.453	18 **Ar** 39.948
19 **K** 39.0983	20 **Ca** 40.078	21 **Sc** 44.9559	22 **Ti** 47.88	23 **V** 50.9415	24 **Cr** 51.996	25 **Mn** 54.9380	26 **Fe** 55.847	27 **Co** 58.9332	28 **Ni** 58.69	29 **Cu** 63.546	30 **Zn** 65.39	31 **Ga** 69.72	32 **Ge** 72.61	33 **As** 74.9216	34 **Se** 78.96	35 **Br** 79.904	36 **Kr** 83.80
37 **Rb** 85.4678	38 **Sr** 87.62	39 **Y** 88.9059	40 **Zr** 91.224	41 **Nb** 92.9064	42 **Mo** 95.94	43 **Tc** (98)	44 **Ru** 101.07	45 **Rh** 102.9055	46 **Pd** 106.42	47 **Ag** 107.8682	48 **Cd** 112.41	49 **In** 114.82	50 **Sn** 118.710	51 **Sb** 121.757	52 **Te** 127.60	53 **I** 126.9045	54 **Xe** 131.29
55 **Cs** 132.9054	56 **Ba** 137.33	57 ***La** 138.9055	72 **Hf** 178.49	73 **Ta** 180.9479	74 **W** 183.85	75 **Re** 186.207	76 **Os** 190.2	77 **Ir** 192.22	78 **Pt** 195.08	79 **Au** 196.9665	80 **Hg** 200.59	81 **Tl** 204.383	82 **Pb** 207.2	83 **Bi** 208.9804	84 **Po** (209)	85 **At** (210)	86 **Rn** (222)
87 **Fr** (223)	88 **Ra** 226.0254	89 **†Ac** 227.0278	104 **Db** (261)	105 **Jl** (262)	106 **Rf** (263)	[107] **Bh** (262)	[108] **Hn** (265)	[109] **Mt** (268)									

*Lanthanide series	58 **Ce** 140.12	59 **Pr** 140.9077	60 **Nd** 144.24	61 **Pm** (145)	62 **Sm** 150.36	63 **Eu** 151.96	64 **Gd** 157.25	65 **Tb** 158.9254	66 **Dy** 162.50	67 **Ho** 164.9304	68 **Er** 167.26	69 **Tm** 168.9342	70 **Yb** 173.04	71 **Lu** 174.967
†Actinide series	90 **Th** 232.0381	91 **Pa** 231.0359	92 **U** 238.0289	93 **Np** 237.048	94 **Pu** (244)	95 **Am** (243)	96 **Cm** (247)	97 **Bk** (247)	98 **Cf** (251)	99 **Es** (252)	100 **Fm** (257)	101 **Md** (258)	102 **No** (259)	103 **Lr** (260)

[a]The larger labels are common American usage. The smaller labels are those recommended by the International Union of Pure and Applied Chemistry.